Computational Methods for the Determination of Formation Constants

MODERN INORGANIC CHEMISTRY

Series Editor: John P. Fackler, Jr.
Texas A&M University

METAL INTERACTIONS WITH BORON CLUSTERS
Edited by Russell N. Grimes

HOMOGENEOUS CATALYSIS
WITH METAL PHOSPHINE COMPLEXES
Edited by Louis H. Pignolet

THE JAHN-TELLER EFFECT AND
VIBRONIC INTERACTIONS IN MODERN CHEMISTRY
I. B. Bersuker

MÖSSBAUER SPECTROSCOPY
APPLIED TO INORGANIC CHEMISTRY, *Volume 1*
Edited by Gary J. Long

CARBON-FUNCTIONAL ORGANOSILICON COMPOUNDS
Edited by Václav Chvalovský and Jon M. Bellama

COMPUTATIONAL METHODS FOR THE DETERMINATION
OF FORMATION CONSTANTS
Edited by David J. Leggett

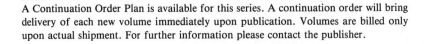

A Continuation Order Plan is available for this series. A continuation order will bring
delivery of each new volume immediately upon publication. Volumes are billed only
upon actual shipment. For further information please contact the publisher.

Computational Methods for the Determination of Formation Constants

Edited by
DAVID J. LEGGETT

Dow Chemical USA
Texas Division
Freeport, Texas

Plenum Press • New York and London

Library of Congress Cataloging in Publication Data

Main entry under title:

Computational methods for the determination of formation constants.

(Modern inorganic chemistry)
Bibliography: p.
Includes index.
1. Phase rule and equilibrium—Data processing. I. Leggett, David J. II. Series.
QD503.C66   1985   541.3′63′0285   85-16991
ISBN 0-306-41957-2

©1985 Plenum Press, New York
A Division of Plenum Publishing Corporation
233 Spring Street, New York, N.Y. 10013

Printed in the United States of America

TO

M. J. S. L.
S. R. L.
K. E. L.

Contributors

Alex Avdeef • Syracuse University, Department of Chemistry, Syracuse, New York 13210

Josef Havel • Department of Analytical Chemistry, J. E. Purkyne University, Kotlarska 2, 611 37 Brno, Czechoslovakia

David J. Leggett • Dow Chemical USA, Texas Division, Freeport, Texas 77541

Peter M. May • Department of Applied Chemistry, University of Wales, Institute of Science and Technology, Cardiff CF1 3NU, United Kingdom

Milan Meloun • Department of Analytical Chemistry, University of Chemical Technology, Leninovo nam. 565, 532 10 Pardubice, Czechoslovakia

I. Nagypal • Computer Center, Institute of Inorganic and Analytical Chemistry, Lajos Kossuth University, H-4010, Debrecen, Hungary

D. D. Perrin • Medical Chemistry Group, The John Curtin School of Medical Research, The Australian National University, P.O. Box 334, Canberra City, A.C.T., 2601, Australia

Antonio Sabatini • Dipartimento di Chimica dell' Università di Firenze and I.S.S.E.C.C. – CNR, Florence, Italy

H. Stunzi • Universite de Neuchatel, Institut de Chimie, Avenue de Bellevaux 51, CH-2000 Neuchatel, Switzerland

Alberto Vacca • Dipartimento di Chimica dell' Università di Firenze and I.S.S.E.C.C. – CNR, Florence, Italy

David R. Williams • Department of Applied Chemistry, University of Wales, Institute of Science and Technology, Cardiff CF1 3NU, United Kingdom

L. Zekany • Computer Center, Institute of Inorganic and Analytical Chemistry, Lajos Kossuth University, H-4010, Debrecen, Hungary

Preface

This volume is concerned with methods that are available for the calculation of formation constants, in particular computational procedures. Although graphical methods have considerable value in the exploration of primary (raw) data they have been overtaken by computational methods, which, for the most part, take primary data and return the refined formation constants. Graphical methods are now considered complementary to these general computational procedures.

This volume brings together programs that span the lifetime of computer-assisted determination of formation constants. On one hand the reader will find listings of programs that are derived from LETAGROP (b.1961) and the GAUSS-G/SCOGS (b.1962) families. On the other hand programs are presented that are the newest members of the SCOGS lineage and from the on-going MINIQUAD series. One program is presented that describes a computational approach to the classical Hedstrom–Osterberg methods; another that takes care of electrode calibration in a simple yet rigorous manner.

Potentiometry and spectrophotometry are the most popular experimental techniques for equilibrium studies, and the programs in this volume reflect this. Four programs handle potentiometric data, two will process spectrophotometric data, and one makes use of both types of data separately or in combination.

This collection of programs also represents a collection of well-tested numerical algorithms for the general problem of data fitting. As such the volume will have value to users not necessarily devoted to full-time solution equilibrium studies. Each program employs a different, and in several instances novel, method for data input.

Every effort has been made to ensure that the codings are accurate. Of course, each program has been run on the author's computer. More importantly the programs have also been tested on a Honeywell 66/60 and on a Vax 11/780, under the FORTRAN 66 compiler. The input data, supplied by each author and used to test the programs, have been included. Minor changes were made to some of the programs to work around local features of the original machines. The program listings were printed out using a daisy-wheel printer after being down-loaded from the Vax 11/780 to a Dec-Mate word processor. The listings, so produced, clearly distinguish between a zero and upper case "oh"; D and "oh" are also obviously different.

I would like to take this opportunity to acknowledge the patience and forebearance of a number of people in what has turned out to be a longer than expected time to produce the volume: the contributors to this volume; Plenum as publishers; my wife.

It would be remiss of me not to acknowledge the positive guidance of Professor W.A.E. (Pete) McBryde during my graduate student days; but for him none of this would have been possible.

David J. Leggett

Contents

3 MAGEC: A Program for the Definite Calibration of the Glass Electrode

PETER M. MAY and DAVID R. WILLIAMS

4 SCOGS2: A Nonlinear Least-Squares Program for the Evaluation of Formation Constants of Metal Complexes

D.D. PERRIN and H. STUNZI

5 MINIQUAD and MIQUV: Two Approaches for the Computation of Formation Constants from Potentiometric Data

ALBERTO VACCA and ANTONIO SABATINI

6 SQUAD: Stability Quotients from Absorbance Data

DAVID J. LEGGETT

7 General Computer Programs for the Determination of Formation Constants from Various Types of Data

JOSEF HAVEL and MILAN MELOUN

8 PSEQUAD: A Comprehensive Program for the Evaluation of Potentiometric and/or Spectrophotometric Equilibrium Data Using Analytical Derivatives

L. ZEKANY and I. NAGYPAL

9 STBLTY: Methods for Construction and Refinement of Equilibrium Models

ALEX AVDEEF

1

The Determination of Formation Constants

An Overview of Computational Methods for Data Processing

DAVID J. LEGGETT

1. INTRODUCTION

The interaction of metal ions and ligands, in solution, giving rise to equilibrium mixtures of coordination complexes has been studied systematically since 1941,[1] although many references atest to 30 or more years of earlier research. Until about 1960 data obtained from these studies were subjected to graphical analysis. Many valuable contributions were made by Hazel and Francis Rossotti (their book[2] remains the definitive text on the subject) and by Lars Gunnar Sillen.[3] The impact of Sillen and co-workers on solution equilibria has not been restricted to innovative graphical procedures. In 1961[4] the first general computer program for the evaluation of formation constants from equilibrium data was announced. Many programs have followed from several independent research groups, including the LETAGROP[5] family of computational techniques.

This review will cover the literature up to the end of 1983. The programs will be classified according to the principal algorithm used. Several reviews, discussing the subject from different viewpoints, have been published.[6-11] Allen and McMeeking[12] have also reviewed the general topic of optimization methods.

The "best fit" of formation constants to the observed data may be expressed in terms of the least-squares criterion. The quantity U, the sum of squares of residuals, is defined according to

$$U = \sum_{k=1}^{\text{NPTS}} (Y_{\text{obs}}^k - Y_{\text{calc}}^k)^2 \tag{1}$$

DAVID J. LEGGETT • Dow Chemical U.S.A., Texas Division, Freeport, Texas 77541.

where Y^k_{obs} is the observed data and Y^k_{calc} is the value of the objective function defined in general terms as $f(\mathbf{x}, \mathbf{t})$. The parameter vector, $\mathbf{x}$, will always comprise the formation constants and may also include additional parameters relating to the model (e.g., molar absorptivities, enthalpy of formation, etc.) and/or parameters relating to the experimental conditions (e.g., analytical concentrations, electrode calibration constants, etc). This vector is commonly known as the dependent variable vector. The vector $\mathbf{t}$ contains the variables determined by the experimenter, known as the independent variables. Thus we seek to locate the set of dependent variables that will minimize U. Inherent to this approach is the need to define the equilibrium model before undertaking the minimization process. A general method that does not require predefinition of the model will be described in Section 4.

2. DIRECT SEARCH AND UNIVARIATE METHODS

Several such methods are available. They are, as the name suggests, techniques wherein one or more parameters are systematically varied. The function may be minimized by monitoring the impact of these parameter variations. The simplest search methods are Grid, Star, and Composite design; Fibonacci, and Golden Section. More elaborate methods include Simplex, Hooke and Jeeves, and interpolation techniques.

2.1. Grid, Star, and Composite Designs

These three related techniques involve the establishment of an n-dimensional figure, the minimum being located within the figure. The objective function is evaluated at each point of the figure. Although the Star and Composite design figures are more often encountered in experimental design protocols,[13] they can serve as useful alternatives to the simple Grid Search. While these methods are slow to converge, they provide a uniform overview of the hypersurface and hence can reveal multiple minima. Izatt et al.[14] employed a Grid Search of calorimetric data and used the coordinates of the located minimum (log β_1 and log β_2) as initial estimates for Davidon's [15] cubic interpolation. Nagano and Metzler adopted the same approach as Izatt in order to locate suitable starting estimates for their program PITMAP.[16] It was noted that the grid method required approximately 60 times more CPU resources than did Davidon's algorithm. Leussing[17] used a contracting grid approach in an ingenious manner to determine three formation constants for Schiff's base complexes with nickel. Natansohn et al.[18] employed the contracting grid method using a steepest descent alogorithm to reposition the grid under certain conditions. This publication describes an interesting approach to the application of experimental design techniques and error analysis to equilibrium modeling. The information given by Natansohn is presented in a manner that is not seen in the majority of publications relating to the estimation of formation constants. SQUAD[19] (described in chapter 6) has an option for a Star design search to obtain initial estimates for the sought parameters.

2.2. Fibonacci and Golden Section

The Fibonacci sequence, F_n, discovered by Leonardo of Pisa (1175–1230), has been used to minimize a unimodal function. The incorporation of this sequence into the minimization procedure is due to Kiefer.[20] The method is reliable for functions with a single minimum but may not converge for general functions.

The Golden Section is based on the fact that for any pair of Fibonacci numbers the relationship

$$\lim_{n \to \infty} \frac{F_n}{F_{n+1}} = \frac{\sqrt{5} - 1}{2} \tag{2}$$

is valid. Consequently the Golden Section algorithm is a limiting form of the Fibonacci search. The Golden Section method has been employed for the determination of a single constant.[21]

2.3. Hooke and Jeeves, Simplex

The method of Hooke and Jeeves involves a two-step procedure. In the first step each parameter axis is explored in order to locate the best parameter vector in the exploratory space. Next, the search is moved in the direction of each vector. These two steps alternate, the search length being reduced as the minimum is approached.

The Simplex algorithm, originated by Spendley *et al.*[22] and improved by Nelder and Mean,[23] is essentially the hypersurface-sensitive adaptive movement of an *n*-dimensional figure through hyperspace. The Simplex technique has been employed in several areas of chemistry, and has been admirably reviewed and demonstrated by Deming.[24] In terms of numerical minimization Nelder and Mead's version has been described by Walsh[25] as "...one of the most efficient pattern search methods currently available...." May and Williams use the Simplex algorithm as the major refinement method in their electrode calibration program MAGEC, described in a later chapter.

Although the computer code requirements for these search methods are small, making them ideal for mini- and microcomputer use, convergence is slow when close to the minimum.

2.4. Quadratic and Cubic Interpolation Methods

Consider the linear search problem as the minimization of the function $f(x)$ along the line

$$x = x_k + \lambda d \tag{3}$$

where x_k is the current point and d is the given search direction. Minimization may be achieved by minimizing the value of λ using quadratic interpolation[26] or cubic interpolation.[15] Powell's[26] method, i.e., quadratic fitting of the objective function at three values of λ followed by location of the turning point of the quadratic equation, provides values of λ that may be the minimum value. The process locates the minimum iteratively. Quadratic interpolation is a one-dimensional search and may be incorporated into other algorithms requiring an efficient linear search, such as Powell 65.[27]

Alternatively, the minimum along a descent line may be located by means of a cubic interpolation procedure.[15] Davidon's method requires that the function value and the gradient are available at two points which bracket the minimum. Similar to Powell's method, Davidon's technique can be used in algorithms that require a linear search. The procedure is an integral part of the variable metric minimization methods, the most widely used of which is the Davidon–Fletcher–Powell (DVP) method.[28]

The principal difference between the two approaches is that while the former uses function values only, the latter requires the evaluation of the function and its derivatives. Thus Davidon's approach is more elaborate than Powell's but usually locates the minimum of $f(x)$ in fewer iterations.

Powell's algorithm has been employed by Sabatini et al.[29] in MINIQUAD, where the normal equation matrix (factorized) is linearly optimized by the subroutine LIMIN.

The application of the DVP algorithm to the computation of force constants was illustrated by Gans.[30, 31] Subsequently, the DVP method has been incorporated into Gans and Vacca's program STEW.[32] The authors noted improved convergence speed and numerical stability compared to SCOGS[33] and LETAGROP[5]; but STEW was comparable to LEAST.[32] It was also demonstrated that the refinement of the logarithm of the formation constant is intrinsically ill conditioned.

In a study of aluminum fluoride interactions, using a fluoride ion-specific electrode, Baumann[35] processed the data using the DVP algorithm.

Christensen et al.[14] have adapted Davidon's basic variable metric algorithm to permit the evaluation of formation constants from calorimetric data.

2.5. Summary

The availability of computers, together with faster CPU, make the search and univariate methods described here increasingly attractive. Sections 2.1, 2.2, and 2.3 briefly describe the more popular search techniques that have the distinct advantage of being straightforward in their methods of operation. Other methods that warrant usage in the area of minimization by search techniques include the secant and dichotomous search methods; the algorithm of Davies, Swann, and Campey, all for single-variable optimizations[36]; and the direct search method of Chandler[37] as implemented by Kankare[38] in the spectrophotometric determination of formation constants.

The linear search techniques of Powell and Davidon are of a more complex nature and correspondingly more powerful. Jacoby, Kowalik, and Pizzo[39] suggest that Davidon's method be used when gradients are available, otherwise that of Davies, Swann, and Campey is recommended for the initial interpolation, switching to Powell's for subsequent moves.

3. LEAST-SQUARES METHODS

The function shown in equation (1) may be minimized by any of several direct search or descent methods. However, to do so takes no account of the fact that equation (1) is a sum of squares.

Examination of the literature shows that the least-squares approach has been commonly used to minimize equation (1).

3.1. Newton–Raphson

Elementary calculus states that by equating the first-order differential of $y = f(x)$, i.e., dy/dx, to zero and solving for x locates the exact minimum (or maximum) of the function. The generalization of this basic premise is the foundation of the gradient method. The n parameters will be manipulated as the column vector x; therefore the transpose of x is given by

$$x^T = [x_1, x_2, \ldots, x_n] \tag{4}$$

Changes in the values of the parameters are denoted by the vector x; its transpose being given by

$$\Delta x^T = [\Delta x_1, \Delta x_2, \ldots, \Delta x_n] \tag{5}$$

The Jacobian gradient vector, g, is given by the transpose of the gradient vector f defined as

$$g^T = \nabla f = \left[\frac{\partial f}{\partial x_1}, \frac{\partial f}{\partial x_2}, \cdots, \frac{\partial f}{\partial x_n} \right] \tag{6}$$

The symmetric matrix of second order partial differentials of f is known as the Hessian. It is denoted by H and defined as

$$H = \begin{bmatrix} \dfrac{\partial^2 f}{\partial x_1^2} & \dfrac{\partial^2 f}{\partial x_1 \, \partial x_2} & \cdots & \dfrac{\partial^2 f}{\partial x_1 \, \partial x_n} \\[2ex] \dfrac{\partial^2 f}{\partial x_2 \, \partial x_1} & & & \\[2ex] \vdots & & & \vdots \\[2ex] \dfrac{\partial^2 f}{\partial x_n \, \partial x_1} & \cdots & & \dfrac{\partial^2 f}{\partial x_n^2} \end{bmatrix} \tag{7}$$

In principle if the first- and second-order derivatives exist and can be evaluated then the Taylor expansion of the function may be employed to minimize the function. The first few terms of a multidimensional Taylor expansion are given by

$$f(x + \Delta x) = f(x) + g^T \Delta x + \tfrac{1}{2} \Delta x^T H \Delta x \ldots \tag{8}$$

At the minimum the Jacobian vector has all zero elements and the Hessian matrix will be positive definite. If the objective function $f(x)$ is everywhere quadratic then equation (8) represents the exact expansion of $f(x)$ and the minimum may therefore be located within the radius of convergence. For higher orders of $f(x)$ it is assumed that the approximation of $f(x)$ to a quadratic improves as the minimum is approached. In these situations an iterative procedure is used to locate the minimum.

For a position near the minimum truncating the Taylor expansion and rewriting in a different form gives

$$f(x_{min}) \cong f(x) + g^T \Delta x + \tfrac{1}{2} \Delta x^T H \Delta x \tag{9}$$

where $x_{min} = x + \Delta x$. Subsequently it may be shown[8, 40] that partial differentiation of equation (9) and setting the result to zero gives

$$\Delta x = -H^{-1} g \tag{10}$$

as the approximation to the required movement toward the minimum. Equation (10), when used directly to adjust initial estimates of x to those at the minimum, is the generalized Newton–Raphson method.[40]

The Newton–Raphson algorithm is used in the well-known LETAGROP family of programs.[5] Since the introduction of the program suite in 1961[4] numerous modifications have been made including the twist-matrix concept (*vrida* = *twist*) to overcome skewed error surfaces[5]; and the ability to refine parameters at two levels.[41] Leggett and McBryde[42] generalized an earlier FORTRAN version of PITMAP[43, 16] by including a general method of solving mass balance equations. PITMAP is based on the LETAGROP minimization strategy. Havel and Meloun present a FORTRAN implementation of the LETAGROP VRID algorithm in a later chapter. Meloun's adaptation of LETAGROP has also been used in SPEKFOT, a program by Suchanek and Sucha[44] that also uses the basic Newton–Raphson algorithm to solve the mass balance equations. The coding for this latter application is taken from Sayce's SCOGS.[33]

Sabatini and Vacca[34] described a new method for the least-squares refinement of formation constants, using either the Newton–Raphson (LEAST NR) or the Gauss–Newton (LEAST GN) algorithm. Hartley's search method[45] was used to optimize the shift calculations in both algorithms. It was found that for the test case LEAST NR version was about three times slower than LEAST GN. However, LEAST NR proved to be twice as fast as SCOGS,[33] which employs the Gauss–Newton algorithm with numerical differentiation. The major point made by Sabatini and Vacca was that the sum of the squared residuals of all mass balance equations be minimized with respect to the unknown parameters. This implies that [M] and [L] are treated as independent variables at the same level as the formation constants. Therefore errors associated with the measurement are carried by all mass balance equations rather than only by the mass balance for acid. Moreover, this approach allows analytical differentiation of the sum of squares function, considerably improving the overall rate of convergence.

The Newton–Raphson algorithm has been used widely to calculate species concentrations that relate to the mass balance equations encountered in formation constant refinement problems.

In 1961 Tobias and Hugus[46] employed the Newton–Raphson algorithm in one of the first Gauss–Newton driven programs for formation constant refinement. Subsequently, Tobias and Yasuda[47] generalized the Newton–Raphson and Gauss–Newton algorithms into a program known as GAUSS G. These authors also used a program, MAPZ, based on the pit mapping procedure of Sillen.[4] Since 1963 numerous applications have been reported of the Newton–Raphson algorithm being used to solve the mass balance equations. The subroutine COGSNR (sometimes refered to, incorrectly, as COGS,[48]) used by Sayce[33] in SCOGS to compute species concentrations may be found in many subsequent programs. The coding of COGSNR owes much to the original work of Tobias and Yasuda.[47] Most recently Motekaitis[49, 50] and Atkins[51] have embodied the Newton–Raphson approach in conventional ways for the solution of mass balance equations.

Ting Po I and Nancollas[52] have provided a detailed examination of various algorithmic variations to the original Newton–Raphson method. They compared the computational efficiencies of each modification concluding that matrix scaling and Hartley's method[46] increased the convergence rate.

The computational efficiency of several variants of the Newton–Raphson algorithm as well as other numerical techniques have been determined on a number of equilibrium systems.[53] It was found that the implementation of the Newton–Raphson algorithm as coded in MINIQUAD[29] was the most efficient and least dependent upon the nature of the equilibrium system.

Despite the work of Sillen *et al.*, Sabatini and Vacca, and Havel the algorithm has not found the popularity that is seen with the Gauss–Newton method. Jacoby, Kowalik, and Pizzo[39] suggest that the algorithm is well suited to situations where the objective function is nearly quadratic and/or good initial estimates are available. They also recommend that analytical expressions for the first and second derivatives should be available, and that some type of acceleration modification be used to avoid the need to update the Hessian every iteration—observations that are similar to those of with Sabatini and Vacca[34] and Nancollas.[52]

3.2. Gauss–Newton

The general definition of the objective function $f(x, t_m)$ allows an alternative definition of the least-squares criterion. The total residual is a function of the individual data errors

$$U = g\{f(x, t_1), f(x, t_2), \ldots, f(x, t_m)\} \tag{11}$$

for m observations. Replacing the function g by a sum of squares and $f(x, t_k)$ by $f_k(x)$ gives

$$U = \sum_{k=1}^{m} [f_k(x)]^2 \tag{12}$$

Let us define a new matrix, G, that gives the variation of each function $f_k(x)$ with each parameter:

$$G = \begin{bmatrix} \dfrac{\partial f_1}{\partial x_1} & \dfrac{\partial f_1}{\partial x_2} & \cdots & \dfrac{\partial f_1}{\partial x_n} \\[2ex] \dfrac{\partial f_2}{\partial x_1} & & & \\[2ex] \vdots & & \vdots & \\[2ex] \dfrac{\partial f_m}{\partial x_1} & \cdots & & \dfrac{\partial f_m}{\partial x_n} \end{bmatrix} = \dfrac{\partial f_k(x)}{\partial x_j} \tag{13}$$

The Jacobian g may be obtained by partial differentiation of equation (12):

$$\frac{\partial E}{\partial x_i} = \sum_{k=1}^{m} 2[f_k(x)] \frac{\partial f_k(x)}{\partial x_i} \tag{14}$$

A second partial differentiation of equation (12) gives

$$\frac{\partial^2 E}{\partial x_i\, \partial x_j} = 2 \sum_{k=1}^{m} \frac{\partial [f_k(x)]}{\partial x_i} \frac{\partial [f_k(x)]}{\partial x_j} + 2 \sum_{k=1}^{m} f_k(x) \frac{\partial^2 [f_k(x)]}{\partial x_i\, \partial x_j} \tag{15}$$

Assuming that the second-order term may be neglected, equation (15) is seen to be an approximation of the Hessian:

$$H = 2A^T A \tag{16}$$

These definitions for the Hessian, equation (16), and for the Jacobian, equation (14), may be substituted into equation (10):

$$\Delta x = [A^T A]^{-1} A^T f \tag{17}$$

This provides the basic algorithm of the Gauss–Newton method.(40)

   This unmodified Gauss–Newton method has been widely used in the area of formation constant determination. The introduction of this algorithm may be traced back to Tobias and Yasuda.[46,47] GAUSS G[47] was the first general program that did not require recoding of sections of the program when different models were fitted to the data. GAUSS G also introduced COGS, the fore-runner of COGSNR.[33] Perrin and Sayce[54] generalized GAUSS G further paying particular attention to the initial processing of the raw data. Subsequently, GAUSS was further generalized and re-named SCOGS.[33] The brute force algorithm used to solve the mass balance equations (COGS) was replaced by the more rapid, generalized Newton–Raphson technique (COGSNR). Straightforward techniques were used in SCOGS to (i) avoid overshifting of the refining parameters during the minimization process; (ii) ensure that numerical differentiation would provide "significant" values for the differentials. The program is still in frequent use worldwide either in its original form (algorithmically speaking) or in variously modified forms. The general routine COGSNR has found its way into a large number of programs that do not necessarily use the Gauss–Newton technique to refine the formation constants. The latest version of SCOGS will be discussed in chapter 4. The original coding of SCOGS was used by Leggett as the basis of SQUAD,[55] which was at that time a hybrid of SCOGS and PITMAP.[43] Further use of that version of SQUAD proved unsatisfactory and the hybrid was abandoned in favor of a spectrophotometric version[19] of SCOGS. The current version of SQUAD is presented in chapter 6. The SCOGS approach was also applied, with encouraging results, to polarographic data, giving rise to the program POLAG.[56]

Probably the most sophisticated of all current programs that use the Gauss–Newton method for minimization is MINIQUAD.[29] Three major points serve to illustrate this statement. First the program is centered around a version of Gauss–Newton wherein convergence is guaranteed by parameter shift optimization using Powell's linear minimization algorithm. Second, all differentials are calculated from analytical derivatives and all unknown quantities (including the unknown free concentrations) are refined. These two properties of the program are linked and derive from earlier work.[34] Third, the formation constants β_j, rather than log β_j, are refined. Citations to this program indicate that it (and a later version MINIQUAD-75) is achieving worldwide use. Sabatini and Vacca present two new programs, related to MINI-QUAD, in chapter 5.

A series of programs by Gaizer [57-60] uses the Gauss–Newton algorithm to process spectrophotometric[57,60] or potentiometric data.[58,59] SPEF-3[57] will handle a limited number of wavelengths but for any general mononuclear complex; MINI-SPEF[60] is designed to run on 16-kbyte computers, is written in BASIC, and can only handle one wavelength. PH-POT[58] is similar in many respects to SPEF-3; MINI-POT[59] is the 16-kbyte version of PH-POT.

Raymond et al.[61] has used the Gauss–Newton algorithm in a weighted least-squares program (unpublished), the weighting scheme originating from Avdeff (see chapter 9).

Seward[62] has employed the Gauss–Newton method in a program that enables the calculation of formation constants from solubility data at elevated temperatures. Analytical differentials are used, thereby ensuring rapid convergence.

Feldberg et al.[63] employed a Gauss–Newton-based algorithm in their analysis of spectrophotometric data obtained from $PdBr_nCl_{4-n}$ data. Encountering highly skewed, narrow error surfaces these authors developed a simple and ingenious linear search method to avoid stalling of the iterative process.

3.3. Levenberg and Marquardt

It has been shown by Levenberg[64] and more recently by Marquardt[65] that the steepest descent algorithm is more rapidly convergent away from the minimum while Gauss–Newton iteration is to be preferred as the minimum is approached. These observations were implemented by incorporation of the term into the Gauss–Newton least-squares increment:

$$x = -[\lambda I + A^T A]^{-1} A^T f \tag{18}$$

where λ is an adjustable parameter and the identity matrix is used to scale the parameters and is successively updated. The procedure ensures that as λ tends to infinity the steepest descent direction is followed; as λ tends to 0 the standard Gauss–Newton direction is followed. Difficulties unresolved by this approach lie mainly in the area of choosing a suitable value for λ. Fletcher[66] has proposed a self-adaptive strategm for the selection of λ. The reduction in the sum of squares when using the Marquardt parameter is compared to a first-order prediction. If the reductions are similar then the minimization progresses with the Gauss–Newton increment; dissimilar reductions imply that a steepest descent technique is more appropriate and λ is modified accordingly.

Zuberbuhler and Kaden have employed the Marquardt modification of the Gauss–Newton algorithm to good effect in a number of programs. Refinement of spectrophotometric data has been achieved using a program designed to fit an HP9821 desk-top computer.[67] The calculation procedure for obtaining molar absorptivities was based on that used in SQUAD.[69] The same group subsequently developed ELORAMA, written in BASIC, to run on an APPLE II computer.[68] This program operates, in principle, in the same manner as their earlier program. However, a number of devices are used to work around the problems associated with single precision arithmetic. The algorithms and coding used for ELORAMA may prove to be important as more research groups add this type of computer to the inventory of "standard" laboratory equipment. MARFIT, designed to process potentiometric data on an HP9821 computer, also employed the Marquardt algorithm.[69] Species concentrations were calculated using the Newton–Raphson technique. However, the program was cumbersome to use as it required recoding for each new model tested. This difficulty was overcome in the latest program published by this group. TITFIT[70] also used the Marquardt algorithm and is coded to use analytical derivatives. The use of analytical derivatives is handled differently from MINIQUAD in that only the titrant volume is minimized, Zuberbuhler and Kaden favoring an approach suggested by Nagypal.[71] A novel addition to the program is its ability to generate in-line code optimized for each new model tested. This facility, it is claimed, is faster than writing a general algorithm.

Arena et al. [72] have developed a program that is capable of refining all relevant parameters in an acid–base titration. Parameters refined include E^0, liquid junction potential, all concentrations, and the Nernst slope. The program, ACBA, also employs analytical derivatives following the same general approach as Zuberhuler and Kaden.[70] However, in contrast, Arena et al. resort to eigenvalues and eigenvectors[73] in order to overcome ill-conditioning of the the normal equation matrix.

Gans et al.[74] tested two separate versions of a successor to MINIQUAD.[29] The first approach removed the unknown free concentrations from the refinement step, calculating them independently at each titration point. The second approach abandoned the linear optimization of shifts replacing that procedure with the Marquardt algorithm. Fletcher's method for selection of the Marquardt parameter is followed.[66] Neither approach was entirely satisfactory and the authors therefore developed a hybrid of the two techniques, known as MINIQUAD 75. This program possesses the initial refinement power of the first approach complemented by the convergence power of the Marquardt algorithm.

Alcock et al.[75] have published a general program, DALSFEK, that is capable of refining both spectrophotometric and potentiometric data. Again the Marquardt algorithm is used, but in this situation the sum of squares of the errors in the absorbances or the measured emf's is minimized. The mass balance equations are solved for each titration point using the same algorithm as is used to refine the formation constants. Similarly to other programs of this type, analytical derivatives are available. The reader is referred to the authors' recent book,[76] in which excellent accounts are given of modeling equilibrium systems, together with a extended description of DALSFEK.

3.4. Objective Function Differentiation

Most gradient techniques require that the objective function be differentialable at least once. Analytical differentiation may, or may not, be possible depending on the exact form of the function. Numerical differentiation is a practical alternative, although it is less accurate and requires more computational effort. It may be achieved using Newton's forward or backward formula or by Sterling's central difference formula.[39] Newton's formulas require $N + 1$ function evaluations for N parameters minimized whereas Sterling's technique requires $2N + 1$ evaluations.

Function evaluation during refinement of formation constants involves the solution of the mass balance equations, usually at a large number of pHs. This is especially true for potentiometric data. Iterative algorithms, such as Newton–Raphson, are employed for this purpose and can account for a significant fraction of the total CPU resources used during formation constant refinement. Consequently, programs that are formulated to permit analytical differentiation of the objective function would be expected to converge more rapidly than equivalent programs using numerical differentiation.

4. OTHER METHODS

4.1. Curve Fitting

Bond[6] has reviewed the application of polynomial procedures to the determination of formation constants, in particular cautioning against the overzealous use of such procedures. Momoki[77] investigated the refinement of polarographic data using both a standard Gaussian approach and a method referred to as the $1/F_2^0$ method, an approach relying on fitting polynomials by least squares. Meites,[78] in his on-going work on multiparametric curve-fitting,[79] describes the fitting of a polynomial function to polarographic data.

4.2. HOSK Method

In 1955 Hedstrom[80] presented a number of relationships that applied to the equilibrium reactions between two reactants, A and B. In describing the principle of the method Hedstrom stated "No *a priori* assumption, whatsoever, is made as to the compositions or abundance of the complexes formed." The method permitted the evaluation of [A] and [B], given the initial concentrations of A and B. The salient equation arises from Maxwell's relations by way of the Jacobian determinant

$$\log [A] = \left[\int_0^B \partial \log (\partial A/[B]/B)_B \, dB \right]_A \tag{19}$$

A graphical approach was employed to integrate the equations.

Osterberg[81] demonstrated that the technique could be used to determine the free ligand concentration in a conventional metal ligand titration. Freeman and Martin[82] applied the principle of Osterberg's approach to copper amino acid equilibrium titrations. In this study relationships were developed for free ligand concentrations and for the variation of the number of species per mole of metal ions. This graphical procedure was used as a prelude to refinement of their data using LETAGROP VRID.

The basic technique was automated by Sarkar and Kruck[83]—hence the acronym HOSK (Hedstrom–Osterberg–Sarkar–Kruck). Sarkar and Kruck derived a valuable extension to the method by showing that the original method could be extended to include the calculation of free metal along with free ligand concentrations. The importance of this development is that all free concentrations are available *before* models are fitted to the data. This is in direct contrast to the majority of other methods, where a model is *first* proposed before fitting to the data.

The HOSK method revived interest in this elegant approach by Hedstrom. McBryde[84] presented a alternative derivation for Osterberg's key relationships revealing inter-relationships among the functions derived and applied in terms of other familiar quantities. Subsequently, Guevremont and Rabenstein[85] and Field and McBryde[86] evaluted the HOSK method from the data gathering and data analysis point of view. They concluded, independently, that the method is quite sensitive to experimental error. Unless data of the highest precision are used the method can give rise to misleading results. Guevremont and Rabenstein described a computer-controlled titration system that allows for the gathering of high-precision data.

Avdeef and Raymond[87] in a general derivation of Hedstrom's original equation have shown that for the system (M, L, H, m, l, h) six Jacobian unit determinants can be derived. It was shown that the elements of the inverse of the Jacobian matrix $J[(M,L,H)/(\ln m, \ln l, \ln h)]$ are the partial derivatives of the mass balance equations. Avdeef presents the program STBLTY, which embodies the principles of the Hedstrom method, in chapter 9.

Vadasdi[88] has shown that the starting point for Hedstrom's method may be used to determine the composition of species present in an equilibrium system, given the free concentration of each component.

5. SUMMARY

This review of computational methods used to extract formation constants from primary data has shown that this area of research remains active and healthy. The Marquardt variation of the Gauss–Newton algorithm seems to be the most favored numerical approach. The trend towards programs that use analytical differentiation is much welcomed. Attempts to down-load programs to laboratory microcomputers have met with success and have spawned some ingenious matrix algebra to work around problems of lack of precision. With the remarkable advances in the computer industry it remains to be seen how long these mathematical maneuvers are necessary.

These advances in various aspects of solution equilibria are paving the way for a totally automatic species selector. Robust algorithms, coding designed for personal computers, and computer-controlled titration equipment are all necessary components of a species selector. It now remains to design the artificial intelligence algorithm to oversee the task of deriving the unique model to fit a set of data. Realistically this is no small problem to solve; however, we now have the raw materials available for the task.

6. REFERENCES

1. J. Bjerrum, *Metal Ammine Formation in Aqueous Solution*, Haase, Copenhagen (1941, reprinted 1957).
2. F. J. C. Rossotti and H. S. Rossotti, *The Determination of Stability Constants*, McGraw-Hill, New York (1961).
3. L. G. Sillen, in *Coordination Chemistry, A.C.S. Monograph* (A. E. Martell, ed.), Vol. 8, pp. 491–541, Van Nostrand Reinhold Co., New York (1972).
4. D. Dyrssen, N. Ingri, and L. G. Sillen, "Pit-Mapping"—A General Approach for Computer Refinement of Equilibrium Constants, *Acta Chem. Scand.* **15**, 694–696 (1961).
5. N. Ingri and L. G. Sillen, High-Speed Computers as a Supplement to Graphical Methods. IV. An ALGOL Version of LETAGROP VRID, *Arkiv Kemi* **23**, 97–121 (1964).
6. A. M. Bond, Some Suggested Calculation Procedures and the Variation in Results Obtained from Different Calculation Methods for Evaluation of Concentration Stability Constants of Metal Ions in Aqueous Solution, *Coord. Chem. Rev.* **6**, 377–405 (1971).
7. F. J. C. Rossotti, H. S. Rossotti, and R. J. Whewell, The Use of Electronic Computing Techniques in the Calculation of Stability Constants, *J. Inorg. Nucl. Chem.* **33**, 2051–2065 (1971).
8. P. Gans, Numerical Methods for Data-Fitting Problems, *Coord. Chem. Rev.* **19**, 99–124 (1976).
9. F. Gaizer, Computer Evaluation of Complex Equilibria, *Coord. Chem. Rev.* **27**, 195–222 (1979).
10. P. Gans, Computer-Assisted Methods for the Investigation of Solution Equilibria, *Adv. Mole. Relax. Interaction Proc.* **18**, 139–148 (1980).
11. D. J. Leggett, The Determination of Stability Constants: A Review, *Am. Lab.* **14**(1), 29–35 (1982).
12. G. C. Allen and R. F. McMeeking, Deconvolution of Spectra by Least-Squares Fitting, *Anal. Chim. Acta.* (Computer Techniques and Optimization) **103**, 73–108 (1978).
13. G. J. Hahn in *Statistics* (R. F. Hirsch, ed.), 1977 Eastern Analytical Symposium, The Franklin Institute Press, Philadelphia, Pennsylvania (1978).

14. R. M. Izatt, D. Eatough, R. L. Snow, and J. J. Christensen, Computer Evaluation of Entropy Titration Data. Calorimetric Determination of Log β_i, ΔH_i, and ΔS_i Values for the Silver(I) and Copper(II) Pyridine Systems, *J. Phys. Chem.* **72**, 1208–1213 (1968).

15. W. C. Davidon, Variable Metric Method for Minimization, AEC(US) Research and Development ANL-5990, Argonne National Laboratory, Argonne, Illinois, 1959; See also ANL-5990 (Rev. 2) (1966).

16. K. Nagano and D. E. Metzler, Machine Computation of Equilibrium Constants and Plotting of Spectra of Individual Ionic Species in the Pyridoxal–Alanine System, *J. Am. Chem. Soc.* **89**, 2891–2900 (1967).

17. D. L. Leussing, Schiff Base Complexes. A Numerical Study of the Nickel(II)-Pyruvate-Glycinate System Using a High Speed Computer, *Talanta* **11**, 189–201 (1964).

18. S. Natansohn, J. I. Krugler, J. E. Lester, M. S. Chagnon, and R. S. Finocchiaro, Stability Constants of Complexes of Molybdate and Tungstate Ions with *o*-Hydroxy Aromatic Ligands, *J. Phys. Chem.* **84**, 2972–2980 (1980)

19. D. J. Leggett and W. A. E. McBryde, General Computer Program for the Computation of Stability Constants from Absorbance Data, *Anal. Chem.* **47**, 1065–1070 (1975).

20. H. J. Kieffer, Sequential Minimax Search for a Maximum, *Proc. Am. Math. Soc.* **4**, 269–282 (1970).

21. P. Gans and H. M. N. H. Irving, The Calculation of Stability Constants of Weak Complexes from Spectrophotometric Data, *J. Inorg. Nucl. Chem.* **34**, 1885–1890 (1972).

22. W. Spendley, G. R. Hext, and F. R. Himsworth, Sequential Applications of Simplex Designs in Optimization and Evolutionary Operation, *Technometrics* **4**, 441–461 (1962).

23. J. A. Nelder and R. Mead, A Simplex Method for Function Mimimization, *Comput. J.* **7**, 308–313 (1965).

24. S. N. Deming and L. R. Parker, A Review of Simplex Optimization in Analytical Chemistry, *Crit. Rev. Anal. Chem.* **7**, 187–202 (1978).

25. G. R. Walsh, *Methods of Optimization*, Wiley, New York (1975).

26. M. J. D. Powell, An Efficient Method for Finding the Minimum of a Function of Several Variables without Calculating Derivatives, *Comput. J.* **7**, 155–162 (1964).

27. M. J. D. Powell, A Method for Minimizing a Sum of Squares of Non-Linear Functions Without Calculating Derivatives, *Comput. J.* **7**, 303–307 (1965).

28. R. Fletcher and M. J. D. Powell, A Rapidly Convergent Descent Method for Minimization, *Comput. J.* **6**, 163–168 (1963).

29. A Sabatini, A. Vacca, and P. Gans, MINIQUAD—A General Computer Programme for the Computation of Formation Constants from Potentiometric Data, *Talanta* **21**, 53–77 (1974).

30. P. Gans, A New Method for Computing Vibrational Force Constants, *Chem. Commun.* 1504–1505 (1970).

31. P. Gans, Force Constant Computations. Part II. Some Model Calculations to Test the Fletcher–Powell Method and to Analyse the Ill-Conditioned Force Constant Refinement, *J. Chem Soc. (A)* 2017–2024 (1971).

32. P. Gans and A. Vacca, Application of the Davidon–Fletcher–Powell Method to the Calculation of Stability Constants, *Talanta* **21**, 45–51 (1974).

33. I. G. Sayce, Computer Calculations of Equilibrium Constants of Species Present in Mixtures of Metal Ions and Complexing Agents, *Talanta* **22**, 1397–1421 (1968).

34. A. Sabatini and A. Vacca, A New Method for Least Squares Refinement of Stability Constants, *J. Chem. Soc. Dalton* 1693–1698 (1972).

35. E. W. Baumann, Determination of the Stability Constants of Hydrogen and Aluminum Fluorides with a Fluoride Selective Electrode, *J. Inorg. Nucl. Chem.* **31**, 3155–3162 (1969).

36. W. H. Swann, Report on the Development of a New Direct Search Method of Optimization, I.C.I. Ltd., Central Instrument Laboratory Research Note 64/3 (1964).

37. J. P. Chandler, Minimum of a Function of Several Variables, Program 66.1, QCPE, Indiana University, Bloomington, Indiana, 1966.

38. J. J. Kankare, Computation of Equilibrium Constants for Multicomponent Systems from Spectrophotometric Data, *Anal. Chem.* **42**, 1322–1326 (1970).

39. S. L. S. Jacoby, J. S. Kowalik, and J. T. Pizzo, *Iterative Methods for Nonlinear Optimization Problems*, Prentice-Hall, Englewood Cliffs, New Jersey (1972).

40. P. R. Adby and M. A. H. Dempster, *Introduction to Optimization Methods*, Chapman and Hall, London (1974).

41. L. G. Sillen and B. Warnqvist, High-speed Computer as a Supplement to Graphical Methods. VI. A Strategy for Two-level Adjustment of Common and "Group" Parameters. Some Features that Avoid Divergence, *Arkiv Kemi* **31**, 315–339 (1968).

42. D. J. Leggett and W. A. E. McBryde, Picoline-2-Aldehyde Thiosemicarbazone: The Dissociation Constants and Reaction with Various Metals, *Talanta* **21**, 781–789 (1974).

43. J. A. Thomson, Computer-Assisted Studies on Quinizarin-2-Sulfonic Acid and its Complexation with Iron(III), Ph.D. thesis, University of Waterloo, Waterloo, Ontario, Canada (1970).

44. M. Suchanek and L. Sucha, The Use of the Program SPEKTFOT for the Computation of the Equilibrium Constants and the Molar Absorption Coefficients of Substances in Solution, *Sbornik Vys. Sk. Chemickotechnol. Praze* **13**, 41–57 (1978).

45. H. O. Hartley, The Modified Gauss–Newton Method for the Fitting of Non-Linear Regression Functions by Least Squares, *Technometrics* **3**, 269–280 (1961).

46. R. S. Tobias and Z. Z. Hugus, Least Squares Computer Calculations of Chloride Complexing of Tin(II), and the Validity of the Ionic Medium Method, *J. Phys. Chem.* **65**, 2165–2169 (1961).

47. R. S. Tobias and M. Yasuda, Computer Analysis of Stability Constants in Three-Component Systems with Polynuclear Complexes, *Inorg. Chem.* **2**, 1307–1310 (1963).

48. D. D. Perrin and I. G. Sayce, Computer Calculations of Equilibrium Concentrations in Mixture of Metal Ions and Complexing Species, *Talanta* **14**, 833–842 (1967).

49. R. J. Motekaitis and A. E. Martell, Program PKAS: A Novel Algorithm for the Computation of Successive Protonation Constants, *Can. J. Chem.* **60**, 168–173 (1982).

50. R. J. Motekaitis and A. E. Martell, BEST—A New Program for Rigorous Calculation of Equilibrium Parameters of Complex Multicomponent Systems, *Can. J. Chem.* **60**, 2403–2409 (1982).

51. C. E. Atkins, S. E. Park, J. A. Blaszak, and D. R. McMillin, A Two-Level Approach to Deconvoluting Absorbance Data Involving Multiple Species. Applications to Copper Systems, *Inorg. Chem.* **23**, 569–572 (1984).

52. Ting-Po I and G. H. Nancollas, EQUIL—A General Computational Method for the Calculation of Solution Equilibria, *Anal. Chem.* **44**, 1940–1950 (1972).

53. D. J. Leggett, Machine Computation of Equilibrium Concentrations-Some Practical Considerations, *Talanta* **24**, 535–542 (1977).

54. D. D. Perrin and I. G. Sayce, Stability Constants of Polynuclear Mercaptoacetate Complexes of Nickel and Zinc, *J. Chem. Soc.(A)* 82–89 (1967).

55. D. J. Leggett and W. A. E. McBryde, Metal Ion Interactions of Picoline-2-Aldehyde Thiosemicarbazone, *Talanta* **22**, 1005–1011 (1975).

56. D. J. Leggett, POLAG—A General Computer Program to Calculate Stability Constants from Polarographic Data, *Talanta* **27**, 787–793 (1980).

57. F. Gaizer and M. Mate, Computerized Calculations of Complex Equilibria, I. A. General Program for the Evaluation of Spectrophotometric Equilibrium Measurements, *Acta Chim. Sci. Hung.* **103**, 355–363 (1980).

58. F. Gaizer, Computerized Calculations of Complex Equilibria, II. The Calculation of the Protonation Constants of Polyfunctional Ligands and the Stability Constants of Metal Complexes from Potentiometric Data, *Acta Chim. Sci. Hung.* **103**, 397–403 (1980).

59. F. Gaizer and A. Puskas, A Desk-Computer Program for Calculation of the Parameters of Acid–Base Titration Curves and Protonation or Metal-Complex Stability Constants from Potentiometric Data, *Talanta* **28**, 565–573 (1981).

60. F. Gaizer and A. Puskas, Desk-Computer Program for Evaluation of Complex Equilibria from Spectrophotometric Data, *Talanta* **28**, 925–929 (1981).

61. W. R. Harris, C. J. Carrano, S. R. Cooper, S. R. Sofen, A. E. Avdeef, J. V. McArdle, and K. N. Raymond, Coordination Chemistry of Microbial Iron Transport Compounds. 19. Stability Constants and Electrochemical Behavior of Ferric Enterobactin and Model Complexes, *J. Am. Chem. Soc.* **101**, 6097–6104 (1979).

62. L. P. Aldridge and T. M. Seward, HACK—A FORTRAN Program for Calculating Equilibrium Con-

stants from Solubility Data at Elevated Temperatures, Department of Scientific and Industrial Research, Chemistry Division, Lower Hutt, New Zealand, Report C.D. 2227 (1976).

63. S. Feldberg, P. Klotz, and L. Newman, Computer Evaluation of Equilibrium Constants from Spectrophotometric Data, *Inorg. Chem.* **11**, 2860–2865 (1972).

64. K. A. Levenberg, A Method for the Solution of Certain Nonlinear Problems in Least Squares, *Quart. Appl. Math.* **2**, 164–168 (1944).

65. D. W. Marquardt, An Algorithm for Least Squares Estimation of Nonlinear Parameters, *J. Soc. Indust. Appl. Math.* **11**, 431–441 (1963).

66. R. Fletcher, A Modified Marquardt Subroutine for Non-Linear Least Squares, A.E.R.E. (U.K.) Harwell Technical Report No. R6799 (1971).

67. A. D. Zuberbuhler and T. A. Kaden, Handling of Electronic Absorption Spectra with a Desk-Top Computer, *Talanta* **26**, 1111–1118 (1979).

68. H. Gampp, M. Maeder, and A. D. Zuberbuhler, General Non-Linear Least-Squares Program for the Numerical Treatment of Spectrophotometric Data on a Single-Precision Game Computer, *Talanta* **27**, 1037–1045 (1980).

69. H. Gampp, M. Maeder, A. D. Zuberbuhler, and T. A. Kaden, Microprocessor-Controlled System for Automatic Acquisition of Potentiometric Data and their Non-Linear Least-Squares Fit in Equilibrium Studies, *Talanta* **27**, 513–518 (1980).

70. A. D. Zuberbuhler and Th. A. Kaden, TITFIT, A Comprehensive Program for Numerical Treatment of Potentiometric Data by Using Analytical Derivatives and Automatically Optimized Subroutines with the Newton-Gauss-Marquardt Algorithm, *Talanta* **29**, 201–206 (1982).

71. I. Nagypal, I. Paka, and L. Zekany, Analytical Evaluation of the Derivatives Used in Equilibrium Calculations, *Talanta* **25**, 549–550 (1978).

72. G. Arena, E. Rizzarelli, and S. Sammartano, A Non-Linear Least-Squares Approach to the Refinement of all Parameters Involved in Acid-Base Titrations, *Talanta* **26**, 1–14 (1979).

73. J. Greenstadt, On the Relative Efficiencies of Gradient Methods, *Math Comput.* **21**, 360–371 (1967).

74. P. Gans, A. Sabatini, and A. Vacca, An Improved Computer Program for the Computation of Formation Constants from Potentiometric Data, *Inorg. Chim. Acta* **18**, 237–239 (1976).

75. R. M. Alcock, F. R. Hartley, and D. E. Rogers, A Damped Non-Linear Least-Squares Program (DALSFEK) for the Evaluation of Equilibrium Constants from Spectrophotometric and Potentiometric Data, *J. C. S. Dalton* 115–123 (1978).

76. F. R. Hartley, C. Burgess, and R. Alcock, *Solution Equilibria*, Ellis Horwood, J. Wiley, London (1980).

77. K. Momoki, H. Sato, and H. Ogawa, Calculation of Successive Formation Constants from Polarographic Data Using a High-Speed Computer, *Anal. Chem.* **39**, 1072–1079 (1967).

78. L. Meites, Automatic Classification of Chemical Behaviour by Sequential Hypothesization and Multiparametric Curve-Fitting. III. Fully Computerized Elucidation of Polarographic Data on Stepwise Complex Formation, *Talanta* **22**, 733–738 (1975).

79. L. Meites, Some New Techniques for the Analysis and Interpretation of Chemical Data, *Crit. Rev. Anal. Chem.* **8**, 1–53 (1979).

80. B. Hedstrom, Equilibria in Systems with Polynuclear Complex Formation, *Acta Chem. Scand.* **9**, 613–621 (1955).

81. R. Osterberg, The Copper(II) Complexity of O-Phosphorylethanolamine, *Acta Chem. Scand.* **14**, 471–485 (1960).

82. H. C. Freeman and R.-P. Martin, Potentiometric Study of Equilibria in Aqueous Solution Between Copper(II) Ions, L (or D)-Histidine and L-Threonine and Their Mixtures, *J. Biol. Chem.* **244**, 4823–4830 (1969).

83. B. Sarkar and T. P. A. Kruck, Theoretical Considerations and Equilibrium Conditions in Analytical Potentiometry. Computer Facilitated Mathematical Analysis of Equilibria in a Multicomponent System, *Can. J. Chem.* **51**, 3541–3548 (1973).

84. W. A. E. McBryde, On An Extension of the Use of pH-Titrations for Determination of Free Metal and Free Ligand Concentrations During Metal Complex Formation, *Can. J. Chem.* **51**, 3572–3576 (1973).

85. R. Guevremont and D. L. Rabenstein, A Study of the Osterberg–Sarkar–Kruck Method for Evaluating

Free Metal and Free Ligand Concentrations in Solutions of Complex Equilibria, *Can. J. Chem.* **55**, 4211–4221 (1977).

86. T. B. Field and W. A. E. McBryde, Determination of Stability Constants by pH Titrations: A Critical Examination of Data Handling, *Can. J. Chem.* **56**, 1202–1211 (1978).

87. A. Avdeef and K. N. Raymond, Free Metal and Free Ligand Concentrations Determined From Titrations Using Only a pH Electrode. Partial Derivatives in Equilibrium Studies, *Inorg. Chem.* **18**, 1605–1611 (1979).

88. K. Vadasdi, On Determining the Composition of Species Present in a System from Potentiometric Data, *J. Phys. Chem.* **78**, 816–820 (1974).

2

Strategies for Solution Equilibria Studies with Specific Reference to Spectrophotometry

JOSEF HAVEL and MILAN MELOUN

1. ELUCIDATION OF THE EQUILIBRIUM MODEL

1.1. Introduction

During equilibrium studies one attempts to solve the problem of designing the best model for the system, i.e., to determine the number of complex species $M_pL_qH_r$ in solution, the stoichiometry, stability constants, and molar absorptivities at selected wavelengths. The overall formation constant of the complex $M_pL_qH_r$ is defined by

$$pM + qL + rH \rightleftharpoons M_pL_qH_r \tag{1}$$

where

$$\beta_{pqr} = [M_pL_qH_r]/[M]^p[L]^q[H]^r \tag{2}$$

Spectrophotometry and potentiometry, as well as other instrumental techniques, have proved useful for this type of investigation. An overall picture of the experimental approach is shown in Figure 1. These techniques are used to assist in (a) the selection of the best of several plausible equilibrium models; (b) the estimation of stability constants and other parameters related to the particular equilibrium model; (c) the calculation of distribution diagrams for all species over the range of the experimental conditions.

JOSEF HAVEL and MILAN MELOUN • Department of Analytical Chemistry, J. E. Purkyne University, Kotlarska 2, 611 37 Brno; Department of Analytical Chemistry, University of Chemical Technology, Leninovo nam. 565, 532 10 Pardubice, Czechoslovakia.

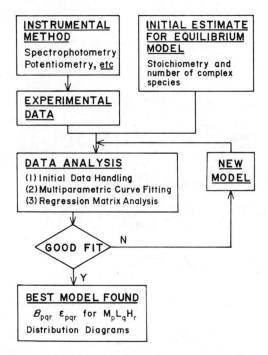

FIGURE 1. Overall scheme for equilibrium model determination.

The first two objectives are sought in a cyclic manner until the equilibrium model, together with all relevant parameters, is at least partially completed. In practice an experimental procedure is adopted and the data obtained are processed in a manner consistent with the initial estimate of the equilibrium model. When a bad fit is noted between the experimental and calculated observable the experimental plan may be modified accordingly. However, computational methods cannot entirely replace the inventiveness and/or judgment derived from the sound chemical reasoning arising from objective experimentation.

The application of spectrophotometric techniques to the elucidation of the equilibrium model usually takes on the following general pattern:

i. Spectra are obtained from solutions containing an excess of the metal ion, an excess of the ligand, and at equimolar ratios of both components.
ii. Absorbance data are obtained as a function pH, for proton-dependent systems, at several wavelengths, with metal excess and then ligand excess solutions, i.e., $A = f(\text{pH})_{C_M, C_L, \lambda}$ curves are produced.
iii. Absorbance data are also obtained where one component concentration is varied. These studies are usually performed at constant pH, i.e., $A = f(C_M)_{C_L, \text{pH}, \lambda}$ or $A = f(C_L)_{C_M, \text{pH}, \lambda}$ curves.
iv. Continuous variation experiments are performed at selected pH's and wavelengths, i.e., $A = f(\chi)_{(C_M + C_L), \text{pH}, \lambda}$ where $\chi = C_L/(C + C_M)$

In each of the above the selected wavelengths should include not only the absorption maxima but also other wavelengths thereby providing an additional mechanism for establishing the veracity of the equilibrium model. For i, ii, and iii a wide range of metal of ligand concentration ratios should be selected to ensure that all possible species are observed.

1.2. Experimental Techniques

In most cases it is necessary to prepare solutions with varying pH or component concentrations. This classical batch approach is tedious and prone to error, especially if a large number of solutions is to be prepared. Several advantages accrue from a combined titration technique with the simultaneous monitoring of absorbance and electrochemical parameters, usually pH. There are two possible approaches:

(i) Internal Titration Performed in the Cuvette.[1] The cell, about 10–30 ml, is placed in the sample compartment and a stirrer, pH electrode, thermometer, and the capillary tip of a microburette are immersed in the solution being measured. The physical arrangement of the equipment should not be disturbed during the course of the titration. If necessary, polyethylene inlet tubes are included to provide an inert gas atmosphere.

(ii) External Titration Performed Outside the Cuvette.[2] A 150-ml double-walled thermostated titration vessel is connected to the photometer cuvette with polyethylene tubes or through glass capillary tubes provided with glass ball joints. One of the capillaries in the cuvette is connected to the titration vessel and the other is connected either to a nitrogen source or to a syringe. The preparation of solutions having different pH values is carried out in the titration vessel under an inert gas atmosphere that also permits solution mixing. After the solution has come to equilibrium in the vessel the pH (or pX, using an ion selectrode) is measured and the solution is then transferred to the cuvette by pressurizing the titration vessel with the inert gas or with the syringe. Once the absorbance at different wavelengths has been measured the solution is returned to the titration vessel. Finally the pH (pX) is measured again. This procedure is repeated as many times as desired.

1.3. Data Analysis

Preliminary analysis of data involves mostly graphical methods. Identification of part of the equilibrium model under certain experimental conditions, such as ligand or metal excess, may lead to simplifying assumptions. These assumptions can, in turn, permit linear transformations of an objective function employing transformed data. Linear regression analysis of the data is then performed, often with a suitable computer program. The second stage of data analysis would be multiparametric curve fitting, employing the nonlinear least-squares approach, to determine the unknown equilibria parameters from the primary experimental data. Initial guesses for the parameters are normally required and may be obtained from the preliminary data treatment. The final phase in data treatment might be a matrix regression analysis based

on the model derived from the second step. Final improvements to the equilibrium
model are often made at this stage.

1.4. Adequacy of the Selected Model

The principal aim of these studies is to select an appropriate equilibrium model
from among many feasible models and to establish whether or not this equilibrium
model adequately represents the data. As to the criteria used to select the "best"
model, one or more of the following tests may be used:

i. The lowest sum of squares of residuals, that is the differences between experimental and calculated vaues;
ii. The minimum variance of the dependent variable, s_y^2;
iii. A statistical examination of the set of residuals.

Since data analysis employs a regression technique (linear, nonlinear, or matrix),
certain underlying assumptions for regression analysis need to be heeded. These are
independence of the nonobservable errors, ϵ, constant variance, and normal (Gauss
ian) distribution for ϵ. If the equilibrium model adequately represents the data, the
residuals should possess characteristics that agree with, or at least do not contradict,
these basic assumptions. The analysis of residuals is one way of checking that the
assumptions are not violated. If the equilibrium model describes the data well, the
residuals should be randomly distributed about the calculated values derived from the
regression equation. Examination of residual plots vs. the independent variable, x, or
a plot of the frequency of the residuals vs. the magnitude of the residuals, have been
suggested as numerical and/or graphical aids to assist in this analysis. The following
five approaches to residual analysis are usually employed:

i. Detection of an outlier.
ii. Detection of a trend in the residuals.
iii. Detection of an abrupt shift in the residuals values.
iv. Detection of changes in the error variance (usually assumed to be constant).
v. Examination of residuals to ascertain if they are normally distributed.

Graphical presentation of the residuals assists in the diagnosis since one extreme
value can simultaneously affect several of the numerical tests. For a more objective
statistical analysis, the residual vector $r = [r_i, i = 1, \cdots, n]$ can be described by
its four moments.

The first moment is the arithmetic mean:

$$m_{r,1} = \left(\sum_{i=i}^{n} r_i \right) \Big/ n \tag{3}$$

which should be zero, for perfect data.

The second moment, the variance:

$$m_{r,2} = \left(\sum r_i^2\right)\!\!\Big/n - \left[\left(\sum r_i\right)\!\!\Big/n\right]^2 \qquad (4)$$

and its square root, the residual standard deviation $s = (m_{r,2})^{1/2}$, should be similar in value to the instrumental error of the dependent variable, y.

The third moment, the coefficient of asymmetry (skewness) provides information about the shape of the error distribution curve:

$$m_{r,3} = \left[\sum (r_i - m_{r,1})^3\right]\!\!\Big/(n \cdot s^3) \qquad (5)$$

and should be zero for normal Gaussian distribution. Finally, the fourth moment, the coefficient of curtosis, also characterizes the shape of distribution curve:

$$m_{r,4} = \left[\sum (r_i - m_{r,1})^4\right]\!\!\Big/(n \cdot s^4) \qquad (6)$$

and attains the value 3 for a normal Gaussian distribution. The residuals may be divided into eight classes within which there will be a "normal" probability of encountering 12.5% of all residuals. Thus the classes are defined by the limits $-\infty$, $-1.15s$, $-0.675s$, $-0.319s$, 0.0, 0.319s, 0.675s, 1.15s, ∞. A goodness-of-fit statistic, χ^2, is then derived from the difference in experimental and calculated probability. Since the residual standard deviation, s, is computed from the residuals χ^2 has six degrees of freedom. A fit can be accepted at the appropriate confidence level if the experimental value of χ^2 is less than the expected value.

Rarely are all these requirements met in an absolute sense when analyzing equilibrium data by multiparametric curve fitting. Nonetheless, the statistical analysis of residuals serves well for the investigation of the goodness-of-fit of the regression curve.

2. DETERMINATION OF pK_a's FROM ABSORBANCE pH STUDIES

2.1. Introduction

The determination of dissociation constants of acids and bases from absorbance–pH curves can be summarized as follows: (a) measurement of A vs. pH data; (b) preliminary data analysis using graphical or simplified numerical methods; (c) rigorous final refinement of dissociation constants by multiparametric curve fitting; (d) regression matrix analysis of A vs. pH spectra.

2.2. Measurement of Absorbance vs. pH Data

Absorbance vs. pH data may be gathered using the conventional batch technique or by the internal (or external) titration procedure described in Section 1.2.

A comparison of the titration methods using data from absorbance pH curves for methyl orange gave a dissociation constant $pK_a = 3.37 \pm 0.01$ with residual standard deviation $s(A)$ in the range 0.022–0.034 for the internal titration and $pK_a = 3.36 \pm 0.01$ with $s(A)$ in the range 0.009–0.015 for the external titration. In other words both procedures lead to the same value for the dissociation constant.

2.3. Preliminary Data Handling

The equilibrium for the monoprotic acid HL, with both species HL and L absorbing, may be expressed by

$$A = dC_L(\epsilon_L K_a + \epsilon_{HL}\{H\})/(K_a + \{H\}) \tag{7}$$

where A is the absorbance of the solution at $p\{H\}$, d is the path length (cm), C_L is the total concentration of a monoprotic acid, ϵ_L and ϵ_{HL} are the molar absorptivities of L and HL, and K_a is a mixed dissociation constant.

There are several mathematical transformations that will permit the evaluation of K_a, ϵ_{HL}, and ϵ_L from the data A and $p\{H\}$. All transformations shown in Table 1 are linear, having the form

$$y = a + bx \tag{8}$$

and consequently estimates of a and b may be obtained from conventional linear regression analysis of y upon x. Further, the extent of the linear relationship between x and y may be estimated from the correlation coefficient, R.

The results of a linear regression analysis of the A and $p\{H\}$ data obtained from an external spectrophotometric titration of methyl orange is shown in Table 2. The final row of results are those obtained by processing the data with the program DCLET.[10] Although the values of pK_a, ϵ_{HL}, and ϵ_L obtained from the ten transfor-

TABLE 1. Ten Transformations of Equation (7) to Forms Suitable for Direct and Logarithmic Analysis of A vs. pH Curve Using Linear Regression

Independent variable	Dependent variable	Intercept	Slope	Reference
$\log a_{H^+}$	$\log[(A_L - A)/(A - A_{HL})]$	$-\log K_a$	1	3
$\log[(A_L - A)/(A - A_{HL})]$	$\log a_{H^+}$	$\log K_a$	1	3
$(A - A_{HL}) \cdot a_{H^+}/A$	$1/A$	$1/A_L$	$1/(K_a \cdot A_L)$	4
$1/[a_{H^+} \cdot (A_{HL} - A)]$	$A/[a_{H^+} \cdot (A_{HL} - A)]$	$1/K_a$	A_L	5
$a_{H^+} \cdot (A - A_{HL})$	A	A_L	$1/(-K_a)$	6
$a_{H^+}/(A_L - A)$	$a_{H^+} \cdot A/(A_L - A)$	K_a	A_{HL}	5
A	$(A_L - A)a_{H^+}$	$-A_{HL}/K_a$	$1/K_a$	7
$(A_L - A)/a_{H^+}$	A	A_{HL}	K_a	8
$a_{H^+}/(A_L - A)$	a_{H^+}	$-K_a$	$(A_L - A_{HL})$	9
$1/a_{H^+}$	$1/(A_L - A)$	$1/(A_L - A_{HL})$	$K_a/(A_L - A_{HL})$	6

TABLE 2. Estimated Values of pK_a, ϵ_{HL}, and ϵ_L for Methyl Orange Together with Various Statistical Parameters Obtained from the Ten Transformations Shown in Table 1

Transform	pK_a	ϵ_{HL}	ϵ_L	$m_{r,1}$	S	R
1	3.358	41,588	11,249	6.0×10^{-4}	0.0050	0.9997
2	3.439	41,588	11,249	-1.7×10^{-2}	0.0122	0.9997
3	3.372	41,588	10,936	7.5×10^{-4}	0.0055	0.9888
4	3.356	41,588	11,195	-2.1×10^{-4}	0.0053	0.9993
5	3.386	41,588	10,483	-1.3×10^{-5}	0.0053	0.0052
6	3.337	42,071	11,249	3.0×10^{-4}	0.0024	0.9999
7	3.338	42,044	11,249	4.0×10^{-4}	0.0024	0.0087
8	3.339	40,006	11,249	2.0×10^{-2}	0.0141	0.9987
9	3.335	42,026	11,249	1.3×10^{-3}	0.0023	0.9999
10	3.350	41,460	11,249	3.5×10^{-3}	0.0051	0.9994
DCLET	3.330	42,198	11,299	5.7×10^{-5}	0.0021	—

mations are comparable, there are some differences to be noted in the values of the statistical parameters s, $m_{r,1}$, etc. This is attributable, in part, to a modification of the random error distribution that occurs as a consequence of the transformations.

2.4. Curve-Fitting Analysis of the Absorbance vs. pH Curve

Nonlinear regression analysis of the A vs. $p\{H\}$ curve may be achieved using the program DCLET.[10] Molar absorptivities for all light-absorbing species, at all wavelengths studied, together with the dissociation constant(s) can be estimated. For the polybasic acid, $H_j L$ the overall mixed protonation constant of the species $H_j L$, i.e., β_{01j}, or its mixed dissociation constant, K_{aj} is given by

$$\beta_{01j} = [H_j L]/[L]\{H\} = 1 \bigg/ \sum_{1}^{j} K_{aj} \tag{9}$$

Assuming that all species absorb radiation at a given wavelength, the absorbance, A, of the solution will be equal to

$$A = dC_{H_j L} \cdot \frac{\epsilon_L + \sum_{j=1}^{J} \epsilon_{H_j L} 10^{(j \log a_H + \log \beta_{H_j L})}}{1 + \sum_{j=1}^{J} 10^{(j \log a_H + \log \beta_{H_j L})}} \tag{10}$$

where ϵ_L, ϵ_{HL}, $\cdots$, $\epsilon_{H_j L}$ are the molar absorptivities for L, HL, $\cdots$ $H_j L$, and $C_{H_j L}$ is the total concentration of the acid $H_j L$. The program DCLET performs a least-squares analysis of the A, $p\{H\}$ data, thereby obtaining values for the molar absorptivities and constant(s) that best describe the data. The functional relationship, equation (10), is used in this context.

2.5. Matrix Regression Analysis

The same parameters may also be evaluated by matrix regression analysis of the spectra obtained at different pH values. One of the most sophisticated programs written to date is SQUAD,[11] described in a later chapter. Consequently another advanced program FA608 + EY608,[12] also used in our laboratories, was selected to demonstrate this aspect of data analysis.

Program FA608[12] determines the number of light-absorbing species in solution by factor analysis, a technique applied to spectrophotometry by Wernimont.[13] The rank of the absorbance matrix A is determined from the corresponding second moment matrix $M = (1/n_s) A A^T$ by examining the eigenvalues and eigenvectors. The Beer–Lambert law may be expressed, in general form, by $A = E \cdot C$ for unit path length, where A is the $n_w \times n_s$ absorbance matrix, E is the $n_w \times n_c$ matrix absorptivities, and C is the $n_c \times n_s$ concentration matrix; n_w, n_s, n_c denote the number of wavelengths, solutions, and complex species, respectively.

Let the eigenvalues of M matrix be g_i and suppose that there are k independent light-absorbing complex species in solution. The residual standard deviation of absorbance s_k is given by

$$s_k = \left\{ \left[\mathrm{tr}\,(M) - \sum_{i=1}^{k} g_i \right] \Big/ (n_w - k) \right\}^{1/2} \tag{11}$$

where $\mathrm{tr}(M)$ is the trace of M. The s_k is then compared with the standard deviation of the absorbance for the spectrophotometer S_A. If the condition $s_k \approx s_A$ is met, the rank of the absorbance matrix represents the number of the light-absorbing species in solution, that is, $n_c \approx k$.

If any erroneous absorbance points are detected they are corrected and the factor analysis repeated.

Subprogram EY608 estimates formation constants (or dissociation constants) for k different components. Individual species spectra of the variously protonated species and the distributed diagram for the species are also computed. The reliability of the computed dissociation constants and molar absorptivities improves as the calculated standard deviation of the absorbance, $s(A)$, approaches the residual standard deviation of the absorbance s_k as determined from the factor analysis.

An absorbance matrix of 22 spectra comprising 13 wavelengths for methyl orange was analyzed by factor analysis. The root, residual trace of the second moment M and the residual standard deviation of the absorbance, s_k, is calculated for $k = 1, 2, \cdots, n$. For the data gathered $s_k = 0.0032$. The closest value of $s(A)$ equal to 0.003 is for $k = 2$. This value compares well with specifications for the spectrophotometer used for this study (VSU2-G, Zeiss, Jena, GDR).

The dissociation constant of methyl orange, estimated using EY608, was 3.332

± 0.003. Values of molar absorptivities of ϵ_L and ϵ_{HL} estimated by the EY608 program are in excellent agreement with those calculated by the DCLET program.

2.6. Thermodynamic Dissociation Constants

Thermodynamic dissociation constants may be estimated by an extrapolation of the $pK_a = f(I)$ dependence to zero ionic strength or, more rigorously, by nonlinear least-squares analysis using the program DHLET.[14] The ionic strength dependency of the mixed dissociation constant may be expressed by the extended Debye–Huckel equation

$$pK_{a,i} = pK_a^T - A \cdot I_i^{1/2} (1 - 2z)/(1 + B\mathring{a}I_i^{1/2}) + CI_i \tag{12}$$

for the equilibrium $HL^z = L^{z-1} + H^+$, assuming that the ion-size parameter $\mathring{a}$ for both ions HL^z and L^{z-1} are approximately equal, and the overall salting-out coefficient, C, is defined by $C = C_{HL} - C_L$. The constants A and B have values 0.5115 and 0.3291, respectively, for aqueous solutions and 25 °C.

Nonlinear least-squares analysis was applied to pK_a vs. I data for selected sulfonephthaleins. The thermodynamic dissociation constants, ion-size parameter, and salting-out coefficient for two indicators were estimated using DHLET, Table 3. The reliability of these parameter estimates is demonstrated by inspection of the first four moments of the residual, also shown in Table 3.

3. INITIAL METHODS FOR INVESTIGATING COMPLEX EQUILIBRIA

3.1. Introduction

Graphical methods, with or without computer programs, play an important diagnostic role in initial data processing. The objective at this stage of the investigation is to identify the major species together with approximate values for the relevant stability constants. Often one or more of the following will be used: (a) A vs. pH or C_M (C_L); (b) continuous variation; (c) mole ratio; (d) corresponding solutions. Experimental conditions are arranged so that only the major species exists (>80% of all complexes). The methods, together with many variations, have been reviewed extensively by Sommer et al. [15]

3.2. Graphical Analysis of Absorbance vs. pH Curves

Graphical linearizations may be used if either a single complex is formed or if one complex predominates. Side reactions, other than ligand dissociation and metal hydrolysis, must not occur. Various linearized transformations[15] may then be made depending on the perceived equilibrium model and resulting algebraic equations.

To illustrate one graphical technique absorbance data for the reaction between 1-(2-thiazolylazo)-2-naphthol-3,6-disulfonic acid and Pr(III) were collected at four

TABLE 3. Nonlinear Regression Estimation of the Thermodynamic Constant, Ion-Size Parameter, and Salting-Out Coefficient of Two Sulfonephthalein Indicators from the Ion Strength Dependency of the Mixed Dissociation Constant; Data Analysis Performed by DHLET

Bromothymol blue			Bromocresol purple		
I	pK_a	r^a	I	pK_a	r^a
0.0025	7.143	11.6	0.010	6.085	6.2
0.011	7.075	2.0	0.022	6.029	−10.3
0.024	7.029	−1.3	0.040	6.009	3.8
0.045	6.979	−11.6	0.060	5.986	5.3
0.067	6.954	−9.8	0.116	5.941	0.4
0.127	6.924	4.1	0.232	5.901	−0.5
0.251	6.878	2.4	0.392	5.861	−16.4
0.423	6.844	−3.2	0.594	5.871	6.8
0.635	6.838	7.1	0.923	5.856	−2.4
1.004	6.822	0.1	1.330	5.863	0.2
1.445	6.826	1.3	2.000	5.894	12.5
2.260	6.843	−2.7	3.720	5.947	−5.6
4.000	6.915	0.2			

$$pK_a^T = 7.199 \pm 0.004 \qquad 6.197 \pm 0.006$$
$$a = 7.763 + 0.209 \qquad 8.807 \pm 0.328$$
$$C = 0.054 \pm 0.003 \qquad 0.055 \pm 0.004$$
$$s(pK_a^T) = 0.0069 \qquad 0.0088$$
$$m_{r,1} = 1.46 \times 10^{-6} \qquad -5.56 \times 10^{-7}$$
$$m_{r,2}^{1/2} = 0.0060 \qquad 0.0076$$
$$m_{r,3} = -0.209 \qquad -0.569$$
$$m_{r,4} = 2.939 \qquad 2.821$$
$$\chi^2 = 3.62 < 12.60 \qquad 4.00 < 12.60$$
$$R \text{ factor} = 0.000861 \qquad 0.001283$$

$^a r = 10^3 (pK_{a,\exp} - pK_{a,\text{calc}})$

wavelengths over the range pH 2.20–4.37. The metal ion was in large excess over the ligand ($C_M = 9.62 \times 10^{-3} M$; $C_L = 4.98 \times 10^{-5} M$). The molar absorptivity of the complex, usually determined from the horizontal part of the absorbance pH curve, was unavailable due to praeseodymium hydrolysis at [H$^+$] less than about 10^{-5} M (Figure 2, solid triangles). Since the ligand (TAN-3,6-S) does not dissociate in this pH range ($pK_a = 7.83$) and $C_M \gg C_L$ the general equation linking experimental data and the unknown stability constants and molar absorptivities can be transformed as

$$A = \epsilon_1 C_L + (\epsilon_2 C_L - A)[H]^{-q} C_M \beta^* = A_{02} + F_1 \beta^* \qquad (13)$$

$$A = \epsilon_2 C_L - (A - \epsilon_1 C_L)[H]^{-q} C_M^{-1} \beta^{*-1} = A_{02} - F_2 \beta^{*-1} \qquad (14)$$

These transformations are based on the assumption that ML is formed, which had been previously established,[16] and that for the reaction

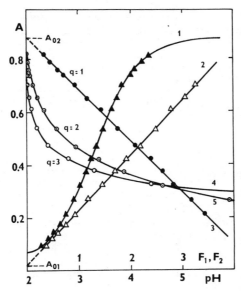

FIGURE 2. Graphical analysis of absorbance–pH curve for 1-(2-thiazolylazo)-2-naphthol-3,6-disulfonic acid and Pr (III). 1, A vs. pH at 580 nm; , $A = f(F_1)$ for $q = 1$; 3–5, $A = f(F_2)$ for $q = 1,2$, and 3.

$$Pr^{3+} + H_qL^{(q-3)} \rightleftharpoons PrL + qH^+$$

with formation constant β^*, the value of q is to be determined.

Equations (13) and (14) will give straight lines only if the correct value of q is used. In this particular application of the transforms, ϵ_1 and ϵ_2 are the molar absorptivities for the ligand and the complex, respectively. Reasonable values are obtained from previous experiments.[16] It can be seen from Figure 2 that straight lines are obtained for only $q = 1$. The intercepts of $A = f(F_1)$ and $A = f(F_2)$ give exact values for ϵ_1 and ϵ_2, which may be then used again in equations (13) and (14) until internal consistency is achieved. Finally a logarithmic combination of equations (13) and (14) yields

$$\log [(A - A_{01})/(A_{02} - A)] = q \cdot pH + \log C_M + \log \beta^* \tag{15}$$

where $A_{01} = \epsilon_1 C_L$ and $A_{02} = \epsilon_2 C_L$. Plotting the left-hand side as a function of pH yields, when the left-hand side becomes zero (pH_0),

$$\log \beta^* = -q \cdot pH_0 - \log C_M \tag{16}$$

and hence β^* for the reaction in question. The results of this graphical approach and the results of processing the same data using PRCEK[3] are shown in Table 4. Variations on these principles, described above, have been reviewed.[15] Applications of the various types of transformations have been presented,[16-20] and in two instances[17,18] a detailed comparison has been made between graphical and computer-aided analyses of absorbance curves.

TABLE 4. Results of $A = f(\text{pH})$ Curve Analysis for the 1-(2-Thiazolylazo)-2-Naphthol-3,6-Disulfonic Acid, Pr(III) System

Wavelength (nm)	Log β^*		
	Direct analysis[a]	Log analysis[b]	PRCEK[c]
540	−1.27	−1.27	−1.271 ± 0.0008
560	−1.29	−1.28	−1.296 ± 0.0008
570	−1.29	−1.28	−1.290 ± 0.0010
580	−1.28	−1.28	−1.293 ± 0.0011

	Molar absorptivities ϵ_1 and ϵ_2			
	Graphical analysis		PRCEK analysis	
540	6730	24,640	6503.7 ± 22.2	24,353 ± 34
560	1770	23,590	1879.8 ± 18.3	23,691 ± 26
570	640	21,800	773.0 ± 26.2	21,967 ± 24
580	330	17,600	369.4 ± 30.7	17,652 ± 31

[a] Transform using equations (13) and (14).
[b] Transform using equation (16).
[c] Average values of direct and logarithmic analysis given. Sum of residuals, U, given by $\Sigma(A_{exp} - A_{calc})$, using log $\beta^* = -1.287$, is 5.39×10^{-4}.

3.3. Analysis of Absorbance vs. pH Curves Using PRCEK

PRCEK performs a rigorous linear least-squares analysis of the absorbance, pH data employing one of the several linear transformations described in previous sections. The program has been coded efficiently so that it occupies the minimum of core. The advantage of using PRCEK is that many transformations[15] may be tested during initial data analysis. A printer plot is included to display the results of the linear least-squares analysis along with several statistical indicators of the goodness of fit. Five systems, analyzed by a variety of techniques, are summarized in Table 5.

3.4. Continuous Variation Techniques

The continuous variation (Job's) method is a familiar and widely used technique to determine the stoichiometry of a principal species and its conditional stability constant. The advantages, disadvantages, and pitfalls of using continuous variation have been discussed by McBryde.[26]. One of the most important restrictions of the method is that no side reactions can occur, other than ligand dissociation and metal ion hydrolysis.

For the general reaction

$$m\text{M} + n\text{L} \rightleftharpoons \text{M}_m\text{L}_n \tag{17}$$

the conditional stability constant, β', is defined as

$$\beta'_{mn} = [\text{M}_m\text{L}'_n]/[\text{M}']^m[\text{L}']^n \tag{18}$$

TABLE 5. A Comparison of Graphical and Computer Analysis of Selected Equilibrium Systems

A. 4-(2-pyridylazo)resorcinol (PAR); 0.1 M (KNO$_3$); 25°C[a]

Constant	Graphical	PRCEK[b]	FA608 + EY608[c]
pK_1	5.46	5.44 ± 0.01	5.57 ± 0.05
pK_2	11.97	12.02 ± 0.03	11.95 ± 0.04
log k_{11H}[d]	−0.16	−0.08	−0.19

B. 2-(2-thiazolylazo)-4-methoxyphenol (TAMP); 0.1 M (KNO$_3$); 25°C[e]

Constant	Graphical	PRCEK[b]	LETAGROP-SPEFO[f]
pK	7.83 ± 0.03	7.827 ± 0.015	7.836 ± 0.009

C. Ni-TAMP; 30% (v/v) ethanol; 0.1 M (KNO$_3$); 25°C[g]

Constant	Graphical	PRCEK	LETAGROP-SPEFO
β_1	7.6	7.60 ± 0.05	7.595 ± 0.024
β_2	15.81	15.805 ± 0.05	15.815 ± 0.03

D. 2-(2-pyridylazo)-1-naphthol-4-sulfonic acid (PANS); 0.1 M, 25°C[h]

Constant	PRCEK[i]	SQUAD[j]	SPEKTFOT[k]
pK_1	2.817 ± 0.006	2.795 ± 0.001	2.8979 ± 0.030

E. Uranyl (V1)-maltol (LH)[l]

Constant[m]	Graphical	LETAGROP-SPEFO
β_1^*	−0.32	−0.36 ± 0.04
β_2^*	−2.24	−2.27 ± 0.04
β_3^*	−7.59	−7.50 ± 0.02

[a]Reference 29.
[b]Reference 3.
[c]Reference 12.
[d]k_{11H} = [MLH][H]/[M][LH$_2$].
[e]Reference 17.
[f]Reference 21.
[g]Reference 18.
[h]Reference 22.
[i]One wavelength.
[j]36 wavelengths.
[k]Reference 23.
[l]References 24 and 25.
[m]β_n^* = [UO$_2$L$_n$][H]^n/[UO$_2$][LH]^n.

The implicit functional relationship for Job's method may be defined as

$$F(x,y) = \beta'_m - [(m + n)/C]^{m+n-1} y E_1^{-m} E_2^{-n} = 0 \tag{19}$$

where

$$C = C_M + C_L \tag{20}$$

$$E_1 = (m + n)x - my \tag{21}$$

$$E_2^{\cdot} = (m + n)(1 - x) - ny \tag{22}$$

$$x = C_M/(C_M + C_L) \tag{23}$$

$$y = (A - A_s)/(A_{max} - A_{max,s}) \tag{24}$$

The quantity $(A - A_s)$ is the absorbance of the measured solution, A, corrected for the absorbance of the ligand solution at the same concentration. The maximum absorbance, assuming no dissociation of the complex, is denoted by $(A_{max} - A_{max,s})$. The primed concentrations are effective concentrations arising through the use of β'_{mn}. The program JOBCON, by Likussar,[27] is used to process the data obtained from a conventional continuous variation experiment. Various combinations of m and n are used to determine values of β'_{mn} and y_{max}. The values of m and n that give the lowest relative standard deviation, as determined by JOBCON, are taken to the correct values.

There are several limitations to the application of JOBCON that are, for the most part, related to continuous variation rather than the program. These are that one complex is formed under the prevailing experimental conditions and data are collected in the mole fraction range 0.1–0.9 thereby ignoring the influence of complexes formed in large excesses of ligand or metal.

The advantage of using JOBCON is mainly in eliminating the manual arithmetic and plotting involved in testing several combinations of m and n. Additionally the program provides a statistical analysis of the experimental data.

3.5. Mole Ratio Techniques

A second familiar technique in investigating principal metal ligand equilibria by spectrophotometry is the mole ratio approach. The original method involves the examination of plots of absorbance vs. the ratio $C_L:C_M$, where C_L is normally kept constant. The technique is frequently used in a titration situation. Under the conditions where only one complex is formed:

$$C_M = [M'] + m[M_mL'_n] = C_M^0 \cdot V_M/V_0 \qquad (25)$$

$$C_M = [L'] + n[M_mL'_n] = C_L^* f_L \qquad (26)$$

where C_M^0 is the initial concentration of the method ion (titrant), C_L^* is the approximate ligand concentration, V_M is the titer volume, V_0 is the initial volume, and f_L is the effective concentration factor. For the mole ratio $q_M (= C_M/C_L)$ and the normalized absorbance, α, defined as

$$\alpha = [A(1 + V_M/V_0) - A_0]/(A_{ext} - A_0) \qquad (27)$$

where A_0 is the initial absorbance before addition of metal and A_{ext} is the absorbance of a solution containing a large excess of metal ion, the functional relationship, F, is derived

$$F = (m/n)\alpha + \alpha^{1/m}[(n\beta'_{mn})^{1/m}C_L{}^{(m+n-1)/m}(1 - \alpha)^{n/m}]^{-1} - C_M/C_L = 0 \quad (28)$$

The dependent variable is A and the independent variable is V_M. The program, MRLET,[28] minimizes F using the least-squares method, by testing various combinations of m and n and determining the unknown parameters β'_{mn}, A_{ext}, and f_L. Again

TABLE 6. Selection of the Most Probable Stoichiometry of ML_n for Semimethyl Thymol Blue with Pb(II) and Zn(II) Using the Program MRLET

	Lead(II) SMTB		Zinc(II) SMTB	
Stoichiometry	1:1	1:2	1:1	1:2
A_{ext}	0.4676 ± 0.0011	0.4699 ± 0.0071	0.4039 ± 0.0012	0.3019 ± 0.0082
$-\log f_L$	0.110 ± 0.002	0.106 ± 0.164	0.127 ± 0.005	0.142 ± 0.152
$\log \beta_{1n}$	6.86 ± 0.04	10.30 ± 0.16	6.14 ± 0.03	11.56 ± 0.21
Error sum square, U	1.09×10^{-4}	1.37×10^{-2}	8.50×10^{-5}	2.62×10^{-1}
$S(A)$	0.0019	0.0295	0.0017	0.0968
Conclusion	Accepted	Rejected	Accepted	Rejected

the best values of m and n are determined by examining the statistical analysis performed by MRLET. Table 6 shows the results obtained from MRLET for two equilibrium systems, distinguishing between ratios of 1:1 and 1:2.

3.6. Method of Corresponding Solutions

Corresponding solutions are defined as having the same degree of complex formation, $\bar{n}$, and hence the same [L]. This situation occurs between solutions having different C_M and C_L. In practice absorbance measurements are made on sets of solutions, each set having a constant metal ion concentration and within each set varying ligand concentration. The absorbances, at selected wavelengths, are plotted for each set, A vs. C_L. Each curve represents the absorbance as a function of C_L for each C_M. Corresponding solutions will have the same absorbance. Consequently, by drawing lines at constant A that intersect the curves the compositions of the corresponding solutions may be identified.

By definition,

$$\bar{n} = (C_L' - C_L'')/(C_M' - C_M'') \tag{29}$$

$$[L] = (C_M'C_L'' - C_M''C_L')/(C_M' - C_M'') \tag{30}$$

The method is applicable for complexes of a stability that will make $(C_L - [L])$ smaller than C_L. Once $\bar{n}$ and [L] have been determined several methods are available for the determination of formation constants from $\bar{n}$, [L] data, once the maximum number of ligands bound to the metal, N, has been determined. This information may often be obtained from $\bar{n}$ vs. pL plots.

For example if $N = 2$, Rossotti and Rossotti[29] have shown that

$$\bar{n}/(1 - \bar{n})[L] = \beta_1 + \beta_2[L](2 - \bar{n})/(1 - \bar{n}) \tag{31}$$

is linear, β_1 and β_2 being obtained from the slope and intercept of an appropriate plot.

Other methods of determining formation constants from $\bar{n}$, pL data include one due to Scatchard,[30] where a new function, Q is defined:

$$Q(L) = \bar{n}/\{(N - \bar{n})[L]\} \tag{32}$$

Extrapolation of the curve $Q(L) = f([L])$ for $L \to 0$ gives the ratio β_1/N while extrapolation for $L \to \infty$ leads to the value for $N\beta_N$.

Other approaches include the elimination method[29] and a matrix method, due to Romary et al.,[31] employing a computer program solves the system of equations

$$\bar{n} + \sum_{n=1}^{N} (\bar{n} - n) [L]^n \beta_n = 0 \tag{33}$$

by gradual elimination.

4. CONCLUSIONS

If the equilibrium system under consideration can be described by only one or two parameters, computers offer only a slight advantage over graph paper. There are two types of graphical methods: linear transformations followed by a regression analysis of the transformed data (with or without the aid of a computer) and curve-fitting techniques, usually involving one normalized variable. However, when more than three species are present curve-fitting procedures are of little value. There are exceptions to this general statement. Once the model has been elucidated it is often instructive to attempt to derive special transformations based on a selected region of the data or by ignoring species that have been shown (or are believed) to be minor in their contribution to the measured quantity.

Graphical methods have the great advantage of allowing the worker to remain in touch with the several steps in the calculations. Additionally graphical methods clearly demonstrate the presence of bad points and systematic errors.

In 1974 H. Rossotti wrote a paper describing, in some detail, the concepts and techniques involved in the design and publication of work on stability constants.[32] We are in total agreement with that publication and can do no better than suggest that the reader study that eloquent discourse on the principles and practice of solution equilibrium investigation.

5. REFERENCES

1. M. Meloun and J. Pancl, Complexation Equilibria of Some Azo Derivatives of 8-Hydroxyquinoline-5-Sulphonic Acid: I. The Least Squares Computer Estimation of Dissociation Constants of SNAZOX and Naphthylazoxine 6 S from Spectrophotometric Data, *Collect. Czech. Chem. Commun.* **41**, 2365–2385 (1976).
2. V. Kubán, L. Sommer, and J. Havel, Ni(II) Metal Chelates with 2(2-Thiazolylazo)-4-Methoxy Phenol in Solution—A Combined Graphical and Numerical Interpretation of Absorbance Curves, *Collect. Czech. Chem. Commun.* **40**, 604–633 (1975).

3. J. Havel and V. Kubán, Simple Programs for the Evaluation of Equilibrium Constants. I. PRCEK—A Linear Least-Squares Program for the Evaluation of Equilibrium Constants from Spectrophotometric Data, *Scripta Fac. Sci. Nat. UJEP Brun. Chem.* **2**, 1:87–98 (1971).

4. A. Ågren, The Complex Formation Between Iron(III) Ion and Some Phenols. IV. The Acidity Constant of the Phenolic Group, *Acta Chem. Scand.* **9**, 49–56 (1955).

5. L. A. Flexser, L. P. Hammett, and A. Dingwall, The Determination of Ionization by Ultraviolet Spectrophotometry: Its Validity and its Application to the Measurement of the Strength of Very Weak Bases, *J. Am. Chem. Soc.* **57**, 2103–2115 (1935).

6. J. P. Maroni and J. P. Calmon, Détermination spectrophotometrique des pK de dissociation des β-dicétones. I. Technique expérimentale et mise au point de mèthodes graphiques, *Bull. Soc. Chim. France* **1964**, 519–524.

7. I. Weil and J. C. Morris, Equilibrium Studies on *N*-Chloro Compounds. II. The Base Strength of *N*-Chloro Dialkylamines and of Monochloramine, *J. Am. Chem. Soc.* **71**, 3123–3126 (1949).

8. I. Iwasaki and S. R. B. Cooke, Dissociation Constant of Xanthic Acid as Determined by Spectrophotometric Method, *J. Phys. Chem.* **63**, 1321–1322 (1959).

9. I. Ya. Bernstein and Yu L. Kaminskij, *Spectrophotometric Analysis in Organic Chemistry* (Russian), Khimiya Publishing House, Leningrad (1975), p. 153.

10. M. Meloun and J. Cermák, Multiparametric Curve Fitting: IV. Computer-Assisted Estimation of Successive Dissociation Constants and of Molar Absorptivities from Absorbance-pH Curves by the DCLET Program *Talanta* **26**, 569–573 (1979).

11. D. J. Leggett and W. A. McBryde, General Computer Program for the Computation of Stability Constants from Absorbance Data, *Anal. Chem.* **47**, 1065–1070 (1975).

12. J. J. Kankare, Computation of Equilibrium Constants for Multicomponent Systems from Spectrophotometric Data, *Anal. Chem.* **42**, 1322–1326 (1970).

13. G. Wernimont, Evaluating Laboratory Performance of Spectrophotometers, *Anal. Chem.* **39**, 554–566 (1967).

14. M. Meloun and S. Kotrlý, Multiparametric Curve Fitting. II. Determination of Thermodynamic Dissociation Constants and Parameters of the Extended Debye–Hückel Expression. Application for Some Sulphonephthalein Indicators, *Collect. Czech. Chem. Commun.* **42**, 2115–2125 (1977).

15. L. Sommer, V. Kubán, and J. Havel, Spectrophotometric Studies of Complexation in Solution. *Folia Fac. Sci. Nat. Univ. Purk. Brunensis, Tom XI, Chemia* 7, Opus 1 (1970).

16. J. Vosta and J. Havel, Spectrophotometric Study of the Complexation of Pr(III) with 4-(2-Thiazolyl-azo) Resorcinol and 1-(2-Thiazolylazo)-2-Naphthol-3,6-Disulphonic Acid, *Collect. Czech. Chem. Commun.* **42**, 2871–2889 (1977).

17. V. Kubán and J. Havel, Some 2-(2-Thiazolylazo)-4-Methoxyphenol (TAMP) Complex Equilibria. I. Acid–Base Properties of TAMP in Water and in Various Mixed Solvents, *Acta Chem. Scand.* **27**, 528–540 (1973).

18. M. Langová, J. Havel, and L. Sommer, Complexation of Mercury(II) with 2-(2-Thiazolylazo)-4-Methoxyphenol, *Chem. Anal. (Warsaw)* **17**, 989–1001 (1972).

19. V. Koblízková, V. Kubán, and L. Sommer, Spectrophotometrisches Studium der komplexbildenden Gleichgewichte des Hg(II) und seine Bestimmung mit 2-(2-Pyridylazo)-1-naphthol-4-sulfonsaure, *Chem. Zvesti* **33**(4), 485–498 (1979).

20. V. Koblízková, V. Kubán, and L. Sommer, Spectrophotometric Study of Complexation Equilibria and Methods of Determining Cu(II) and Ni(II) with 2-(2-Pyridylazo)-1-Naphthol-4-Sulphonic Acid *Collect. Czech. Chem. Commun.* **43**, 2711–2731 (1978).

21. L. G. Sillén and B. Warnqvist, High-Speed Computers as a Supplement to Graphical Methods. 10. Application of LETAGROP to Spectrophotometric Data, for Testing Models and Adjusting Equilibrium Constants, *Ark. Kemi* **31**, 377–390 (1969).

22. L. Jancár, J. Havel, V. Kubán, and L. Sommer, Effective Spectrophotometric Study of the Complexation Reactions of 2-(2-Pyridylazo)-1-Naphthol-4-Sulphonic Acid with Zn(II) and Cd(II) using the SQUAD-G Program, *Collect. Czech. Chem. Commun.* **47**, 2654–2675 (1982).

23. M. Suchánek and L. Sucha, The Use of Program SPEKTFOT for the Computation of the Equilibrium Constants and the Molar Absorption Coefficients of Substances in Solutions, *Sci. Papers Prague Inst. Chem. Technol. H* **13**, 41–57 (1978).

24. E. Chiacchierini, J. Havel, and L. Sommer, UO_2 Complexes with Phenolic Ligands. XI. On the Reaction of Uranium(VI) with Maltol, *Collect. Czech. Chem. Commun.* **33**, 4215–4247 (1968).
25. J. Havel, in *Contributions to Coordination Chemistry in Solution* (E. Hogfeldt. ed.), Treatment of Spectrophotometric Data on the Uranyl–Maltol–H^+ Ternary System by Means of a Generalized Least Squares Program LETAGROP-SPEFO, *Trans. Royal Inst. Technol. Stockholm*, No 277, *Pure Appl. Chem.* **34**, 371–383 (1972).
26. W. A. E. McBryde, Spectrophotometric Determination of Equilibrium Constants in Solution, *Talanta* **21**, 979–1004 (1974).
27. W. Likussar, Computer Approach to the Continuous Variations Method for Spectrophotometric Determination of Extraction and Formation Constants, *Anal. Chem.* **45**, 1926–1931 (1973).
28. M. Meloun and M. Javurek, Multiparametric Curve Fitting. 5. MRFIT, MRLET Computer Programs to Calculate the Stability Constant of $M_m L_n$ Complex from Photometric Titration Curve *Talanta*, in print (1983).
29. F. J. C. Rossotti and H. S. Rossotti, Graphical Methods for Determining Equilibrium Constants. I. System of Mononuclear Complexes, *Acta Chem. Scand.* **9**, 1166–1176 (1955).
30. G. Scatchard, unpublished results., taken from Ref. 9.
31. J. K. Romary, D. L. Donnelly, and A. C. Andrews, Computer Programs for the Rigorous analysis of Potentiometric Complex Formation and Proton Dissociation Data, *J. Inorg. Nucl. Chem.* **29**, 1805–1806 (1967).
32. H. S. Rossotti, Design and Publication of Work on Stability Constants, *Talanta* **21**, 809–829 (1974).

3

MAGEC

A Program for the Definitive Calibration of the Glass Electrode

PETER M. MAY and DAVID R. WILLIAMS

1. INTRODUCTION

"The glass electrode is to solution chemistry what silicon chips are to computers—the central controlling component."[1] It is clear, however, that an uncalibrated electrode is not worth much more than a stirring rod.

Several decades ago, glass electrodes were almost invariably calibrated using buffers of specific pH values. This approach is still used in work where only a relative scale of acidity is required, but since IUPAC has introduced an operational definition of pH,[2] it can no longer be directly associated with hydrogen ion acitivity.

Even when a specific mathematical relationship between the pH of a buffer and the hydrogen ion activity was defined, difficulties remained in inferring hydrogen ion concentration values of working solutions, such as those used to measure formation constants. First, the hydrogen ion activity coefficients are not the same in the buffer and in the test solutions because the two solutions have differing ionic compositions. All "adjustments" made to allow for this difference, be they either theoretical or empirical, clearly introduce inaccuracy.[1] Moreover, differences in liquid junction potentials applicable to the buffer and to the test solution impose further difficulty.[3, 4]

These problems are worsened when high concentrations of background electrolytes are used for controlling the ionic strength of test solutions. Thus, researchers who employ potentiometric methods to measure formation constants have instead long used solutions of known hydrogen ion concentration.[5] This is usually achieved by titrating strong acid solutions with strong alkali and plotting the emf at each point against the corresponding values of $\log [H^+]$, where $[H^+]$ denotes the free hydrogen ion *concentration*. Ideally, a linear calibration curve for the electrode system ought to be obtained.

PETER M. MAY and DAVID R. WILLIAMS • Department of Applied Chemistry, University of Wales, Institute of Science and Technology, Cardiff CF1, 3NU, United Kingdom.

There are many reasons why an ideal response is not, in fact, observed in practice. Unless there is a sufficient excess of acid or alkali to ensure that the solution is fairly well concentration buffered, small errors become significant. Such errors can arise from, for example, the presence of glass itself[6] and the imperfect behavior of glass electrodes in alkaline solution due to increasing sensitivity to alkali metal ions above $-\log [H^+] = 11.0$.[7] Liquid junction potentials can also cause curvature at high concentrations of hydrogen ion.

From the foregoing, it may be concluded that there is a restricted range of free hydrogen ion concentration over which strong acid versus strong base titration data are suitable for calibration purposes. In our laboratory, the best response is found to occur when $[H^+]$ or $[OH^-]$ lie in the range $2\text{--}10 \times 10^{-3}$. Data collected inside these limits give a precision that is adequate for the determination of formation constants, but as measurements for most complexing systems lie outside these calibration ranges the data depend on extrapolation and are thus sensitive to small errors in the analytical concentrations of the acid or alkali.

Further, it is well known that the standard potential of the glass membrane tends to vary from day to day (owing to asymmetry effects) and that liquid junction potentials are not easily reproduced.[4, 7-10]

These difficulties can only be overcome by performing internal calibrations of the electrodes from measurements on the test solution itself. There are two means of achieving this: the first and simplest is to calibrate by a series of constant additions. Alternatively, for titrations of weak acids or bases, one can calculate free hydrogen ion concentrations at various points from the protonation constants, provided these are known accurately. Neither approach is, however, applicable generally because a number of parameters, including the glass electrode intercept and one or more equilibrium constants, often have to be determined simultaneously. Straightforward solutions are exceptional so general optimization methods must be found instead.

This chapter describes a general approach to the optimization of parameters required for calibrating glass electrode systems. It also shows how protonation constants of a ligand can also be determined simultaneously.[11] The relevant algorithms are incorporated into a Fortran program named MAGEC (*M*ultiple *A*nalysis of titration data for *G*lass *E*lectrode *C*alibration).

2. DESCRIPTION OF THE PROGRAM MAGEC

2.1. Theory

The classical electrochemical cell has a test solution surrounding a glass electrode which is in electrical contact with a reference electrode via a salt bridge:

first reference half-cell (in glass electrode)	test solution	second reference half-cell (external)
g		l

Boundaries, g and l, respectively, indicate the glass membrane and the liquid junction at the interface between the salt bridge and the test solution. There are four contributions to the potential difference between the two reference electrodes.[7-10] Two arise from the reference electrodes themselves. Their contributions are independent of the composition of the test solution and may be represented as a fixed, combined potential, E_r. On the other hand, the potential differences generated across the boundaries of g and l (E_g and E_l) depend on the activities of all the chemical species on either side of them. Thus,

$$E_{cell} = E_r + E_l + E_g \qquad (1)$$

Glass electrodes, in general, are found experimentally to exhibit a Nernstian response over a wide range of concentration[9] and so equation (1) can be rewritten as

$$E = E_r + E_l + E_g^0 + \frac{RT}{F} \ln \{H^+\} \qquad (2)$$

E_g^0 is the standard glass electrode potential at unit activity of hydrogen ions, R is the universal gas constant, T is the absolute temperature, F is the Faraday constant, and $\{H^+\}$ represents the hydrogen ion *activity*.

Provided the ionic strength of the test solution remains constant, the free hydrogen ion activity, $\{H^+\}$, can be expressed in terms of concentration. Hence, one obtains equation (3) by collecting together all the constants as E_{const} and setting $s = 2.303RT/F$:

$$E_{cell} = E_{const} + s \log [H^+] \qquad (3)$$

2.2. Use of MAGEC for Electrode Calibration

In the circumstances where strong acid versus strong base titrations are to be used in the calibration of the electrode system, MAGEC uses a subroutine entitled CALIBT. CALIBT first analyzes the data by the method of Gran.[12, 13] This gives a good indication of glass electrode performance and also yields an end point that is independent of the slope and intercept used in equation (3). If extrapolation of the data from before the end point produces a value significantly lower than that obtained from data after the end point, this suggests that an alkaline titrant may have become contaminated with carbon dioxide from the atmosphere. The end point obtained from the Gran extrapolations provides an independent check at the CALIBT optimization of the strong acid concentration referred to later.

Additional processing of strong acid versus strong base titration is in three stages:

i. The input concentration values are used to calculate free hydrogen ion concentrations at each point and a linear least-squares fit is performed on the data before and after the end point and over the entire range. (This first analysis is used mainly for comparison with subsequent output. Owing to relatively small errors in the concentration of titrant and titrand, the least-squares straight line does not normally possess a Nernstian slope).

ii. The concentration of the titrand is varied slightly until the slope from the data before the end point coincides with the Nernstian values. A correction factor, corresponding to a selectivity coefficient for the electrode, is calculated and applied to the data after the end point in order to establish the most Nernstian responses in this region also.

iii. The titrant concentration is adjusted in a similar manner to that described above, but on the basis of the whole range of data. Naturally, to maintain the same end point, a corresponding change in the titrand concentration then needs to be made. Very close agreements, of the order of 0.1 mV, can thus be achieved between the calculated and observed values for the emf at each point.

A factor that critically affects the refinement of the titrant concentrations in the final stage is the value used for the dissociation constant of water, K_w. This parameter is very sensitive to correlations with the concentration of alkali in the buret. In other words, either the base titrant concentration or K_w (but not both) can be determined by finding the value that yields the best least-squares slope. In those situations when K_w is uncertain, CALIBT permits the user to systematically vary the estimate.

When attempting to measure E_{const} and the ligand protonation constants at the same time, two further equations are applicable. These are the mass balance equations (4) and (5) for ligand and for protons:

$$C_L = [L] + \sum_{p=1}^{P} \sum_{r=1}^{R} p\beta_{pr}[L]^p[H^+]^r \tag{4}$$

$$C_H = [H^+] + \sum_{p=1}^{P} \sum_{r=1}^{R} r\beta_{pr}[L]^p[H^+]^r \tag{5}$$

Here, C_L and C_H represent the total concentration of ligand and protons, respectively; [L] and [H$^+$] represent the concentrations of free ligand and protons, respectively; β_{pr} represents the formation constant for the general reaction,

$$pL + rH^+ \rightleftharpoons L_pH_r$$

It is clear that (3), (4), and (5) constitute just two independent equations for each titration point. C_L and C_H are calculated from the initial volume of the titrant, the volume of each added increment of titrant and K_w. E is measured at each titration point. It follows that for n titration points there are $2n$ independent equations containing $(pr + n + 2)$ unknowns, i.e., the electrode parameters, E_{const} and s, pr β values, and n free ligand concentrations. Thus, in principle, the $2n$ equations may be solved simultaneously to yield E_{const} and the β values from a single set of titration data, provided a sufficiently large number of titration points is used.

C_L and C_H values contain analytical errors and the value of K_w is sometimes uncertain, so these may also be regarded as unknowns.

Accordingly, the main analysis applied by MAGEC to all titrations involving ligands is one of general optimization of parameter values by minimizing an objective function based on titer volumes. This is based on the simplex method introduced by Nelder and Mead.[14] Any of the parameters, E_{const}, s, the β values, C_L, C_H, and K_w can be flagged for refinement. Usually, the requirement is to find the value for E_{const} in each of a series of titrations with the ultimate objective of determining the protonation constants of a ligand.

2.3. Conclusions

Numerical analysis of titration data has the potential to convert the "art" of electrode calibration into a quantitative "science," even for relatively inexpert researchers. This provides access to the measurement of exact concentrations of hydrogen ions, which can be important in many areas of chemistry. Hopefully, solution chemists will soon agree on a standardized approach, based directly upon hydrogen ion con-

TABLE 1. Results of Glass Electrode Calibration by Titration of Hydrochloric Acid (ca. 0.0167 M) with Sodium Hydroxide (0.100 M) at 37°C in 0.15 M NaCl Background Electrolyte.[a]

Vol.	E_{obs}	E_{calc}	Residual
0.00	265.15	265.07	0.08
0.20	263.75	263.80	−0.05
0.40	262.45	262.48	−0.03
0.60	261.10	261.12	−0.02
0.80	259.70	259.70	0.00
1.00	258.25	258.22	0.03
1.20	256.65	256.68	−0.03
1.40	255.00	255.06	−0.06
1.60	253.40	253.36	0.04
1.81	251.50	251.48	0.02
2.00	249.70	249.67	0.03
2.20	247.60	247.66	−0.06
2.40	245.50	245.51	−0.01
2.60	243.20	243.20	0.00
2.80	240.70	240.71	−0.01
3.00	238.00	238.00	0.00
3.20	235.00	235.01	−0.01
3.40	231.65	231.70	−0.05
3.60	228.00	227.96	0.04
3.80	223.70	223.67	0.03
4.00	218.65	218.62	0.03
4.20	212.50	212.48	0.02
5.80	−282.95	−282.93	0.08
6.00	−288.85	−288.72	0.00
6.20	−293.70	−293.43	−0.11
6.40	−297.65	−297.40	−0.07
6.60	−301.05	−300.81	−0.03
6.80	−304.10	−303.81	−0.06
7.00	−306.75	−306.47	−0.02
7.20	−309.10	−308.87	0.05
7.40	−311.20	−311.05	0.15

[a]Thirty-seven titration points were collected. Six of these were found to be unsuitable (unbuffered concentrations) in the range 4.40–5.60 cm³ of titrant. The end point was determined by MAGEC to be 4.997 ± 0.001 cm³. (This compares with values determined by Gran plot of 4.998 ± 0.002 (data before the endpoint) and 5.006 ± 0.006 (data after the end point). The apparent $pK_w = 13.333$. Using the data before the end point, the best fit involving 22 points gave $E_0 = 374.5 ± 0.07$ and $s = 61.53 ± 0.03$. Using the data after the end point, the best fit involving nine points with an alkaline error correction factor of 1.49 × 10^{-13} gave $E_0 = 373.6 ± 9.0$ and $s = 61.50 ± 0.83$. Using all the buffered data, the best fit involving 31 points gave $E_0 = 374.5 ± 0.01$ and $s = 61.55 ± 0.005$.

TABLE 2. Results of Determination of the Protonation Constants of Cysteine at 37°C and in 0.150 M [NaCl] Background Electrolyte: 348 Data Points Were Determined in Five Potentiometric Titrations.

Parameter	Initial value	After first MINIQUAD	After first MAGEC	After second MINIQUAD	After second MAGEC	After third MINIQUAD	After third MAGEC
E_0 titration 1	423.0		422.6		422.5		422.5
E_0 titration 2	423.0		421.4		421.4		421.4
E_0 titration 3	423.0		422.7		422.7		422.7
E_0 titration 4	423.0		423.7		423.7		423.7
E_0 titration 5	423.0		423.3		423.4		423.4
$\log \beta_{11}$	10.0	10.103 ± 0.006		10.098 ± 0.002		10.097 ± 0.002	
$\log \beta_{12}$	18.0	18.023 ± 0.009		18.022 ± 0.003		18.223 ± 0.003	
$\log \beta_{13}$	20.0	19.968 ± 0.011		19.964 ± 0.004		19.966 ± 0.004	
Sum of squared residuals MINIQUAD		3.46×10^{-6}		4.18×10^{-7}		4.02×10^{-7}	
R factor		0.0045		0.0016		0.0015	

centrations. This would facilitate both the comparison of formation constant values and the elimination of errors inherent in the use of the pH scale. We believe that the numerical procedures outlined in this volume will be seen to have many advantages over traditional methods. This is particularly true in the context of pH scale measurements and the associated "adjustments" which they require.

The examples, shown in Tables 1 and 2, illustrate the use of MAGEC in calibrating a glass electrode system both in the strong acid versus strong base titration range as well as in the hydrogen ion concentration range when significant ligand–proton–metal ion interactions occur. The major advantage lies in the ability of MAGEC, together with MAGEC–MINIQUAD cycling, to solve this problem even when the protonation constants of the ligand are not known *a priori*. Thus, E_{const} becomes optimized for a given titration and any change in the liquid junction potential is incorporated into the particular E_{const} value obtained. Small variations in the liquid junction potential arising, for example, from poor reproducibility or concentration effects thus cease to be a problem.

Few limitations are imposed by the program itself. Any of the parameters E_{const}, s, β values, total concentrations of ligand and proton and K_w, can be determined provided there are sufficient titration points. However, when small differences in the values of different parameters have similar effects on the objective function, difficulties with correlation can arise. Thus the usual precautions which are adopted during any optimization procedure are to be heeded in this situation. Difficulties arising from this source may be minimized by using the MAGEC–MINIQUAD cycling procedure, in which only a few of the parameters are estimated simultaneously.

3. REFERENCES

1. P. W. Linder, P. M. May, R. G. Torrington, and D. R. Williams, *Analysis Using Glass Electrodes*, Open University Press, Milton Keynes, (1984).
2. *Manual of Symbols and Terminology for Physiochemical Quantities and Units*, 1973 Edition, International Union of Pure and Applied Chemistry (1975).
3. H. M. Irving, M. G. Miles, and L. D. Pettit, A Study of Some Problems in Determining the Stoicheiometric Proton Dissociation Constants of Complexes by Potentiometric Titrations Using a Glass Electrode, *Anal. Chim. Acta* **38**, 475–488 (1967).
4. G. Mattock and D. M. Band, in *Glass Electrodes For Hydrogen Ion and Other Cations* G. Eisenman, ed., Dekker, New York (1967), Chaps. 2 and 9.
5. G. Anderegg, in *Proc. Summer School on Stability Constants* (P. Paoletti, R. Barbucci, and L. Fabbrizzi, eds.), Bivigliano, Florence, Italy (1974), pp.11.
6. D. R. Williams, in *Proc. Summer School on Stability Constants* (P. Paoletti, R. Barbucci, and L. Fabbrizzi, eds.), Bivigliano, Florence, Italy (1974), pp. 125.
7. K. Schwabe, in *Electroanalytical Chemistry*, Vol. 10 in *Advances in Analytical Chemistry and Instrumentation* (H. W. Nurnberg, ed.), Wiley, London (1974), Chap. 7, pp. 495.
8. M. Filomena, G. F. G. Camoes, and A. K. Covington, New Procedure for Calibrating Glass Electrodes, *Anal. Chem.* **46**, 1547 (1974).
9. H. Rossotti, *The Study of Ionic Equilibria*, Longman, London (1978).
10. A. Albert and E. P. Serjeant, *The Determination of Ionisation Constants*, Chapman Hall, London (1971).
11. P. M. May, D. R. Williams, P. W. Linder, and R. G. Torrington, The Use of Glass Electrodes for the Determination of Formation Constants. Part I. A Definitive Method for Calibration, *Talanta* **29**, 249–256 (1982).

12. G. Gran, Determination of the Equivalence Point in Potentiometric Titrations, Pt I, *Acta Chem. Scand.* **4,** 559–567 (1950); Pt. II, *Analyst* **77,** 661–671 (1952).
13. F. J. C. Rossotti and H. Rossotti, Potentiometric Titrations Using Gran Plots, *J. Chem. Ed.* **42,** 375–378 (1965).
14. J. A. Nelder and R. Mead, A Simplex Method for Function Minimization, *Comput. J.* **7,** 308–313 (1965).
15. A. Sabatini, A. Vacca, and P. Gans, MINIQUAD—A General Computer Programme for the Computation of Formation Constants from Potentiometric Data, *Talanta* **21,** 53–57 (1974).
16. P. Gans, A. Sabatini, and A. Vacca, An Improved Computer Program for the Computation of Formation Constants from Potentiometric Data, *Inorg. Chim. Acta* **18,** 237–239 (1976).

4. INSTRUCTIONS FOR THE USE OF MAGEC

4.1. Basic Principles

In our experience, the calibration of an electrode system is best carried as follows.

1. An initial estimate of E_{const} is obtained by applying a CALIBT analysis to a strong acid versus a strong base titration. An additional advantage of this step stems from the ability to use the CALIBT analysis for checking the reliability of an individual glass electrode. In the MAGEC analysis of strong acid versus strong base titration data, the appearance of sudden and marked increases in the titre volume residuals indicates unreliability of the electrode. Figure 1 illustrates the recommended procedure for using CALIBT.

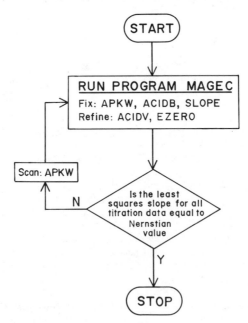

FIGURE 1. Use of the subroutine CALIBT. Flow diagram of the usual procedure with strong acid versus strong base titrations. APKW, apparent dissociation constant of water (K_w); ACIDV, titrand acid concentration (negative for alkali); ACIDB, titrant acid concentration (negative for alkali); EZERO, electrode intercept (E_{const}); SLOPE, electrode slope (s).

2. The E_{const} value obtained in step 1 is used in conjunction with titration data obtained for a ligand (usually acidified with mineral acid) versus a strong base, in a suitable program to refine the values of the protonation constants for the ligand. MINIQUAD[15, 16] (see Chapter 4) is most suitable for this purpose. Individual titrations should be processed together to yield global values for each protonation constant.
3. The ligand titration data and the protonation constants of 2 are then processed by the main routine of MAGEC to yield refined E_{const} values.
4. Steps 2 and 3 are repeated in a cyclical fashion until convergence is obtained.
5. If desired, values of E_{const} and protonation constants may be improved further by refining the values of C_H, C_L, and K_w in a series of subsequent cycles. Care must be taken, however, to ensure that the calculation is not performed using more degrees of freedom than are warranted by the accuracy of the data.

The flow diagram of Figure 2 illustrates steps 2–4.

4.2. Data Input Requirements

MAGEC analyzes potentiometric glass electrode data from titrations where the titrand and titrant may be a combination of strong acid, strong base, and/or ligand

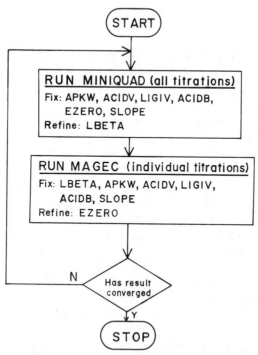

FIGURE 2. Use of the main routine of MAGEC. Flow diagram of the usual procedure with titrations involving ligands. LBETA, ligand protonation constants; LIGIV, ligand concentration in vessel; APKW, ACIDV, ACIDB, EZERO, SLOPE, as for Figure 1.

solutions. There are a variety of processing options that have been discussed in previous sections.

Assembling the data deck is straightforward, and includes a simple method of specifying which combination of nine parameters may be refined.The details of the input data will now be described. The FORMAT specifications of the input data are in square brackets. Only boldfaced variable names require values for that particular item.

The input file consists of the following lines:

1. 1 line [20A4]: Descriptive title of the system under consideration.
2. 1 line [A3,2X,I1,1X,I1,2X,3G10.3]: **KEY,IVAL,JVAL,(PK(I,J),J=1,3)**

    KEY: PKW is the key word used to indicate that the remaining data on this card relate to the pK_w of water.

    IVAL: Refinement key for this parameter. The allowed values are 0 (=no refinement); 1 (=refinement without boundary constraints); 2 (=refinement within boundary constraints).

    PK(I,1): pK_w of water, read in as a negative logarithmic value.

    PK(I,2): Lower refinement limit. Set to zero unless IVAL = 2.

    PK(I,2): Upper refinement limit. Set to zero unless IVAL = 2.
3. 1 line [as item 2]: **KEY,IVAL,JVAL,(PK(I,J),J=1,3)**

    KEY: Key word is PK1, i.e., first stepwise protonation constant.

    IVAL: As item 2.

    PK(I,1): Value of first protonation constant, read in as a logarithmic value.

    PK(I,2): $\left.\begin{array}{l}\text{PK(I,2):}\\ \text{PK(I,3):}\end{array}\right\}$ As item 2.
4. *N* lines [as item 2]: **KEY,IVAL,JVAL,(PK(I,J),J=1,3)**

    KEY: Key word is PK2, i.e., second *stepwise* protonation constant.

    IVAL: As item 2.

    $\left.\begin{array}{l}\text{PK(I,1):}\\ \text{PK(I,2):}\\ \text{PK(I,3):}\end{array}\right\}$ As item 3.

    Note: There will be a total of *N* cards having key words PK1, PK2, . . . , PK*N*, for *N* protonation constants, A maximum of six protonation constants may be read in. If protonation constants are inappropriate *omit* item(s) 4.
5. 1 line [as item 2]: **KEY,IVAL,JVAL,(VZERO(J),J=1,3)**

    KEY: Key word is VZ0, i.e., the initial volume in the vessel.

    IVAL: As item 2.

    VZERO(1),VZERO(2),VZERO(3): The initial volume value, lower, and upper boundary constraints (if **IVAL** = 2), respectively.
6. 1 line [as item 2]: **KEY,REFINE(2),JVAL,(HVESSL(J),J=1,3)**

    KEY: Key word is H+V, i.e., mineral acid in titration vessel.

    REFINE(2): Alternative name for IVAL, same as item 2.

    HVESSL(1),HVESSL(2),HVESSL(3): Concentration of mineral acid in titration vessel, lower and upper boundary constraints. Use negative values for alkali concentrations.
7. 1 line [as item 2]: **KEY,REFINE(3),JVAL,(LVESSL(J),J=1,3)**

KEY: Key word is LGV, i.e., ligand in titration vessel.

REFINE(3): As item 5.

JVAL: Number of dissociable protons on ligand.

LVESSL(1),LVESSL(2),LVESSL(3): Concentration of ligand in titration vessel, lower and upper boundary constraints.

Note: This card must be present, even if no ligand is present. In this situation set LVESSL(1) to zero.

8. 1 line [as item 2]:**KEY,REFINE(4),JVAL,(HBURET(J),J=1,3)**

KEY: Key word is H + B, i.e., mineral acid in buret.

REFINE(4): As item 5.

HBURET(1),HBURET(2),HBURET(3): Concentration of mineral acid in buret, lower and upper boundary constraints. Use negative values for alkali concentrations.

9. 1 line [as item 2]: **KEY,REFINE(5),NDPB,(LBURET(J),J=1,3)**

KEY: Key word is LGB, i.e., ligand in buret.

REFINE(5): As item 5.

NDPB: Number of dissociable protons on ligand.

LBURET(1),LBURET(2),LBURET(3): Concentration of ligand in buret, lower and upper boundary constraints. This card must be present even if no ligand is present in the buret. In this sitaution set LBURET(1) to zero.

10. 1 line [as item 2]: **KEY,REFINE(6),JVAL,(EZERO(J),J=1,3)**

KEY: Key word is EZO, i.e., glass electrode intercept.

REFINE(6): As item 5.

EZERO(1),EZERO(2),EZERO(3):Glass electrode intercept parameters, lower and upper boundary constraints.

11. 1 line [as item 2]: **KEY,REFINE(7),JVAL,(SLOPE(J),J=1,3)**

KEY: Key word is SLP, i.e., glass electrode slope.

REFINE(7): As item 5.

SLOPE(1),SLOPE(2),SLOPE(3): Glass electrode slope parameters, lower and upper boundary constraints.

12. N line [2G10.3]: **V(I),E(I)**

V(I): The ith volume of titrant added, in ml.

E(I): The resulting ith potential, in mV.

Note: There will be $(N - 1)$ data pairs. The Nth card has V(N) equal to -1.00 signifying the end of the data set. A maximum of 100 points are allowed.

This completes the description of the input file for one set of data. A second set of data may be processed by adding items 1–12 after the final card of the previous data set.

5. PRESENTATION OF THE PROGRAM

5.1. Listing of MAGEC

The following is a listing of the program MAGEC.

```
1    C
2    C
3    C                         PROGRAM MAGEC.
4    C                         -------------
5    C
6    C
7    C
8    C                         MULTIPLE ANALYSIS
9    C                        OF TITRATION DATA FOR
10   C                     GLASS ELECTRODE CALIBRATION.
11   C
12   C
13   C
14   C    ****************************************************************
15   C
16   C
17   C
18   C     THE PROGRAM ANALYSES POTENTIOMETRIC GLASS ELECTRODE DATA
19   C     FROM TITRATIONS IN WHICH THE TITRAND AND THE TITRANT,
20   C     RESPECTIVELY, MAY BE    (1) A STRONG ACID OR A STRONG BASE,
21   C     (2) A LIGAND OR (3) A COMBINATION OF (1) AND (2).
22   C
23   C     ESTIMATES OF THE CONCENTRATIONS, THE EQUILIBRIUM CONSTANTS AND
24   C     THE PARAMETERS FOR THE NERNSTIAN EQUATION OF THE GLASS ELECTRODE
25   C     RESPONSE TO HYDROGEN ION CONCENTRATION CAN BE REFINED.    THE USER
26   C     SPECIFIES WHICH ARE TO BE SIMULTANEOUSLY OPTIMISED.    UPPER AND
27   C     LOWER LIMITS CAN BE IMPOSED ON ALL PARAMETERS BEING REFINED.
28   C
29   C     THE OPTIMISATION IS PERFORMED BY SUBROUTINE NELM AND GENERALLY,
30   C     THE EQUILIBRIUM CONCENTRATIONS ARE EVALUATED BY SUBROUTINE ML AND
31   C     SUBROUTINE FUNCT.
32   C
33   C     SUBPROGRAM CALIBT ONLY ACCEPTS DATA PERTAINING TO MONOBASIC
34   C     REACTANTS.    IT IS MAINLY USED TO ANALYSE STRONG ACID VERSUS
35   C     STRONG BASE TITRATIONS.    THE PARAMETERS OF THE NERNSTIAN
36   C     EQUATION ARE FOUND BY LINEAR LEAST SQUARES BEST FIT.
37   C     IN ADDITION IT PERFORMS GRAN PLOT CALCULATIONS, SCANS VALUES OF
38   C     PKW AND PK1 AND ADJUSTS THE CONCENTRATIONS OF REACTANTS IF THIS
39   C     IMPROVES THE AGREEMENT BETWEEN THE THEORETICAL AND THE OBSERVED
40   C     SLOPE FOR THE GLASS ELECTRODE RESPONSE.
41   C
42   C
43   C
44   C     THE PROGRAM WAS DEVELOPED AT THE UNIVERSITY OF WALES INSTITUTE
45   C     OF SCIENCE AND TECHNOLOGY IN 1978 AND 1979.
46   C
47   C     IT IS WRITTEN IN FORTRAN IV.
48   C
49   C
50   C
51   C    ****************************************************************
52   C
53   C
54   C
55   C                         DATA INPUT
56   C                         ----------
57   C
58   C
59   C     The following information is required for each titration.
60   C
61   C
62   C
63   C     Item    Flag                        Description
64   C
65   C
66   C     1                   The title
67   C     2       PKW         Ionisation constant of water (as negative log)
68   C     3       PK1         First stepwise protonation constant (as log)
69   C     3       PK2          - up to six acid PK values can be accepted
70   C     4       VZO         Initial volume in vessel
71   C     5       H+V         Mineral acid conc. in vessel
72   C     6       LGV         Ligand conc. in vessel
73   C     7       H+B         Mineral conc. in burette
74   C     8       LGB         Ligand conc. in burette
75   C     9       EZO         Glass electrode intercept parameter
```

```
76   C    10       SLP          Glass electrode slope parameter
77   C    11                    Titration data pairs (vol,emf).
78   C                          Up to 100 points. Volume in ml, emf in mV.
79   C                          USE -1.00 AFTER LAST VOL;EMF CARD TO
80   C                          INDICATE END OF DATA SET.
81   C
82   C
83   C
84   C
85   C                             Format requirements
86   C
87   C
88   C       ITEM  1  : (20A4)
89   C       ITEMS 2-11: (A3,2X,I1,1X,I1,2X,3G10.3)
90   C       ITEM  12 : (2G10.3)
91   C
92   C
93
94   C
95   C                             Data requirements
96   C
97   C
98   C       ITEMS 2-10:    A3  -  Flag as defined above
99   C                      I1  -  Refine key
100  C                      I1  -  Ligand NDP (only for ITEMS 6 and 8)
101  C                      G10.3 - Value or estimate of parameter
102  C                      G10.3 - Lower refinement limit of value
103  C                      G10.3 - Upper refinement limit of value
104  C
105  C
106  C       Refine keys:   0 = no refinement
107  C                      1 = normal refinement
108  C                      2 = refine between upper and lower limits
109  C
110  C
111  C
112  C       NDP:           Number of dissociable protons on the ligand
113  C       Mineral acid:  Use negative values for alkali concentrations
114  C
115  C
116  C
117  C   ****************************************************************
118  C
119  C
120        INTEGER  TITLE, LIT(15)
121        DIMENSION  X(7), Y(7), H(7), XS(7), FP(8), ID(7)
122        INTEGER  OUT, REFINE
123        REAL LVESSL, LBURET
124        LOGICAL TITEND
125        COMMON /ONE/  PK(7,3), VZERO(3), HVESSL(3), LVESSL(3), HBURET(3),
126       &    LBURET(3), EZERO(3), SLOPE(3), REFINE(15), NDPV, NDPB
127        COMMON /TWO/ CI(7),CX(2),TT(2),HX(2),TOLC(2),DT(2),DDT(2,2)
128        COMMON /THREE/  V(100), E(100), NP
129        COMMON /FOUR/  BETA(7), ARRAY(3), TOL, NCONST, NBETAH, JQR(2,7)
130        COMMON /FIVE/  IN, OUT, IFAIL, JFAIL, AL10, KOUNT
131        COMMON /SIX/  TITLE(20), P(100)
132  C
133        DATA  LIT /'PKW','PK1','PK2','PK3','PK4','PK5','PK6',
134       &    'VZO','H+V','LGV','H+B','LGB','EZO','SLP','   '/
135  C
136  C
137  C
138  9200 FORMAT(20A4)
139  9201 FORMAT(8G10.3)
140  9202 FORMAT(A3,2X,I1,1X,I1,2X,3G10.3)
141  9203 FORMAT('0')
142  9204 FORMAT(//'0',1H*,3X,1H*,6X,3(1H*),7X,4(1H*),5X,5(1H*),6X,
143       &    4(1H*),/' ',2(2(1H*),1X),4X,1H*,3X,1H*,5X,3(1H*,9X),
144       &    /' ',1H*,1X,1H*,1X,1H*,5X,5(1H*),5X,1H*,1X,3(1H*),
145       &    5X,4(1H*),6X,1H*,/' ',3(1H*,3X,1H*,5X),1H*,9X,1H*,
146       &    /' ',2(1H*,3X,1H*,5X),1X,4(1H*),5X,5(1H*),6X,4(1H*),
147       &    /' ',45(1H-),////)
148  9205 FORMAT(///'0',30X,20A4)
149  9206 FORMAT('0',8X,A3,12X,I1,13X,F17.3,17X,2F15.3)
150  9207 FORMAT('0',8X,A3,12X,I1,13X,1PE17.4,17X,2E15.4)
```

```
151      9208 FORMAT('Ø',8X,A3,12X,I1,12X,I1,1PE17.4,20X,2E15.4)
152      9209 FORMAT('+',89X,'N/A')
153      9210 FORMAT(////'ØDIMENSION LIMITS OF TITRATION DATA ARRAYS EXCEEDED')
154      9211 FORMAT(////'Ø','SEQUENCE ERROR DETECTED IN THE DATA')
155      9212 FORMAT(////'Ø','ERROR IN THE LIMITS FOR REFINEMENT')
156      9213 FORMAT(////'Ø','ACID DISSOCIATION CONSTANT ERROR')
157      9214 FORMAT('Ø','EXECUTION TERMINATED.',///)
158      9215 FORMAT(////'Ø',4X,'V(OBS.) V(CALC.) EMF(OBS.) EMF(CALC.)',
159         &      7X,'RESIDUALS',12X,'PH',9X,'FL',8X,
160         &      'EZO(CALC.)   SLP(CALC.)',///)
161      9216 FORMAT(/////'Ø',2X,'REFINED VALUES FOR THE PARAMETERS ARE NOW:',///
162         &      'Ø',2X,'OBJ. FUNCT.',5X,A4,6(8X,A4))
163      9217 FORMAT('+',102X,'OPTIMIZATION RECORD',//)
164      9219 FORMAT('+',102X,7(1H*),3X,'CONVERGED',//)
165      9220 FORMAT(/////'Ø',4X,'INPUT VALUES FOR THE TITRATION PARAMETERS ARE:'
166         &      ,///'Ø',4X,'IDENTIFIER',4X,'REFINE KEY',7X,'NDP',10X,'VALUE',
167         &      26X,'LOWER AND UPPER LIMITS',//)
168      9221 FORMAT('Ø',1P8E12.3)
169    C
170    C
171    C      SECTION ONE:    INITIALISATION.
172    C
173    C
174           IN = 5
175           OUT = 6
176           IFAIL = Ø
177           JFAIL = Ø
178           EPS = 1.ØE-6
179           TOL = 1.ØE-4
180           AL1Ø = ALOG(1Ø.ØØ)
181           TITEND = .FALSE.
182    C
183    C
184    C      SECTION TWO:    DATA INPUT
185    C
186    C
187      200 WRITE(OUT,92Ø3)
188           READ(IN,9200,END=299)   TITLE
189           WRITE(OUT,92Ø4)
190           WRITE(OUT,92Ø5)    TITLE
191           WRITE(OUT,922Ø)
192           M = 1
193           DO 207 I=1,7
194           READ(IN,92Ø2)  KEY, IVAL, JVAL, (PK(I,J), J=1,3)
195           N = 1
196           IF(IVAL.EQ.2)  N = 3
197           WRITE(OUT,92Ø6)  KEY, IVAL, (PK(I,J), J=1,N)
198           IF(N.EQ.1)  WRITE(OUT,92Ø9)
199           IF(KEY.EQ.LIT(8))  GO TO 21Ø
200           IF(KEY.NE.LIT(I))  GO TO 25Ø
201           DO 2Ø5 L=1,3
202      205 ARRAY(L) = PK(I,L)
203           IF(IVAL.GT.Ø)  CALL SETUP(ID,LIT(I),M,X,H,IVAL,ARRAY)
204           IF(I.LE.2)  GO TO 207
205           IF(PK(I,1).LT.PK(I-1,1))  GO TO 27Ø
206      207 REFINE(I+7) = IVAL
207           READ(IN,92Ø2)  KEY, IVAL, JVAL, (VZERO(J),J=1,3)
208           IF(KEY.NE.LIT(8))  GO TO 29Ø
209           IF(IVAL.EQ.2.AND.(PK(I,2).GT.PK(I,1).OR.PK(I,3).LT.PK(I,1)))
210         &    GO TO 26Ø
211           GO TO 215
212    C
213    C      REFINE(1)  -  VZERO
214    C      REFINE(2)  -  HVESSL
215    C      REFINE(3)  -  LVESSL
216    C      REFINE(4)  -  HBURET
217    C      REFINE(5)  -  LBURET
218    C      REFINE(6)  -  EZERO
219    C      REFINE(7)  -  SLOPE
220    C      REFINE(8)  -  PKW
221    C      REFINE(9) TO REFINE(14)  -  PK1 TO PK6
222    C
223      210 NCONST = I - 1
224           DO 212 J=1,3
225      212 VZERO(J) = PK(I,J)
226      215 REFINE(1) = IVAL
```

```
227            IF(REFINE(1).EQ.2.AND.(VZERO(2).GT.VZERO(1).OR.VZERO(3)
228          &   .LT.VZERO(1)))  GO TO 260
229            IF(REFINE(1).GT.0)  CALL SETUP(ID,LIT(8),M,X,H,REFINE(1),VZERO)
230            READ(IN,9202)  KEY, REFINE(2), JVAL, (HVESSL(J),J=1,3)
231            N = 1
232            IF(REFINE(2).EQ.2)  N = 3
233            WRITE(OUT,9207)  KEY, REFINE(2), (HVESSL(J), J=1,N)
234            IF(N.EQ.1)  WRITE(OUT,9209)
235            IF(KEY.NE.LIT(9))  GO TO 250
236            IF(REFINE(2).EQ.2.AND.(HVESSL(2).GT.HVESSL(1).OR.HVESSL(3)
237          &   .LT.HVESSL(1))  GO TO 260
238            IF(REFINE(2).GT.0)  CALL SETUP(ID,LIT(9),M,X,H,REFINE(2),HVESSL)
239            READ(IN,9202)  KEY, REFINE(3), NDPV, (LVESSL(J),J=1,3)
240            N = 1
241            IF(REFINE(3).EQ.2)  N = 3
242            WRITE(OUT,9208)  KEY, REFINE(3), NDPV, (LVESSL(J), J=1,N)
243            IF(N.EQ.1)  WRITE(OUT,9209)
244            IF(KEY.NE.LIT(10))  GO TO 250
245            IF(REFINE(3).EQ.2.AND.(LVESSL(2).GT.LVESSL(1).OR.LVESSL(3)
246          &   .LT.LVESSL(1)))  GO TO 260
247            IF(REFINE(3).GT.0)  CALL SETUP(ID,LIT(10),M,X,H,REFINE(3),LVESSL)
248            READ(IN,9202)  KEY, REFINE(4), JVAL, (HBURET(J),J=1,3)
249            N = 1
250            IF(REFINE(4).EQ.2)  N = 3
251            WRITE(OUT,9207)  KEY, REFINE(4), (HBURET(J), J=1,N)
252            IF(N.EQ.1)  WRITE(OUT,9209)
253            IF(KEY.NE.LIT(11))  GO TO 250
254            IF(REFINE(4).EQ.2.AND.(HBURET(2).GT.HBURET(1).OR.HBURET(3)
255          &   .LT.HBURET(1)))  GO TO 260
256            IF(REFINE(4).GT.0)  CALL SETUP(ID,LIT(11),M,X,H,REFINE(4),HBURET)
257            READ(IN,9202)  KEY, REFINE(5), NDPB, (LBURET(J),J=1,3)
258            N = 1
259            IF(REFINE(5).EQ.2)  N = 3
260            WRITE(OUT,9208)  KEY, REFINE(5), NDPB, (LBURET(J), J=1,N)
261            IF(N.EQ.1)  WRITE(OUT,9209)
262            IF(KEY.NE.LIT(12))  GO TO 250
263            IF(REFINE(5).EQ.2.AND.(LBURET(2).GT.LBURET(1).OR.LBURET(3)
264          &   .LT.LBURET(1)))  GO TO 260
265            IF(REFINE(5).GT.0)  CALL SETUP(ID,LIT(12),M,X,H,REFINE(5),LBURET)
266            READ(IN,9202)  KEY, REFINE(6), JVAL, (EZERO(J),J=1,3)
267            N = 1
268            IF(REFINE(6).EQ.2)  N = 3
269            WRITE(OUT,9206)  KEY, REFINE(6), (EZERO(J), J=1,N)
270            IF(N.EQ.1)  WRITE(OUT,9209)
271            IF(KEY.NE.LIT(13))  GO TO 250
272            IF(REFINE(6).EQ.2.AND.(EZERO(2).GT.EZERO(1).OR.EZERO(3)
273          &   .LT.EZERO(1)))  GO TO 260
274            IF(REFINE(6).GT.0)  CALL SETUP(ID,LIT(13),M,X,H,REFINE(6),EZERO)
275            READ(IN,9202)  KEY, REFINE(7), JVAL, (SLOPE(J),J=1,3)
276            N = 1
277            IF(REFINE(7).EQ.2)  N = 3
278            WRITE(OUT,9206)  KEY, REFINE(7), (SLOPE(J), J=1,N)
279            IF(N.EQ.1)  WRITE(OUT,9209)
280            IF(KEY.NE.LIT(14))  GO TO 250
281            IF(REFINE(7).EQ.2.AND.(SLOPE(2).GT.SLOPE(1).OR.SLOPE(3)
282          &   .LT.SLOPE(1)))  GO TO 260
283            IF(REFINE(7).GT.0)  CALL SETUP(ID,LIT(14),M,X,H,REFINE(7),SLOPE)
284     C
285            DO 231 I=1,100
286            READ(IN,9201,END=235)  V(I), E(I)
287            IF(V(I).LT.0.000)  GO TO 239
288     231 CONTINUE
289            I = 101
290            READ(IN,9202,END=235)  KEY
291            WRITE(OUT,9210)
292            GO TO 290
293     235 TITEND = .TRUE.
294     239 NP = I - 1
295            GO TO 300
296     C
297     250 WRITE(OUT,9211)
298            GO TO 290
299     260 WRITE(OUT,9212)
300            GO TO 290
301     270 WRITE(OUT,9213)
302     290 WRITE(OUT,9214)
```

```
303     299 STOP
304   C
305   C
306   C      SECTION THREE:    REFINEMENT BY SUBROUTINE NELM.
307   C
308   C
309     300 NBETAH = NCONST - 1
310         M = M - 1
311         IF(NBETAH.LT.1)   GO TO 400
312         DO 301 I=1,NBETAH
313         JQR(1,I) = I
314     301 JQR(2,I) = 1
315         JQR(1,NCONST) = -1
316         JQR(2,NCONST) = Ø
317   C     IF(M-1)  350, 320, 340
318   C 320 KOUNT = -50
319   C     X(1) = Y(1) + H(1)
320   C     WRITE(OUT,9215)
321   C     CALL CALC(M,X,F,ICALC)
322   C 325 X(1) = Y(1) - H(1)
323   C     WRITE(OUT,9215)
324   C     CALL CALC(M,X,F,ICALC)
325   C 327 X(1) = Y(1)
326   C     GO TO 350
327         IF(M.EQ.Ø)   GO TO 350
328     340 KOUNT = -25
329         WRITE(OUT,9215)
330         CALL CALC(Ø,X,F,ICALC)
331     341 KOUNT = Ø
332         WRITE(OUT,9216)   (ID(I), I=1,M)
333         WRITE(OUT,9217)
334         DO 344 I=1,M
335     344 Y(I) = X(I)
336         IF(M.LT.7)   Y(M+1) = SLOPE(1)
337         M1 = M + 1
338         MP = M * (M + 2)
339         CALL NELM(X,F,EPS,P,FP,XS,H,Ø,M,M1,MP)
340         IF(KOUNT.LE.M)   GO TO 355
341         WRITE(OUT,9221)   F, (X(I), I=1,M)
342         IF(IFAIL.EQ.Ø)        WRITE(OUT,9219)
343         IFAIL = Ø
344   C
345     350 KOUNT = -100
346         WRITE(OUT,9215)
347         CALL CALC(M,X,F,ICALC)
348   C     RESET ORIGINAL PARAMETER VALUES.
349     355 REFINE(7) = 2
350         SLOPE(3) = SLOPE(1) - ABS(SLOPE(1)) / 10.ØØ
351         SLOPE(2) = SLOPE(1)
352     360 CALL CALC(M,Y,F,ICALC)
353   C
354   C
355   C      SECTION FOUR
356   C
357   C
358     400 IF(NBETAH.GT.1.OR.ABS(LBURET(1)).GT.1.ØE-8)   GO TO 999
359         IF(LVESSL(1).GT.1.E-8.AND.HBURET(1).LE.Ø..AND.NDPV.NE.1) GOTO 999
360         IF(LVESSL(1).GT.1.E-8.AND.HBURET(1).GT.Ø..AND.NDPV.NE.Ø) GOTO 999
361         PKWMIN = PK(1,1)
362         PKWMAX = Ø.ØØØ
363         BLGMIN = PK(2,1)
364         BLGMAX = Ø.ØØØ
365         IF(REFINE(8).NE.2)   GO TO 401
366         PKWMIN = PK(1,2)
367         PKWMAX = PK(1,3)
368         GO TO 450
369     401 IF(REFINE(9).NE.2)   GO TO 450
370         BLGMIN = PK(2,2)
371         BLGMAX = PK(2,3)
372     450 CALL CALIBT(VZERO(1),EZERO(1),SLOPE(1),HVESSL(1),LVESSL(1),
373       &  HBURET(1),PKWMIN,PKWMAX,BLGMIN,BLGMAX)
374   C
375     999 IF(.NOT.TITEND)   GO TO 200
376         STOP
377         END
```

```
 1    C
 2    C
 3    C       ******************************************************************
 4            SUBROUTINE NELM(X,F,EPS,P,FP,XS,H,IX,N,N1,NP)
 5    C
 6    C       REFERENCE:- NELDER J.A. AND MEAD R., COMPUT. J., 1965, 7, 308.
 7    C
 8            DIMENSION X(N), P(NP), FP(N1), H(N), XS(N)
 9       9000 FORMAT('+',100X,3I3,3X,'TEST')
10       9001 FORMAT('+',100X,3I3,3X,'REFLECTION')
11       9002 FORMAT('+',100X,3I3,3X,'EXPANSION')
12       9003 FORMAT('+',100X,3I3,3X,'CONTRACTION')
13            ISH = 1
14            IS = 0
15            NN = N * (N + 1)
16            CALL CALC(N,X,F,ICALC)
17            IF(ICALC.EQ.1)GO TO 100
18            FP(1) = F
19            IF(IX.NE.0)  GO TO 2
20            DO 1 I=1,N
21            K = I
22            DO 1 J=1,N1
23            P(K) = X(I)
24            IF(I-J+1.NE.0)  GO TO 1
25            P(K) = X(I) + H(I)
26          1 K = K + N
27          2 K = 1 + N
28            DO 3 I=2,N1
29            DO 4 J=1,N
30            X(J) = P(K)
31          4 K = K + 1
32            CALL CALC(N,X,F,ICALC)
33            IF(ICALC.EQ.1)GO TO 100
34          3 FP(I) = F
35            IF(FP(1)-FP(2).GT.0.0)  GO TO 5
36            IH = 2
37            IL = 1
38            GO TO 6
39          5 IH = 1
40            IL = 2
41          6 IF(N1.LE.2) GO TO 49
42            DO 7 I=3,N1
43            IF(FP(I)-FP(IH).GT.0.0)  GO TO 8
44            IF(FP(I)-FP(IL).GE.0.0)  GO TO 7
45            IL = I
46            GO TO 7
47          8 IH = I
48          7 CONTINUE
49         49 XN = N
50         50 K1 = NN
51            DO 9 I=1,N
52            K = I
53            S = 0.0
54            DO 10 J=1,N1
55            IF(J-IH.EQ.0)  GO TO 10
56            S = S + P(K)
57         10 K = K + N
58            K1 = K1 + 1
59          9 P(K1) = S / XN
60            K = NN + 1
61            DO 11 I=1,N
62            X(I) = P(K)
63         11 K = K + 1
64            CALL CALC(N,X,F0,ICALC)
65            IF(ICALC.EQ.1)GO TO 100
66            WRITE(6,9000)  IL, IH, IS
67            S = 0.0
68            DO 12 I=1,N1
69         12 S = S + (FP(I) - F0) ** 2
70            S = S / XN
71            IF(S-EPS.LE.0.00)  GO TO 100
72            IF((IH-1).EQ.0)  GO TO 13
73            IS = 1
74            GO TO 14
75         13 IS = 2
```

```
76      14 DO 15 I=1,N1
77         IF(I-IH.EQ.Ø)  GO TO 15
78         IF(FP(I)-FP(IS).LE.Ø.Ø)  GO TO 15
79         IS = I
80      15 CONTINUE
81    C**** REFLECTION
82         K = (IH - 1) * N + 1
83         KØ = NN + 1
84         DO 16 I=1,N
85         X(I) = 2.Ø * P(KØ) - P(K)
86         K = K + 1
87      16 KØ = KØ + 1
88         K = K - N
89         CALL CALC(N,X,F,ICALC)
90         IF(ICALC.EQ.1)GO TO 100
91         WRITE(6,9001)  IL, IH, IS
92         IF(F-FP(IL).GE.Ø.Ø)  GO TO 20
93    C**** EXPANSION
94         KØ = NN + 1
95         DO 17 I=1,N
96         XS(I) = 2.Ø * X(I) - P(KØ)
97      17 KØ = KØ + 1
98         CALL CALC(N,XS,FS,ICALC)
99         IF(ICALC.EQ.1)GO TO 100
100        WRITE(6,9002)  IL, IH, IS
101        IF(FS-FP(IL).GE.Ø.Ø)  GO TO 18
102        DO 19 I=1,N
103        P(K) = XS(I)
104     19 K = K + 1
105        FP(IH) = FS
106        IL = IH
107        IH = IS
108        GO TO 50
109     18 IL = IH
110        IH = IS
111        FP(IL) = F
112     21 DO 22 I=1,N
113        P(K) = X(I)
114     22 K= K + 1
115        GO TO 50
116     20 IF(F-FP(IS).GE.Ø.Ø)  GO TO 23
117        FP(IH) = F
118        IH = IS
119        GO TO 21
120     23 IF(F-FP(IH).GE.Ø.Ø)  GO TO 25
121        DO 24 I=1,N
122        P(K) = X(I)
123     24 K = K + 1
124        FP(IH) = F
125    C**** CONTRACTION
126        K = K - N
127     25 KØ = NN + 1
128        DO 26 I=1,N
129        XS(I) = Ø.5 * (P(K) + P(KØ))
130        K = K + 1
131     26 KØ = KØ + 1
132        K = K - N
133        CALL CALC(N,XS,FS,ICALC)
134        IF(ICALC.EQ.1)GO TO 100
135        WRITE(6,9003)  IL, IH, IS
136        IF(FS-FP(IH).GE.Ø.Ø)  GO TO 40
137        DO 27 I=1,N
138        P(K) = XS(I)
139     27 K = K + 1
140        FP(IH) = FS
141        IF(FP(1)-FP(2).GT.Ø.Ø)  GO TO 28
142        IH = 2
143        GO TO 29
144     28 IH = 1
145     29 DO 31 I=3,N1
146        IF(FP(I)-FP(IH).LE.Ø.Ø)  GO TO 31
147        IH = I
148     31 CONTINUE
149        GO TO 50
150     40 FP(1) = FP(IL)
151        KKK = MOD(ISH,1Ø)
```

```
152              ISH = ISH + 1
153              IF((IL-1).EQ.Ø)  GO TO 43
154              K = (IL - 1) * N
155              DO 41 I=1,N
156              K = K + 1
157              X(I) = P(K)
158              P(K) = P(I)
159           41 P(I) = X(I)
160              IL = 1
161           43 K = N
162              DO 42 I=2,N1
163              DO 42 J=1,N
164              K = K + 1
165              P(K) = Ø.5 * (P(K) + P(J))
166              ISH = ISH + 1
167           42 CONTINUE
168              GO TO 2
169          100 IL = 1
170              DO 101 I=2,N1
171              IF(FP(I)-FP(IL).GE.Ø.ØØ)  GO TO 101
172              IL = I
173          101 CONTINUE
174    C         WRITE(6,2349)  IL, FØ, (FP(I), I=1,N1)
175    C2349 FORMAT('Ø','NELM EXIT',I5,1P7E15.4,//)
176              F = FØ
177              IF(F.LT.FP(IL))  RETURN
178              K = (IL - 1) * N
179              DO 102 I=1,N
180              K = K + 1
181              X(I) = P(K)
182          102 CONTINUE
183              F = FP(IL)
184              RETURN
185              END




  1    C
  2    C         *****************************************************************
  3    C
  4              SUBROUTINE CALC(M,X,F,ICALC)
  5    C
  6    C         THIS ROUTINE IS USED TO CALCULATE THE OBJECTIVE FUNCTION
  7    C         VALUES FOR SUBROUTINE NELM.  IT ALSO PRINTS AN ANALYSIS
  8    C         OF THE TITRATION DATA FOR A GIVEN SET OF PARAMETER VALUES.
  9    C
 10              INTEGER  OUT, REFINE
 11              REAL LVESSL, LBURET
 12              COMMON /ONE/  PK(7,3), VZERO(3), HVESSL(3), LVESSL(3), HBURET(3),
 13           &    LBURET(3), EZERO(3), SLOPE(3), REFINE(15), NDPV, NDPB
 14              COMMON /TWO/ CI(7),CX(2),TT(2),HX(2),TOLC(2),DT(2),DDT(2,2)
 15              COMMON /THREE/  V(100), E(100), NP
 16              COMMON /FOUR/  BETA(7), ARRAY(3), TOL, NCONST, NBETAH, JQR(2,7)
 17              COMMON /FIVE/  IN, OUT, IFAIL, JFAIL, AL10, KOUNT
 18              DIMENSION X(7)
 19     9000 FORMAT(' ',2F9.3,2F10.2,F12.3,F9.3,F13.3,1PE13.2,ØPF16.3,F11.3)
 20     9001 FORMAT(' ','FAILURE IN SUBROUTINE CALC')
 21     9002 FORMAT(' ','CAUSED BY NON-CONVERGENCE IN SUBROUTINE ML')
 22     9003 FORMAT(' ','THE MAXIMUM NUMBER OF ITERATIONS HAS BEEN EXCEEDED.')
 23     9004 FORMAT('+',118X,'?')
 24     9005 FORMAT('Ø','A QUESTION MARK INDICATES A')
 25     9006 FORMAT('+',120X,'UNBUFFERED')
 26     9007 FORMAT(' ',1P8E12.3)
 27     9008 FORMAT(' ',2F9.3,F10.2,10X,F12.3,5X,'FAILURE TO SOLVE FOR FREE '
 28           & ,'HYDROGEN ION CONCENTRATION')
 29    C
 30    C
 31              ILOAD=Ø
 32              ICALC=Ø
 33              KFAIL = Ø
 34              IF(M.EQ.Ø)  GO TO 20
 35              KOUNT = KOUNT + 1
 36              IF(KOUNT.GE.M*50+80)  GO TO 60
 37    C
```

```
38    C       LOAD THE PARAMETERS BEING REFINED.
39    C
40            I = 0
41            J = 7
42            DO 15 K=1,NCONST
43            J = J + 1
44            IF(REFINE(J).LT.1)  GO TO 15
45            DO 10 L=1,3
46      10 ARRAY(L) = PK(K,L)
47            CALL LOAD(ILOAD,F,I,M,X,REFINE(J),ARRAY)
48            IF(ILOAD.EQ.1)GO TO 50
49            PK(K,1) = ARRAY(1)
50      15 CONTINUE
51            IF(REFINE(1).GT.0)  CALL LOAD(ILOAD,F,I,M,X,REFINE(1),VZERO)
52            IF(ILOAD.EQ.1)GO TO 56
53            IF(REFINE(2).GT.0)  CALL LOAD(ILOAD,F,I,M,X,REFINE(2),HVESSL)
54            IF(ILOAD.EQ.1)GO TO 56
55            IF(REFINE(3).GT.0)  CALL LOAD(ILOAD,F,I,M,X,REFINE(3),LVESSL)
56            IF(ILOAD.EQ.1)GO TO 56
57            IF(REFINE(4).GT.0)  CALL LOAD(ILOAD,F,I,M,X,REFINE(4),HBURET)
58            IF(ILOAD.EQ.1)GO TO 56
59            IF(REFINE(5).GT.0)  CALL LOAD(ILOAD,F,I,M,X,REFINE(5),LBURET)
60            IF(ILOAD.EQ.1)GO TO 56
61            IF(REFINE(6).GT.0)  CALL LOAD(ILOAD,F,I,M,X,REFINE(6),EZERO)
62            IF(ILOAD.EQ.1)GO TO 56
63            IF(REFINE(7).GT.0)  CALL LOAD(ILOAD,F,I,M,X,REFINE(7),SLOPE)
64            IF(ILOAD.EQ.1)GO TO 56
65    C
66    C       CALCULATION OF THE SUM OF SQUARED RESIDUALS.
67    C
68      20 I = 1
69            J = NCONST
70            ALFH = 0.000
71      21 ALFH = ALFH + PK(J,1) * AL10
72            BETA(I) = ALFH
73            I = I + 1
74            J = J - 1
75            IF(J.GT.1)  GO TO 21
76            BETA(I) = -PK(1,1) * AL10
77    C
78    C       IF(KOUNT.GE.4)  GO TO 60
79    C       WRITE(OUT,9007)  VZERO(1),HVESSL(1),LVESSL(1),HBURET(1),
80    C     * LBURET(1), EZERO(1), SLOPE(1)
81    C       WRITE(OUT,9007)  BETA
82    C
83            THZERO = (HVESSL(1) + LVESSL(1) * NDPV) * VZERO(1)
84            THBC = HBURET(1) + LBURET(1) * NDPB
85            TLZERO = LVESSL(1) * VZERO(1)
86            ALSLP = AL10 / SLOPE(1)
87    C
88            F = 0.000
89            DO 40 I=1,NP
90            VOL = VZERO(1) + V(I)
91            TT(1) = (THZERO + THBC * V(I)) / VOL
92            TT(2) = (TLZERO + LBURET(1) * V(I)) / VOL
93            ALFH = (E(I) - EZERO(1)) * ALSLP
94            CX(1) = EXP(ALFH)
95            ALFL = 1.0000
96            DO 31 K=1,NBETAH
97      31 ALFL = ALFL + EXP(BETA(K) + ALFH * K)
98            CX(2) = TT(2) / ALFL
99            ALFL = ALOG(CX(2))
100           DO 32 K=1,NBETAH
101     32 CI(K) = EXP(BETA(K)+ALFL+ALFH*K)
102           CI(NCONST) = EXP(BETA(NCONST)-ALFH)
103           THCALC = CX(1)
104           DO 33 J=1,NCONST
105     33 THCALC = THCALC + CI(J) * JQR(1,J)
106           VCALC = (THZERO - THCALC * VZERO(1)) / (THCALC - THBC)
107           RESID = VCALC - V(I)
108           IF(KOUNT.GT.0)  GO TO 40
109   C       WRITE(OUT,9007)  TT, CX, TOL, ALFL
110   C       WRITE(OUT,9007)  CI
111           CALL ML(NCONST,2,TOL,2,BETA,CI,CX,TT,HX,TOLC,DT,DDT,JQR)
```

```
112            IF(IFAIL.LE.0.AND.CX(1).GT.0.00)  GO TO 34
113            KFAIL = KFAIL + 1
114            IF(KFAIL.GT.8)  GO TO 60
115            WRITE(OUT,9008)  V(I), VCALC, E(I), RESID
116            IF(IFAIL.GT.0)  IFAIL = JFAIL
117            JFAIL = IFAIL
118            GO TO 40
119         34 ECALC = EZERO(1) + ALOG(CX(1)) / ALSLP
120            ALFH = -ALOG10(CX(1))
121            ALFL = ECALC - E(I)
122            THCALC = E(I) + SLOPE(1) * ALFH
123            SCALC = (EZERO(1) - E(I)) / ALFH
124            WRITE(OUT,9000)   V(I), VCALC, E(I), ECALC, RESID, ALFL, ALFH,
125          &  CX(2), THCALC, SCALC
126            IF(IFAIL.LT.JFAIL)  WRITE(OUT,9004)
127            JFAIL = IFAIL
128            TT(1) = (THZERO + THBC*(V(I)+0.004)) / VOL
129            CALL ML(NCONST,2,TOL,2,BETA,CI,CX,TT,HX,TOLC,DT,DDT,JQR)
130            IF(IFAIL.LE.0.AND.CX(1).GT.0.00)  GO TO 35
131            WRITE(OUT,9004)
132            IF(IFAIL.GT.0)  IFAIL = JFAIL
133            JFAIL = IFAIL
134            GO TO 40
135         35 ALFH = ALFH + ALOG10(CX(1))
136            ALFH = ABS(ALFH) * 60.0
137            IF(ALFH.GT.0.3.AND.KOUNT.LE.0)  WRITE(OUT,9006)
138         40 F = F + RESID ** 2
139         50 IF(KOUNT.GT.0)  GO TO 55
140            IF(IFAIL.NE.0)  GO TO 58
141            RETURN
142         55 WRITE(OUT,9007)  F, (X(I), I=1,M)
143            IFAIL = 0
144            JFAIL = 0
145         56 RETURN
146         58 WRITE(OUT,9005)
147         60 WRITE(OUT,9001)
148            IF(IFAIL.NE.0)  WRITE(OUT,9002)
149            IF(IFAIL.LT.1)  WRITE(OUT,9003)
150            IFAIL = 10
151            JFAIL = 0
152            ICALC=1
153            RETURN
154            END




  1   C
  2   C
  3   C      ****************************************************************
  4          SUBROUTINE SETUP(ID,LIT,M,X,H,IREF,ARRAY)
  5   C
  6   C      THIS SUBROUTINE INITIALISES THE ARRAYS USED BY NELM FOR REFINEMENT
  7   C
  8          DIMENSION  X(7), H(7), ARRAY(3), ID(7)
  9          INTEGER OUT
 10          COMMON /FIVE/  IN, OUT, IFAIL, JFAIL, AL10, KOUNT
 11          IF(M.GE.8)  GO TO 20
 12          IF(IREF.EQ.2)  GO TO 10
 13          IXA = 1
 14          XA = ARRAY(1) / 12.000
 15          IF(XA.LT.0.0000)  IXA = -1
 16          ARRAY(2) = ARRAY(1) - XA * IXA
 17          ARRAY(3) = ARRAY(1) + XA * IXA
 18       10 H(M) = (ARRAY(2) + (ARRAY(3) - ARRAY(2)) / 10.0) - ARRAY(1)
 19          X(M) = ARRAY(1)
 20          ID(M) = LIT
 21          M = M + 1
 22          RETURN
 23       20 WRITE(OUT,99)
 24       99 FORMAT(//'0','ATTEMPT TO REFINE TOO MANY PARAMETERS',/' ',
 25          & 'ERROR TERMINATION',///)
 26          STOP
 27          END
```

```
 1   C
 2   C   ******************************************************************
 3   C
 4       SUBROUTINE ML(NK,NMBE,TOL,NC,HLNB,CI,CX,TT,HX,TOLC,DT,DDT,JQR)
 5   C
 6   C   TAKEN, WITH PERMISSION, FROM MINIQUAD.
 7   C   REFERENCE:- SABATINI A., VACCA A. AND GANS P., TALANTA, 1974, 21,
 8   C
 9       DIMENSION HLNB(NK),CI(NK)
10       DIMENSION JQR(NMBE,NK)
11       DIMENSION CX(NMBE),TT(NMBE),HX(NMBE)
12       DIMENSION TOLC(NC),DT(NC),DDT(NC,NC)
13       COMMON /FIVE/  JINP, JOUT, IFAIL, JFAIL, AL10, KOUNT
14   C
15   C       THIS ROUTINE CALCULATES ESTIMATES OF THE FREE CONCENTRATIONS O
16   C       LIGAND ETC.  USING A NUMBER OF MASS-BALANCE EQUATIONS EQUAL TO
17   C       NUMBER OF UNKNOWNS (THOSE FOR WHICH THERE IS NO POTENTIAL), TH
18   C       'NEWTON-RAPHSON' METHOD IS USED, WITH FIRST DERIVATIVES ONLY.
19   C       ESTIMATES ARE ALSO REQUIRED FOR THIS ROUTINE BUT 1.E-07 WILL
20   C       SUFFICE IF A MORE ACCURATE VALUE IS NOT AVAILABLE.
21   C
22       NEMF = NMBE - NC
23       IF(NEMF.EQ.0)  GO TO 103
24       DO 102 I=1,NEMF
25       IPNC = I + NC
26   102 HX(IPNC)=ALOG(CX(IPNC))
27   103 NCICL = 0
28   C       A CYCLE COUNTER. 100 CYCLES ARE PERMITTED AS MAXIMIMUM.
29       DO 105 I=1,NC
30   C       TOLC(I) PROVIDES A RELATIVE TOLERANCE FOR USE WITH THE
31   C       CONVERGENCE CRITERION
32   105 TOLC(I)=ABS(TT(I))*TOL
33   121 NCICL=NCICL+1
34       DO 125 J=1,NC
35   C       XC(J) IS ONE OF THE UNKNOWN CONCENTRATIONS THAT ARE BEING CALC
36   C       AS IT CANNOT TAKE A NEGATIVE VALUE, THE STEP LENGTH OF THE COR
37   C       VECTOR HX IS REDUCED SO THAT NONE OF THEM TAKES A NEGATIVE VAL
38   122 IF(CX(J))123,123,125
39   123 DO 124 I=1,NC
40       HX(I)=0.5*HX(I)
41   124 CX(I)=CX(I)-HX(I)
42       GO TO 122
43   125 CONTINUE
44       DO 126 I=1,NC
45       HX(I)=ALOG(CX(I))
46   C
47   C       DT(I) IS THE DIFFERENCE BETWEEN T OBSERVED AND T CALCULATED
48   C       FOR THE MASS-BALANCE EQUATION (I), I.E. IT IS THE RESIDUAL.
49   C
50   126 DT(I)=CX(I)-TT(I)
51   C
52   C   CHANGES AS RECOMMENDED BY D. LEGGETT  (TALANTA, 1977,24,535)
53   C
54       DO 128 J=1,NK
55       W=HLNB(J)
56       DO 127 I=1,NMBE
57   127 W=W+HX(I)*JQR(I,J)
58   C       CI(J) IS THE CONCENTRATION OF THE SPECIES (J) DEFINED BY
59   C       THE INDICES IN JQR
60       CI(J)=EXP(W)
61       DO 128 I=1,NC
62   128 DT(I)=DT(I)+JQR(I,J)*CI(J)
63       IF(KOUNT.GT.1000)  RETURN
64       DO 129 I=1,NC
65   C       CONVERGENCE CRITERION. WHEN ALL THE MASS-BALANCE EQUATIONS
66   C       SATISFIED TO THE REQUIRED RELATIVE TOLERANCE, CONTROL IS
67   C       PASSED BACK TO THE CALLING PROGRAM.
68       IF(ABS(DT(I))-TOLC(I))129,129,131
69   129 CONTINUE
70       GO TO 190
71   131 DO 152 I=1,NC
72       DO 151 J=I,NC
73   C       DDT IS THE JACOBIAN FOR THE SYSTEM, AND IT IS SYMMETRICAL AND
74   C       SQUARE.  ITS ELEMENTS ARE THE RELATIVE DERIVATIVES, SO THAT
75   C       THEY ARE OBTAINED DIRECTLY FROM THE CONCENTRATION TERMS
```

```
76   C               PREVIOUSLY CALCULATED.
77           DDT(I,J)=0.
78           DO 151 L=1,NK
79           IF(JQR(I,L))149,151,149
80      149  IF(JQR(J,L))150,151,150
81      150  W=JQR(I,L)*JQR(J,L)*CI(L)
82           DDT(I,J)=DDT(I,J)+W
83      151  CONTINUE
84      152  DDT(I,I)=DDT(I,I)+CX(I)
85           CALL LINEQ(DDT,NC,DT,4)
86           IF (IFAIL) 160,160,190
87   C               DT CONTAINS THE RELATIVE CORRECTIONS TO THE PARAMETERS.
88   C               HX WILL CONTAIN THE ABSOLUTE CORRECTIONS.
89      160  DO 165 I=1,NC
90           HX(I)=-DT(I)*CX(I)
91      165  CX(I)=CX(I)+HX(I)
92   C               IF 100 CYCLES HAVE BEEN EXCEEDED CONTROL IS RETURNED TO CALC.
93           IF(NCICL.LT.101)  GO TO 121
94           IFAIL = IFAIL - 1
95      190  RETURN
96           END




1    C
2    C
3    C   ****************************************************************
4            SUBROUTINE LINEQ(A,N,B,KFAIL)
5    C
6    C   TAKEN, WITH PERMISSION, FROM MINIQUAD (REFERENCED ABOVE).
7    C
8            DIMENSION A(N,N), B(N)
9            COMMON /FIVE/  IN, OUT, IFAIL, JFAIL, AL10, KOUNT
10   C        SOLVES THE N SIMULTANEOUS LINEAR EQUATIONS A*X=B WITH M RIGHT-
11   C        SIDES IN B. THE SOLUTION VECTORS ARE LEFT IN B AND THE MATRIX
12   C        REPLACED BY ITS INVERSE. AFTER CHOLESKI FACTORING OF A TO GIVE
13   C        THE FORWARD SUBSTITUTIONS L*Y=B AND L*Z=E AND THE BACKWARD SUB
14   C        LT*X=Y AND LT*AINV=Z ARE PERFORMED
15           IF (N-1) 455,5,9
16        5  T=A(1,1)
17           IF(T.LE.0)  GO TO 455
18        6  A(1,1)=1./T
19           B(1) = B(1) / T
20           RETURN
21        9  DO 80 I=1,N
22           I1=I-1
23           DO 70 J=I,N
24           S=A(I,J)
25           IF (I1) 10,30,10
26       10  DO 20 K=1,I1
27       20  S=S-A(I,K)*A(J,K)
28       30  X=S
29           IF (J-I) 60,40,60
30       40  IF (X) 45,45,50
31       45  IFAIL=KFAIL
32           GO TO 400
33       50  A(I,I)=1./SQRT(X)
34           GO TO 70
35       60  A(J,I)=X*A(I,I)
36       70  CONTINUE
37       80  CONTINUE
38   C        FORWARD SUBSTITUTION ON RIGHT HAND SIDES
39           B(1) = B(1) * A(1,1)
40           DO 120 I=2,N
41           I1=I-1
42           S = B(I)
43           DO 110 K=1,I1
44      110  S = S - A(I,K) * B(K)
45      120  B(I) = S * A(I,I)
46   C        FORWARD SUBSTITUTION FOR INVERSION
47           DO 170 J=1,N
48           J1=J+1
49           IF (J1-N) 140,140,170
```

```
50      140 DO 160 I=J1,N
51          I1=I-1
52          S=0.
53          DO 150 K=J,I1
54      150 S=S-A(I,K)*A(J,K)
55      160 A(J,I)=S*A(I,I)
56      170 CONTINUE
57  C           BACKWARD SUBSTITUTION
58          B(N) = B(N) * A(N,N)
59          DO 220 J=1,N
60      220 A(J,N)=A(J,N)*A(N,N)
61          DO 290 II=2,N
62          I=N-II+1
63          T=A(I,I)
64          I1=I+1
65          S = B(I)
66          DO 240 K=I1,N
67      240 S = S - A(K,I) * B(K)
68      245 B(I) = S * T
69          DO 280 J=1,I
70          S=A(J,I)
71          DO 270 K=I1,N
72      270 S=S-A(K,I)*A(J,K)
73          A(J,I)=S*T
74      280 CONTINUE
75      290 CONTINUE
76          DO 300 I=2,N
77          I1=I-1
78          DO 300 J=1,I1
79      300 A(I,J)=A(J,I)
80      400 RETURN
81  C
82      455  IFAIL=KFAIL
83          RETURN
84          END
```

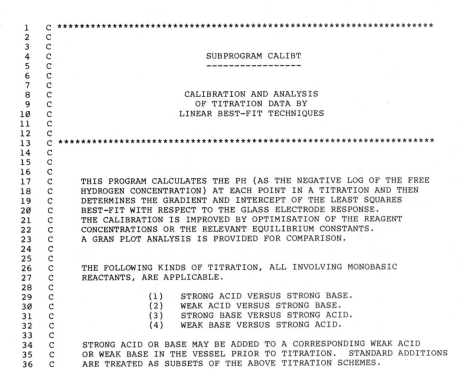

```
1   C ***********************************************************************
2   C
3   C
4   C                        SUBPROGRAM CALIBT
5   C                        -----------------
6   C
7   C
8   C                    CALIBRATION AND ANALYSIS
9   C                      OF TITRATION DATA BY
10  C                    LINEAR BEST-FIT TECHNIQUES
11  C
12  C
13  C ***********************************************************************
14  C
15  C
16  C
17  C      THIS PROGRAM CALCULATES THE PH (AS THE NEGATIVE LOG OF THE FREE
18  C      HYDROGEN CONCENTRATION) AT EACH POINT IN A TITRATION AND THEN
19  C      DETERMINES THE GRADIENT AND INTERCEPT OF THE LEAST SQUARES
20  C      BEST-FIT WITH RESPECT TO THE GLASS ELECTRODE RESPONSE.
21  C      THE CALIBRATION IS IMPROVED BY OPTIMISATION OF THE REAGENT
22  C      CONCENTRATIONS OR THE RELEVANT EQUILIBRIUM CONSTANTS.
23  C      A GRAN PLOT ANALYSIS IS PROVIDED FOR COMPARISON.
24  C
25  C
26  C      THE FOLLOWING KINDS OF TITRATION, ALL INVOLVING MONOBASIC
27  C      REACTANTS, ARE APPLICABLE.
28  C
29  C               (1)    STRONG ACID VERSUS STRONG BASE.
30  C               (2)    WEAK ACID VERSUS STRONG BASE.
31  C               (3)    STRONG BASE VERSUS STRONG ACID.
32  C               (4)    WEAK BASE VERSUS STRONG ACID.
33  C
34  C      STRONG ACID OR BASE MAY BE ADDED TO A CORRESPONDING WEAK ACID
35  C      OR WEAK BASE IN THE VESSEL PRIOR TO TITRATION.   STANDARD ADDITIONS
36  C      ARE TREATED AS SUBSETS OF THE ABOVE TITRATION SCHEMES.
```

```
37    C
38    C
39    C
40    C        THE NERNST SIGN CONVENTION IS TAKEN TO BE POSITIVE
41    C        (I.E. INCREASING PH GIVES DECREASING EMF VALUES.)
42    C        TITRATION VOLUMES MUST INCREASE MONOTONICALLY.
43    C
44    C
45    C
46    C        WITH STRONG ACID / STRONG BASE TITRATIONS, A SCAN OF PKW CAN BE
47    C        IMPLEMENTED TO ESTIMATE AN OPTIMUM VALUE.
48    C        IT IS IMPORTANT TO OBTAIN AGREEMENT BETWEEN:
49    C        (1) THE OBSERVED AND THEORETICAL SLOPE FOR THE ELECTRODE RESPONSE
50    C        AND (2) THE VALUES OF EZERO FOUND FOR DATA CORRESPONDING TO THE
51    C        ACID RANGE COMPARED WITH THAT COVERING ALL BUFFERED POINTS.
52    C        THESE ARE BETTER CRITERIA THAN A MINIMUM STANDARD DEVIATION.
53    C        A SCAN OF THE LIGAND PROTONATION CONSTANT IS ALSO POSSIBLE.
54    C
55    C*********************************************************************
56    C
57    C
58    C
59    C
60             SUBROUTINE CALIBT(VZERO,EZERO,SLOPE,THZERO,TLZERO,AHZERO,
61           &  PKWMIN,PKWMAX,BLGMIN,BLGMAX)
62    C
63    C
64    C
65             DIMENSION  W(100),H(100),PH(100),PHA(100),PHB(100),G(100),VG(100),
66           &  VOL(100),X(100),Y(100),ET(100),EA(100),EB(100),EC(100),THT(100)
67             INTEGER  IN, OUT, CARD
68             REAL NERNST
69             LOGICAL  LIGAND, ALKALI, ADJUST
70    C
71    C
72    C
73             COMMON /TWO/ CI(7),CX(2),TT(2),HX(2),TOLC(2),DT(2),DDT(2)
74             COMMON /THREE/  V(100), E(100), N
75             COMMON /FOUR/  BETA(7), ARRAY(3), TOL, NCONST, NBETAH, JQR(2,7)
76             COMMON /FIVE/  IN, OUT, IFAIL, JFAIL, AL10, KOUNT
77             COMMON /SIX/  CARD(20), P(100)
78    C
79             EQUIVALENCE (P(1),W(1))
80    C
81    C
82    C
83       200 FORMAT('1')
84       201 FORMAT('0')
85       210 FORMAT(/'0',44X,'SUBPROGRAM CALIBT',/45X,17(1H-),////'0',20A4,//)
86       211 FORMAT('0','THE INITIAL VOLUME IS',F7.2,/
87           &' ','THE EXPECTED ELECTRODE INTERCEPT IS',F8.2,/
88           &' ','THE EXPECTED NERNSTIAN SLOPE IS',F7.2)
89       212 FORMAT('0','THE ACID CONCENTRATION IN THE TITRATION VESSEL IS',
90           &1PE11.3,/' THE LIGAND CONCENTRATION IN THE TITRATION VESSEL IS',
91           &1PE11.3,/' THE ACID CONCENTRATION IN THE BURETTE IS',1PE11.3)
92       213 FORMAT(' ','THE END POINT IS EXPECTED AT',F7.3)
93       215 FORMAT(' ','THE VALUE OF PKW IS',F7.3,/' ',26(1H-),/)
94       216 FORMAT(' ','THE VALUE OF LOG BETAH IS',F7.3,/' ',31(1H-),/)
95       220 FORMAT('0','TITRATION POINT NUMBER',I3,' IS IN ERROR',15X,2F12.2)
96       222 FORMAT('0','INITIAL CONCENTRATION DATA PROBABLY INCORRECT',//)
97       231 FORMAT('0','THE GRAN-PLOT VALUES AT EACH POINT ARE:',11X,'I',7X,
98           &  'VOL',6X,'EMF',8X,'V(I)',12X,'G(I)',///)
99       235 FORMAT('0','PKW IS IN ERROR')
100      236 FORMAT('0','THE VALUE OF LOG BETAH IS IN ERROR')
101      240 FORMAT('0','THE SEARCH FOR UNSUITABLE POINTS HAS REMOVED ')
102      241 FORMAT(' ',45X,I6,2F10.2,7(1PE15.4))
103      245 FORMAT('+',45X,'NONE')
104      246 FORMAT('0','TOO FEW POINTS REMAIN FOR LEAST SQUARES ANALYSIS',//)
105      250 FORMAT(' (50% OF THE POINTS, TAKEN ABOUT THE MIDDLE OF THE SET)')
106      251 FORMAT(' ','BEST-FIT INVOLVING',I3,' POINTS GIVES EZERO =',
107          &  F7.1,' (',F4.2,') AND A SLOPE =',F7.2,' (',F4.2,')',/
108          &  ' ','OVERALL STANDARD DEVIATION =',1PE11.3)
109      252 FORMAT(///'0','USING DATA BEFORE THE ENDPOINT,')
110      253 FORMAT('0','USING DATA AFTER THE ENDPOINT,')
111      254 FORMAT('0','USING ALL THE BUFFERED DATA,')
112      255 FORMAT(///'0',8X,'VOL.'9X,'TH',11X,'PH',9X,'E(OBS.)',6X,'E(CORR).'
```

```
113        &,6X,'E(CALC.)',7X,'RESIDUAL',5X,'PKW(CALC.)',3X,'EZERO(CALC.)',//)
114    256 FORMAT(' ',I3,F9.2,1PE14.3,0PF11.3,6F14.2)
115    257 FORMAT('0','PKW(CALC.) IS OBTAINED USING THE CURRENT VALUE FOR',
116        & ' EZERO (',F6.1,') AND THE NERNSTIAN SLOPE.  AVERAGE =',F7.3)
117    258 FORMAT('0','EZERO(CALC.) IS OBTAINED USING THE CURRENT VALUE FOR',
118        & ' PKW (',F6.2,') AND THE NERNSTIAN SLOPE.  AVERAGE =',F7.1,///)
119    259 FORMAT(' ','WITH AN ALKALI ERROR CORRECTION FACTOR OF',1PE10.2)
120    260 FORMAT('1',/'0',10X,35(1H*),5X,'NEW SCAN ITERATION',5X,35(1H*),/)
121    261 FORMAT('+','NOT REQUIRED.',/
122        & 'THE RESULTS ARE INDEPENDENT OF THE VALUE OF PKW.')
123    262 FORMAT(//////'0',44X,'CALIBT CONCENTRATION ADJUSTMENT')
124    263 FORMAT(' ',43X,'AND TO DETERMINE A VALUE FOR PKW.')
125    264 FORMAT(///'0','WITH A VESSEL-ACID CONCENTRATION OF',1PE11.3,/' ',
126        & 'AND A BURETTE-ACID CONCENTRATION OF',1PE11.3,/' ',
127        & 'THE FOLLOWING RESULTS ARE OBTAINED',////)
128    265 FORMAT(' ','THE ENDPOINT NOW OCCURS AT',F7.3,///)
129    266 FORMAT(' ',36X,'TO DETERMINE THE BURETTE-ACID CONCENTRATION.',//)
130    267 FORMAT(' ',37X,'TO DETERMINE THE VESSEL-ACID CONCENTRATION')
131  C
132  C
133  C    SECTION ONE:   INPUT PROCEDURES
134  C
135  C
136     10 WRITE(OUT,200)
137        LIGAND = .FALSE.
138        ALKALI = .FALSE.
139        ADJUST = .FALSE.
140        NADJ = -10
141  C
142        WRITE(OUT,210)   CARD
143        NERNST = SLOPE
144        WRITE(OUT,211)   VZERO, EZERO, NERNST
145        IF(AHZERO.GT.0.000)  ALKALI = .TRUE.
146        DUMMY = TLZERO
147        IF(ALKALI)  DUMMY = -DUMMY
148        ENDPT = -VZERO * (THZERO + DUMMY) / AHZERO
149        ARESET = AHZERO
150        TRESET = THZERO
151        ERESET = ENDPT
152        WRITE(OUT,212)   THZERO, TLZERO, AHZERO
153        WRITE(OUT,213)   ENDPT
154        IF(TLZERO.GT..1E-7)  LIGAND = .TRUE.
155        PKW = PKWMIN
156        WRITE(OUT,215)   PKW
157        PKWINC = (PKWMAX - PKWMIN) / 5.000
158        IF(PKWINC.LT.0.000)  PKWINC = 0.000
159        BLGINC = 0.000
160        IF(.NOT.LIGAND)  GO TO 16
161        BLG = BLGMIN
162        WRITE(OUT,216)   BLG
163        IF(BLGMAX.GT.BLGMIN)  BLGINC = (BLGMAX - BLGMIN) / 5.000
164        IF(BLGINC.GT.0.000)  PKWINC = 0.000
165     16 CONTINUE
166  C
167  C
168  C    SECTION TWO:   CHECKING PROCEDURES
169  C
170  C
171        DO 20 I=2,N
172        IF(V(I).LT.V(I-1))  GO TO 21
173     20 CONTINUE
174        GO TO 30
175     21 WRITE(OUT,220) I, V(I), E(I)
176        IF(I.EQ.2)  WRITE(OUT,222)
177        GO TO 65
178  C
179  C
180  C    SECTION THREE:   GRAN PLOT ANALYSIS
181  C
182  C
183     30 DO 31 I=1,N
184     31 H(I) = 10.000 ** ((E(I) - EZERO) / SLOPE)
185        WRITE(OUT,231)
186        IG = 1
187        DO 32 I=1,N
188        IF(LIGAND.AND.V(I).LT.0.00099)  GO TO 32
```

```
189            IF(V(I).GT.ENDPT)  GO TO 33
190            G(IG) = VZERO + V(I)
191            IF(LIGAND)  G(IG) = V(I)
192            HOLD = H(I)
193            IF(ALKALI)  HOLD = 1.000 / HOLD
194            G(IG) = G(IG) * HOLD
195            VG(IG) = V(I)
196            WRITE(OUT,241)  I, V(I), E(I), VG(IG), G(IG)
197            IG = IG + 1
198      32 CONTINUE
199      33 IG = IG - 1
200            INIT = 1
201            IF(IG.GT.5)  CALL STRAIT(VG,G,W,IG,ENDPT,INIT,IFIN,X,Y,OUT)
202            NFLAG = IG + 1
203            IF(NFLAG.GT.N-3)  GO TO 35
204            IG = N - NFLAG + 2
205            IF(NFLAG.GT.1)  WRITE(OUT,231)
206            DO 34 I=NFLAG,N
207            IG = IG - 1
208            G(IG) = VZERO + V(I)
209            HOLD = H(I)
210            IF(.NOT.ALKALI)  HOLD = 1.000 / HOLD
211            G(IG) = G(IG) * HOLD
212            VG(IG) = V(I)
213      34 WRITE(OUT,241)  I, V(I), E(I), VG(IG), G(IG)
214            IG = N - NFLAG + 1
215            INIT = N
216            IF(IG.GT.5)  CALL STRAIT(VG,G,W,IG,ENDPT,INIT,IFIN,X,Y,OUT)
217    C
218      35 IF(PKW.GT.12.0.AND.PKW.LT.15.00)  GO TO 36
219            IF(PKW.GT.0.5.OR.PKW.LT.-0.5)  WRITE(OUT,235)
220            GO TO 65
221      36 IF(.NOT.LIGAND)  GO TO 37
222            IF(BLG.GT.2.00.AND.BLG.LT.12.00)  GO TO 40
223            IF(BLG.GT.0.5.OR.BLG.LT.-0.5)  WRITE(OUT,236)
224            GO TO 65
225      37 TH2 = THZERO * 0.1
226            IF(NADJ.GT.15)  TH2 = TH2 / 5.00
227            TH1 = THZERO + TH2
228            TH2 = THZERO - TH2
229    C
230    C
231    C     SECTION FOUR:   CALCULATION OF FREE CONCENTRATIONS
232    C
233    C
234      40 IT = 0
235            IA = 0
236            IB = 0
237            NFLAG = 0
238            KWSCAN = 1
239            THMOL = THZERO * VZERO
240            IF(.NOT.(LIGAND.OR.ADJUST))  WRITE(OUT,215)  PKW
241            IF(LIGAND)  WRITE(OUT,216)  BLG
242            IF(.NOT.ADJUST)  WRITE(OUT,240)
243            IF(.NOT.LIGAND)  GO TO 45
244    C
245    C     WEAK ACID OR BASE
246    C
247            TLMOL = TLZERO * VZERO
248            IF(.NOT.ALKALI)  THMOL = THMOL + TLMOL
249            DUMMY = -PKW
250            WK = 10.00 ** DUMMY
251            BETAH = 10.000 ** BLG
252            DO 44 I=1,N
253            W(I) = 1.000
254            VOLUME = VZERO + V(I)
255            TL = TLMOL / VOLUME
256            TH = (THMOL + AHZERO * V(I)) / VOLUME
257            HLO = -1.000
258            HHI = -1.000
259            HFREE = -1.000
260            IF(TH.LE.0.000)  GO TO 41
261    C
262    C     ACID APPROXIMATION
263    C
264            HOLD = 1.0 + BETAH * (TL - TH)
```

```
265              HLO = HOLD ** 2 + 4.0 * BETAH * TH
266              HLO = (SQRT(HLO) - HOLD) / (2.0 * BETAH)
267              HFREE = HLO
268         41 IF(TL.LE.TH) GO TO 42
269   C
270   C        ALKALI APPROXIMATION
271   C
272              HOLD = -(TH + WK * BETAH)
273              HHI = HOLD ** 2 + 4.0 * BETAH * (TL - TH) * WK
274              HHI = (SQRT(HHI) - HOLD) / (2.0 * BETAH * (TL - TH))
275              IF(HFREE.LE.0.000)  HFREE = HHI
276              IF(HLO.LT.1.00E-12.OR.HHI.LT.1.00E-12)  GO TO 42
277   C
278              DUMMY = SQRT(WK)
279              HFREE = -1.000
280              IF(HLO.GT.DUMMY*100.0)  HFREE = HLO
281              IF(HHI.LT.DUMMY/100.0)  HFREE = HHI
282              IF(HFREE.GT.0.000)  GO TO 42
283              HOLD = HLO * HHI / WK
284              DUMMY = 1.000 / HOLD
285              IF(HOLD.LT.1.000)  HOLD = 1.000
286              IF(DUMMY.LT.1.000)  DUMMY = 1.000
287              HLO = ALOG(HLO) * HOLD
288              HHI = ALOG(HHI) * DUMMY
289              HFREE = (HLO + HHI) / (HOLD + DUMMY)
290              HFREE = EXP(HFREE)
291         42 IF(HFREE.LT.1.0E-12)  GO TO 43
292              FREEL = TL / (1.0 + BETAH * HFREE)
293              FREELH = TL - FREEL
294              IF((FREEL.LT.1.0E-3.OR.FREELH.LT.1.0E-3).AND.
295            &    (HFREE.GT.1.0E-11.AND.HFREE.LT.1.0E-3))  GO TO 43
296              IT = IT + 1
297              PH(IT) = -ALOG10(HFREE)
298              VOL(IT) = V(I)
299              THT(IT) = TH
300              ET(IT) = E(I)
301              IF(V(I).GT.ENDPT)  GO TO 421
302              IA = IA + 1
303              EA(IA) = E(I)
304              PHA(IA) = PH(IT)
305              GO TO 44
306        421 IB = IB + 1
307              EB(IB) = E(I)
308              PHB(IB) = PH(IT)
309              GO TO 44
310         43 IT = IT + 1
311              IF(IT.EQ.I)  WRITE(OUT,201)
312              IT = IT - 1
313              WRITE(OUT,241)  I, V(I), E(I)
314         44 CONTINUE
315              IF(IT.EQ.N)  WRITE(OUT,245)
316              IF(IT.GE.5)  GO TO 53
317              WRITE(OUT,246)
318              GO TO 65
319   C
320   C        STRONG ACID OR BASE
321   C
322         45 DO 48 I=1,N
323              W(I) = 1.0000
324              VOLUME = VZERO + V(I)
325              TH = (THMOL + AHZERO * V(I)) / VOLUME
326              IF(TH.LT.0.00175)  GO TO 46
327              KWSCAN = 1
328              IT = IT + 1
329              IA = IA + 1
330              ET(IT) = E(I)
331              EA(IA) = E(I)
332              EC(IT) = E(I)
333              PH(IT) = -ALOG10(TH)
334              PHA(IA) = PH(IT)
335              VOL(IT) = V(I)
336              THT(IT) = TH
337              GO TO 48
338         46 IF(TH.LE.-0.00175.AND.TH.GT.-0.01)  GO TO 47
339              IF(ADJUST)  GO TO 48
340              IT = IT + 1
```

```
341              IF(IT.EQ.I)  WRITE(OUT,201)
342              IT = IT - 1
343              WRITE(OUT,241)  I, V(I), E(I)
344              GO TO 48
345        47 IT = IT + 1
346              IB = IB + 1
347              ET(IT) = E(I)
348              EB(IB) = E(I)
349              EC(IT) = E(I)
350              PH(IT) = PKW + ALOG10(-TH)
351              PHB(IB) = PH(IT)
352              VOL(IT) = V(I)
353              THT(IT) = TH
354        48 CONTINUE
355    C
356    C
357    C     SECTION FIVE:    LINEAR LEAST SQUARES ANALYSIS
358    C
359    C
360        50 IF(.NOT.ADJUST.AND.IT.EQ.N)  WRITE(OUT,245)
361              IF(IT.LE.0)   STOP 50
362              IF(IT.LE.5)   GO TO 54
363              IF(IA.LT.5.OR.IB.LT.5)  GO TO 53
364              CALL LINFIT(PHA,EA,W,IA,SLOPE,EZERO,SXSLP,SXINT,STDDEV)
365              SLOPE = -SLOPE
366              ACIDSL = SLOPE
367              IF(ADJUST)  GO TO 53
368              IF(.NOT.ALKALI.OR.LIGAND)  WRITE(OUT,252)
369              IF(.NOT.LIGAND.AND.ALKALI)  WRITE(OUT,253)
370              IF(IB.EQ.0)  WRITE(OUT,250)
371              WRITE(OUT,251)   IA, EZERO, SXINT, SLOPE, SXSLP, STDDEV
372    C
373              CALL LINFIT(PHB,EB,W,IB,SLOPE,EZERO,SXSLP,SXINT,STDDEV)
374              SLOPE = -SLOPE
375              IF(.NOT.ALKALI.OR.LIGAND)  WRITE(OUT,253)
376              IF(.NOT.LIGAND.AND.ALKALI)  WRITE(OUT,252)
377              IF(IA.EQ.0)  WRITE(OUT,250)
378              WRITE(OUT,251)   IB, EZERO, SXINT, SLOPE, SXSLP, STDDEV
379              PKWCLC = SLOPE
380              HOLD = 0.0
381              DUMMY = 1.0E-12
382    C
383              DO 52 J=1,2
384              IF(ABS(SLOPE-NERNST).LE.0.005)  GO TO 52
385              IG = IT
386              IIB=IB
387              DO 51 I=1,IB
388              HFREE = 10.000 ** (-PH(IG))
389              HHI = HFREE + DUMMY * 1.0E-12
390              HFREE = HFREE + DUMMY
391              HFREE = (ALOG10(HFREE) + ALOG10(HHI)) / 2.0
392              EB(IIB) = ET(IG) + NERNST * (PH(IG) + HFREE)
393              EC(IG) = EB(IIB)
394              IIB=IIB-1
395        51 IG = IG - 1
396              CALL LINFIT(PHB,EB,W,IB,SLOPE,EZERO,SXSLP,SXINT,STDDEV)
397              SLOPE = -SLOPE
398              PKWCLC = (DUMMY - HOLD) / (SLOPE - PKWCLC)
399              HOLD = DUMMY
400              DUMMY = PKWCLC * NERNST - PKWCLC * SLOPE + DUMMY
401              PKWCLC = SLOPE
402        52 CONTINUE
403              IF(.NOT.ALKALI.OR.LIGAND)  WRITE(OUT,253)
404              IF(.NOT.ALKALI.OR.LIGAND)  WRITE(OUT,259) DUMMY
405              IF(.NOT.LIGAND.AND.ALKALI)  WRITE(OUT,252)
406              IF(IA.EQ.0)  WRITE(OUT,250)
407              WRITE(OUT,251)   IB, EZERO, SXINT, SLOPE, SXSLP, STDDEV
408    C
409        53 IF(ADJUST.AND.NADJ.LT.10.AND.IA.GE.5.AND.IB.GE.5)  GO TO 59
410              CALL LINFIT(PH,EC,W,IT,SLOPE,EZERO,SXSLP,SXINT,STDDEV)
411              SLOPE = -SLOPE
412              IF(IA.LT.5.OR.IB.LT.5)  ACIDSL = SLOPE
413              BASESL = SLOPE
414              IF(ADJUST)  GO TO 59
415              WRITE(OUT,254)
416              WRITE(OUT,251)   IT, EZERO, SXINT, SLOPE, SXSLP, STDDEV
```

```
417   C
418   C
419   C
420     54 WRITE(OUT,255)
421        HOLD = 0.000
422        DUMMY = 0.000
423        PKWCLC = 0.000
424        IF(IA.EQ.0)  IA = IT + 1
425        DO 55 I=1,IT
426   C    X(I) = PH(I)
427   C    Y(I) = ET(I)
428        ECALC = EZERO - SLOPE * PH(I)
429        ERESID = EC(I) - ECALC
430        EZOCLC = EC(I) + NERNST * PH(I)
431        HOLD = HOLD + EZOCLC
432        IF(I.LE.IA)  GO TO 55
433        IF(I.EQ.IA+1)  WRITE(OUT,201)
434        PKWCLC = ((EZERO - EC(I)) / NERNST) - PH(I) + PKW
435        DUMMY = DUMMY + (1.0 / 10.00 ** PKWCLC)
436     55 WRITE(OUT,256)  I, VOL(I), THT(I), PH(I), ET(I), EC(I), ECALC,
437      &    ERESID, PKWCLC, EZOCLC
438   C
439        WRITE(OUT,201)
440        IF(IT.LE.5.OR.IA.LT.5)  GO TO 65
441        IF(IB.LT.3)  GO TO 56
442        PKWCLC = DUMMY / FLOAT(IB)
443        PKWCLC = -ALOG10(PKWCLC)
444        WRITE(OUT,257)  EZERO, PKWCLC
445     56 EZOCLC = HOLD / FLOAT(IT)
446        WRITE(OUT,258)  PKW, EZOCLC
447   C
448   C
449   C
450     59 IF(LIGAND)  GO TO 60
451        NADJ = NADJ + 1
452        IF(NADJ-11)  591, 592, 594
453    591 ADJUST = .TRUE.
454        ACIDSL = ACIDSL - NERNST
455        IF(NERNST.LT.0.00)  ACIDSL = -ACIDSL
456        IF(ACIDSL.GT.0.000)  TH1 = THZERO
457        IF(ACIDSL.LT.0.000)  TH2 = THZERO
458        THZERO = (TH1 + TH2) / 2.000
459        GO TO 40
460    592 IF(.NOT.ADJUST)  GO TO 593
461        ENDPT = -VZERO * THZERO / AHZERO
462        WRITE(OUT,262)
463        WRITE(OUT,267)
464        IF(IB.GT.3)  WRITE(OUT,263)
465        WRITE(OUT,201)
466        WRITE(OUT,264)  THZERO, AHZERO
467        WRITE(OUT,265)  ENDPT
468        ADJUST = .FALSE.
469        NADJ = 10
470        GO TO 40
471    593 ADJUST = .TRUE.
472        TH2 = AHZERO * 0.1
473        IF(NADJ.GT.15)  TH2 = TH2 / 5.00
474        TH1 = AHZERO + TH2
475        TH2 = AHZERO - TH2
476        NADJ = NADJ + 1
477    594 IF(NADJ-21)  597, 37, 595
478    595 IF(NADJ-31)  591, 593, 596
479    596 IF(NADJ-40)  597, 598, 60
480    597 BASESL = BASESL - NERNST
481        IF(NERNST.LT.0.00)  BASESL = -BASESL
482        IF(BASESL.GT.0.000)  TH2 = AHZERO
483        IF(BASESL.LT.0.000)  TH1 = AHZERO
484        AHZERO = (TH1 + TH2) / 2.000
485        THZERO = -AHZERO * ENDPT / VZERO
486        GO TO 40
487    598 IF(IA.LT.5.OR.IB.LT.5)  GO TO 65
488        WRITE(OUT,262)
489        WRITE(OUT,266)
490        WRITE(OUT,264)  THZERO, AHZERO
491        WRITE(OUT,265)  ENDPT
492        ADJUST = .FALSE.
```

```
493              GO TO 40
494    C
495    C
496    C
497         60 IF(PKWINC.LE.0.0001)  GO TO 62
498              NADJ = 0
499              PKW = PKW + PKWINC
500              IF(PKW.GT.PKWMAX)  GO TO 62
501              WRITE(OUT,260)
502              IF(KWSCAN.EQ.0)  WRITE(OUT,260)
503              IF(KWSCAN.EQ.0)  GO TO 62
504              THZERO = TRESET
505              AHZERO = ARESET
506              ENDPT = ERESET
507              WRITE(OUT,264)  THZERO, AHZERO
508              WRITE(OUT,213)  ENDPT
509              GO TO 35
510         62 IF(.NOT.LIGAND.OR.BLGINC.LE.0.0001)  GO TO 65
511              BLG = BLG + BLGINC
512              IF(BLG.GT.BLGMAX)  GO TO 65
513              WRITE(OUT,260)
514              GO TO 35
515    C
516         65 WRITE(OUT,200)
517              RETURN
518              END




  1    C
  2    C     ****************************************************************
  3    C
  4              SUBROUTINE STRAIT(V,G,W,N,ENDPT,INIT,IFIN,X,Y,OUT)
  5    C
  6    C     THIS SUBROUTINE SEARCHES DATA STORED IN THE ARRAYS
  7    C     NAMED V AND G TO FIND THE SEGMENT GIVING THE STRAIGHTEST LINE.
  8    C     ONLY THOSE LINES WITH ABSSICA INTERCEPTS WITHIN PTOL PERCENT OF
  9    C     THE ESTIMATED END POINT VALUE ARE ACCEPTED;  IF THIS CONDITION IS
 10    C     SATISFIED, THE SEGMENT GIVING THE LOWEST STANDARD DEVIATION IS
 11    C     LOCATED;  OTHERWISE, THE LINE FOUND IS THAT GIVING THE CLOSEST
 12    C     AGREEMENT WITH THE END POINT ESTIMATE.    SUBROUTINE LINFIT IS
 13    C     CALLED BETWEEN 20 AND 50 TIMES;  ITER DETERMINES HOW FINELY
 14    C     THE DATA IS DIVIDED INTO LINE SEGMENTS.
 15    C
 16              DIMENSION  V(N), G(N), W(N), X(N), Y(N)
 17              INTEGER OUT
 18    C
 19        201 FORMAT(///'0','GRAN-PLOT EXTRAPOLATIONS',/' ',24(1H*),///'0',
 20          &  2X,'SEGMENT',8X,'END PT.',5X,'STD. DEV.',//)
 21        202 FORMAT(' ',I3,2X,'-',I4,4X,F10.4,3(1PE15.4))
 22        203 FORMAT(//'0','THE STRAIGHTEST SEGMENT GIVES',3X,F10.4,/'0',
 23          & 'THE MINIMUM STD. DEVIATION GIVES',3X,F10.4)
 24        205 FORMAT('+',48X,'OMITTED BECAUSE THE ESTIMATED ENDPOINT IS ',
 25          & 'OUTSIDE THE PERMITTED RANGE OF', F5.1,'%')
 26        207 FORMAT('0','THE WEIGHTED AVERAGE GIVES',3X,F8.4,/'0',
 27          & 'THE ACTUAL AVERAGE GIVES',3X,F8.4,////)
 28        210 FORMAT(' FAILED TO FIND THE ENDPOINT',//)
 29        211 FORMAT(' CLOSEST AGREEMENT WITH THE ESTIMATED VALUE IS WITHIN',
 30          & F6.1,'%',//)
 31        215 FORMAT('0ANOTHER ATTEMPT WITH INCREASED ENDPOINT RANGE',///)
 32        218 FORMAT('0',I5,' POINTS HAVE BEEN OMITTED BECAUSE THEIR ESTIMATED
 33          &ENDPOINTS ARE OUTSIDE A RANGE OF',F6.1,'%',///)
 34    C
 35    C
 36    C
 37              PTOL = 5.0
 38              ISTART = INIT
 39          5 ITER = 5
 40              IF(ISTART.EQ.1)  ISTART = -1
 41              INIT = 0
 42              ENDOLD = 1.000E19
 43              STDOLD = 1.000E19
 44              STDSTD = 1.000E19
 45              SAVINT = 0.000
 46              SAVSLP = 0.000
```

```
47            KOUNT = 0
48            WGTAVE = 0.000
49            ENDAVE = 0.000
50            STDAVE = 0.000
51            NDUD = 0
52    C
53            IF(N.LT.6)  STOP
54            MMIN = 5
55            IF(N.GT.12)  MMIN = 6
56            IF(N.GT.25)  MMIN = 7
57            IF(N.GT.50)  MMIN = 9
58            NM1 = N - 1
59            MINC = (N - MMIN) / ITER
60            IF(MINC.LT.1)  MINC = 1
61            DO 10 I=1,N
62     10  W(I) = 1.000
63            IF(PTOL.LT.7.00)  WRITE(OUT,201)
64            DO 35 M=MMIN,NM1,MINC
65            MP1 = M + 1
66            IINC = ((N - M) / ITER)
67            IF(IINC.LT.1)  IINC = 1
68    C
69            DO 35 I=M,N,IINC
70            DO 20 J=1,M
71            K = I + J - M
72            X(J) = V(K)
73     20  Y(J) = G(K)
74            J = I - M + 1
75            CALL LINFIT(X,Y,W,M,SLOPE,XINT,SXSLP,SXINT,STDDEV)
76            ENDTRY = -XINT / SLOPE
77            ENDAVE = ENDAVE + ENDTRY
78            WGTAVE = WGTAVE + ENDTRY / STDDEV
79            STDAVE = STDAVE + 1.000 / STDDEV
80            KOUNT = KOUNT + 1
81            K = IABS(ISTART-J+1)
82            L = IABS(ISTART-I+1)
83            IF(PTOL.LT.7.00)  WRITE(OUT,202)  K, L, ENDTRY, STDDEV
84            ENDNEW = (ABS(ENDTRY-ENDPT) / ENDPT) * 100.0
85            IF(ENDNEW.LT.PTOL)  GO TO 30
86            IF(PTOL.LT.7.00)  WRITE(OUT,205)  PTOL
87            ENDAVE = ENDAVE - ENDTRY
88            KOUNT = KOUNT - 1
89            WGTAVE = WGTAVE - ENDTRY / STDDEV
90            STDAVE = STDAVE - 1.000 / STDDEV
91            NDUD = NDUD + 1
92            IF(ENDNEW.GT.ENDOLD)  GO TO 35
93            ENDOLD = ENDNEW
94            GO TO 32
95     30  ENDOLD = 0.0000
96            IF(STDDEV/FLOAT(I-J+2).GT.STDOLD)  GO TO 34
97     32  INIT = J
98            IFIN = I
99            SAVINT = XINT
100           SAVSLP = SLOPE
101           STDOLD = STDDEV / FLOAT(I-J+2)
102    34  IF(STDDEV.GT.STDSTD)  GO TO 35
103           STDSTD = STDDEV
104           SAVEND = ENDTRY
105    35  CONTINUE
106           IF(KOUNT.GE.1)  GO TO 38
107           WRITE(OUT,210)
108           IF(ENDOLD.LT.1.0E2)  WRITE(OUT,211)  ENDOLD
109    36  IF(PTOL.GT.7.0)  GO TO 39
110           PTOL = PTOL * 2
111           IF(ENDOLD.GT.PTOL) RETURN
112           WRITE(OUT,215)
113           GO TO 5
114    38  STDDEV = STDOLD * FLOAT(IFIN-INIT+2)
115           ENDTRY = -SAVINT / SAVSLP
116           ENDAVE = ENDAVE / FLOAT(KOUNT)
117           WGTAVE = WGTAVE / STDAVE
118           WRITE(OUT,203)  ENDTRY, SAVEND
119           WRITE(OUT,207)  WGTAVE,  ENDAVE
120           IF(FLOAT(NDUD).GT.FLOAT(KOUNT)/2.0)  GO TO 36
121           RETURN
122    39  WRITE(OUT,218)  NDUD, PTOL
```

```
123          RETURN
124          END




  1    C
  2    C     ******************************************************************
  3    C
  4          SUBROUTINE LINFIT(X,Y,W,N,XSLOPE,XINT,SXSLP,SXINT,STDDEV)
  5    C
  6    C
  7    C     DEFINITION OF VARIABLE NAMES:
  8    C
  9    C
 10    C     X, Y            = DATA ARRAYS
 11    C     W              = ARRAY FOR WEIGHTING FACTORS
 12    C     N              = NUMBER OF DATA POINTS
 13    C     NW             = FLAG FOR WEIGHTING
 14    C     XINT, XSLOPE   = INTERCEPT AND SLOPE OF LEAST SQUARES LINE
 15    C     SXINT, SXSLP   = STANDARD DEVIATIONS OF XINT AND XSLOPE
 16    C     STDDEV         = STANDARD DEVIATION OF THE LINE FIT
 17    C
 18    C
 19          DIMENSION X(N), Y(N), W(N)
 20          WW = 0.000
 21          WX = 0.000
 22          WY = 0.000
 23          WXY = 0.000
 24          WXX = 0.000
 25          WYY = 0.000
 26          DO 10 I=1,N
 27          WW = WW + W(I)
 28          WY = WY + W(I) * Y(I)
 29          WX = WX + W(I) * X(I)
 30          WXY = WXY + W(I) * X(I) * Y(I)
 31          WXX = WXX + W(I) * X(I) ** 2
 32       10 WYY = WYY + W(I) * Y(I) ** 2
 33          DENOM = WW * WXX - WX ** 2
 34          XSLOPE = (WW * WXY - WX * WY) / DENOM
 35          XINT = (WXX * WY - WX * WXY) / DENOM
 36          VSUM = 0.000
 37          DO  20  I=1,N
 38       20 VSUM = VSUM + (Y(I) - XINT - XSLOPE * X(I)) ** 2
 39          SS = VSUM / FLOAT(N-2)
 40          STDDEV = SQRT(SS)
 41          SXINT = (SS/WW)*(1.0+((WX**2)/DENOM))
 42          IF(SXINT.GT.0.000)  SXINT = SQRT(SXINT)
 43          SXSLP = SS*WW/DENOM
 44          IF(SXSLP.GT.0.000)  SXSLP = SQRT(SXSLP)
 45          RETURN
 46          END




  1    C
  2    C     ******************************************************************
  3    C
  4          SUBROUTINE LOAD(ILOAD,F,I,M,X,IREF,ARRAY)
  5    C
  6    C     THIS SUBROUTINE IS USED TO APPLY UPPER AND LOWER LIMITS
  7    C     TO THE OPTIMISATION BY RETURNING LARGE OBJECTIVE FUNCTION
  8    C     VALUES WHENEVER A TRIAL VALUE IS OUTSIDE THE PERMITTED RANGE.
  9    C
 10          DIMENSION  X(M), ARRAY(3)
 11          I = I + 1
 12          ARRAY(1) = X(I)
 13          IF(IREF.EQ.1)  RETURN
 14          IF(ARRAY(2).LT.X(I).AND.ARRAY(3).GT.X(I))  RETURN
 15          XA = (ARRAY(3) + ARRAY(2)) / 2.0
 16          XA = (X(I) - XA) / (ARRAY(3) - ARRAY(2))
 17          IF(XA.LT.0.000)  XA = - XA
 18          F = XA * 1.00E10
 19          ILOAD=1
 20          RETURN
 21          END
```

5.2. Input Data

The following is the listing of data for an electrode calibration by titration of hydrochloric acid with sodium hydroxide. This data was used to generate the results shown in Table 1.

```
|<--  Column 1

PKW   2      1.331E+01 1.325E+01 1.335E+01      3.000   238.000
VZO   0      3.000E+01 0.000E+00 0.000E+00      3.200   235.000
H+V   1      1.667E-02 0.000E+00 0.000E+00      3.400   231.650
LGV   0 0    0.000E+00 0.000E+00 0.000E+00      3.600   228.000
H+B   0     -1.000E-01 0.000E+00 0.000E+00      3.800   223.700
LGB   0 0    0.000E+00 0.000E+00 0.000E+00      4.000   218.650
EZO   1      3.600E+02 0.000E+00 0.000E+00      4.200   212.500
SLP   1      6.154E+01 0.000E+00 0.000E+00      4.400   204.650
        0.000    265.150                        4.600   193.800
        0.200    263.750                        4.800   175.450
        0.400    262.450                        5.200  -241.700
        0.600    261.100                        5.400  -263.250
        0.800    259.700                        5.600  -274.800
        1.000    258.250                        5.800  -282.950
        1.200    256.650                        6.000  -288.850
        1.400    255.000                        6.200  -293.700
        1.600    253.400                        6.400  -297.650
        1.810    251.500                        6.600  -301.050
        2.000    249.700                        6.800  -304.100
        2.200    247.600                        7.000  -306.750
        2.400    245.500                        7.200  -309.100
        2.600    243.200                        7.400  -311.200
        2.800    240.700                       -1.000
```

4

SCOGS2

A Nonlinear Least-Squares Program for the Evaluation of Formation Constants of Metal Complexes

D. D. PERRIN and H. STUNZI

1. INTRODUCTION

Bjerrum's *enbar* ($\bar{n}$) approach,[1] provided a useful tool in studying quantitatively the stepwise formation of mononuclear complexes, but this method failed if polynuclear, protonated, or hydrolyzed species were present. The use of electronic digital computers overcame this barrier: programs were developed in which formation constants were sought by minimizing the sums of squares of residuals in "analytical hydrogen ion concentrations." The residual is the difference between the experimental value and the amount calculated from the pH and the concentrations of all species using computed constants. (a general discussion of least-squares calculations has been presented by Wentworth.[2])

Two of the earliest noteworthy programs were LETAGROP,[3-6] and GAUSS.[7] The former uses a "pit-mapping" approach in which the error-square sum function of N variables in a system is defined as a "paraboloid in *n*-dimensional space" and the calculation seeks the minimum or "pit" in such a surface.[3,5] GAUSS G[7] is a nonlinear least-squares program which calculates shifts in the formation constants, partial derivatives being obtained numerically by incrementing the constants. A modified GAUSS method was developed for all types of equilibria in a system comprising one kind of metal ion and one kind of ligand.[8-10]

This was subsequently replaced by a more general program, SCOGS ("*S*tability *C*onstants *o*f *G*eneral *S*pecies),[11-13] which applied to systems containing up to two

---

D. D. PERRIN and H. STUNZI • Medical Chemistry Group, The John Curtin School of Medical Research, The Australian National University, P.O. Box 334, Canberra City, A.C.T., 2601, Australia. Dr. Stunzi's present address is Universite de Neuchatel, Institut de Chimie, Avenue de Bellevaux 51, CH-2000 Neuchatel, Switzerland.

71

kinds of metal ions (M′ and M″) and two kinds of ligands (L′ and L″). The general complex can be written as $(M'_{m1}M''_{m2}L'_{l1}L''_{l2}H_h)$ and the cumulative formation constant is

$$\beta_{m1,m2,l1,l2,h} = \frac{[M'_{m1}M''_{m2}L'_{l1}L''_{l2}H_h]}{[M']^{m1}[M'']^{m2}[L']^{l1}[L'']^{l2}[H]^h}$$

where $m1$, $m2$, $l1$, and $l2$, are positive integers or zero and h is a positive integer for protonated species, a negative integer for hydroxo complexes, or zero. For simple species, several of these subscripts will be zero:

$$\log \beta_{0,0,1,0,1} = pK_a \text{ of HL}'$$

$$\log \beta_{0,0,1,0,2} = pK_a \text{ of HL}' + pK_a \text{ of H}_2\text{L}'$$

$$\beta_{1,0,n,0,0,} = \beta_n \text{ of ML}_n$$

$$\log \beta_{1,0,0,0,-1} = -pK_a \text{ of } M'_{aq} \text{ (formation of M'OH)}$$

More complicated species include mixed ligand complexes M′L′L″ $(\beta_{1,0,1,1,0})$, protonated complexes M′L′$_2$H $(\beta_{1,0,2,0,1})$, hydroxo complexes M′L′(OH)$_2$ $(\beta_{1,0,1,0,-2})$, polymeric complexes M′$_2$L′$_2$(OH)$_2$ $(\beta_{2,0,2,0,-2})$, mixed metal complexes M′M″L′ $(\beta_{1,1,1,0,0})$, or even complexes containing two metal ions and two ligands M′M″L′L″ $(\beta_{1,1,1,1,0})$. Note that the original program SCOGS defined the constants with the number of OH instead of number of H (h).

In SCOGS2, an estimate of the value of f is needed to calculate free [H$^+$] from pH, for insertion in the ionic balance equation. In equilibria where proton gain or loss is not involved, "practical" and "concentration" constants are the same. For an equilibrium involving addition or removal of protons SCOGS2 computes "practical," or Bronsted, constants made up of hydrogen ion activity and molar concentrations of the other species. These can be converted to concentration constants, β_c, by the relation

$$\log \beta_c = \log \beta_p + h \log f$$

SCOGS has been widely used as a "workhorse" since it was first written in this department in 1968. It has also been adapted in other laboratories. We have made several modifications and corrections. The present version, SCOGS2, will process either (volume of titrant, pH) or (pM, pH) data; give a full or limited printout for either type of data; has a simplified input and self-explanatory variable names; decides on the size of the increments in the numerical differentiation and when to end iteration; and is successful in cases where the original SCOGS failed.

2. DESCRIPTION OF THE PROGRAM SCOGS2

2.1. General

SCOGS2 is a general computer program that accepts data from pH (option SCOGSH) or pM (option SCOGSM) titrations to refine a set of estimated formation constants. For the SCOGSH option the quantity minimized is the sum of squares of differences between experimental and calculated volume of titrant added, over all experimental values. The SCOGSM option minimizes pM rather than titer. The main points of the program SCOGS2 will be discussed using the SCOGSH option. Differences between this option and SCOGSM are detailed in Section 2.3.

The total concentrations of each metal ion, M' and M'', the ligands, L' and L'', and hydrogen ion, H, are known from the quantities used to make up the test solution (analytical total concentrations). C_H is calculated from the number of displaceable protons on each ligand and the initial excess of inorganic acid, or inorganic base. It decreases on each addition of titrant, or increases for each addition of titrant acid. The equilibria in a given solution are defined by a set of N formation constants β_j (see above), some of which may be known, some to be determined. For each species J, the equilibrium concentration $[C]_j$ can be calculated from the formation constant β_j, the free metal and ligand concentrations, and the activity of hydrogen ion:

$$\beta_j = \beta_{m1j, m2j, l1j, l2j, hj}$$

$$[C] = \beta_j [M']^{m1j} [M'']^{m2j} [L']^{l1j} [L'']^{l2j} \{H\}^{hj}$$

The total concentrations are also from the sum over all species:

$$C_{M'} = \sum_{J=1}^{N} m1_j [C]_j$$

$$C_{M''} = \sum_{J=1}^{N} m2_j [C]_j$$

$$C_{L'} = \sum_{J=1}^{N} l1_j [C]_j$$

$$C_{L''} = \sum_{J=1}^{N} l2_j [C]_j$$

$$C_H = \{H\}/f - \{OH\}/f + \sum_{J=1}^{N} h_j [C]_j$$

(f is the mean activity coefficient of H^+ and OH^-: $\{H^+\} = [H^+] \cdot f$).

For each set of constants, the free metal ion and ligand concentrations are those that fulfil

$$C_{tot, calculated} = C_{tot, experimental}$$

whereas the "calculated volume of titrant added" is derived from C_H.

The desired result from calculations with SCOGS2 is to refine estimated formation constants so that the "calculated volume of titrant added" best represents the experimental values.

SCOGS2 comprises a MAIN program, the subroutine COGSNR and a standard matrix inversion subroutine SUB760 which is called from MAIN and from COGSNR.

2.2. MAIN Program of SCOGS2—SCOGSH Option

The input data are made up of the main parameters valid for the set of experiments to be evaluated together, including the number of metal ions, ligands, and species; the constants comprise those known and estimates for those to be determined, and a key word denoting which option is to be employed.

For each titration, the input data comprise the analytical concentrations of the various components, and the readings; volume of titrant added (ml), pH. The readings from all titrations are arranged sequentially ranging from number 1 to TNR. The variable IFR keeps track of the numbers where each new titration starts.

For each point k, initial estimates of the free metal ion and free ligand concentrations are sought. For the first point in each titration the former are assumed to be equal to the total metal ion concentrations, whereas the latter are calculated from the pK_a values, assuming no metal complex formation. For the kth point, these initial estimates are taken as equal to the final values for the $(k\text{-}1)$th point, except when the determination of free ligand or metal ion concentration did not converge properly for the point $(k\text{-}1)$. These calculations are performed within subroutine COGSNR. C_H^{calc} is also calculated by COGSNR and is returned to the MAIN program where the "calculated titer" is determined from it. The residual, R_k, is the difference between experimental and calculated titer, R_k being the quantity minimized.

By incrementing and decrementing each constant to be determined, the differential $dR_k/d(\log \beta_j)$ is calculated (the concentrations are again found with COGSNR). The Jacobian and Hessian matrices for the least-squares equations are progressively built up. After the final point, the Hessian matrix is inverted (SUB760) and solved to yield the shifts that are to be applied to each constant. Thus, a new set of constants is obtained, and printed in the output. The goodness of the fit is expressed as "standard deviation in volume of titrant." (Note that since the experimental error in titer is similar over all points, no weighting of the readings has to be applied.) If the standard deviation in volume of titrant is smaller with the improved formation constants, another cycle of refinement is made. However, if the shifts to the constants are all small and the standard deviation has not been improved, the final values of the formation constants are printed together with their estimated standard deviations. After the refinement is complete a table is printed listed all residuals, total concentrations, and the concentration of each species expressed as its negative logarithm (pC_j) for each titration point. The residuals are listed under "C–E," which stands for "volume of titrant added: calculated–experimental." The pC_j values are listed under "PCONC" in order corresponding to the input order of formation constants.

The SCOGSDMH option suppresses all but the titration titles and progress of refinement output.

Further features of the MAIN program are:

  a. The selection, at execution time, of the increments in $\log \beta_j$ for numerical differentiation.
  b. If the concentration of a species J at point K is smaller than $ca.$ $10^{-9}\ M$, $dR_k/d(\log \beta_j)$ is not calculated($= 0$).
  c. If the numerical differentiation gives $dR_k/d(\log \beta_j) = 0$, the increment for

log β_j is increased. Limited experience with a version of MAIN without lines 340 to 351 suggests that this is not necessary, i.e., $dR_k/d(\log \beta_j) = 0$ is acceptable as the result of the differentiation.

d. If the calculated shifts to be applied to the sought stability constants are too large, smaller shifts are used. These will be 0.98 or 0.49 depending on the circumstances.

e. If the goodness of the fit is too poor and the improvement between cycles is not significant, the calculation is terminated.

f. If the subroutine COGSNR fails more than 20 times to converge, calculation is terminated.

2.3. SCOGSM Option

The SCOGSM option accepts pM, pH pairs of readings and minimizes the "standard deviation in pM" (pM $= -\log [M]$). 10^{-pM} is taken as an initial estimate to be refined by COGSNR. C_H is not relevant in this program. The set of data given below as an example for calculations with SCOGSM is from Stunzi.[14] The SCOGSDMM option provides only the titration titles and process of refinement as printout.

2.4. Subroutine COGSNR

This subroutine accepts the total concentration of metal ions and ligands at one experimental point and the set of formation constants with their subscript-indices. It refines the estimated free metal ion and ligand concentrations from the MAIN program, and then calculates the concentrations of all species. The final values for the free metal ion and ligand concentrations are reached when the calculated total concentrations and the analytical total concentrations agree to within 0.00002%.

The main algorithm of COGSNR (*Concentration of Generalized Species by the Newton–Raphson method*) has been discussed by Sayce.[10] In a few cases, especially when very stable complexes $M_xL_xH_{-x}$ ($x = 1, 2, 3,$) were formed, the original version of COGSNR failed to converge. This was traced to overshooting in the "shifts" applied to the estimates for the free metal ion or ligand concentrations. Limitation of these "shifts" to a factor of less than 4 and more than 0.1 has alleviated this problem. Furthermore, if the iterations do not converge after 90 cycles, one of two different algorithms is used: COGS from the program COMICS,[9] for complicated systems or KONZ[15] for simple systems (1 metal ion, 1 ligand, and all $m1_j < 3$). The latter finds a new estimate for [L] by dividing the previous estimate by 2 until the new C_L is too small. The iterative calculation then proceeds with a weighted logarithmic mean between the current estimates of [L] that are too large and too small. For each iteration, the free metal ion concentration is found by solving the equation for C_M which is linear or quadratic in [M].

2.5. Limits of SCOGS2

The present version of SCOGS2 allows for 10 titrations, a total of 200 readings and 24 species. This was found adequate for all uses in this department. However, an increase of these limits can easily be made by changing the dimensions in lines 51 to 60 in the MAIN program and lines 15 and 17 in COGSNR. (Note that the dimension of the variable IFR is maximum number of titrations + 1.) To allow for more than two kinds of metal ions or ligands would involve more substantial alterations of the program, primarily in COGSNR.

Some limitations of the program SCOGS2 can be overcome by unconventional definition of the input constants. Although SCOGS2 is not set up to deal with a Z–E equilibrium, formation constants where the ligand undergoes such isomerization have been obtained by assuming the presence of a "ghost" metal ion, M''.[16] (For example, L_E is treated as L' and L_Z as $M''L'$.)

2.6. Some Examples of the Use of SCOGS

As the following recent examples indicate, SCOGS is currently used internationally.

Yamauchi et al.[17] applied SCOGS to titration data from which they refined formation constants for histidine-containing ternary amino acid copper (II) complexes. Davidenko et al.[18, 19] used this program to obtain formation constants of ternary complexes of $3d$ transition metal ions with nucleotides and glycine or histidine. The behavior of UO_2^{2+} with some cryptates in aqueous solution was studied quantitatively by Spiess et al.[20] Potentiometric titration data for the ligand 2,2,4,-trimethyl-1,5,9-triazacyclodecane with nickel(II), copper(II), and zinc(II) ions yielded the same formation constants when either of the computer programs SCOGS or MINIQUAD was used.[21] Colin and Pinart[22] applied SCOGS to evaluate formation constants of mercury(II) chelates with pyridyl-2,6-dimethanol. Similarly, Lajunen et al.[23] used SCOGS to calculate the formation constants of aluminum complexes with 3-hydroxy-7-sulfo-2-naphthoic acid.

There are significant interrelationships among published computer programs. SQUAD (Stability QUotients from Absorbance Data), which calculates formation constants for the same range of species as SCOGS, using absorbance data,[24] is based on the algorithm used in SCOGS. Nagypal[25] extended SCOGS so that any fragment (such as ML) of a complex can be treated as a component of larger complexes, giving a program that is suitable for calculating formation constants from polarographic, distribution, or spectrophotometric measurements. POLAG, a general computer program to calculate formation constants from polarographic data, derives, in turn, from SQUAD.[26]

Sabatini and Vacca[27] computed three complex formation equilibria constants using SCOGS, LG/3 (a program having the same basic approach as LETAGROP) and their program LEAST (for use in one-metal, one-ligand situations). SCOGS took longest to iterate but all methods gave essentially the same constants (the variation in log K was in the second decimal place). They conclude that the Gauss–Newton method is to be preferred for the computer refinement of formation constants. The minor disadvantage of the relative slowness of SCOGS is more than compensated for by the greater generality of this program.

Field and McBryde[28] state that "experience with simulated titration data has revealed that the program SCOGS consistently achieves good recovery of the starting equilibrium constants and also the smallest values of standard deviation in pM and

TABLE 1. Refinement of the Constants from Data of Cu^{2+}/Phosphonoacetate Titrations

Stab. const: Subscripts[b]: Complex:	$\log K_1$ 1,0,1,0,0 ML	$\log \beta_2$ 1,0,2,0,0 ML_2	$\log \beta_{111}$ 1,0,1,0,1 MLH	$\log \beta_{121}$ 1,0,2,0,1 ML_2H	$\log\beta_{122}$ 1,0,2,0,2 ML_2H_2	$\log \beta_{11-1}$ 1,0,1,0,-1 MLOH	$\log \beta_{201}$ 2,0,1,0,0 M_2L	Std. dev.[a]
Estimates	1.0	3.0						
Refined, 2 cy.[c]	2.0[d]	4.0[d]						0.06
Estimates	4.0	7.0						
Refined, 8 cy.	7.02(4)[e]	10.7(2)						0.016
Estimates	7.0	10.7	11.0	15.7	19.7			
Refined, 5 cy.	7.16(0)	11.02(1)	11.78(1)	16.2(3)	21.5(1)			0.00075
Estimates	7.16	11.0	11.8	16.2	21.5	−0.3	10.0	
Refined, 5 cy.	7.14(0)	10.99(1)	11.76(0)	16.36(8)	21.53(6)	−1.65(9)	8.98(2)	0.00032

[a]Standard deviation in titer (ml titrant added).
[b]ml, $m2$, $l1$, $l2$, h.
[c]Number of refinement cycles for SCOGS2.
[d]Estimates rejected by SCOGS2 as being too low.
[e]Standard deviation $(1 - \sigma)$ of last digit given for $\log \beta$.

pL." SCOGS is relied on to indicate "correct" formation constants derived from real titrations, and hence serves as a basis for comparison with other existing computer programs.

2.7. Using SCOGS2 to Build the Equilibrium Model

SCOGS would not refine reliably if *bad* estimates of the constants were used as starting values.[29] This could be avoided by using, as preliminary estimates, values of formation constants for similar ligands and metals, if available. Formation constant compilations are useful as source material. The recommended strategy for applying SCOGS or SCOGS2 is to begin with estimates of log β values for the most likely complexes. These constants are given a preliminary "refinement" and additional complex species are then added to the model, guided by their effect on the computed standard deviation in titre, the trends in residuals "C–E," and estimated standard deviation of each constant. An example serves to illustrate these principles.

Phosphonoacetic acid (H_3L) has three pK_a values and was expected to form a range of copper complexes. To test the ability of SCOGS2 to refine "bad" constants, improbably small starting values of log K_1 (1.0 and log β_2 (3.0) were used in treating appropriate pH titration data. In two cycles of refinement, these values increased by the maximum allowed shift per cycle, to 2.0 and 4.0. Calculation terminated without any improvement in the large standard deviation in "ml titrant added" (0.06). This indicated that the estimates of log K_1 and log β_2 were much too small. New estimates (log K_1 = 4.0 and log β_2 = 7.0) were refined in 8 cycles to log K_1 = 7.02 and log β_2 = 10.7. The large standard deviation in ml titrant (0.016 ml) was indicative of further species being important. From the buffer regions of the test solutions (pH 3.5–5.5) protonated complexes such as CuLH, CuL$_2$H, and CuL$_2$H$_2$ were also postulated and expected to have pK_a values of 4–5. This gave initial estimates for log $\beta_{1,0,1,0,1}$, log $\beta_{1,0,2,0,2}$ as listed in Table 1. Five cycles of refinement gave a new set of constants with a standard deviation of 0.00075 ml. It decreased to 0.00032 ml when log $\beta_{1,0,1,0,-1}$ and log $\beta_{2,0,1,0,0}$ were included. (Starting estimates were based on pK_a *ca.* 7.5 for CuL = CuLOH + H and log K *ca.* 3 for CuL + Cu = Cu$_2$L.) The final values are also given in Table 1 and the maximal concentrations of each species are listed in Table 2. For the last run, seven stability constants were varied simultaneously, while the three pK_a values of the ligand and the two constants for the hydrolysis of copper ion (log $\beta_{1,0,0,0,-1}$ and log $\beta_{2,0,0,0,-2}$) were kept constant.

TABLE 2. Computed Maximal Concentrations of Complex Species for Final Run of Cu^{2+}/Phosphonoacetate (Percent of $[M]_{tot}$ in Species)

Metal: ligand ratio	ML (%)	ML$_2$ (%)	MLH (%)	ML$_2$H (%)	ML$_2$H$_2$ (%)	MLOH (%)
1:1	88	1	44	0.4	0.6	0.1
1:2	82	51	50	1.3	1.3	3.1
3:4	87	7	54	1.4	1.5	0.2
2:5	80	69	61	3.6	3.5	1.2
1:1	77	0.5	49	0.4	0.8	0
2:1	10	0	32	0	0.2	0

The other example given below (Section 5.2.2) will show the refinement of two rather poor estimates for constants for mixed ligand complexes, where 17 constants for the protonation of ligands, hydrolysis of metal ions, and the binary complexes are held constant.[31]

3. REFERENCES

1. J. Bjerrum, *Metal Ammine Formation in Aqueous Solution*, Haase, Copenhagen (1941, reprinted 1957).
2. W. E. Wentworth, Rigorous Least Squares Adjustment. Application to Some Non-Linear Equations, I., *J. Chem. Educ.* **42**, 96–103; II, 162–167 (1965).
3. L. G. Sillén, High-Speed Computer as a Supplement to Graphical Methods. I. Functional Behaviour of the Error Square Sum, *Acta Chem. Scand.* **16**, 159–172 (1962).
4. N. Ingri and L. G. Sillén, High-Speed Computers as a Supplement to Graphical Methods. II. Some Computer Programs for Studies of Complex Formation Equilibria, *Acta Chem. Scand.* **16**, 173–191 (1962).
5. L. G. Sillén, High-Speed Computers as a Supplement to Graphical Methods. III. Twist Matrix Methods for Minimising the Error-square Sum in Problems with Many Unknown Constants, *Acta Chem. Scand.* **18**, 1085–1098 (1964).
6. N. Ingri and L. G. Sillén, High-Speed Computers as a Supplement to Graphical Methods. IV. An ALGOL version of LETAGROP VRID, *Ark. Kemi* **23**, 97–121 (1964).
7. R. S. Tobias and M. Yasuda, Computer Analysis of Stability Constants in Three-Component Systems with Polynuclear Complexes, *Inorg. Chem.* **2**, 1307–1310 (1963).
8. D. D. Perrin and I. G. Sayce, Stability Constants of Polynuclear Mercaptoacetate Complexes of Nickel and Zinc, *J. Chem. Soc.(A)* **1967**, 82–89.
9. D. D. Perrin and I. G. Sayce, Computer Calculation of Equilibrium Concentrations in Mixtures of Metal Ions and Complexing Species, *Talanta* **14**, 833–842 (1967).
10. D. D. Perrin, I. G. Sayce, and V. S. Sharma, Mixed Ligand Complex Formation by Copper (II) Ion, *J. Chem. Soc (A)* **1967**, 1755–1759.
11. I. G. Sayce, Computer Calculation of Equilibrium Constants of Species Present in Mixtures of Metal Ions and Complexing Agents, *Talanta* **15**, 1397–1411 (1968).
12. I. G. Sayce, Computer Calculation of Equilibrium Constants by Use of the Program SCOGS: A Correction, *Talanta* **18**, 653 (1971).
13. I. G. Sayce and V. S. Sharma, Computer Calculation of Equilibrium Constants using Program SCOGS. A Further Modification, *Talanta* **19**, 831 (1972).
14. H. Stunzi, Copper Complexation by Isatin β-Thiosemicarbazones in Aqueous Solution, *Aust. J. Chem.* **34**, 2549–2561 (1981).
15. G. Anderegg, Die Anwendung der elektronishchen Rechenmaschine zur Lösung einiger Probleme der Komplexchemie, *Helv. Chim. Acta* **45**, 901–907 (1962).
16. H. Stunzi, Can Chelation be Important in the Antiviral Activity of Isatin β-Thiosemicarbazones?, *Aust. J. Chem.*, **35**, 1145–1155 (1982).
17. N. K. Davidenko and P. A. Manorik, Mixed-ligand Complexes of Copper(II) and Nickel (II) with Nucleotides and Glycine, *Zh. Neorg. Khim.* **25**, 437–444 (1980).
18. O. Yamauchi, T. Takaba, and T. Sakurai, Solution Equilibria of Histidine-containing Ternary Amino Acid-Copper(II) Complexes in 20% Dioxane-Water, *Bull. Chem. Soc. Jpn* **53**, 106–111 (1980).
19. N. K. Davidenko, P. A. Manorik, and K. B. Yatsimirskii, Ternary Complexes of the Ions of 3d Transition Metals with Adenine Nucleotides and Histidine, *Zh. Neorg. Khim.* **25**, 883–890 (1980).
20. B. Spiess, F. Arnaud-Neu, and M. J. Schwing-Weil, Behaviour of Uranium (VI) with some Cryptands in Aqueous Solution, *Inorg. Nucl. Chem. Lett.* **15**, 13–16 (1979).
21. R. W. Renfrew, R. S. Jamison, and D. C. Weatherburn, Aqueous Solution Equilibria Involving the Ligand 2,2,4-Trimethyl-1,5,9-triazacyclododecane and Nickel(II), Copper(II), and Zinc(II), *Inorg. Chem.* **18**, 1584–1589 (1979).
22. J. L. Colin and J. Pinart, Etude des chélates mercuriques de pyridyl 2,6-dimethanol, *Bull. Soc. Chim. France* **1**, 7–10 (1979).
23. L. H. J. Lajunen, R. Petrola, P. Schildt, O. Korppi-Tominola, and O. Makitie, Complex Formation between Aluminium and 3-Hydroxy-7-sulpho-2-naphthoic Acid, *Talanta* **27**, 75–78 (1980).

24. D. J. Leggett and W. A. E. McBryde, General Computer Program for the Computation of Stability Constants from Absorbance Data, *Anal. Chem.* **47**, 1065–1070 (1975).

25. I. Nagypal, General Method for the Calculation of the Concentration, Distribution and Formation Constants in Chemical Equilibrium Systems, *Magy. Kem. Foly.* **80**, 49–55 (1974).

26. D. J. Leggett, POLAG—A General Computer Program to Calculate Stability Constants from Polaro-graphic Data, *Talanta* **27**, 787–793 (1980).

27. A. Sabatini and A. Vacca, A New Method for Least-Squares Refinement of Stability Constants, *J. Chem. Soc. (Dalton)* **1972**, 1693–1698.

28. T. B. Field and W. A. E. McBryde, Determination of Stability Constants by pH Titrations: A Critical Examination of Data Handling, *Can. J. Chem.* **56**, 1202–1211 (1978).

29. P. Gans and A. Vacca, Application of the Davidon–Fletcher–Powell Method to the Calculation of Stability Constants, *Talanta* **21**, 45–51 (1974).

30. H. Stunzi and D. D. Perrin, Stability Constants of Metal Complexes of Phosphonoacetic Acid, *J. Inorg. Biochem.* **10**, 309–316 (1979).

31. H. Stunzi, R. L. N. Harris, D. D. Perrin, and T. Teitei, Stability Constants for Metal Complexation by Isomers of Mimosine and Related Compounds, *Aust. J. Chem.* **33**, 2207–2220 (1980).

4. INSTRUCTIONS FOR SCOGS2

The input data requirements for SCOGS2 are displayed in Table 3. Free format has been used throughout the data input phase of SCOGS2. Consequently multiple

TABLE 3. Input Data Requirements for SCOGS2

Item	Description	Card No.	Entry/Explanation
1	Option	1	SCOGSH/Refine volume of titrant, pH data, full output. Other options are SCOGSDMM, SCOGSM, or SCOGSDMM. See text for full details.
2	Number of titrations to be evaluated together, maximum: 10	2	2/2 titrations
3	Number of kinds of metal ions ($\leqslant 2$), ligands ($\leqslant 2$), number of species ($\leqslant 24$) including protonated ligands, hydrolyzed metal ions, and all complexes	3	1,2,19/1 metal ion (Cu^{2+}), 2 ligands (his, M*). 19 constants are given
4	For each species a card with $m1$, $m2$, $l1$, $l2$, h, $\log \beta_{m1,m2,l1,l2,h}$	4	0.0,1,0,1, 8.906/pK_a of histidine (HL)
		5	0,0,1,0,2, 14.84/pK_a of HL + pK_a of H_2L of histidine
	(If the last constant of the list is 1,0,0,0, -2, $\log \beta$, then $\beta = [M'] \cdot \{H\}^{-2}$, i.e. the solubility product of $M'(OH)_2$. Note that the same constant within the list relates to soluble $M'(OH)_2$, i.e., $\beta = [M'(OH)_2]/[M']\{H\}^{-2}$	6	0,0,1,0,3, 16.55/pK_a of HL + pK of H_2L + pK_a of H_3L of histidine
		7	0,0,0,1,1, 8.8/pK_a of M* (HL)
		8–22	/the other constants: the third integer always refers to histidine, the fourth to M*

TABLE 3. Continued

Item	Description	Card No.	Entry/Explanation
5	Number of displaceable protons on ligand 1, on ligand 2	23	1,1/both ligands were added to the solution in the monoprotonated form (HL)
6	Burette correction factor (if the readings of the burette are correct type 1)	24	1./no burette corection required
7	Title for following titration (gives a heading on the output)	25	4 CU HIS M* ....
8	C_M, $C_{M''}$, $C_{L'}$, $C_{L''}$ inorganic acid, strength of titrant base (mol/liter), initial volume (ml), correction of pH (The first five entries are initial total molar concentrations; the correction of pH applies to each reading of this titration: pH = pH + PHCOR) For titration with strong acid instead of strong base treat H^+ as negative OH^-: for "concentration of titrant base" give "negative concentration of titrant acid." If there is a certain amount of inorganic base present give its negative concentration instead of inorganic acid.	26	0.001967, 0, 0.002005, 0.001935, 0.003879, 0.993, 50.237 [Cu]$_{tot}$, no second metal ion, [his]$_{tot}$, [M*]$_{tot}$, molar concentration of HNO_2 present, molar concentration of titrant KOH, volume in ml, no pH correction required
9	For each experimental reading ("point") a card with "ml titrant base added," pH	27	0.21, 3.803/the first reading: 0.21 ml KOH added, pH = 3.803
	[in SCOGSM: pM, pH]	28–47	/the subsequent readings
	For the last point in the current titration give "ml titrant base added,"-pH (the negative sign is used as a marker; it is ignored in the calculations)	48	0.375, −5.401/0.375 ml KOH added, this is the last point of this titration, pH = 5.401
10	If more titrations belong to this set (item 1) go to item 7, else go to item 11	49–71	/these cards give the title, concentrations, and readings of the second titration
11	Log K_w ($-pK_w$, activity coefficient f of H^+ and OH [K_w = (H) · (OH), f = 1. for concentration constants]	72	−13.62, 0.81/$-pK_w$; activity coeff. f = 0.81
12	Number of constants to be varied (determined), followed by their ranking position in item 4 (if none: give 0,0; for a listing of concentrations)	73	2,18,19/2 constants are to be refined, they are number 18 (card No. 20) and number 19 (card No. 21)
13	Go to item 2. This gives the number of titrations for the next calculation. Or 0 (zero) if execution of SCOGS2 is to be terminated.	74	0/last card, terminate execution.

entries, per card, are separated by commas. In Table 3 a general description of the
input data is given, item by item, and may be found under Description. The right-
hand side of the table provides specific examples of each item in the form Card No.,
Entry and Explanation (of entry). A complete listing of this data set, for the system
copper (II)-histidine-(3-hydroxy-1-methylpyridin-4(1*H*)-one) (alias M*), is given in
Section 5.2.2.

5. PRESENTATION OF THE PROGRAM

5.1. Listing of SCOGS2

The following is a listing of the program SCOGS2.

```
 1   C
 2   C      PROGRAM SCOGS2, MAIN PROGRAM      HST NOV. 81
 3   C                                                   SCOGS2.MAIN
 4   C
 5   C      THIS PROGRAM IS WRITTEN IN FORTRAN-V
 6   C
 7   C      USED ON A UNIVAC 1100/82 COMPUTER
 8   C              COMPUTER SERVICES CENTRE
 9   C              AUSTRALIAN NATIONAL UNIVERSITY
10   C              CANBERRA 2600  AUSTRALIA
11   C
12   C      INPUT
13   C      *****
14   C              FREE FORMAT, SEPARATE VALUES BY A COMMA
15   C
16   C      1  CONTROL WORD TO INDICATE WHICH CALCULATION OPTION
17   C         IS TO BE USED.
18   C         SCOGSH = PROCESS VOL. OF TITRANT;PH DATA.
19   C         SCOGSM = PROCESS PM;PH DATA.
20   C         SCOGSDMH = DEMAND VERSION OF SCOGSH
21   C         SCOGSDMM = DEMAND VERSION OF SCOGSM
22   C      2  NO OF TITRATIONS, INEXP.  (0 IF INPUT IS COMPLETED)
23   C      3  NO OF METAL IONS, NM (0,1,2), NO OF LIGANDS, NL (0,1,2)
24   C         NO OF CONSTANTS, N. (1 - 24)
25   C      4  FOR EACH COMPLEX A CARD: MM1,MM2,ML1,ML2,MH,LOG BETA
26   C            MM(1,J),  MM(2,J) = NO OF METAL IONS 1 AND 2 IN JTH COMPLEX
27   C            ML(1,J),  ML(2,J) = NO OF LIGANDS 1 AND 2 IN JTH COMPLEX
28   C            MH(J) = NO OF H+ IN THE COMPLEX (MH<0 FOR OH-)
29   C            (IF LAST CONSTANT IS (1,0,0,0,-2,X) THEN X = LOG M*H**-2)
30   C      5  NO OF DISPLACEABLE PROTONS ON LIGAND 1, ON LIGAND 2, NDP(I).
31   C      6  BURETTE CORRECTION FACTOR, BURCAL. (IF NONE: 1.)
32   C      7  TITLE FOR ITH TITRATION, TITLE(ITH,J).
33   C      8  INITIAL TOTAL CONCENTRATIONS OF M1, MT(1,NEXP), M2, MT(2,NEXP),
34   C         L1, LT(1,NEXP), L2, LT(2,NEXP), ACID, INORGANIC ACID ADDED,
35   C         ACID, CONCENTRATION OF TITRANT BASE, BA(NEXP), INITIAL VOLUME,
36   C         V(NEXP), CORRECTION TO BE APPLIED TO ALL PH VALUES, PHCOR,
37   C         WHERE NEXP REFERS TO THE NEXPTH TITRATION.
38   C      9  FOR EACH READING A CARD :  ML,TITR(K), PH, PH(K).
39   C            LAST READING: ML, -PH
40   C      10 IF MORE TITRATONS: GO TO ITEM 6
41   C         ELSE GO TO ITEM 11
42   C
43   C      11 -PKW, F  (F = ACTIVITY COEFF.; PH = -LOG (H) = -LOG H1*F )
44   C                  (F = 1.  IF  PH = -LOG H1 )
45   C      12 NO OF CONSTANTS TO BE VARIED, NCV. INDICES ACCORDING
46   C         TO ORDER UNDER ITEM 4 IG(I) FOR THE ITH CONSTANT TO BE VARIED.
47   C
48   C      13 GO TO ITEM 1
49   C
50   C*****************************************************************
51         REAL PH(200),LOGKW,KW,ILBO(24),ILB(24),MT(2,24),LT(2,24)
52         REAL*8 TYPE,MODE(4)
```

```
53            DATA MODE/'SCOGSH','SCOGSM','SCOGSDMH','SCOGSDMM'/
54            INTEGER TNR
55            DOUBLE PRECISION CK(24),CC(24,24),MTOT(2),LTOT(2),
56           &LOGB(24),TITLE(30,8),BETA(24),LF(2),MF(2),C(24),
57           &DE(24),SHIFT(24),HTOT,CTITR,DTITR,R,RO,CT2,CT3,HTC
58            DIMENSION ML(2,24),MM(2,24),MH(24),NDP(2),BA(30),HT(30),V(30),
59           &IG(24),INCIT(200),NPTSET(31),ICH(24),DMY(24),IFR(31),TITR(200),
60           &RES(200)
61    C
62            COMMON C,MTOT,LTOT,LF,MF,HTC,BETA,IWRT,
63           &ML,MM,MH,NL,NM,N,HF,F,KW,NIT,NNCI,KJ
64    C
65       99 FORMAT()
66      102 FORMAT(8A8)
67      104 FORMAT(//9X,'M1  M2  L1  L2  H     LOG.BETA'/)
68      108 FORMAT(1X,I3,2X,5(2X,I2),3X,F8.4)
69      114 FORMAT(2X,I2,7X,1PD12.4,3D12.4,0PF11.6,F11.6,F8.3,F14.4,10X,I3)
70      123 FORMAT(/' -PKW=',F7.3,5X,'F=',F5.2/)
71      124 FORMAT(/' NUMBER OF PARAMETERS  ',I2,9X,'TOTAL NUMBER OF READINGS
72           & ',I3/)
73      173 FORMAT(/' INCREMENT FOR  CONSTANT ',I2,' RAISED')
74      175 FORMAT(/' INCREMENTS CHANGED FOR ',I3,' POINT(S)')
75      182 FORMAT(/' OVERSHIFT,  CONSTANT NO. ',I2,' SHIFT(I)=',F8.3)
76      183 FORMAT(/' EXTREME OVERSHIFT, CONSTANT NO. ',I2,' SHIFT(I)=',F8.3)
77      196 FORMAT(/' STANDARD DEVIATION IN ML TITRANT WITH THE INPUT CONSTA'
78           &,'NTS :',1PE10.3/)
79      197 FORMAT(/,17X,' STANDARD DEVIATION IN ML TITRANT : ',1PE10.3,/)
80      195 FORMAT(/' STANDARD DEVIATION IN   PM  WITH THE INPUT CONSTANTS :'
81           &1PE10.3/)
82      199 FORMAT(/,17X,' STANDARD DEVIATION IN   PM  : ',1PE10.3,/)
83      201 FORMAT(/' STDLB NEGATIVE')
84      202 FORMAT(1X,I2,5X,5I2,F12.4,'  +-',F8.4,2X,'SHIFT=',F7.4)
85      206 FORMAT(/'  1/4 SHIFTS APPLIED FOR NEXT CYCLE')
86      207 FORMAT(10X,I2,' CYCLES CALCULATED '/)
87      210 FORMAT(//,8X,'PH    PM    RES C-E   M1 TOT    M2 TOT    L1 TOT',
88           &3X,'L2 TOT    PCONC ',/,30X,4('  FREE '))
89      209 FORMAT(//,8X,'PH    ML    RES C-E   M1 TOT    M2 TOT    L1 TOT',
90           &3X,'L2 TOT    PCONC ',/,30X,4('  FREE '))
91      212 FORMAT(/,I4,F8.3,F7.4,F9.5,2X,4(1PE9.3),/,30X,4(1PE9.3),1X,
92           &9(0PF7.3),/,3(67X,9(F7.3)/))
93      218 FORMAT('0','BURETTE CORRECTION FACTOR : ',F8.5)
94      240 FORMAT('0',13X,I2,' METAL IONS     ',I2,' LIGANDS',10X,
95           &I2,' COMPLEXES')
96      241 FORMAT(34X,I2,' H ',I2,' H')
97      400 FORMAT(//2X,'TITRATION',5X,'TITLE'/4X,'NUMBER')
98      401 FORMAT(7X,I2,7X,8A8)
99      402 FORMAT(//2X,'COMPOSITION OF SOLUTIONS USED IN EACH TITRATION',
100          &1X,'TITRATION',3X,'METAL 1',5X,'METAL 2',5X,'LIGAND 1',
101          &4X,'LIGAND 2',5X,'ACID',7X,'BASE',6X,'INITIAL',
102          &9X,'PH',11X,'NO. OF'/2X,'NUMBER',6X,25('-'),' MOLES PER LITER '
103          &25('-'),2X,'VOLUME',6X,'CORRECTION',7X,'POINTS')
104      999 FORMAT(1H1)
105      799 FORMAT('  STD. DEV. = ',F10.5,/)
106   C
107   C    SET READ UNIT NUMBER, IRDR, AND WRITE UNIT NUMBER, IWRT.
108   C
109           IRDR=4
110           IWRT=6
111   C
112   C    READS IN GENERAL DATA
113   C
114           READ(IRDR,102)TYPE
115        1 WRITE(IWRT,999)
116           READ(IRDR,*)INEXP
117           IF(INEXP)2222,1000,2
118     2222 READ(IRDR,*)J,(MM(I,J),I=1,2),(ML(I,J),I=1,2),MH(J),LOGB(J)
119           DO 2223 I=1,2
120           MF(I)=0.
121     2223 LF(I)=0.
122           IFINAL=0
123           GO TO 2229
124        2 IFINAL=0
125           READ(IRDR,*)NM,NL,N
126           DO 109 J=1,N
127      109 READ(IRDR,*)(MM(I,J),I=1,2),(ML(I,J),I=1,2),MH(J),LOGB(J)
128           READ(IRDR,*)(NDP(I),I=1,2)
```

```
129            DO 112 I=1,2
130            LF(I)=0.
131      112 MF(I)=0.
132            K=0
133   C
134   C       READ BURETTE CORRECTION FACTOR( 1.0 IF NO CORRECTION REQUIRED   )
135   C
136            READ(IRDR,*)BURCAL
137            DO 121 NEXP=1,INEXP
138   C
139   C       READS IN TITRATION DATA FOR ALL TITRATIONS. SET CONVERGENCE
140   C       FOR CONVERGENCE IN COGSNR, INCIT(K).
141   C       K = THE TITRATION POINT NUMBER COUNTER, RUNNING CONSECUTIVELY
142   C       FOR ALL TITRATIONS.
143   C       NOR = THE NUMBER OF POINTS IN EACH TITRATION.
144   C       IFR(NEXP) =  FIRST POINT NUMBER FOR EACH NEW TITRATION.
145   C       TNR = TOTAL NUMBER OF READINGS.
146   C
147            READ(IRDR,102)(TITLE(NEXP,J),J=1,8)
148            READ(IRDR,*)MT(1,NEXP),MT(2,NEXP),LT(1,NEXP),LT(2,NEXP),ACID,
149          &BA(NEXP),V(NEXP),PHCOR
150            IFR(NEXP)=K+1
151            IF(TYPE.EQ.MODE(2).OR.TYPE.EQ.MODE(4))GO TO 115
152            HT(NEXP)=LT(1,NEXP)*FLOAT(NDP(1))+LT(2,NEXP)*FLOAT(NDP(2))+ACID
153      115 K=K+1
154            READ(IRDR,*)TITR(K),PH(K)
155            PH(K)=PH(K)+PHCOR
156            TITR(K)=TITR(K)*BURCAL
157            INCIT(K)=0
158            IF(PH(K))117,115,115
159      117 NOR=K-IFR(NEXP)+1
160            NPTSET(NEXP)=NOR
161            PH(K)=-PH(K)+2*PHCOR
162      121 CONTINUE
163            READ(IRDR,*)LOGKW,F
164            READ(IRDR,*)NCV,(IG(I),I=1,NCV)
165            TNR=K
166            IFR(INEXP+1)=K+1
167            WRITE(IWRT,400)
168            DO 300 NEXP=1,INEXP
169      300 WRITE(IWRT,401)NEXP,(TITLE(NEXP,J),J=1,8)
170            IF(TYPE.EQ.MODE(3).OR.TYPE.EQ.MODE(4))GO TO 2229
171            WRITE(IWRT,402)
172            DO 301 NEXP=1,INEXP
173      301 WRITE(IWRT,114)NEXP,MT(1,NEXP),MT(2,NEXP),LT(1,NEXP),LT(2,NEXP),
174          &ACID,BA(NEXP),V(NEXP),PHCOR,NPTSET(NEXP)
175            WRITE(IWRT,218)BURCAL
176            WRITE(IWRT,240)NM,NL,N
177            WRITE(IWRT,241)(NDP(I),I=1,NL)
178            WRITE(IWRT,104)
179            DO 131 J=1,N
180      131 WRITE(IWRT,108)J,(MM(I,J),I=1,2),(ML(I,J),I=1,2),MH(J),LOGB(J)
181            WRITE(IWRT,123)LOGKW,F
182            WRITE(IWRT,124)N,TNR
183   C
184   C       SETS INCREMENTS FOR NUMERICAL DIFFERENTIATION
185   C       ILB(I) IS THE INCREMENT FOR THE ITH CONSTANT TO BE REFINED.
186   C       ILBO(I) HOLDS THE ORIGINAL VALUES OF ILB(I).
187   C
188     2229 DO 127 I=1,NCV
189            IS=IG(I)
190            ILB(I)=MM(2,IS)**2+MM(1,IS)**2+1
191            ILB(I)=ILB(I)*0.0004
192            IF(ML(1,IS).GE.3.OR.ML(2,IS).GE.3)ILB(I)=ILB(I)*5.
193      127 ILBO(I)=ILB(I)
194   C
195   C       BEGINS REFINEMENT
196   C
197   C       NCC=NUMBER OF REFINEMENT CYCLES PERFORMED
198   C       NNCI=NUMBER OF NONCONVERGENT CYCLES IN COGSNR.
199   C       SQRO=SUM OF SQUARES OF RESIDUALS (SSR) FOR INITIAL VALUES
200   C            OF CONSTANTS.
201   C       SQR=SSR FOR EACH NEW SET OF REFINIG CONSTANTS.
202   C       INCCH=INDICATES IF DIFFERENTIATION INCREMENT HAS BEEN CHANGED
203   C            (=1) OR NOT (=0)
```

```
204   C        ICH(J)=INDICATOR FOR WHICH CONSTANT HAVING INCREMENT RAISED (=1)
205   C           OR NOT (=0).
206   C        NICH=NUMBER OF TIMES INCREMENT HAS CHANGED.
207   C        SQRSI=STANDARD DEVIATION FOR EACH TITRATION.
208   C
209    88 NCC=0
210       KW=10.**LOGKW
211       NNCI=0
212   130 NCC=NCC+1
213       SQRO=0.0
214       INCCH=0
215       DO 133 J=1,24
216       ICH(J)=0
217       CK(J)=0.0
218       DO 133 J1=1,24
219   133 CC(J,J1)=0.0
220       IPR=0
221       IF(NCV)208,208,134
222   134 K=0
223       SQR=0.
224       NICH=0
225       SQRSI=0
226       KKK=0
227       NEXP=0
228   C
229   C     COME HERE FOR POINT BY POINT ANALYSIS OF THE TITRATION DATA.
230   C
231   C     VCOR=CORRECTED DILUTION FACTOR FOR ADDED TITRANT.
232   C     HTOT=EXPERIMENTAL TOTAL HYDROGEN ION CONCENTRATION.
233   C     HTC=CALCULATED TOTAL HYDROGEN ION CONCENTRATION.
234   C
235   135 K=K+1
236       KKK=KKK+1
237       IF(K.EQ.IFR(NEXP+1))NEXP=NEXP+1
238       VCOR=1.0
239       IF(TYPE.EQ.MODE(2).OR.TYPE.EQ.MODE(4))GO TO 137
240       VCOR=V(NEXP)/(V(NEXP)+TITR(K))
241       HTOT=HT(NEXP)*VCOR
242   137 DO 136 I=1,2
243       MTOT(I)=MT(I,NEXP)*VCOR
244   136 LTOT(I)=LT(I,NEXP)*VCOR
245       IF(TYPE.EQ.MODE(2).OR.TYPE.EQ.MODE(4))GO TO 138
246       VCOR=(V(NEXP)+TITR(K))/BA(NEXP)
247   138 DTITR=TITR(K)
248       J=0
249       INUM=0
250   C
251   C     DETERMINES INITIAL ESTIMATE FOR FREE METAL AND FREE LIGAND CONC.
252   C
253   C     HF=FREE HYDROGEN ION CONCENTRATION.
254   C     MF=FREE METAL ION CONCENTRATION.
255   C     DMY=FRACTION OF TOTAL LIGAND PRESENT AS FREE LIGAND, ASSUMING NO
256   C         COMPLEXATION.
257   C
258       HF=10.**(-PH(K))
259       IF(K-IFR(NEXP))1421,143,1421
260   1421 IF(NM)141,141,142
261   142 IF(INCIT(K-1))140,140,141
262   143 IF(TYPE.EQ.MODE(1).OR.TYPE.EQ.MODE(3))INCIT(K)=INCIT(K)-1
263   141 DO 144 I=1,NM
264   144 MF(I)=MTOT(I)
265       IF(TYPE.EQ.MODE(2).OR.TYPE.EQ.MODE(4))MF(1)=10.**(-TITR(K))
266       DO 146 I=1,NL
267       DMY(I)=1.0
268       DO 146 J1=1,N
269       IF(ML(I,J1))146,146,147
270   147 DO 148 K1=1,NM
271       IF(MM(K1,J1))146,148,146
272   148 CONTINUE
273       DMY(I)=DMY(I)+10.**LOGB(J1)*HF**MH(J1)
274   146 CONTINUE
275       DO 145 I=1,NL
276       LF(I)=LTOT(I)/DMY(I)
277   145 CONTINUE
278   140 IF(INCIT(K).GT.0.AND.K.LT.TNR)GO TO 135
```

```
279              DO 149 I=1,N
280         149 BETA(I)=10.**LOGB(I)
281              KJ=K
282  C
283  C         CALLS SUBROUTINE FOR CALCULATION OF ALL FREE METAL, FREE LIGAND
284  C         AND ALL SPECIES CONCENTRATIONS
285  C
286       1491 CALL COGSNR
287  C
288  C         PERFORMS SEVERAL CHECKS TO SEE IF MASS-BALANCE EQUATIONS
289  C         FOR THE KTH POINT HAVE BEEN SOLVED SUCCESSFULLY.
290  C
291              IF(NNCI.GT.20.OR.NNCI.GT.(TNR-NCV-2))GO TO 1
292              CTITR=(HTOT-HTC)*VCOR
293              IF(TYPE.EQ.MODE(2).OR.TYPE.EQ.MODE(4))CTITR=-DLOG10(MF(1))
294              IF(TYPE.EQ.MODE(1).OR.TYPE.EQ.MODE(3))CTITR=(HTOT-HTC)*VCOR
295              R=CTITR-DTITR
296              IF(NIT-100)220,220,221
297          221 R=0.0
298              IF(TYPE.EQ.MODE(1).OR.TYPE.EQ.MODE(3))GO TO 222
299              INCIT(K)=1
300              GO TO 220
301          222 INCIT(K)=INCIT(K)+1
302              IF(INCIT(K).LT.1)GO TO 1491
303          220 IF(IFINAL)152,152,153
304          152 IF(IPR-1)154,216,216
305          154 IF(NIT.GT.100)GO TO 1641
306              INUM=INUM+1
307              GO TO(155,156,157),INUM
308          155 SQRO=SQRO+R*R
309              RO=R
310  C
311  C         BEGINS NUMERICAL DIFFERENTIATION
312  C         CT2 = CALCULATED TITER FOR INCREMENTED CONSTANT.
313  C         CT3 = CALCULATED TITER FOR DECREMENTED CONSTANT.
314  C
315       1601 J=J+1
316              DE(J)=0.
317              IS=IG(J)
318              IF(C(IS)*(MM(1,IS)+MM(2,IS)+ML(1,IS)+ML(2,IS)).LT.1.E-10)
319             &GO TO 1581
320  C
321  C         INCREMENT, BY ILB, THE JTH CONSTANT.
322  C
323          160 BETA(IS)=10.**(LOGB(IS)+ILB(J))
324              GO TO 1491
325          156 CONTINUE
326              CT2=CTITR
327  C
328  C         DECREMENT, BY ILB, THE JTH CONSTANT.
329  C
330              BETA(IS)=10.**(LOGB(IS)-ILB(J))
331              GO TO 1491
332          157 CONTINUE
333              CT3=CTITR
334  C
335  C         RESTORE THE JTH CONSTANT TO THE ORIGINAL VALUE.
336  C
337  C         CHECK TO SEE IF ORIGINAL INCREMENT FOR DIFFERENTIATION WAS
338  C         SUFFICIENT.
339  C
340              BETA(IS)=10.**LOGB(IS)
341              IF(CT2-CT3)158,159,158
342          159 INUM=1
343  C
344  C         INCREASES INCREMENT USED IN NUMERICAL DIFFERENTIATION, IF THIS
345  C         WAS TOO SMALL
346  C
347              IF(ILB(J).GT.0.2)GO TO 1581
348              ILB(J)=ILB(J)*5.0
349              INCCH=1
350              ICH(J)=1
351              GO TO 160
352  C
353  C         COMPUTE THE JACOBIAN, DE(J), FOR THE JTH CONSTANT.
354  C
```

```
355        158 DE(J)=(CT2-CT3)/(2.*ILB(J))
356       1581 ILB(J)=ILBO(J)
357            INUM=1
358            IF(J.LT.NCV)GO TO 1601
359     C
360     C      SETS UP AN APPROXIMATION TO THE HESSIAN MATRIX, CC(II,JJ)
361     C
362            DO 163 II=1,NCV
363            CK(II)=CK(II)-RO*DE(II)
364            DO 163 JJ=1,NCV
365            CC(II,JJ)=CC(II,JJ)+DE(II)*DE(JJ)
366        163 CONTINUE
367            NICH=NICH+INCCH
368       1641 INCCH=0
369            IF(K-TNR)135,167,167
370        167 CONTINUE
371            IF(NICH)169,169,170
372        170 DO 171 I=1,NCV
373            IF(ICH(I))171,171,172
374        172 WRITE(IWRT,173)IG(I)
375        171 CONTINUE
376            WRITE(IWRT,175)NICH
377     C
378     C       CALLS MATRIX INVERSION SUBROUTINE, AND SOLVES FOR CHANGES,
379     C      SHIFT(I), IN THE ITH CONSTANT.
380     C      IOS = INDICATOR TO DENOTE LARGE SHIFTS.
381     C
382        169 CALL SUB760(CC,NCV,24)
383     C
384            IOS=-1
385            DO 178 I=1,NCV
386            SHIFT(I)=0.0
387            DO 177 J=1,NCV
388        177 SHIFT(I)=SHIFT(I)+CC(I,J)*CK(J)
389     C
390     C      IF SIZE OF A SHIFT WAS GREATER THAN 1 LOG UNIT REDUCES SHIFT TO
391     C      0.98. IF >1 SHIFTS ARE GREATER THAN 0.5
392     C          REDUCES ALL SHIFTS TO HALF VALUE. PRINTS APPROPRIATE MESSAGE
393     C
394            IF(DABS(SHIFT(I)).GT.200.0.AND.NCC.GT.1)NCC=99
395            IF(1.-DABS(SHIFT(I)))179,180,180
396        180 IF(0.5-DABS(SHIFT(I)))181,178,178
397        181 IOS=IOS+1
398            WRITE(IWRT,182)IG(I),SHIFT(I)
399            GO TO 178
400        179 CONTINUE
401            WRITE(IWRT,183)IG(I),SHIFT(I)
402            SHIFT(I)=(SHIFT(I)/DABS(SHIFT(I)))*0.98
403            IOS=IOS+1
404        178 CONTINUE
405            IF(IOS)190,190,191
406        191 DO 192 I=1,NCV
407        192 SHIFT(I)=0.5*SHIFT(I)
408     C
409     C      CALCULATE THE STANDARD DEVIATION FOR THE DATA AND
410     C      UPDATE THE CONSTANTS.
411     C
412        190 SQRO=SQRO/FLOAT(TNR-NNCI-NCV)
413            DO 193 I=1,NCV
414            IS=IG(I)
415        193 LOGB(IS)=LOGB(IS)+SHIFT(I)
416            IF((NCC-1).GT.0)GO TO 194
417            STDML=SQRT(SQRO)
418     C
419     C      PRINTS STANDARD DEVIATION IN ML TITRANT WITH INITIAL CONSTANTS
420     C
421            IF(TYPE.EQ.MODE(1).OR.TYPE.EQ.MODE(3))WRITE(IWRT,196)STDML
422            IF(TYPE.EQ.MODE(2).OR.TYPE.EQ.MODE(4))WRITE(IWRT,195)STDML
423        194 IPR=1
424            GO TO 134
425        216 SQR=SQR+R*R
426            IF(K-TNR)135,217,217
427        217 SQRN=SQR/FLOAT(TNR-NNCI-NCV)
428            STDML=SQRT(SQRN)
429     C
430     C      PRINTS STANDARD DEVIATION IN ML TITRANT WITH NEW CONSTANTS
```

```
431  C
432        IF(TYPE.EQ.MODE(1).OR.TYPE.EQ.MODE(3))WRITE(IWRT,197)STDML
433        IF(TYPE.EQ.MODE(2).OR.TYPE.EQ.MODE(4))WRITE(IWRT,199)STDML
434        DO 198 I=1,NCV
435        IS=IG(I)
436        STDLB=CC(I,I)*SQRN
437        IF(STDLB.LT.0.0)WRITE(IWRT,201)
438        STDLB=SQRT(ABS(STDLB))
439        IF(STDLB.GT.200.0.AND.NCC.GT.1)NCC=99
440        IF(IPR.EQ.2)STDLB=99999.
441  C
442  C    PRINTS IMPROVED CONSTANTS, ESTIMATED STANDARD DEVIATIONS AND
443  C    SHIFTS FROM FORMER VALUES
444  C
445    198 WRITE(IWRT,202)IS,MM(1,IS),MM(2,IS),ML(1,IS),ML(2,IS),MH(IS),
446       &LOGB(IS),STDLB,SHIFT(I)
447        WRITE(IWRT,207)NCC
448  C
449  C    CHECKS CONVERGENCES, DECIDES IF MORE ITERATIONS
450  C
451        DO 297 I=1,NCV
452        IF(ABS(SHIFT(I))-0.0001499)297,299,299
453    297 CONTINUE
454        GO TO 208
455    299 IF(SQRN.LT.SQRO*0.97.AND.NCC.LT.8.AND.SQRN.LT.0.02)GOTO 130
456        IF(ABS((SQRO-SQRN)/SQRO).LT.0.0001)GO TO 208
457        IF(SQRO-SQRN)203,204,204
458    203 DO 205 I=1,NCV
459        IS=IG(I)
460        SHIFT(I)=0.25*SHIFT(I)
461    205 LOGB(IS)=LOGB(IS)-3*SHIFT(I)
462        WRITE(IWRT,206)
463        IPR=2
464        GO TO 134
465    204 IF(NCC-5)130,208,208
466  C
467  C    AFTER FINAL CYCLE OF REFINEMENT CALCULATES AND PRINTS TABULATION
468  C    OF QUANTITIES OF INTEREST USING FINAL VALUES OF CONSTANTS
469  C
470    208 IFINAL=1
471        NOR=-NCV
472        IF(TYPE.EQ.MODE(3).OR.TYPE.EQ.MODE(4))GO TO 134
473        IF(TYPE.EQ.MODE(1))WRITE(IWRT,209)
474        IF(TYPE.EQ.MODE(2))WRITE(IWRT,210)
475        GO TO 134
476  C
477  C    CALCULATE LOG(1/C(J)) FOR EACH SPECIES CONCENTRATION.
478  C
479    153 DO 777 J=1,N
480    777 C(J)=-DLOG10(C(J))
481        ANIT=FLOAT(NIT)/1000.
482        IF(TYPE.EQ.MODE(3).OR.TYPE.EQ.MODE(4))GO TO 778
483        WRITE(IWRT,212)K,PH(K),TITR(K),R,MTOT(1),MTOT(2),LTOT(1),LTOT(2),
484       &MF(1),MF(2),LF(1),LF(2),(C(J),J=1,N),ANIT
485    778 SQRSI=SQRSI+R*R
486        RES(KKK)=R
487        NOR=NOR+1
488  C
489  C     CALCULATE STANDARD DEVIATION FOR EACH TITRATION OF THE SET,
490  C     IF END OF INDIVIDUAL TITRATION HAS BEEN REACHED.
491  C
492        IF(K-IFR(NEXP+1)+1)7761,7762,7761
493   7762 IF(NOR)7763,7763,7764
494   7764 SQRSI=SQRT(SQRSI/FLOAT(NOR))
495        WRITE(IWRT,799)SQRSI
496   7763 SQRSI=0
497        IF(TYPE.EQ.MODE(3).OR.TYPE.EQ.MODE(4))GO TO 7765
498  C
499  C     PRODUCE SCATTER PLOT OF RESIDUALS AT END OF EACH SET
500  C     OF TITRATION DATA.
501  C
502        IF(NEXP.EQ.INEXP)GO TO 7766
503        IF(NPTSET(NEXP).LT.5)GO TO 7765
504   7766 CALL SCATER(RES,IWRT,KKK)
505        KKK=0
506   7765 NOR=-NCV
```

```
507    7761 CONTINUE
508         IF(K-TNR)135,1001,1001
509    1001 IF(TYPE.EQ.MODE(1).OR.TYPE.EQ.MODE(2))GO TO 1
510         IF(MT(1,NEXP).LE.0..AND.MT(2,NEXP).LE.0..AND.INEXP.EQ.1)
511        &GO TO 1003
512         GO TO 1
513    1003 READ(IRDR,*)HX,Y
514         IF(HX.EQ.0..AND.Y.EQ.0.)GO TO 1
515         IF(HX.NE.0.)LT(1,NEXP)=HX
516         IF(Y.NE.0.)ACID=Y
517         HT(NEXP)=LT(1,NEXP)*FLOAT(NDP(1))+LT(2,NEXP)*FLOAT(NDP(2))+ACID
518         IFINAL=0
519         GO TO 88
520    1000 STOP
521         END
```

```
  1    C
  2    C
  3    C
  4          SUBROUTINE COGSNR
  5    C                        24 CPLX,  2 MET,  2 LIG        1 M(OH)2
  6    C      SCOGS2.COGSNR
  7    C                                    HST NOV. 81
  8    C
  9    C      INSOLUBLE METAL HYDROXIDE IF LAST CONSTANT IS 1,0,0,0,-2,X
 10    C
 11    C      THIS SUBROUTINE CALCULATES THE CONCENTRATIONS OF EACH
 12    C      SPECIES USING THE NEWTON RAPHSON ALGORITHM.
 13    C
 14          REAL KW
 15          DOUBLE PRECISION TERM(24),CONVM(2),CONVL(2),C(24),
 16         &MTOT(2),LTOT(2),LF(2),MF(2),MTC(2),LTC(2),SEM(4,4),AA,
 17         &SEV(4),SHFT,BETA(24),HTC,A1,B1,C1,D1,M1,L1,LMI,LMA,WMI,WMA,MLIM
 18          EQUIVALENCE(SEV(1),A1),(SEV(2),B1),(SEV(3),C1),
 19         &(SEV(4),D1),(L1,LF(1)),(M1,MF(1)),(SEM(1,1),WMA),(SEM(1,2),WMI)
 20    C
 21          DIMENSION ML(2,24),MM(2,24),MH(24)
 22    C
 23          COMMON C,MTOT,LTOT,LF,MF,HTC,BETA,IWRT,
 24         &ML,MM,MH,NL,NM,N,HF,F,KW,NIT,NNCI,KJ
 25     998 FORMAT(' ITERATION DID NOT CONVERGE, POINT NO ',I3,I9,2(1PD11.4))
 26     996 FORMAT(5X,7(1PD11.4))
 27    C
 28    C      ESTABLISH CONVERGENCE CRITERION OF TERMINATION OF COGSNR.
 29    C
 30          DO 690 I=1,NM
 31     690 CONVM(I)=MTOT(I)*2.0D-08
 32          DO 691 I=1,NL
 33     691 CONVL(I)=LTOT(I)*2.0D-08
 34    C
 35    C      DECIDE IF INSOLUBE HYDROXIDE IS PRESENT.
 36    C
 37          IHYD=0
 38          M2L1L2=MM(2,N)+ML(1,N)+ML(2,N)
 39          IF(MM(1,N).EQ.1.AND.M2L1L2.EQ.0.AND.MH(N).EQ.-2)IHYD=1
 40          MLIM=100.
 41          IF(IHYD.EQ.1)MLIM=BETA(N)*HF*HF
 42          N=N-IHYD
 43          KOPO=1
 44          IF(NM.GT.1.OR.NL.GT.1)GO TO 699
 45          DO 698 K=1,N
 46          IF(MM(1,K).GT.2)GO TO 699
 47     698 CONTINUE
 48          KOPO=0
 49     699 ILARGE=0
 50          ISMALL=0
 51          NIT=-190
 52          DO 1 K=1,N
 53       1 TERM(K)=BETA(K)*HF**MH(K)
 54       2 CONTINUE
 55          IF(MF(1).LE.MLIM)GO TO 22
 56          MF(1)=MLIM
```

```
57              IF(NIT.LT.-99.AND.NIT.GT.-188)NIT=-99
58         22 DO 3 K=1,N
59          3 C(K)=TERM(K)
60              IF(NM)42,42,41
61         41 DO 4 K=1,N
62            DO 4 J=1,NM
63          4 C(K)=C(K)*MF(J)**MM(J,K)
64         42 DO 5 K=1,N
65              IF(NL)5,5,50
66         50 DO 6 J=1,NL
67          6 C(K)=C(K)*LF(J)**ML(J,K)
68          5 CONTINUE
69        100 NIT=NIT+1
70    C
71    C       CALCULATES EACH TOTAL METAL AND TOTAL LIGAND CONCENTRATION
72    C       FROM CURRENT VALUES OF FREE METAL AND LIGAND CONCENTRATIONS.
73    C
74            DO 8 I=1,NM
75            MTC(I)=MF(I)
76            DO 8 K=1,N
77            AA=MM(I,K)
78          8 MTC(I)=MTC(I)+AA*C(K)
79            IF(IHYD.EQ.0)GOTO 77
80            C(N+1)=1.E-20
81            IF(MF(1).EQ.MLIM)C(N+1)=MTOT(1)-MTC(1)
82         77 DO 10 I=1,NL
83            LTC(I)=LF(I)
84            DO 10 K=1,N
85            AA=ML(I,K)
86         10 LTC(I)=LTC(I)+AA*C(K)
87            IF(NIT-100)11,11,999
88    C
89    C       CHECKS DEGREE OF CONVERGENCE
90    C
91         11 DO 12 I=1,NM
92            IF(DABS(MTC(I)-MTOT(I)).GT.CONVM(I))GO TO 14
93         12 CONTINUE
94            DO 13 I=1,NL
95            IF(DABS(LTC(I)-LTOT(I)).GT.CONVL(I))GO TO 14
96         13 CONTINUE
97            N=N+IHYD
98            HTC=(HF-KW/HF)/F
99            DO 2001 J=1,N
100           AA=MH(J)
101      2001 HTC=HTC+AA*C(J)
102           RETURN
103   C
104   C       IF CONVERGENCE INSUFFICIENT SETS UP, AND SOLVES, MATRIX FOR IMP-
105   C       ROVED VALUES OF EACH FREE METAL AND FREE LIGAND CONCENTRATION.
106   C
107   C       IF NO CONVERGENCE AFTER 190 ITERATIONS, USES COMICS-ALGORITHM.
108   C       OR KONZ(1 METAL AND 1 LIGAND ONLY)
109   C
110        14 CONTINUE
111           IF(NIT.LT.-99)GO TO 1401
112           IF(LF(1).GT.LTOT(1))LF(1)=LTOT(1)/3.
113           IF(KOPO)601,640,2222
114   C
115   C        COMICS ALGORITHM
116   C
117      2222 DO 3333 I=1,NM
118      3333 MF(I)=MF(I)*SQRT(MTOT(I)/MTC(I))
119           DO 4444 I=1,NL
120      4444 LF(I)=LF(I)*SQRT(LTOT(I)/LTC(I))
121           GO TO 2
122   C
123   C       NORMAL ALGORITHM - SETS UP NEWTON RAPHSON ALGORITHM.
124   C
125      1401 M9=NM+1
126           M2=NM+NL
127           DO 1001 I=1,M2
128           DO 1001 J=1,M2
129      1001 SEM(I,J)=0.0
130           DO 1002 I=1,NM
131           SEV(I)=-MTOT(I)+MTC(I)
132      1002 SEM(I,I)=-MF(I)
```

```
133              DO 1003 I=M9,M2
134              SEV(I)=-LTOT(I-NM)+LTC(I-NM)
135         1003 SEM(I,I)=-LF(I-NM)
136              DO 1007 K=1,N
137              DO 1004 I=1,NM
138              DO 1004 J=1,NM
139              AA=MM(I,K)*MM(J,K)
140         1004 SEM(I,J)=SEM(I,J)-C(K)*AA
141              DO 1006 I=M9,M2
142              DO 1006 J=1,NM
143              AA=MM(J,K)*ML(I-NM,K)
144              SEM(I,J)=SEM(I,J)-C(K)*AA
145         1006 SEM(J,I)=SEM(I,J)
146              DO 1007 I=M9,M2
147              DO 1007 J=M9,M2
148              AA=ML(I-NM,K)*ML(J-NM,K)
149              SEM(I,J)=SEM(I,J)-C(K)*AA
150         1007 CONTINUE
151    C
152    C         CALLS MATRIX INVERSION SUBROUTINE
153    C
154              CALL SUB760(SEM,M2,4)
155              DO 1013 I=1,M2
156              SHFT=0.0
157              DO 1012 J=1,M2
158         1012 SHFT=SHFT+SEM(I,J)*SEV(J)
159              IF(SHFT.LT.-0.9)SHFT=-0.9
160              IF(SHFT.GT.3.)SHFT=3.
161              IF(I-NM)1015,1015,1016
162         1015 MF(I)=MF(I)+MF(I)*SHFT
163              GOTO 1013
164         1016 LF(I-NM)=LF(I-NM)+LF(I-NM)*SHFT
165         1013 CONTINUE
166              GO TO 2
167    C
168    C         KONZ ALGORITHM
169    C
170          601 SHFT=LTC(1)-LTOT(1)
171              IF(SHFT)620,640,610
172          610 LMA=L1
173              WMA=1./DABS(SHFT)
174              ILARGE=1
175              IF(ISMALL)612,612,631
176          612 L1=LMA/2
177              GO TO 640
178          620 LMI=L1
179              WMI=1./DABS(SHFT)
180              ISMALL=1
181              IF(ILARGE)630,630,631
182          630 L1=LMI*2
183              GO TO 640
184          631 L1=DEXP((WMA*DLOG(LMA)+WMI*DLOG(LMI))/(WMA+WMI))
185          640 B1=1.
186              A1=0.
187              C1=-MTOT(1)
188              DO 641 K=1,N
189              IF(MM(1,K).EQ.1)B1=B1+TERM(K)*L1**ML(1,K)
190              IF(MM(1,K).EQ.2)A1=A1+2*TERM(K)*L1**ML(1,K)
191          641 CONTINUE
192              IF(A1)643,643,642
193          643 M1=MTOT(1)/B1
194              GO TO 660
195          642 D1=B1**2-4*A1*C1
196              IF(D1)997,997,650
197          650 D1=SQRT(D1)
198              M1=(-B1+D1)/(2*A1)
199              IF(M1.LE.10.**(-25))GO TO 997
200          660 SHFT=LTC(1)
201              KOPO=-1
202              GO TO 2
203    C
204    C         EXITS IF NO CONVERGENCE AFTER ADDITIONAL 100 ITERATIONS
205    C
206          997 WRITE(IWRT,996)A1,B1,C1,D1,M1,L1,SHFT
207              M1=MTOT(1)
208          999 WRITE(IWRT,998)KJ,NIT,LTC(1),SHFT
```

```
209          NNCI=NNCI+1
210          N=N+IHYD
211          RETURN
212          END




  1    C
  2    C
  3    C
  4          SUBROUTINE SUB760(A7,N7,ND7)
  5    C                                        SCOGS2.SUB760
  6          DOUBLE PRECISION A7(ND7,ND7)
  7    C
  8    C     STANDARD, UNPROTECTED MATRIX INVERSION ROUTINE, USED
  9    C     IN COGSNR AND BY GAUSS-NEWTON ALGORITHM.
 10    C
 11          DO 4 K7=1,N7
 12          CM7=A7(K7,K7)
 13          A7(K7,K7)=1.0
 14          DO 5 J7=1,N7
 15        5 A7(K7,J7)=A7(K7,J7)/CM7
 16          DO 4 I7=1,N7
 17          IF(I7-K7)2,4,2
 18        2 CM7=A7(I7,K7)
 19          A7(I7,K7)=0.0
 20          DO 3 J7=1,N7
 21        3 A7(I7,J7)=A7(I7,J7)-CM7*A7(K7,J7)
 22        4 CONTINUE
 23          RETURN
 24          END




  1          SUBROUTINESCATER(R,IWRT,ITN)
  2    C
  3    C     THIS ROUTINE PRODUCES A SCATTER PLOT OF THE RESIDUALS
  4    C     FOR THE CALCULATION TITRE(CALC.) - TITRE(OBS.)
  5    C     THE SCATTER PLOT HAS A SELF  EXPANDING Y-AXIS.
  6    C     THE SCATTER PLOT IS USEFUL FOR DETERMINING WHETHER THE
  7    C     THE FITTED MODEL REASONABLY DESCRIBES THE DATA.
  8    C
  9          INTEGERLINE(100),BLANK/' '/,STAR/'*'/,POINT/'.'/,CROSS/'+'/,
 10         &II(9)
 11          DATAII/10,20,30,40,50,60,70,80,90/
 12          DIMENSIONR(200)
 13      600 FORMAT(2X,'.',2X,1PE12.2,'+',100A1,'+')
 14      601 FORMAT(2X,25('.'),9('+(',I2,').....'))
 15      602 FORMAT('0')
 16    C
 17    C     FINDS MAXIMUM RESIDUAL VALUE IN VECTOR R.
 18    C
 19          RMAX=0.0
 20          DO11K=1,ITN
 21          IF(ABS(R(K)).GT.RMAX)RMAX=ABS(R(K))
 22       11 CONTINUE
 23    C
 24    C     ESTABLISHES A SUITABLE YMAX AND YMIN (=-YMAX) FOR Y-AXIS
 25    C     OF SCATTER PLOT, ALWAYS MAINTAINING A SCATTER PLOT Y-AXIS
 26    C     OF 31 PRINTER LINES.
 27    C
 28          ICOUNT=0
 29          IITN=ITN/10
 30          TEST=7.5E-04
 31       10 IF(RMAX.LT.TEST)GOTO20
 32          RMAX=RMAX/2.0
 33          ICOUNT=ICOUNT+1
 34          GOTO10
 35       20 RANGE=5.0E-05*(2.0**ICOUNT)*15.0
 36          RINC=RANGE/15.0
 37          WRITE(IWRT,602)
 38          WRITE(IWRT,601)(II(I),I=1,IITN)
 39          ISET=ITN/100
```

```
40              IST=-99
41              IF(ISET.EQ.0)ISET=1
42      C
43      C       START PRINTING OUT PRINTER PLOT LINE-BY-LINE.
44      C
45              DO50ILINE=1,ISET
46              IST=IST+100
47              IFIN=IST+99
48              IF(IFIN.GT.ITN)IFIN=ITN
49              DO30I=1,31
50              IF(I.NE.16)GOTO31
51      C
52      C       COME HERE ONLY ONCE TO DRAW X-AXIS IN MIDDLE OF PLOT.
53      C
54              DO42K=IST,IFIN
55           42 LINE(K)=POINT
56              GOTO32
57      C
58      C       COME HERE TO LOAD X-VECTOR, LINE, FOR CURRENT Y VALUE WITH
59      C       BLANKS.
60      C
61      31      DO40K=IST,IFIN
62      40      LINE(K)=BLANK
63      32      CONTINUE
64              DO43J=1,IITN
65              JJ=II(J)
66      43      LINE(JJ)=CROSS
67      C
68      C       LOAD LINE WITH STAR IF RESIDUAL VALUE IS WITHIN THE RANGE
69      C       OF THE CURRENT Y-VALUE LINE UNDER CONSIDERATION.
70      C
71              DO41K=1,ITN
72              IF(R(K).LT.RANGE)GOTO41
73              LINE(K)=STAR
74              R(K)=-99.9
75      41      CONTINUE
76              IF(I.EQ.16)RANGE=0.E0
77              WRITE(IWRT,600)RANGE,(LINE(K),K=IST,IFIN)
78      30      RANGE=RANGE-RINC
79              WRITE(IWRT,601)(II(I),I=1,IITN)
80              WRITE(IWRT,602)
81      50      CONTINUE
82              RETURN
83              END
```

5.2. Input Data

5.2.1. Data for the System Copper(II)–Phosphonoacetate.

The following is a listing of a data set obtained from the copper(II)-phosphono-acetate system. Six titrations have been included and processed as one complete set using the SCOGSH option.

```
¶<-- Column 1              ¶<--  Column 1

SCOGSH                     2,0,0,0,-2, -10.6

6                          1,0,1,0,0, 7.14

1,1,12                     1,0,2,0,0, 10.99

0,0,1,0,1, 7.931           1,0,1,0,1, 11.76

0,0,1,0,2, 12.735          1,0,2,0,1, 16.42

0,0,1,0,3, 13.97           1,0,2,0,2, 21.5

1,0,0,0,-1, -7.4           1,0,1,0,-1, -1.66

                           2,0,1,0,0, 9.
```

¶<-- Column 1

1,0

1.

CU PAA 1:1   .002   21.3.78   08//1   ** PAA.CU **

.001770, 0, .001969, 0, .00385, .993, 50.237, 0

.11, 3.498

.12, 3.606

.13, 3.72

.14, 3.837

.15, 3.954

.16, 4.069

.17, 4.1835

.18, 4.297

.19, 4.409

.2, 4.52

.21, 4.634

.22, 4.75

.23, 4.875

.24, 5.01

.245, 5.085

.25, 5.166

.255, 5.256

.26, 5.359

.265, 5.479

.27, 5.625

.275, -5.815

CU PAA   1:2   .002 23.3.78   11//

.000983, 0, .001967, 0, .00387, .993, 50.237, 0

.11, 3.638

.12, 3.776

.13, 3.92

.14, 4.059

.15, 4.198

.16, 4.336

.17, 4.47

.18, 4.605

.19, 4.7405

.195, 4.811

.2, 4.885

.205, 4.963

¶<-- Column 1

.21, 5.045

.215, 5.134

.22, 5.233

.225, 5.345

.23, 5.476

.26, 6.781

.265, 6.984

.2675, 7.085

.27, 7.19

.2725, 7.292

.275, 7.395

.2775, 7.51

.28, -7.619

CU PAA   3:4   .004   18.4.78   29//1

.002949, 0, .003968, 0, .007856, 1.985, 50.237, 0

.12, 3.373

.13, 3.508

.14, 3.651

.15, 3.797

.16, 3.943

.17, 4.081

.18, 4.218

.19, 4.35

.2, 4.482

.21, 4.612

.22, 4.745

.23, 4.89

.235, 4.966

.24, 5.051

.245, 5.141

.25, 5.242

.255, 5.365

.265, 5.66

.27, 5.853

.275, -6.11

CU PAA   2:5   .005   28.4.78   40//2

.001967, 0, .004811, 0, .009567, 1.985, 50.18, 0

.17, 3.92

.18, 4.072

¶<-- Column 1

.19, 4.219

.195, 4.2895

.2, 4.357

.21, 4.49

.215, 4.552

.22, 4.617

.225, 4.683

.23, 4.747

.24, 4.88

.245, 4.95

.25, 5.023

.26, 5.182

.265, 5.27

.27, 5.366

.28, 5.592

.285, 5.724

.29, 5.878

.3, 6.217

.305, 6.38

.31, 6.552

.315, 6.717

.32, 6.883

.325, 7.053

.33, 7.226

.335, -7.406

CU PAA 1:1   .004   60//2   12.5.78

.003933, 0, .003905, 0, .006588, 1.985, 50.237, 0

.09, 3.297

¶<-- Column 1

.1, 3.417

.11, 3.543

.12, 3.675

.13, 3.808

.14, 3.941

.15, 4.0675

.16, 4.193

.17, 4.317

.18, 4.438

.19, 4.559

.2, 4.683

.21, 4.814

.22, 4.955

.23, 5.118

.24, 5.318

.245, -5.449

CU PAA 2:1   .004   61//   15.5.78

.007866, 0, .003905, 0, .00659, 1.985, 50.237, -.005

.12, 3.4035

.13, 3.514

.14, 3.633

.15, 3.755

.16, 3.881

.17, 4.007

.18, -4.13

-13.62,.81

7,12,6,8,11,7,9,10

0

5.2.2. Data for the System Copper(II)–Histidine-(3-Hydroxy-1-Methylpyridin-4-(1H)-one)

The following is a listing of a data set for the copper–histidine-(3-hydroxy-1-methylpyridin-4-(1*H*)-one) ternary system.[31] Two titrations have been included and processed as one complete set using the SCOGSDMH option.

¶<-- Column 1	¶<-- Column 1
SCOGSDMH	0,0,1,0,2, 14.84
2	0,0,1,0,3, 16.55
1,2,19	0,0,0,1,1, 8.8
0,0,1,0,1, 8.906	0,0,0,1,2, 12.15

¶<-- Column 1

1,0,0,0,-1, -7.4

2,0,0,0,-2, -10.

1,0,1,0,0, 9.8

1,0,1,0,1, 13.86

1,0,2,0,0, 17.45

1,0,2,0,1, 23.2

1,0,2,0,2, 26.59

1,0,1,0,-1, 2.1

2,0,2,0,-2, 7.5

1,0,0,1,0, 9.35

1,0,0,1,1, 10.94

1,0,0,2,0, 16.93

1,0,1,1,0, 20.

1,0,1,1,1, 22.

1,1

1.

CU HIS M* 1:1:1 .002   8.6.78   68//1   ** CUHISMS **

.001967, 0, .002005, .001935, .003879, .993, 50.237, 0

.21, 3.803

.22, 3.871

.23, 3.939

.24, 4.007

.25, 4.071

.26, 4.139

.27, 4.208

.28, 4.279

.29, 4.352

.3, 4.427

.31, 4.506

.32, 4.593

.33, 4.688

.335, 4.74

.34, 4.796

.345, 4.8565

¶<-- Column 1

.35, 4.922

.355, 4.995

.36, 5.076

.365, 5.167

.37, 5.274

.375, -5.401

CU HIS MS 1:1:1 .004 73//1

   14.6.78

.003933, 0, .004006, .003845,

   .00391, 1.985, 50.1, 0

.11, 3.602

.12, 3.674

.13, 3.746

.14, 3.8185

.15, 3.892

.16, 3.964

.17, 4.035

.18, 4.109

.19, 4.185

.2, 4.264

.21, 4.347

.22, 4.435

.23, 4.532

.24, 4.64

.245, 4.7

.25, 4.7655

.255, 4.837

.26, 4.914

.265, 5.003

.27, 5.104

.275, -5.2255

-13.62, .81

2,18,19

0

5.2.3. Data for the System Copper(II)–*p*-Sulfonatobezaldehyde Thiosemicarbazone

The following is a listing of a data set for the copper(II)–*p*-sulfonatobenzaldehyde thiosemicarbazone system.[14] Twelve single-point batch titrations have been included and processed as one complete set using the SCOGSM option.

```
¶<-- Column 1

SCOGSM

12

1,1,4

1,0,1,0,1, 19.

1,0,2,0,2, 34.4

0,0,1,0,1, 10.92

1,0,2,0,1, 27.8

0,0

1.

SBAT.LPH31     THE LOW PH VALUE CU SBAT     11.9.80  1:3  .001

.001016, 0, .00267, 0, 0, 1., 50., 0,

9.1, -3.0

1:4 .001  3.9.

.00101, 0, .00362, 0, 0, 1., 50., 0,

9.79, -3.2

1:6  12.9.  .001

.00101, 0, .00563, 0, 0, 1., 50., 0,

10.48, -2.7

1:4 .0005, 1.9.

.0005, 0, .00189, 0, 0, 1., 50., 0,

9.6, -2.7

1:10, 4.9.,  .0004

.000404, 0, .00381, 0, 0, 1., 50., 0,

10.8, -2.
```

```
¶<-- Column 1

1:4, .0002, 5.9.

.000203, 0, .000708, 0, 0, 1., 125., 0,

9.19, -3.

 1:10, .00008, 10.9.

.000081, 0, .000777, 0,  0, 1., 125., 0,

10., -2.5

 1:5 .00004,  15.9.

.0000407, 0, .000193, 0, 0, 1., 125., 0,

9., -3.4

1:50, .00002,  10.9.

.000016, 0, .000799, 0, 0, 1., 125., 0,

10.89, -4.9

 9.12.  1:40    .0001

.0001, 0, .00397, 0, 0, 1., 50., 0

11.53, -4.34

11.12,   1:20    .00008

.00008, 0, .00157, 0, 0, 1., 125., 0

10.8, -3.94

15.12.   1:6.5    .0003

.0003, 0, .00185, 0, 0, 1., 50., 0

10.1, -3.73

-13.58, .83

2,1,2

0
```

5

MINIQUAD and MIQUV

Two Approaches for the Computation of Formation Constants from Potentiometric Data

ALBERTO VACCA and ANTONIO SABATINI

1. INTRODUCTION

The first version of the program MINIQUAD, published in 1974,[1] was based on an original algorithm previously described.[2] The program permitted the calculation of formation constants of complex species in solution and was applicable to most kinds of potentiometric titration data. Two major changes have been made in MINIQUAD in order to increase speed and reliability,[3] and to simultaneously treat titration curves with different numbers of reactants.[4]

Recently MIQUV, a new program based on a different algorithm, has been developed and extensively tested. MINIQUAD and MIQUV are compatible as far as data input is concerned, even though completely different algorithms have been employed.

2. DESCRIPTION OF THE PROGRAMS

MINIQUAD and MIQUV refine, using the Gauss–Newton least-squares method, cumulative formation constants of reactions:

$$a\mathrm{A} + b\mathrm{B} + c\mathrm{C} + \cdots \rightleftharpoons \mathrm{A}_a\mathrm{B}_b\mathrm{C}_c \cdots$$

where A, B, C, ... are the uncomplexed reactant species, $\mathrm{A}_a\mathrm{B}_b\mathrm{C}_c$ ... is the formed species, and $a, b, c, \ldots$ are the stoichiometric reaction coefficients. Therefore, the

ALBERTO VACCA and ANTONIO SABATINI • Dipartimento di Chimica dell' Università di Firenze and I.S.S.E.C.C.–CNR, Florence, Italy.

99

cumulative formation constants are defined as follows:

$$\beta_{abc...} = \frac{[A_aB_bC_c \cdots]}{[A]^a[B]^b[C]^c \cdots}$$

where $[A_aB_bC_c \ldots]$ is the concentration of the complex species, and $[A]$, $[B]$, $[C]$, ... are the concentrations of the uncomplexed reactant species at the equilibrium.

Using initial estimates of the formation constants, the set of simultaneous normal equations is built up and then solved in order to obtain the corrections to be applied to the parameters β. The new values of the formation constants are a better approximation to the final values and are employed in the next refinement cycle. Such an iterative procedure provides β values which give the best agreement between calculated and experimental data.

Array dimensions for both programs permit the treatment of data from systems containing a maximum of four independent reactants and two potentiometric electrodes. These limits can, in principle, be increased to deal with more complicated systems. In general limitations due to the array sizes may be easily overcome since execution-time dimensioning has been used in all the subprograms.

Experimental data from potentiometric titrations, carried out using a different number of reactant species and/or potentiometric electrodes, may be processed simultaneously. This might be the case for measurements arising from the protonation of the ligand(s), the hydrolysis of the metal ion(s), the formation of simple and/or mixed complexes using only one electrode (for example, a glass electrode) for some measurements, and two electrodes (for example, glass and amalgam electrode) for others.

2.1. Program MINIQUAD

For each experimental point of a titration curve the following system of mass balance equations must be valid:

$$C_A = [A] + \Sigma\, a\, \beta_{abc...}\, [A]^a\, [B]^b\, [C]^c \cdots$$

$$C_B = [B] + \Sigma\, b\, \beta_{abc...}\, [A]^a\, [B]^b\, [C]^c \cdots \tag{1}$$

$$C_C = [C] + \Sigma\, c\, \beta_{abc...}\, [A]^a\, [B]^b\, [C]^c \cdots$$

etc.

where C_A, C_B, C_C, ... are the total (analytical) concentrations of the independent reactants A, B, C, ...

The program MINIQUAD computes the values of the cumulative formation constants which minimize the sum of the squared residuals between observed and calculated analytical concentrations:

$$U = \sum_i (C_i^{obs} - C_i^{calc})^2 \tag{2}$$

where the sum covers all the mass balance equations for all the experimental points.

The main program starts by defining the maximum values of the array dimensions, the input/output unit numbers, tolerances, and accuracies. It then reads a descriptive title and a set of integers related to the system under consideration. Subsequently the subprograms DINP, MINIM, DOUT, and, optionally, STATS are called successively. Finally, an indicator is read which terminates the data set and initiates job execution. The major subroutines will now be described.

Subroutine DINP. This subroutine reads the initial estimates of the values and the data relevant to each titration. At each experimental point one or two free reactant concentrations have been measured potentiometrically. The unknown concentrations of all free reactants are calculated point by point by means of the subroutines ML and MQ. ML performs a preparatory function: it calculates, using the Newton–Raphson algorithm, the unknown concentrations by solving the system of the mass balance equations relative to the reactants whose free concentration is unknown. These concentrations are then used in MQ as starting values for the minimization of the sum of the squared residuals on all the mass balance equations for each experimental point. All the derivatives, needed to construct the normal equations, are evaluated analytically.

Subroutine MINIM. This routine performs the refinement process of the β values. The derivatives of all the mass balance equations, with respect to the unknown free concentrations and to the formation constants to be refined, are evaluated analytically. This calculation is performed, point by point, using the unknown free concentrations obtained in MQ and the current β values. Using the values of the derivatives and the residuals between observed and calculated total concentrations, the matrix of normal equation coefficients and the gradient vector relative to the parameters β is obtained. The shifts to be applied to the formation constants are computed in the subroutine LINEQ by solving the system of normal equations. The subprogram MON1 is optionally called in order to monitor the progress of minimization. The values of the shifts are then tested in order to check if any formation constant will become negative. In this situation, all the shifts are proportionally reduced so that the β value corresponding to the most negative relative shift becomes one hundredth of the previous value. If this happens for the same formation constant in four consecutive refinement cycles, this constant is set equal to zero and refined no further. The shifts are then applied to the values before starting a new refinement cycle.

Two convergence criteria must be fulfilled before control is returned to the main program. First, the ratios between corresponding shifts and standard deviations must be less than TOLB (0.1). Second the relative decrease of the error square sum must be less than TOLU (0.0001).

Subroutine DOUT. This routine initiates printout of all relevant information concerning the refinement procedure: values of the error square sum, formation constants, relative standard deviations, and correlation coefficients. If the refinement procedure fails a concise message is printed that describes the reason for the premature termination.

Subroutine STATS. This routine is optionally called from the main program and performs a statistical analysis[5] of the residuals on the mass-balance equations. Additionally, the total concentrations of the reactants, the residuals, and the formation percentages of the complex species with respect to the total concentration of one or more reactants may be printed, point by point.

Use of MINIQUAD.   The program MINIQUAD has been extensively used by our research group over the last ten years, many equilibrium systems having been studied and elucidated. Two such examples serve to illustrate the application of the program. The first[6] describes in detail the procedure used in the choice of the most probable chemical model. The second[4] illustrates a method for calculating the errors associated with the stepwise constants from the values obtained for the cumulative equilibria.

MINIQUAD has been used worldwide by many research groups. Sylva and Davidson[7] used MINIQUAD to investigate the hydrolysis of copper (II) and dioxouranium (VI). The authors introduced two changes in the program to permit the refinement of the initial amount of acid and the calibration coefficient, λ, of the pH meter. Williams *et al.*[8] have applied MINIQUAD to the study of ternary complexes (four mass-balance equations) in the metal–ligand interactions in biofluids. The performance of MINIQUAD and SCOGS[9] on the same systems were compared in this study.

Pettit *et al*[10] used MINIQUAD to refine formation constants of silver (1) complexes from titrations carried out with two indicator electrodes.

A listing of the latest version of MINIQUAD is given in Section 5.1. As an example of the use of the program, the input data which are needed for the selection of the chemical model in the system nickel(II)-1,1,1,-tris-(aminomethyl) methane,[6] are also reported.

2.2. Program MIQUV

The program MIQUV computes the β values which minimize the sum of the squared residuals between observed and calculated emf values:

$$U = \sum_i w_i (E_i^{obs} - E_i^{calc})^2 \tag{3}$$

where the sum is over all the potentiometric readings, and w_i is the weighting factor assigned to the ith observation, calculated from

$$w_i = 1/\sigma_i^2 \tag{4}$$

where σ_i^2 is the estimated variance associated with the ith observation.

Only instrumental uncertainties in the emf measurements, σ_E, and in the added titrant volume, σ_v, are taken into account as a source of error. Since the fluctuations in emf and volume measurements are clearly uncorrelated, the estimated variance may be calculated from

$$\sigma_i^2 = \sigma_E^2 + (\partial E_i/\partial v_i)^2 \cdot \sigma_v^2 \tag{5}$$

MIQUV can be used to refine "special" parameters in addition to formation constants. These "special" parameters are quantities which are characteristic of each titration such as the initial amounts of reactants, their concentration in the titrant solution, the standard cell potentials, the coefficients of the correction terms for the effects of the liquid-junction, and the initial volume of the solution. This option was included in Sillen's LETAGROP,[11] and was used to detect the eventual presence of systematic errors. In our opinion caution must be exercised using the program for this purpose. However, the refinement of "special" parameters is very important when no other possibility exists to determine, with sufficient accuracy, one or more of the quantities mentioned above.

It may happen that different "special" parameters must have the same value, as a consequence of the experimental conditions. For example, if the same titrant solution has been used in different titrations, the value of the relevant concentration will be the same for all these titrations. When such "special" parameters are refined, proper constraints are invoked so that equal shifts are applied to these parameters during the minimization process.

In the main program the array dimensions, the input–output unit numbers, the tolerances, and the accuracies are first defined and then the minimization process is started.

Subroutine DINS performs the input of the data, which are a descriptive title of the experiment; a set of integers related to the particular system; the temperature of the cell solution; σ_v; σ_E; the initial estimates of the β values and the potentiometric titration data. The weighting factors, w_i, are required by this routine and obtained from subroutine WEIGHTS. Equation (4) defines w_i and equation (5) shows the equation used to calculate w_i. The partial differential $(\partial E_i/\partial v_i)$ is estimated from the slope of the potentiometric curve (E against v). Approximate values of the partial differential are calculated within subroutine SPLINE, called from WEIGHTS, using cubic spline interpolation.[12] As described by Stanton and Hoskins,[12] two boundary conditions must be set in order to allow "exact" interpolation. In the routine SPLINE the second derivatives of the spline function are assumed to be zero at the first and last point of the titration curves, implying that at least two points per curve as necessary. The approximation for the partial derivative improves as the experimental points become closer together. The array dimensions of the subprogram SPLINE permit a maximum of 101 points per titration curve. The subroutine SNTPV is called, after DINS, by the main program. In this routine all the parameter values to be refined are transferred to the vector PARAM from the proper variables bearing special names.

The subroutine FUNV, which is called after SNTPV by the main program, is used to calculate the emf values, via the Nernst equation, relative to the potentiometric electrodes used in the measurements. The current values of the parameters are employed at each cycle. The values of the free concentrations [A], [B], [C], ... are calculated, in the subroutine CCFR, by solving the simultaneous equations (1) using the Newton–Raphson algorithm. Occasionally the convergence criteria in CCFR are not fulfilled after 100 interactions. In this situation a warning message is printed out and the calculation continues on to the next experimental point. We know, however, from experience that the free reactant concentrations so calculated are still reliable and such an event does not affect the minimization process. FUNV also computes the residuals between observed and calculated emf values, the weighted error square sum, the standard deviations for each titration curve and for the complete data set.

At this point control is passed back to the main program where the calculation of the design matrix is performed. The elements of this matrix are the derivatives $(\partial E_i / \partial p_j)$ for each point, with respect to the parameters p_j to be refined. Approximate values of these quantities are obtained by numerical differentiation. The parameters to be refined are incremented one at a time, subroutine FUNZ being used to compute the emf values. Subroutine PTSNV transfers the terms from the vector PARAM into the corresponding variables (the reverse of the operation performed by subroutine SNTPV). The differences, ΔE_i, between the emf values obtained in FUNZ (using incremented parameters) and those obtained in FUNV (using nonincremented parameters) are divided by the incrementing quantity Δp_j, for the jth parameter. The increment Δp_j is set equal to 0.002 times the value of the parameter for the first refinement cycle; in the subsequent cycles Δp_j is calculated as 0.02 times the standard deviation of the parameter.

After the terms of the gradient vector and the coefficient matrix for the normal equations have been calculated, the latter is inverted by the subroutine MATIN. The shifts to be applied to the parameters and their standard deviations are then obtained.

A test, analogous to that used by MINIQUAD, is then performed to check if the application of the shifts causes negative values of the formation constants. In this case the routine BETANEG, which performs the same task as NEG in MINIQUAD, is called.

The shifts obtained are then applied to the terms of the vector PARAM employing subroutine PTSNV at the begining of each subsequent refinement cycle.

The minimization is complete when both criteria of convergence (identical with those for MINIQUAD) are fulfilled. The subroutine DOUS is used to print out all the relevant information on the refined parameters, and is essentially the same as DOUT in MINIQUAD.

After DOUS, the subroutine STANS may be optionally called. STANS is similar to STATS (MINIQUAD) and includes a graphical display of formation percentages.

A listing of MIQUV is given in Section 5.2. A hypothetical chemical model consisting of four reactants forming 12 complex species was chosen as a test mode for MIQUV and arbitrary values were assigned to the relevant formation constants. Theoretical titration curves were calculated with varying numbers of reactants and/or potentiometric electrodes. The emf values so obtained were then suitably altered to account for the random fluctuations in the measurement of both emf and titrant volume. Section 5.2.1 contains these input data.

MIQUV has only been recently developed. Since that time, it has been used by our research group for all the systems which have been treated with MINIQUAD. It has been possible therefore to compare the performances of the two programs.

2.3. Comments

As already reported,[3] MINIQUAD allows the refinement of formation constants even if the initial estimates are incorrect by more than one order of magnitude. Under these conditions, convergence is still rapid.

The following limitations of the program should be mentioned:

i. execution is terminated for titration curves in which all the free reactant concentrations are known;

ii. neither corrections for the effect of liquid-junction potential, nor dependence of activity coefficients on the ionic strength are considered.

Moreover it should be noted that the statistical analysis of the residuals performed in MINIQUAD may be subject to criticism. The least-squares method has statistical significance only if the sum of the squared residuals of physical quantities affected by random errors is minimized *and* if a correct weighting procedure is used. This is not the case for MINIQUAD, where the residuals are calculated on the total concentrations and unit weights are employed throughout. Undoubtedly the random errors on the measurements of the emf and volume of the titrant solution influence the residuals of the mass balance equations. It is, however, extremely complicated, if not impossible, to calculate the propagation of these random fluctuations on the values of the different analytical concentrations of the reactants (C_A, C_B, C_C, ...).

For the above reason we decided to develop an alternative algorithm, on which MIQUV is now based, where the sum of the weighted residuals between observed and calculated emf values is minimized. Estimates of the fluctuations in the measurements of emf and volume allow a correct weighting of the experimental observations.

It is worthwhile mentioning the following limitations of MIQUV:

i. the decimal cologarithm of the free species concentrations (e.g., pH) cannot be used as input data;

ii. it is not possible to treat individual potentiometric points not belonging to a titration curve, i.e., a "batch" titration; furthermore the points of the curve need to be in sequence;

iii. activity coefficients have been ignored, as in MINIQUAD;

iv. it is possible to take into account corrections for the effect of the liquid-junction potential in both acid and basic solutions; these corrections can be applied for only one electrode per titration curve.

Although these limitations may be easily overcome by making small code changes in the programs, this necessity has never arisen for the treatment of experimental data collected in our laboratory.

MIQUV has two major points in its favor. As described above, it calculates the emf values as a function of the volume of titrant. These quantities are experimental data and hence directly measured. Therefore the deviations of the calculated values from the experimental ones are easily interpretable. Moreover MIQUV allows the refinement of "special" parameters. This facility is a very useful one, and sometimes a necessity.

On the other hand MINIQUAD is much faster than MIQUV and convergence is attained even if poor estimates of the initial parameters are available. On the basis of the above considerations it is advisable to adopt the following procedure: preliminary refinement of the equilibrium constants with MINIQUAD, if good estimates of the β values are not available, followed by final refinement with MIQUV using the β values obtained by MINIQUAD as starting values, together with adjustment of "special" parameters, if required.

3. REFERENCES

1. A. Sabatini, A. Vacca, and P. Gans, MINIQUAD—A General Computer Programme for the Computation of Formation Constants from Potentiometric Data, *Talanta* **21**, 53–77 (1974).
2. A. Sabatini and A. Vacca, A New Method for Least-Squares Refinement of Stability Constants, *J. Chem. Soc., Dalton Trans.* **1972**, 1693–1698.
3. P. Gans, A. Sabatini, and A. Vacca, An Improved Computer Program for the Computation of Formation Constants from Potentiometric Data, *Inorg. Chim. Acta* **18**, 237–239 (1976).
4. M. Micheloni, A. Sabatini, and A. Vacca, Nickel(II), Copper(II) and Zinc(II) Complexes of 1,1,1-Tris(aminomethyl)propane. A Calculation Procedure of Stepwise Formation Constants and Their Standard Errors from the Values Obtained for the Cumulative Equilibria, *Inorg. Chim. Acta* **25**, 41–48 (1977).
5. M. R. Spiegel, *Theory and Problems of Statistics,* Schaum Publishing Co., New York (1961).
6. A. Sabatini and A. Vacca, Complex Formation Equilibria Between 1,1,1-Tris(aminomethyl)ethane and Divalent Transition Metal Ions: The Chemical Model and Thermodynamic Quantities, *Coord. Chem. Rev.* **16**, 161–169 (1975).
7. R. N. Sylva and M. R. Davidson, The Hydrolysis of Metal Ions. Part 1. Copper(II), *J. Chem. Soc., Dalton Trans.* **1979**, 232–235; Hydrolysis of Metal Ions. Part 2. Dioxouranium(VI), *J. Chem. Soc., Dalton Trans.* **1979**, 465–471.
8. G. Berton, P. M. May, and D. R. Williams, Computer Simulation of Metal-ion Equilibria in Biofluids. Part 2. Formation Constants for Zinc(II)-Citrate-Cysteinate Binary and Ternary Complexes and Improved Models of Low-Molecular-Weight Zinc Species in Blood Plasma, *J. Chem. Soc., Dalton Trans.* **1978**, 1433–1438.
9. I. G. Sayce, Computer Calculation of Equilibrium Constants of Species Present in Mixtures of Metal Ions and Complexing Agents, *Talanta* **15**, 1397–1411 (1968).
10. L. D. Pettit, K. F. Siddiqui, H. Kozlowski, and T. Kowalik, Potentiometric and ^{1}H NMR Studies on Silver(I) Interaction with S-Methyl-L-Cysteine, L-Methionine and L-Ethionine, *Inorg. Chim. Acta* **55**, 87–91 (1981).
11. N. Ingri and L. G. Sillén, High Speed Computers as a Supplement to Graphical Methods. IV. An ALGOL Version of LETAGROP VRID, *Ark. Kemi* **23**, 47–71 (1964).
12. R. G. Stanton and W. D. Hoskins, in *Physical Chemistry. An Advanced Treatise* (H. Eyring, D. Henderson, and W. Jost, eds.), Academic, New York (1975), Vol. XIA, pp. 346–352.
13. A. Vacca, A. Sabatini, and M. A. Gristina, Two Problems Involved in Solving Complex Formation Equilibria: The Selection of Species and the Calculation of Stability Constants, *Coord. Chem. Rev.* **8**, 45–53 (1972).

4. INSTRUCTIONS FOR THE PROGRAMS

4.1. Details of the Input File

Both programs may use the same input file if only formation constants are to be refined. A few additional data are needed in the input file for MIQUV if "special" parameters are to be refined. The FORMAT specifications of the input data are in square brackets. Boldfaced variable names are required for both programs; regular typeface variable names are for MIQUV only.

The input file consists of the following lines.

1. 1 line [18A4]: descriptive title of the system under consideration.
2. 1 line [7I5]: **LARS, NK, N, MAXIT, IPRIN, NMBEO,** NCONS

> LARS: an indicator for the data points to be considered in the refinement: with LARS = 1 all the data points are used, with LARS = 2 alternate points, with

LARS = 3 every third point, etc. (the last point of each titration curve is always used).

NK: the total number of formation constants.

N: the number of formation constants to be refined in the input file for MIN-IQUAD; the number of "special" parameters, only, to be refined in the input file for MIQV.

MAXIT: the maximum number of iterative cycles to be performed. If, for some reason, convergence is not attained within MAXIT cycles the job is terminated. If MAXIT = 0 no refinement is performed; however, information about the system under consideration may be obtained depending on the values of IPRIN, JPRIN, and JP(I) (*vide infra*).

IPRIN: an indicator which determines the amount of output. If IPRIN = 0 the progress of the refinement is not monitored. IPRIN = 1 produces the printout of iterative cycle number; error square sum; values of the parameters to be refined; relative shifts; updated parameters and relative standard deviations for each cycle. With MIQUV a listing of the volumes, measured and calculated emf values, weighted and unweighted residuals are obtained after the last cycle. With IPRIN = 2, the potentiometric input data are also listed before the first refinement cycle. With MIQUV the weights w_i, calculated in the subprogram WEIGHTS, are printed for each potentiometric point.

NMBEO: the total number of independent reactants used in the system. The number of reactants may be less than NMBEO, for any particular titration.

NCONS: the number of constraints established between "special" parameters to be refined. One constraint is needed to keep two "special" parameters equal during the refinement process. If N "special" parameters are to be kept equal, $N - 1$ constraints are needed.

NCONS is not used by MINIQUAD.

3. 1 line [4F10.6]: **TEMP, SIGMAV, SIGMAE(1), SIGMAE(2)**

TEMP: the temperature (°C) at which the measurements have been carried out.

SIGMAV: the estimated instrumental uncertainty in the measurement of the titrant volume.

SIGMAE(1), SIGMAE(2): the estimated instrumental uncertainties in the measurements of the potential of the first and, if present, the second electrode. If only one electrode is employed, SIGMAE(2) is unused.

SIGMAV, SIGMAE(1), and SIGMAE(2) are not used by MINIQUAD.

4. NK lines [F10.6, 5I5]: **BETA(I), JPOT(I), PQRO(J,I),** (NMBEO values), **KEY(I)**

BETA(I) and JPOT(I): define the values of the ith formation constant. The relationship is $\beta_i = \text{BETA(I)} \times 10^{\text{JPOT(I)}}$.

PQRO(1,I), PQRO(2,I),..., PQRO(NMBEO,I), are stoichiometric coefficients of the species with formation constant β_i. The order of the coefficients for the first formation constant is arbitrary except that those relative to a reactant, whose free concentration is determined potentiometrically, must come last. The stoichiometric coefficients of the other constants must be given using the same order. Moreover, this choice implies that a progressive integer number, from 1 to NMBEO, is attributed to each independent reactant. These reactant indicators are used in the successive input to specify which reactants are present in each titration experiment, and are associated with the stoichiometric coefficient order established by PQRO(1,I).

KEY(I): the refinement key of the ith formation constant: 1 for constants to be refined, otherwise 0. The number of β values to be refined is determined by the number of unitary keys.

In MINIQUAD, if this number is not equal to the value of N previously read in, a message is printed, the number of unitary refinement keys is assigned to N, and the execution continues.

5. 1 or 2 lines [14I5]: IVAR(I) (N values)

IVAR(I): is present in the input file for MIQUV, *only* when $N > 0$; *it must be absent for* MINIQUAD. IVAR(I) is the sequence number of the ith "special" parameter to be refined. The "special" parameters are organized, curve by curve, in the following order: initial total amounts of the reactants, their concentrations in the titrant, the standard cell potential(s), two coefficents for liquid-junction correction, and initial volume. This is also the order in which these "special" quantities are read in from the input file.

6. 1 to 3 lines [14I5]: KCONS(I,1), KCONS(I,2) (NCONS pairs of values)

This item must be absent in the input file for MINIQUAD *and is present for* MIQUV, *only if* $N > 0$ *and* NCONS > 0.

KCONS(I,1), KCONS(I,2): two sequence numbers which specify the ith pair of "special" parameters constrained to keep the same value in the refinement procedure.

7. The following sets of lines for each titration curve:

a. 1 line [10I5]: **NMBE, JNMB(I)** (NMBE values), **NEMF, JP(I)** (NMBE values)

NMBE: the number of reactants present in the titration.

JNMB(I): reactant indicator corresponding to the ith reagent present in the titration experiment.

NEMF: number of potentiometric electrodes used in the titration curve.

JP(I): reactant indicator used to specify the reactants for which the formation percentages are to be calculated in the statistical analysis subroutines.

b. 1 line [3I5]: **JEL(I)** (NEMF values), **JCOUL**

JEL(I): number of electrons transferred at the ith electrode. If the decimal cologarithms of concentration (e.g., pH), instead of emf values, are used then (JEL(I) = 0. This option is valid *only* for MINIQUAD.

JCOUL: equal to zero is normal. If JCOUL = 1 the total volume of the solution is constant during the titration (e.g., coulometric experiments).

c. 1 line [4F10.6]: **TOTC(I)** (NMBE values)

TOTC: total amount (mmol) of the ith reactant initially present in the solution. The order of reactants is the same as that specified by the NMBE values of JNMB.

d. 1 line [4F10.6]: **ADDC(I)** (NMBE values)

ADDC(I): concentration (mmol cm^{-3}) of the ith reactant in the titrant solution; the order is dictated by JNMB.

e. 1 line [4F10.6]: **EZERO(I)** (NEMF values), AJ, BJ

EZERO(I): standard potential of the ith electrode as defined in the Nernst equation:

EMF(I) = EZERO(I) + [(TEMP + 273.16)/11.6048 JEL(I)] ln (CX).

AJ, BJ: coefficients of the correction terms for the effect of liquid potentials in acid and basic solution, respectively. The emf values [EMF(I)] are corrected according to the following equation:

$$EMF(I)_{corr} = EMF(I) + AJ.CH + BJ.HKW/CH$$

where HKW is the ionic product of water and CH is the hydrogen ion concentration. These corrections are not performed by MINIQUAD. If no correction for liquid-junction potential is to be performed by MIQUV, the two coefficients are set to zero. Note that the emf values of only *one* electrode can be corrected by MIQUV. If more than one electrode is used in the titration, the correction is applied to the emf values measured with the *last* electrode.

f. 1 line [F10.6]: **VINIT**

VINIT: is the initial volume (cm³) of the solution.

g. For each point of the titration curve:

1 line [I5,3F8.3]: **LUIGI, TITRE, EMF(I)** (NEMF values)

LUIGI: different from zero only for last points of the titrations;
LUIGI > 0 indicates the end of a titration curve; LUIGI < 0 indicates the end of the last titration curve.

TITRE: volume (cm³) of the titrant solution added. It is assumed that only one titrant solution is used which may contain more than one reactant.

EMF(I): potential (mV) measured by the ith electrode. With MINIQUAD the decimal cologarithm of the concentration of the free species is to be read, instead of the potential, if the corresponding value of JEL(I) is equal to zero.

8. 1 line [I5]: **JPRIN**

JPRIN: controls the output produced by the subprograms STATS and STANS as follows:

JPRIN	Statistical analysis	Tables	Graphs
0	no	no	no
1	yes	no	no
2	yes	yes	no
3	yes	no	yes
4	yes	yes	yes

If JPRIN > 1 the number of tables and/or graphs is determined by the values previously assigned to the reactant indicators JP(I) for each titration curve.

9. 1 line [I5]: **NSET**

NEST: for another set of formation constants set NSET to one, using the same experimental points, followed by items 1–4, 8, and 9 *only,* changed as appropriate; NSET = 0 for another complete set of data; NSET = −1 for the termination of the job.

4.2. Description of the Output

The outputs of both programs are quite similar and every effort has been made to render them self-explanatory. As described in Section 4.1, the amount and type of output is determined by the values of the indicators IPRIN, JPRIN, and JP(I) contained in the input file.

The first part of the output gives information about the system under consideration, including the starting values of the formation constants and the initial conditions of each titration curve, etc. Optionally (IPRIN = 2), a list of all titration points may be obtained. Such a list, when produced by MIQUV, also reports the weights assigned to each observation. These values should vary regularly within a titration experiment, depending on the slope of the titration curve. Any departure from a regular trend may be indicative of error in the data entry of the potentiometric points.

When the refinement is terminated, the results of the minimization are reported: the value of the error square sum, the best values of the parameters, their standards deviations, and the correlation matrix.

If JPRIN > 0, the residuals are subjected to a statistical analysis and the relevant results are shown. This output has been described elsewhere.[1] As a measure of the agreement between observed and calculated values, the crystallographic R factor, used in MINIQUAD,[1, 13] is replaced by the estimated standard deviation in MIQUV.

The additional output, which is obtained if JPRIN > 1, consists of tables and/or graphs as described in Section 4.1.

The residuals plot produced by MINIQUAD needs some explanation. Each line, which refers to a different titration point, shows the residuals for the mass balance equations, in graphical form. Alphabetic symbols are used for this purpose: A refers to the first reactant, B to the second, etc. The whole line spans six times σ, the estimate standard deviation, from -3σ (left margin) to $+3\sigma$ (right margin). A residual less than -3σ or greater $+3\sigma$ is represented by the corresponding character at the left or right margin, respectively. If two or more symbols are to be written in the same position, only the last one is printed.

5. PRESENTATION OF THE PROGRAMS

5.1. Listing of MINIQUAD

The following is a listing of the program MINIQUAD.

```
 1    C*************************************************************
 2    C
 3    C      ** MINIQUAD 82 A   **
 4    C
 5    C         PROGRAM FOR THE CALCULATION OF CUMULATIVE FORMATION CONSTANTS
 6    C         FROM POTENTIOMETRIC DATA.
 7    C
 8    C         AUTHORS: A. SABATINI, A. VACCA AND P. GANS
 9    C
10    C         LATEST UPDATED VERSION: JAN. 1982.
11    C
12    C         BIBLIOGRAPHY:  A. SABATINI AND A. VACCA, J. CHEM. SOC. DALTON,
13    C         1972,1963;
```

```
14   C       A. SABATINI, A. VACCA AND P. GANS, TALANTA, 1974, 21,45;
15   C       TALANTA, 1974, 21, 53.
16   C       P. GANS, A. SABATINI AND A. VACCA, INORG. CHIM. ACTA, 1976,
17   C       18, 237.
18   C
19   C
20           IMPLICIT REAL*8(A-H,O-Z)
21           INTEGER PQR(4,16),PQRØ(4,16),PQRV(4,16,2Ø)
22           INTEGER TITLE(18)
23           DIMENSION JTP(2Ø),NMBEV(2Ø),NCV(2Ø)
24           DIMENSION BETA(16),BETAV(16),SB(16),JPOT(16),HLNB(16),
25          &KEY(16),IVAR(16),SIGMA(16),JS(16),CS(16),DY(16),GB(16)
26           DIMENSION CX(4),ADDC(4),DEPS(4),HX(4),TT(4),TOTC(4)
27           DIMENSION GC(3),TOL(3),DC(3)
28           DIMENSION JEL(2),EZERO(2),EMF(2)
29           DIMENSION CI(16,4ØØ)
30           DIMENSION B3(16,16),BB(16,16)
31           DIMENSION T(4,4ØØ),CONC(4,4ØØ),EPS(4,4ØØ)
32           DIMENSION JP(4,2Ø),JNMB(4,2Ø)
33           DIMENSION A(4,19)
34           DIMENSION B2(3,16),D1(3,16)
35           DIMENSION B1(3,3),B(3,3)
36           COMMON JINP,JOUT,AL1Ø,LARS,MAXIT,TOLB,TOLU,ACCM,IPRIN,NRUN,
37          &RELAC,IFAIL
38         1 FORMAT(18A4)
39         2 FORMAT(//'1MINIQUAD 82 A   ',18A4/)
40         3 FORMAT(10I5)
41   C
42   C           MAXIMUM VALUES FOR ARRAY DIMENSIONS
43   C
44   C       NOTE:  THIS PARTICULAR VERSION CAN HANDLE UP TO:
45   C       1)   4ØØ TITRATION POINTS, MAXTP.
46   C       2)   2Ø TITRATION CURVES, MAXTC.
47   C       3)   16 FORMATION CONSTANTS, MAXK.
48   C       4)   4 INDEPENDENT REACTANTS, MAXMB.
49   C       5)   2 POTENTIOMETRIC ELECTRODES, SIMULTANEOUSLY, MAXEL.
50   C
51   C       EXPANSION OF THESE LIMITS MAYBE OBTAINED BY SUITABLE ENLARGEMENT
52   C       OF THE CORRESPONDING ARRAYS AND BY CHANGING THE FOLLOWING
53   C       MAXIMUM SIZE PARAMETERS.
54   C
55   C
56           MAXTP=4ØØ
57           MAXTC=2Ø
58           MAXK=16
59           MAXMB=4
60           MAXCX=3
61           MAXEL=2
62           MAXP=MAXK+MAXCX
63   C
64   C           DEFINITION OF I/O CHANNELS,TOLERANCES, AND ACCURACIES
65   C
66           JINP=5
67           JOUT=6
68           TOLB=Ø.1
69           TOLU=1.E-Ø4
70           ACCM=1.E-37
71           AL1Ø=DLOG(1Ø.D+ØØ)
72           RELAC=1.E-Ø4
73   C
74   C       READ IN TITLE, INITIAL PARAMETERS AND COMMENCE REFINEMENT.
75   C
76           NSET=Ø
77       100 READ(JINP,1)TITLE
78           WRITE(JOUT,2)TITLE
79           IFAIL=Ø
80           READ(JINP,3)LARS,NK,N,MAXIT,IPRIN,NMBEØ,NCONS
81   C
82   C           CALL THE DATA INPUT ROUTINE.
83   C
84           CALL DINP(NK,N,NMBEV,NCV,BETA,JPOT,PQRØ,KEY,HLNB,T,CONC,BETAV,
85          &IVAR,JEL,EZERO,EMF,TOTC,ADDC,DEPS,CX,TT,CI,HX,TOL,EPS,B1,B,
86          &NP,IP,NSET,A,GC,DC,NMBEØ,JTP,PQRV,NTC,JP,JNMB,PQR,MAXTP,MAXTC,
87          &MAXK,MAXMB,MAXCX,MAXEL,MAXP)
88   C
89   C       CHECK THAT DINP HAS OPERATED SUCCESSFULLY.
```

```
 90    C
 91            IF(IFAIL)116,116,200
 92      116 NRUN=0
 93    C
 94    C       CALL THE MINIMISATION ROUTINE.
 95    C
 96      125 CALL MINIM(NK,N,NMBEV,NCV,NP,BETAV,JPOT,PQRV,HLNB,T,CONC,IVAR,
 97           &A,GC,DC,SB,EPS,DEPS,HX,TT,CI,CX,B3,SIGMA,BETA,U,KEY,B2,DY,B1,
 98           &B,D1,TOL,GB,JTP,NTC,BB,PQR,MAXTP,MAXTC,MAXK,MAXMB,MAXCX,
 99           &MAXEL,MAXP)
100            IF(IFAIL)130,150,150
101    C
102    C       THE NUMBER OF PARAMETERS TO BE REFINED IS REDUCED BY ONE
103    C       IF A SPECIES HAS BEEN REJECTED BY SUBROUTINE NEG.  THE
104    C       INDICATOR IFAIL IS THEN RESTORED TO THE NORMAL VALUE.
105    C
106      130 N=N-1
107            IFAIL=0
108    C
109    C       CHECK THAT AT LEAST ONE REFINEABLE PARAMETER SURVIVES.
110    C
111            IF(N)135,135,125
112      135 IFAIL=4
113            GO TO 150
114    C
115    C       CALL THE DATA OUTPUT ROUTINE
116    C
117      150 CALL DOUT(NK,N,NMBE0,BETAV,JPOT,PQR0,IVAR,SIGMA,U,KEY,B3,HLNB,
118           &NP,TITLE,MAXMB,MAXK)
119    C
120    C       JPRIN IS USED TO CONTROL THE SUBSEQUENT OUTPUT OF THE PROGRAM.
121    C
122            READ(JINP,3)JPRIN
123            IF(IFAIL.GT.0.OR.JPRIN.EQ.0)GO TO 158
124            NT=NMBE0
125    C
126    C       CALL THE STATISTICAL ANALYSIS ROUTINE
127    C
128            CALL STATS(NK,NMBEV,PQRV,T,EPS,CI,NP,U,CONC,JPRIN,NT,JP,JTP,
129           &NTC,TITLE,PQR,JS,CS,MAXMB,MAXK,MAXTC,MAXTP)
130    C
131    C       NSET IS USED FOR THE TERMINATION OF THE JOB OR THE CONTINUATION
132    C       OF THE JOB ON THE SAME OR A NEW SET OF DATA.
133    C
134      158 READ(JINP,3)NSET
135            IF(NSET)200,100,100
136      200 STOP
137            END
```

```
  1    C
  2    C       SUBROUTINE DINP
  3    C
  4    C       DATA INPUT ROUTINE. THE INPUT DATA ARE READ IN. THE UNKNOWN
  5    C    FREE CONCENTRATIONS FOR EACH TITRATION POINT ARE FIRST CALCULATED
  6    C    IN THE ROUTINE MQ, VIA GAUSS-NEWTON LEAST SQUARES METHOD, USE
  7    C    THE STARTING VALUES OF THE FORMATION CONSTANTS.  FOR DETAILS
  8    C    OF THE INPUT STATEMENTS SEE THE ACCOMPANYING INSTRUCTIONS.
  9    C
 10         SUBROUTINE DINP(NK,N,NMBEV,NCV,BETA,JPOT,PQR0,KEY,HLNB,T,CONC,
 11        &BETAV,IVAR,JEL,EZERO,EMF,TOTC,ADDC,DEPS,CX,TT,CI,HX,TOL,EPS,B1,B,
 12        &NP,IP,NSET,A,GC,DC,NMBE0,JTP,PQRV,NTC,JP,JNMB,PQR,MAXTP,MAXTC,
 13        &MAXK,MAXMB,MAXCX,MAXEL,MAXP)
 14         IMPLICIT REAL*8(A-H,O-Z)
 15         INTEGER PQR(MAXMB,MAXK),PQR0(MAXMB,MAXK),PQRV(MAXMB,MAXK,MAXTC)
 16         DIMENSION JTP(MAXTC),NMBEV(MAXTC),NCV(MAXTC)
 17         DIMENSION BETA(MAXK),JPOT(MAXK),HLNB(MAXK),KEY(MAXK),
 18        &          BETAV(MAXK),IVAR(MAXK)
 19         DIMENSION CX(MAXMB),HX(MAXMB),TT(MAXMB),TOTC(MAXMB),
 20        &          ADDC(MAXMB),DEPS(MAXMB)
 21         DIMENSION GC(MAXCX),TOL(MAXCX),DC(MAXCX)
 22         DIMENSION JEL(MAXEL),EZERO(MAXEL),EMF(MAXEL)
 23         DIMENSION CI(MAXK,MAXTP)
```

```
24              DIMENSION T(MAXMB,MAXTP),CONC(MAXMB,MAXTP),EPS(MAXMB,MAXTP)
25              DIMENSION JP(MAXMB,MAXTC),JNMB(MAXMB,MAXTC)
26              DIMENSION A(MAXMB,MAXP)
27              DIMENSION B1(MAXCX,MAXCX),B(MAXCX,MAXCX)
28              COMMON JINP,JOUT,AL10,LARS,MAXIT,TOLB,TOLU,ACCM,IPRIN,NRUN,
29             &RELAC,IFAIL
30            1 FORMAT('(',I2,')',F11.4,'E',I3,I8,7X,5I4)
31            2 FORMAT(//10X,'FORMATION    REFINEMENT    STOICHEIOMETRIC'/
32             &10X,'CONSTANTS        KEYS        COEFFICIENTS'/)
33            3 FORMAT(12I5)
34            4 FORMAT(F10.6,7I5)
35            5 FORMAT(7F10.6)
36            6 FORMAT(I5,13F8.3)
37            7 FORMAT(1H0,'FAILURE TO CALCULATE INITIAL FREE CONCENTRATIONS '
38             &'AT POINT :', I4)
39            8 FORMAT(/5X,'LARS',4X,'MAXIT',3X,'IPRIN',7X,'TOLB',11X,'TOLU',11X,
40             & 'ACCM',10X,'RELAC'/3I8,4E15.3//6X,'THERE ARE',I3,' EQUILIBRIUM CO
41             &NSTANTS,',I3,' OF WHICH ARE TO BE REFINED'//6X,'THE MAXIMUM NUMBER
42             & OF MASS-BALANCE EQUATIONS IS',I2//6X,'REACTION TEMPERATURE',F8.2,
43             &' DEGREES CENTIGRADE'/)
44            9 FORMAT(/' THEREFORE RUN ABANDONED BEFORE REFINEMENT')
45           10 FORMAT(/I5,' DATA POINTS HAVE BEEN READ IN AND',I5,
46             &' MAXIMUM WERE EXPECTED'/)
47           11 FORMAT(/'    THE NUMBER N OF CONSTANTS TO BE REFINED DOES NOT AGR
48             &EE WITH THE NUMBER OF POSITIVE KEYS'/'    THE LATTER IS TAKEN:',
49             &' REFINEMENT BEGINS'/)
50           12 FORMAT(///2X,'CURVE',I2,4X,'THERE ARE',I2,' MASS-BALANCE EQUATION
51             &S AND',I2,' UNKNOWN FREE CONCENTRATION(S)PER DATA POINT'//22X,'IN
52             &ITIAL QUANTITIES',4X,'TITRANT CONCENTRATION',4X,'STANDARD POTENTIA
53             &L'/25X,'(MILLIMOLES)',11X,'(MOLES/LITRE)',11X,'(MILLIVOLTS)'/)
54           13 FORMAT(6X,'REACTANT *',I1,'*',8X,F7.4,17X,F7.4,17X,F7.2)
55           14 FORMAT(6X,'REACTANT *',I1,'*',8X,F7.4,17X,F7.4,15X,'NO ELECTRODE
56             &')
57           15 FORMAT(/6X,'INITIAL VOLUME=',F8.2,' MILLILITERS')
58           16 FORMAT(/I8,' DATA POINTS IN CURVE',I3)
59           17 FORMAT(/)
60              READ(JINP,5)TEMP,SIGMAV,SIGMAE
61  C
62  C         NOTE THAT SIGMAV AND SIGMAE ARE NOT USED BY THIS VERSION.
63  C
64              WRITE(JOUT,8)LARS,MAXIT,IPRIN,TOLB,TOLU,ACCM,RELAC,
65             &NK,N,NMBE0,TEMP
66              J=0
67              WRITE(JOUT,2)
68              DO 105 I=1,NK
69              READ(JINP,4)BETA(I),JPOT(I),(PQR0(K,I),K=1,NMBE0),KEY(I)
70              WRITE(JOUT,1)I,BETA(I),JPOT(I),KEY(I),(PQR0(K,I),K=1,NMBE0)
71              IF(KEY(I))105,105,103
72          103 J=J+1
73              BETAV(J)=BETA(I)
74              IVAR(J)=I
75          105 HLNB(I)=DLOG(BETA(I))+AL10*JPOT(I)
76              IF(N-J)106,107,106
77          106 WRITE(JOUT,11)
78              N=J
79          107 IP=0
80              NTC=0
81          108 NTC=NTC+1
82              IF(NSET)109,110,109
83          109 NMBE=NMBEV(NTC)
84              NC=NCV(NTC)
85              DO 111 I=1,NMBE
86              DO 111 J=1,NMBE
87              IF(JNMB(I,NTC).EQ.JP(J,NTC))JP(J,NTC)=I
88          111 CONTINUE
89              NEMF=NMBE-NC
90              GO TO 2110
91          110 READ(JINP,3)NMBE,(JNMB(I,NTC),I=1,NMBE),NEMF,
92             &(JP(I,NTC),I=1,NMBE0)
93              NC=NMBE-NEMF
94              WRITE(JOUT,12)NTC,NMBE,NC
95              NMBEV(NTC)=NMBE
96              DO 1110 I=1,NC
97         1110 CX(I)=1.E-07
98              NCV(NTC)=NC
99              READ(JINP,3)(JEL(L),L=1,NEMF),JCOUL
```

```
100      2110 DO 1111 K=1,NMBE
101           KK=JNMB(K,NTC)
102           DO 1111 I=1,NK
103      1111 PQRV(K,I,NTC)=PQR0(KK,I)
104           KK=1
105           DO 1116 K=1,NMBE0
106           IF(K.EQ.JNMB(KK,NTC))GO TO 1115
107      1112 DO 1114 I=1,NK
108           IF(PQR0(K,I).EQ.0)GO TO 1114
109           DO 1113 J=1,NMBE
110      1113 PQRV(J,I,NTC)=0
111      1114 CONTINUE
112           GO TO 1116
113      1115 IF(KK.EQ.NMBE)GO TO 1116
114           KK=KK+1
115      1116 CONTINUE
116           DO 1117 I=1,NK
117           DO 1117 K=1,NMBE
118      1117 PQR(K,I)=PQRV(K,I,NTC)
119           IF(NSET.EQ.0)GO TO 2117
120      2109 IP=IP+1
121           DO 3109 L=1,NMBE
122           TT(L)=T(L,IP)
123      3109 CX(L)=CONC(L,IP)
124           GO TO 125
125      2117 READ(JINP,5)(TOTC(K),K=1,NMBE)
126           READ(JINP,5)(ADDC(K),K=1,NMBE)
127           READ(JINP,5)(EZERO(L),L=1,NEMF)
128           READ(JINP,5)VINIT
129           DO 1119 K=1,NMBE
130           IF(K.GT.NC)GO TO 1118
131           WRITE(JOUT,14)JNMB(K,NTC),TOTC(K),ADDC(K)
132           GO TO 1119
133      1118 I=K-NC
134           WRITE(JOUT,13)JNMB(K,NTC),TOTC(K),ADDC(K),EZERO(I)
135      1119 CONTINUE
136           WRITE(JOUT,15)VINIT
137           IF(IPRIN.GT.1)WRITE(JOUT,17)
138       112 IP=IP+1
139           IS=LARS
140       113 READ(JINP,6)LUIGI,TITRE,(EMF(L),L=1,NEMF)
141           IS=IS-1
142           IF(LUIGI)115,114,115
143       114 IF(IS)115,115,113
144       115 IF(IP-MAXTP)2115,2115,140
145      2115 DO 118 L=1,NEMF
146           NCPL=NC+L
147           IF(JEL(L))116,117,116
148       116 CX(NCPL)=DEXP((EMF(L)-EZERO(L))*JEL(L)*11.6049/(TEMP+273.16))
149           GO TO 118
150       117 CX(NCPL)=DEXP(-EMF(L)*AL10)
151       118 CONTINUE
152           VOL=VINIT
153           IF(JCOUL .EQ. 0)VOL=VOL+TITRE
154      1120 DO 120 K=1,NMBE
155           TT(K)=(TOTC(K)+TITRE*ADDC(K))/VOL
156       120 T(K,IP)=TT(K)
157  C
158  C          CALL THE ROUTINE FOR CALCULATION OF STARTING VALUES OF
159  C          UNKNOWN FREE REACTANT CONCENTRATIONS
160  C
161       125 CALL ML(NK,NMBE,NEMF,NC,HLNB,CI,CX,TT,HX,EPS,B1,PQR,
162          &KEY,N,S,TOL,A,IP,DC,MAXMB,MAXK,MAXCX,MAXTP,MAXP)
163  C
164  C          CALL THE ROUTINE FOR LEAST-SQUARES REFINEMENT OF UNKNOWN
165  C          FREE REACTANT CONCENTRATIONS
166  C
167           CALL MQ(NK,NMBE,NEMF,NC,HLNB,CI,CX,TT,HX,5,EPS,DEPS,B1,B,PQR,
168          &KEY,N,S,TOL,A,GC,DC,IP,MAXMB,MAXK,MAXCX,MAXTP,MAXP)
169           IF(NSET)128,126,128
170       126 IF((IPRIN-1).GT.0)WRITE(JOUT,6)IP,TITRE,(EMF(L),L=1,NEMF)
171  C
172  C          CHECK THAT MQ HAS OPERATED SUCCESSFULLY.
173  C
174       128 IF(IFAIL)129,129,370
175       129 DO 130 K=1,NMBE
```

```
176       130 CONC(K,IP)=CX(K)
177       135 IF(NSET)136,140,136
178       136 IF(IP.LT.JTP(NTC))GO TO 2109
179           IF(IP.EQ.NP)RETURN
180           GO TO 108
181       140 IF(LUIGI.EQ.0)GO TO 112
182           JTP(NTC)=IP
183           IF(NTC.NE.1)GO TO 145
184           NPC=IP
185           GO TO 150
186       145 NTC1=NTC-1
187           NPC=IP-JTP(NTC1)
188       150 WRITE(JOUT,16)NPC,NTC
189           IF(LUIGI.GT.0)GO TO 108
190           WRITE(JOUT,10)IP,MAXTP
191   C
192   C         CHECK THAT THE MAXIMUM ALLOWABLE NUMBER OF TITRATION POINTS
193   C         IS NOT EXCEEDED.
194   C
195           IF(IP-MAXTP)300,300,350
196       300 NP=IP
197           RETURN
198       350 IFAIL=5
199           GO TO 400
200       370 WRITE(JOUT,7)IP
201       400 WRITE(JOUT,9)
202           RETURN
203           END
```

```
1    C
2    C         SUBROUTINE DOUT
3    C
4    C         DATA OUTPUT ROUTINE. ALL THE RELEVANT INFORMATION ON THE
5    C         REFINED FORMATION CONSTANTS IS PRINTED OUT. IF REFINEMENT
6    C         HAS FAILED, A CONCISE FAILURE MESSAGE IS PRINTED.
7    C
8    C         IN MANY FORMAT STATEMENTS THE FIRST CHARACTER OF THE RECORD TO
9    C         BE PRINTED IS A CARRIAGE CONTROL CHARACTER SPECIFIED AS 1HX
10   C         WHERE 'X' HAS ONE OF THE FOLLOWING MEANINGS:-
11   C
12   C             X                 MEANING
13   C          (BLANK)           ADVANCE ONE LINE BEFORE PRINTING
14   C             0              ADVANCE TWO LINES BEFORE PRINTING
15   C             1              ADVANCE TO FIRST LINE OF NEW PAGE
16   C             +              NO ADVANCE
17   C
18           SUBROUTINE DOUT(NK,N,NMBE,BETAV,JPOT,PQR,IVAR,SIGMA,U,KEY,B3,
19          &HLNB,NP,TITLE,MAXMB,MAXK)
20           IMPLICIT REAL*8(A-H,O-Z)
21           INTEGER PQR(MAXMB,MAXK)
22           INTEGER TITLE(18)
23           DIMENSION BETAV(MAXK),JPOT(MAXK),HLNB(MAXK),KEY(MAXK),
24          &SIGMA(MAXK),IVAR(MAXK)
25           DIMENSION B3(MAXK,MAXK)
26           COMMON JINP,JOUT,AL10,LARS,MAXIT,TOLB,TOLU,ACCM,IPRIN,NRUN,
27          &RELAC,IFAIL
28         1 FORMAT(1H0,'FAILURE IN FACTORIZATION PROCESS.'/
29          &' REFINEMENT ABANDONED')
30         2 FORMAT(1H0,'MAXIMIMUM NO. OF ITERATIONS PERFORMED, REFINEMENT TER
31          &MINATED')
32         3 FORMAT(1H0,'MATRIX OF NORMAL EQUATIONS NOT POSITIVE DEFINITE, REF
33          &INEMENT ABANDONED')
34         4 FORMAT(1H0,'ALL THE FORMATION CONSTANTS BECAME NEGATIVE. REFINEME
35          &NT TERMINATED')
36         5 FORMAT(1H0,'REFINEMENT CONVERGED SUCCESSFULLY')
37         6 FORMAT(I4,' ITERATIONS'/' SUM OF SQUARES=',E17.9//
38          &35X,' VALUE     STD. DEVIATION     LOG BETA     STD. DEVIATION'/)
39         7 FORMAT(1H ,'BETA(',I2,') ',5I2)
40       100 FORMAT(1H1,3X,18A4)
41       110 FORMAT(1H+,23X,'NEGATIVE AT CYCLE',I3)
42       111 FORMAT(1H+,23X,'CONSTANT')
43       112 FORMAT(1H+,23X,'REFINED ')
44       113 FORMAT(1H+,32X,F10.5,'E',I3,14X,F14.5)
```

```
45        114 FORMAT(1H+,32X,2(F10.5,'E',I3),F14.5,6X,F7.5)
46        115 FORMAT(///' MATRIX OF CORRELATION COEFFICIENTS RHO I,J'//4H     I/)
47        116 FORMAT(I4,11F10.3/5X,11F10.3)
48        117 FORMAT(6H0 J   ,I5,10I10/11I10)
49            WRITE(JOUT,100)TITLE
50    C
51    C        IFAIL IS POSITIVE IF THE REFINEMENT HAS FAILED
52    C
53            IF(IFAIL)15,15,10
54        10 GO TO(11,12,13,14),IFAIL
55        11 WRITE(JOUT,1)
56            GO TO 23
57        12 WRITE(JOUT,2)
58            GO TO 16
59        13 WRITE(JOUT,3)
60            GO TO 23
61        14 WRITE(JOUT,4)
62            GO TO 23
63        15 WRITE(JOUT,5)
64        16 WRITE(JOUT,6)NRUN,U
65            K=1
66            DO 20 I=1,NK
67            WRITE(JOUT,7)I,(PQR(J,I),J=1,NMBE)
68            IF(KEY(I))17,18,19
69        17 KK=IABS(KEY(I))
70            WRITE(JOUT,110)KK
71            GO TO 20
72        18 WRITE(JOUT,111)
73            XC=DEXP(HLNB(I)-AL10*JPOT(I))
74            XL=HLNB(I)/AL10
75            WRITE(JOUT,113)XC,JPOT(I),XL
76            GO TO 20
77        19 WRITE(JOUT,112)
78            IV=IVAR(K)
79            SDC=SIGMA(K)*BETAV(K)/100.
80            XL=DLOG(BETAV(K))/AL10+JPOT(IV)
81            IF(1.-SIGMA(K)/100.)191,191,192
82       191 SDL=1.E+30
83            GO TO 193
84       192 SDL=DABS(DLOG(1.-SIGMA(K)/100.))/AL10
85       193 WRITE(JOUT,114)BETAV(K),JPOT(IV),SDC,JPOT(IV),XL,SDL
86            K=K+1
87        20 CONTINUE
88    C
89    C        IF MORE THAN ONE PARAMETER HAS BEEN REFINED, THE LOWER
90    C        TRIANGULAR PART OF THE MATRIX OF THE CORRELATION MATRIX
91    C        IS PRINTED OUT.
92    C
93            IF(N-1)23,23,21
94        21 WRITE(JOUT,115)
95            DO 22 I=2,N
96            IM=I-1
97            WRITE(JOUT,116)IVAR(I),(B3(I,J),J=1,IM)
98        22 CONTINUE
99            WRITE(JOUT,117)(IVAR(J),J=1,IM)
100       23 RETURN
101           END




 1    C
 2    C        SUBROUTINE LINEQ
 3    C
 4    C        GENERAL PURPOSE LINEAR EQUATION SOLVER BASED ON CHOLESKI
 5    C        FACTORIZATION
 6    C
 7    C        THIS ROUTINE SOLVES THE N LINEAR EQUATIONS   A * X = B
 8    C        THE SOLUTION VECTOR IS STORED IN B AND THE MATRIX IS
 9    C        REPLACED BY ITS INVERSE.
10    C
11            SUBROUTINE LINEQ(A,N,B,JFAIL,MAXK)
12            IMPLICIT REAL*8(A-H,O-Z)
13            DIMENSION A(MAXK,MAXK),B(MAXK)
14            COMMON JINP,JOUT,AL10,LARS,MAXIT,TOLB,TOLU,ACCM,IPRIN,NRUN,
```

```
15              &RELAC,IFAIL
16               IF(N-1)46,5,9
17           5  T=A(1,1)
18               IF(T-ACCM)46,46,6
19           6  A(1,1)=1./T
20               B(1)=B(1)/T
21               RETURN
22          46  IFAIL=JFAIL
23               RETURN
24           9  DO 80 I=1,N
25               I1=I-1
26               DO 70 J=I,N
27               S=A(I,J)
28               IF(I1)10,30,10
29          10  DO 20 K=1,I1
30          20  S=S-A(I,K)*A(J,K)
31          30  X=S
32               IF(J-I)60,40,60
33          40  IF(X)45,45,50
34          45  IFAIL=JFAIL
35               RETURN
36          50  A(I,I)=1./ DSQRT(X)
37               GOTO 70
38          60  A(J,I)=X*A(I,I)
39          70  CONTINUE
40          80  CONTINUE
41  C
42  C               FORWARD SUBSTITUTION
43  C
44               B(1)=B(1)*A(1,1)
45               DO 120 I=2,N
46               I1=I-1
47               S=B(I)
48               DO 110 K=1,I1
49         110  S=S-A(I,K)*B(K)
50         120  B(I)=S*A(I,I)
51  C
52  C           FORWARD SUBSTITUTION FOR THE INVERSE
53  C
54               DO 170 J=1,N
55               J1=J+1
56               IF(J1-N)140,140,170
57         140  DO 160 I=J1,N
58               I1=I-1
59               S=0.
60               DO 150 K=J,I1
61         150  S=S-A(I,K)*A(J,K)
62         160  A(J,I)=S*A(I,I)
63         170  CONTINUE
64  C
65  C           BACKWARD SUBSTITUTION
66  C
67               B(N)=B(N)*A(N,N)
68               DO 220 J=1,N
69         220  A(J,N)=A(J,N)*A(N,N)
70               DO 290 II=2,N
71               I=N-II+1
72               T=A(I,I)
73               I1=I+1
74               S=B(I)
75               DO 240 K=I1,N
76         240  S=S-A(K,I)*B(K)
77         245  B(I)=S*T
78               DO 280 J=1,I
79               S=A(J,I)
80               DO 270 K=I1,N
81         270  S=S-A(K,I)*A(J,K)
82               A(J,I)=S*T
83         280  CONTINUE
84         290  CONTINUE
85               DO 300 I=2,N
86               I1=I-1
87               DO 300 J=1,I1
88         300  A(I,J)=A(J,I)
89         400  RETURN
90               END
```

```
 1    C
 2    C          SUBROUTINE MINIM
 3    C
 4    C          MINIMISATION ROUTINE
 5    C     REFINEMENT OF THE FORMATION CONSTANTS IS PERFORMED USING THE
 6    C     GAUSS-NEWTON LEAST SQUARES METHOD.  ALL DIFFERENTIAL COEFFICIENTS
 7    C     ARE CALCULATED BY ANALYTICAL EXPRESSIONS.
 8    C
 9    C
10          SUBROUTINE MINIM(NK,N,NMBEV,NCV,NP,BETAV,JPOT,PQRV,HLNB,T,CONC,
11         &IVAR,A,GC,DC,SB,EPS,DEPS,HX,TT,CI,CX,B3,SIGMA,BETA,U,KEY,B2,DY,
12         &B1,B,D1,TOL,GB,JTP,NTC,BB,PQR,MAXTP,MAXTC,MAXK,MAXMB,MAXCX,
13         &MAXEL,MAXP)
14          IMPLICIT REAL*8(A-H,O-Z)
15          INTEGER PQR(MAXMB,MAXK),PQRV(MAXMB,MAXK,MAXTC)
16          DIMENSION JTP(MAXTC),NMBEV(MAXTC),NCV(MAXTC)
17          DIMENSION BETA(MAXK),JPOT(MAXK),HLNB(MAXK),KEY(MAXK),
18         &BETAV(MAXK),IVAR(MAXK),SB(MAXK),SIGMA(MAXK),
19         &DY(MAXK),GB(MAXK)
20          DIMENSION CX(MAXMB),HX(MAXMB),TT(MAXMB),DEPS(MAXMB)
21          DIMENSION GC(MAXCX),TOL(MAXCX),DC(MAXCX)
22          DIMENSION CI(MAXK,MAXTP)
23          DIMENSION B3(MAXK,MAXK),BB(MAXK,MAXK)
24          DIMENSION T(MAXMB,MAXTP),CONC(MAXMB,MAXTP),EPS(MAXMB,MAXTP)
25          DIMENSION A(MAXMB,MAXP)
26          DIMENSION B2(MAXCX,MAXK),D1(MAXCX,MAXK)
27          DIMENSION B1(MAXCX,MAXCX),B(MAXCX,MAXCX)
28          COMMON JINP,JOUT,AL10,LARS,MAXIT,TOLB,TOLU,ACCM,IPRIN,NRUN,
29         &RELAC,IFAIL
30        1 FORMAT(/'  FAILURE TO CALCULATE FREE CONCENTRATIONS AT CYCLE'
31         &,I3,', AT POINT',I4/)
32          KONT=0
33          IBN=0
34          IND=-1
35          PERC=0.
36    C
37    C          U DENOTES THE SUM OF THE SQUARED RESIDUALS.
38    C
39      100 U=0.
40          IND=IND+1
41    C
42    C          IEXIT IS USED TO TEST THE TWO CONVERGENCE CRITERIA.
43    C
44          IEXIT=2
45          DO 101 I=1,N
46          DY(I)=0.
47          DO 101 J=1,N
48      101 B3(I,J)=0.
49          DO 102 I=1,N
50          IV=IVAR(I)
51      102 HLNB(IV)=DLOG(BETAV(I))+AL10*JPOT(IV)
52    C
53    C          LOOP FOR EACH EXPERIMENTAL POINT.
54    C
55          JGRLB=-N
56          JTC=0
57          DO 200 IP=1,NP
58          IF(JTC.NE.0)GO TO 1104
59     1003 JTC=JTC+1
60          NMBE=NMBEV(JTC)
61          NC=NCV(JTC)
62          NEMF=NMBE-NC
63          NC1=NC+1
64          NPNC=N+NC
65          DO 1103 J=1,NK
66          DO 1103 K=1,NMBE
67     1103 PQR(K,J)=PQRV(K,J,JTC)
68     1104 IF(IP.GT.JTP(JTC))GO TO 1003
69          JGRLB=JGRLB+NEMF
70          DO 104 K=1,NMBE
71          CX(K)=CONC(K,IP)
72      104 TT(K)=T(K,IP)
73    C
74    C          CALL THE ROUTINE FOR LEAST-SQUARES REFINEMENT OF UNKNOWN
75    C          FREE REACTANT CONCENTRATIONS
76    C
```

```
 77              IPCOPY=IP
 78              CALL MQ(NK,NMBE,NEMF,NC,HLNB,CI,CX,TT,HX,1,EPS,DEPS,B1,B,PQR,
 79             &KEY,N,DU,TOL,A,GC,DC,IPCOPY,MAXMB,MAXK,MAXCX,MAXTP,MAXP)
 80              IF(IFAIL)107,107,105
 81         105 WRITE(JOUT,1)NRUN,IP
 82              RETURN
 83         107 DO 108 K=1,NC
 84         108 CONC(K,IP)=CX(K)
 85              DO 109 I=1,NMBE
 86              DO 109 J=1,N
 87              IV=IVAR(J)
 88              K=J+NC
 89         109 A(I,K)=PQR(I,IV)*CI(IV,IP)
 90   C
 91   C          THE DERIVATIVES ARE EVALUATED. POINT BY POINT, USING THE
 92   C          CONCENTRATIONS OBTAINED IN THE MQ ROUTINE WITH THE CURRENT
 93   C          VALUES OF THE FORMATION CONSTANTS AND STORED IN THE DESIGN
 94   C          MATRIX.
 95   C
 96              DO 110 I=1,NC
 97              DO 110 L=1,N
 98              J=NC+L
 99              B2(I,L)=0.
100              DO 110 K=1,NMBE
101         110 B2(I,L)=B2(I,L)+A(K,I)*A(K,J)
102              DO 132 I=1,NC
103              DO 132 J=1,N
104              D1(I,J)=0.
105              DO 132 K=1,NC
106         132 D1(I,J)=D1(I,J)-B1(I,K)*B2(K,J)
107              DO 190 I=NC1,NPNC
108              L=I-NC
109              DO 180 J=I,NPNC
110              M=J-NC
111              DO 170 K=1,NMBE
112         170 B3(L,M)=B3(L,M)+A(K,I)*A(K,J)
113              DO 180 K=1,NC
114         180 B3(L,M)=B3(L,M)+D1(K,L)*B2(K,M)
115              DO 185 K=1,NMBE
116         185 DY(L)=DY(L)+A(K,I)*EPS(K,IP)
117         190 CONTINUE
118   C
119   C          B3 AND DY CONTAIN THE COEFFICIENTS OF THE NORMAL EQUATION AND
120   C          THE ELEMENTS OF THE GRADIENT VECTOR, RESPECTIVELY.
121   C
122              U=U+DU
123         200 CONTINUE
124              IF(PERC)204,210,204
125   C
126   C          CHECK THAT THE SUM OF SQUARES IS NOT DECREASED.
127   C
128         204 IF(U-(1.+TOLU)*U0)205,208,208
129         205 IF(IND)206,206,210
130   C
131   C          FIRST CONVERGENCE CRITERION:  RELATIVE DECREASE OF U LESS
132   C          THAN TWOFOLD.
133   C
134         206 IF((1.-TOLU)*U0-U)207,210,210
135         207 IEXIT=1
136              GO TO 210
137         208 IF(IND-7)209,210,210
138   C
139   C          IN CASE OF DIVERGENCE THE SHIFT VECTOR IS HALVED.
140   C
141         209 PERC=-0.5*DABS(PERC)
142              GO TO 263
143         210 U0=U
144              IND=-1
145              DO 211 I=1,N
146              SB(I)=DY(I)
147              DO 211 J=I,N
148         211 B3(J,I)=B3(I,J)
149              DO 212 I=1,N
150              DO 212 J=1,N
151         212 BB(I,J)=B3(I,J)
152         213 DO 1213 I=1,N
```

```
153              SB(I)=DY(I)
154              DO 1212 J=1,N
155      1212 B3(I,J)=BB(I,J)
156      1213 B3(I,I)=BB(I,I)
157   C
158   C          CALL THE LINEAR EQUATION SOLVER
159   C
160   C          THIS ROUTINE REPLACES THE MATRIX B3 BY ITS INVERSE AND THE
161   C          SOLUTION VECTOR IS STORED IN SB.
162   C
163              CALL LINEQ(B3,N,SB,3,MAXK)
164   C
165   C          IF THE MATRIX B3 IS NOT POSITIVE DEFINITE CONTROL IS PASSED BACK
166   C          TO THE MAIN PROGRAM
167   C
168              IF(IFAIL.GT.0)RETURN
169       214 W=JGRLB
170              W=DSQRT(U/W)
171              DO 215 I=1,N
172              GB(I)=DSQRT(B3(I,I))
173       215 SIGMA(I)=GB(I)*W
174   C
175   C          THE RELATIVE STANDARD DEVIATIONS OF THE PARAMETERS ARE STORED IN
176   C          THE VECTOR SIGMA.
177   C
178              GO TO(220,240),IEXIT
179   C
180   C          SECOND CONVERGENCE CRITERION:  ALL RELATIVE SHIFTS ARE LESS
181   C          THAN TOLB*SIGMA.
182   C
183       220 DO 223 I=1,N
184              IF(DABS(SB(I)/SIGMA(I))-TOLB)223,223,222
185       222 IEXIT=2
186              GO TO 240
187       223 CONTINUE
188   C
189   C          CALCULATION OF THE CORRELATION COEFICIENTS.
190   C
191              DO 235 I=1,N
192              DO 235 J=1,I
193       235 B3(I,J)=B3(I,J)/GB(I)/GB(J)
194       240 DO 245 I=1,N
195       245 SIGMA(I)=SIGMA(I)*100.
196              IF(IPRIN)256,256,255
197   C
198   C          CALL MONITOR OF PROGRESS OF MINIMIZATION
199   C
200       255 CALL MON1(N,NK,U,PERC,BETAV,JPOT,SB,SIGMA,IVAR,MAXK)
201   C
202   C          IF BOTH CONVERGENCE CRITERIA ARE FULFILLED CONTROL IS PASSED
203   C          BACK TO THE MAIN PROGRAM
204   C
205       256 IF(IEXIT.EQ.1)RETURN
206       257 NRUN=NRUN+1
207   C
208   C          CHECK ON THE NUMBER OF ITERATIONS PERFORMED.
209   C
210              IF(NRUN-MAXIT)259,259,390
211       259 PERC=1.
212   C
213   C          CHECK FOR NEGATIVE EQUILIBRIUM CONSTANTS
214   C
215              CALL NEG(N,NK,BETAV,PERC,SB,KEY,IVAR,HLNB,JPOT,CI,KONT,IBN,NPNC,
216          &NP,MAXK,MAXTP)
217              IF(IFAIL.NE.0)RETURN
218   C
219   C          CALCULATION OF THE SHIFT VECTOR AND OF THE NEW VALUES OF THE
220   C          PARANETERS.
221   C
222       260 DO 262 I=1,N
223       262 SB(I)=BETAV(I)*SB(I)
224       263 DO 265 I=1,N
225       265 BETAV(I)=BETAV(I)+PERC*SB(I)
226              GO TO 100
227       390 IFAIL=2
228       400 RETURN
229              END
```

```
 1    C
 2    C         SUBROUTINE ML
 3    C
 4    C         THE UNKNOWN FREE CONCENTRATIONS ARE CALCULATED FOR USE AS
 5    C         INITIAL ESTIMATES IN SUBSEQUENT REFINEMENT(SUBROUTINE MQ)
 6    C         USING A NUMBER OF MASS BALANCE EQUATIONS EQUAL TO THE NUMBER
 7    C         OF UNKNOWNS (THOSE FOR WHICH THERE IS NO POTENTIAL).  THE
 8    C         NEWTON-RAPHSON METHOD IS USED, WITH FIRST DERIVATIVES ONLY -
 9    C         INITIAL ESTIMATES ARE ALSO REQUIRED FOR THIS ROUTINE, BUT
10    C         1.0E-07 FOR THE FIRST POINT IN THE FIRST TITRATION CURVE
11    C         IS SATISFACTORY. INITIAL ESTIMATES FOR THE OTHER POINTS ARE
12    C         TAKEN AS THE VALUE OBTAINED FOR THE PREVIOUS POINT.
13    C
14    C
15              SUBROUTINE ML(NK,NMBE,NEMF,NC,HLNB,CI,CX,TT,HX,EPS,B1,PQR,
16             &KEY,N,S,TOL,A,IP,DC,MAXMB,MAXK,MAXCX,MAXTP,MAXP)
17              IMPLICIT REAL*8(A-H,O-Z)
18              INTEGER PQR(MAXMB,MAXK)
19              DIMENSION HLNB(MAXK),KEY(MAXK)
20              DIMENSION CX(MAXMB),HX(MAXMB),TT(MAXMB)
21              DIMENSION DC(MAXCX),TOL(MAXCX)
22              DIMENSION CI(MAXK,MAXTP)
23              DIMENSION EPS(MAXMB,MAXTP)
24              DIMENSION A(MAXMB,MAXP)
25              DIMENSION B1(MAXCX,MAXCX)
26              COMMON JINP,JOUT,AL10,LARS,MAXIT,TOLB,TOLU,ACCM,IPRIN,NRUN,
27             &RELAC,IFAIL
28    C
29    C         ITER IS A CYCLE COUNTER.  100 CYCLES ARE PERMITTED AS MAX.
30    C
31              ITER=-1
32              DO 101 I=1,NEMF
33              J=I+NC
34        101 HX(J)=DLOG(CX(J))
35    C
36    C         TOL(I) PROVIDES A RELATIVE TOLERANCE FOR USE WITH CONVERGENCE
37    C         CRITERIA.
38    C
39              DO 103 I=1,NC
40        103 TOL(I)=TT(I)*RELAC
41        110 ITER=ITER+1
42              IF(ITER-100)116,116,330
43        116 DO 118 I=1,NC
44              W=CX(I)
45        118 HX(I)=DLOG(W)
46    C
47    C         LOOP FOR THE CALCULATION OF CONCENTRATIONS.
48    C
49              DO 150 I=1,NK
50              IF(KEY(I))150,120,120
51        120 W=HLNB(I)
52              DO 140 J=1,NMBE
53              IF(PQR(J,I))130,140,130
54        130 W=W+PQR(J,I)*HX(J)
55        140 CONTINUE
56              CI(I,IP)=DEXP(W)
57        150 CONTINUE
58              S=0.
59    C
60    C         LOOP FOR THE CALCULATION OF RESIDUALS AND SUM OF SQUARES.
61    C
62              DO 175 I=1,NC
63              EPS(I,IP)=TT(I)-CX(I)
64              DO 170 J=1,NK
65              IF(PQR(I,J))160,170,160
66        160 EPS(I,IP)=EPS(I,IP)-PQR(I,J)*CI(J,IP)
67        170 CONTINUE
68        175 S=S+EPS(I,IP)*EPS(I,IP)
69              IF(ITER)186,186,176
70        176 GO TO(178,189),IDIM
71    C
72    C         CHECK THAT SUM OF SQUARES IS NOT INCREASED.
73    C
74        178 IF(S-(1.+TOLU)*S0)186,180,180
75    C
76    C         IN CASE OF DIVERGENCE SHIFT IS HALVED.
77    C
```

```
78      180 DO 182 I=1,NC
79      182 CX(I)=CX(I)-0.5*DC(I)
80          IDIM=2
81          GO TO 116
82    C
83    C       CONVERGENCE CRITERION:  ALL THE MASS BALANCE EQUATIONS MUST
84    C       SATISFY THE REQUIRED TOLERANCE.
85    C
86      186 DO 188 I=1,NC
87          IF(DABS(EPS(I,IP))-TOL(I))188,189,189
88      188 CONTINUE
89          GO TO 330
90      189 SØ=S
91          IDIM=1
92    C
93    C       CALCULATION OF THE JACOBIAN.
94    C
95          DO 210 I=1,NC
96          DO 210 J=1,NC
97          A(I,J)=0.
98          DO 210 K=1,NK
99          IF(PQR(J,K))190,210,190
100     190 IF(PQR(I,K))200,210,200
101     200 A(I,J)=A(I,J)+PQR(I,K)*PQR(J,K)*CI(K,IP)
102     210 CONTINUE
103         DO 215 I=1,NC
104     215 A(I,I)=A(I,I)+CX(I)
105     220 DO 221 I=1,NC
106         DO 221 J=1,NC
107     221 B1(I,J)=A(I,J)
108   C
109   C         CALL THE MATRIX INVERSION ROUTINE
110   C
111         CALL INVER(B1,NC,ACCM,IFAIL,MAXCX)
112         IF(IFAIL)230,230,312
113   C
114   C       LOOP FOR THE CALCULATION OF THE RELATIVE SHIFT
115   C
116     230 DO 240 I=1,NC
117         DC(I)=0.
118         DO 240 J=1,NC
119     240 DC(I)=DC(I)+B1(I,J)*EPS(J,IP)
120     250 SCN=0.
121         K=0
122         W=1.
123   C
124   C       CX CONTAINS THE CONCENTRATIONS OF THE FREE SPECIES.  AS THEY
125   C       CANNOT TAKE NEGATIVE VALIES THE STEP LENGTH OF THE CORRECTION
126   C       VECTOR IS REDUCED SO THAT NONE OF THEM WILL BECOME NEGATIVE.
127   C
128         DO 254 I=1,NC
129         IF(DC(I)+0.99)252,254,254
130     252 IF(DC(I)-SCN)253,254,254
131     253 K=I
132         SCN=DC(I)
133     254 CONTINUE
134         IF(K)256,256,255
135     255 W=-0.99/SCN
136     256 DO 257 I=1,NC
137         DC(I)=CX(I)*DC(I)*W
138     257 CX(I)=CX(I)+DC(I)
139         GO TO 110
140     312 IFAIL=0
141         SCN=0.
142         DO 316 I=1,NC
143         DO 316 J=1,NC
144         W=A(I,J)/EPS(I,IP)
145         IF(DABS(SCN)-DABS(W))315,316,316
146     315 SCN=W
147         L=J
148     316 CONTINUE
149         DO 320 I=1,NC
150     320 DC(I)=0.
151         DC(L)=1./SCN
152         GO TO 250
153     330 RETURN
154         END
```

```
 1    C
 2    C          SUBROUTINE INVER
 3    C
 4    C          MATRIX INVERSION ROUTINE
 5    C
 6               SUBROUTINE INVER(B,N,ACCM,IFAIL,MAXCX)
 7               IMPLICIT REAL*8(A-H,O-Z)
 8               DIMENSION B(MAXCX,MAXCX)
 9               DO 104 K=1,N
10               DIS=B(K,K)
11               IF(DABS(DIS)-ACCM)110,110,100
12         100 B(K,K)=1.
13               DO 101 J=1,N
14         101 B(K,J)=B(K,J)/DIS
15               DO 104 I=1,N
16               IF(I-K)102,104,102
17         102 DIS=B(I,K)
18               B(I,K)=0.
19               DO 103 J=1,N
20         103 B(I,J)=B(I,J)-DIS*B(K,J)
21         104 CONTINUE
22               RETURN
23         110 IFAIL=1
24               RETURN
25               END
```

```
 1    C
 2    C          SUBROUTINE MQ
 3    C
 4    C          GAUSS-NEWTON LEAST SQUARES REFINEMENT OF THE UNKNOWN FRE
 5    C          REACTANT CONCENTRATIONS
 6    C
 7               SUBROUTINE MQ(NK,NMBE,NEMF,NC,HLNB,CI,CX,TT,HX,JFAIL,EPS,DEPS,
 8              &B1,B,PQR,KEY,N,S,TOL,A,GC,DC,IP,MAXMB,MAXK,MAXCX,MAXTP,MAXP)
 9               IMPLICIT REAL*8(A-H,O-Z)
10               INTEGER PQR(MAXMB,MAXK)
11               DIMENSION HLNB(MAXK),KEY(MAXK)
12               DIMENSION CX(MAXMB),HX(MAXMB),TT(MAXMB),DEPS(MAXMB)
13               DIMENSION GC(MAXCX),TOL(MAXCX),DC(MAXCX)
14               DIMENSION CI(MAXK,MAXTP)
15               DIMENSION EPS(MAXMB,MAXTP)
16               DIMENSION A(MAXMB,MAXP)
17               DIMENSION B1(MAXCX,MAXCX),B(MAXCX,MAXCX)
18               COMMON JINP,JOUT,AL10,LARS,MAXIT,TOLB,TOLU,ACCM,IPRIN,NRUN,
19              &RELAC,IFAIL
20               ITER=0
21    C
22    C          ITER IS A CYCLE COUNTER.  100 CYCLES ARE PERMITTED AS MAX.
23    C
24               PMARQ=0.
25               S0=1./ACCM
26               DO 100 I=1,NC
27         100 DC(I)=0.
28               DO 101 I=1,NEMF
29               J=I+NC
30         101 HX(J)=DLOG(CX(J))
31         110 CONTINUE
32         116 DO 118 I=1,NC
33               W=CX(I)+DC(I)
34         118 HX(I)=DLOG(W)
35    C
36    C          LOOP FOR THE CALCULATION OF CONCENTRATIONS.
37    C
38               DO 150 I=1,NK
39               IF(KEY(I))150,120,120
40         120 W=HLNB(I)
41               DO 140 J=1,NMBE
42               IF(PQR(J,I))130,140,130
43         130 W=W+PQR(J,I)*HX(J)
44         140 CONTINUE
45               CI(I,IP)=DEXP(W)
46         150 CONTINUE
47               W=0.
48    C
49    C          LOOP FOR THE CALCULATION OF RESIDUALS AND SUM OF SQUARES.
```

```
 50    C
 51          DO 175 I=1,NMBE
 52          DEPS(I)=TT(I)-CX(I)
 53          IF(I-NC)158,158,159
 54     158  DEPS(I)=DEPS(I)-DC(I)
 55     159  DO 170 J=1,NK
 56          IF(PQR(I,J))160,170,160
 57     160  DEPS(I)=DEPS(I)-PQR(I,J)*CI(J,IP)
 58     170  CONTINUE
 59          EPS(I,IP)=DEPS(I)
 60     175  W=W+DEPS(I)*DEPS(I)
 61          S=W
 62          IF(ITER)179,179,1170
 63    1170  GO TO(1171,1175), IRET
 64    1171  IF(DABS(S0-S)/S -0.01*TOLU)330,330,1175
 65    1175  R=(S0-S)/DS
 66          IF(R-0.25)1176,1180,1180
 67    1176  IF(PMARQ)1179,1177,1179
 68    1177  CUTOF=0.
 69          DO 1178 I=1,NC
 70    1178  CUTOF=CUTOF+B1(I,I)*B(I,I)
 71          CUTOF=1./CUTOF
 72          CUTUP=CUTOF*1.E+07
 73          PMARQ=CUTOF
 74    1179  PMARQ=PMARQ*2.
 75          IF(PMARQ.LT.CUTUP)GO TO 176
 76          PMARQ=CUTUP
 77          GO TO 177
 78    1180  IF(R-0.75)176,176,1181
 79    1181  PMARQ=PMARQ*0.5
 80          IF(PMARQ-CUTOF)1182,1182,176
 81    1182  PMARQ=0.
 82     176  IF(S-(1.+TOLU)*S0)177,180,180
 83     177  DO 178 I=1,NC
 84     178  CX(I)=CX(I)+DC(I)
 85     179  ITER=ITER+1
 86          IF(ITER-100)190,190,322
 87     180  DO 183 I=1,NC
 88          DO 182 J=1,NC
 89     182  B1(I,J)=B(I,J)
 90     183  B1(I,I)=B1(I,I)*(1.+PMARQ)
 91          GO TO 248
 92     190  IF(S.LT.S0) S0=S
 93          DO 210 I=1,NMBE
 94          DO 210 J=1,NC
 95          A(I,J)=0.
 96          DO 210 K=1,NK
 97          IF(PQR(J,K))195,210,195
 98     195  IF(PQR(I,K))200,210,200
 99     200  A(I,J)=A(I,J)+PQR(I,K)*PQR(J,K)*CI(K,IP)
100     210  CONTINUE
101          DO 215 I=1,NC
102     215  A(I,I)=A(I,I)+CX(I)
103          DO 225 I=1,NC
104          W=0.
105          DO 220 J=1,NMBE
106     220  W=W+A(J,I)*DEPS(J)
107     225  GC(I)=W
108          DO 235 I=1,NC
109          DO 235 J=I,NC
110          B1(I,J)=0.
111          DO 230 L=1,NMBE
112     230  B1(I,J)=B1(I,J)+A(L,I)*A(L,J)
113     235  B1(J,I)=B1(I,J)
114          DO 239 I=1,NC
115          DO 238 J=1,NC
116     238  B(I,J)=B1(I,J)
117     239  B1(I,I)=B1(I,I)*(1.+PMARQ)
118     248  NOT=0
119    C
120    C          CALL THE MATRIX INVERSION ROUTINE
121    C
122     249  CALL INVER(B1,NC,ACCM,IFAIL,MAXCX)
123          DO 251 I=1,NC
124          DC(I)=0.
125          DO 251 J=1,NC
```

```
126      251 DC(I)=DC(I)+B1(I,J)*GC(J)
127          IRET=2
128      253 DO 255 I=1,NC
129          IF(DABS(DC(I))-0.001*DSQRT(B1(I,I)*S/NEMF))255,255,260
130      255 CONTINUE
131          IRET=1
132      260 SCN=0.
133          W=1.
134          K=0
135          DO 264 I=1,NC
136          IF(DC(I)+0.99)262,264,264
137      262 IF(DC(I)-SCN)263,264,264
138      263 K=I
139          SCN=DC(I)
140      264 CONTINUE
141          IF(K)266,266,265
142      265 W=-0.99/SCN
143      266 DO 267 I=1,NC
144      267 DC(I)=DC(I)*W
145          DS=0.
146          DO 280 I=1,NC
147          DO 278 J=1,NC
148      278 DS=DS-B(I,J)*DC(I)*DC(J)
149      280 DS=DS+2.*DC(I)*GC(I)
150          DO 285 I=1,NC
151      285 DC(I)=CX(I)*DC(I)
152          GO TO 110
153      322 IFAIL=JFAIL
154          RETURN
155      330 DO 331 I=1,NC
156          DO 331 J=1,NC
157      331 B1(I,J)=B(I,J)
158    C
159    C          CALL THE MATRIX INVERSION ROUTINE
160    C
161          CALL INVER(B1,NC,ACCM,IFAIL,MAXCX)
162          IFAIL=0
163          RETURN
164          END
```

```
  1    C
  2    C          SUBROUTINE NEG
  3    C
  4    C          CALCULATED SHIFTS ARE CHECKED TO SEE IF ANY NEGATIVE
  5    C          FORMATION CONSTANT MAY RESULT
  6    C
  7          SUBROUTINE NEG(N,NK,BETAV,PERC,SB,KEY,IVAR,HLNB,JPOT,CI,KONT,
  8         &IBN,NPNC,NP,MAXK,MAXTP)
  9          IMPLICIT REAL*8(A-H,O-Z)
 10          DIMENSION BETAV(MAXK),JPOT(MAXK),IVAR(MAXK),HLNB(MAXK),
 11         &SB(MAXK),KEY(MAXK)
 12          DIMENSION CI(MAXK,MAXTP)
 13          COMMON JINP,JOUT,AL10,LARS,MAXIT,TOLB,TOLU,ACCM,IPRIN,NRUN,
 14         &RELAC,IFAIL
 15        1 FORMAT('0BETA(',I2,')NEGATIVE')
 16          SBN=0.
 17          J=0
 18          DO 30 I=1,N
 19          IF(SB(I)+0.99)10,30,30
 20       10 IF(SB(I)-SBN)20,30,30
 21       20 J=I
 22          SBN=SB(J)
 23       30 CONTINUE
 24          IF(J)120,120,40
 25       40 PERC=-0.99/SBN
 26          IF(KONT)50,50,60
 27       50 IBN=J
 28          GO TO 70
 29       60 IF(IBN-J)130,70,130
 30       70 KONT=KONT+1
 31          IF(KONT-3)140,80,80
 32       80 IV=IVAR(J)
```

```
33    C
34    C         IV IS THE INDEX OF THE PARAMETER TO BE REJECTED.  A NEGATIVE
35    C         VALUE IS ASSIGNED TO ITS REFINEMENT KEY AND THE CONCENTRATION
36    C         OF THE CORRESPONDING COMPLEX IS SET TO ZERO.  THE RELEVANT
37    C         VECTORS ARE THEN COMPACTED.
38    C
39              KEY(IV)=-NRUN
40              DO 90 I=1,NP
41       90 CI(IV,I)=0.
42              IF(J-N)100,105,105
43      100 DO 110 I=J,N
44              IP=I+1
45              IVAR(I)=IVAR(IP)
46              SB(I)=SB(IP)
47      110 BETAV(I)=BETAV(IP)
48      105 WRITE(JOUT,1)IV
49              IFAIL=-1
50      120 KONT=0
51              GO TO 140
52      130 KONT=1
53              IBN=J
54      140 RETURN
55              END
```

```
1     C
2     C         SUBROUTINE MON1
3     C
4     C         OPTIONAL MONITOR OF THE PROGRESS OF MINIMIZATION
5     C
6              SUBROUTINE MON1(N,NK,U,PERC,BETAV,JPOT,SB,SIGMA,IVAR,MAXK)
7              IMPLICIT REAL*8(A-H,O-Z)
8              DIMENSION BETAV(MAXK),JPOT(MAXK),IVAR(MAXK),SIGMA(MAXK),
9             &SB(MAXK)
10             COMMON JINP,JOUT,AL10,LARS,MAXIT,TOLB,TOLU,ACCM,IPRIN,NRUN,
11            &RELAC,IFAIL
12           1 FORMAT(7H0 CYCLE,I4,5H  U=,E14.7/)
13           2 FORMAT(11X,'FRACTION OF STEP LENGTH TAKEN',F10.4/)
14           3 FORMAT(1H ,21X,'   VALUE    REL. SHIFT  STD.DEVN.(REL.PERCENT)')
15           4 FORMAT(17H            BETA(,I2,1H),F10.5,'E',I3,E13.5,F14.2)
16             FRACT=DABS(PERC)
17             WRITE(JOUT,1)NRUN,U
18             IF(FRACT)7,7,5
19    C
20    C         THE FRACTION OF THE STEP LENGTH IS PRINTED OUT IF LESS THAN 0.9.
21    C
22           5 IF(FRACT-0.9)6,7,7
23           6 WRITE(JOUT,2)FRACT
24           7 WRITE(JOUT,3)
25             DO 8 I=1,N
26             K=IVAR(I)
27           8 WRITE(JOUT,4)K,BETAV(I),JPOT(K),SB(I),SIGMA(I)
28             RETURN
29             END
```

```
1     C
2     C         SUBROUTINE STATS
3     C
4     C         A STATISITICAL ANALYSIS OF THE RESIDUALS OF ALL THE MASS
5     C         BALANCE EQUATIONS. FOR FURTHER DETAILS SEE M.R. SPIEGEL,
6     C         THEORY AND PROBLEMS OF STATISTICS, MC GRAW HILL.
7     C
8              SUBROUTINE STATS(NK,NMBEV,PQRV,T,EPS,CI,NP,U,CONC,JPRIN,NT,JP,
9             &JTP,NTC,TITLE,PQR,JS,CS,MAXMB,MAXK,MAXTC,MAXTP)
10             IMPLICIT REAL*8(A-H,O-Z)
11             INTEGER PQR(MAXMB,MAXK),PQRV(MAXMB,MAXK,MAXTC)
12             INTEGER SYM(26),TITLE(18)
13             INTEGER PLUS,BLANK,STAR,PT(102)
14             DIMENSION JPOP(8),CLIM(8),EXFR(8)
15             DIMENSION JTP(MAXTC),NMBEV(MAXTC)
```

```
16              DIMENSION JS(MAXK),CS(MAXK)
17              DIMENSION CI(MAXK,MAXTP)
18              DIMENSION T(MAXMB,MAXTP),CONC(MAXMB,MAXTP),EPS(MAXMB,MAXTP)
19              DIMENSION JP(MAXMB,MAXTC)
20              COMMON JINP,JOUT,AL10,LARS,MAXIT,TOLB,TOLU,ACCM,IPRIN,NRUN,
21             &RELAC,IFAIL
22              DATA PLUS,BLANK,STAR,SYM/'+',' ','*','A','B','C','D','E','F',
23             &'G','H','I','J','K','L','M','N','O','P','Q','R','S','T',
24             &'U','V','W','X','Y','Z'/
25            1 FORMAT ('1STATISTICS ON', 18A4//)
26            2 FORMAT(I5,5X,10E11.3)
27            3 FORMAT(7I5)
28            4 FORMAT(//' R FACTOR=', F12.6)
29            6 FORMAT(I5,E13.4,4X,12F8.2/18X, 8F8.2)
30            7 FORMAT(1H1,' TABLE OF FORMATION PERCENTAGES'///)
31            9 FORMAT(1H1)
32           10 FORMAT(//E15.5,'      ARITHMETIC MEAN'/
33             &E15.5,'      MEAN DEVIATION'/
34             &E15.5,'      STANDARD DEVIATION'/
35             &E15.5,'      VARIANCE'/
36             &E15.5,'      MOMENT COEFFICIENT OF SKEWNESS'/
37             &E15.5,'      MOMENT COEFFICIENT OF KURTOSIS'//)
38           11 FORMAT(//' CONCENTRATIONS AND RESIDUALS TABLE - CURVE',I3/
39             &' COLS 1 -',I2,' TOTAL REACTANT CONCENTRATIONS'/
40             &' COLS',I2,2H -,I2,' RESIDUALS ON MASS BALANCE EQUATIONS'/)
41           12 FORMAT(16X,'CLASS  LIMITS',13X,'PROBABILITY',10X,'FREQUENCY',8X,
42             &'PARTIAL'/12X,'LOWER',10X,'HIGHER', 7X,'CALC', 7X,'OBS', 7X,'CALC',
43             &5X, 'OBS',5X,'CHI-SQUARE'//)
44           17 FORMAT(I5,2E15.5,2F10.4,F10.1,I8,F12.3)
45           18 FORMAT(//' OBSERVED CHI SQUARE IS', F10.2/
46             &' CHI SQUARE(6,0.95)SHOULD BE ', F10.2/)
47           20 FORMAT(//' FORMATION PERCENTAGES RELATIVE TO TOTAL CONCENTRATION
48             &OF REACTANT',I2,' - CURVE',I3)
49           21 FORMAT(I4,116A1)
50           22 FORMAT('1PLOTS OF FORMATION PERCENTAGES')
51           23 FORMAT('0POINT  FREE CONCN. ',12I8/16X, 8I8/)
52           24 FORMAT('1RESIDUALS PLOT'/)
53           25 FORMAT(/5X,A1,2(49X,A1)/5X,A1,10(9X,A1)/1X,109A1)
54           26 FORMAT(/' CURVE',I3)
55              NR=8
56              WRITE(JOUT,1)TITLE
57              JTP1=0
58              NTOT=0
59              DO 100 JTC=1,NTC
60              NTOT=NTOT+NMBEV(JTC)*(JTP(JTC)-JTP1)
61          100 JTP1=JTP(JTC)
62              W=NTOT
63              SD=DSQRT(U/W)
64     C
65     C        SD IS THE STANDARD DEVIATION OF THE FIT.  THE RESIDUALS ARE TO BE
66     C        DIVIDED IN EIGHT CLASSES WITH THE UPPER LIMITS CONTAINED IN
67     C        THE VECTOR CLIM.
68     C
69              CLIM(1)=-1.150*SD
70              CLIM(2)=-0.675*SD
71              CLIM(3)=-0.319*SD
72              CLIM(4)=0.000
73              CLIM(5)=0.319*SD
74              CLIM(6)=0.675*SD
75              CLIM(7)=1.150*SD
76              CLIM(8)=SD/ACCM
77              DO 102 I=1,NR
78          102 JPOP(I)=0
79              AM=0.
80              DM=0.
81              VAR=0.
82              COSQ=0.
83              COKU=0.
84              RDEN=0.
85              JTC=1
86              NMBE=NMBEV(1)
87              DO 112 IP=1,NP
88          104 IF(IP.LE.JTP(JTC))GO TO 4104
89              JTC=JTC+1
90              NMBE=NMBEV(JTC)
91              GO TO 104
```

```
92     4104 DO 110 I=1,NMBE
93          DO 106 K=1,NR
94          IF(EPS(I,IP)-CLIM(K))105,105,106
95      105 JPOP(K)=JPOP(K)+1
96          GO TO 108
97      106 CONTINUE
98          JPOP(NR)=JPOP(NR)+1
99      108 AM=AM+EPS(I,IP)
100         RDEN=RDEN+T(I,IP)*T(I,IP)
101         DM=DM+ DABS(EPS(I,IP))
102         VAR=VAR+EPS(I,IP)**2
103         COSQ=COSQ+EPS(I,IP)**3
104     110 COKU=COKU+EPS(I,IP)**4
105     112 CONTINUE
106         AM=AM/W
107         DM=DM/W
108         VAR=VAR/W
109         COSQ=COSQ/(W*VAR*SD)
110         COKU=COKU/(W*VAR*VAR)
111         WRITE(JOUT,10)AM,DM,SD,VAR,COSQ,COKU
112  C
113  C          COMPUTATION OF THE CHI-SQUARE STATISTIC.
114  C
115         DO 120 I=1,NR
116     120 EXFR(I)=0.125
117         OBSCH=0.
118         WRITE(JOUT,12)
119         R1=-CLIM(NR)
120         DO 130 K=1,NR
121         EXPOP=EXFR(K)*W
122         OBFR=JPOP(K)/W
123         RAPP=((JPOP(K)-EXPOP)**2)/EXPOP
124         WRITE(JOUT,17)K,R1,CLIM(K),EXFR(K),OBFR,EXPOP,JPOP(K),RAPP
125         R1=CLIM(K)
126     130 OBSCH=OBSCH+RAPP
127         EXPCH=12.6
128         WRITE(JOUT,18)OBSCH,EXPCH
129  C
130  C          COMPUTATION OF THE AGREEMENT FACTOR R
131  C
132         RFACT=DSQRT(U/RDEN)
133         WRITE(JOUT,4)RFACT
134         GO TO(201,140,201,140),JPRIN
135  C
136  C          LISTING OF TOTAL CONCENTRATIONS AND RESIDUALS OF ALL POINTS.
137  C
138     140 IP=0
139         WRITE(JOUT,9)
140         DO 200 JTC=1,NTC
141         NMBE=NMBEV(JTC)
142         N1=NMBE+1
143         N2=NMBE+NMBE
144         WRITE(JOUT,11)JTC,NMBE,N1,N2
145     150 IP=IP+1
146         WRITE(JOUT,2)IP,(T(I,IP),I=1,NMBE),(EPS(I,IP),I=1,NMBE)
147         IF(IP.LT.JTP(JTC))GO TO 150
148     200 CONTINUE
149     201 GO TO(350,350,202,202),JPRIN
150  C
151  C          GRAPH OF RESIDUALS FOR ALL POINTS
152  C
153     202 WRITE(JOUT,24)
154         IP=0
155         DO 260 JTC=1,NTC
156         WRITE(JOUT,26)JTC
157         NMBE=NMBEV(JTC)
158     204 IP=IP+1
159         DO 210 I=1,92
160     210 PT(I)=BLANK
161         DO 220 I=2,92,15
162     220 PT(I)=PLUS
163         DO 250 I=1,NMBE
164         X=EPS(I,IP)/SD*15.+47.5
165         J=IDINT(X)
166         IF(J-1)230,230,235
167     230 J=2
168         GO TO 250
```

```
169        235 IF(J-92)250,250,240
170        240 J=92
171        250 PT(J)=SYM(I)
172            WRITE(JOUT,21)IP,(PT(I),I=1,92)
173            IF(IP.LT.JTP(JTC))GO TO 204
174        260 CONTINUE
175        350 GO TO(500,351,366,351),JPRIN
176     C
177     C       LISTING OF ALL FREE REACTANT CONCENTRATIONS AND PERCENTAGES
178     C       OF FORMATION FOR ALL POINTS.
179     C
180        351 WRITE(JOUT,7)
181            DO 365 I=1,NT
182            IP=0
183            DO 360 JTC=1,NTC
184            JT=JP(I,JTC)
185            IF(JT .NE.0)GO TO 4511
186            IP=JTP(JTC)
187            GO TO 360
188       4511 NMBE=NMBEV(JTC)
189            DO 3511 J=1,NK
190            DO 3511 K=1,NMBE
191       3511 PQR(K,J)=PQRV(K,J,JTC)
192            WRITE(JOUT,20)JT,JTC
193            NS=0
194            DO 353 J=1,NK
195            IF(PQR(JT,J))352,353,352
196        352 NS=NS+1
197            JS(NS)=J
198        353 CONTINUE
199            WRITE(JOUT,23)(JS(J),J=1,NS)
200        354 IP=IP+1
201            TS=T(JT,IP)-EPS(JT,IP)
202            DO 357 J=1,NS
203            JJ=JS(J)
204        357 CS(J)=CI(JJ,IP)*PQR(JT,JJ)*100./TS
205            WRITE(JOUT,6)IP,CONC(JT,IP),(CS(J),J=1,NS)
206            IF(IP.LT.JTP(JTC))GO TO 354
207        360 CONTINUE
208        365 CONTINUE
209            GO TO(500,500,366,366),JPRIN
210     C
211     C       GRAPH OF PERCENTAGES OF FORMATION FOR ALL POINTS
212     C
213        366 WRITE(JOUT,22)
214            DO 480 I=1,NT
215            IP=0
216            DO 480 JTC=1,NTC
217            JT=JP(I,JTC)
218            IF(JT.NE.0)GO TO 4661
219            IP=JTP(JTC)
220            GO TO 480
221       4661 NMBE=NMBEV(JTC)
222            DO 3661 J=1,NK
223            DO 3661 K=1,NMBE
224       3661 PQR(K,J)=PQRV(K,J,JTC)
225            WRITE(JOUT,20)JT,JTC
226            WRITE(JOUT,25)(PLUS,K=1,123)
227        400 IP=IP+1
228            TS=T(JT,IP)-EPS(JT,IP)
229            DO 440 K=1,102
230        440 PT(K)=BLANK
231            PT(2)=PLUS
232            PT(102)=PLUS
233            DO 460 J=1,NK
234            X=CI(J,IP)*PQR(JT,J)*100./TS+0.5
235            K=IDINT(X)
236            IF(K-100)450,457,445
237        445 K=100
238        450 IF(K)460,460,455
239        455 K=K+2
240            IF(PT(K)-BLANK)456,458,456
241        456 PT(K)=STAR
242            GO TO 460
243        457 K=102
244        458 PT(K)=SYM(J)
245        460 CONTINUE
```

```
246          WRITE(JOUT,21)IP,PT
247          IF(IP.LT.JTP(JTC))GO TO 400
248      480 CONTINUE
249      500 RETURN
250          END
```

5.1.1. Input Data

The following is a listing of a data set obtained from the nickel(II)-1,1,1-tris(aminomethyl) methane[6] system that has been processed by MINIQUAD.

```
¶<--  Column 1                                    0.80 -  54.2

  NICKEL(II)-TAME SYSTEM : "BEST" MODEL          0.84 -  62.4
    1    8    0   10    0    3    0
                                                 0.88 -  72.2
   25.     0.002      0.2
                                                 0.93 -  87.2
  1.4        10   1    0    1    1
                                                 0.96 -  97.8
  2.5        18   1    0    2    1
                                                 1.00 -111.9
  1.8        24   1    0    3    1
                                                 1.04 -123.7
  1.8       -14   0    0   -1    1
                                                 1.08 -133.6
  1.2        10   1    1    0    1
                                                 1.12 -141.8
  2.0        17   2    1    0    1
                                                 1.16 -149.4
  0.3        16   1    1    1    1
                                                 1.20 -156.3
  1.5        24   2    1    1    1
                                                 1.24 -162.5
    2    1    3    1
                                                 1.28 -168.9
    1    0
                                                 1.32 -174.6
  0.12109   0.42549
                                                 1.36 -180.8
  0.        -.18828
                                                 1.40 -187.0
  316.5   0.      0.
                                                 1.44 -193.2
  93.10
                                                 1.48 -200.0
           0.32    57.4
                                                 1.52 -206.9
           0.37    30.0
                                                 1.56 -214.6
           0.40    18.8
                                                 1.60 -222.7
           0.44     8.2
                                                 1.64 -230.5
           0.48 -   0.9
                                                 1.68 -238.0
           0.52 -   8.3
                                                 1.72 -244.7
           0.56 -  14.9
                                                 1.76 -251.1
           0.61 -  22.8
                                                 1.80 -256.6
           0.64 -  27.5
                                                 1.84 -261.8
           0.68 -  33.7
                                                 1.88 -266.4
           0.72 -  39.9
                                                 1.92 -270.5
           0.76 -  46.7
```

	1.96	-274.7	0.80 - 31.5
	2.00	-278.5	0.84 - 36.8
	2.04	-282.1	0.88 - 42.2
	2.08	-285.6	0.92 - 47.8
	2.12	-288.7	0.96 - 54.0
	2.16	-291.7	1.00 - 60.8
	2.20	-294.7	1.04 - 68.4
	2.24	-297.4	1.08 - 77.6
	2.28	-300.1	1.12 - 88.0
	2.32	-302.6	1.16 - 99.5
	2.36	-304.9	1.20 -111.5
	2.40	-307.1	1.24 -121.8
	2.44	-309.3	1.28 -130.9
	2.48	-311.3	1.32 -138.3
	2.52	-313.2	1.36 -144.9
	2.57	-315.7	1.41 -152.4
	2.60	-316.9	1.44 -156.3
	2.64	-318.6	1.48 -161.4
	2.68	-320.2	1.52 -166.3
	2.72	-321.7	1.56 -171.3
	2.76	-323.3	1.60 -176.3
	2.80	-324.6	1.64 -181.3
1	2.84	-326.0	1.68 -186.4
2	1 3	1	1.72 -191.4
1	0		1.76 -197.1
0.14553	0.51138		1.80 -203.0
0.	-.18828		1.84 -209.1
315.92	0.	0.	1.88 -215.3
101.65			1.92 -221.7
	0.38	59.4	1.96 -228.3
	0.42	40.3	2.00 -234.8
	0.46	25.1	2.04 -240.8
	0.50	14.3	2.08 -246.5
	0.54	6.2	2.12 -251.8
	0.60 -	4.2	2.16 -256.6
	0.64 -	10.4	2.20 -261.2
	0.68 -	15.8	2.25 -266.3
	0.72 -	21.1	2.30 -270.9
	0.76 -	26.4	2.35 -275.4

```
      2.40 -279.4                   0.96 - 16.3
      2.45 -283.4                   1.03 - 19.1
      2.50 -286.9                   1.10 - 21.8
      2.55 -290.5                   1.20 - 25.8
      2.60 -293.6                   1.30 - 29.8
      2.65 -296.9                   1.47 - 37.0
      2.70 -299.8                   1.60 - 43.1
      2.76 -303.0                   1.70 - 48.9
      2.80 -305.1                   1.80 - 55.9
      2.85 -307.5                   1.90 - 65.2
      2.90 -310.0                   1.98 - 75.8
      2.95 -312.2                   2.04 - 86.7
      3.00 -314.4                   2.08 - 95.2
      3.05 -316.4                   2.12 -103.8
      3.10 -318.4                   2.16 -112.5
      3.14 -319.8                   2.21 -126.6
      3.18 -321.2            1      2.26 -162.7
      3.22 -322.7            3   1   2   3   1   2
      3.26 -323.9            1   0
      3.31 -325.6        0.12273   0.05931   0.43176
      3.36 -327.0        0.        0.        -.18710
      3.41 -328.5        315.45    0.        0.
1     3.45 -329.6        100.41
3   1   2   3   1   2             0.40     27.5
1   0                            0.44     18.0
0.12119   0.10025   0.42667       0.48     10.5
0.        0.        -.18710       0.52      5.7
315.99    0.        0.            0.56      1.5
102.82                            0.64 -  5.2
          0.38     33.6           0.72 - 10.1
          0.41     25.2           0.80 - 14.7
          0.44     19.1           0.90 - 20.2
          0.48     12.7           1.00 - 24.9
          0.52      8.5           1.10 - 30.0
          0.60      1.7           1.20 - 35.7
          0.68 -  3.4             1.30 - 42.0
          0.77 -  8.2             1.40 - 49.9
          0.85 - 11.8             1.50 - 60.9
```

```
        1.60 -  76.7                      1.13 -123.4
        1.70 -  92.8                      1.18 -140.0
        1.80 -105.5                       1.23 -176.2
        1.90 -116.4                       1.28 -265.4
        2.00 -127.9                       1.30 -275.8
        2.10 -142.0                       1.32 -284.0
        2.14 -149.3                       1.34 -290.8
        2.18 -158.6                       1.38 -299.8
        2.22 -171.4                       1.41 -304.9
        2.24 -191.1                       1.44 -309.5
        2.28 -220.9                       1.49 -315.1
  1     2.32 -246.8                       1.52 -318.0
  3  1   2   3   1   2                    1.55 -320.6
  1  0                                    1.58 -323.0
0.12164  0.07479  0.42805                 1.62 -325.8
0.       0.       -.34365       -1        1.66 -328.5
314.27   0.       0.             1
102.11                           1
        0.23    22.5      NICKEL(II)-TAME SYSTEM : FIRST THREE COMPLEX SPECIES MODEL
        0.25    14.7         1    7   0   10   1   3   0
        0.27     9.1        25.     0.002      0.2
        0.31     1.6        1.4     10   1   0   1   1
        0.36 -   5.6        2.5     18   1   0   2   1
        0.41 -  11.2        1.8     24   1   0   3   1
        0.46 -  15.9        1.8    -14   0   0  -1   1
        0.51 -  20.1        1.2     10   1   1   0   1
        0.55 -  23.4        0.3     16   1   1   1   1
        0.61 -  28.1        1.      21   1   1   2   1
        0.68 -  34.1         1
        0.73 -  38.5         1
        0.77 -  42.9      NICKEL(II)-TAME SYSTEM : SECOND THREE COMPLEX SPECIES MODEL
        0.81 -  47.4         1    7   0   10   1   3   0
        0.86 -  54.8        25.     0.002      0.2
        0.91 -  64.6        1.4     10   1   0   1   1
        0.94 -  72.8        2.5     18   1   0   2   1
        0.98 -  85.2        1.8     24   1   0   3   1
        1.04 -102.2         1.8    -14   0   0  -1   1
        1.08 -111.3         1.2     10   1   1   0   1
```

```
2.0       17   2   1   0   1
0.3       16   1   1   1   1
 1
 1
```

NICKEL(II)-TAME SYSTEM : ANOTHER
FOUR COMPLEX SPECIES MODEL

```
 1   8   0   10   1   3   0
25.    0.002     0.2
1.4       10   1   0   1   1
2.5       18   1   0   2   1
1.8       24   1   0   3   1
1.8      -14   0   0  -1   1
1.2       10   1   1   0   1
2.0       17   2   1   0   1
0.3       16   1   1   1   1
1.        31   2   1   2   1
 1
 1
```

NICKEL(II)-TAME SYSTEM : FIVE
COMPLEX SPECIES MODEL

```
 1   9   0   10   1   3   0
25.    0.002     0.2
1.4       10   1   0   1   1
2.5       18   1   0   2   1
1.8       24   1   0   3   1
1.8      -14   0   0  -1   1
1.2       10   1   1   0   1
2.0       17   2   1   0   1
0.3       16   1   1   1   1
1.5       24   2   1   1   1
1.        31   2   1   2   1
 1
-1
```

5.2. Listing of MIQUV

The following is a listing of the program MIQUV.

```
 1    C   * MIQUV 82 *
 2    C
 3    C       PROGRAM FOR THE CALCULATION OF CUMULATIVE FORMATION CONSTANTS
 4    C       FROM POTENTIOMETRIC DATA.
 5    C
 6    C       AUTHORS:  A. SABATINI AND A. VACCA
 7    C       LATEST UPDATE VERSION:  JAN. 1982
 8    C
 9            IMPLICIT REAL*8(A-H,O-Z)
10            INTEGER PQR(4,16),PQRØ(4,16),PQRV(4,16,15)
11            INTEGER TITLE(18)
12            DIMENSION TITRE(400)
13            DIMENSION PARAM(42),PARI(42),IVAR(42),JTYP(42),JCURV(42)
14            DIMENSION PARØ(22),RESH(22),SD(22)
15            DIMENSION C(22),DX(22)
16            DIMENSION VINIT(15),JTP(15),AJ(10),BJ(15),NEMFV(15)
17            DIMENSION NMBEV(15),SIG(15),JCOUL(15)
18            DIMENSION BETA(16),JPOT(16),HLNB(16),CI(16),KNEG(16),KEY(16)
19            DIMENSION JS(16),CS(16)
20            DIMENSION TT(4),CX(4),HX(4),DT(4),DCX(4),TOLTC(4)
21            DIMENSION SIGMAE(2),DW(2)
22            DIMENSION H(101),X(101),Y(101),Y2(101)
23            DIMENSION DM(600,22)
24            DIMENSION FREEC(400,4)
25            DIMENSION EMF(400,2),EMFC(400,2),EMFD(400,2),WS(400,2)
26            DIMENSION DY(400,2)
27            DIMENSION B(22,22)
28            DIMENSION KCONS(20,2)
29            DIMENSION TOTC(15,4),ADDC(15,4)
```

```
30            DIMENSION EZERO(15,2),JEL(15,2)
31            DIMENSION JNMB(4,15),JP(4,15)
32            DIMENSION DMC(4,4)
33            COMMON RTF,TOL,ACCM,DLNAM,AL1Ø
34            COMMON NK,NKV,N,NSP,NCONS,NPAR,NE,NTC,NMBEØ,NKW,NSET,
35         &LARS,MAXIT,IPRIN,JINP,JOUT
36       1  FORMAT(I5)
37       2  FORMAT(/1ØX,'OLD PARAMETERS   REL.SHIFTS   NEW PARAMETERS',
38         &'   REL.STD.DEV.')
39       3  FORMAT(' (',I2,')',4E15.5)
40       4  FORMAT(I5,2X,4E15.5)
41       5  FORMAT(//' CYCLE',I3,4X,'SUM OF SQUARES =',E12.6,4X,
42         &'STANDARD DEVIATION =',E1Ø.4)
43       6  FORMAT(' SINGULAR MATRIX FOUND IN SOLVING NORMAL EQUATIONS'/
44         &' CALCULATION ABANDONED')
45    C
46    C          MAXIMUM VALUES FOR ARRAY DIMENSIONS
47    C      NOTE: THIS PARTICULAR VERSION CAN HANDLE UP TO:
48    C      1) 6ØØ EMF VALUES, MAXEMF.
49    C      2) 4ØØ TITRATION POINTS, MAXTP.
50    C      3) 1Ø1 POINTS PER TITRATION CURVE, MAXSP.
51    C      4) 42 REFINABLE PARAMETERS, MAXPAR.
52    C      5) 22 INDEPENDENT REFINABLE PARAMETERS, MAXPA.
53    C      6) 2Ø CONSTRAINTS (EQUALITY RELATIONSHIPS) BETWEEN PAIRS OF
54    C            PARAMETERS, MAXCON.
55    C      7) 15 TITRATION CURVES, MAXTC.
56    C      8) 16 FORMATION CONSTANTS, MAXK.
57    C      9) 4 INDEPENDENT REACTANTS, MAXCX.
58    C      1Ø) 2 POTENTIOMETRIC ELECTRODES, SIMULTANEOUS, MAXEL.
59    C
60    C      EXPANSION OF THESE LIMITS MAY BE ACHIEVED BY SUITABLE
61    C      ENLARGEMENT OF THE CORRESPONDING ARRAYS AND CHANGING THE
62    C      APPROPRIATE STATEMENTS.
63    C
64            MAXEMF=6ØØ
65            MAXTP=4ØØ
66            MAXSP=1Ø1
67            MAXPAR=42
68            MAXPA=22
69            MAXCON=2Ø
70            MAXTC=15
71            MAXK=16
72            MAXCX=4
73            MAXEL=2
74    C
75    C          DEFINITION OF I/O CHANNELS, TOLERANCES AND ACCURACIES
76    C
77            JINP=5
78            JOUT=6
79            TOL=Ø.ØØØ1
80            TOLU=Ø.ØØØ1
81            TOLB=Ø.1
82            ACCM=1.D-37
83            DLNAM=DLOG(ACCM)
84            AL1Ø=DLOG(1Ø.D+ØØ)
85            NSET=Ø
86    C
87    C          CALL THE DATA INPUT ROUTINE
88    C
89      100 CALL DINS(TITRE,IVAR,JTP,NEMFV,NMBEV,VINIT,AJ,BJ,BETA,JPOT,
90         &JEL,H,X,Y,Y2,EMF,WS,KCONS,TOTC,ADDC,EZERO,JNMB,JP,PQRØ,PQRV,
91         &SIGMAE,KEY,TITLE,KNEG,JCOUL,MAXCX,MAXK,MAXTC,MAXTP,MAXPAR,
92         &MAXEL,MAXSP,MAXCON,MAXPA,MAXEMF)
93            DO 1Ø5 I=1,NPAR
94      105 RESH(I)=Ø.ØØ2
95            INEG=Ø
96            IEXIT=Ø
97            NCICL=Ø
98            UØP=1./ACCM
99    C
100   C          CALL THE ROUTINE FOR SPECIAL NAMES TO PARAMETERS CONVERSION
101   C
102     120 CALL SNTPV(NMBEV,NEMFV,IVAR,PARAM,BETA,TOTC,ADDC,EZERO,VINIT,
103        &AJ,BJ,JTYP,JCURV,KEY,MAXPAR,MAXK,MAXTC,MAXEL,MAXCX)
104            DO 121 I=1,NPAR
105     121 PARI(I)=PARAM(I)
```

```
106              GO TO 124
107    C
108    C          CALL THE ROUTINE FOR PARAMETERS TO SPECIAL NAMES CONVERSION
109    C
110        123 CALL PTSNV(NMBEV,NEMFV,IVAR,PARAM,BETA,TOTC,ADDC,EZERO,VINIT,
111           &AJ,BJ,KCONS,KEY,MAXPAR,MAXK,MAXTC,MAXEL,MAXCX,MAXCON)
112    C
113    C          CALL THE ROUTINE FOR THE CALCULATION OF E.M.F.VALUES AND
114    C          OF ERROR SQUARE SUM
115    C
116        124 CALL FUNV (JTP,HLNB,BETA,JPOT,VINIT,TITRE,TOTC,ADDC,EZERO,JEL,
117           &EMF,EMFC,WS,AJ,BJ,TT,DY,PQR,PQRV,FREEC,CX,HX,DT,DMC,DCX,CI,
118           &TOLTC,U,NCICL,NEMFV,NMBEV,KEY,SIGMA,SIG,JCOUL,MAXCX,MAXK,MAXTC,
119           &MAXEL,MAXTP)
120            IF(IPRIN.EQ.0)GO TO 125
121            WRITE(JOUT,5)NCICL,U,SIGMA
122        125 IF(U0P.GT.U)GO TO 127
123            DO 126 I=1,NPAR
124        126 PARAM(I)=(PARAM(I)+PAR0(I))*0.5
125            GO TO 123
126        127 IF(IEXIT.EQ.0)GO TO 128
127            IF(INEG.EQ.1)GO TO 128
128    C
129    C          SECOND CONVERGENCE CRITERION: RELATIVE DECREASE OF U
130    C          LESS THAN TOLU
131    C
132            IF((U0-U)/U0.LT.TOLU)GO TO 480
133        128 IF(NCICL.EQ.MAXIT)GO TO 480
134            U0=U
135            NCICL=NCICL+1
136            DO 150 I=1,NPAR
137        150 PAR0(I)=PARAM(I)
138    C
139    C          CONSTRUCTION OF THE DESIGN MATRIX DM
140    C
141            DO 200 JC=1,NPAR
142            PARAM(JC)=PAR0(JC)*(1.+RESH(JC))
143    C
144    C          CALL THE ROUTINE FOR PARAMETERS TO SPECIAL NAMES CONVERSION
145    C
146            CALL PTSNV(NMBEV,NEMFV,IVAR,PARAM,BETA,TOTC,ADDC,EZERO,VINIT,
147           &AJ,BJ,KCONS,KEY,MAXPAR,MAXK,MAXTC,MAXEL,MAXCX,MAXCON)
148    C
149    C          CALL THE ROUTINE FOR THE CALCULATION OF E.M.F.VALUES
150    C          USING THE INCREMENTED VALUE OF ONE PARAMETER(PARAM(JC))
151    C
152            CALL FUNZ (JTP,HLNB,BETA,JPOT,VINIT,TITRE,TOTC,ADDC,EZERO,JEL,
153           &EMFC,EMFD,JTYP,AJ,BJ,TT,PQR,PQRV,JC,FREEC,CX,HX,DT,DMC,DCX,CI,
154           &TOLTC,JCURV,KCONS,NEMFV,NMBEV,KEY,JCOUL,MAXCX,MAXK,MAXTC,MAXPAR,
155           &MAXEL,MAXTP,MAXCON)
156            K=0
157            IPF=0
158            DO 160 IC=1,NTC
159            IPI=IPF+1
160            IPF=JTP(IC)
161            NEMF=NEMFV(IC)
162            DO 160 J=1,NEMF
163            DO 160 I=IPI,IPF
164            K=K+1
165        160 DM(K,JC)=(EMFD(I,J)-EMFC(I,J))/RESH(JC)
166        200 PARAM(JC)=PAR0(JC)
167    C
168    C          CONSTRUCTION OF THE NORMAL EQUATIONS
169    C
170            DO 410 K=1,NPAR
171            C(K)=0.
172            DO 408 L=1,NPAR
173            B(K,L)=0.
174            M=0
175            IPF=0
176            DO 408 IC=1,NTC
177            IPI=IPF+1
178            IPF=JTP(IC)
179            NEMF=NEMFV(IC)
180            DO 408 J=1,NEMF
181            DO 408 I=IPI,IPF
```

```
182              M=M+1
183          408 B(K,L)=B(K,L)+DM(M,K)*DM(M,L)*WS(I,J)
184              M=0
185              IPF=0
186              DO 410 IC=1,NTC
187              IPI=IPF+1
188              IPF=JTP(IC)
189              NEMF=NEMFV(IC)
190              DO 410 J=1,NEMF
191              DO 410 I=IPI,IPF
192              M=M+1
193          410 C(K)=C(K)+DM(M,K)*DY(I,J)*WS(I,J)
194  C
195  C              CALL THE MATRIX INVERSION ROUTINE
196  C
197              CALL MATIN(B,NPAR,DET,MAXPA,JOUT,ACCM)
198  C
199  C              COMPUTATION OF THE RELATIVE SHIFTS
200  C
201              DO 430 I=1,NPAR
202              DX(I)=0.
203              DO 420 L=1,NPAR
204          420 DX(I)=DX(I)+B(I,L)*C(L)
205              SD(I)=DSQRT(DABS(B(I,I)*SIGMA))
206          430 RESH(I)=.02*SD(I)
207              IEXIT=0
208              INEG=0
209              FRACT=1.D+00
210  C
211  C              CHECK FOR NEGATIVE FORMATION CONSTANTS
212  C
213              DO 460 I=1,NKV
214              IF(1.D+00+DX(I))440,440,460
215          440 FRAC=-0.99D+00/DX(I)
216              K=I
217              DO 450 J=1,NPAR
218          450 DX(J)=FRAC*DX(J)
219              FRACT=FRACT*FRAC
220          460 CONTINUE
221              IF(FRACT.EQ.1.D+00)GO TO 468
222  C
223  C              CALL THE ROUTINE FOR EVENTUAL ELIMINATION OF THE
224  C              NEGATIVE FORMATION CONSTANT
225  C
226              CALL BETNEG(K,FRACT,PARAM,PAR0,PARI,KCONS,RESH,DX,SD,KEY,
227             &KNEG,JTYP,JCURV,MAXPAR,MAXPA,MAXK,MAXCON)
228              INEG=1
229          468 IF(IPRIN.GT.0)WRITE(JOUT,2)
230              J=0
231  C
232  C              THE SHIFTS ARE APPLIED TO THE PARAMETERS
233  C
234              DO 471 I=1,NPAR
235              PARAM(I)=PARAM(I)*(1.D+00+DX(I))
236              IF(IPRIN.EQ.0)GO TO 471
237              IF(I.GT.NKV)GO TO 470
238          469 J=J+1
239              IF(KEY(J).NE.1)GO TO 469
240              WRITE(JOUT,3)J,PAR0(I),DX(I),PARAM(I),SD(I)
241              GO TO 471
242          470 J=IVAR(I-NKV)
243              WRITE(JOUT,4)J,PAR0(I),DX(I),PARAM(I),SD(I)
244          471 CONTINUE
245              U0P=U0*(1.+TOL)
246              IF(INEG.EQ.1)GO TO 123
247  C
248  C              FIRST CONVERGENCE CRITERION: ALL RATIOS BETWEEN SHIFTS
249  C              AND STANDARD DEVIATIONS LESS THAN TOLB
250  C
251              DO 475 I=1,NPAR
252              IF(DABS(DX(I)/SD(I)).GT.TOLB)GO TO 123
253          475 CONTINUE
254              IEXIT =1
255              GO TO 123
256  C
257  C              CALL THE DATA OUTPUT ROUTINE
```

```
258   C
259     480 CALL DOUS(IEXIT,KEY,HLNB,BETA,JPOT,PARAM,SD,PARØ,DX,U,UØ,NCICL,
260         &TITLE,PQRØ,IVAR,PARI,SIGMA,JTP,NEMFV,DW,DY,WS,TITRE,EMF,EMFC,
261         &SIG,B,MAXCX,MAXK,MAXTC,MAXEL,MAXPAR,MAXPA,MAXTP)
262   C
263   C       JPRIN IS USED TO CONTROL THE SUBSEQUENT OUTPUT OF THE PROGRAM.
264   C
265           READ(JINP,1)JPRIN
266           IF(MAXIT.EQ.Ø.AND.JPRIN.NE.Ø)GO TO 485
267           IF(IEXIT.EQ.Ø.OR.JPRIN.EQ.Ø)GO TO 490
268   C
269   C       CALL THE STATISTICAL ANALISYS ROUTINE
270   C
271     485 CALL STANS(TITLE,JTP,DY,WS,EMF,NEMFV,NMBEV,PQRV,PQR,TOTC,ADDC,
272         &VINIT,TITRE,KEY,HLNB,BETA,JPOT,CX,HX,CI,JP,FREEC,U,JPRIN,JCOUL,
273         &JS,CS,MAXCX,MAXK,MAXTC,MAXTP,MAXEL)
274   C
275   C       NSET IS USED FOR THE TERMINATION OF THE JOB OR THE
276   C       CONTINUATION OF THE JOB ON THE SAME OR A NEW SET OF DATA.
277   C
278     490 READ(JINP,1)NSET
279           IF(NSET)500,100,100
280     500 STOP
281           END
282   C
283   C     SUBROUTINE BETNEG
284   C         THE NEGATIVE FORMATION CONSTANT IS ELIMINATED IF IT
285   C         RESULTED NEGATIVE IN THE LAST FOUR REFINEMENT CYCLES
286   C
287           SUBROUTINE BETNEG(K,FRACT,PARAM,PARØ,PARI,KCONS,RESH,DX,SD,
288         &KEY,KNEG,JTYP,JCURV,MAXPAR,MAXPA,MAXK,MAXCON)
289           IMPLICIT REAL*8(A-H,O-Z)
290           DIMENSION PARAM(MAXPAR),PARI(MAXPAR),JCURV(MAXPAR)
291           DIMENSION JTYP(MAXPAR)
292           DIMENSION RESH(MAXPA),PARØ(MAXPA),SD(MAXPA),DX(MAXPA)
293           DIMENSION KNEG(MAXK),KEY(MAXK)
294           DIMENSION KCONS(MAXCON,2)
295           COMMON RTF,TOL,ACCM,DLNAM,AL1Ø
296           COMMON NK,NKV,N,NSP,NCONS,NPAR,NE,NTC,NMBEØ,NKW,NSET,
297         &LARS,MAXIT,IPRIN,JINP,JOUT
298       1 FORMAT(/' BETA',I3,' NEGATIVE - FRACTION OF STEP LENGTH TAKEN'
299         &,F7.5)
300       2 FORMAT(/' BETA',I3,' NEGATIVE',I5,' FORMATION CONSTANTS TO ',
301         &'BE REFINED')
302       3 FORMAT(' NO MORE PARAMETERS TO BE REFINED')
303       4 FORMAT(' IONIC PRODUCT OF WATER NEGATIVE')
304       5 FORMAT(' REFINEMENT ABANDONED')
305           L=Ø
306           DO 70 I=1,NK
307           IF(KEY(I).LT.1)GO TO 70
308           L=L+1
309           IF(L.NE.K)GO TO 70
310           KNEG(I)=KNEG(I)+1
311           DO 10 J=1,NK
312           IF(I.EQ.J)GO TO 10
313           KNEG(J)=Ø
314      10 CONTINUE
315           IF(KNEG(I).EQ.4)GO TO 20
316           IF(IPRIN.GT.Ø)WRITE(JOUT,1)I,FRACT
317           RETURN
318      20 KEY(I)=-1
319           KNEG(I)=Ø
320           NKV=NKV-1
321           NPAR=NPAR-1
322           IF(NKW.NE.I)GO TO 25
323           WRITE(JOUT,4)
324           WRITE(JOUT,5)
325           STOP
326      25 IF(NKW.GT.I)NKW=NKW-1
327           IF(NPAR.GT.Ø)GO TO 30
328           WRITE(JOUT,3)
329           WRITE(JOUT,5)
330           STOP
331      30 DO 40 J=K,NPAR
332           J1=J+1
333           PARØ(J)=PARØ(J1)
```

```
334          RESH(J)=RESH(J1)
335          DX(J)=DX(J1)
336       40 SD(J)=SD(J1)
337          DO 50 J=1,NCONS
338          DO 50 L=1,2
339       50 KCONS(J,L)=KCONS(J,L)-1
340          NN=N+NKV
341          DO 60 J=K,NN
342          J1=J+1
343          PARAM(J)=PARAM(J1)
344          PARI(J)=PARI(J1)
345          JTYP(J)=JTYP(J1)
346       60 JCURV(J)=JCURV(J1)
347          IF(IPRIN.GT.Ø)WRITE(JOUT,2)I,NKV
348          RETURN
349       70 CONTINUE
350          RETURN
351          END
```

```
1     C
2     C     SUBROUTINE SPLINE
3     C
4     C        CUBIC SPLINE INTERPOLATING ROUTINE
5     C        TAKEN WITH MODIFICATIONS FROM R.G.STANTON AND W.D.HOSKINS,
6     C        'NUMERICAL ANALYSIS', IN 'PHYSICAL CHEMISTRY, AN ADVANCED
7     C        TREATISE',  ACADEMIC PRESS INC.(1975)
8     C
9            SUBROUTINE SPLINE(H,X,Y,Y2,N,P)
10           IMPLICIT REAL*8(A-H,O-Z)
11           DIMENSION P(6),X(N),Y(N),Y2(N),H(N),A(101),C(101),D(101)
12           M=N-1
13           DO 6 J=1,M
14         6 H(J)=X(J+1)-X(J)
15           M=M-1
16           DO 7 J=1,M
17           Y2(J+1)=(H(J+1)+H(J))/3.
18           A(J)=H(J)/6.
19           C(J)=A(J)
20         7 D(J+1)=(Y(J+2)-Y(J+1))/H(J+1)-(Y(J+1)-Y(J))/H(J)
21           M=M+1
22           Y2(1)=P(2)-H(1)*P(1)/3.
23           C(1)=-P(1)*C(1)
24           D(1)=P(3)-P(1)*(Y(2)-Y(1))/H(1)
25           D(N)=P(6)+P(4)*(Y(M)-Y(N))/H(M)
26           Y2(N)=P(5)+H(M)*P(4)/3.
27           C(M)=H(M)/6.
28           A(M)=P(4)*C(M)
29           DO 1 K=2,N
30           I=N+2-K
31           J=I-1
32           Z=C(J)/Y2(I)
33           Y2(J)=Y2(J)-Z*A(J)
34         1 D(J)=D(J)-Z*D(I)
35           Y2(1)=D(1)/Y2(1)
36           DO 2 I=2,N
37         2 Y2(I)=(D(I)-A(I-1)*Y2(I-1))/Y2(I)
38           RETURN
39           END
```

```
1     C
2     C     SUBROUTINE CCFR
3     C        THE CONCENTRATIONS OF FREE REACTANTS AT EQUILIBRIUM ARE
4     C        CALCULATED BY SOLVING SIMULTANEOUS MASS BALANCE EQUATIONS
5     C
6            SUBROUTINE CCFR(NX,NK,NITER,JOUT,CX,TC,HX,EPS,DM,DX,HLNB,CI,PQR
7          &,TOLTC,TOL,NKMAX,NXMAX,ACCM,DLNAM)
8            IMPLICIT REAL*8(A-H,O-Z)
9            INTEGER PQR(NXMAX,NKMAX)
10           DIMENSION CX(NXMAX),TC(NXMAX),HX(NXMAX),EPS(NXMAX),DX(NXMAX),
```

```
11          &DM(NXMAX,NXMAX),TOLTC(NXMAX),HLNB(NKMAX),CI(NKMAX)
12        1 FORMAT(I4,   ' ITERATIONS PERFORMED IN CCFR: CALCULATION CONTINUES
13          &ON NEXT EXPERIMENTAL POINT'/)
14        2 FORMAT(' SINGULAR MATRIX FOUND IN SUBROUTINE CCFR'/
15          &' FAILURE TO CALCULATE FREE REACTANT CONCENTRATIONS'/
16          &' CALCULATION ABANDONED')
17          ITMAX=NITER
18          NITER=0
19          TOL1=TOL+1.D+00
20          JCK=2
21          DO 90 IX=1,NX
22          W=DABS(TC(IX))
23          IF(CX(IX).LE.ACCM)CX(IX)=W*1.D-03+1.D-07
24          IF(W.LT.1.D-08)TC(IX)=1.D-08
25          TOLT=W*TOL
26          IF(TOLT.LT.1.D-08)TOLT=1.D-08
27       90 TOLTC(IX)=TOLT*1.D-03
28      100 NITER=NITER+1
29      110 CALL CCCS(HLNB,PQR,CX,HX,NKMAX,NXMAX,NK,NX,DLNAM,CI)
30          SUM=0.D+00
31          DO 120 IX=1,NX
32          W=CX(IX)-TC(IX)
33          DO 118 IK=1,NK
34      118 W=W+PQR(IX,IK)*CI(IK)
35          EPS(IX)=W
36      120 SUM=SUM+W*W
37          GO TO(130,150), JCK
38      130 IF(SUM.LE.SUMP)GO TO 140
39          DO 132 IX=1,NX
40          DX(IX)=0.5D+00*DX(IX)
41      132 CX(IX)=CX(IX)-DX(IX)
42          JCK=2
43          GO TO 110
44      140 DO 145 IX=1,NX
45          IF(DABS(EPS(IX)).GT.TOLTC(IX))GO TO 150
46      145 CONTINUE
47          GO TO 500
48      150 SUMP=SUM*TOL1
49          JCK=1
50          DO 170 IX=1,NX
51          DO 170 JX=1,NX
52          W=0.D+00
53          IF(IX.EQ.JX)W=CX(IX)
54          DO 160 IK=1,NK
55      160 W=W+PQR(IX,IK)*PQR(JX,IK)*CI(IK)
56      170 DM(IX,JX)=W
57          CALL MATIN(DM,NX,DET,NXMAX,JOUT,ACCM)
58          IF(DET.EQ.0)GO TO 205
59          DO 180 IX=1,NX
60          W=0.D+00
61          DO 175 JX=1,NX
62      175 W=W-DM(IX,JX)*EPS(JX)
63          DX(IX)=W*CX(IX)
64      180 CX(IX)=CX(IX)+DX(IX)
65      190 DO 195 IX=1,NX
66          IF(CX(IX).LE.ACCM)GO TO 198
67      195 CONTINUE
68          GO TO 210
69      198 FRACT=-0.5D+00*(CX(IX)/DX(IX)-1.D+00)
70          DO 200 IX=1,NX
71          CX(IX)=CX(IX)-DX(IX)
72          DX(IX)=FRACT*DX(IX)
73      200 CX(IX)=CX(IX)+DX(IX)
74          JCK=2
75          GO TO 190
76      205 WRITE(JOUT,2)
77          STOP
78      210 IF(NITER.LT.ITMAX)GO TO 100
79          WRITE(JOUT,1)ITMAX
80      500 RETURN
81          END



1    C
2    C    SUBROUTINE DINS
```

```
 3     C
 4     C       DATA INPUT SUBROUTINE.
 5     C       THE INPUT DATA ARE READ IN.  THE WEIGHTS FOR ALL OBSERVATIONS
 6     C       ARE CALCULATED IN SUBROUTINE WEIGHTS.  FOR DETAILS
 7     C       REGARDING THE WEIGHTIG PROCEDURE AND INPUT STATEMENTS SEE
 8     C       PRECCEDING SECTIONS.
 9     C
10             SUBROUTINE DINS(TITRE,IVAR,JTP,NEMFV,NMBEV,VINIT,AJ,BJ,BETA,
11            &JPOT,JEL,H,X,Y,Y2,EMF,WS,KCONS,TOTC,ADDC,EZERO,JNMB,JP,PQRØ,PQRV,
12            &SIGMAE,KEY,TITLE,KNEG,JCOUL,MAXCX,MAXK,MAXTC,MAXTP,MAXPAR,MAXEL,
13            &MAXSP,MAXCON,MAXPA,MAXEMF)
14             IMPLICIT REAL*8(A-H,O-Z)
15             INTEGER PQRØ(MAXCX,MAXK),PQRV(MAXCX,MAXK,MAXTC)
16             INTEGER TITLE(18)
17             DIMENSION TITRE(MAXTP)
18             DIMENSION IVAR(MAXPAR)
19             DIMENSION JTP(MAXTC),NEMFV(MAXTC),NMBEV(MAXTC)
20             DIMENSION VINIT(MAXTC),AJ(MAXTC),BJ(MAXTC),JCOUL(MAXTC)
21             DIMENSION BETA(MAXK),JPOT(MAXK),KEY(MAXK),KNEG(MAXK)
22             DIMENSION SIGMAE(MAXEL)
23             DIMENSION H(MAXSP),X(MAXSP),Y(MAXSP),Y2(MAXSP)
24             DIMENSION EMF(MAXTP,MAXEL),WS(MAXTP,MAXEL)
25             DIMENSION KCONS(MAXCON,2)
26             DIMENSION TOTC(MAXTC,MAXCX),ADDC(MAXTC,MAXCX)
27             DIMENSION EZERO(MAXTC,MAXEL),JEL(MAXTC,MAXEL)
28             DIMENSION JNMB(MAXCX,MAXTC),JP(MAXCX,MAXTC)
29             COMMON RTF,TOL,ACCM,DLNAM,AL1Ø
30             COMMON NK,NKV,N,NSP,NCONS,NPAR,NE,NTC,NMBEØ,NKW,NSET,
31            &LARS,MAXIT,IPRIN,JINP,JOUT
32           1 FORMAT('(',I2,')',F11.4,'E',I3,I7,7X,4I4)
33           2 FORMAT(//1ØX,'FORMATION    REFINEMENT    STOICHEIOMETRIC'/
34            &1ØX,'CONSTANTS       KEYS         COEFFICIENTS'/)
35           3 FORMAT(14I5)
36           4 FORMAT(F1Ø.6,7I5)
37           5 FORMAT(7F1Ø.6)
38           6 FORMAT(I5,F8.3,8F8.2)
39           7 FORMAT(/I3,' FORMATION CONSTANTS TO BE REFINED')
40           8 FORMAT(6X,'REACTION TEMPERATURE ',F7.2,' DEGREES CENTIGRADE'
41            &//6X,'STANDARD DEVIATION ON VOLUME IS',F5.3)
42           9 FORMAT(/' THEREFORE RUN ABANDONED BEFORE REFINEMENT')
43          10 FORMAT(/I5,' DATA POINTS HAVE BEEN READ IN ')
44          11 FORMAT(//2X,'CURVE',I2,4X,'THERE ARE',I2,' MASS-BALANCE EQUA',
45            &'TIONS AND',I2,' UNKNOWN FREE CONCENTRATION(S)PER DATA POINT'
46            &//22X,'INITIAL QUANTITIES',4X,'TITRANT CONCENTRATION',4X,
47            &' STANDARD POTENTIAL'/25X,'(MILLIMOLES)',11X,'(MOLES/LITRE)',
48            &12X,'(MILLIVOLTS)')
49          12 FORMAT (6X,'REACTANT *',I1,'*',8X,F7.4,17X,F7.4,17X,F7.2)
50          13 FORMAT (6X,'REACTANT *',I1,'*',8X,F7.4,17X,F7.4,15X,'NO ELE',
51            &'CTRODE')
52          14 FORMAT(/6X,'INITIAL VOLUME =',F8.2,' MILLILITERS')
53          15 FORMAT(I8,' DATA POINTS IN CURVE',I3)
54          16 FORMAT(/6X,'AJ(CORRECTION TERM FOR LIQUID-JUNCTION POTENTIAL'
55            &,' IN ACID   SOLUTION)IS',F6.1,
56            &//6X,'BJ(CORRECTION TERM FOR LIQUID-JUNCTION POTENTIAL'
57            &,' IN BASIC SOLUTION)IS',F6.1)
58          17 FORMAT(/6X,'STANDARD DEVIATIONS ON E.M.F.ARE',F4.2,' AND'
59            &,F4.2)
60          18 FORMAT(/6X,'STANDARD DEVIATION ON E.M.F.IS',F4.2)
61          19 FORMAT(/I3,' SPECIAL PARAMETERS TO BE REFINED:',2ØI4/29X,2ØI4)
62          20 FORMAT(/' CONSTRAINTS:',1Ø(I2,'=',I2,3X)/13X,1Ø(I2,'=',I2,3X))
63          21 FORMAT(' THE NUMBER OF INDEPENDENT PARAMETERS TO BE REFINED '
64            &,' IS GREATER THAN',I3)
65          22 FORMAT(' THE TOTAL NUMBER OF PARAMETERS TO BE REFINED IS'
66            &,' GREATER THAN',I3)
67          23 FORMAT(' THE NUMBER OF CONSTRAINTS IS GREATER THAN',I3)
68          24 FORMAT(' NO PARAMETERS TO BE REFINED')
69          25 FORMAT(' THE NUMBER OF TITRATION CURVES IS GREATER THAN',I3)
70          26 FORMAT(' THE NUMBER OF DATA POINTS IS GREATER THAN',I4)
71          27 FORMAT(' WRONG CONSTRAINTS SETTING')
72          28 FORMAT(' THE NUMBER OF EXPERIMENTAL E.M.F.VALUES(',I3,
73            &')IS GREATER THAN MAXEMF(',I3,')')
74          29 FORMAT(' IN CURVE',I3,' THERE ARE MORE THAN ',I4,' POINTS'
75            &,'(THE MAXIMUM ALLOWED IN SUBROUTINE SPLINE)')
76          30 FORMAT(18A4)
77          31 FORMAT('1E.M.F.-VOLUME TITRATION CURVE ',18A4)
78          32 FORMAT(/5X,'LARS',4X,'MAXIT',3X,'IPRIN',7X,'TOL',11X,'ACCM'/
79            &3I8,2E15.3//6X,'THE MAXIMUM NUMBER OF MASS BALANCE ',
```

```
80             &'EQUATIONS IS',I2)
81          33 FORMAT(' THE NUMBER OF FORMATION CONSTANTS IS GREATER THAN',I3)
82             READ(JINP,30)TITLE
83             WRITE(JOUT,31)TITLE
84             READ(JINP,3)LARS,NK,N,MAXIT,IPRIN,NMBEØ,NCONS
85             WRITE(JOUT,32)LARS,MAXIT,IPRIN,TOL,ACCM,NMBEØ
86             IF(NK.GT.MAXK)GO TO 290
87             IF(NCONS.GT.MAXCON)GO TO 220
88             READ(JINP,5)TEMP,SIGMAV,SIGMAE
89             WRITE(JOUT,8)TEMP,SIGMAV
90             RTF=(TEMP+273.16)/11.6048
91             IF(SIGMAE(2).EQ.Ø)GO TO 101
92             WRITE(JOUT,17)(SIGMAE(I),I=1,2)
93             GO TO 102
94         101 WRITE(JOUT,18)SIGMAE(1)
95         102 SIGMAV=SIGMAV*SIGMAV
96             SIGMAE(1)=SIGMAE(1)*SIGMAE(1)
97             SIGMAE(2)=SIGMAE(2)*SIGMAE(2)
98             WRITE(JOUT,2)
99             NKV=Ø
100            DO 103 I=1,NK
101            READ(JINP,4)BETA(I),JPOT(I),(PQRØ(K,I),K=1,NMBEØ),KEY(I)
102            WRITE(JOUT,1)I,BETA(I),JPOT(I),KEY(I),(PQRØ(K,I),K=1,NMBEØ)
103            KNEG(I)=Ø
104            IF(KEY(I).EQ.Ø)GO TO 103
105            NKV=NKV+1
106        103 CONTINUE
107            WRITE(JOUT,7)NKV
108            NPAR=NKV+N
109            IF(NPAR.GT.MAXPA)GO TO 200
110            IF(NPAR+NCONS.GT.MAXPAR)GO TO 210
111            IF(NPAR.LT.1)GO TO 230
112            N1=NMBEØ-1
113    C
114    C       SEARCH FOR THE SEQUENCE NUMBER OF THE DISSOCIATION CONSTANT
115    C       OF WATER.
116    C
117            DO 1104 I=1,NK
118            L=Ø
119            M=Ø
120            DO 1103 K=1,NMBEØ
121            IF(PQRØ(K,I).EQ.Ø)L=L+1
122            IF(PQRØ(K,I).EQ.-1)M=M+1
123       1103 CONTINUE
124            IF(L.EQ.N1.AND.M.EQ.1)NKW=I
125       1104 CONTINUE
126            NSP=N
127            IF(NSP.EQ.Ø)GO TO 109
128            READ(JINP,3)(IVAR(I),I=1,N)
129    C
130    C       IVAR CONTAINS THE IDENTIFICATION NUMBERS OF THE 'SPECIAL'
131    C       PARAMETERS TO BE REFINED.
132    C
133            WRITE(JOUT,19)N,((IVAR(K)),K=1,N)
134            IF(NCONS.EQ.Ø)GO TO 109
135    C
136    C       THE CONSTRAINTS ARE READ IN.
137    C
138            READ(JINP,3)((KCONS(I,J),J=1,2),I=1,NCONS)
139            WRITE(JOUT,20)((KCONS(IC,I),I=1,2),IC=1,NCONS)
140    C
141    C       THE IDENTIFICATION NUMBERS OF THE CONSTRAINED PARAMETERS
142    C       ARE STORED IN THE VECTOR IVAR AFTER THE IDENTIFICATION
143    C       NUMBERS OF THE INDEPENDENT 'SPECIAL' PARAMETERS.
144    C
145            DO 108 I=1,NCONS
146            DO 104 K=1,2
147            DO 104 J=1,NSP
148            IF(KCONS(I,K).EQ.IVAR(J))GO TO 105
149        104 CONTINUE
150            GO TO 260
151        105 N=N+1
152            IF(K.EQ.2)GO TO 106
153            IVAR(N)=KCONS(I,2)
154            GO TO 107
155        106 IVAR(N)=KCONS(I,1)
```

```
156        107 KCONS(I,1)=N+NKV
157            KCONS(I,2)=J+NKV
158        108 CONTINUE
159        109 IF(NSET.EQ.1)GO TO 170
160            NE=Ø
161            IP=Ø
162            NTC=Ø
163   C
164   C          LOOP FOR EACH TITRATION CURVE.
165   C
166        110 NTC=NTC+1
167            IF(NTC.GT.MAXTC)GO TO 240
168            READ(JINP,3)NMBE,(JNMB(I,NTC),I=1,NMBE),NEMF,
169           &(JP(I,NTC),I=1,NMBEØ)
170            NC=NMBE-NEMF
171            WRITE(JOUT,11)NTC,NMBE,NC
172            NMBEV(NTC)= NMBE
173            NEMFV(NTC)=NEMF
174            READ(JINP,3)(JEL(NTC,L),L=1,NEMF),JCOUL(NTC)
175            DO 111 K=1,NMBE
176            KK=JNMB(K,NTC)
177            DO 111 I=1,NK
178        111 PQRV(K,I,NTC)= PQRØ(KK,I)
179            DO 112 I=1,NMBE
180            DO 112 J=1,NMBE
181            IF(JNMB(I,NTC).EQ.JP(J,NTC))JP(J,NTC)=I
182        112 CONTINUE
183            KK=1
184            DO 116 K=1,NMBEØ
185            IF(K.EQ.JNMB(KK,NTC))GO TO 115
186            DO 114 I=1,NK
187            IF(PQRØ(K,I).EQ.Ø)GO TO 114
188            DO 113 J=1,NMBE
189        113 PQRV(J,I,NTC)= Ø
190        114 CONTINUE
191            GO TO 116
192        115 IF(KK.EQ.NMBE)GO TO  116
193            KK=KK+1
194        116 CONTINUE
195            READ(JINP,5)(TOTC(NTC,K),K=1,NMBE)
196            READ(JINP,5)(ADDC(NTC,K),K=1,NMBE)
197            READ(JINP,5)(EZERO(NTC,K),K=1,NEMF),AJ(NTC),BJ(NTC)
198            READ(JINP,5)VINIT(NTC)
199            DO 119 K=1,NMBE
200            IF(K.GT.NC)GO TO 118
201            WRITE(JOUT,13)JNMB(K,NTC),TOTC(NTC,K),ADDC(NTC,K)
202            GO TO 119
203        118 I=K-NC
204            WRITE(JOUT,12)JNMB(K,NTC),TOTC(NTC,K),ADDC(NTC,K),EZERO(NTC,I)
205        119 CONTINUE
206            WRITE(JOUT,16)AJ(NTC),BJ(NTC)
207            WRITE(JOUT,14)VINIT(NTC)
208   C
209   C          LOOP FOR EACH EXPERIMENTAL POINT.
210   C
211        120 IP=IP+1
212            IF(IP.GT.MAXTP)GO TO 250
213            IS=LARS
214        125 READ(JINP,6)LUIGI,TITRE(IP),(EMF(IP,K),K=1,NEMF)
215            IS=IS-1
216            IF(LUIGI.NE.Ø)GO TO 140
217            IF(IS.GT.Ø)GO TO 125
218            GO TO 120
219        140 JTP(NTC)=IP
220            IF(NTC.NE.1)GO TO 145
221            NPC=IP
222            GO TO 150
223        145 NTC1=NTC-1
224            NPC=IP-JTP(NTC1)
225        150 WRITE(JOUT,15)NPC,NTC
226            IF(NPC.GT.MAXSP)GO TO 280
227            NE=NE+NPC*NEMF
228            IF(NE.GT.MAXEMF)GO TO 270
229   C
230   C          CALL THE ROUTINE FOR WEIGHTING THE EXPERIMENTAL DATA
231   C
```

```
232            CALL WEIGHT(IP,NPC,NEMF,TITRE,EMF,SIGMAV,SIGMAE,WS,H,X,Y,Y2,
233           &MAXTP,MAXEL,MAXSP)
234            IF(IPRIN.LT.2)GO TO 160
235            K=IP-NPC+1
236            DO 155 I=K,IP
237        155 WRITE(JOUT,6)I,TITRE(I),(EMF(I,K),K=1,NEMF),
238           &(WS(I,K),K=1,NEMF)
239        160 IF(LUIGI.GT.0)GO TO 110
240            NP=IP
241            WRITE(JOUT,10)IP
242            GO TO 190
243        170 DO 188 IC=1,NTC
244            NMBE=NMBEV(K)
245            DO 175 K=1,NMBE
246            KK=JNMB(K,IC)
247            IF(KK.EQ.0)GO TO 175
248            DO 174 I=1,NK
249        174 PQRV(K,I,IC)= PQR0(KK,I)
250        175 CONTINUE
251            KK=1
252            DO 186 K=1,NMBE0
253            IF(K.EQ.JNMB(KK,IC))GO TO 182
254            DO 180 I=1,NK
255            IF(PQR0(K,I).EQ.0)GO TO 180
256            DO 178 J=1,NMBE
257        178 PQRV(J,I,IC)= 0
258        180 CONTINUE
259            GO TO 186
260        182 IF(KK.EQ.NMBE)GO TO   186
261            KK=KK+1
262        186 CONTINUE
263        188 CONTINUE
264        190 RETURN
265        200 WRITE(JOUT,21)MAXPA
266            GO TO 300
267        210 WRITE(JOUT,22)MAXPAR
268            GO TO 300
269        220 WRITE(JOUT,23)MAXCON
270            GO TO 300
271        230 WRITE(JOUT,24)
272            GO TO 300
273        240 WRITE(JOUT,25)MAXTC
274            GO TO 300
275        250 WRITE(JOUT,26)MAXTP
276            GO TO 300
277        260 WRITE(JOUT,27)
278            GO TO 300
279        270 WRITE(JOUT,28)NE,MAXEMF
280            GO TO 300
281        280 WRITE(JOUT,29)NTC,MAXSP
282            GO TO 300
283        290 WRITE(JOUT,33)MAXK
284        300 WRITE(JOUT,9)
285            STOP
286            END
```

```
  1   C
  2   C      SUBROUTINE DOUS
  3   C          DATA OUTPUT ROUTINE.ALL THE RELEVANT INFORMATION ON THE
  4   C          REFINED PARAMETERS IS PRINTED OUT.OPTIONALLY, A LIST OF
  5   C          ALL TITRATION DATA MAY BE OBTAINED.
  6   C
  7   C      IN MANY FORMAT STATEMENTS THE FIRST CHARACTER OF THE RECORD TO
  8   C      BE PRINTED IS A CARRIAGE CONTROL CHARACTER SPECIFIED AS 1HX
  9   C      WHERE 'X' HAS ONE OF THE FOLLOWING MEANINGS:-
 10   C
 11   C         X                    MEANING
 12   C      (BLANK)        ADVANCE ONE LINE BEFORE PRINTING
 13   C         0           ADVANCE TWO LINES BEFORE PRINTING
 14   C         1           ADVANCE TO FIRST LINE OF NEW PAGE
 15   C         +           NO ADVANCE
 16   C
```

```
17              SUBROUTINE DOUS(IEXIT,KEY,HLNB,BETA,JPOT,PARAM,SD,PARØ,DX,U,UØ,
18         &NCICL,TITLE,PQRØ,IVAR,PARI,SIGMA,JTP,NEMFV,DW,DY,WS,TITRE,EMF,
19         &EMFC,SIG,B,MAXCX,MAXK,MAXTC,MAXEL,MAXPAR,MAXPA,MAXTP)
20              IMPLICIT REAL*8(A-H,O-Z)
21              INTEGER PQRØ(MAXCX,MAXK)
22              INTEGER TITLE(18)
23              DIMENSION TITRE(MAXTP)
24              DIMENSION JTP(MAXTC),NEMFV(MAXTC),SIG(MAXTC)
25              DIMENSION BETA(MAXK),JPOT(MAXK),HLNB(MAXK),KEY(MAXK)
26              DIMENSION DW(MAXEL)
27              DIMENSION PARAM(MAXPAR),IVAR(MAXPAR),PARI(MAXPAR)
28              DIMENSION PARØ(MAXPA),SD(MAXPA),DX(MAXPA),B(MAXPA,MAXPA)
29              DIMENSION EMF(MAXTP,MAXEL),EMFC(MAXTP,MAXEL),DY(MAXTP,MAXEL),
30         &WS(MAXTP,MAXEL)
31        ·     COMMON RTF,TOL,ACCM,DLNAM,AL1Ø
32              COMMON NK,NKV,N,NSP,NCONS,NPAR,NE,NTC,NMBEØ,NKW,NSET,
33         &LARS,MAXIT,IPRIN,JINP,JOUT
34            1 FORMAT(1H1,3X,18A4)
35            2 FORMAT(1HØ,'REFINEMENT CONVERGED SUCCESSFULLY'//
36         &I4,' ITERATIONS'/' SUM OF SQUARES =',E12.6/
37         &' STANDARD DEVIATION =',E1Ø.4//35X,
38         &'     VALUE     STD.DEVIATION     LOG BETA     STD.DEVIATION'/)
39            3 FORMAT(1H ,'BETA(',I2,') ',4I2)
40            4 FORMAT(1H+,22X,'NEGATIVE ')
41            5 FORMAT(1H+,22X,'CONSTANT')
42            6 FORMAT(1H+,22X,'REFINED ')
43            7 FORMAT(1H+,32X,F1Ø.5,'E',I3,14X,F14.5)
44            8 FORMAT(1H+,32X,2(F1Ø.5,'E',I3),F14.5,6X,F7.5)
45            9 FORMAT(//' PARAMETER   INITIAL VALUE   FINAL VALUE   STD.DEV.'
46         &/)
47           10 FORMAT(I6,6X,F12.5,3X,2F12.5)
48           11 FORMAT(1HØ,' MAXIMUM NUMBER(',I2,')OF ITERATIONS PERFORMED'/
49         &' REFINEMENT ABANDONED')
50           12 FORMAT(' SUM OF SQUARES =',E12.6,' AT CYCLE',I3/
51         &' SUM OF SQUARES =',E12.6,' AT CYCLE',I3/)
52           13 FORMAT(' VALUES AT CYCLE',I3,'   REL.SHIFTS   VALUES AT CYCLE',
53         &I3,'   STD.DEV.')
54           14 FORMAT(F14.5,8X,F1Ø.5,F17.5,F15.5)
55           15 FORMAT(/' CURVE',I3,6X,'SIGMA =',F6.2)
56           16 FORMAT(I5,F8.3,2(2F9.2,2F7.2))
57           17 FORMAT(//'  CORRELATION MATRIX:'/)
58           18 FORMAT(I3,1X,22F5.2)
59           19 FORMAT(/3X,22I5)
60           20 FORMAT(//)
61              WRITE(JOUT,1)TITLE
62              IF(MAXIT.EQ.Ø)GO TO 22Ø
63              IF(IEXIT.EQ.Ø)GO TO 26Ø
64              WRITE(JOUT,2)NCICL,U,SIGMA
65              K=Ø
66              DO 16Ø I=1,NK
67              WRITE(JOUT,3)I,(PQRØ(J,I),J=1,NMBEØ)
68              IF(KEY(I).GT.-1)GO TO 12Ø
69              WRITE(JOUT,4)
70              GO TO 16Ø
71          12Ø XL=HLNB(I)/AL1Ø
72              IF(KEY(I).EQ.1)GO TO 13Ø
73              WRITE(JOUT,5)
74              WRITE(JOUT,7)BETA(I),JPOT(I),XL
75              GO TO 16Ø
76          13Ø WRITE(JOUT,6)
77              K=K+1
78              SDC=SD(I)*BETA(I)
79              IF(SD(I).LT.1.D+ØØ)GO TO 14Ø
80              SDL=1.E+3Ø
81              GO TO 15Ø
82          14Ø SDL= DABS(DLOG(1.-SD(I)))/AL1Ø
83          15Ø WRITE(JOUT,8)BETA(I),JPOT(I),SDC,JPOT(I),XL,SDL
84          16Ø CONTINUE
85              IF(NSP.EQ.Ø)GO TO 19Ø
86              WRITE(JOUT,9)
87              DO 18Ø I=1,NSP
88              L=K+I
89              SD(L)=SD(L)*PARAM(L)
90          18Ø WRITE(JOUT,1Ø)IVAR(I),PARI(L),PARAM(L),SD(L)
91          19Ø IF(IPRIN.EQ.Ø)GO TO 22Ø
92              WRITE(JOUT,17)
```

```
93              DO 210 I=2,NPAR
94              K=I-1
95              DO 200 J=1,K
96      200 B(I,J)=B(I,J)/DSQRT(B(I,I)*B(J,J))
97      210 WRITE(JOUT,18)I,(B(I,J),J=1,K)
98              WRITE(JOUT, 19)(I,I=1,K)
99              WRITE(JOUT,20)
100     220 IPI=1
101             DO 250 JTC=1,NTC
102             WRITE(JOUT,15)JTC,SIG(JTC)
103             IF(IPRIN.EQ.0)GO TO 250
104             IPF=JTP(JTC)
105             NEMF=NEMFV(JTC)
106             DO 240 I=IPI,IPF
107             DO 230 J=1,NEMF
108     230 DW(J)=DY(I,J)*DSQRT(WS(I,J))
109     240 WRITE(JOUT,16)I,TITRE(I),(EMF(I,J),EMFC(I,J),DY(I,J),DW(J),
110             &J=1,NEMF)
111             IPI=IPF+1
112     250 CONTINUE
113             RETURN
114     260 WRITE(JOUT,11)MAXIT
115             NCICL1=NCICL-1
116             WRITE(JOUT,12)U,NCICL,U0,NCICL1
117             WRITE(JOUT,13)NCICL1,NCICL
118             DO 270 I=1,NPAR
119     270 WRITE(JOUT,14)PAR0(I),DX(I),PARAM(I),SD(I)
120             RETURN
121             END


1   C
2   C    SUBROUTINE FUNV
3   C
4   C    FOR EACH EXPERIMENTAL POINT THE FREE REACTANT CONCENTRATIONS
5   C    ARE CALCULATED IN CCFR.  THE EMF VALUES ARE COMPUTED VIA THE
6   C    NERNST EQUATION USING THE RELEVANT FREE CONCENTRATION VALUES.
7   C    THE RESIDUALS AND WEIGHTED ERROR SQUARE SUM IS COMPUTED.
8   C
9            SUBROUTINE FUNV(JTP,HLNB,BETA,JPOT,VINIT,TITRE,TOTC,ADDC,EZERO,
10           &JEL,EMF,EMFC,WS,AJ,BJ,TT,DY,PQR,PQRV,FREEC,CX,HX,DT,DM,DX,CI,
11           &TOLTC,U,NCICL,NEMFV,NMBEV,KEY,SIGMA,SIG,JCOUL,MAXCX,MAXK,MAXTC,
12           &MAXEL,MAXTP)
13           IMPLICIT REAL*8(A-H,O-Z)
14           INTEGER PQR(MAXCX,MAXK),PQRV(MAXCX,MAXK,MAXTC)
15           DIMENSION BETA(MAXK),JPOT(MAXK),HLNB(MAXK),CI(MAXK),KEY(MAXK)
16           DIMENSION TOTC(MAXTC,MAXCX),ADDC(MAXTC,MAXCX)
17           DIMENSION EZERO(MAXTC,MAXEL),JEL(MAXTC,MAXEL)
18           DIMENSION VINIT(MAXTC),JTP(MAXTC),AJ(MAXTC),BJ(MAXTC)
19           DIMENSION TITRE(MAXTP),FREEC(MAXTP,MAXCX)
20           DIMENSION DY(MAXTP,MAXEL),EMF(MAXTP,MAXEL),EMFC(MAXTP,MAXEL)
21           DIMENSION WS(MAXTP,MAXEL)
22           DIMENSION TT(MAXCX),CX(MAXCX),HX(MAXCX),DT(MAXCX),DX(MAXCX),
23           &TOLTC(MAXCX),DM(MAXCX,MAXCX)
24           DIMENSION NEMFV(MAXTC),NMBEV(MAXTC),SIG(MAXTC),JCOUL(MAXTC)
25           COMMON RTF,TOL,ACCM,DLNAM,AL10
26           COMMON NK,NKV,N,NSP,NCONS,NPAR,NE,NTC,NMBE0,NKW,NSET,
27           &LARS,MAXIT,IPRIN,JINP,JOUT
28           K=0
29           DO 100 IK=1,NK
30           IF(KEY(IK).EQ.-1)GO TO 100
31           K=K+1
32           HLNB(K)=DLOG(BETA(IK))+AL10*JPOT(IK)
33      100 CONTINUE
34           NKP=K
35           HKW=DEXP(HLNB(NKW))
36           U=0.D+00
37           IPF=0
38           DO 240 JTC=1,NTC
39           IPI=IPF+1
40           IPF=JTP(JTC)
41           NEMF=NEMFV(JTC)
42           NMBE=NMBEV(JTC)
```

```
43              NMB1=NMBE-1
44              SIG(JTC)=0.D+00
45              IF(JCOUL(JTC).EQ.1)VOL=VINIT(JTC)
46              DO 130 J=1,NMBE
47          130 CX(J)=0.D+00
48              DO 135 I=1,NMBE
49              J=0
50              DO 135 L=1,NK
51              IF(KEY(L).EQ.-1)GO TO 135
52              J=J+1
53              PQR(I,J)=PQRV(I,L,JTC)
54          135 CONTINUE
55              DO 230 IP=IPI,IPF
56              IF(JCOUL(JTC).EQ.0)VOL=VINIT(JTC)+TITRE(IP)
57              DO 160 J=1,NMBE
58          160 TT(J)=(TOTC(JTC,J)+TITRE(IP)*ADDC(JTC,J))/VOL
59              IF(NCICL.EQ.0)GO TO 175
60              DO 170 J=1,NMBE
61          170 CX(J)=FREEC(IP,J)
62              GO TO 180
63          175 IF(IP.GT.IPI)GO TO 180
64              DO 178 J=1,NMBE
65          178 CX(J)=0.D+00
66          180 NITER=100
67  C
68  C           CALL THE ROUTINE FOR THE CALCULATION OF FREE REACTANT
69  C           CONCENTRATIONS
70  C
71              CALL CCFR(NMBE,NKP,NITER,JOUT,CX,TT,HX,DT,DM,DX,HLNB,
72             &CI,PQR,TOLTC,TOL,MAXK,MAXCX,ACCM,DLNAM)
73              DO 190 J=1,NMBE
74          190 FREEC(IP,J)=CX(J)
75              NM=NMBE-NEMF
76              DO 210 I=1,NEMF
77              EMFC(IP,I)=EZERO(JTC,I)+(RTF/JEL(JTC,I))*DLOG(CX(NM+I))
78              IF(I.NE.NEMF)GO TO 210
79              IF(AJ(JTC).EQ.0.AND.BJ(JTC).EQ.0)GO TO 210
80              EMFC(IP,I)=EMFC(IP,I)+AJ(JTC)*CX(NMBE)+BJ(JTC)*HKW/CX(NMBE)
81          210 CONTINUE
82              DO 220 J=1,NEMF
83              S=EMF(IP,J)-EMFC(IP,J)
84              DY(IP,J)=S
85              S=S*S*WS(IP,J)
86              SIG(JTC)=SIG(JTC)+S
87          220 U=U+S
88          230 CONTINUE
89              SIG(JTC)=DSQRT(SIG(JTC)/((IPF-IPI+1)*NEMF))
90          240 CONTINUE
91              SIGMA=DSQRT(U/(NE-NPAR))
92              RETURN
93              END




1   C
2   C       SUBROUTINE FUNZ
3   C
4   C           THE E.M.F.VALUES ARE COMPUTED USING THE INCREMENTED VALUE
5   C           OF ONE PARAMETER
6   C
7           SUBROUTINE FUNZ(JTP,HLNB,BETA,JPOT,VINIT,TITRE,TOTC,ADDC,EZERO,
8          &JEL,EMFC,EMFD,JTYP,AJ,BJ,TT,PQR,PQRV,JC,FREEC,CX,HX,DT,DM,DX,
9          &CI,TOLTC,JCURV,KCONS,NEMFV,NMBEV,KEY,JCOUL,MAXCX,MAXK,MAXTC,
10         &MAXPAR,MAXEL,MAXTP,MAXCON)
11          IMPLICIT REAL*8(A-H,O-Z)
12          INTEGER PQR(MAXCX,MAXK),PQRV(MAXCX,MAXK,MAXTC)
13          DIMENSION BETA(MAXK),JPOT(MAXK),HLNB(MAXK),CI(MAXK),KEY(MAXK)
14          DIMENSION JTYP(MAXPAR),JCURV(MAXPAR)
15          DIMENSION TOTC(MAXTC,MAXCX),ADDC(MAXTC,MAXCX)
16          DIMENSION EZERO(MAXTC,MAXEL),JEL(MAXTC,MAXEL)
17          DIMENSION VINIT(MAXTC),JTP(MAXTC),AJ(MAXTC),BJ(MAXTC)
18          DIMENSION TITRE(MAXTP),FREEC(MAXTP,MAXCX)
19          DIMENSION EMFD(MAXTP,MAXEL),EMFC(MAXTP,MAXEL)
20          DIMENSION TT(MAXCX),CX(MAXCX),HX(MAXCX),DT(MAXCX),DX(MAXCX),
```

```
21          &TOLTC(MAXCX),DM(MAXCX,MAXCX),KCONS(MAXCON,2)
22           DIMENSION NEMFV(MAXTC),NMBEV(MAXTC),JCOUL(MAXTC)
23           COMMON RTF,TOL,ACCM,DLNAM,AL10
24           COMMON NK,NKV,N,NSP,NCONS,NPAR,NE,NTC,NMBE0,NKW,NSET,
25          &LARS,MAXIT,IPRIN,JINP,JOUT
26           K=0
27           DO 100 IK=1,NK
28           IF(KEY(IK).EQ.-1)GO TO 100
29           K=K+1
30           HLNB(K)=DLOG(BETA(IK))+AL10*JPOT(IK)
31      100 CONTINUE
32           NKP=K
33           HKW=DEXP(HLNB(NKW))
34           IPF=0
35           DO 240 JTC=1,NTC
36           IPI=IPF+1
37           IPF=JTP(JTC)
38           NEMF=NEMFV(JTC)
39           IF(JCURV(JC).EQ.0)GO TO 160
40           IF(JCURV(JC).EQ.JTC)GO TO 160
41           DO 140 IC=1,NCONS
42           IF(KCONS(IC,2).NE.JC)GO TO 140
43           K=KCONS(IC,1)
44           IF(JCURV(K).EQ.JTC)GO TO 160
45      140 CONTINUE
46           DO 150 IF=1,NEMF
47           DO 150 IP=IPI,IPF
48      150 EMFD(IP,IF)=EMFC(IP,IF)
49           GO TO 240
50      160 NMBE=NMBEV(JTC)
51           NMB1=NMBE-1
52           IF(JCOUL(JTC).EQ.1)VOL=VINIT(JTC)
53           DO 165 I=1,NMBE
54           J=0
55           DO 165 L=1,NK
56           IF(KEY(L).EQ.-1)GO TO 165
57           J=J+1
58           PQR(I,J)=PQRV(I,L,JTC)
59      165 CONTINUE
60           DO 230 IP=IPI,IPF
61           IF(JCOUL(JTC).EQ.0)VOL=VINIT(JTC)+TITRE(IP)
62           DO 190 J=1,NMBE
63      190 CX(J)=FREEC(IP,J)
64           IF(JTYP(JC).GT.1)GO TO 220
65           DO 210 J=1,NMBE
66      210 TT(J)=(TOTC(JTC,J)+TITRE(IP)*ADDC(JTC,J))/VOL
67           NITER=100
68    C
69    C           CALL THE ROUTINE FOR THE CALCULATION OF FREE REACTANT
70    C           CONCENTRATIONS
71    C
72           CALL CCFR(NMBE,NKP,NITER,JOUT,CX,TT,HX,DT,DM,DX,HLNB,
73          &CI,PQR,TOLTC,TOL,MAXK,MAXCX,ACCM,DLNAM)
74      220 NM=NMBE-NEMF
75           DO 225 I=1,NEMF
76           EMFD(IP,I)=EZERO(JTC,I)+(RTF/JEL(JTC,I))*DLOG(CX(NM+I))
77           IF(I.NE.NEMF)GO TO 225
78           IF(AJ(JTC).EQ.0.AND.BJ(JTC).EQ.0)GO TO 225
79           EMFD(IP,I)=EMFD(IP,I)+AJ(JTC)*CX(NMBE)+BJ(JTC)*HKW/CX(NMBE)
80      225 CONTINUE
81      230 CONTINUE
82      240 CONTINUE
83           RETURN
84           END
```

```
1     C
2     C     SUBROUTINE MATIN
3     C
4     C         MATRIX INVERSION ROUTINE
5     C
6           SUBROUTINE MATIN(ARRAY,NORDER,DET,NMAX,JOUT,ACCM)
7           IMPLICIT REAL*8(A-H,O-Z)
```

```
8               DIMENSION ARRAY(NMAX,NMAX),IK(22),JK(22)
9         10    DET=1.
10        11    DO 100 K=1,NORDER
11              AMAX=0.
12        21    DO 30 I=K,NORDER
13              DO 30 J=K,NORDER
14        23    IF(DABS(AMAX)-DABS(ARRAY(I,J)))24,24,30
15        24    AMAX=ARRAY(I,J)
16              IK(K)=I
17              JK(K)=J
18        30    CONTINUE
19        31    IF(DABS(AMAX)-ACCM)32,41,41
20        32    DET=0.
21              GO TO 140
22        41    I=IK(K)
23              IF(I-K)21,51,43
24        43    DO 50 J=1,NORDER
25              SAVE=ARRAY(K,J)
26              ARRAY(K,J)=ARRAY(I,J)
27        50    ARRAY(I,J)=-SAVE
28        51    J=JK(K)
29              IF(J-K)21,61,53
30        53    DO 60 I=1,NORDER
31              SAVE=ARRAY(I,K)
32              ARRAY(I,K)=ARRAY(I,J)
33        60    ARRAY(I,J)=-SAVE
34        61    DO 70 I=1,NORDER
35              IF(I-K)63,70,63
36        63    ARRAY(I,K)=-ARRAY(I,K)/AMAX
37        70    CONTINUE
38        71    DO 80 I=1,NORDER
39              DO 80 J=1,NORDER
40              IF(I-K)74,80,74
41        74    IF(J-K)75,80,75
42        75    ARRAY(I,J)=ARRAY(I,J)+ARRAY(I,K)*ARRAY(K,J)
43        80    CONTINUE
44        81    DO 90 J=1,NORDER
45              IF(J-K)83,90,83
46        83    ARRAY(K,J)=ARRAY(K,J)/AMAX
47        90    CONTINUE
48              ARRAY(K,K)=1./AMAX
49        100   DET=DET*AMAX
50        101   DO 130 L=1,NORDER
51              K=NORDER-L+1
52              J=IK(K)
53              IF(J-K)111,111,105
54        105   DO 110 I=1,NORDER
55              SAVE=ARRAY(I,K)
56              ARRAY(I,K)=-ARRAY(I,J)
57        110   ARRAY(I,J)=SAVE
58        111   I=JK(K)
59              IF(I-K)130,130,113
60        113   DO 120 J=1,NORDER
61              SAVE=ARRAY(K,J)
62              ARRAY(K,J)=-ARRAY(I,J)
63        120   ARRAY(I,J)=SAVE
64        130   CONTINUE
65        140   RETURN
66              END
```

```
1    C
2    C        SUBROUTINE CCCS
3    C            THE EQUILIBRIUM CONCENTRATIONS OF COMPLEX SPECIES
4    C            AT A GIVEN EXPERIMENTAL POINT ARE CALCULATED
5    C
6              SUBROUTINE CCCS(HLNB,PQR,CX,HX,NKMAX,NXMAX,NK,NX,DLNAM,CI)
7              IMPLICIT REAL*8(A-H,O-Z)
8              INTEGER PQR(NXMAX,NKMAX)
9              DIMENSION HLNB(NKMAX),CI(NKMAX),CX(NXMAX),HX(NXMAX)
10             DO 101 IX=1,NX
11       101   HX(IX)=DLOG(CX(IX))
12             DO 131 IK=1,NK
```

```
13              W=HLNB(IK)
14              DO 111 IX=1,NX
15         111  W=W+PQR(IX,IK)*HX(IX)
16              IF(W.GT.DLNAM)GO TO 121
17              CI(IK)=Ø.
18              GO TO 131
19         121  CI(IK)=DEXP(W)
2Ø         131  CONTINUE
21              RETURN
22              END
```

```
1   C
2   C        SUBROUTINE PTSNV
3   C
4   C            PARAMETERS TO SPECIAL NAMES CONVERSION ROUTINE
5   C
6   C        THE VALUES OF THE PARAMETERS TO BE REFINED, STORED IN THE
7   C        VECTOR PARAM, ARE TRANSFERRED INTO THE CORRESPONDING
8   C        VARIABLE NAMES.
9   C
1Ø          SUBROUTINE PTSNV(NMBEV,NEMFV,IVAR,PARAM,BETA,TOTC,ADDC,EZERO,
11         &VINIT,AJ,BJ,KCONS,KEY,MAXPAR,MAXK,MAXTC,MAXEL,MAXCX,MAXCON)
12          IMPLICIT REAL*8(A-H,O-Z)
13          DIMENSION IVAR(MAXPAR),PARAM(MAXPAR),BETA(MAXK),KEY(MAXK)
14          DIMENSION EZERO(MAXTC,MAXEL),VINIT(MAXTC),AJ(MAXTC),BJ(MAXTC)
15          DIMENSION NMBEV(MAXTC),NEMFV(MAXTC)
16          DIMENSION ADDC(MAXTC,MAXCX),TOTC(MAXTC,MAXCX)
17          DIMENSION KCONS(MAXCON,2)
18          COMMON RTF,TOL,ACCM,DLNAM,AL1Ø
19          COMMON NK,NKV,N,NSP,NCONS,NPAR,NE,NTC,NMBEØ,NKW,NSET,
2Ø         &LARS,MAXIT,IPRIN,JINP,JOUT
21          IF(NCONS.EQ.Ø)GO TO 1Ø1
22          DO 1ØØ IC=1,NCONS
23         1ØØ  PARAM(KCONS(IC,1))=PARAM(KCONS(IC,2))
24         1Ø1  J=Ø
25          DO 1Ø5 IK=1,NK
26          IF(KEY(IK).NE.1)GO TO 1Ø5
27          J=J+1
28          BETA(IK)=PARAM(J)
29         1Ø5  CONTINUE
3Ø          DO 14Ø K=1,N
31          J=Ø
32          L=NKV+K
33          DO 135 IC=1,NTC
34          NMBE=NMBEV(IC)
35          NEMF=NEMFV(IC)
36          DO 11Ø IMBE=1,NMBE
37          J=J+1
38          IF(J.NE.IVAR(K))GO TO 11Ø
39          TOTC(IC,IMBE)=PARAM(L)
4Ø          GO TO 14Ø
41         11Ø  CONTINUE
42          DO 12Ø IMBE=1,NMBE
43          J=J+1
44          IF(J.NE.IVAR(K))GO TO 12Ø
45          ADDC(IC,IMBE)=PARAM(L)
46          GO TO 14Ø
47         12Ø  CONTINUE
48          DO 128 IEMF=1,NEMF
49          J=J+1
5Ø          IF(J.NE.IVAR(K))GO TO 128
51          EZERO(IC,IEMF)=PARAM(L)
52          GO TO 14Ø
53         128  CONTINUE
54          J=J+1
55          IF(J.NE.IVAR(K))GO TO 13Ø
56          AJ(IC)=PARAM(L)
57          GO TO 14Ø
58         13Ø  J=J+1
59          IF(J.NE.IVAR(K))GO TO 133
6Ø          BJ(IC)=PARAM(L)
61          GO TO 14Ø
```

```
62        133 J=J+1
63            IF(J.NE.IVAR(K))GO TO 135
64            VINIT(IC)=PARAM(L)
65            GO TO 140
66        135 CONTINUE
67        140 CONTINUE
68            RETURN.
69            END
```

```
 1    C
 2    C     SUBROUTINE SNTPV
 3    C
 4    C         SPECIAL NAMES TO PARAMETERS CONVERSION ROUTINE
 5    C
 6    C         THE VALUES OF THE VARIABLES TO BE REFINED ARE TRANSFERRED
 7    C         FROM THEIR SPECIAL NAMES INTO THE VECTOR PARAM.
 8    C
 9            SUBROUTINE SNTPV(NMBEV,NEMFV,IVAR,PARAM,BETA,TOTC,ADDC,EZERO,
10           &VINIT,AJ,BJ,JTYP,JCURV,KEY,MAXPAR,MAXK,MAXTC,MAXEL,MAXCX)
11            IMPLICIT REAL*8(A-H,O-Z)
12            DIMENSION IVAR(MAXPAR),PARAM(MAXPAR),BETA(MAXK),KEY(MAXK)
13            DIMENSION NMBEV(MAXTC),NEMFV(MAXTC),VINIT(MAXTC),AJ(MAXTC)
14            DIMENSION BJ(MAXTC)
15            DIMENSION JTYP(MAXPAR),JCURV(MAXPAR)
16            DIMENSION EZERO(MAXTC,MAXEL)
17            DIMENSION TOTC(MAXTC,MAXCX),ADDC(MAXTC,MAXCX)
18            COMMON RTF,TOL,ACCM,DLNAM,AL10
19            COMMON NK,NKV,N,NSP,NCONS,NPAR,NE,NTC,NMBE0,NKW,NSET,
20           &LARS,MAXIT,IPRIN,JINP,JOUT
21            K=0
22            DO 105 IK=1,NK
23            IF(KEY(IK).NE.1)GO TO 105
24            K=K+1
25            PARAM(K)=BETA(IK)
26            JTYP(K)=0
27            JCURV(K)=0
28        105 CONTINUE
29            DO 140 K=1,N
30            J=0
31            L=NKV+K
32            DO 135 IC=1,NTC
33            NMBE=NMBEV(IC)
34            NEMF=NEMFV(IC)
35            DO 110 IMBE=1,NMBE
36            J=J+1
37            IF(J.NE.IVAR(K))GO TO 110
38            PARAM(L)=TOTC(IC,IMBE)
39            GO TO 138
40        110 CONTINUE
41            DO 120 IMBE=1,NMBE
42            J=J+1
43            IF(J.NE.IVAR(K))GO TO 120
44            PARAM(L)=ADDC(IC,IMBE)
45            GO TO 138
46        120 CONTINUE
47            DO 125 IEMF=1,NEMF
48            J=J+1
49            IF(J.NE.IVAR(K))GO TO 125
50            PARAM(L)=EZERO(IC,IEMF)
51            GO TO 136
52        125 CONTINUE
53            J=J+1
54            IF(J.NE.IVAR(K))GO TO 130
55            PARAM(L)=AJ(IC)
56            GO TO 136
57        130 J=J+1
58            IF(J.NE.IVAR(K))GO TO 133
59            PARAM(L)=BJ(IC)
60            GO TO 136
61        133 J=J+1
62            IF(J.NE.IVAR(K))GO TO 135
63            PARAM(L)=VINIT(IC)
```

```
64            GO TO 138
65       135 CONTINUE
66       136 JTYP(L)=2
67            GO TO 139
68       138 JTYP(L)=1
69       139 JCURV(L)=IC
70       140 CONTINUE
71            RETURN
72            END
```

```
 1   C
 2   C      SUBROUTINE STANS
 3   C
 4   C         A STATISTICAL ANALYSIS OF THE WEIGHTED RESIDUALS OF ALL
 5   C            THE E.M.F.VALUES
 6   C      TABLES AND/OR GRAPHS OF THE FORMATION PERCENTAGES OF THE
 7   C      COMPLEX SPECIES RELATIVE TO PRE-ESTABLISHED REACTANTS
 8   C      CAN BE OBTAINED.   THE REACTANT INDICATORS ARE CONTAINED IN THE
 9   C      VECTOR JP.
10   C
11          SUBROUTINE STANS(TITLE,JTP,EPS,WS,EMF,NEMFV,NMBEV,PQRV,PQR,
12         &TOTC,ADDC,VINIT,TITRE,KEY,HLNB,BETA,JPOT,CX,HX,CI,JP,FREEC,U,
13         &JPRIN,JCOUL,JS,CS,MAXCX,MAXK,MAXTC,MAXTP,MAXEL)
14          IMPLICIT REAL*8(A-H,O-Z)
15          INTEGER TITLE(18),PQRV(MAXCX,MAXK,MAXTC),PQR(MAXCX,MAXK)
16          INTEGER PLUS,BLANK,STAR,SYM(20),PT(102)
17          DIMENSION JPOP(8),CLIM(8)
18          DIMENSION TITRE(MAXTP)
19          DIMENSION JTP(MAXTC),NEMFV(MAXTC),NMBEV(MAXTC),VINIT(MAXTC)
20          DIMENSION JCOUL(MAXTC)
21          DIMENSION KEY(MAXK),HLNB(MAXK),BETA(MAXK),JPOT(MAXK),CI(MAXK)
22          DIMENSION JS(MAXK),CS(MAXK)
23          DIMENSION CX(MAXCX),HX(MAXCX)
24          DIMENSION TOTC(MAXTC,MAXCX),ADDC(MAXTC,MAXCX)
25          DIMENSION FREEC(MAXTP,MAXCX)
26          DIMENSION EPS(MAXTP,MAXEL),WS(MAXTP,MAXEL),EMF(MAXTP,MAXEL)
27          DIMENSION JP(MAXCX,MAXTC)
28          COMMON RTF,TOL,ACCM,DLNAM,AL10
29          COMMON NK,NKV,N,NSP,NCONS,NPAR,NE,NTC,NMBEØ,NKW,NSET,
30         &LARS,MAXIT,IPRIN,JINP,JOUT
31          DATA PLUS,BLANK,STAR,SYM/'+',' ','*','A','B','C','D','E','F','G',
32         &'H','I','J','K','L','M','N','O','P','Q','R','S','T'/
33        1 FORMAT  ('1STATISTICS ON', 18A4//)
34        2 FORMAT('1FORMATION DIAGRAMS ',18A4)
35        6 FORMAT(I5,E13.4,4X,12F8.2/18X, 8F8.2)
36        7 FORMAT('1TABLE OF FORMATION PERCENTAGES ',18A4//)
37       10 FORMAT(//E15.5,'      ARITHMETIC MEAN'/
38         &E15.5,'      MEAN DEVIATION'/
39         &E15.5,'      STANDARD DEVIATION'/
40         &E15.5,'      VARIANCE'/
41         &E15.5,'      MOMENT COEFFICIENT OF SKEWNESS'/
42         &E15.5,'·      MOMENT COEFFICIENT OF KURTOSIS'//)
43       12 FORMAT(16X,'CLASS  LIMITS',13X,'PROBABILITY',10X,'FREQUENCY',8X,
44         &'PARTIAL'/12X,'LOWER',10X,'HIGHER',7X,'CALC',7X,'OBS',7X,'CALC'·
45         &5X,'OBS',5X,'CHI-SQUARE'/)
46       17 FORMAT(I5,2E15.5,2F10.4,F10.1,I8,F12.3)
47       18 FORMAT(//' OBSERVED CHI SQUARE IS', F10.2/
48         &' CHI SQUARE(6,Ø.95)SHOULD BE ', F10.2/)
49       20 FORMAT(//' FORMATION PERCENTAGES RELATIVE TO TOTAL ',
50         &'CONCENTRATION OF REACTANT',I2,'-CURVE',I3)
51       21 FORMAT(I4,116A1)
52       23 FORMAT('ØPOINT  FREE CONCN.',12I8/16X, 8I8)
53       25 FORMAT(/5X,A1,2(49X,A1)/5X,A1,10(9X,A1)/1X,109A1)
54          NT=NMBEØ
55          HNTOT=NE
56          NR=8
57          WRITE(JOUT,1)TITLE
58          SD= DSQRT(U/(HNTOT-NPAR))
59          CLIM(1)=-1.150*SD
60          CLIM(2)=-0.675*SD
61          CLIM(3)=-Ø.319*SD
62          CLIM(4)= 0.000
```

```
 63            CLIM(5)= 0.319*SD
 64            CLIM(6)= 0.675*SD
 65            CLIM(7)= 1.150*SD
 66            CLIM(8)=SD/ACCM
 67            DO 102 I=1,NR
 68        102 JPOP(I)=0
 69            AM=0.
 70            DM=0.
 71            VAR=0.
 72            COSQ=0.
 73            COKU=0.
 74            RDEN=0.
 75            JTC=1
 76            NP0=1
 77            DO 112 JTC=1,NTC
 78            NEMF=NEMFV(JTC)
 79            NP=JTP(JTC)
 80            DO 110 IP=NP0,NP
 81            DO 110 JE=1,NEMF
 82            EPS(IP,JE)=EPS(IP,JE)*DSQRT(WS(IP,JE))
 83            DEMF=EPS(IP,JE)
 84            DO 106 K=1,NR
 85            IF(DEMF-CLIM(K))105,105,106
 86        105 JPOP(K)=JPOP(K)+1
 87            GO TO 108
 88        106 CONTINUE
 89            JPOP(NR)=JPOP(NR)+1
 90        108 AM=AM+DEMF
 91            DM=DM+ DABS(DEMF)
 92            VAR=VAR+DEMF**3
 93            COSQ=COSQ+DEMF**3
 94        110 COKU=COKU+DEMF**4
 95            NP0=NP+1
 96        112 CONTINUE
 97            AM=AM/HNTOT
 98            DM=DM/HNTOT
 99            VAR=VAR/HNTOT
100            COSQ=COSQ/(HNTOT*VAR*SD)
101            COKU=COKU/(HNTOT*VAR*VAR)
102            WRITE(JOUT,10)AM,DM,SD,VAR,COSQ,COKU
103            EXFR= 0.125
104            OBSCH=0.
105            WRITE(JOUT,12)
106            R1=-CLIM(NR)
107            DO 130 K=1,NR
108            EXPOP=EXFR*HNTOT
109            OBFR=JPOP(K)/HNTOT
110            RAPP=((JPOP(K)-EXPOP)**2)/EXPOP
111            WRITE(JOUT,17)K,R1,CLIM(K),EXFR,OBFR,EXPOP,JPOP(K),RAPP
112            R1=CLIM(K)
113        130 OBSCH=OBSCH+RAPP
114            EXPCH=12.6
115            WRITE(JOUT,18)OBSCH,EXPCH
116            IF(JPRIN.EQ.1)GO TO 500
117            NKP=0
118            DO 150 IK=1,NK
119            IF(KEY(IK).EQ.-1)GO TO 150
120            NKP=NKP+1
121            HLNB(NKP)=DLOG(BETA(IK))+AL10*JPOT(IK)
122        150 CONTINUE
123        350 GO TO(500,351,366,351),JPRIN
124        351 WRITE(JOUT,7)TITLE
125            DO 365 I=1,NT
126            IP=0
127            DO 360 JTC=1,NTC
128            IF(JCOUL(JTC).EQ.1)VOL=VINIT(JTC)
129            JT=JP(I,JTC)
130            IF(JT.NE.0)GO TO 352
131            GO TO 360
132        352 NMBE=NMBEV(JTC)
133            L=0
134            DO 1353 J=1,NK
135            IF(KEY(J).LT.0)GO TO 1353
136            L=L+1
137            DO 353 K=1,NMBE
138        353 PQR(K,L)=PQRV(K,J,JTC)
```

```
139     1353 CONTINUE
140          WRITE(JOUT,20)JT,JTC
141          NS=0
142          DO 355 J=1,NK
143          IF(PQR(JT,J))354,355,354
144      354 NS=NS+1
145          JS(NS)=J
146      355 CONTINUE
147          WRITE(JOUT,23)(JS(J),J=1,NS)
148      356 IP=IP+1
149          IF(JCOUL(JTC).EQ.0)VOL=VINIT(JTC)+TITRE(IP)
150          TS=(TOTC(JTC,JT)+TITRE(IP)*ADDC(JTC,JT))/VOL
151          DO 357 K=1,NMBE
152      357 CX(K)=FREEC(IP,K)
153   C
154   C          CALL THE ROUTINE FOR THE CALCULATION OF THE EQUILIBRIUM
155   C          CONCENTRATIONS OF COMPLEX SPECIES
156   C
157          CALL CCCS(HLNB,PQR,CX,HX,MAXK,MAXCX,NKP,NMBE,DLNAM,CI)
158          DO 358 J=1,NS
159          JJ=JS(J)
160      358 CS(J)=CI(JJ)*PQR(JT,JJ)*100./TS
161          WRITE(JOUT,6)IP,CX(JT),(CS(J),J=1,NS)
162          IF(IP.LT.JTP(JTC))GO TO 356
163      360 CONTINUE
164      365 CONTINUE
165          IF(JPRIN.LT.3)GO TO 500
166      366 WRITE(JOUT,2)TITLE
167          DO 480 I=1,NT
168          IP=0
169          DO 480 JTC=1,NTC
170          IF(JCOUL(JTC).EQ.1)VOL=VINIT(JTC)
171          JT=JP(I,JTC)
172          IF(JT.NE.0)GO TO 380
173          GO TO 480
174      380 NMBE=NMBEV(JTC)
175          DO 390 J=1,NK
176          DO 390 K=1,NMBE
177      390 PQR(K,J)= PQRV(K,J,JTC)
178          WRITE(JOUT,20)JT,JTC
179          WRITE(JOUT,25)(PLUS,K=1,123)
180      400 IP=IP+1
181          IF(JCOUL(JTC).EQ.0)VOL=VINIT(JTC)+TITRE(IP)
182          TS=(TOTC(JTC,JT)+TITRE(IP)*ADDC(JTC,JT))/VOL
183          DO 420 K=1,NMBE
184      420 CX(K)=FREEC(IP,K)
185   C
186   C          CALL THE ROUTINE FOR THE CALCULATION OF THE EQUILIBRIUM
187   C          CONCENTRATIONS OF COMPLEX SPECIES
188   C
189   C
190          CALL CCCS(HLNB,PQR,CX,HX,MAXK,MAXCX,NKP,NMBE,DLNAM,CI)
191          DO 440 K=1,102
192   C
193      440 PT(K)=BLANK
194          PT(2)=PLUS
195          PT(102)=PLUS
196          DO 460 J=1,NK
197          X=CI(J)*PQR(JT,J)*100./TS+0.5
198          K=IDINT(X)
199          IF(K-100)450,457,445
200      445 K=100
201      450 IF(K)460,460,455
202      455 K=K+2
203          IF(PT(K)-BLANK)456,458,456
204      456 PT(K)=STAR
205          GO TO 460
206      457 K=102
207      458 PT(K)=SYM(J)
208      460 CONTINUE
209          WRITE(JOUT,21)IP,PT
210          IF(IP.LT.JTP(JTC))GO TO 400
211      480 CONTINUE
212      500 RETURN
213          END
```

```
 1    C
 2    C       SUBROUTINE WEIGHT
 3    C          WEIGHTS FOR ALL E.M.F.EXPERIMENTAL VALUES ARE COMPUTED
 4    C
 5            SUBROUTINE WEIGHT(IP,NPC,NEMF,TITRE,EMF,SIGMAV,SIGMAE,
 6           &WS,H,X,Y,Y2,MAXTP,MAXEL,MAXSP)
 7            IMPLICIT REAL*8(A-H,O-Z)
 8            DIMENSION TITRE(MAXTP)
 9            DIMENSION H(MAXSP),X(MAXSP),Y(MAXSP),Y2(MAXSP)
10            DIMENSION P(6)
11            DIMENSION SIGMAE(MAXEL)
12            DIMENSION EMF(MAXTP,MAXEL),WS(MAXTP,MAXEL)
13            DATA P/0.0D0,1.0D0,0.0D0,0.0D0,1.0D0,0.0D0/
14            JPI=IP-NPC
15            DO 50 JE=1,NEMF
16            DO 10 I=1,NPC
17            JP=JPI+I
18            X(I)=TITRE(JP)
19         10 Y(I)=EMF(JP,JE)
20    C
21    C          CALL THE ROUTINE FOR CUBIC SPLINE INTERPOLATION
22    C
23            CALL SPLINE(H,X,Y,Y2,NPC,P)
24            WS(JPI+1,JE)=(Y(2)-Y(1))/H(1)-H(1)*(Y2(1)+Y2(2)/2.D+00)/3.D+00
25            DO 20 J=2,NPC
26            K=JPI+J
27            I=J-1
28         20 WS(K,JE)=(Y(J)-Y(I))/H(I)+H(I)*(Y2(I)/2.D+00+Y2(J))/3.D+00
29            Y2(1)=WS(JPI+1,JE)
30            IP1=NPC-1
31            DO 30 J=2,IP1
32            I=J-1
33            K=J+1
34            W1=WS(JPI+I,JE)
35            W2=WS(JPI+J,JE)
36            W3=WS(JPI+K,JE)
37         30 Y2(J)=.1*(W3+W1)+.4*W2+.2*((Y(J)-Y(I))/H(I)+(Y(K)-Y(J))/H(J))
38            Y2(NPC)=WS(JPI+NPC,JE)
39            DO 40 I=1,NPC
40            JP=JPI+I
41         40 WS(JP,JE)=1./(SIGMAE(JE)+Y2(I)*Y2(I)*SIGMAV)
42         50 CONTINUE
43            RETURN
44            END
```

5.2.1. Input Data

The following is a listing of a data set generated for a hypothetical chemical model of four reactants forming 12 complex species. Random errors, to simulate instrumental fluctuations, were applied to the emf values.

¶<-- Column 1						
"PERFECT" TITRATIONS FOR PROGRAM TESTING						
1 12	0	20	2	4	0	
25.00000	.00200	.10000	.10000			
1.20000	4	0	1	1	0	1
3.90000	-3	0	1	1	-1	1
2.50000	1	0	2	1	-1	1
5.20000	-7	0	2	1	-2	1
5.70000	-7	0	1	2	-2	1
2.20000	-13	0	1	2	-3	1
1.20000	8	1	1	1	0	1
3.50000	1	1	1	1	-1	1
4.00000	0	2	2	1	-2	1
2.00000	6	1	0	0	1	1
1.30000	8	0	1	0	1	1
1.75000	-14	0	0	0	-1	0
4	1	2	3	4	1	3

1	0	0	
.10000	.10000	.05000	.30000
.00000	.00000	.00000	-.20000
350.00000	.00000	.00000	
80.00000			

0	.00	178.26
0	.10	172.66
0	.20	165.00
0	.30	153.98
0	.40	137.24
0	.50	81.11
0	.60	25.48
0	.70	9.14
0	.80	-2.23
0	.90	-11.79
0	1.00	-21.26
0	1.10	-29.36
0	1.20	-39.29
0	1.30	-50.45
0	1.40	-63.53
0	1.50	-78.75
0	1.60	-93.72
0	1.70	-108.48
0	1.80	-125.81
0	1.90	-150.46
1	2.00	-212.63

4	1	2	3	4	2	3
2	1	0			
.05000	.05000	.05000	.10000		
.00000	.00000	.00000	-.20000		
150.00000	350.00000	.00000	.00000		
80.00000					

0	.00	55.22	69.37
0	.04	54.90	28.18
0	.08	54.26	11.58
0	.12	53.69	.79
0	.16	52.79	-6.29
0	.20	51.99	-13.26
0	.24	50.83	-19.31
0	.28	50.00	-24.79
0	.32	48.81	-30.56
0	.36	47.43	-35.17
0	.40	46.05	-40.41
0	.44	44.26	-44.44
0	.48	42.03	-50.20
0	.52	39.91	-54.69
0	.56	37.40	-60.30
0	.60	34.31	-65.82
0	.64	30.19	-72.22
0	.68	24.38	-80.71
0	.72	17.34	-89.98
0	.76	6.17	-103.79
0	.80	-7.64	-121.79
0	.84	-22.15	-141.65
0	.88	-41.96	-170.39
0	.92	-66.92	-207.82
0	.96	-84.62	-234.37
1	1.00	-93.50	-247.66

3	2	3	4	2	2
2	1	0			
.80000	.40000	1.00000			
.00000	.00000	-.20000			
150.00000	350.00000	.00000	.00000		
80.00000					

0	.00	81.91	196.09
0	.30	81.80	186.79
0	.60	81.79	172.67
0	.90	81.93	137.75
0	1.20	80.47	49.92
0	1.50	78.62	25.69
0	1.80	76.58	13.20
0	2.10	74.63	4.72
0	2.40	72.59	-2.13
0	2.70	70.88	-7.86
0	3.00	68.56	-12.92
0	3.30	66.03	-17.81

0	3.60	63.45	-22.50		0	1.60	-250.04		
0	3.90	60.78	-27.28		0	1.70	-268.12		
0	4.20	57.50	-32.01		0	1.80	-278.28		
0	4.50	53.95	-37.18		0	1.90	-285.60		
0	4.80	49.62	-42.75		1	2.00	-291.36		
0	5.10	45.05	-48.33		2	2	4	1	0
0	5.40	39.65	-54.84		1	0	0		
0	5.70	33.57	-61.87		.20000	.25000			
0	6.00	27.04	-69.42		.00000	-.20000			
0	6.30	19.90	-77.09	350.00000	.00000	.00000			
0	6.60	12.29	-85.38	75.00000					
0	6.90	4.88	-93.72		0	.00	161.82		
1	7.20	-2.86	-102.01		0	.10	149.26		
2	1	4	1	0	0	.20	121.32		
1	0	0			0	.30	-53.77		
.20000	.30000				0	.40	-85.40		
.00000	-.20000				0	.50	-101.64		
350.00000	.00000	.00000			0	.60	-113.78		
80.00000					0	.70	-124.70		
0	.00	178.12			0	.80	-134.81		
0	.10	172.51			0	.90	-145.77		
0	.20	165.20			0	1.00	-157.60		
0	.30	154.65			0	1.10	-174.20		
0	.40	138.03			0	1.20	-199.83		
0	.50	88.55			0	1.30	-240.52		
0	.60	33.63			0	1.40	-262.45		
0	.70	12.89			0	1.50	-275.92		
0	.80	-1.62			0	1.60	-284.27		
0	.90	-12.22			0	1.70	-290.76		
0	1.00	-22.96			0	1.80	-295.65		
0	1.10	-32.84			0	1.90	-299.87		
0	1.20	-44.80			-1	2.00	-303.44		
0	1.30	-58.27			4				
0	1.40	-79.29			-1				
0	1.50	-164.12							

6

SQUAD

Stability Quotients from Absorbance Data

DAVID J. LEGGETT

1. INTRODUCTION

Compared to potentiometric methods, the spectrophotometric determination of formation constants, aided by computer-based data analysis, has not been extensively developed. The reasons are several, including the apparently lower accuracy of spectrophotometric data; the much larger amount of data generated by digitizing the spectra; numerical problems associated with data processing; lack of precision (compared to potentiometric investigations) of the refined constants.

The advent of inexpensive mini- and micro-computers, together with a wide range of on-board computer-controlled spectrophotometers, answer the first two difficulties with this technique. For example, Zuberbuhler and Kaden have demonstrated the solution to these problems,[1] including full processing of the spectral data.[2] The third problem is being addressed now that more research efforts are directed toward this area of solution equilibria. The fourth difficulty remains largely unresolved. Nonetheless, spectrophotometric investigation of complex formation, when used in conjunction with potentiometry, provides a reliable and independent means to verify the equilibrium model.[3,4]

The program described in this chapter is SQUAD. It is designed to calculate the best values for the stability constants of the proposed equilibrium model by employing a nonlinear least-squares approach. The program is completely general in scope, having the capability to refine stability constants for the general complex $M_m M'_l H_j L_n L'_q$, where $m, l, n, q \geq 0$ and j is positive (protons), negative (hydroxide ions), or zero. Therefore, the same program may be used to study acid-base equilibria for ligands that are weak acids (or bases); $M_m L_n$; mixed-ligand (or mixed-metal) complexes; protonated or hydroxo complexes. SQUAD was originally designed to process absorbance data from aqueous solutions. Recently extensive modifications have been made to the

---

DAVID J. LEGGETT • Dow Chemical U.S.A., Texas Division, Freeport, Texas 77541.

program that now permit the analysis of data from any type of solution. The current version of SQUAD is, in many respects, a distillation of a number of earlier programs. Nagano and Metzler[5] developed PITMAP implementing a FORTRAN coding of Sillen's twist matrix algorithm[6] and using the concept of "two-level adjustment of common and group parameters."[7] PITMAP was extensively modified by Thompson[8] but still required recoding of the species concentration subroutine for each new model tested. This represented a considerable handicap when many models were fitted to the data, and the problem was alleviated by incorporating subrouting COGSNR, from SCOGS,[9] into PITMAP.[10] Experiences with this version of PITMAP were not entirely satisfactory and a new approach was therefore sought to extract formation constants from spectrophotometric data. Consequently, SCOGS was rewritten in subroutine form, one set of routines for the Gauss–Newton least-squares algorithm and another set to solve the mass balance equations. A sum of squares of residuals subroutine was added, which incorporated Cramer's method for solving a system of linear equations, as coded in PITMAP.[7, 8] Thus the original version of SQUAD was born.[11] Extensive use of this version of SQUAD uncovered the problem of the calculation of negative molar absorptivities. This difficulty was overcome by replacing Cramer's algorithm with a multiple regressive (MR) and a nonnegative linear least-squares algorithm (NNLS);[12] MR or NRLS being used, at the user's discretion, to solve the Beer's law equations.

The final major modification to SQUAD has involved the method of data entry. Up to this point the model had been defined, within the input data file, by a large number of integers. The various calculations and output options were also chosen using numerical values. It was found that new users of SQUAD rapidly became disenchanted with the intricacies of data input and as a result did not fully utilize the program. The current version of SQUAD, described below, is distinctly user-friendly. This has resulted in an increase in the size of the program, though execution times have not been adversely affected.

SQUAD has been distributed internationally over the past four years. Kadish *et al.* have used the program, in conjunction with cyclic voltammetry, to study axial binding reactions and redox properties of ruthenium (II) and ruthenium (III) tetraphenylporphyrin axial ligation investigations.[14] Cruywagen and Heyns[15] employed SQUAD to evaluate the pK_a's of oxalic acid using UV data in the range 200–230 nm.[15] Vlckova *et al.* have investigated the reactions between Cd and 4-(2-pyridylazo)resorcine with SQUAD.[16] Podlahova and Podlaha determined constants for reactions between twelve metal ions and ethylenediphosphinetetraacetic acid.[17] O'Mara, Walsh, and Hynes performed a SQUAD-assisted study of the interactions of Ni^{2+} and Co^{2+} with Murexide.[18]

2. DESCRIPTION OF THE PROGRAM SQUAD

2.1. Mathematical Algorithms

SQUAD employs the two-level adjustment of parameters suggested by Sillen.[7] The strategy used in shown in Figure 1.

The subroutines DIFF, RESID, ECOEF, CCSCC (and COGSNR), and SOLVE

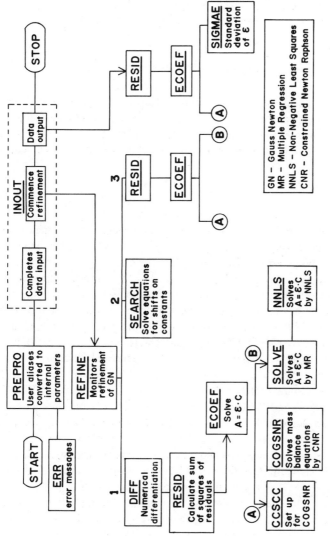

FIGURE 1. A block diagram for the computer program SQUAD showing the interrelationships of the various subroutines.

are called successively by subroutine REFINE. Route 1, route 2, and then route 3 are followed, in order, during a single refinement cycle. Once convergence has been achieved control passes back to the data output section of subroutine INOUT for the postrefinement printouts. The algorithms in SQUAD are well-tried versions of familiar procedures, that have been used in many chemically oriented computational applications. Nonetheless, the coding of SQUAD is structured such that any of the algorithms may be replaced by newer, more efficient numerical methods.

For each absorbance value $A_{i,k}$ the equation

$$A_{i,k} = \sum_{1}^{J} [\text{species}]_{i,j} \times \epsilon_{j,k} \tag{1}$$

where $[\text{species}]_{i,j}$ is the concentration of the jth species in the ith solution (spectrum) and $\epsilon_{j,k}$ is the molar absorptivity of the jth species at the kth wavelength. SQUAD computes the values of the overall formation constant(s) which minimize the sum of the squared residentials between observed and calculated absorbance values:

$$U = \sum_{1}^{K} \sum_{1}^{I} (A_{i,k}^{\text{obs}} - A_{i,k}^{\text{calc}})^2 \tag{2}$$

The main program starts by initializing I/O unit numbers and maximum sizes for all execution-time dimensioned arrays. Subroutine PREPRO is then called to perform the majority of the data-input and input file error checking. This routine will be described in Section 2.2. Control then passes to INOUT, completes data input, and then commences the refinement of the formation constants by a single call to REFINE. The major subroutines invoked during the refinement process will now be described.

Subroutine REFINE. This routine controls the principal refinement algorithm, which uses the classic Gauss–Newton approach. However, prior to commencing the refinement process a grid search is performed to determine if better initial estimates for the formation constants can be located. This procedure may be omitted by removing the call to subroutine GRID. During a single refinement cycle REFINE calls, in turn, routines DIFF and SEARCH. Once the new set of constants have been found RESID is called again and the fit to the data determined. Convergence is deemed complete if all shifts are less than 0.1% of each constant. However the process can be terminated by a user-selected maximum number of iterations, as indicated in Section 4.2. Finally, the standard deviations of the molar absorptivities are determined before control returns to INOUT for the final printout of the results.

Subroutines DIFF and SEARCH. Numerical differentiation, employing Stirling's central difference algorithm, is performed by DIFF. The increment for differentiation is fixed at 0.5% of each constant. The Jacobian and Hessian matrices are developed within DIFF. They are solved by SEARCH yielding the shifts to be applied to those formation constants being refined. The correlation matrix is also calculated at this time. The reader's attention is directed to Gans' review[19] for a lucid description and discussion of the Gauss–Newton algorithm.

Subroutine RESID. This routine acts as a service routine, during the numerical differentiation step in the Gauss–Newton algorithm, and to accumulate U, the squared sum of residuals. The routine also controls the solution of Beer's law for the current set of formation constants, calling subroutine ECOEF for this purpose.

Subroutines ECOEF and SIGMAE. For a given set of constants ECOEF first calls CCSCC, which will return the concentrations of complexed and uncomplexed species in solution. At this point ECOEF prepares arrays for the solution of the linear equations (Beer's law) by SOLVE. Prior to calling SOLVE the absorbance array is "corrected" for those species having known molar absorptivities. Essentially this entails the formation of a new absorbance matrix, $A_{i,k}^{\text{unknown}}$ from $(A_{i,k}^{\text{total}} - A_{i,k}^{\text{known}})$, where $A_{i,k}^{\text{known}}$ is the absorbance calculated from known molar absorptivities and the appropriate species concentrations calculated by CCSCC. This process results in a procedurally simpler problem for SOLVE to handle. Once the molar absorptivities have been calculated they are inserted back into the total molar absorptivity array; $A_{i,k}^{\text{calc}}$ is then obtained and hence U. When convergence has been achieved a final call is made to ECOEF, at the entry point SIGMAE, to obtain the standard deviations of the calculated molar absorptivities.

Subroutines CCSCC and COGSNR. These two routines are used to calculate the concentrations of all species in solution, as dictated by the model under investigation. The total concentration of each component, the pH of the solution (if applicable), and the current set of formation constants are input data for these routines. [Note: The phrase "current set of formation constants" means all sets encountered during the refinement process including those temporary values generated during the numerical differentiation procedure.] CCSCC operates as a bookkeeping and control routine for COGSNR. The latter routine solves the mass balance equations, solution by solution, using a constrained Newton–Raphson procedure. The coding of COGSNR used in SQUAD is a generalized version of that originally presented by Tobias and Yasuda in GAUSS-G.[20] As such it represents a well-tried and robust implementation of the Newton–Raphson algorithm, failing only in the most extreme conditions.[21]

Subroutine SOLVE. Once the concentrations of all species for the current set of constants have been calculated, and the absorbance matrix corrected, control passes to subroutine SOLVE. At this point one of two algorithms is selected to solve the overdetermined set of simultaneous equations. For most data sets the traditional approach of multiple regression (MR) may be employed. However, in certain situations the nonnegative linear least squares (NNLS) may be invoked, where the solution vector is constrained to be equal to or greater than zero. The numerical aspects of MR and NNLS have been previously described.[12] Subroutine NNLS uses routines H12, G1, and G2. Further discussion on the use of NNLS and MR will be presented in Section 2.3.

Subroutine SEARCH. When the matrix of nonlinear least-squares equations is complete control returns to REFINE and then to SEARCH, where these equations are solved, according to the Gauss–Newton algorithm, to yield the shifts to be applied to the refining constants. Checks are made to ensure that these shifts do not exceed preset limits, and if so, the shifts are adjusted accordingly. The covariance matrix is also calculated at this point.

Finally control passes back to REFINE, where the progress of refinement is determined. If convergence has been attained the program returns to INOUT and begins the output. Otherwise another refinement cycle is commenced.

Subroutine INOUT. As described earlier INOUT completes the data input and calls subroutine REFINE. Once convergence has been achieved INOUT controls the output of the results. First tables of all concentrations for all species and solutions are printed, followed by the calculated molar absorptivities together with their standard deviations. A number of printer plots are produced, the quantity being controlled by an input keyword (see Section 4.2).

2.2. Principles of Data Input

One important aspect of computer programs of the size of SQUAD is that the method of data input be as straightforward as possible. This is particularly true for first-time users and users with little programming experience. Considerable attention has been paid to this aspect of SQUAD. The method of data input will now be described.

The data file consists of seven sections, some of which may be omitted. Each section starts with a keyword and most finish with the keyword **END.** Table I gives an overview of the basic components of data. Subroutine PREPRO performs the task of parsing the "text" of the first two sections of the input file. The first section, **DICTIONARY,** provides SQUAD with the alias names that will be used to identify the components of the complexes. These names are used to "connect" the user to the program variables MTL1 (first metal in complex), MTL2 (second metal), LIG1 (first ligand in complex), LIG2 (second ligand), PROT (the proton), HYDR (hydroxide ion as ligand).

The second section of the data deck, **SPECIES,** is used to define the model. Abbreviations are included, when defining each complex, to indicate if the formation constants are to be refined (VB) or held constant (FB), and if molar absorptivities are to be calculated (VE) or are already known (FE). Thus, PREPRO acts as a simple compiler enabling species definitions and alias names to be converted into internal vectors and arrays that define the model being fitted to the data. Section 4.2 gives the rules that pertain to **DICTIONARY** and **SPECIES,** together with an example.

These two sections have proven to be valuable in allowing first-time users to employ SQUAD with little or no difficulty. Although the method of data input is straightforward, errors can occur, such as misspelling component names or omitting a semicolon or parenthesis. Extensive error checks are performed within PREPRO and appropriate error messages, via subroutine ERR, will be displayed indicating the likely source and probable location of the error. Consequently, an incorrectly constructed data set will be detected by the program rather than by the compiler.

The third section, **OTHER,** is used to indicate if the molar absorptivities of any of the *components* are to be calculated, or if the molar absorptivities of more than one species are the same. The fourth section, **DATA,** provides information relating to the

TABLE 1. Overview of Data Deck Layout

A Suitable Title	
A Suitable Sub-title	
DICTIONARY:	
.	
.	
.	Section 1 (required)
.	
END:	
SPECIES.	
.	
.	Section 2 (required)
.	
END:	
OTHER:	
.	
.	Section 3 (optional)
.	
DATA:	
.	
.	Section 4 (required)
.	
MOL.ABS.:	
.	
.	
END:	
.	
.	Section 5 (required)
.	
BASELINE:	
.	
.	Section 6 (optional)
.	
SPECTRA:	
.	
.	Section 7 (required)
.	
−1.0	Last Card (required)

basic parameters of the system under study. These data include the wavelength range of interest, whether β's or log β's are to be refined, which algorithm to use when solving the system of linear equations arising from Beer's law, etc. The fifth section, **MOL.ABS.**, contains the known molar absorptivities, previously determined or known to be zero, for any component and/or complex species. The data are prefaced by the formulas of the component and/or complex species.

Section 6, **BASELINE,** permits baseline corrections to be made, if needed. Section 7, **SPECTRA,** holds the total concentrations of each component, the pH of the solution (if applicable), the pathlength, and the absorbance data for each spectrum.

Full details and explanatory examples of the data input requirements are presented in Section 4.2.

2.3. Suggestions for the Use of SQUAD

Assuming that spectral data have been gathered and a satisfactory model has been fitted to the data, the question "How do we know the answers are correct?" must be addressed. Clearly the model is only applicable to the current set of data. Nonetheless, provided that no gross errors have been made during the data gathering stage it *may* be safe to assume that the refined constants are representative of the metal–ligand equilibrium system. It is *wiser*, though, to prepare fresh stock solutions and obtain spectral data for different $C_M : C_L$ ratios, thereby building confidence in the deduced model and associated constants.

SQUAD calculates a number of statistical parameters that may be used to judge the validity of a particular model. The correlation matrix, obtained from the inverted Hessian matrix, indicates which constants are not separately defined by the current data. An analysis of the correlations between constants may be employed to suggest additional $C_M : C_L$ ratios to investigate.

Two other parameters are calculated by SQUAD. These are σ_{DATA} and σ_{CONST}. The "standard deviation in the absorbance data," σ_{DATA}, provides an overall measure of the fit of the model to the data. Assuming that the correct model has been found, then residuals due to experimental error alone would lead to a value of σ_{DATA} of about ± 0.001 to ± 0.004, depending on the type of spectrophotometer used. Experience has shown that $\sigma_{\text{DATA}} >$ approx. 0.01 is indicative of a poor fit. Estimates for the standard deviation of each refined constant, σ_{CONST}, approximately 1% of that constant, are indicative of a good fit. The reader's attention is drawn to the comments of Braibanti and Brushci[22] concerning acceptable levels of standard deviations of formation constants. Several situations may give rise to larger than expected values for σ_{DATA} and σ_{CONST}. For example, the model ML is being fitted to the data but in reality both ML and ML_2 are formed. As a result both σ_{DATA} and σ_{CONST} will be high and inspection of σ_{SPECT} (the standard deviation of *each* spectrum) would reveal increasing values as $C_L : C_M$ increases. The reverse situation, where ML *and* ML_2 are being fitted but the data is only representative of ML, would cause σ_{DATA} to be high but show no distinct trend in σ_{SPECT}. The value of the constant for ML_2 would be successively reduced eventually eliminating the contribution of ML_2 to the refinement process. This would be shown by increasing values for σ_{SPECT} for ML_2 and subsequent termination of the refinement.

Gross errors in preparing stock solutions or preparing solutions for a specific spectrum will also lead to large values for σ_{DATA} and σ_{CONST}. The former error is often difficult to uncover until new stock solutions are prepared. However, the latter problem is easily detected by inspecting the values of σ_{SPECT}. Removing the data related to the "bad" solution should reduce the values of σ_{DATA} and σ_{CONST} but *not* substantially alter the values of the refined constants. Finally, careless analytical technique will give rise to numerous, small sources of experimental error resulting in high values for σ_{DATA}. This type of problem may also be apparent in a slow rate of convergence. Obviously the samples of metal and ligand should be as pure as possible. However, efforts devoted to laborious purifications followed by inadequate precautions in solution preparation gain little but needless frustration.

Examination of the calculated molar absorptivities for the complex(es) may also provide an indication of the quality of the data and the model. Plots of these values, for each species, should give smooth continuous curves usually bearing some resemblance to portions of the observed spectra for various $C_M : C_L$ ratios. Severely disjointed plots are strongly indicative of a less than adequate fit to the data. It is also possible that negative molar absorptivities will be calculated by SQUAD. Although this situation arises most frequently in conjunction with problems described earlier in this section, there are instances where the occurrence of negative values may be traced to numerical problems. These can occur due to small spectral contributions of one or more species to a part of the spectral region caused by low concentration *or* weak absorption by the species in that part of the spectrum.[12]

SQUAD provides the user with two algorithms for solving the system of linear equations arising from Beer's law. The multiple regression algorithm is used during the initial data refinement. If negative molar absorptivities are detected the data should be first checked for data-entry and/or for experimental errors. All plausible models are then tested to ascertain that the negative values are not due to fitting the wrong model. However, should all these strategies fail to remove the negative values, then the user would switch to the nonnegative least-squares algorithm (NNLS). A practical demonstration of the validity and use of NNLS has been provided in a previous publication.[12]

2.4. Modifications to SQUAD

SQUAD has been written in discrete subroutines, each one or group, representing the implementation of a specific algorithm. Consequently, the user may replace COGSNR, for example, with a more effective routine that will solve the mass balance equations, at a specified pH. Alternatively, those users who feel more comfortable with the Marquardt algorithm would replace subroutines DIFF and SEARCH by an appropriate block of code while leaving REFINE virtually intact and not disturbing RESID, CCSCC, COGSNR, ECOEF, NNLS, etc.

The program uses execution-time dimensioning and consequently there are no reasonable restrictions on the amount of data or size of model that may be processed. However, experience has shown that attempts to refine six or more constants simultaneously in a model are likely to lead to nonconvergence unless the initial estimates are within one log unit of the final values. In practice, the normal procedure of building the model, starting with the major species, will avoid this problem. The main program provides a list of those arrays that need to be increased (or decreased) in size to fit the user's particular needs.

3. REFERENCES

1. G. Hanisch, T. A. Kaden, and A. D. Zuberbuhler, Handling of Electronic Absorption Spectra with a Desk Top Computer—I. A Fully Automatic Spectrophotometric Titration System with On-line Data Acquisition, *Talanta* **26**, 563–567 (1979).
2. A. D. Zuberbuhler and T. A. Kaden, Handling of Electronic Absorption Spectra with a Desk Top Computer—II. Calculation of Stability Constants from a Spectrophotometric Titrator, *Talanta* **26**, 1111–1118 (1979).
3. F. R. Hartley, C. Burgess, and R. Alcock, *Solution Equilibria*, Chap. 5, Halstead Press, J. Wiley & Sons, New York (1980).
4. W. A. E. McBryde, Spectrophotometric Determination of Equilibrium Constants in Solution, *Talanta* **21**, 979–1004 (1974).
5. K. Nagano and D. E. Metzler, Machine Computation of Equilibrum Constants and Plotting of Spectra of Individual Ionic Species in the Pyridoxal-Alanine System, *J. Am. Chem. Soc.* **89**, 2891–2900 (1967).
6. N. Ingri and L. G. Sillen, High-Speed Computer as a Supplement to Graphical Methods III. Twist Matrix Methods for Minimizing the Error-Square Sum in Problems with Many Unknown Constants, *Acta Chem. Scand.* **18**, 1085–1098 (1964).
7. L. G. Sillen and B. Warnqvist, High-Speed Computer as a Supplement to Graphical Methods. 6. A Strategy for Two-level LETAGROP Adjustment of Common and "Group" Parameters. Some Features that Avoid Divergence, *Acta Chem. Scand.* **31**, 315–339 (1968).
8. J. A. Thompson, Ph.D. thesis, University of Waterloo, Department of Chemistry, 1970.
9. I. G. Sayce, Computer Calculations of Equilibrium Constants of Species Present in Mixtures of Metal Ions and Complexing Agents, *Talanta* **15**, 1397–1411. (1968).
10. D. J. Leggett and W. A. E. McBryde, Metal Ion Interactions of Picoline-2-Aldehyde Thiosemicarbazone, *Talanta* **22**, 781–789 (1975).
11. D. J. Leggett and W. A. E. McBryde, General Computer Program for the Computation of Stability Constants for the Absorbance Data, *Anal. Chem.* **47**, 1065–1070 (1975).
12. D. J. Leggett, Numerical Analysis of Multicomponent Spectra, *Anal. Chem.* **49**, 276–281 (1977).
13. K. M. Kadish, D. J. Leggett, and D. Chang, Investigation of the Electrochemical Reactivity and Axial Legand Binding Reactions of Tetraphenylporphyrin Carbonyl Complexes of Ruthenium (II), *Inorg. Chem.* **21**, 3618–3622 (1982).
14. D. J. Leggett, S. L. Kelly, L. R. Shiue, Y. T. Wu, D. Chang, and K. M. Kadish, A Computational Approach to the Spectrophotometric Determination of Stability Constants—II. Application to Metalloporphyrin–Axial Ligand Interactions in Non-Aqueous Solvents, *Talanta* **30**, 579–586 (1983).
15. J. J. Cruywagen and J. B. B. Heyns, Determination of the Dissociation Constants of Oxalic Acid and the Ultraviolet Spectra of the Oxalate Species in 3M Perchlorate Medium, *Talanta* **30**, 197–200 (1983).
16. S. Vlchova, L. Jancar, V. Kuban, and J. Havel, Spectrophotometric Study of the Complex Equilibria of Cadmium (II) Ions with 4-(2-Pyridylazo)Resorcine (PAR) Using the SQUAD-G Program and the Method of Determining Cd(II) Ions with PAR, *Coll. Czech. Chem. Commun.* **47**, 1086–1099 (1982).
17. J. Podlahova and J. Podlaha, The Stability Constants of Ethylenediphosphinetetraacetate Complexes, *Coll. Czech. Chem. Commun.* **47**, 1078–1085 (1982).
18. C. O'Mara, J. Walsh, and M. J. Hynes, The Interaction of Ni^{2+} and Co^{2+} with Murexide, *Inorg. Chim. Acta*, **92**, L1–L2 (1984).
19. P. Gans, Numerical Methods for Data-Fitting Problems, *Coord. Chem. Rev.* **19**, 99–124 (1976).
20. R. S. Tobias and M. Yasuda, Computer Analysis of Stability Constants in Three-Component Systems with Polynuclear Complexes, *Inorg. Chem.* **2**, 1307–1310 (1963).
21. D. J. Leggett, Programs to Compute Equilibrium Concentrations—Some Practical Considerations, *Talanta* **24**, 535–542 (1977).
22. A. Braibanti and C. Bruschi, Evaluation of Errors in Concentrations of Complexes in Systems at Equilibrium, *Ann. Chim.* **67**, 471–480 (1977).

4. INSTRUCTIONS FOR SQUAD

4.1. Introduction

Section 4.2 describes, in detail, the required and optional parts of a data input file for SQUAD; a general overview of the file structure has been given in Section 2.2. The FORMAT specification for each input item is given in square brackets. A knowledge of FORMAT codes for character, integer, and floating point data is assumed.

4.2. Details of the Data Input

The input file consists of the following lines. All entries begin in column one, unless specified.

Section 0 1 line [20A4] a suitable title
         1 line [20A4] a suitable subtitle, or blank card

Section 1 1 line [80A1] **DICTIONARY:**
         1 line [80A1] **MTL1**=aaaa;**MTL2**=bbbb;**LIG1**=cccc;
                                       LIG2=dddd;**PROT**=eeee;**HYDX**=ffff:
         1 line [80A1] **END:**

Not all the key words, **MTL1**=,**MTL2**=, *etc.*, need be present. An example illustrates the point. For the system copper/alanine in water,

    MTL1=Copper=**CU**   **LIG1**=Alanine=**ALA**

    PROT=Proton=**H**   **HYDX**=Hydroxide=**OH**

Therefore Section 1 would be

    DICTIONARY:
    MTL1=CU;LIG1=ALA;PROT=H;HYDX=OH:
    END:

Rules   1. User-supplied aliases are limited to no more than four characters.
         2. Each alias is separated by a semicolon (;), except at the end of a line, where a colon (:) is used.
         3. All component aliases used in the SPECIES section must have been defined within **DICTIONARY**.
         4. Component aliases may appear in any order.
         5. Embedded blanks will *not* be removed.

Section 2 1 lines [80A1] **SPECIES:**
        N1 lines [80A1] Species descriptor, one line per complex, N1 species.
        1 line [80A1] **END:**

The species descriptor is built up as shown in the following example. For the copper/alanine system the complexes, known to exist, are HL, H_2L, $Cu_2(OH)_2$. The complexes ML, ML_2, and MHL are believed to exist. Their presence is to be confirmed from the data to be processed. The species descriptor cards are set up as follows:

SPECIES:
CU(1)ALA(1);8.20;VB;VE:
CU(1)ALA(2);15.60;VB;VE:
CU(1)H(1)ALA(1);10.9;VB;VE:
H(1)ALA(1);9.50;FB;FE:
H(2)ALA(1);12.30;FB;FE:
CU(2)OH(2); -10.86;FB;FE:
END:

Rules   1. The stoichiometric coefficients of each complex are enclosed in parentheses.
2. The formula of the complex is terminated by a semicolon.
3. The constant follows the formula and is read in as the logarithm of the overall formation constant. IUPAC convention is used.
4. The constant may be either fixed (**FB**) or refined (**VB**) by SQUAD. Molar absorptivities may be either read in as part of the input file (**FE**), as assumed to be zero (**FE**), or calculated by SQUAD (**VE**).
5. The end of each species descriptor is signified by a colon.

Section 3   1   line   [80A1]   **OTHER**:
N2   lines   [80A1]   OTHER options, N2 possibilities.
1   line   [80A1]   **END**:

This section is optional and may be omitted if none of the **OTHER** options are needed. Two calculation possibilities may be invoked using this section. First, the molar absorptivities of the free uncomplexed species may be calculated. This is achieved by using the name of the component followed by **VE**. Second, the molar absorptivities of two, or more, species may be forced to be equal. Consider species A and B. Beer's law for these species would be

$$A = \epsilon_A [A] + \epsilon_B [B]$$

If $\epsilon_A = \epsilon_B$ for all wavelengths considered then

$$A = \epsilon_{AB} ([A] + [B])$$

This calculation option is included by using the formula of each species, separated by a plus (+) sign. For example,

OTHER:
CU;VE:
CU(1)H(1)ALA(1)+CU(1)H(2)ALA(1):
END:

Section 4   1   line   [80A1]   **DATA:**

1   line   [3F10.3]   Start wavelength, λ_{min}, stop wavelength, λ_{max}, wavelength increment, λ_{inc}, for the absorbance data

1   line   [A4]   **LOGB** Refine log β's $\Big\}$ Choose one

                     EXPB Refine β's

Note: All β's in the **SPECIES** section must be either log β or β—they cannot be mixed.

1   line   [A4]   **PRIN** Initial conditions printed $\Big\}$ Choose one

                     NOPR Reduced output

1   line   [A4]   **CARD** Direct portion of output to card punch $\Big\}$ Choose one

                     NOCD No cards produced

Note: The use of **CARD** will provide output for species concentrations and molar absorptivities of each species at each wavelength for all spectra. The unit number of the output device is specified in the main program as the integer variable PUNIT.

1   line   [A4]   **MR** Multiple regression algorithm $\Big\}$ Choose one

                     NNLS Nonlinear least-squares algorthim

1   line   [A4,I2]   **NOPL** No printer plots produced $\Big\}$ Choose one

                     PLOT 1 Printer plot produced for every spectrum

                     PLOT 2 Printer plot produced for every second spectrum.

1   line   [A4]   **CRT** 80 column printer plots $\Big\}$ Choose one

                     PAGE 132 column printer plots

Note: If **NOPL** has been choosen either **CRT** or **PAGE** *must* be included at this point. The line cannot be left blank or omitted.

1   line   [I3]   Number of iterative cycles to be performed before quitting. SQUAD will stop automatically if convergence has been achieved before the maximum number has been reached.

1   line   [F10.3]   Gamma H value. Used to convert 10.0^{**} ($-pH$ reading) to [Hydrogen Ion]. (See W. A. E. McBryde, *Analyst (London)* **94,** 337 (1969) for details.) If unknown or inapplicable set 1.0.

An example of the setup for this section is shown below:

```
Col. 1            Col. 11   Col. 21  Col. 31
↓                 ↓         ↓        ↓
DATA:
                  360.0     500.0    5.0
LOGB
PRIN
NOCD
MR
PLOT 1
CRT
 10
                  0.82
```

Note: This section does not end with the **END:** keyword.

Section 5 1   line   [80A1]   **MOL.ABS.**:
             1   line   [80A1]   Species formulas
             1   line   [80A1]   **END**:
             N3 lines [8D10.4] Molar absorptivities

This section is optional. The input is in two parts. The first comprises the species (or free components) which have known molar absorptivities. Each species is separated by a semicolon. The input line is terminated by a colon. The second part consists of the values of the molar absorptivities. They are entered in the order determined by the species formulas card. There will be as many sets as wavelengths, i.e., N3 sets. The layout is shown below:

```
Col. 1            Col. 11          Col. 21 . . . . . . .
↓                 ↓                ↓
MOL.ABS.:
CU; CU(2)OH(2):
END:
value (Cu)        value (Cu₂(OH)₂), i.e., value at λ_min
value (Cu)        value (Cu₂(OH)₂), i.e., value at λ_min + λ_inc
     .                 .
     .                 .
     .                 .
value (Cu)        value (Cu₂(OH)₂), i.e., value at λ_max
```

The layout above uses: $value\ (Cu_2(OH)_2)$, i.e., value at λ_{min}; $\lambda_{min} + \lambda_{inc}$; λ_{max}.

Section 6 1   line   [80A1]   **BASELINE**:
             N3 lines [8F10.3] Base-line absorbances, eight per line, one value per wavelength. Data entered from λ_{min} to λ_{max} at λ_{inc} intervals.

This section is optional. There is no **END:** card.

Section 7   1   line   [80A1]   **SPECTRA**:

           1   line   [8D10.4]   TM1, TM2, TL1, TL2, pH, pathlength; where TM1, TM2, TL1, TL2 are the total concentrations of the components MTL1, MTL2, LIG1, LIG2. If MTL2 is absent zero must be entered. For non-aqueous measurements or where pH has not been measured, use zero. Otherwise include the measured value in columns 41–50. Finally, the path length is entered in columns 51–60.

        N4   lines   [8D10.4]   Enter absorbance values, eight per card, from λ_{min} to λ_{max} at λ_{inc} intervals.

        1   line   [F10.4]   −1.0. Termination marker for end of spectra.

The total concentrations of components and absorbance values for one spectrum are entered as a set. There will be as many sets as spectra. The final card (−1.0) allows the user to remove spectra from numerical consideration yet keep them as physical part of the data file.

5. PRESENTATION OF THE PROGRAM

5.1. Listing of SQUAD

There follows a listing of the program SQUAD.

```
 1   C*******************
 2   C
 3   C
 4   C      $ $ $ $ $   S Q U A D   $ $ $ $ $
 5   C
 6   C
 7   C
 8   C      THE CALCULATION OF STABILITY CONSTANTS FROM ABSORBANCE DATA.
 9   C      AUTHOR:  D.LEGGETT.   ANAL. CHEM.,47,1065(1975)
10   C
11   C      LATEST UPDATE VERSION JANUARY 1982.  THIS VERSION USES
12   C      A RADICALLY DIFFERENT METHOD OF DATA INPUT.
13   C
14   C      THE ACCOMPANYING MANUAL SHOULD BE CONSULTED FOR DETAILS
15   C      OF THE DATA SET MAKEUP.  THE FOLLOWING GIVES THE BASIC
16   C      COMPOSITION OF THE DATA DECK.
17   C
18   C
19   C      ITEM 1     A SUITABLE TITLE.
20   C      ITEM 2     SUITABLE SUB-TITLE.
21   C      ITEM 3     DICTIONARY:
22   C                 ..........
23   C                 END:
24   C      ITEM 4     SPECIES:
25   C                 ..........
26   C                 ..........
27   C                 END:
28   C      ITEM 5     OTHER:
29   C                 ..........
30   C                 ..........
31   C                 END:
32   C      ITEM 6     DATA:
33   C      ITEM 7     LAMBDA START;LAMBDA FINISH;LAMBDA INC.(3F10.4 FORMAT)
```

```
34   C        ITEM 8    LOGB    OR    EXPB
35   C        ITEM 9    PRIN    OR    NOPR
36   C        ITEM 10   CARD    OR    NOCD
37   C        ITEM 11   MR      OR    NNLS
38   C        ITEM 12   NOPL    OR    PLOT 1   OR   PLOT 1   OR   PLOT 3  ETC.
39   C        ITEM 13   CRT     OR    PAGE
40   C        ITEM 14   NUMBER OF ITERATIVE CYCLES  (I3 FORMAT)
41   C        ITEM 15   GAMMA H VALUE  (F10.4 FORMAT)
42   C        ITEM 16   MOL.ABS.:
43   C                  .........
44   C                  END:
45   C                  IF THIS SECTION IS USED THERE MUST BE VALUES FOR
46   C                  THE MOLAR ABSORPTIVITIES OF EACH SPECIES, ONE
47   C                   WAVELENGTH PER ITEM .
48   C        ITEM 17   BASELINE:
49   C                  BASELINE DATA IS READ IN HERE (8F10.4 FORMAT)
50   C        ITEM 18   SPECTRA:
51   C                  THE SPECTRAL DATA AND ANALYTICAL COMPOSITION OF EACH
52   C                  SOLUTION IS READ IN HERE, TOGETHER WITH THE PATHLENGTH
53   C                  AND PH OF EACH SOLUTION (IF APPROPRIATE).
54   C                  (4D10.4,2F10.4 FORMAT) FOR SOLUTION COMPOSITION,
55   C                  (8F10.4 FORMAT) FOR ABSORBANCE VALUES.
56   C        FINAL     -1.0
57   C
58   C*******************************************************************
59   C
60   C
61   C
62   C        THIS PROGRAM USES EXECUTION-TIME DIMENSIONING.
63   C        SET MAXIMUM ARRAY SIZES WHERE:-
64   C        NWVLS = MAXIMUM NUMBER OF WAVELENGTHS
65   C        NSOLNS = MAXIMUM NUMBER OF SOLUTIONS
66   C        NCOMP = MAXIMUM NUMBER OF COMPLEXES
67   C        NBETA = MAXIMUM NUMBER OF STABILITY CONSTANTS TO BE REFINED
68   C        NETACL = MAXIMUM NUMBER OF MOLAR ABSORPTIVITIES OF SPECIES
69   C                    TO BE CALCULATED
70   C
71   C
72   C        THE FOLLOWING LIST IS FOR ALL EXECUTION TIME DIMENSIONED
73   C        ARRAYS (ARRANGED ALPHABETICALLY).
74   C
75   C        B1(NSOLNS),CC(NBETA,NBETA),CK(NETACL),COMPLX(30,NCOMP4),
76   C        CONC(NSOLNS,NETACL),C1(NSOLNS),DE(NETACL),E(NCOMP),
77   C        EC(NCOMP6,NWVLS),EQ(NCOMP4,NWVLS),IPO(NBETA),
78   C        IWORK(NETACL),JCOMB(NCOMP,5),JEQ(NETACL),JJNO(NCOMP),
79   C        ML(2,NCOMP),MM(2,NCOMP),MN(NCOMP),PATH(NSOLNS),PH(NSOLNS),
80   C        Q(NSOLNS,NCOMP4),SPEC(NSOLNS,NWVLS),SP2(NWVNSL,NBETA),
81   C        SSRSOL(NWVLS),STD(NETACL,NWVLS),TL(2,NSOLNS),TM(2,NSOLNS),
82   C        WORK(NSOLNS,NWVLS),WORK1(NETACL,NETACL),WORK2(NSOLNS,NETACL),
83   C        X(NETACL)
84   C
85   C        THE FOLLOWING TABLE ARRANGES ARRAYS BY DIMENSION PARAMETERS.
86   C
87   C        NSOLNS NCOMP NCOMP4 NCOMP6 NWVLS   NWVNSL NBETA NETACL
88   C        ****** ***** ****** ****** *****   ****** ***** ******
89   C        B1     E     COMPLX EC     EC      SP2    CC    CK
90   C        C1     JCOMB EQ            EQ             IPO   CONC
91   C        CONC   JJNO  Q            SPEC            SP2   DE
92   C        PATH   ML                 SSRSOL                IWORK
93   C        PH     MM                 STD                   JEQ
94   C        Q      MN                 WORK                  STD
95   C        SPEC                                            WORK1
96   C        TL                                              WORK2
97   C        TM                                              X
98   C        WORK
99   C        WORK2
100  C
101  C        REMEMBER:  WHEN CHANGING NCOMP DON'T FORGET NCOMP4 AND NCOMP6
102  C                   WHEN CHANGING NSOLNS OR NWVLS DON'T FORGET NWVNSL
103  C
104  C
105  C*******************************************************************
106           IMPLICIT REAL*8 (A-H,O-Z)
107           CHARACTER*13 DOC
108           COMMON/INTEGR/NCV,NBA,JQ,JNO,NL,NM,NTS,NPASS,NUMPH,NCD,M,
109          &NBANUM,JJA,IBETA,NCC,NPAGE,IPLOT,IPRIN,ICARD ,NLMR
```

```
110          COMMON/CONST/WVLST,WVLFIN,WVLINC,S1,F,XDF,XSD,SQRA,SSQR,WT
111          COMMON/UNIT/RUNIT,WUNIT,PUNIT
112          DIMENSION Q(25,24),TM(2,25),TL(2,25),PH(25)
113          DIMENSION SPEC(25,50),PATH(25),WORK(25,50),SP2(1250,6),
114         &SSRSOL(50),WORK2(25,10)
115          DIMENSION JEQ(10),JJNO(20),JCOMB(20,5),IWORK(10)
116          DIMENSION ML(2,20),MM(2,20),MN(20)
117          DIMENSION E(20),CK(10),X(10),IPO(6),B1(25),C1(25),
118         &CC(6,6),WORK1(10,10),DE(10)
119          DIMENSION EQ(24,50),EC(26,50),STD(10,50),CONC(25,10)
120          INTEGER COMPLX(30,24),CARD,TITLE(20),TITLE1(20),
121         &BETA,ALGOR,PLOT,DISP,PRINT,RUNIT,PUNIT,WUNIT
122          DATA COMPLX/
123         &'M','T','L','1',' ',' ','I','S',' ',' ','A','B','S','E','N','T',16*'.',
124         &'M','T','L','2',' ',' ','I','S',' ',' ','A','B','S','E','N','T',16*'.',
125         &'L','I','G','1',' ',' ','I','S',' ',' ','A','B','S','E','N','T',16*'.',
126         &'L','I','G','2',' ',' ','I','S',' ',' ','A','B','S','E','N','T',16*'.',
127         &600*'.'/
128      601 FORMAT(3F10.4)
129      602 FORMAT(20A4)
130      603 FORMAT(A4,I2)
131      604 FORMAT(I3)
132      605 FORMAT(/3X,20A4)
133      606 FORMAT('1',7X,' SS    QQQQ    U   U   AAA    DDDD'/
134         &7X,'S S    Q    Q   U   U   A   A   D   D'/
135         &7X,'S      Q    Q   U   U   A   A   D   D'/
136         &7X,' SS    Q    Q   U   U   AAAAA   D   D'/
137         &7X,'   S   Q    Q   U   U   A   A   D   D'/
138         &7X,'   S   Q    Q   U   U   A   A   D   D'/
139         &7X,'S S   Q Q Q  U   U   A   A   D   D'/
140         &7X,' SS    QQQQ    UUU    A   A   DDDD '/
141         &18X,'Q'///)
142      607 FORMAT(3X,'LOGARITHMIC STABILITY CONSTANTS WILL BE REFINED')
143      608 FORMAT(3X,'ANTILOGARITHM OF LOG(BETA) WILL BE REFINED')
144      609 FORMAT(3X,'ABBREVIATED INPUT DATA WILL PRINTED OUT')
145      610 FORMAT(3X,'FULL INPUT DATA WILL BE PRINTED OUT')
146      611 FORMAT(3X,'NO CARDS WILL BE PRODUCED')
147      612 FORMAT(3X,'FINAL MOLAR ABSORPTIVITIES AND ALL SPECIES',
148         &' CONCENTRATIONS WILL BE PUNCHED ON CARDS')
149      613 FORMAT(3X,'MOLAR ABSORPTIVITIES WILL BE CALCULATED USING',
150         &' THE MULTIPLE REGRESSION ALGORITHM')
151      614 FORMAT(3X,'MOLAR ABSORPTIVITIES WILL BE CALCULATED USING ',
152         &'THE NON-NEGATIVE LINEAR LEAST SQUARES ALGORITHM')
153      615 FORMAT(3X,'NO PRINTER PLOTS WILL BE PRODUCED')
154      616 FORMAT(3X,'EACH SPECTRUM WILL BE DISPLAYED AS A PRINTER PLOT')
155      617 FORMAT(//3X,'THE FOLLOWING PRINT-OUT AND CALCULATION OPTIONS',
156         &' ARE IN EFFECT:-'/)
157      618 FORMAT(3X,'EVERY OTHER SPECTRUM WILL BE DISPLAYED AS A ',
158         &'PRINTER PLOT.')
159      619 FORMAT(3X,'EVERY THIRD SPECTRUM WILL BE DISPLAYED AS A ',
160         &'PRINTER PLOT.')
161      620 FORMAT(3X,'EVERY',I2,'TH SPECTRUM WILL BE DISPLAYED AS A ',
162         &'PRINTER PLOT.')
163          TYPE *,' ENTER FILE NAME FOR SQUAD RUN'
164          READ(*,800)DOC
165      800 FORMAT(A13)
166          OPEN(UNIT=4,FILE=DOC,STATUS='OLD')
167   C      OPEN(UNIT=6,FILE='OUT.DAT',STATUS='OLD')
168   C################################################################
169   C################################################################
170   C##                                                          ##
171   C##      SET THE UNIT NUMBERS FOR CARD READER (RUNIT),       ##
172   C##      LINE PRINTER (WUNIT) AND CARD PUNCH (PUNIT)         ##
173   C##                                                          ##
174          RUNIT=4
175          WUNIT=6
176          PUNIT=7
177   C##                                                          ##
178   C##                                                          ##
179   C################################################################
180   C################################################################
181   C
182   C      SET EXECUTION TIME DIMENSIONS.
183   C
184          NWVLS=50
185          NSOLNS=25
```

```
186          NCOMP=20
187          NBETA=6
188          NETACL=10
189          NWVNSL=NWVLS*NSOLNS
190          NCOMP4=NCOMP+4
191          NCOMP6=NCOMP+6
192          WRITE(WUNIT,606)
193    C
194    C     READ AND PRINT OUT SUITABLE TITLES
195    C
196          READ(RUNIT,602)(TITLE(I),I=1,20)
197          READ(RUNIT,602)(TITLE1(I),I=1,20)
198          WRITE(WUNIT,605)(TITLE(I),I=1,20)
199          WRITE(WUNIT,605)(TITLE1(I),I=1,20)
200    C
201    C     CALL PREPROCESSING ROUTINE, PREPRO, TO ESTABLISH THE
202    C     MODEL TO BE TESTED.  EXTENSIVE ERROR CHECKING OF
203    C     THE DATA DECK IS PERFORMED IN THIS ROUTINE.
204    C
205          CALL PREPRO(E,COMPLX,JCOMB,MM,ML,MN,IPO,JEQ,JJNO,NCOMP,NCOMP4,
206         &NBETA,NETACL)
207          READ(RUNIT,601)WVLST,WVLFIN,WVLINC
208          READ(RUNIT,602)BETA
209          CALL DATCHK(BETA,IBETA,1,0,1)
210          READ(RUNIT,602)PRINT
211          CALL DATCHK(PRINT,IPRIN,7,1,0)
212          READ(RUNIT,602)CARD
213          CALL DATCHK(CARD,ICARD,3,1,0)
214          READ(RUNIT,602)ALGOR
215          CALL DATCHK(ALGOR,NLMR,9,0,1)
216          READ(RUNIT,603)PLOT,NPLOT
217          CALL DATCHK(PLOT,IPLOT,5,NPLOT,0)
218          READ(RUNIT,602)DISP
219          CALL DATCHK(DISP,NPAGE,11,0,1)
220          READ(RUNIT,604)NCD
221          READ(RUNIT,601)F
222          WRITE(WUNIT,617)
223          IF(IBETA.EQ.0)WRITE(WUNIT,607)
224          IF(IBETA.EQ.1)WRITE(WUNIT,608)
225          IF(IPRIN.EQ.0)WRITE(WUNIT,609)
226          IF(IPRIN.EQ.1)WRITE(WUNIT,610)
227          IF(ICARD.EQ.0)WRITE(WUNIT,611)
228          IF(ICARD.EQ.1)WRITE(WUNIT,612)
229          IF(NLMR.EQ.0)WRITE(WUNIT,613)
230          IF(NLMR.EQ.1)WRITE(WUNIT,614)
231          IF(IPLOT.EQ.0)WRITE(WUNIT,615)
232          IF(IPLOT.EQ.1)WRITE(WUNIT,616)
233          IF(IPLOT.EQ.2)WRITE(WUNIT,618)
234          IF(IPLOT.EQ.3)WRITE(WUNIT,619)
235          IF(IPLOT.GT.3)WRITE(WUNIT,620)IPLOT
236    C
237    C        CALL INOUT TO PRINT OUT THE INPUT DATA AND INITIATE
238    C        THE DATA PROCESSING.
239    C
240          CALL INOUT(CC,WORK,WORK1,WORK2,SP2,SPEC,Q,TM,TL,EQ,EC,STD,CONC,
241         &DE,B1,PATH,C1,E,X,CK,PH,SSRSOL,COMPLX,JCOMB,MM,ML,MN,IPO,
242         &JEQ,JJNO,IWORK,TITLE,TITLE1,NWVLS,NSOLNS,NETACL,NBETA,NCOMP,
243         &NWVNSL,NCOMP4,NCOMP6)
244          STOP
245          END
```

```
 1    C###############################################################
 2    C###############################################################
 3    C##                                                         ##
 4    C##                                                         ##
 5    C##        $ $ $ $ $   D A T C H K   $ $ $ $ $              ##
 6    C##                                                         ##
 7    C##                                                         ##
 8    C###############################################################
 9    C###############################################################
10    C
```

```
11   C         THIS ROUTINE IS USED TO CHECK THAT LEGAL CONTROL WORDS
12   C         HAVE BEEN USED IN DATA: SECTION (ITEMS 8 TO 15).
13   C
14             SUBROUTINE DATCHK(WORD,IWORD,TYPE,I1,I2)
15             INTEGER CODE(12),TEST(2),WORD,RUNIT,WUNIT,PUNIT,TYPE,HEAD(4)
16             COMMON/UNIT/RUNIT,WUNIT,PUNIT
17             DATA CODE/'LOGB','EXPB','CARD','NOCD','PLOT','NOPL','PRIN','NOPR',
18            &'MR','NNLS','CRT','PAGE'/,HEAD/'*** ',' ERR','OR  ','***'/
19    607 FORMAT(2X,4A4,'UNKNOWN CONTROL WORD FOUND.  I.E....',2A4/
20            &18X,'EXPECTING...... ',A4,' OR ',A4/
21            &2X,'CORRECT ERRORS. EXECUTION TERMINATED.  BYE-BYE.'////)
22             TEST(1)=WORD
23             IWORD=-1
24             IF(WORD.EQ.CODE(TYPE))IWORD=I1
25             IF(WORD.EQ.CODE(TYPE+1))IWORD=I2
26             IF(IWORD.LT.0)GO TO 500
27             RETURN
28    500 WRITE(WUNIT,607)(HEAD(J),J=1,4),(TEST(I),I=1,2),
29            &CODE(TYPE),CODE(TYPE+1)
30             STOP
31             END
```

```
 1   C#################################################################
 2   C#################################################################
 3   C##                                                           ##
 4   C##                                                           ##
 5   C##        $ $ $ $ $   P R E P R O   $ $ $ $ $                ##
 6   C##                                                           ##
 7   C##                                                           ##
 8   C#################################################################
 9   C#################################################################
10   C
11   C     THIS ROUTINE TAKES THE USER-ORIENTED MODEL DICTIONARY AND
12   C     MODEL DESCRIPTION AND DETERMINES THE STOICHIOMETRY OF EACH
13   C     COMPLEX; THE INITIAL VALUES FOR STABILITY CONSTANTS ETC.
14   C     SEE USER MANUAL FOR FULL DETAILS CONCERNING DATA DECK SET-UP.
15   C     EXTENSIVE ERROR CHECKING OF THE DATA DECK IS PERFORMED AND
16   C     THE USER IS INFORMED OF ANY ERRORS FOUND AND THEIR LOCATION.
17   C
18             SUBROUTINE PREPRO(E,COMPLX,JCOMB,MM,ML,MN,IPO,JEQ,JJNO,
19            &NCOMP,NCOMP4,NBETA,NETACL)
20             IMPLICIT INTEGER*4(A-Z)
21             COMMON/INTEGR/NCV,NBA,JQ,JNO,NL,NM,NTS,NPASS,NUMPH,NCD,M,
22            &NBANUM,JJA,IBETA,NCC,NPAGE,IPLOT,IPRIN,ICARD,NLMR
23             COMMON/UNIT/RUNIT,WUNIT,PUNIT
24             DIMENSION ML(2,NCOMP),MM(2,NCOMP),MN(NCOMP)
25             DIMENSION JEQ(NETACL),JJNO(NCOMP),JCOMB(NCOMP,5),IPO(NBETA)
26             REAL*8 E(NCOMP)
27             DIMENSION DICT(6,6),TEST(4,7),CHAR(23),CARD(80),COMPON(6),
28            &SAVE(20),NUMBER(8),JJEQ(25),TYPEE(5),ORDER(4),TYPEB(5),FIX(5),
29            &ABSENT(6),COMPLX(30,NCOMP4),SETUP(40,80),VARY(5),DIRECT(10,4)
30             DATA TEST/'M','T','L','1','M','T','L','2','L','I','G','1',
31            &'L','I','G','2','P','R','O','T','H','Y','D','X','E','N',
32            &'D',':'/
33             DATA FIX/'F','I','X','E','D'/,VARY/'V','A','R','Y',' '/
34             DATA CHAR /'0','1','2','3','4','5','6','7','8','9',
35            &';','=','(',')',',','.','F','V',' ',':','B','E','+','-'/
36             DATA COMPON/6*0/,SETUP/3200*' '/
37             DATA DIRECT/'D','I','C','T','I','O','N','A','R','Y',
38            &'S','P','E','C','I','E','S',':',2*' ',
39            &'O','T','H','E','R',':',4*' ',
40            &'D','A','T','A',':',5*' '/
41             DATA ABSENT/'A','B','S','E','N','T'/
42    100 FORMAT(80A1)
43    601 FORMAT(//8X,'DICTIONARY...'/7X,' METAL1  METAL2  LIGAND1 ',
44            &' LIGAND2  PROTON  HYDROXO '/6X,2(2X,6('*')),2(2X,7('*')),
45            &2X,6('*'),2X,7('*'),/7X,2(1X,6A1,1X),2(2X,6A1,1X),
46            &1X,6A1,1X,2X,6A1,1X/)
47    604 FORMAT(5X,I2,5X,30A1,F8.4,5X,5A1,7X,5A1)
48    607 FORMAT(//2X,'SPECIES',5X,'FORMULA',22X,'LOG BETA',4X,
49            &'F OR V',6X,'MOL.ABS.')
```

```
50      610 FORMAT(//2X,'NUMBER OF METALS =',I2,3X,'NUMBER OF LIGANDS =',
51          &I2,3X,'NUMBER OF COMPLEXES =',I2//2X,'NUMBER OF ',
52          &'CONSTANTS TO BE VARIED =',I2)
53      611 FORMAT(/2X,'NUMBER OF SETS OF MOLAR ABSORPTIVITIES',
54          &' TO BE FOUND =',I3)
55      612 FORMAT(5X,30A1)
56      613 FORMAT(/5X,'COMBINATION SET NUMBER',I3)
57      615 FORMAT(3X,30A1)
58      616 FORMAT(//3X,'THE FOLLOWING SPECIES AND/OR COMPLEXES HAVE',
59          &' FIXED (ZERO OR READ IN) MOL. ABS.')
60      620 FORMAT(3X,'I.E...  ',30A1)
61      621 FORMAT(/3X,'THE MOL. ABS. OF ALL OTHER COMPONENTS WILL BE',
62          &' FIXED'//)
63      622 FORMAT(3X,'THE FOLLOWING MOL. ABS. OF THE COMPONENT(S)',
64          &' WILL BE VARIED.')
65      623 FORMAT(5X,'SOME SPECIES HAVE THE SAME MOLAR ABSORPTIVITIES.',
66          &' THEY ARE COMBINED AS FOLLOWS:-')
67    C
68    C       INITIALIZE SEVERAL PARAMETERS INCLUDING THE DICTIONARY ARRAY, DICT
69    C
70          DO 198 J=1,4
71      198 ORDER(J)=0
72          DO 193 I=1,NCOMP
73          DO 192 J=1,2
74          MM(J,I)=0
75      192 ML(J,I)=0
76          DO 191 J=1,5
77      191 JCOMB(I,J)=0
78          JJNO(I)=0
79      193 MN(I)=0
80          JJNO(1)=1
81          JJNO(2)=2
82          JJNO(3)=3
83          JJNO(4)=4
84          EQUALS=CHAR(12)
85          SEMI=CHAR(11)
86          BLANK=CHAR(18)
87          COLON=CHAR(19)
88          NCARD=0
89          NCV=0
90          NCOMB=0
91          JQ=0
92          JNO=4
93          NSET=0
94          DO 194 J=1,6
95          DO 194 K=1,6
96      194 DICT(K,J)=ABSENT(K)
97    C
98    C       COMMENCE READING IN THE CARDS FOLLOWING THE TITLE CARDS.
99    C       EACH CARD IS COPIED INTO THE HOLDING ARRAY SETUP AS AN
100   C       80 CHARACTER VECTOR.
101   C
102      12 NCARD=NCARD+1
103          READ(RUNIT,100)(SETUP(NCARD,I),I=1,80)
104   C
105   C       CHECK THAT ALL CARDS HAVE A ':' AS THE LAST ELEMENT.
106   C
107          JJ=81
108          DO 199 J=1,80
109          JJ=JJ-1
110          CHECK=SETUP(NCARD,JJ)
111          IF(CHECK.EQ.BLANK)GO TO 199
112          IF(CHECK.NE.COLON)GO TO 196
113          GO TO 197
114      199 CONTINUE
115      196 DO 195 J=1,80
116      195 CARD(J)=SETUP(NCARD,J)
117          CALL ERR(JJ+1,3,CARD)
118   C
119   C         CONTINUE IF ":" IS FOUND.
120   C         CARD CONTAINS A HEADING SUCH AS "DICTIONARY"; "SPECIES";
121   C         "OTHER"; OR "DATA".  IF "DATA" IS FOUND THEN PROCESSING
122   C         OF THE MODEL IS COMPLETE.  IF NOT READ IN ANOTHER CARD.
123   C
124      197 DO 10 I=1,5
125          IF(SETUP(NCARD,I).NE.DIRECT(I,4))GO TO 12
```

```
126        10 CONTINUE
127     C
128     C     THERE ARE NCARD SETUP CARDS.
129     C     DETERMINE THE ORDER TO PROCESS THE SETUP CARDS.
130     C     THE ORDERING OF THE DATA DECK IS NOT IMPORTANT.  HOWEVER
131     C     THE ORDERING OF THE PROCESSING IS IMPORTANT, AND IS SETUP IN
132     C     THE VECTOR ORDER. NAMELY
133     C        (1)   ESTABLISH NAME OF THE COMPONENT METAL AND LIGAND(S).
134     C              SAVE THESE IN DICT VECTOR.....DICTIONARY
135     C        (2)   SET UP THE STOICHIOMETRY OF THE COMPLEXES.....SPECIES
136     C        (3)   CHECK TO SEE IF ANY SPECIAL DATA PROCESSING IS
137     C              REQUIRED.....OTHER
138     C        (4)   READ IN REMAINING DATA.....DATA
139     C
140           DO 24 I=1,NCARD
141           DO 22 K=1,4
142           DO 23 J=1,10
143           IF(SETUP(I,J).NE.DIRECT(J,K))GO TO 22
144        23 CONTINUE
145           ORDER(K)=I
146           GO TO 24
147        22 CONTINUE
148        24 CONTINUE
149           DO 27 J=1,4
150           IF(ORDER(J).GT.0)GO TO 27
151           DO 28 JJ=J,3
152        28 ORDER(JJ)=ORDER(JJ+1)
153           ORDER(4)=0
154        27 CONTINUE
155     C
156     C     COME HERE TO DETERMINE THE ORDER IN WHICH THE SUB-SECTIONS
157     C     OF THE DATA DECK ARE TO BE PROCESSED.
158     C     THE VALUE OF IJK DIRECTS CONTROL TO THE PARTICULAR
159     C     PREPROCESSING SECTION.
160     C
161           IJK=0
162        20 IJK=IJK+1
163           ICARD=ORDER(IJK)
164           IF(ICARD.EQ.NCARD)GO TO 998
165        25 GO TO(200,300,500,9999),IJK
166     C********************************
167     C
168     C     DICTIONARY CARD ANALYSIS.
169     C
170     C********************************
171       200 ICARD=ICARD+1
172           DO 26 I=1,80
173        26 CARD(I)=SETUP(ICARD,I)
174     C
175     C     CHECK THAT THE NUMBER OF = EQUALS THE NUMBER OF ;
176     C
177           JJ=0
178           JJJ=1
179           DO 208 J=1,80
180           IF(CARD(J).EQ.EQUALS)JJ=JJ+1
181           IF(CARD(J).EQ.SEMI)JJJ=JJJ+1
182       208 CONTINUE
183           IF(JJ.EQ.JJJ)GO TO 209
184           CALL ERR(0,1,CARD)
185     C
186     C     COMMENCE DICTIONARY CARD ANALYSIS.
187     C
188       209 JJ=81
189           DO 190 J=1,80
190           JJ=JJ-1
191           IF(CARD(JJ).EQ.COLON)GO TO 189
192       190 CONTINUE
193       189 CARD(JJ)=SEMI
194           CARD(JJ+1)=COLON
195           J=1
196           JST=1
197       211 IF(CARD(J).EQ.SEMI)GO TO 201
198           J=J+1
199           GO TO 211
200     C
201     C     CHECK FOR COMPONENT DEFINITION NAME.
```

```
202    C
203        201 DO 202 K=1,6
204            DO 204 L=1,4
205            J=JST+L-1
206            IF(CARD(J).NE.TEST(L,K))GO TO 202
207        204 CONTINUE
208            IF(CARD(J+1).EQ.EQUALS)GO TO 212
209            CALL ERR(J,6,CARD)
210    C
211    C      LOAD COMPON(1) WITH ONE FOR FIRST METAL PRESENT; (2) WITH ONE
212    C      FOR SECOND METAL PRESENT; (3) WITH ONE FOR FIRST LIGAND
213    C      PRESENT; (4) WITH ONE FOR SECOND LIGAND PRESENT.
214    C
215        212 COMPON(K)=1
216    C
217    C      EQUATE USER NAMED COMPONENT WITH COMPONENT DEFINITION NAME
218    C
219            J=J+2
220            DICT(1,K)=BLANK
221            DO 205 M=1,5
222            TST=CARD(J)
223            IF(TST.EQ.SEMI)GO TO 206
224            IF(M.EQ.5)GO TO 203
225            DICT(M+1,K)=TST
226            COMPLX(M,K)=TST
227        205 J=J+1
228        202 CONTINUE
229            CALL ERR(J,2,CARD)
230        203 CALL ERR(J,6,CARD)
231        206 IF(K.GT.4)GO TO 227
232            DO 226 MJ=M,30
233        226 COMPLX(MJ,K)=BLANK
234        227 DO 228 MJ=M,5
235        228 DICT(MJ+1,K)=BLANK
236            JST=J+1
237            J=J+1
238            IF(CARD(J).NE.COLON)GO TO 211
239    C
240    C      DEFINITION CARD ANALYSIS COMPLETE IF ":" FOUND.
241    C      INTERNAL SCALARS NM (#OF METALS) AND NL (#OF LIGANDS)
242    C      ARE CALCULATED.
243    C
244            NM=COMPON(1)+COMPON(2)
245            NL=COMPON(3)+COMPON(4)
246            WRITE(WUNIT,601)((DICT(K,J),K=1,6),J=1,6)
247            DO 222 K=1,4
248            JJ=31
249            DO 221 J=1,30
250            JJ=JJ-1
251            IF(COMPLX(JJ,K).NE.CHAR(15))GO TO 222
252        221 COMPLX(JJ,K)=BLANK
253        222 CONTINUE
254            WRITE(WUNIT,607)
255            DO 210 J=1,6
256            DO 220 K=1,5
257        220 DICT(K,J)=DICT(K+1,J)
258        210 DICT(6,J)=BLANK
259            ICARD=ICARD+1
260            GO TO 20
261    C********************************
262    C
263    C      ANALYSIS OF INDIVIDUAL SPECIES CARDS.
264    C
265    C********************************
266        300 N=0
267        399 ICARD=ICARD+1
268            DO 350 I=1,80
269        350 CARD(I)=SETUP(ICARD,I)
270            DO 351 I=1,4
271            IF(CARD(I).NE.TEST(I,7))GO TO 352
272        351 CONTINUE
273            ICARD=ICARD+1
274            GO TO 20
275        352 N=N+1
276            IF(N.GT.NCOMP)CALL ERR(0,18,CARD)
277            I=0
```

```
278   C
279   C          LOOK FOR LEFT PARENTHESIS IN COMPLEX FORMULA.
280   C
281       318 DO 305 J=1,20
282       305 SAVE(J)=BLANK
283           DO 314 J=1,5
284           I=I+1
285           IF(CARD(I).EQ.CHAR(13))GO TO 301
286       314 SAVE(J)=CARD(I)
287           CALL ERR(I,7,CARD)
288   C
289   C          LOOK FOR RIGHT PARENTHESIS IN COMPLEX FORMULA.
290   C
291       301 DO 302 J=5,8
292           I=I+1
293           IF(CARD(I).EQ.CHAR(14))GO TO 303
294       302 SAVE(J)=CARD(I)
295           CALL ERR(I,8,CARD)
296       303 DO 304 K=1,6
297           IF(COMPON(K).EQ.0)GO TO 304
298           DO 315 J=1,4
299           IF(SAVE(J).NE.DICT(J,K))GO TO 304
300       315 CONTINUE
301           GO TO 319
302       304 CONTINUE
303           CALL ERR(I-3,5,CARD)
304   C
305   C          EXTRACT STOICHIOMETRIC COEFFICIENT FROM WITHIN LOCATED PARENTHESIS
306   C
307       319 DO 310 J=1,3
308       310 NUMBER(J)=99
309   C
310   C          ISOLATE COEFFICIENT.
311   C
312           DO 307 J=5,8
313           NUMRAL=SAVE(J)
314           IF(NUMRAL.EQ.BLANK)GO TO 307
315           DO 306 JJ=1,10
316           IF(NUMRAL.EQ.CHAR(JJ))GO TO 308
317       306 CONTINUE
318       308 NUMBER(J-4)=JJ-1
319       307 CONTINUE
320   C
321   C          CONVERT CHARACTER REPRESENTATION OF COEFFICIENT TO INTEGER
322   C
323           DO 311 J=1,3
324           IF(NUMBER(J).EQ.99)GO TO 312
325       311 CONTINUE
326       312 SUM=0
327           KK=1
328       313 SUM=SUM*10+NUMBER(KK)
329           KK=KK+1
330           J=J-1
331           IF(J.GT.1)GO TO 313
332   C
333   C          TRANSFER STOICHIOMETRIC COEFFICIENT TO ML AND MM ARRAYS
334   C          OR THE MN VECTOR.
335   C
336           GO TO (320,320,321,321,322,322),K
337       320 MM(K,N)=SUM
338           GO TO 323
339       321 K=K-2
340           ML(K,N)=SUM
341           GO TO 323
342       322 IF(K.EQ.5)SUM=-SUM
343           MN(N)=SUM
344       323 IF(CARD(I+1).NE.SEMI)GO TO 318
345           DO 332 IJ=1,I
346       332 COMPLX(IJ,N+4)=CARD(IJ)
347           I=I+1
348   C
349   C          EXTRACT THE STABILITY CONSTANT FROM THE CHARACTER STRING.
350   C
351           DO 417 J=1,20
352       417 SAVE(J)=BLANK
353           JJ=0
```

```
354            I=I+1
355            ISIGN=1
356            IF(CARD(I).NE.CHAR(23))GO TO 440
357            I=I+1
358            ISIGN=-1
359       440  I=I-1
360            JPER=0
361            DO 400 J=1,20
362            I=I+1
363            IF(CARD(I).EQ.SEMI)GO TO 402
364            IF(CARD(I).EQ.CHAR(15))GO TO 401
365            JJ=JJ+1
366            SAVE(JJ)=CARD(I)
367            GO TO 400
368       401  JPER=J
369       400  CONTINUE
370            CALL ERR(I,9,CARD)
371       402  IF(JPER.NE.0)GO TO 441
372            CALL ERR(I,9,CARD)
373  C
374  C       SAVE CONTAINS THE CHARACTER REPRESENTATION OF THE FORMATION
375  C       CONSTANT, WITHOUT THE DECIMAL POINT.
376  C       LOCATION OF THE DECIMAL POINT IS HELD IN JPER
377  C       JJ = NUMBER OF DIGITS IN THE NUMBER.
378  C
379       441  CONTINUE
380            DO 408 J=1,JJ
381            NUMRAL=SAVE(J)
382            DO 404 JJ1=1,10
383            IF(NUMRAL.EQ.CHAR(JJ1))GO TO 408
384       404  CONTINUE
385            CALL ERR(I-1,15,CARD)
386       408  NUMBER(J)=JJ1-1
387  C
388  C       CONVERT THE STABILITY CONSTANT TO FLOATING POINT NUMBER.
389  C
390            SUM=0
391            KK=1
392            JJJ=JJ+1
393       403  SUM=SUM*10+NUMBER(KK)
394            KK=KK+1
395            JJ=JJ-1
396            IF(JJ.GT.0)GO TO 403
397            SUM=SUM*ISIGN
398            E(N)=FLOAT(SUM)/10.0**(JJJ-JPER)
399  C
400  C       CHECK TO FIND WHETHER CONSTANT AND/OR THE MOLAR
401  C       ABSORPTIVITY IS TO BE VARIED (V) OR FIXED (F).
402  C
403            NPASS=0
404       421  IF(NPASS.EQ.2)GO TO 407
405            I=I+1
406            SWITCH=CARD(I)
407            IF(SWITCH.NE.CHAR(16))GO TO 412
408            GO TO 411
409       412  IF(SWITCH.EQ.CHAR(17))GO TO 411
410            CALL ERR(I,4,CARD)
411       411  I=I+1
412            IF(CARD(I).EQ.CHAR(21))GO TO 420
413            IF(CARD(I).EQ.CHAR(20))GO TO 422
414            CALL ERR(I,11,CARD)
415       422  DO 406 J=1,5
416            TYPEB(J)=FIX(J)
417            IF(SWITCH.EQ.CHAR(17))TYPEB(J)=VARY(J)
418       406  CONTINUE
419            IF(SWITCH.EQ.CHAR(16))GO TO 435
420            NCV=NCV+1
421            IF(NCV.GT.NBETA)CALL ERR(0,19,CARD)
422            IPO(NCV)=N
423       435  I=I+1
424            NPASS=NPASS+1
425            GO TO 421
426       420  DO 423 J=1,5
427            TYPEE(J)=FIX(J)
428            IF(SWITCH.EQ.CHAR(17))TYPEE(J)=VARY(J)
429       423  CONTINUE
```

```
430          IF(SWITCH.EQ.CHAR(16))GO TO 436
431          JQ=JQ+1
432          IF(JQ.GT.NETACL)CALL ERR(0,20,CARD)
433          JJEQ(JQ)=N+4
434          GO TO 439
435      436 JNO=JNO+1
436          JJNO(JNO)=N+4
437      439 I=I+1
438          NPASS=NPASS+1
439          GO TO 421
440      407 N4=N+4
441          WRITE(WUNIT,604)N,(COMPLX(J,N4),J=1,30),E(N),(TYPEB(J),J=1,5),
442         &(TYPEE(J),J=1,5)
443          JJ=31
444          DO 450 J=1,30
445          JJ=JJ-1
446          IF(COMPLX(JJ,N4).NE.CHAR(15))GO TO 399
447      450 COMPLX(JJ,N4)=BLANK
448          GO TO 399
449 C********************************
450 C
451 C        COME HERE TO DEAL WITH ANY "OTHER" REQUIREMENTS.
452 C
453 C********************************
454      500 ICARD=ICARD+1
455          DO 501 I=1,80
456      501 CARD(I)=SETUP(ICARD,I)
457          DO 502 I=1,4
458          IF(CARD(I).NE.TEST(I,7))GO TO 503
459      502 CONTINUE
460          ICARD=ICARD+1
461          GO TO 540
462      503 DO 504 K=1,2
463          DO 505 I=1,80
464          IF(CARD(I).EQ.COLON)GO TO 504
465          IF(CARD(I).EQ.CHAR(11*K))GO TO 506
466      505 CONTINUE
467      504 CONTINUE
468          CALL ERR(0,12,CARD)
469      506 GO TO(510,520),K
470 C
471 C        COME HERE TO DETERMINE WHICH UNCOMPLEXED SPECIES' MOLAR
472 C        ABSORPTIVITY IS TO BE VARIED.
473 C
474      510 DO 511 K=1,4
475          DO 512 I=1,5
476          IF(CARD(I).EQ.SEMI)GO TO 513
477          IF(CARD(I).NE.DICT(I,K))GO TO 511
478      512 CONTINUE
479      511 CONTINUE
480          CALL ERR(I,5,CARD)
481 C
482 C        CHECK THAT THE NEXT THREE CHARACTERS ARE ";VE".
483 C
484      513 I=0
485      514 I=I+1
486          IF(CARD(I).NE.SEMI)GO TO 514
487          IF(CARD(I+1).NE.CHAR(17))GO TO 515
488          IF(CARD(I+2).NE.CHAR(21))GO TO 515
489          JQ=JQ+1
490          JJEQ(JQ)=K
491          JNO=JNO-1
492          DO 516 J=K,JNO
493      516 JJNO(J)=JJNO(J+1)
494          GO TO 500
495      515 CALL ERR(0,10,CARD)
496 C
497 C        COME HERE TO DETERMINE WHICH SPECIES WILL BE CONSIDERED TO
498 C        HAVE THE SAME MOLAR ABSORPTIVITIES.
499 C
500      520 IST=1
501          IFIN=30
502          NCOMB=0
503          NSET=NSET+1
504      521 DO 522 J=1,JQ
505          II=0
```

```
506              DO 523 I=IST,IFIN
507              II=II+1
508              TST=CARD(I)
509              IF(TST.EQ.CHAR(22))GO TO 524
510              IF(TST.NE.COMPLX(II,J+4))GO TO 522
511          523 CONTINUE
512          522 CONTINUE
513              CALL ERR(I,13,CARD)
514          524 IST=I+1
515              IFIN=IST+30
516          525 NCOMB=NCOMB+1
517              JCOMB(NSET,NCOMB)=J+4
518              IF(TST.EQ.COLON)GO TO 525
519              GO TO 521
520    C
521    C         COME HERE TO TRANSFER MOLAR ABSORPTIVITY CALCULATION
522    C         REQUIREMENTS TO INTERNAL WORK ARRAYS.
523    C
524          540 IF(NSET.EQ.0)GO TO 20
525              DO 541 ISET=1,NSET
526              DO 542 ICOMB=1,5
527              IF(JCOMB(ISET,ICOMB).EQ.0)GO TO 541
528              JCOM=JCOMB(ISET,ICOMB)
529              DO 543 JJQ=1,JQ
530              IF(JCOM.NE.JJEQ(JJQ))GO TO 543
531              JJEQ(JJQ)=0
532          543 CONTINUE
533          542 CONTINUE
534          541 JEQ(ISET)=ICOMB-1
535              DO 544 J=1,JQ
536              IF(JJEQ(J).EQ.0)GO TO 544
537              NSET=NSET+1
538              JEQ(NSET)=1
539              JCOMB(NSET,1)=JJEQ(J)
540          544 CONTINUE
541              JQ=NSET
542              GO TO 20
543    C********************************
544    C
545    C         COME HERE TO PRINT OUT THE REMAINDER OF THE INFORMATION.
546    C
547    C********************************
548          998 NTS=N
549              WRITE(WUNIT,610)NM,NL,NTS,NCV
550              IF(NCOMB.GT.0)GO TO 702
551              DO 703 J=1,JQ
552              JEQ(J)=1
553          703 JCOMB(J,1)=JJEQ(J)
554          702 WRITE(WUNIT,611)JQ
555              IWRTRG=0
556              DO 704 J=1,JQ
557              JN=JEQ(J)
558              DO 705 JJ=1,JN
559              JCOM=JCOMB(J,JJ)
560              IF(JCOM.GT.4)GO TO 704
561              IWRTRG=IWRTRG+1
562              IF(IWRTRG.EQ.1)WRITE(WUNIT,622)
563          705 WRITE(WUNIT,620)(COMPLX(L,JCOM),L=1,30)
564          704 CONTINUE
565              IF(IWRTRG.NE.0)WRITE(WUNIT,621)
566              LL=0
567              DO 552 L=1,JQ
568          552 LL=LL+JEQ(L)
569              IF(LL.NE.JQ)WRITE(WUNIT,623)
570              LM=0
571              DO 550 L=1,JQ
572              IF(JEQ(L).EQ.1)GO TO 550
573              LM=LM+1
574              LL=JEQ(L)
575              WRITE(WUNIT,613)LM
576              DO 551 M=1,LL
577              MMM=JCOMB(L,M)-4
578          551 WRITE(WUNIT,612)(COMPLX(J,MMM+4),J=1,30)
579          550 CONTINUE
580              IF(JNO.EQ.0)RETURN
581              WRITE(WUNIT,616)
```

```
582          DO 553 J=1,JNO
583          JJ=JJNO(J)
584          DO 554 I=1,6
585          IF(COMPLX(I+8,JJ).NE.ABSENT(I))GO TO 555
586     554 CONTINUE
587          GO TO 553
588     555 WRITE(WUNIT,615)(COMPLX(L,JJ),L=1,30)
589     553 CONTINUE
590    9999 RETURN
591          END
```

```
  1  C################################################################
  2  C################################################################
  3  C##                                                            ##
  4  C##                                                            ##
  5  C##           $ $ $ $ $   E R R   $ $ $ $ $                    ##
  6  C##                                                            ##
  7  C##                                                            ##
  8  C################################################################
  9  C################################################################
 10          SUBROUTINE ERR(JCOL,ERROR,CARD)
 11  C
 12  C       THIS ROUTINE PRINTS OUT ERROR MESSAGES IF AN ERROR IN THE
 13  C       DATA DECK SETUP HAS BEEN DETECTED.
 14  C
 15          INTEGER ERROR,CARD(80),RUNIT,PUNIT,WUNIT,HEAD(4)
 16          COMMON/INTEGR/NCV,NBA,JQ,JNO,NL,NM,NTS,NPASS,NUMPH,NCD,M,
 17         &NBANUM,JJA,IBETA,NCC,NPAGE,IPLOT,IPRIN,ICARD,NLMR
 18          COMMON/UNIT/RUNIT,WUNIT,PUNIT
 19          DATA HEAD/'*** ',' ERR','OR  ','*** '/,BLANK/' '/,STAR/'*'/
 20      601 FORMAT(5(/),2X,4A4,'ERROR IN DICTIONARY CARD - PROBABLE',
 21         &' CAUSE MISSING "=" OR ";"'/)
 22      602 FORMAT(5(/),2X,4A4,'ILLEGAL COMPONENT DEFINITION NAME ON LEFT',
 23         &' HAND SIDE OF "="'/)
 24      603 FORMAT(5(/),2X,4A4,'THIS CARD IS NOT TERMINATED WITH A ":".'/)
 25      604 FORMAT(5(/),2X,4A4,'VARY OR FIXED CONSTANT?  USE "V" OR',
 26         &' "F" AFTER CONSTANT.'/)
 27      605 FORMAT(5(/),2X,4A4,'COMPONENT NAME IS NOT PART OF THE ',
 28         &' DICTIONARY.'/)
 29      606 FORMAT(5(/),2X,4A4,'USER DEFINED COMPONENT NAME CONTAINS MORE ',
 30         &'THAN FOUR CHARACTERS.'/)
 31      607 FORMAT(5(/),2X,4A4,'LEFT PARENTHESIS NOT FOUND.'/)
 32      608 FORMAT(5(/),2X,4A4,'RIGHT PARENTHESIS NOT FOUND.'/)
 33      609 FORMAT(5(/),2X,4A4,'DECIMAL POINT OR SEMICOLON, ASSOCIATED',
 34         &' WITH THE STABILITY CONSTANT IS MISSING.'/)
 35      610 FORMAT(5(/),2X,4A4,'IN SECTION "OTHER"  WHAT IS TO BE VARIED.'/)
 36      611 FORMAT(5(/),2X,4A4,'CONSTANTS OR MOLAR ABSORPTIVITY?',
 37         &'  USE "B" OR "E" TO DENOTE WHICH.'/)
 38      612 FORMAT(5(/),2X,4A4,'IN SECTION "OTHER" THE SYMBOL "+" OR ";" IS',
 39         &' MISSING.'/)
 40      613 FORMAT(5(/),2X,4A4,'SPECIES FOUND IN "OTHER" SECTION IS NOT A',
 41         &' LEGAL COMPLEX.'/)
 42      614 FORMAT(5(/),2X,4A4,'PREVIOUS ASSIGNMENT REQUIRES THE MOL. ABS.',
 43         &' OF THIS SPECIES TO BE CALCULATED BY SQUAD.'/)
 44      615 FORMAT(5(/),2X,4A4,'EXPECTING STAB. CONST. BUT LETTER FOUND'/)
 45      616 FORMAT(5(/),2X,4A4,'NUMBER OF SPECTRA EXCEEDS',
 46         &' EXECUTION-TIME DIMENSION PARAMETER NSOLNS')
 47      617 FORMAT(5(/),2X,4A4,'NUMBER OF WAVELENGTHS EXCEEDS',
 48         &' EXECUTION-TIME DIMENSION PARAMETER NWVLS')
 49      618 FORMAT(5(/),2X,4A4,'NUMBER OF COMPLEXES EXCEEDS',
 50         &' EXECUTION-TIME DIMENSION PARAMETER NCOMP')
 51      619 FORMAT(5(/),2X,4A4,'NUMBER OF STABILITY CONSTANTS EXCEEDS',
 52         &' EXECUTION-TIME DIMENSION PARAMETER NBETA')
 53      620 FORMAT(5(/),2X,4A4,'NUMBER OF MOL. ABS. SETS EXCEEDS',
 54         &' EXECUTION-TIME DIMENSION PARAMETER NETACL')
 55      695 FORMAT(2X,'INCREASE APPROPRIATE MAXIMUM ARRAY SIZE',
 56         &' PARAMETER IN MAIN PROGRAM')
 57      696 FORMAT(2X,'ERROR OCCURS AT OR NEAR STAR'//)
 58      697 FORMAT(1X,100A1)
 59      698 FORMAT(2X,'EXECUTION TERMINATED.  CORRECT ERRORS.  BYE-BYE')
 60      699 FORMAT(2X,'ERROR FOUND IN THIS CARD...'/2X,75A1)
 61          GO TO(101,102,103,104,105,106,107,108,109,110,111,112,113,114,
```

```
62          &115,116,117,118,119,120),ERROR
63      101 WRITE(WUNIT,601)(HEAD(J),J=1,4)
64          GO TO 999
65      102 WRITE(WUNIT,602)(HEAD(J),J=1,4)
66          GO TO 999
67      103 WRITE(WUNIT,603)(HEAD(J),J=1,4)
68          GO TO 999
69      104 WRITE(WUNIT,604)(HEAD(J),J=1,4)
70          GO TO 999
71      105 WRITE(WUNIT,605)(HEAD(J),J=1,4)
72          GO TO 999
73      106 WRITE(WUNIT,606)(HEAD(J),J=1,4)
74          GO TO 999
75      107 WRITE(WUNIT,607)(HEAD(J),J=1,4)
76          GO TO 999
77      108 WRITE(WUNIT,608)(HEAD(J),J=1,4)
78          GO TO 999
79      109 WRITE(WUNIT,609)(HEAD(J),J=1,4)
80          GO TO 999
81      110 WRITE(WUNIT,610)(HEAD(J),J=1,4)
82          GO TO 999
83      111 WRITE(WUNIT,611)(HEAD(J),J=1,4)
84          GO TO 999
85      112 WRITE(WUNIT,612)(HEAD(J),J=1,4)
86          GO TO 999
87      113 WRITE(WUNIT,613)(HEAD(J),J=1,4)
88          GO TO 999
89      114 WRITE(WUNIT,614)(HEAD(J),J=1,4)
90          GO TO 999
91      115 WRITE(WUNIT,615)(HEAD(J),J=1,4)
92          GO TO 999
93      116 WRITE(WUNIT,616)(HEAD(J),J=1,4)
94          GO TO 997
95      117 WRITE(WUNIT,617)(HEAD(J),J=1,4)
96          GO TO 997
97      118 WRITE(WUNIT,618)(HEAD(J),J=1,4)
98          GO TO 997
99      119 WRITE(WUNIT,619)(HEAD(J),J=1,4)
100         GO TO 997
101     120 WRITE(WUNIT,620)(HEAD(J),J=1,4)
102     997 WRITE(WUNIT,695)
103         STOP
104     999 WRITE(WUNIT,699)(CARD(I),I=1,75)
105         IF(JCOL.EQ.0)GO TO 998
106         WRITE(WUNIT,697)(BLANK,J=1,JCOL),STAR
107         WRITE(WUNIT,696)
108     998 WRITE(WUNIT,698)
109         STOP
110         END
```

```
1   C###############################################################
2   C###############################################################
3   C##                                                         ##
4   C##                                                         ##
5   C##        $ $ $ $ $  I N O U T  $ $ $ $ $                  ##
6   C##                                                         ##
7   C##                                                         ##
8   C###############################################################
9   C###############################################################
10         SUBROUTINE INOUT(CC,WORK,WORK1,WORK2,SP2,SPEC,Q,TM,TL,EQ,EC,STD,
11        &C11,DE,B1,PATH,C1,E,X,CK,PH,SSRSOL,COMPLX,JCOMB,MM,ML,MN,IPO,
12        &JEQ,JJNO,IWORK,TITLE,TITLE1,NWVLS,NSOLNS,NETACL,NBETA,NCOMP,
13        &NWVNSL,NCOMP4,NCOMP6)
14   C
15   C
16   C       THIS ROUTINE COMPLETES THE DATA INPUT AND CONTROLS THE OUTPUT
17   C       FROM SQUAD.
18   C
19         IMPLICIT REAL*8 (A-H,O-Z)
20         COMMON/INTEGR/NCV,NBA,JQ,JNO,NL,NM,NTS,NPASS,NUMPH,NCD,M,
21        &NBANUM,JJA,IBETA,NCC,NPAGE,IPLOT,IPRIN,ICARD,NLMR
22         COMMON/UNIT/RUNIT,WUNIT,PUNIT
```

```
23        COMMON/CONST/WVLST,WVLFIN,WVLINC,S1,F,XDF,XSD,SQRA,SSQR,WT
24        REAL*8 DEXP,DABS,DSQRT
25        DIMENSION Q(NSOLNS,NCOMP4),PH(NSOLNS),TM(2,NSOLNS),TL(2,NSOLNS)
26        DIMENSION SPEC(NSOLNS,NWVLS),SP2(NWVNSL,NBETA),PATH(NSOLNS),
27       &WORK(NSOLNS,NWVLS),SSRSOL(NWVLS)
28        DIMENSION JEQ(NETACL),JJNO(NCOMP),JCOMB(NCOMP,5),IWORK(NETACL)
29        DIMENSION C1(NSOLNS),E(NCOMP),IPO(NBETA),X(NETACL),CK(NETACL),
30       &WORK1(NETACL,NETACL),CC(NBETA,NBETA),DE(NETACL),B1(NSOLNS)
31        DIMENSION EQ(NCOMP4,NWVLS),EC(NCOMP6,NWVLS),STD(NETACL,NWVLS),
32       &C11(NSOLNS,NETACL),WORK2(NSOLNS,NETACL)
33        DIMENSION ML(2,NCOMP),MM(2,NCOMP),MN(NCOMP)
34        DIMENSIONJNOP(25),APER(100),BASELN(100)
35        INTEGER TITLE(20),HEAD(4),TITLE1(20),CODE(6),CARD(80),
36       &RUNIT,IVECT(100),COMPLX(30,NCOMP4),IBLAN/' '/,COPY(80),
37       &WUNIT,PUNIT,END(4),WORD1,SEMI,COLON,ABSENT(6),WORD2,
38       &CHAR(41)
39        DATA END/'E','N','D',':'/,SEMI/';'/,COLON/':'/,
40       &ABSENT/'A','B','S','E','N','T'/
41        DATA CHAR/'E','T','M','N','L','Q','1','2','3','4',
42       &'5','6','7','8','9','A','B','C','D','E','F','G','H','I','J',
43       &'K','L','M','N','O','P','Q','R','S','T','U','V','W','X','Y',
44       &'Z'/,HEAD/'*** ','  ERR','OR ','*** '/
45        DATA CODE/'MOL.','ABS.','BASE','LINE','SPEC','TRA:'/
46   600  FORMAT(20A4)
47   601  FORMAT(2A4)
48   602  FORMAT(80A1)
49   606  FORMAT(8F10.4)
50   607  FORMAT(4E10.4,2F10.4)
51   612  FORMAT(//2X,'GAMMA FOR H+ = ',F6.4/)
52   613  FORMAT(//2X,'*** BRONSTED CONSTANTS WILL BE CALCULATED ***'/)
53   614  FORMAT(///2X,'SPECTRAL REGION COVERED IS ',F5.1,' TO ',F5.1,
54       &' AT',F5.1,' INTERVALS'/)
55   615  FORMAT(//2X,'THERE ARE NO KNOWN MOLAR ABSORPTIVITIES',
56       &' FOR THIS SYSTEM.'/)
57   616  FORMAT(//2X,'COMPOSITION OF SOLUTIONS USED TO OBTAIN SPECTRA'/
58       &1X,'SPECTRUM',5X,'METAL 1',4X,'METAL 2',5X,'LIGAND 1',
59       &4X,'LIGAND 2',5X,'PH',3X,'PATHLENGTH'/2X,'NUMBER',
60       &4X,15('-'),' MOLES PER LITER ',14('-'),11X,'(CMS.)')
61   617  FORMAT(//2X,'NUMBER OF CYCLES DESIRED =',I3//2X,'TEMPERATURE ',
62       &'IS ',F4.1,'DEGREE C'/)
63   618  FORMAT(2X,I3,7X,1PD10.4,2X,3(D10.4,2X),0PF6.3,3X,F6.3)
64   619  FORMAT('1',2X,5('* '),'INTERMEDIATE CALCULATIONS',5(' *'))
65   620  FORMAT(12X,'RESIDUALS FOR SPECTRUM NUMBER',I3,
66       &' POINT-BY-POINT'/)
67   621  FORMAT(//7X,'NSPECIES', 8(5X,I2,5X)/25X,8(5X,I2,5X)/)
68   622  FORMAT(5X,'SOLN.',I3, 8(1PE12.4)/23X,8(E12.4))
69   623  FORMAT(2X,'WAVELENGTH',11F10.2)
70   624  FORMAT(5X,'SOLN.',I3, 4(1PE12.4))
71   625  FORMAT(//20X,'STANDARD DEVIATION OF ABSORBANCE FOR SOLUTION ',
72       &I2,' = ',1PD11.4//)
73   626  FORMAT(5X,'USING MULTIPLE REGRESSION ALGORITHM.'/)
74   627  FORMAT(5X,'USING NON-NEGATIVE LINEAR LEAST SQUARES',
75       &' ALGORITHM.'/)
76   628  FORMAT(/5X,'WAVELENGTH',4X,4(5X,I2,8X)/)
77   629  FORMAT(//5X,'MOLAR ABSORPTIVITIES OF INDIVIDUAL SPECIES'
78       &,' CALCULATED BY PROGRAM'/)
79   630  FORMAT(5X,F8.2,2X,1PD15.4,5D15.4/18X,(6D15.4))
80   632  FORMAT(8(E10.4))
81   633  FORMAT('1',10X,'CONCENTRATION MATRIX Q(NUMPH,NSPECIES)')
82   634  FORMAT(//15X,4('FREE ',5A1,2X))
83   635  FORMAT( 1X,12F10.3/41X,7F10.3)
84   636  FORMAT(//15X,'PERCENT CONTRIBUTION TO CALCULATED SPECTRUM',
85       &' NUMBER',I3,' FROM EACH SPECIES PRESENT IN SOLUTION'//
86       &13X,4('FREE ',5A1),7(6X,I2,2X)/40X,8(6X,I2,2X)/)
87   637  FORMAT(4X,'RESIDUAL',1PD10.2,10D10.2/)
88   639  FORMAT(8F10.4)
89   640  FORMAT(5X,'HEADING',I3,' REFERS TO THE SPECIES  ',30A1)
90   641  FORMAT(//8X,8I12/8X,8I12)
91   642  FORMAT('1',3X,'THE PRINTER PLOTS BELOW USE VARIOUS CHARACTERS',
92       &' TO SPECIFY THE FOLLOWING:-'/3X,'CHARACTER E IS USED TO REPR',
93       &'ESENT THE OBSERVED ABSORBANCES'/3X,'CHARACTER T IS USED TO ',
94       &'REPRESENT THE CALCULATED ABSORBANCES'/3X,'CHARACTER S IS ',
95       &'USED TO INDICATE THAT A(OBS.) EQUALS A(CALC.)')
96   643  FORMAT(3X,'CHARACTER ',A2,'IS USED TO REPRESENT THE',
97       &' SPECIES ',30A1)
98   644  FORMAT(//2X,'STANDARD DEVN. OF CALCULATED MOLAR ABSORPTIVITIES')
```

```
 99         645 FORMAT(1ØX,'NO CONSTANTS WILL BE VARIED'/1ØX,'MOLAR ABSORPTIVI',
1ØØ            &'TIES WILL BE CALCULATED USING INPUT CONSTANTS'//)
1Ø1         646 FORMAT(5X,F6.2,1PD12.4,7D12.4/4(11X,8D12.4))
1Ø2         647 FORMAT(//2ØX,'KNOWN AND FIXED MOLAR ABSORPTIVITIES'//)
1Ø3         649 FORMAT(8(D1Ø.4))
1Ø4         65Ø FORMAT(//2X,'BASELINE CORRECTION NOT APPLIED'//)
1Ø5         651 FORMAT(//2X,'BASELINE CORRECTION APPLIED'//)
1Ø6         652 FORMAT(5(/),2X,4A4,2X,'UNKNOWN CONTROL WORD FOUND'/18X,'I.E.....',
1Ø7            &2A4//18X,'CORRECT ERRORS. EXECUTION TERMINATED.  BYE-BYE'//)
1Ø8         653 FORMAT(3X,'PRINTER PLOTS WILL BE PRODUCED FOR SPECTRA',
1Ø9            &' NUMBERS',I2,8(',',I2),15(',',I3))
11Ø         654 FORMAT('1',3X,'PRINTER PLOT OF ALL MOLAR ABSORPTIVITIES')
111         655 FORMAT('1',3X,'PRINTER PLOT OF ABSORBANCE DATA'/3X,'SPECTRA 1',
112            &' TO',I3,' ARE REPRESENTED BY SYMBOLS:-'/3X,4Ø(A1,',',1X))
113         699 FORMAT(//2ØX,'*** ERROR  ***',3X,'END OF FILE',
114            &' FOUND.  PROBABLE CAUSE, THE LAST CARD IS NOT NEGATIVE')
115   C
116   C       INITIALIZE INTERNAL PARAMETERS AND ARRAYS.
117   C
118            WT=1.
119            NPASS=Ø
12Ø            IBSL=Ø
121            NBA=(WVLFIN-WVLST)/WVLINC+1.1
122            IF(NBA.GT.NWVLS)CALL ERR(Ø,17,COPY)
123            JJA=NTS+4
124            JJB=JNO+1
125            DO 1Ø1 J=1,JJA
126            DO 1Ø2 K=1,NBA
127       1Ø2 EQ(J,K)=Ø.Ø
128            DO 1Ø1 I=1,NSOLNS
129       1Ø1 Q(I,J)=Ø.Ø
13Ø   C
131   C       DETERMINE IF MOLAR ABSORPTIVITIES, BASELINE AND/OR
132   C       SPECTRAL DATA ARE TO BE READ-IN.
133   C
134       172 READ(RUNIT,6Ø1)WORD1,WORD2
135            DO 17Ø I=1,3
136            I2=2*I
137            I2M1=I2-1
138            IF(WORD1.NE.CODE(I2M1))GO TO 17Ø
139            IF(WORD2.EQ.CODE(I2))GO TO 171
14Ø       17Ø CONTINUE
141            WRITE(WUNIT,652)(HEAD(J),J=1,4),WORD1,WORD2
142            STOP
143       171 GO TO (181,2Ø2,2Ø3),I
144   C********************************
145   C
146   C       READ IN FIXED MOLAR ABSORPTIVITIES.
147   C
148   C********************************
149       181 IJ=Ø
15Ø       193 READ(RUNIT,6Ø2)(CARD(I),I=1,8Ø)
151            DO 182 I=1,4
152            IF(CARD(I).EQ.END(I))GO TO  194
153       182 CONTINUE
154   C
155   C       CHECK THAT LAST CHARACTER IN STRING IS A COLON
156   C
157            IJI=81
158            DO 187 J=1,8Ø
159            IJI=IJI-1
16Ø            IF(CARD(IJI).EQ.COLON)GO TO 188
161       187 CONTINUE
162            CALL ERR(IJI,3,CARD)
163   C
164   C       SAVE THE FORMULA OF EACH SPECIES IN COPY
165   C
166       188 IJI=Ø
167            ICOLON=Ø
168       189 DO 184 J=1,8Ø
169       184 COPY(J)=IBLAN
17Ø            DO 185 J=1,3Ø
171            IJI=IJI+1
172            IF(CARD(IJI).EQ.COLON)GO TO 183
173            IF(CARD(IJI).EQ.SEMI)GO TO 186
174       185 COPY(J)=CARD(IJI)
```

```
175        183 ICOLON=1
176        186 DO 190 J=1,JJA
177            DO 175 L=1,30
178            IF(COPY(L).EQ.IBLAN)GO TO 178
179            IF(COPY(L).NE.COMPLX(L,J))GO TO 190
180        175 CONTINUE
181        178 JNOPOS=J
182            DO 191 JJ=1,JNO
183            IF(JNOPOS.EQ.JJNO(JJ))GO TO 192
184        191 CONTINUE
185            CALL ERR(0,14,COPY)
186        190 CONTINUE
187            CALL ERR(0,13,COPY)
188        192 IJ=IJ+1
189            JNOP(IJ)=JNOPOS
190            IF(ICOLON.EQ.1)GO TO 193
191            GO TO 189
192      C
193      C       READ IN MOLS. ABS. AND PLACE IN CORRECT ELEMENTS OF EQ.
194      C
195        194 CONTINUE
196            DO 195 K=1,NBA
197            READ(RUNIT,649)(EC(J,1),J=1,IJ)
198            DO 173 J=1,IJ
199            JJ=JNOP(J)
200            EQ(JJ,K)=EC(J,1)
201        173 CONTINUE
202        195 CONTINUE
203            GO TO 172
204      C********************************
205      C
206      C       COME HERE TO PERFORM BASELINE CORRECTION, IF NEEDED.
207      C
208      C********************************
209        202 IBSL=1
210            READ(RUNIT,639)(BASELN(K),K=1,NBA)
211            GO TO 172
212      C********************************
213      C
214      C       COME HERE TO READ IN SPECTRA
215      C
216      C********************************
217        203 I=0
218        138 I=I+1
219            IF(I.GT.NSOLNS)CALL ERR(0,16,COPY)
220            READ(RUNIT,607)TM1,TM2,TL1,TL2,PHI,PATHI
221            IF(TM1.LT.0.0)GO TO 127
222            TM(1,I)=TM1
223            TM(2,I)=TM2
224            TL(1,I)=TL1
225            TL(2,I)=TL2
226            PH(I)=PHI
227            PATH(I)=PATHI
228            READ(RUNIT,606,END=5432)(SPEC(I,K),K=1,NBA)
229            GO TO 138
230        127 NUMPH=I-1
231            JJ=0
232            DO 179 I=1,NUMPH
233            ICHK=I-(I/IPLOT)*IPLOT
234            IF(ICHK.NE.0)GO TO 179
235            JJ=JJ+1
236            COPY(JJ)=I
237        179 CONTINUE
238            IF(IPLOT.GT.2)WRITE(WUNIT,653)(COPY(I),I=1,JJ)
239      C
240      C
241      C       THE FOLLOWING DO-LOOP INITIALIZES ANY NEGATIVE ABSORBANCES
242      C       TO ZERO AFTER THE DATA HAS BEEN CORRECTED.
243      C
244      C
245            IF(IBSL.EQ.0)GO TO 106
246            DO 306 I=1,NUMPH
247            DO 306 K=1,NBA
248            A=SPEC(I,K)-BASELN(K)
249            IF(A.LT.0.0E0) A=0.0E0
250        306 SPEC(I,K)=A
```

```
251   C
252   C         COMPLETE PRINTING INPUT DATA AND CALCULATION OPTIONS.
253   C
254     106 IF(ABS(F-1.E0).LT.1.E-06)WRITE(WUNIT,613)
255         IF(ABS(F-1.E0).GT.1.E-06)WRITE(WUNIT,612)F
256         IF(NCV.EQ.0)WRITE(WUNIT,645)
257         TEMP=25.
258         WRITE(WUNIT,617)NCD,TEMP
259         WRITE(WUNIT,614)WVLST,WVLFIN,WVLINC
260         IF((NM+NL+NTS).EQ.JQ)GO TO 142
261         IF(IPRIN.EQ.0)GO TO 9876
262         WRITE(WUNIT,647)
263         DO 314 J=1,JNO
264         JJ=JJNO(J)
265     314 WRITE(WUNIT,640)JJ,(COMPLX(L,JJ),L=1,30)
266         WRITE(WUNIT,641)(JJNO(J),J=1,JNO)
267         WVL=WVLST
268         DO 140 K=1,NBA
269         DO 103 J=1,JNO
270         JJ=JJNO(J)
271     103 EC(J,1)=EQ(JJ,K)
272         WRITE(WUNIT,646)WVL,(EC(J,1),J=1,JNO)
273     140 WVL=WVL+WVLINC
274         GO TO 141
275     142 WRITE(WUNIT,615)
276     141 WRITE(WUNIT,616)
277         DO 108 I=1,NUMPH
278     108 WRITE(WUNIT,618)I,TM(1,I),TM(2,I),TL(1,I),TL(2,I),PH(I),PATH(I)
279    9876 IF(IBSL.EQ.0)WRITE(WUNIT,650)
280         IF(IBSL.EQ.1)WRITE(WUNIT,651)
281         WRITE(WUNIT,619)
282         XDF=NBA*(NUMPH-JQ)-NCV
283         XSD=NBA-NCV
284   C*********************************
285   C
286   C         CALL SUBROUTINE REFINE TO COMMENCE THE REFINEMENT
287   C         PROCESS.
288   C
289   C*********************************
290         CALL REFINE(CC,WORK,WORK1,WORK2,C11,EQ,EC,STD,Q,TM,TL,
291        &SPEC,SP2,PATH,SSRSOL,PH,E,B1,C1,CK,X,DE,COMPLX,IPO,JCOMB,JEQ,
292        &JJNO,IWORK,MM,ML,MN,NWVLS,NSOLNS,NETACL,NBETA,NCOMP,NWVNSL,
293        &NCOMP4,NCOMP6)
294   C*********************************
295   C
296   C         FINAL PRINT OUT STARTS HERE CONSISTING OF ALL CONCENTRATIONS AND
297   C         MOLAR ABSORPTIVITIES BASED ON FINAL SET OF STABILITY CONSTANTS.
298   C
299   C*********************************
300   C
301   C         PRINT OUT CONCENTRATIONS OF ALL SPECIES
302   C
303         WRITE(WUNIT,633)
304         WRITE(WUNIT,634)((COMPLX(L,J),L=1,5),J=1,4)
305         DO 110 I=1,NUMPH
306     110 WRITE(WUNIT,624)I,(Q(I,J),J=1,4)
307         WRITE(WUNIT,621)(I,I=1,NTS)
308         DO 111 I=1,NUMPH
309     111 WRITE(WUNIT,622)I,(Q(I,J),J=5,JJA)
310   C
311   C         PRINT OUT ALL CALCULATED MOLAR ABSORPTIVITIES.
312   C
313         IF(JQ.EQ.0)GO TO 132
314         WRITE(WUNIT,629)
315         IF(NLMR.EQ.0)WRITE(WUNIT,626)
316         IF(NLMR.EQ.1)WRITE(WUNIT,627)
317         DO 206 J=1,JQ
318         JN=JEQ(J)
319         DO 204 JJ=1,JN
320         JCOM=JCOMB(J,JJ)
321     204 WRITE(WUNIT,640)J,(COMPLX(L,JCOM),L=1,30)
322     206 CONTINUE
323         WRITE(WUNIT,628)(I,I=1,JQ)
324         WVL=WVLST
325         DO 112 K=1,NBA
326         WRITE(WUNIT,630)WVL,(EC(J,K),J=1,JQ)
```

```
327        112 WVL=WVL+WVLINC
328            WRITE(WUNIT,644)
329            WRITE(WUNIT,628)(I,I=1,JQ)
330            WVL=WVLST
331            DO 135 K=1,NBA
332            WRITE(WUNIT,630)WVL,(STD(J,K),J=1,JQ)
333        135 WVL=WVL+WVLINC
334    C
335    C          START PLOTTING THE DATA RETURNED FROM REFINE.
336    C          THE MOLAR ABSORPTIVITIES, ABSORBANCE DATA AND A(OBS.)
337    C          VS A(CALC.) TOGETHER WITH EACH SPECIES CONTRIBUTION TO
338    C          THE OBSERVED SPECTRA ARE PLOTTED AS PRINTER PLOTS.
339    C          THE VECTOR SSRSOL (FOR THE X-AXIS) AND THE ARRAY
340    C          EC (FOR THE Y-AXIS) AND STD (FOR POINT-BY-POINT RESIDUALS)
341    C          ARE REUSED.
342    C
343            IF(IPLOT.EQ.0)GO TO 132
344    C
345    C          PLOT THE MOLAR ABSORPTIVITIES.
346    C
347            WRITE(WUNIT,654)
348            DO 160 J=1,4
349        160 JEQ(J)=0
350            IF(NM.NE.0)JEQ(1)=1
351            IF(NM.EQ.2)JEQ(2)=1
352            IF(NL.NE.0)JEQ(3)=1
353            IF(NL.EQ.2)JEQ(4)=1
354            DO 161 J=1,4
355            IF(JEQ(J).EQ.0)GO TO 161
356            JP2=J+2
357            WRITE(WUNIT,643)CHAR(JP2),(COMPLX(L,J),L=1,30)
358        161 CONTINUE
359            DO 162 J=5,JJA
360            JP2=J+2
361        162 WRITE(WUNIT,643)CHAR(JP2),(COMPLX(L,J),L=1,30)
362            WVL=WVLST
363            DO 119 K=1,NBA
364            SSRSOL(K)=WVL
365        119 WVL=WVL+WVLINC
366            DO 118 K=1,NBA
367            DO 118 J=1,JJA
368        118 EC(J,K)=EQ(J,K)
369            CALL PRTPLT(SSRSOL,EC,NWVLS,NCOMP6,NBA,JJA,NPAGE,2)
370    C
371    C          PLOT THE ABSORBANCE DATA.
372    C
373            NP6=NUMPH+6
374            WRITE(WUNIT,655)NUMPH,(CHAR(J),J=7,NP6)
375            J=0
376            ISHIFT=2
377            DO 116 I=1,NUMPH
378            J=J+1
379            DO 117 K=1,NBA
380        117 EC(J,K)=SPEC(I,K)
381            IF(I.EQ.NUMPH)GO TO 115
382            JCK=J-(J/4)*4
383            IF(JCK.NE.0)GO TO 116
384        115 ISHIFT=ISHIFT+4
385            CALL PRTPLT(SSRSOL,EC,NWVLS,NCOMP6,NBA,J,NPAGE,ISHIFT)
386            J=0
387        116 CONTINUE
388    C
389    C          PLOT OBSERVED, CALCULATED AND SPECIES CONTRIBUTIONS TO
390    C          EACH SPECTRUM, OR TO THE NTH SPECTRUM.
391    C
392            WRITE(WUNIT,642)
393            DO 163 J=1,4
394            IF(JEQ(J).EQ.0)GO TO 163
395            JP2=J+2
396            WRITE(WUNIT,643)CHAR(JP2),(COMPLX(L,J),L=1,30)
397        163 CONTINUE
398            DO 164 J=5,JJA
399            JP2=J+2
400        164 WRITE(WUNIT,643)CHAR(JP2),(COMPLX(L,J),L=1,30)
401            DO 114 I=1,NUMPH
402            ICHK=I-(I/IPLOT)*IPLOT
```

```
403          IF(ICHK.NE.Ø)GO TO 114
404          SSQ=Ø.Ø
405          PATHI=PATH(I)
406          DO 196 K=1,NBA
407          EC(1,K)=SPEC(I,K)
408          ABSORB=Ø.Ø
409          DO 197 J=1,JJA
410          ASPEC=Q(I,J)*EQ(J,K)*PATHI
411          ABSORB=ABSORB+ASPEC
412          JP2=J+2
413      197 EC(JP2,K)=ASPEC
414          R=ABSORB-SPEC(I,K)
415          SSQ=SSQ+R*R*WT
416          BASELN(K)=R
417      196 EC(2,K)=ABSORB
418          SQRSOL=DSQRT(SSQ/XSD)
419          CALL PRTPLT(SSRSOL,EC,NWVLS,NCOMP6,NBA,JJA+2,NPAGE,Ø)
420          WRITE(WUNIT,625)I,SQRSOL
421          WRITE(WUNIT,620)I
422          NTIM=Ø
423          NTIMES=1+NBA/11
424          IREM=NBA-NTIMES*11
425      131 NTIM=NTIM+1
426          NST=1+(NTIM-1)*11
427          NFIN=NTIM*11
428          IF(NFIN.GT.NBA)NFIN=NBA
429          WRITE(WUNIT,623)(SSRSOL(K),K=NST,NFIN)
430          WRITE(WUNIT,637)(BASELN(K),K=NST,NFIN)
431          IF(NFIN.NE.NBA)GO TO 131
432          CALL SCATER(BASELN,NBA)
433          WRITE(WUNIT,636)I,((COMPLX(L,J),L=1,5),J=1,4),(J,J=1,NTS)
434          WVL=WVLST
435          DO 122 K=1,NBA
436          PATHI=PATH(I)
437          DO 121 J=1,JJA
438          ABSORB=ABSORB+Q(I,J)*EQ(J,K)
439          APER(J)=Ø.Ø
440          IF(ABSORB.GT.Ø.Ø)APER(J)=EQ(J,K)*Q(I,J)*100.ØDØ/ABSORB
441      121 CONTINUE
442          WRITE(WUNIT,635)WVL,(APER(J),J=1,JJA)
443      122 WVL=WVL+WVLINC
444      114 CONTINUE
445  C
446  C          PUNCH OUT CONCENTRATION AND MOLAR ABSORPTIVITIES ARRAYS IF
447  C          NEEDED, ICARD=1.
448  C
449      132 IF(ICARD.EQ.Ø)RETURN
450          WRITE(PUNIT,600)TITLE
451          WRITE(PUNIT,600)TITLE1
452          DO 124 K=1,NBA
453      124 WRITE(PUNIT,632)(EQ(J,K),J=1,JJA)
454          DO 125 K=1,NUMPH
455      125 WRITE(PUNIT,632)(Q(K,J),J=1,JJA)
456          RETURN
457     5432 WRITE(WUNIT,699)
458          STOP
459          END
```

```
1    C###############################################################
2    C###############################################################
3    C##                                                         ##
4    C##                                                         ##
5    C##        $ $ $ $ $  S O L V E  $ $ $ $ $                  ##
6    C##                                                         ##
7    C##                                                         ##
8    C###############################################################
9    C###############################################################
10          SUBROUTINE SOLVE(CC,C,WORK2,EC,STD,SPEC,PATH,B1,C1,B,W,X,JCOMB,
11        &JEQ,JJNO,INDEX,NWVLS,NSOLNS,NETACL,NCOMP,NCOMP6)
12   C
13   C          THIS SUBROUTINE SOLVES A SET OF OVER DETERMINED LINEAR EQUATIONS
14   C          USING MULTIPLE LINEAR REGRESSION OR BY NON-NEGATIVE LINEAR
```

```
15    C         LEAST-SQUARES ALGORITHM.
16    C         THE ALGORITHM CHOSEN IS CONTROLLED BY THE SCALAR NLMR:-
17    C         NLMR = 0  USE MULTIPLE REGRESSION
18    C         NLMR = 1  USE NON-NEGATIVE LINEAR LEAST-SQUARES.
19    C
20              IMPLICIT REAL*8(A-H,O-Z)
21              COMMON/INTEGR/NCV,NBA,JQ,JNO,NL,NM,NTS,NPASS,NUMPH,NCD,M,
22             &NBANUM,JJA,IBETA,NCC,NPAGE,IPLOT,IPRIN,ICARD,NLMR
23              DIMENSION EC(NCOMP6,NWVLS),STD(NETACL,NWVLS),C(NSOLNS,NETACL)
24              DIMENSION PATH(NSOLNS),SPEC(NSOLNS,NWVLS)
25              DIMENSION JEQ(NETACL),JJNO(NCOMP),JCOMB(NCOMP,5),INDEX(NETACL)
26              DIMENSIONCC(NETACL,NETACL),X(NETACL),W(NETACL),B(NWVLS),
27             &B1(NSOLNS),C1(NSOLNS),WORK2(NSOLNS,NETACL)
28    C
29    C         SEVERAL ARRAYS ARE INITIALIZED TO ZERO.
30    C
31              DO 11 J=1,JQ
32              DO 18 I=1,NUMPH
33           18 STD(J,I)=0.0
34              DO 24 K=1,NBA
35           24 EC(J,K)=0.0
36              DO 11 JJ=1,JQ
37           11 CC(J,JJ)=0.0
38    C
39    C         THE MATRIX CC IS OBTAINED AND SUBSEQUENTLY INVERTED
40    C         THIS PROCESS INVOLVES THE USE OF THE TRANSPOSE OF
41    C         THE ARRAY C.  SINCE C(TRANSPOSE) (I,K) IS THE SAME AS
42    C         C (K,I)  THE TRANSPOSE IS NOT CALCULATED EXPLICITLY.
43    C
44              DO 12 JJ=1,JQ
45              DO 12 J=1,JQ
46              DO 12 I=1,NUMPH
47           12 CC(JJ,J)=C(I,JJ)*C(I,J)+CC(JJ,J)
48              DO 13 K=1,JQ
49              CM760=CC(K,K)
50              CC(K,K)=1.0
51              DO 14 J=1,JQ
52           14 CC(K,J)=CC(K,J)/CM760
53              DO 13 I=1,JQ
54              IF(I-K)15,13,15
55           15 CM760=CC(I,K)
56              CC(I,K)=0.0
57              DO 16 J=1,JQ
58           16 CC(I,J)=CC(I,J)-CM760*CC(K,J)
59           13 CONTINUE
60    C
61    C         CHECK TO SEE IF MULTIPLE REGRESSION (MR) OR NON-NEGATIVE
62    C         LINEAR LEAST SQUARES (NNLS) IS TO BE USED TO SOLVE FOR
63    C         THE MOLAR ABSORPTIVITIES.
64    C
65              IF(NLMR.EQ.1)GO TO 50
66    C
67    C         THE SOLUTION TO THE EQUATIONS IS FOUND BY MULTIPLICATION OF
68    C         (C*C(TRANS))-1*C(TRANS) BY THE ABSORBANCE DATA, FOR MR ONLY.
69    C
70              DO 17 I=1,NUMPH
71              DO 17 JJ=1,JQ
72              DO 17 J=1,JQ
73           17 STD(JJ,I)=CC(JJ,J)*C(I,J)+STD(JJ,I)
74              DO 19 K=1,NBA
75              DO 19 J=1,JQ
76              DO 19 I=1,NUMPH
77           19 EC(J,K)=STD(J,I)*(SPEC(I,K)/PATH(I))+EC(J,K)
78              RETURN
79    C
80    C         COME HERE TO SOLVE FOR MOL. ABS. BY NNLS
81    C
82           50 DO 51 K=1,NBA
83              DO 52 I=1,NUMPH
84              DO 53 J=1,JQ
85           53 WORK2(I,J)=C(I,J)
86           52 B(I)=SPEC(I,K)/PATH(I)
87              CALL NNLS(WORK2,NUMPH,NUMPH,JQ,B,X,RNORM,W,B1,C1,INDEX,MODE,
88             &NSOLNS,NETACL,NWVLS)
89              DO 54 J=1,JQ
90           54 EC(J,K)=X(J)
```

```
91        51 CONTINUE
92           RETURN
93    C
94    C
95    C      ENTRY POINT SIGMAE IS USED TO CALCULATE THE STANDARD DEVIATIONS
96    C      STORED IN STD(I,NBA), FOR THE CALCULATED MOLAR ABSORPTIVITIES
97    C      AFTER THE ITERATIVE CYCLE.
98    C
99    C###############################################################
100   C###############################################################
101   C##                                                         ##
102   C##                                                         ##
103   C##        $ $ $ $ $   S I G M A E   $ $ $ $ $              ##
104   C##                                                         ##
105   C##                                                         ##
106   C###############################################################
107   C###############################################################
108   C
109          ENTRY SIGMAE(CC,C,WORK2,EC,STD,SPEC,PATH,B1,C1,B,W,X,JCOMB,
110         &JEQ,JJNO,INDEX,NWVLS,NSOLNS,NETACL,NCOMP,NCOMP6)
111          NDEGS=NUMPH-JQ
112          DO 31 K=1,NBA
113          S=0.0
114          DO 20 I=1,NUMPH
115          ABSORB=0.0
116          DO 23 J=1,JQ
117       23 ABSORB=ABSORB+C(I,J)*EC(J,K)
118          R=SPEC(I,K)-ABSORB*PATH(I)
119       20 S=S+R*R
120   C
121   C      CALCULATE STD OF THE MOL. ABS.
122   C
123          DO 22 I=1,JQ
124       22 STD(I,K)=DSQRT(CC(I,I)*S/NDEGS)
125       31 CONTINUE
126          RETURN
127          END
```

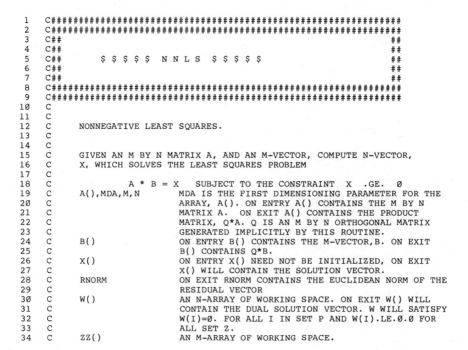

```
1     C###############################################################
2     C###############################################################
3     C##                                                         ##
4     C##                                                         ##
5     C##        $ $ $ $ $   N N L S   $ $ $ $ $                  ##
6     C##                                                         ##
7     C##                                                         ##
8     C###############################################################
9     C###############################################################
10    C
11    C
12    C      NONNEGATIVE LEAST SQUARES.
13    C
14    C
15    C      GIVEN AN M BY N MATRIX A, AND AN M-VECTOR, COMPUTE N-VECTOR,
16    C      X, WHICH SOLVES THE LEAST SQUARES PROBLEM
17    C
18    C             A * B = X    SUBJECT TO THE CONSTRAINT  X  .GE.  0
19    C      A(),MDA,M,N      MDA IS THE FIRST DIMENSIONING PARAMETER FOR THE
20    C                       ARRAY, A(). ON ENTRY A() CONTAINS THE M BY N
21    C                       MATRIX A.  ON EXIT A() CONTAINS THE PRODUCT
22    C                       MATRIX, Q*A. Q IS AN M BY N ORTHOGONAL MATRIX
23    C                       GENERATED IMPLICITLY BY THIS ROUTINE.
24    C      B()              ON ENTRY B() CONTAINS THE M-VECTOR,B. ON EXIT
25    C                       B() CONTAINS Q*B.
26    C      X()              ON ENTRY X() NEED NOT BE INITIALIZED, ON EXIT
27    C                       X() WILL CONTAIN THE SOLUTION VECTOR.
28    C      RNORM            ON EXIT RNORM CONTAINS THE EUCLIDEAN NORM OF THE
29    C                       RESIDUAL VECTOR
30    C      W()              AN N-ARRAY OF WORKING SPACE. ON EXIT W() WILL
31    C                       CONTAIN THE DUAL SOLUTION VECTOR. W WILL SATISFY
32    C                       W(I)=0. FOR ALL I IN SET P AND W(I).LE.0.0 FOR
33    C                       ALL SET Z.
34    C      ZZ()             AN M-ARRAY OF WORKING SPACE.
```

```
 35   C        INDEX()            AN INTEGER WORKING ARRAY OF LENGTH AT LEAST N.ON
 36   C                           EXIT THE CONTENTS OF THIS ARRAY DEFINE THE SETS
 37   C                           AS FOLLOWS:
 38   C                           INDEX(1)    THRU    INDEX(NSETP) = SET P
 39   C                           INDEX(IZ1)  THRU    INDEX(IZ2) = SET Z.
 40   C                           IZ1 = NSETP+1 = NPP1
 41   C                           IZ2 = N
 42   C        MODE               THIS IS A SUCCESS-FAILURE FLAG WITH THE
 43   C                           FOLLOWING MEANINGS:
 44   C                           1   THE SOLUTION HAS BEEN COMPUTED SUCCESSFULLY.
 45   C                           2   THE DIMENSIONS OF THE PROBLEM ARE BAD
 46   C                               EITHER M.LE.Ø OR N.LE.Ø
 47   C                           3   ITERATION COUNT EXCEEDED. MORE THAN 3*N
 48   C                               ITERATIONS.
 49   C
 50   C        CHARLES L. LAWSON/RICHARD J. HANSON,
 51   C        SOLVING LEAST SQUARES PROBLEMS, COPYRIGHT 1974, PP 304-309,
 52   C        REPRINTED BY PERMISSION OF PRENTICE-HALL, INC.,
 53   C        ENGLEWOOD CLIFFS, N.J.
 54   C
 55            SUBROUTINENNLS(A,MDA,M,N,B,X,RNORM,W,ZZ,DUMMY,INDEX,MODE,
 56           &NSOLNS,NETACL,NWVLS)
 57            IMPLICIT REAL*8(A-H,O-Z)
 58            DIMENSIONA(NSOLNS,NETACL),INDEX(NETACL),W(NETACL),ZZ(NSOLNS),
 59           &X(NETACL),DUMMY(NSOLNS),B(NWVLS)
 60            ZERO=Ø.Ø
 61            ONE=1.Ø
 62            TWO=2.Ø
 63            FACTOR=Ø.Ø1
 64            MODE=1
 65            IF(M.GT.Ø.AND.N.GT.Ø)GO TO 1Ø
 66            MODE=2
 67            RETURN
 68        1Ø  ITER=Ø
 69            ITMAX=3*N
 70   C
 71   C        INITIALIZE THE ARRAYS INDEX() AND X()
 72   C
 73            DO2ØI=1,N
 74            X(I)=ZERO
 75        2Ø  INDEX(I)=I
 76            IZ2=N
 77            IZ1=1
 78            NSETP=Ø
 79            NPP1=1
 80   C
 81   C        MAIN LOOP BEGINS HERE.
 82   C
 83        3Ø  CONTINUE
 84   C
 85   C        QUIT IF ALL COEFICIENTS ARE ALREADY IN THE SOLUTION VECTOR
 86   C        OR IF M COLUMNS OF "A" HAVE BEEN TRIANGULARIZED.
 87   C
 88            IF(IZ1.GT.IZ2.OR.NSETP.GE.M)GO TO 35Ø
 89   C
 90   C        COMPUTE COMPONENTS OF THE DUAL (NEGATIVE GRADIENT) VECTOR W().
 91   C
 92            DO5ØIZ=IZ1,IZ2
 93            J=INDEX(IZ)
 94            SM=ZERO
 95            DO4ØL=NPP1,M
 96        4Ø  SM=SM+A(L,J)*B(L)
 97        5Ø  W(J)=SM
 98   C
 99   C        FIND LARGEST POSITIVE W(J).
100   C
101        6Ø  WMAX=ZERO
102            DO7ØIZ=IZ1,IZ2
103            J=INDEX(IZ)
104            IF(W(J)-WMAX)7Ø,7Ø,71
105        71  WMAX=W(J)
106            IZMAX=IZ
107        7Ø  CONTINUE
108   C
109   C        IF WMAX .LE. Ø. GO TO TERMINATION.
110   C        THIS INDICATES SATISFACTION OF THE KUHN-TUCKER CONDITIONS.
```

```
111    C
112            IF(WMAX)350,350,80
113        80 IZ=IZMAX
114           J=INDEX(IZ)
115    C
116    C      THE SIGN OF W(J) IS OK FOR J TO BE MOVED TO SET P
117    C      BEGIN THE TRANSFORMATION AND CHECK NEW DIAGONAL ELEMENT TO AVOID
118    C      NEAR LINEAR DEPENDENCE.
119    C
120           ASAVE=A(NPP1,J)
121           CALLH12(1,NPP1,NPP1+1,M,A(1,J),1,UP,DUMMY,1,1,0,NSOLNS,NETACL,
122          &NSOLNS)
123           UNORM=ZERO
124           IF(NSETP.EQ.0)GO TO 100
125           DO90L=1,NSETP
126        90 UNORM=UNORM+A(L,J)**2
127       100 UNORM=SQRT(UNORM)
128           DIFF1=UNORM+DABS(A(NPP1,J))*FACTOR-UNORM
129           IF(DIFF1)130,130,110
130    C
131    C      COL J IS SUFFICIENTLY INDEPENDENT. COPY B INTO ZZ. UPDATE ZZ AND
132    C      SOLVE FOR ZTEST ( = PROPOSED NEW VALUE FOR X(J) ).
133    C
134       110 DO120L=1,M
135       120 ZZ(L)=B(L)
136           CALLH12(2,NPP1,NPP1+1,M,A(1,J),1,UP,ZZ,1,1,1,NSOLNS,NETACL,
137          &NSOLNS)
138           ZTEST=ZZ(NPP1)/A(NPP1,J)
139    C
140    C      SEE IF ZTEST IS POSITIVE.
141    C
142           IF(ZTEST)130,130,140
143    C
144    C      REJECT J AS A CANDIDATE TO BE MOVED FROM SET Z TO SET P.
145    C      RESTORE A(NPP1,J). SET W(J)=0., AND LOOP BACK TO TEST DUAL
146    C      COEFICIENTS AGAIN.
147    C
148       130 A(NPP1,J)=ASAVE
149           W(J)=ZERO
150           GO TO 60
151    C
152    C      THE INDEX J=INDEX(IZ) HAS BEEN SELECTED TO BE MOVED FROM SET Z
153    C      TO SET P. UPDATE B, UPDATE INDICES. APPLY HOUSEHOLDER
154    C      TRANSFORMATION TO COLUMNS IN NEW SET Z. ZERO SUBDIAGONAL
155    C      ELEMENTS IN COLUMN J. SET W(J)=0.
156    C
157       140 DO150L=1,M
158       150 B(L)=ZZ(L)
159           INDEX(IZ)=INDEX(IZ1)
160           INDEX(IZ1)=J
161           IZ1=IZ1+1
162           NSETP=NPP1
163           NPP1=NPP1+1
164           IF(IZ1.GT.IZ2)GO TO 170
165           DO160JZ=IZ1,IZ2
166           JJ=INDEX(JZ)
167       160 CALLH12(2,NSETP,NPP1,M,A(1,J),1,UP,A(1,JJ),1,MDA,1,NETACL,
168          &NETACL,NSOLNS)
169       170 CONTINUE
170           IF(NSETP.EQ.M)GO TO 190
171           DO180L=NPP1,M
172       180 A(L,J)=ZERO
173       190 CONTINUE
174           W(J)=ZERO
175    C
176    C      SOLVE THE TRIANGULAR SYSTEM.  STORE THE SOLUTON TEMPORARILY
177    C      IN ZZ().
178    C
179           ASSIGN200TONEXT
180           GO TO 400
181       200 CONTINUE
182    C
183    C      SECONDARY LOOP BEGINS HERE.
184    C
185       210 ITER=ITER+1
186           IF(ITER.LE.ITMAX)GO TO 220
```

```
187              MODE=3
188              WRITE(WUNIT,440)
189              GO TO 350
190      220 CONTINUE
191    C
192    C        SEE IF ALL NEW CONSTRAINED COEFICIENTS ARE FEASIBLE.
193    C        IF NOT COMPUTE ALPHA.
194    C
195              ALPHA=TWO
196              DO240IP=1,NSETP
197              L=INDEX(IP)
198              IF(ZZ(IP))230,230,240
199      230 T=-X(L)/(ZZ(IP)-X(L))
200              IF(ALPHA-T)240,240,241
201      241 ALPHA=T
202              JJ=IP
203      240 CONTINUE
204    C
205    C        IF ALL NEW CONSTRAINED COEFICIENTS ARE FEASIBLE THEN ALPHA WILL
206    C        STILL EQUAL 2.   IF SO EXIT FROM SECONDARY LOOP TO MAIN LOOP.
207    C
208              IF(ALPHA-TWO)242,330,242
209    C
210    C        OTHERWISE USE ALPHA, WHICH WILL BE BETWEEN 0. AND 1. TO
211    C        TO INTERPOLATE BETWEEN THE OLD X AND THE NEW X.
212    C
213      242 DO250IP=1,NSETP
214              L=INDEX(IP)
215      250 X(L)=X(L)+ALPHA*(ZZ(IP)-X(L))
216    C
217    C        MODIFY A AND B AND THE INDEX ARRAYS TO MOVE COEFICIENTS I
218    C        FROM SET P TO SET Z.
219    C
220              I=INDEX(JJ)
221      260 X(I)=ZERO
222              IF(JJ.EQ.NSETP)GO TO 290
223              JJ=JJ+1
224              DO280J=JJ,NSETP
225              II=INDEX(J)
226              INDEX(J-1)=II
227              CALLG1(A(J-1,II),A(J,II),CC,SS,A(J-1,II))
228              A(J,II)=ZERO
229              DO270L=1,N
230              IF(L.NE.II)CALLG2(CC,SS,A(J-1,L),A(J,L))
231      270 CONTINUE
232      280 CALLG2(CC,SS,B(J-1),B(J))
233      290 NPP1=NSETP
234              NSETP=NSETP-1
235              IZ1=IZ1-1
236              INDEX(IZ1)=I
237    C
238    C        SEE IF THE REMAINING COEFICIENTS IN SET P ARE FEASIBLE.
239    C        THEY SHOULD BE BECAUSE OF THE WAY ALPHA WAS DETERMINED.
240    C        IF ANY ARE INFEASIBLE IT IS DUE TO ROUND-OFF ERROR.   ANY
241    C        THAT ARE NONPOSITIVE WILL BE SET TO ZERO AND MOVED FROM
242    C        SET P TO SET Z.
243    C
244              DO300JJ=1,NSETP
245              I=INDEX(JJ)
246              IF(X(I))260,260,300
247      300 CONTINUE
248    C
249    C        COPY B() INTO ZZ(). THEN SOLVE AGAIN AND LOOP BACK.
250    C
251              DO310I=1,M
252      310 ZZ(I)=B(I)
253              ASSIGN320TONEXT
254              GO TO 400
255      320 CONTINUE
256              GO TO 210
257    C
258    C        END OF SECONDARY LOOP.
259    C
260      330 DO340IP=1,NSETP
261              I=INDEX(IP)
262      340 X(I)=ZZ(IP)
```

```
263   C
264   C        ALL NEW COEFICIENTS ARE POSITIVE.  LOOP BACK TO BEGINING.
265   C
266            GO TO 30
267   C
268   C        END OF MAIN LOOP.
269   C
270   C
271   C        COME HERE FOR TERMINATION.  COMPUTE THE NORM OF THE FINAL
272   C        RESIDUAL VECTOR.
273   C
274      350 SM=ZERO
275            IF(NPP1.GT.M)GO TO 370
276            DO360I=NPP1,M
277      360 SM=SM+B(I)**2
278            GO TO 390
279      370 DO380J=1,N
280      380 W(J)=ZERO
281      390 RNORM=SQRT(SM)
282            RETURN
283   C
284   C        THE FOLLOWING BLOCK OF CODE IS USED AS AN INTERNAL SUBROUTINE.
285   C        TO SOLVE THE TRIANGULAR SYSTEM.  PUTTING THE SOLUTION IN ZZ().
286   C
287      400 DO430L=1,NSETP
288            IP=NSETP+1-L
289            IF(L.EQ.1)GO TO 420
290            DO410II=1,IP
291      410 ZZ(II)=ZZ(II)-A(II,JJ)*ZZ(IP+1)
292      420 JJ=INDEX(IP)
293      430 ZZ(IP)=ZZ(IP)/A(IP,JJ)
294            GO TO NEXT,(200,320)
295      440 FORMAT(//2X,' NNLS QUITTING ON ITERATION COUNT.')
296            END
```

```
 1   C##################################################################
 2   C##################################################################
 3   C##                                                              ##
 4   C##                                                              ##
 5   C##        $ $ $ $ $   H 1 2   $ $ $ $ $                         ##
 6   C##                                                              ##
 7   C##                                                              ##
 8   C##################################################################
 9   C##################################################################
10   C
11   C
12   C        CONSTRUCTION AND/OR APPLICATION OF A SINGLE HOUSEHOLDER
13   C        TRANSFORMATION    Q = I + U*(U**T)/B
14   C
15   C        MODE  =  1 OR 2 TO SELECT ALGORITHM H1 OR H2
16   C        LPIVOT = INDEX OF THE PIVOT ELEMENT
17   C        L1,M    IF L1.LE.M THE TRANSFORMATION WILL BE CONSTRUCTED TO ZERO
18   C                ELEMENTS INDEXED FROM L1 THROUGH M. IF L1.GT.M THE
19   C                SUBROUTINE AN INDENTITY TRANSFORMATION.
20   C        U(),IUE,UP  ON ENTRY TO H1 U() CONTAINS THE PIVOT VECTOR. IUE IS
21   C                THE STORAGE INCREMENT BETWEEN ELEMENTS. ON EXIT FROM
22   C                H1 U() AND UP CONTAIN QUANTITES DEFINING THE VECTOR U
23   C                OF THE HOUSHOLDER TRANSFORMATION. ON ENTRY TO H2 U()
24   C                AND UP SHOULD CONTAIN QUANTITIES PREVIOUSLY COMPUTED
25   C                BY H1.
26   C                THESE WILL NOT BE MODIFIFED BY H2.
27   C        C()  ON ENTRY TO H1 OR H2 C() CONTAINS A MATRIX WHICH WILL BE
28   C                REGARDED AS A SET OF VECTORS TO WHICH THE HOUSHOLDER
29   C                TRANSFORMATION IS APPLIED. ON EXIT C() CONTAINS THE SET OF
30   C                TRANSFORMED VECTORS.
31   C        ICE = STORAGE INCREMENT BETWEEN ELEMENTS OF VECTORS IN C()
32   C        ICV = STORAGE INCREMENT BETWEEN VECTORS IN C()
33   C        NCV = NUMBER OF VECTORS IC C() TO BE TRANSFORMED. IF NCV.LE.0 NO
34   C                OPERATIONS WILL BE DONE ON C()
35   C
36   C        CHARLES L. LAWSON/RICHARD J. HANSON,
37   C        SOLVING LEAST SQUARES PROBLEMS, COPYRIGHT 1974, PP 304-309,
```

```
38   C        REPRINTED BY PERMISSION OF PRENTICE-HALL, INC.,
39   C        ENGLEWOOD CLIFFS, N.J.
40   C
41            SUBROUTINEH12(MODE,LPIVOT,L1,M,U,IUE,UP,C,ICE,ICV,NCV,NDIM,
42           &NETACL,NSOLNS)
43            IMPLICIT REAL*8(A-H,O-Z)
44            DIMENSIONU(1,NETACL),C(NSOLNS)
45            ONE=1.
46            IF(0.GE.LPIVOT.OR.LPIVOT.GE.L1.OR.L1.GT.M)RETURN
47            CL=DABS(U(1,LPIVOT))
48            IF(MODE.EQ.2)GO TO 60
49   C
50   C        CONSTRUCT THE TRANSFORMATION.
51   C
52            DO10J=L1,M
53    10   CL=DMAX1(DABS(U(1,J)),CL)
54            IF(CL)130,130,20
55    20   CLINV=ONE/CL
56            SM=((U(1,LPIVOT))*CLINV)**2
57            DO30J=L1,M
58    30   SM=SM+((U(1,J))*CLINV)**2
59            SM1=SM
60            CL=CL*SQRT(SM1)
61            IF(U(1,LPIVOT))50,50,40
62    40   CL=-CL
63    50   UP=U(1,LPIVOT)-CL
64            U(1,LPIVOT)=CL
65            GO TO 70
66   C
67   C        APPLY THE TRANSFORM I+U*(U**T)/B TO C
68   C
69    60   IF(CL)130,130,70
70    70   IF(NCV.LE.0)RETURN
71            B=(UP)*U(1,LPIVOT)
72   C
73   C        B MUST BE NONPOSITIVE HERE.  IF B=0. RETURN.
74   C
75            IF(B)80,130,130
76    80   B=ONE/B
77            I2=1-ICV+ICE*(LPIVOT-1)
78            INCR=ICE*(L1-LPIVOT)
79            DO120J=1,NCV
80            I2=I2+ICV
81            I3=I2+INCR
82            I4=I3
83            SM=C(I2)*(UP)
84            DO90I=L1,M
85            SM=SM+C(I3)*(U(1,I))
86    90   I3=I3+ICE
87            IF(SM)100,120,100
88   100   SM=SM*B
89            C(I2)=C(I2)+SM*(UP)
90            DO110I=L1,M
91            C(I4)=C(I4)+SM*(U(1,I))
92   110   I4=I4+ICE
93   120   CONTINUE
94   130   RETURN
95            END
```

```
 1   C############################################################
 2   C############################################################
 3   C##                                                        ##
 4   C##                                                        ##
 5   C##      $ $ $ $ $   G 1   $ $ $ $ $                       ##
 6   C##                                                        ##
 7   C##                                                        ##
 8   C############################################################
 9   C############################################################
10            SUBROUTINEG1(A,B,COS,SIN,SIG)
11   C
12   C        COMPUTE ORTHOGONAL ROTATION MATRIX
13   C        COMPUTE MATRIX ( C,S) SO THAT ( C,S)(A)= (SQRT(A**2+B**2))
```

```
14    C                       (-S,C)              (-S,C)(B)   (          Ø      )
15    C       COMPUTE SIG = SQRT(A**2+B**2)
16    C          SIG IS COMPUTED LAST TO ALLOW FOR THE POSSIBILITY THAT
17    C          SIG MAY BE IN THE SAME LOCATION AS A OR B.
18    C
19    C          CHARLES L. LAWSON/RICHARD J. HANSON,
20    C          SOLVING LEAST SQUARES PROBLEMS, COPYRIGHT 1974, PP 304-309,
21    C          REPRINTED BY PERMISSION OF PRENTICE-HALL, INC.,
22    C          ENGLEWOOD CLIFFS, N.J.
23    C
24            IMPLICIT REAL*8(A-H,O-Z)
25            ZERO=Ø.Ø
26            ONE=1.Ø
27            IF(DABS(A).LE.DABS(B))GO TO 10
28            XR=B/A
29            YR=SQRT(ONE+XR*XR)
30            COS=SIGN(ONE/YR,A)
31            SIN=COS*XR
32            SIG=DABS(A)*YR
33            RETURN
34      10  IF(B)20,30,20
35      20  XR=A/B
36            YR=SQRT(ONE+XR*XR)
37            SIN=SIGN(ONE/YR,B)
38            COS=SIN*XR
39            SIG=DABS(B)*YR
40            RETURN
41      30  SIG=ZERO
42            COS=ZERO
43            SIN=ONE
44            RETURN
45            END
```

```
1     C###################################################################
2     C###################################################################
3     C##                                                             ##
4     C##                                                             ##
5     C##        $ $ $ $ $  G 2  $ $ $ $ $                            ##
6     C##                                                             ##
7     C##                                                             ##
8     C###################################################################
9     C###################################################################
10            SUBROUTINEG2(COS,SIN,X,Y)
11    C
12    C       APPLY THE ROTATION COMPUTED BY G1 TO (X,Y)
13    C
14    C          CHARLES L. LAWSON/RICHARD J. HANSON,
15    C          SOLVING LEAST SQUARES PROBLEMS, COPYRIGHT 1974, PP 304-309,
16    C          REPRINTED BY PERMISSION OF PRENTICE-HALL, INC.,
17    C          ENGLEWOOD CLIFFS, N.J.
18    C
19            IMPLICIT REAL*8(A-H,O-Z)
20            XR=COS*X+SIN*Y
21            Y=-SIN*X+COS*Y
22            X=XR
23            RETURN
24            END
```

```
1     C###################################################################
2     C###################################################################
3     C##                                                             ##
4     C##                                                             ##
5     C##        $ $ $ $ $  R E F I N E  $ $ $ $ $                    ##
6     C##                                                             ##
7     C##                                                             ##
8     C###################################################################
9     C###################################################################
10            SUBROUTINE REFINE(CC,WORK,WORK1,WORK2,C11,EQ,EC,STD,Q,TM,TL,
11           &SPEC,SP2,PATH,SSRSOL,PH,E,B1,C1,CK,X,DE,COMPLX,IPO,JCOMB,JEQ,
```

```
12           &JJNO, IWORK, MM, ML, MN, NWVLS, NSOLNS, NETACL, NBETA, NCOMP, NWVNSL,
13           &NCOMP4, NCOMP6)
14    C
15    C        THIS ROUTINE CONTROLS THE PROGRESS OF MINIMISATION BY THE LEAST
16    C        SQUARES TECHNIQUE USING NUMERICAL DIFFERENTIATION. THE
17    C        IMPLEMENTATION OF THE GAUSS-NEWTON ALGORITHM IS DERIVED FROM
18    C        SCOGS (I.G.SAYCE) ORIGINALY FROM GAUSS (R.S.TOBIAS)
19    C
20             IMPLICIT REAL*8(A-H,O-Z)
21             INTEGER COMPLX(30,NCOMP4),RUNIT,WUNIT,PUNIT
22             COMMON/UNIT/RUNIT,WUNIT,PUNIT
23             COMMON/INTEGR/NCV,NBA,JQ,JNO,NL,NM,NTS,NPASS,NUMPH,NCD,M,
24           &NBANUM,JJA,IBETA,NCC,NPAGE,IPLOT,IPRIN,ICARD,NLMR
25             COMMON/CONST/WVLST,WVLFIN,WVLINC,S1,F,XDF,XSD,SQRA,SSQR,WT
26             DIMENSION SPEC(NSOLNS,NWVLS),SP2(NWVNSL,NBETA),PATH(NSOLNS),
27           &WORK(NSOLNS,NWVLS),SSRSOL(NWVLS)
28             DIMENSION E(NCOMP),IPO(NBETA),X(NETACL),CK(NETACL),C1(NSOLNS),
29           &WORK1(NETACL,NETACL),CC(NBETA,NBETA),DE(NETACL),B1(NSOLNS)
30             DIMENSION JEQ(NETACL),JJNO(NCOMP),JCOMB(NCOMP,5),IWORK(NETACL)
31             DIMENSION EQ(NCOMP4,NWVLS),EC(NCOMP6,NWVLS),STD(NETACL,NWVLS)
32             DIMENSION Q(NSOLNS,NCOMP4),TM(2,NSOLNS),TL(2,NSOLNS),
33           &PH(NSOLNS),WORK2(NSOLNS,NETACL),C11(NSOLNS,NETACL)
34             DIMENSION ML(2,NCOMP),MM(2,NCOMP),MN(NCOMP)
35    196 FORMAT(/' THE STANDARD DEVIATION IN THE ABSORBANCE DATA WITH THE'
36           &,' INPUT CONSTANTS IS ',1PD11.4/)
37    197 FORMAT(/10X,' THE STANDARD DEVIATION IN THE ABSORBANCE DATA IS',
38           &1PD11.4)
39    201 FORMAT(/' X1 NEGATIVE')
40    202 FORMAT(10X,F12.4,F9.4,2X,'SHIFT=',F10.4,' FOR ',30A1)
41    203 FORMAT(10X,1PD12.4,D15.4,2X,'SHIFT=',D13.4,' FOR ',30A1)
42    207 FORMAT(/10X,I2,' CYCLE(S) CALCULATED')
43    218 FORMAT(//10X,'SUM OF SQUARES OF (OBSVD. - CALC.) = ',1PD12.4//)
44    219 FORMAT(//10X,'*** CONVERGENCE ACHIEVED IN',I3,' CYCLES ***'/)
45    220 FORMAT(///2X,'****  REFINEMENT TERMINATED DUE TO ',
46           &'PERSISTENTLY HIGH'/2X,'****   STANDARD DEVIATION OF ONE OR',
47           &' MORE CONSTANTS'///)
48   1007 FORMAT(10X,'STAND. DEV. OF FIT FOR SPECTRUM NUMBER ',I3,' = ',
49           &1PE10.4)
50   1006 FORMAT(//)
51             IF(NCV.EQ.0)GO TO 195
52    C
53    C        GRID IS CALLED TO SEE IF A BETTER ESTIMATE OF THE STARTING
54    C        VALUES FOR THE CONSTANTS CAN BE FOUND USING A STAR DESIGN.
55    C
56             CALL GRID(WORK1,WORK2,C11,EQ,EC,STD,Q,TM,TL,SPEC,DE,
57           &PH,E,B1,C1,PATH,SSRSOL,CK,X,JCOMB,JEQ,JJNO,IWORK,MM,ML,MN,IPO,
58           &NWVLS,NSOLNS,NETACL,NBETA,NCOMP,NCOMP4,NCOMP6)
59    C
60    C        SEVERAL COUNTERS ARE INITIALISED.
61    C
62             ISTDCK=0
63             NCC=0
64    130 NCC=NCC+1
65             NNCI=0
66    C
67    C        THE NUMERICAL DIFFERENTIATION ROUTINE IS CALLED AND RETURNS
68    C        THE COMPLETED LEAST SQUARES EQUATIONS.
69    C
70             CALL DIFF(WORK,WORK1,WORK2,CC,C11,EQ,EC,STD,Q,TM,TL,SPEC,SP2,PH,
71           &E,CK,X,PATH,SSRSOL,B1,C1,DE,IPO,JCOMB,JEQ,JJNO,IWORK,MM,ML,MN,
72           &NWVLS,NSOLNS,NETACL,NBETA,NCOMP,NWVNSL,NCOMP4,NCOMP6)
73    C
74    C        CONTROL PASSES TO THE S/R SEARCH WHERE THE LEAST SQUARES EQUATIONS
75    C        ARE SOLVED AND THE ADJUSTABLE CONSTANTS UPDATED.
76    C
77             CALL SEARCH(CC,WORK1,E,IPO,DE,CK,NETACL,NBETA,NCOMP)
78             IF(NCC-1)194,194,195
79    194 WRITE(WUNIT,196)SSQR
80    C
81    C        STATISTICAL PARAMETERS SUCH AS SSQR ETC. ARE RECALCULATED WITH THE
82    C        UPDATED SET OF CONSTANTS.
83    C
84    195 CALL RESID(WORK1,WORK2,C11,EQ,EC,STD,Q,TM,TL,SPEC,PATH,
85           &PH,E,B1,C1,SSRSOL,CK,X,JCOMB,JEQ,JJNO,IWORK,MM,ML,MN,
86           &NWVLS,NSOLNS,NETACL,NCOMP,NCOMP4,NCOMP6)
87             WRITE(WUNIT,197)SSQR
```

```
 88          IF(NCV.EQ.Ø)GO TO 2Ø8
 89          DO 198 I=1,NCV
 9Ø          IS=IPO(I)
 91          X1=CC(I,I)*SQRA
 92          IF(X1)199,2ØØ,2ØØ
 93      199 X1=-X1
 94          WRITE(WUNIT,2Ø1)
 95      2ØØ X1=DSQRT(X1)
 96  C
 97  C      PRINTS IMPROVED CONSTANTS, ESTIMATED STANDARD DEVIATIONS, X1, AND
 98  C      SHIFTS, DE(I).
 99  C
1ØØ          IF(X1.GT.1ØØ.)ISTDCK=ISTDCK+1
1Ø1          IF(IBETA.EQ.1)WRITE(WUNIT,2Ø3)E(IS),X1,DE(I),(COMPLX(L,(IS+4)),
1Ø2         &L=1,3Ø)
1Ø3      198 IF(IBETA.EQ.Ø)WRITE(WUNIT,2Ø2)E(IS),X1,DE(I),(COMPLX(L,(IS+4)),
1Ø4         &L=1,3Ø)
1Ø5          WRITE(WUNIT,2Ø7)NCC
1Ø6  C
1Ø7  C      IF ALL SHIFTS ARE LESS THAN E(IS)/1.ØD-Ø4 CONVERGENCE IS
1Ø8  C      CONSIDERED TO HAVE BEEN ACHIEVED AND CONTROL IS RETURNED TO
1Ø9  C      INOUT. OTHERWISE PROVIDED THAT THE MAXIMUM NUMBER OF
11Ø  C      ITERATIONS DESIRED, NCD, HAS NOT BEEN EXCEEDED, A NEW
111  C      ITERATIVE CYCLE IS COMMENCED.
112  C
113          IF(ISTDCK.GT.3)GO TO 189
114          ISTOP=Ø
115          DO 214 I=1,NCV
116          IS=IPO(I)
117          CRIT=DABS(E(IS)/1.DØ4)
118          CHEKA=DABS(DE(I))
119          IF(CHEKA.LT.CRIT)ISTOP=ISTOP+1
12Ø      214 CONTINUE
121          IF(ISTOP.EQ.NCV)GO TO 222
122          IF(NCC-NCD)13Ø,2Ø8,2Ø8
123      222 WRITE(WUNIT,219) NCC
124      2Ø8 WRITE(WUNIT,1ØØ6)
125          DO 1ØØ5 I=1,NUMPH
126     1ØØ5 WRITE(WUNIT,1ØØ7)I,SSRSOL(I)
127          WRITE(WUNIT,1ØØ6)
128          WRITE(WUNIT,218)S1
129  C
13Ø  C      AFTER REFINEMENT IS COMPLETE OR NCD HAS BEEN REACHED
131  C      THE STANDARD DEVIATIONS OF THE MOLAR ABSORPTIVITIES ARE CALCULATED
132  C
133          M=1ØØ
134          CALL RESID(WORK1,WORK2,C11,EQ,EC,STD,Q,TM,TL,SPEC,PATH,
135         &PH,E,B1,C1,SSRSOL,CK,X,JCOMB,JEQ,JJNO,IWORK,MM,ML,MN,
136         &NWVLS,NSOLNS,NETACL,NCOMP,NCOMP4,NCOMP6)
137          RETURN
138      189 WRITE(WUNIT,22Ø)
139          STOP
14Ø          END
```

```
 1  C##################################################################
 2  C##################################################################
 3  C##                                                            ##
 4  C##                                                            ##
 5  C##         $ $ $ $ $  D I F F  $ $ $ $ $                      ##
 6  C##                                                            ##
 7  C##                                                            ##
 8  C##################################################################
 9  C##################################################################
1Ø          SUBROUTINE DIFF(WORK,WORK1,WORK2,CC,C11,EQ,EC,STD,Q,TM,TL,SPEC,
11         &SP2,PH,E,CK,X,PATH,SSRSOL,B1,C1,DE,IPO,JCOMB,JEQ,JJNO,IWORK,MM,ML,
12         &MN,NWVLS,NSOLNS,NETACL,NBETA,NCOMP,NWVNSL,NCOMP4,NCOMP6)
13  C
14  C      NUMERICAL DIFFERENTIATION IS PERFORMED BY THIS
15  C      ROUTINE. STERLING'S CENTRAL DIFFERENCE METHOD IS APPLIED.
16  C
17          IMPLICIT REAL*8 (A-H,O-Z)
18          INTEGER RUNIT,WUNIT,PUNIT
```

```
19           COMMON/UNIT/RUNIT,WUNIT,PUNIT
20           COMMON/CONST/WVLST,WVLFIN,WVLINC,S1,F,XDF,XSD,SQRA,SSQR,WT
21           COMMON/INTEGR/NCV,NBA,JQ,JNO,NL,NM,NTS,NPASS,NUMPH,NCD,M,
22          &NBANUM,JJA,IBETA,NCC,NPAGE,IPLOT,IPRIN,ICARD,NLMR
23           DIMENSION SPEC(NSOLNS,NWVLS),SP2(NWVNSL,NBETA),PATH(NSOLNS),
24          &WORK(NSOLNS,NWVLS),WORK2(NSOLNS,NETACL),SSRSOL(NWVLS)
25           DIMENSION E(NCOMP),X(NETACL),IPO(NBETA),CK(NETACL),
26          &CC(NBETA,NBETA),DE(NETACL),B1(NSOLNS),C1(NSOLNS)
27           DIMENSION EQ(NCOMP4,NWVLS),EC(NCOMP6,NWVLS),STD(NETACL,NWVLS),
28          &C11(NSOLNS,NETACL),PH(NSOLNS),WORK1(NETACL,NETACL)
29           DIMENSION Q(NSOLNS,NCOMP4),TM(2,NSOLNS),TL(2,NSOLNS)
30           DIMENSION JEQ(NETACL),JJNO(NCOMP),JCOMB(NCOMP,5),IWORK(NETACL)
31           DIMENSION ML(2,NCOMP),MM(2,NCOMP),MN(NCOMP)
32       101 FORMAT(//10X,'HESSIAN MATRIX APPROXIMATION')
33       102 FORMAT(10X,'ELEMENTS OF CONSTANT VECTOR')
34       104 FORMAT(10X,1PD13.6)
35       110 FORMAT(10X,1PD15.6,6D15.6)
36     C
37     C     THE PARTIAL DIFFERENTIALS ARE COMPUTED AT THIS POINT, ONE CONSTANT
38     C     AT A TIME.
39     C
40           DO 163 J=1,NCV
41           IS=IPO(J)
42           EINC=DABS(E(IS))*0.005
43           E(IS)=E(IS)+EINC
44           CALL RESID(WORK1,WORK2,C11,EQ,EC,STD,Q,TM,TL,SPEC,PATH,
45          &PH,E,B1,C1,SSRSOL,CK,X,JCOMB,JEQ,JJNO,IWORK,MM,ML,MN,
46          &NWVLS,NSOLNS,NETACL,NCOMP,NCOMP4,NCOMP6)
47           DO 162 I=1,NUMPH
48           PATHI=PATH(I)
49           DO 162 K=1,NBA
50           ABSORB=0.0
51           DO 160 JJ=1,JJA
52       160 ABSORB=ABSORB+Q(I,JJ)*EQ(JJ,K)
53       162 WORK(I,K)=ABSORB*PATHI
54           E(IS)=E(IS)-2.0*EINC
55           CALL RESID(WORK1,WORK2,C11,EQ,EC,STD,Q,TM,TL,SPEC,PATH,
56          &PH,E,B1,C1,SSRSOL,CK,X,JCOMB,JEQ,JJNO,IWORK,MM,ML,MN,
57          &NWVLS,NSOLNS,NETACL,NCOMP,NCOMP4,NCOMP6)
58           E(IS)=E(IS)+EINC
59           K1=0
60           DO 170 I=1,NUMPH
61           PATHI=PATH(I)
62           DO 170 K=1,NBA
63           ABSORB=0.0
64           DO 161 JJ=1,JJA
65       161 ABSORB=ABSORB+Q(I,JJ)*EQ(JJ,K)
66           K1=K1+1
67       170 SP2(K1,J)=WORK(I,K)-ABSORB*PATHI
68       163 CONTINUE
69     C
70     C     THE PARTIAL DIFFERENTIALS FOR EACH CONSTANT, DE(J), AND THE
71     C     CONSTANT VECTOR, CK(J), ARE NOW DEVELOPED.
72     C
73           M=1
74           CALL RESID(WORK1,WORK2,C11,EQ,EC,STD,Q,TM,TL,SPEC,PATH,
75          &PH,E,B1,C1,SSRSOL,CK,X,JCOMB,JEQ,JJNO,IWORK,MM,ML,MN,
76          &NWVLS,NSOLNS,NETACL,NCOMP,NCOMP4,NCOMP6)
77           DO 133 J=1,NCV
78           CK(J)=0.0
79           DO 133 J1=1,NCV
80       133 CC(J,J1)=0.0
81           K1=0
82           DO 165 I=1,NUMPH
83           PATHI=PATH(I)
84           DO 165 K=1,NBA
85           K1=K1+1
86           ABSORB=0.0
87           DO 164 J=1,JJA
88       164 ABSORB=ABSORB+Q(I,J)*EQ(J,K)
89           ABSORB=ABSORB*PATHI
90           DO 166 J=1,NCV
91           IS=IPO(J)
92           EINC=DABS(E(IS))*0.005
93           DE(J)=(SP2(K1,J))/(2.*EINC)
94           CK(J)=CK(J)-(ABSORB-SPEC(I,K))*DE(J)
```

```
 95    C
 96    C        FINALLY THE FULL LEAST SQUARES MATRIX, CC(I,J) IS DEVELOPED,
 97    C        AS THE LOWER TRIANGULAR MATRIX.
 98    C
 99             DO 166 JJ=1,J
100        166 CC(JJ,J)=CC(JJ,J)+DE(J)*DE(JJ)
101        165 CONTINUE
102    C
103    C         FILL IN THE UPPER DIAGONAL ELEMENTS OF CC.
104    C
105             DO 168 J=1,NCV
106             JP1=J+1
107             DO 168 I=JP1,NCV
108        168 CC(I,J)=CC(J,I)
109             WRITE(WUNIT,101)
110             DO 112 J=1,NCV
111        112 WRITE(WUNIT,110)(CC(J,I),I=1,NCV)
112             WRITE(WUNIT,102)
113             DO 103 I=1,NCV
114        103 WRITE(WUNIT,104) CK(I)
115         22 RETURN
116             END




  1    C################################################################
  2    C################################################################
  3    C##                                                            ##
  4    C##                                                            ##
  5    C##        $ $ $ $ $   R E S I D   $ $ $ $ $                   ##
  6    C##                                                            ##
  7    C##                                                            ##
  8    C################################################################
  9    C################################################################
 10             SUBROUTINE RESID(WORK1,WORK2,C11,EQ,EC,STD,Q,TM,TL,SPEC,PATH,
 11            &PH,E,B1,C1,SSRSOL,CK,X,JCOMB,JEQ,JJNO,IWORK,MM,ML,MN,
 12            &NWVLS,NSOLNS,NETACL,NCOMP,NCOMP4,NCOMP6)
 13    C
 14    C        THIS ROUTINE CALCULATES THE FOLLOWING QUANTITIES:-
 15    C        THE RESIDUAL OF EACH POINT, R, WHERE
 16    C
 17    C            R = CALC. ABSORBANCE - OBS. ABSORBANCE
 18    C
 19    C        THE SUM OF SQUARES OF R
 20    C        THE VARIANCE AND STANDARD DEVIATION FOR EACH SOLUTION OVER ALL
 21    C        WAVELENGTHS
 22    C        THE STANDARD DEVIATION OF THE COMPLETE DATA SET, SSQR.
 23    C
 24             IMPLICIT REAL*8 (A-H,O-Z)
 25             COMMON/INTEGR/NCV,NBA,JQ,JNO,NL,NM,NTS,NPASS,NUMPH,NCD,M,
 26            &NBANUM,JJA,IBETA,NCC,NPAGE,IPLOT,IPRIN,ICARD,NLMR
 27             COMMON/CONST/WVLST,WVLFIN,WVLINC,S1,F,XDF,XSD,SQRA,SSQR,WT
 28             DIMENSION Q(NSOLNS,NCOMP4),TM(2,NSOLNS),TL(2,NSOLNS),
 29            &PH(NSOLNS),E(NCOMP),WORK2(NSOLNS,NETACL)
 30             DIMENSION SPEC(NSOLNS,NWVLS),PATH(NSOLNS),SSRSOL(NWVLS)
 31             DIMENSION EQ(NCOMP4,NWVLS),EC(NCOMP6,NWVLS),STD(NETACL,NWVLS),
 32            &C11(NSOLNS,NETACL),WORK1(NETACL,NETACL),CK(NETACL),X(NETACL)
 33             DIMENSION C1(NSOLNS),B1(NSOLNS),ML(2,NCOMP),MM(2,NCOMP),MN(NCOMP)
 34             DIMENSION JEQ(NETACL),JJNO(NCOMP),JCOMB(NCOMP,5),IWORK(NETACL)
 35             NBANUM=0
 36             S1=0.
 37             CALL ECOEF(WORK1,WORK2,C11,EQ,EC,STD,Q,TM,TL,SPEC,PH,E,
 38            &B1,C1,PATH,SSRSOL,CK,X,JCOMB,JEQ,JJNO,IWORK,MM,ML,MN,
 39            &NWVLS,NSOLNS,NETACL,NCOMP,NCOMP4,NCOMP6)
 40    C
 41    C        ONCE CONTROL HAS RETURNED FROM ECOEF   CALCULATE
 42    C        THE RESIDUAL R AND THE SUM OF SQUARES OF ALL THE RESIDUALS S1.
 43    C
 44             DO55 I=1,NUMPH
 45             SSQ=0.0
 46             PATHI=PATH(I)
 47             DO 56 K=1,NBA
 48             NBANUM=NBANUM+1
 49             ABSORB=0.0
```

```
50              DO 54 J=1,JJA
51           54 ABSORB=ABSORB+Q(I,J)*EQ(J,K)
52              R=ABSORB*PATHI-SPEC(I,K)
53           56 SSQ=SSQ+R*R*WT
54              S1=S1+SSQ
55              SQRSSQ=SSQ/XSD
56      C
57      C       THE VARIANCE AND STANDARD DEVIATION FOR EACH SOLUTION, SSRSOL(I),
58      C       IS CALCULATED.
59      C
60           55 SSRSOL(I)=DSQRT(SQRSSQ)
61      C
62      C       THE VARIANCE, SQRA, AND THE STANDARD DEVIATION, SSQR, FOR THE
63      C       WHOLE DATA SET IS CALCULATED.
64      C
65              SQRA=S1/XDF
66              SSQR=DSQRT(SQRA)
67              RETURN
68              END


 1      C###############################################################
 2      C###############################################################
 3      C##                                                          ##
 4      C##                                                          ##
 5      C##        $ $ $ $ $   E C O E F   $ $ $ $ $                 ##
 6      C##                                                          ##
 7      C##                                                          ##
 8      C###############################################################
 9      C###############################################################
10              SUBROUTINE ECOEF(WORK1,WORK2,C,EQ,EC,STD,Q,TM,TL,SPEC,PH,E,
11             &B1,C1,PATH,B,CK,X,JCOMB,JEQ,JJNO,IWORK,MM,ML,MN,
12             &NWVLS,NSOLNS,NETACL,NCOMP,NCOMP4,NCOMP6)
13      C
14      C       THIS SUBROUTINE CALCULATES FROM BEER'S LAW EQUATIONS THE MOLAR
15      C       ABSORPTIVITIES OF SELECTED SPECIES.
16      C        NOTE THAT ONLY Q AND EQ ARRAYS, BASED ON THE ENTERING SET
17      C        OF CONSTANTS, E, ARE RETURNED.
18      C
19              IMPLICIT REAL*8 (A-H,O-Z)
20              COMMON/INTEGR/NCV,NBA,JQ,JNO,NL,NM,NTS,NPASS,NUMPH,NCD,M,
21             &NBANUM,JJA,IBETA,NCC,NPAGE,IPLOT,IPRIN,ICARD,NLMR
22              DIMENSION E(NCOMP),WORK2(NSOLNS,NETACL),B1(NSOLNS),C1(NSOLNS)
23              DIMENSION SPEC(NSOLNS,NWVLS),PH(NSOLNS),PATH(NSOLNS),B(NWVLS)
24              DIMENSION EQ(NCOMP4,NWVLS),EC(NCOMP6,NWVLS),STD(NETACL,NWVLS),
25             &C(NSOLNS,NETACL),WORK1(NETACL,NETACL),CK(NETACL),X(NETACL)
26              DIMENSION Q(NSOLNS,NCOMP4),TM(2,NSOLNS),TL(2,NSOLNS)
27              DIMENSION JEQ(NETACL),JJNO(NCOMP),JCOMB(NCOMP,5),IWORK(NETACL)
28              DIMENSION ML(2,NCOMP),MM(2,NCOMP),MN(NCOMP)
29      C
30      C       ALL CONCENTRATIONS ARE CALCULATED IN CCSCC AND RETURNED IN
31      C       Q(NUMPH,JJA).
32      C
33              CALL CCSCC(Q,TM,TL,PH,E,B1,C1,MM,ML,MN,NCOMP,NCOMP4,NSOLNS)
34              IF(JQ.EQ.0)GO TO 66
35      C
36      C        THE CALCULATION:-
37      C
38      C       ABSORBANCE(UNKNOWN)=ABSORBANCE(OBSERVED)-ABSORBANCE(CALCULATED)
39      C
40      C       IS PERFORMED SO THAT THE SOLUTION OF THE LINEAR EQUATIONS IS
41      C       REDUCED TO:
42      C       ABSORBANCE(UNKNOWN)=C(I,J)*EQ(J,K)
43      C
44      C       WHERE EQ(J,K) WILL BE CALCULATED. THE SIGNIFICANCE OF C(I,J) IS
45      C       EXPLAINED BELOW.
46      C
47              DO 20 I=1,NUMPH
48              IF(JNO.EQ.0)GO TO 40
49              DO 21 K=1,NBA
50              ABSORB=0.0
51              DO 22 J=1,JNO
52              JK=JJNO(J)
```

```
 53        22 ABSORB=ABSORB+Q(I,JK)*EQ(JK,K)
 54        21 SPEC(I,K)=SPEC(I,K)-ABSORB*PATH(I)
 55     C
 56     C     THE CONCENTRATIONS OF SPECIES WHOSE MOLAR ABSORPTIVITIES ARE TO BE
 57     C     CALCULATED ARE COPIED INTO THE C ARRAY. THE EVENTUALITY OF MORE
 58     C     THAN ONE SPECIES HAVING THE SAME SET OF MOLAR ABSORPTIVITIES IS
 59     C     DEALT WITH AT THIS TIME. (SEE MANUAL FOR EXPLANATION)
 60     C
 61        40 DO 23 JJQ=1,JQ
 62           A=0.0
 63           IQ=JEQ(JJQ)
 64           DO 24 IIQ=1,IQ
 65           II=JCOMB(JJQ,IIQ)
 66        24 A=A+Q(I,II)
 67        23 C(I,JJQ)=A
 68        20 CONTINUE
 69     C
 70     C     THE VALUES OF CALCULATED MOLAR ABSORPTIVITIES ARE RETURNED FROM
 71     C     SOLVE IN THE ARRAY EC.
 72     C
 73           CALL SOLVE(WORK1,C,WORK2,EC,STD,SPEC,PATH,B1,C1,B,CK,X,JCOMB,
 74          &JEQ,JJNO,IWORK,NWVLS,NSOLNS,NETACL,NCOMP,NCOMP6)
 75     C
 76     C      AT THE END OF THE REFINEMENT PROCESS (M=100) THE STANDARD
 77     C      DEVIATIONS OF THE MOLAR ABSORBITIVITIES ARE CALCULATED
 78     C      BY SIGMAE, AN ENTRY POINT IN SOLVE.
 79     C
 80           IF(M.EQ.100)CALL SIGMAE(WORK1,C,WORK2,EC,STD,SPEC,PATH,B1,C1,B,
 81          &CK,X,JCOMB,JEQ,JJNO,IWORK,NWVLS,NSOLNS,NETACL,NCOMP,NCOMP6)
 82     C
 83     C     THE CALCULATED VALUES OF MOLAR ABSORPTIVITIES ARE COPIED FROM EC
 84     C     TO THE EQ ARRAY.
 85     C
 86        31 DO 32 K=1,NBA
 87           DO 32 JJQ=1,JQ
 88           A=EC(JJQ,K)
 89           IQ=JEQ(JJQ)
 90           DO 33 IIQ=1,IQ
 91           II=JCOMB(JJQ,IIQ)
 92        33 EQ(II,K)=A
 93        32 CONTINUE
 94     C
 95     C      THE ORIGINAL ABSORBANCES ARE REGENERATED AND STORED IN SPEC.
 96     C
 97        66 DO 54 I=1,NUMPH
 98           PATHI=PATH(I)
 99           DO 56 K=1,NBA
100           ABSORB=0.0
101           DO 55 J=1,JNO
102           JK=JJNO(J)
103        55 ABSORB=ABSORB+Q(I,JK)*EQ(JK,K)
104        56 SPEC(I,K)=SPEC(I,K)+ABSORB*PATHI
105        54 CONTINUE
106           RETURN
107           END
```

```
  1     C################################################################
  2     C################################################################
  3     C##                                                            ##
  4     C##                                                            ##
  5     C##       $ $ $ $ $   C C S C C   $ $ $ $ $                    ##
  6     C##                                                            ##
  7     C##                                                            ##
  8     C################################################################
  9     C################################################################
 10           SUBROUTINE CCSCC(Q,TM,TL,PH,E,B,C,MM,ML,MN,NCOMP,NCOMP4,NSOLNS)
 11     C
 12     C
 13     C     THIS ROUTINE IS USED PRIMARILY FOR BOOK-KEEPING FOR COGSNR.
 14     C     THESE TWO ROUTINES CALCULATE THE CONCENTRATIONS OF ALL SPECIES
 15     C     IN SOLUTION, GIVEN ALL RELEVANT STABILITY CONSTANTS, ANALYTICAL
 16     C     CONCENTRATIONS AND HYDROGEN ION CONCENTRATIONS ARE KNOWN. THE
```

```
17  C      CALCULATIONS ARE NOT RESTRICTED TO ONLY ONE METAL AND LIGAND. AT
18  C      PRESENT THE ROUTINES ARE SET UP TO HANDLE A MAXIMUM OF TWO
19  C      METALS AND TWO LIGANDS.
20  C
21  C      NOTE:- IF PH(I) HAS BEEN SET EQUAL TO ZERO IN THE CALLING ROUTINE
22  C      THEN THE CALCULATIONS PERFORMED BY CCSCC AND COGSNR ASSUME THAT
23  C      COMPLEX FORMATION IS PH INDEPENDENT.
24  C
25  C      THE CONCENTRATIONS OF ALL SPECIES IN SOLUTION ARE CONTAINED IN
26  C      THE ARRAY  Q(NUMPH,NTS+4) .
27  C
28         IMPLICIT REAL*8 (A-H,O-Z)
29         COMMON/INTEGR/NCV,NBA,JQ,JNO,NL,NM,NTS,NPASS,NUMPH,NCD,M,
30        &NBANUM,JJA,IBETA,NCC,NPAGE,IPLOT,IPRIN,ICARD,NLMR
31         COMMON/CONST/WVLST,WVLFIN,WVLINC,S1,F,XDF,XSD,SQRA,SSQR,WT
32         COMMON/ACOGS/Y1,Y3,BTOT,CLTOT,TX,VX,UX,NIT,KJ
33         DIMENSIONBTOT(2),CLTOT(2),DMY(2),TX(2),VX(2),Y1(2),Y3(2),DM(2)
34         DIMENSION B(NSOLNS),PH(NSOLNS),E(NCOMP),C(NSOLNS)
35         DIMENSION Q(NSOLNS,NCOMP4),TM(2,NSOLNS),TL(2,NSOLNS)
36         DIMENSION ML(2,NCOMP),MM(2,NCOMP),MN(NCOMP)
37  C
38  C      INITIALISATION OF VARIOUS PARAMETERS AND THE EVALUATION OF
39  C      CONVERGENCE CRITERIA Y1(I) AND Y3(I) WITHIN COGSNR FOR THE
40  C      ITH SOLUTION.
41  C
42         DO 109 I=1,2
43         TX(I)=0.0
44     109 VX(I)=0.0
45         NPASS=NPASS+1
46     302 DO 1000 K=1,NUMPH
47         KJ=K
48         DO 1290 I=1,2
49         BTOT(I)=TM(I,K)
50         CLTOT(I)=TL(I,K)
51         Y1(I)=BTOT(I)*0.0000001
52    1290 Y3(I)=CLTOT(I)*0.0000001
53         IF(PH(K).LT.0.0001)GO TO 1901
54         UX=10.**PH(K)*F
55    1901 CONTINUE
56  C
57  C      WHEN  K=1 INITIAL GUESS FOR FREE METAL, VX(I), AND LIGAND, TX(I)
58  C      CONCENTRATIONS (REQUIRED BY COGSNR) ARE CALCULATED. IF NPASS>1
59  C      (SECOND CALL) THESE INITIAL GUESSES ARE TAKEN AS THOSE
60  C      CALCULATED FOR THE (NPASS-1)TH CYCLE AND CONTROL PASSES TO 140.
61  C
62         IF(NPASS.GT.1)GO TO 140
63     141 IF(NM.EQ.0)GO TO 608
64     609 DO 144 I=1,NM
65     144 VX(I)=BTOT(I)
66         IF(PH(K).LT.0.0001)GO TO 1902
67     608 DO 146 I=1,NL
68         DMY(I)=1.0
69         DO 146 J1=1,NTS
70         IF(ML(I,J1).EQ.0)GO TO 146
71     147 DO 148 K1=1,NM
72         IF(MM(K1,J1).NE.0)GO TO 146
73     148 CONTINUE
74         DM(I)=10.**E(J1)*UX**MN(J1)
75         DMY(I)=DMY(I)+DM(I)
76     146 CONTINUE
77         DO 145 I=1,NL
78     145 TX(I)=CLTOT(I)/DMY(I)
79         GO TO 1903
80    1902 IF(K.GT.1)GO TO 1903
81         DO 1904 I=1,NL
82         TX(I)=CLTOT(I)-4.*BTOT(I)
83         IF(TX(I).LE.0.0)TX(I)=-TX(I)
84    1904 CONTINUE
85    1903 IF(PH(K).GT.0.0001)GO TO 140
86         UX=1.0
87     140 DO 149 I=1,NTS
88         B(I)=E(I)
89         IF(IBETA.EQ.0)B(I)=10.D0**E(I)
90     149 CONTINUE
91         IF(NPASS.EQ.1)GO TO 500
92         VX(1)=Q(K,1)
```

```
93            VX(2)=Q(K,2)
94            TX(1)=Q(K,3)
95            TX(2)=Q(K,4)
96     C
97     C      CALLS SUBROUTINE FOR CALCULATION OF ALL FREE METAL, FREE LIGAND,
98     C      AND SPECIES CONCENTRATIONS
99     C
100      500 CALL COGSNR(B,C,MM,ML,MN,NSOLNS,NCOMP)
101    C
102    C      FREE METAL LIGAND AND SPECIES CONCENTRATIONS ARE COPIED INTO THE
103    C      ARRAY Q FROM THE VECTORS TX, VX, AND C.
104    C
105           Q(K,1)=VX(1)
106           Q(K,2)=VX(2)
107           Q(K,3)=TX(1)
108           Q(K,4)=TX(2)
109           DO 160 I=5,JJA
110      160 Q(K,I)=C(I-4)
111     1900 CONTINUE
112           RETURN
113           END


1      C###################################################################
2      C###################################################################
3      C##                                                             ##
4      C##                                                             ##
5      C##          $ $ $ $ $   C O G S N R   $ $ $ $ $                ##
6      C##                                                             ##
7      C##                                                             ##
8      C###################################################################
9      C###################################################################
10            SUBROUTINE COGSNR(B,C,MM,ML,MN,NSOLNS,NCOMP)
11     C
12     C
13     C      THIS SUBROUTINE SOLVES BY A NEWTON RAPHSON TECHNIQUE THE MASS
14     C      BALANCE EQUATIONS FOR THE PARTICULAR SYSTEM AS DESCRIBED BY
15     C      TM, TL, PH AND THE SET OF SPECIES DEFINITION CARDS.
16     C
17     C
18            IMPLICIT REAL*8 (A-H,O-Z)
19            INTEGER RUNIT,WUNIT,PUNIT
20            COMMON/UNIT/RUNIT,WUNIT,PUNIT
21            COMMON/INTEGR/NCV,NBA,JQ,JNO,NL,NM,N,NPASS,NUMPH,NCD,M,
22           &NBANUM,JJA,IBETA,NCC,NPAGE,IPLOT,IPRIN,ICARD,NLMR
23            COMMON/ACOGS/Y1,Y3,BTOT,CLTOT,TX,VX,UX,NIT,KJ
24            DIMENSION ML(2,NCOMP),MM(2,NCOMP),MN(NCOMP)
25            DIMENSION B(NSOLNS),C(NSOLNS)
26            DIMENSION BTOT(2),CLTOT(2),ALO(2),BO(2),SEI(4,4),SEM(4,4),
27           &SEV(4),Y1(2),Y2(2),Y3(2),Y4(2),SHFT(4),TX(2),VX(2)
28            REAL*8 DABS
29        998 FORMAT(' ITERATION DID NOT CONVERGE,  SOLN NUMBER ',I3)
30     C
31     C      CALCULATES CONCENTRATION OF EACH SPECIES
32     C       THE ACTUAL CALCULATIONS ARE PEROFRMED USING LOGS .
33     C      FOR EXAMPLE THE NORMAL EXPRESSION:-
34     C       TERM(K) = B(K)*UX**MN(K)
35     C      IS REPLACED BY:-
36     C       TERM(K) = DLOG10(B(K)) + MN(K)*DLOG10(UX)
37     C      THIS METHOD OF CALCULATION IS DEMANDED BY THE LIMITATIONS
38     C      OF THE HONEYWELL 66/60 COMPUTER HAVING A DYNAMIC RANGE OF
39     C      FROM 10.0**(-37.0) TO 10.0**(+37.0).
40     C
41            NIT=0
42          2 DO 1 K=1,N
43          1 C(K)=DLOG10(B(K))+MN(K)*DLOG10(UX)
44            IF(NM)42,42,41
45         41 DO 4 K=1,N
46            DO 4 J=1,NM
47          4 C(K)=C(K)+MM(J,K)*DLOG10(VX(J))
48         42 DO 5 K=1,N
49            IF(NL)52,52,50
```

```
50          50 DO 6 J=1,NL
51           6 C(K)=C(K)+ML(J,K)*DLOG10(TX(J))
52          52 TEST=10.0**(-37.0)
53             IF(C(K).LT.-37.0)C(K)=TEST
54           5 C(K)=10.0**C(K)
55         100 NIT=NIT+1
56    C
57    C          CALCULATES EACH TOTAL METAL AND TOTAL LIGAND CONCENTRATION FROM
58    C          MASS BALANCE EQUATIONS.
59    C
60             IF(NM.EQ.0)GO TO 20
61          21 DO 7 I=1,NM
62             BO(I)=VX(I)
63             DO 8 K=1,N
64           8 BO(I)=BO(I)+MM(I,K)*C(K)
65           7 Y2(I)=DABS(BO(I)-BTOT(I))
66          20 IF(NL.EQ.0)GO TO 22
67          23 DO 9 I=1,NL
68             ALO(I)=TX(I)
69             DO 10 K=1,N
70          10 ALO(I)=ALO(I)+ML(I,K)*C(K)
71           9 Y4(I)=DABS(ALO(I)-CLTOT(I))
72          22 IF(NIT.GT.100)GO TO 999
73    C
74    C          CHECKS DEGREE OF CONVERGENCE
75    C
76             IF(NM.EQ.0)GO TO 24
77          11 DO 12 I=1,NM
78             YY=Y1(I)-Y2(I)
79             IF(YY.LT.0.0)GO TO 14
80          12 CONTINUE
81          24 IF(NL.EQ.0)GO TO 25
82          26 DO 13 I=1,NL
83             YY=Y3(I)-Y4(I)
84             IF(YY.LT.0.0)GO TO 14
85          13 CONTINUE
86          25 CONTINUE
87             RETURN
88    C
89    C          IF CONVERGENCE INSUFFICIENT SETS UP, AND SOLVES, MATRIX FOR IMP-
90    C          ROVED VALUES OF EACH FREE METAL AND FREE LIGAND CONCENTRATION
91    C
92          14 M1=NM+1
93             M2=NM+NL
94             IF(NL.EQ.0)M1=1
95             IF(NL.EQ.0)M2=NM
96        1070 DO 1001 I=1,M2
97             DO 1001 J=1,M2
98        1001 SEM(I,J)=0.0
99             IF(NM.EQ.0)GO TO 34
100         28 DO 1002 I=1,NM
101       1002 SEM(I,I)=-VX(I)
102            IF(NL.EQ.0)GO TO 33
103         34 DO 1003 I=M1,M2
104            IMNM=I-NM
105       1003 SEM(I,I)=-TX(IMNM)
106            IF(NM.EQ.0)GO TO 27
107         33 DO 1004 I=1,NM
108            DO 1004 J=1,NM
109            DO 1004 K=1,N
110       1004 SEM(I,J)=SEM(I,J)-C(K)*MM(I,K)*MM(J,K)
111            IF(NL.EQ.0)GO TO 30
112         36 DO 1005 I=M1,M2
113            DO 1005 J=1,NM
114            DO 1005 K=1,N
115            IMNM=I-NM
116       1005 SEM(I,J)=SEM(I,J)-C(K)*MM(J,K)*ML(IMNM,K)
117            DO 1006 I=1,NM
118            DO 1006 J=M1,M2
119            DO 1006 K=1,N
120            JMNM=J-NM
121       1006 SEM(I,J)=SEM(I,J)-C(K)*MM(I,K)*ML(JMNM,K)
122         27 DO 1007 I=M1,M2
123            DO 1007 J=M1,M2
124            DO 1007 K=1,N
```

```
125              IMNM=I-NM
126              JMNM=J-NM
127      1007 SEM(I,J)=SEM(I,J)-C(K)*ML(IMNM,K)*ML(JMNM,K)
128              IF(NM.EQ.0)GO TO 29
129       30 DO 1008 I=1,NM
130      1008 SEV(I)=-BTOT(I)+BO(I)
131       29 IF(NL.EQ.0)GO TO 37
132       38 DO 1009 I=M1,M2
133              IMNM=I-NM
134      1009 SEV(I)=-CLTOT(IMNM)+ALO(IMNM)
135       37 DO 1010 I=1,M2
136              DO 1010 J=1,M2
137      1010 SEI(I,J)=SEM(I,J)
138    C
139    C           INVERT NORMAL EQUATIONS ARRAY SEI
140    C
141              DO 764 K=1,M2
142              CM760=SEI(K,K)
143              SEI(K,K)=1.0
144              DO 765 J=1,M2
145       765 SEI(K,J)=SEI(K,J)/CM760
146              DO 764 I=1,M2
147              I1=I-K
148              IF(I1.EQ.0)GO TO 764
149       762 CM760=SEI(I,K)
150              SEI(I,K)=0.0
151              DO 763 J=1,M2
152       763 SEI(I,J)=SEI(I,J)-CM760*SEI(K,J)
153       764 CONTINUE
154    C
155    C           CALCULATE SHIFTS TO FREE CONCENTRATIONS
156    C
157              DO 1011 I=1,M2
158      1011 SHFT(I)=0.0
159              DO 1012 I=1,M2
160              DO 1012 J=1,M2
161      1012 SHFT(I)=SHFT(I)+SEI(I,J)*SEV(J)
162    C
163    C           LIMITS SHIFTS TO +/- 1.00 AND THEN UPDATES THE FREE
164    C           CONCENTRATIONS.
165    C
166              DO 1013 I=1,M2
167              IF(SHFT(I)+0.9999)1014,1013,1013
168      1014 SHFT(I)=-0.9999
169      1013 CONTINUE
170              IF(NM.EQ.0)GO TO 31
171       32 DO 1015 I=1,NM
172      1015 VX(I)=VX(I)+VX(I)*SHFT(I)
173              IF(NL.EQ.0)GO TO 39
174       31 DO 1016 I=M1,M2
175              K=I-NM
176      1016 TX(K)=TX(K)+TX(K)*SHFT(I)
177       39 GO TO 2
178    C
179    C           EXITS IF NO CONVERGENCE AFTER 100 ITERATIONS
180    C
181       999 WRITE(WUNIT,998) KJ
182              RETURN
183              END
```

```
1    C###############################################################
2    C###############################################################
3    C##                                                         ##
4    C##                                                         ##
5    C##          $ $ $ $ $   S E A R C H   $ $ $ $ $            ##
6    C##                                                         ##
7    C##                                                         ##
8    C###############################################################
9    C###############################################################
10            SUBROUTINE SEARCH(CC,CORR,E,IPO,X,CK,NETACL,NBETA,NCOMP)
11   C
12   C      THIS SUBROUTINE PERFORMS THE SOLUTION OF THE LEAST SQUARES
```

```
13   C      EQUATIONS, CALCULATION OF THE OPTIMUM SHIFT VECTORS (TOGETHER WITH
14   C      SOME PRECAUTIONARY CHECKING), THE IMPROVED CONSTANTS AND THE
15   C      CORRELATION MATRIX FOR THIS CYCLE.
16   C
17          IMPLICIT REAL*8 (A-H,O-Z)
18          INTEGER RUNIT,WUNIT,PUNIT
19          COMMON/UNIT/RUNIT,WUNIT,PUNIT
20          COMMON/INTEGR/NCV,NBA,JQ,JNO,NL,NM,NTS,NPASS,NUMPH,NCD,M,
21         &NBANUM,JJA,IBETA,NCC,NPAGE,IPLOT,IPRIN,ICARD,NLMR
22          DIMENSION E(NCOMP),IPO(NBETA),X(NETACL),CK(NETACL),
23         &CORR(NETACL,NETACL),CC(NBETA,NBETA)
24      182 FORMAT(/' OVERSHIFT, VARIABLE CONSTANT NO. ',I2,' X(I)=',1PE10.3)
25      183 FORMAT(/' EXTREME OVERSHIFT, VARIABLE CONSTANT NO. ',I2,' X(I)=',
26         &1PE10.3)
27      901 FORMAT(10X,1PD15.6,6D15.6)
28      902 FORMAT(/10X,'CORRELATION MATRIX')
29      903 FORMAT(10X,4D15.6)
30      904 FORMAT(10X,'HESSIAN (INVERTED)')
31   C
32   C      AFTER VARIOUS BOOKEEPING OPERATIONS MATRIX BC((I,J) IS INVERTED.
33   C      AN UNPROTECTED IBM MATRIX INVERSION TECHNIQUE IS USED.
34   C
35          DO 764 K=1,NCV
36          CM760=CC(K,K)
37          CC(K,K)=1.0
38          DO 765 J=1,NCV
39      765 CC(K,J)=CC(K,J)/CM760
40          DO 764 I=1,NCV
41          IF(I-K)762,764,762
42      762 CM760=CC(I,K)
43          CC(I,K)=0.0
44          DO 763 J=1,NCV
45      763 CC(I,J)=CC(I,J)-CM760*CC(K,J)
46      764 CONTINUE
47   C
48   C      THE INVERTED MATRIX CC(I,J) IS WRITTEN OUT.
49   C
50          WRITE(WUNIT,904)
51          DO 900 K=1,NCV
52      900 WRITE(WUNIT,901)(CC(K,J),J=1,NCV)
53          DO 176 K=1,NCV
54      176 X(K)=0.0
55   C
56   C      THE SOLUTION VECTOR, X(I), WHICH WILL CONTAIN THE OPTIMUM SHIFT
57   C      PARAMETERS, IS OBTAINED BY MULTIPLICATION OF CC(I,J) INVERT AND
58   C      THE CONSTANT VECTOR CK (J)
59   C
60          DO 177 I=1,NCV
61          DO 177 J=1,NCV
62      177 X(I)=X(I)+CC(I,J)*CK(J)
63   C
64   C      CHECKS SIZE OF SHIFT AND IF GREATER THAN 1 LOG UNIT REDUCES SHIFT
65   C      TO 0.98. IF ANY SHIFT WAS GREATER THAN 0.5 BUT NONE WERE GREATER
66   C      THAN 1 REDUCES  SHIFTS TO HALF VALUE. PRINTS APPROPRIATE MESSAGE
67   C
68   C
69   C      IF IBETA = 1 THEN SHIFTS ARE COMPARED AGAINST 10% AND 20% OF THE
70   C      VALUES OF EACH CONSTANT AS OUTLINED ABOVE.
71   C
72          DO 178 I=1,NCV
73          XAB=DABS(X(I))
74          IS=IPO(I)
75          IOS=0
76          IEOS=0
77          CHEK10=E(IS)*0.1
78          IF(IBETA.EQ.0)CHEK10=0.98
79          CHEK20=CHEK10*2.0
80          SIGN=1.0
81          IF(X(I).LT.0.D0)SIGN=-1.0
82          IF((CHEK20-XAB).LT.0.0)IEOS=1
83          IF((CHEK10-XAB).LT.0.0.AND.(CHEK20-XAB).GT.0.0)IOS=1
84          IF(IOS.EQ.0.AND.IEOS.EQ.0)GO TO 179
85          IF(IOS.EQ.1)WRITE(WUNIT,182)I,X(I)
86          IF(IEOS.EQ.1)WRITE(WUNIT,183)I,X(I)
87          IF(IEOS.EQ.1)X(I)=CHEK10*SIGN
88          IF(IOS.EQ.1)X(I)=X(I)*0.5
```

```
89   C
90   C       APPLIES THE SHIFT PARAMETER TO YIELD UPDATED ADJUSTABLE CONSTANTS.
91   C
92     179 E(IS)=E(IS)+X(I)
93     178 CONTINUE
94   C
95   C       CALCULATES THE CORRELATION MATRIX.
96   C
97         WRITE(WUNIT,902)
98         DO 150 I=1,NCV
99         DO 150 J=1,I
100        CORR(I,J)=CC(I,I)*CC(J,J)
101        IF(CORR(I,J).LT.0.0)CORR(I,J)=1.0
102    150 CONTINUE
103        DO 151 I=1,NCV
104        DO 152 J=1,I
105    152 CORR(I,J)=CC(I,J)/DSQRT(CORR(I,J))
106    151 WRITE(WUNIT,903)(CORR(I,J),J=1,I)
107        RETURN
108        END


1    C#################################################################
2    C#################################################################
3    C##                                                             ##
4    C##                                                             ##
5    C##          $ $ $ $ $   G R I D   $ $ $ $ $                    ##
6    C##                                                             ##
7    C##                                                             ##
8    C#################################################################
9    C#################################################################
10         SUBROUTINE GRID(WORK1,WORK2,C11,EQ,EC,STD,Q,TM,TL,SPEC,EBEST,
11        &PH,E,B1,C1,PATH,SSRSOL,CK,X,JCOMB,JEQ,JJNO,IWORK,MM,ML,MN,IPO,
12        &NWVLS,NSOLNS,NETACL,NBETA,NCOMP,NCOMP4,NCOMP6)
13   C
14   C       THIS ROUTINE, THROUGH A STAR DESIGN, ATTEMPTS TO OBTAIN
15   C       A BETTER STARING SET OF INITIAL GUESSES WITH TO START
16   C       THE REFINEMENT PROCESS.
17   C
18         IMPLICIT REAL*8 (A-H,O-Z)
19         INTEGER RUNIT,WUNIT,PUNIT
20         COMMON/UNIT/RUNIT,WUNIT,PUNIT
21         COMMON/CONST/WVLST,WVLFIN,WVLINC,S1,F,XDF,XSD,SQRA,SSQR,WT
22         COMMON/INTEGR/NCV,NBA,JQ,JNO,NL,NM,NTS,NPASS,NUMPH,NCD,M,
23        &NBANUM,JJA,IBETA,NCC,NPAGE,IPLOT,IPRIN,ICARD,NLMR
24         DIMENSION ML(2,NCOMP),MM(2,NCOMP),MN(NCOMP)
25         DIMENSION SPEC(NSOLNS,NWVLS),PATH(NSOLNS),
26        &SSRSOL(NWVLS)
27         DIMENSION EQ(NCOMP4,NWVLS),EC(NCOMP6,NWVLS),STD(NETACL,NWVLS),
28        &C11(NSOLNS,NETACL),WORK1(NETACL,NETACL),CK(NETACL),X(NETACL)
29         DIMENSION E(NCOMP),IPO(NBETA),B1(NSOLNS),C1(NSOLNS)
30         DIMENSION Q(NSOLNS,NCOMP4),TM(2,NSOLNS),TL(2,NSOLNS),
31        &PH(NSOLNS),WORK2(NSOLNS,NETACL),EBEST(NETACL)
32         DIMENSION JEQ(NETACL),JJNO(NCOMP),JCOMB(NCOMP,5),IWORK(NETACL)
33     600 FORMAT(//5X,'USER-SUPPLIED INITIAL ESTIMATES ARE GOOD'//)
34     601 FORMAT(//5X,'USER-SUPPLIED INITIAL ESTIMATES HAVE BEEN',
35        &' REPLACED'//)
36   C
37   C       ESTABLISH CENTRAL POINT OF STAR DESIGN
38   C
39         CALL RESID(WORK1,WORK2,C11,EQ,EC,STD,Q,TM,TL,SPEC,PATH,
40        &PH,E,B1,C1,SSRSOL,CK,X,JCOMB,JEQ,JJNO,IWORK,MM,ML,MN,
41        &NWVLS,NSOLNS,NETACL,NCOMP,NCOMP4,NCOMP6)
42         SSQRO=SSQR
43   C
44   C       CALCULATE DEGREE OF FIT FOR EACH ARM OF THE STAR.
45   C
46         DO 160 J=1,NCV
47         IS=IPO(J)
48     160 EBEST(J)=E(IS)
49         DO 163 J=1,NCV
50         IS=IPO(J)
51         HINC=0.15*E(IS)
52         E(IS)=E(IS)+HINC
```

```
53            CALL RESID(WORK1,WORK2,C11,EQ,EC,STD,Q,TM,TL,SPEC,PATH,
54       &PH,E,B1,C1,SSRSOL,CK,X,JCOMB,JEQ,JJNO,IWORK,MM,ML,MN,
55       &NWVLS,NSOLNS,NETACL,NCOMP,NCOMP4,NCOMP6)
56            IF(SSQR.GT.SSQRO)GO TO 171
57            DO 161 JJ=1,NCV
58            JS=IPO(JJ)
59      161 EBEST(JJ)=E(JS)
60            SSQRO=SSQR
61      171 E(IS)=E(IS)-2.0*HINC
62            CALL RESID(WORK1,WORK2,C11,EQ,EC,STD,Q,TM,TL,SPEC,PATH,
63       &PH,E,B1,C1,SSRSOL,CK,X,JCOMB,JEQ,JJNO,IWORK,MM,ML,MN,
64       &NWVLS,NSOLNS,NETACL,NCOMP,NCOMP4,NCOMP6)
65            IF(SSQR.GT.SSQRO)GO TO 163
66            DO 162 JJ=1,NCV
67            JS=IPO(JJ)
68      162 EBEST(JJ)=E(JS)
69            SSQRO=SSQR
70      163 E(IS)=E(IS)+HINC
71   C
72   C        SELECT SET OF CONSTANTS THAT GIVES THE BEST DEGREE OF FIT
73   C        AND USE THEM FOR THE INITIAL ESTIMATES.
74   C
75            DO 164 J=1,NCV
76            JS=IPO(J)
77            IF(EBEST(J).NE.E(JS))GO TO 165
78      164 CONTINUE
79            WRITE(WUNIT,600)
80            RETURN
81      165 DO 174 J=1,NCV
82            IS=IPO(J)
83      174 E(IS)=EBEST(J)
84            WRITE(WUNIT,601)
85            RETURN
86            END




1    C################################################################
2    C################################################################
3    C##                                                            ##
4    C##                                                            ##
5    C##        $ $ $ $ $   P R T P L T   $ $ $ $ $                  ##
6    C##                                                            ##
7    C##                                                            ##
8    C################################################################
9    C################################################################
10           SUBROUTINE PRTPLT(X,Y,NWVLS,NCOMP6,NPTS,N,NPAGE,ISHIFT)
11   C
12   C        THIS ROUTINE PRODUCES A PRINTER PLOT OF USER CONTROLLED
13   C        DATA, CONSISTING OF THE COMPLETE SET OF EXPERIMENTAL DATA;
14   C        THE CALCULATED MOLAR ABSORBTIVITIES AND THE
15   C        OBSERVED VS. CALCULATED SPECTRA, TOGETHER WITH THE
16   C        THE INDIVIDUAL SPECIES SPECTRA FOR EACH SOLUTION.
17   C
18           DIMENSION ICHAR(100),X(NWVLS),XLAB(13),YLAB(13),Y(NCOMP6,NWVLS),
19       &LINE(10),LINE1(10),LINE2(10),YMN(20),YMX(20)
20           COMMON/UNIT/RUNIT,WUNIT,PUNIT
21           INTEGER CHAR(41),RUNIT,WUNIT,SHIFT,XLABEL(10),VERT,STAR,S,
22       &PUNIT
23           REAL*8 X,Y
24           DATA LINE1/'I',9*'-'/,VERT/'I'/,LINE2/'I',9*' '/
25           DATA CHAR/'E','T','M','N','L','Q','1','2','3','4',
26       &'5','6','7','8','9','A','B','C','D','E','F','G','H','I','J',
27       &'K','L','M','N','O','P','Q','R','S','T','U','V','W','X','Y',
28       &'Z'/,MINUS/'-'/,IBLAN/' '/,STAR/'*'/,S/'S'/
29           DATA XLABEL/' ',2*'* * ','WAVE','LENG','TH (','NM)',2*' * *',
30       &' '/
31      602 FORMAT(4X,11F10.2)
32      606 FORMAT(30X,10A4)
33      608 FORMAT(1X,A1,8X,102A1,8X,A1)
34      609 FORMAT(1X,A1,1PG8.2,102A1,G8.2,A1)
35      611 FORMAT('1',30X,10A4)
36   C
37   C        LOCATE THE SINGLE LARGEST AND SMALLEST ABSORBANCE VALUES
38   C        FOR ALL DATA SETS TO BE PROCESSED.
```

```
39    C
40          XMAX=X(NPTS)
41          XMIN=X(1)
42          DO 100 J=1,N
43          JPT=J
44          CALL SIZE(Y,YMAX,YMIN,NPTS,JPT,NCOMP6,NWVLS)
45          YMX(J)=YMAX
46          YMN(J)=YMIN
47    100 CONTINUE
48          YMAX=YMX(1)
49          YMIN=YMN(1)
50          DO 102 J=1,N
51          IF(YMX(J).GT.YMAX)YMAX=YMX(J)
52          IF(YMN(J).LT.YMIN)YMIN=YMN(J)
53    102 CONTINUE
54          DIFF=YMAX-YMIN
55          YMAX=YMAX+0.10*DIFF
56          YMIN=YMIN-0.10*DIFF
57          IF(YMIN.LT.0.0)YMIN=0.0
58    C
59    C     SET UP PARAMETERS DEPENDING ON THE PAGE SIZE NEEDED.
60    C
61          IPAGE=NPAGE
62          IF(NPAGE.EQ.0)GO TO 400
63          NX=100
64          NY=60
65          GO TO 401
66    400 NX=68
67          NY=40
68          NPAGE=1
69    401 CALL LABEL(XMAX,XMIN,XFACT,NX,XLAB,10,NN)
70          CALL LABEL(YMAX,YMIN,YFACT,NY,YLAB,5,NN)
71    C
72    C     RELOAD YAXIS LABEL TO READ FROM THE BOTTOM UP.
73    C
74          NND2=NN/2
75          DO 402 J=1,NND2
76          IBOT=NN+1-J
77          YSAVE=YLAB(J)
78          YLAB(J)=YLAB(IBOT)
79    402 YLAB(IBOT)=YSAVE
80    C
81    C     DUMP THE PLOT LINE-BY-LINE
82    C
83          MJ1=0
84          MJ2=0
85          MJ3=1
86          MB=1
87          ME=11
88          IF(NX.LT.100)ME=7
89          DO 300 KJ=1,NPAGE
90          WRITE(WUNIT,611)(XLABEL(J),J=1,10)
91          WRITE(WUNIT,602)(XLAB(I),I=MB,ME)
92    C
93    C     NOW SET UP THE ICHAR VECTOR, LINE BY LINE.
94    C
95          J1=0
96    212 DO 211 JJ=1,10
97          J1=J1+1
98    211 ICHAR(J1)=LINE1(JJ)
99          IF(J1.LT.NX)GO TO 212
100         ICHAR(NX)=VERT
101         IF(NX.LT.80)WRITE(WUNIT,609)VERT,YLAB(MJ3),(ICHAR(J),J=1,NX)
102         IF(NX.GE.80)WRITE(WUNIT,609)VERT,YLAB(MJ3),(ICHAR(J),J=1,NX),
103        &YLAB(MJ3),VERT
104         NYM1=NY-1
105         DO 302 K=1,NYM1
106   C
107   C     SET UP THE GRID FOR THE PLOTTING SURFACE.
108   C
109         DO 203 J=1,10
110   203 LINE(J)=LINE1(J)
111         KK=K-K/5*5
112         IF(KK.EQ.0)GO TO 205
113         DO 204 J=1,10
114   204 LINE(J)=LINE2(J)
```

```
115       205 J1=0
116       202 DO 201 JJ=1,10
117           J1=J1+1
118       201 ICHAR(J1)=LINE(JJ)
119           IF(J1.LT.NX)GO TO 202
120           ICHAR(NX)=VERT
121    C
122    C       START PLOTTING THE DATA.
123    C
124           IXSAVE=1
125           DO 110 J=1,N
126           DO 110I=1,NPTS
127           YY=Y(J,I)
128           IF(YY.LT.-1.0E34)GO TO 110
129           IY=1+IFIX(((YY-YMIN)/YFACT)
130           KTEST=NY-K+1
131           IF(IY.NE.KTEST)GO TO 110
132           XX=X(I)
133           IX=1+IFIX(((XX-XMIN)/XFACT)
134           IF(IX.EQ.0)IX=1
135           L=ICHAR(IX)
136           IF(L.EQ.S)GO TO 126
137           IF(L.EQ.IBLAN.OR.L.EQ.VERT.OR.L.EQ.MINUS)GO TO 127
138           IF(ISHIFT.EQ.0.AND.J.LT.3)GO TO 128
139           ICHAR(IX)=STAR
140           GO TO 126
141       128 ICHAR(IX)=S
142           GO TO 126
143       127 ICHAR(IX)=CHAR(J+ISHIFT)
144       126 Y(J,I)=-1.0E35
145       110 CONTINUE
146    C
147    C       SET UP TO PRINT THE PLOT IN SECTIONS AS DICTATED BY
148    C       THE PAGE PARAMETER, NPAGE.
149    C
150       301 IF(KK.EQ.0)GO TO 308
151           IF(NX.LT.100)GO TO 320
152           WRITE(WUNIT,608)VERT,VERT,(ICHAR(J),J=1,NX),VERT,VERT
153           GO TO 302
154       320 WRITE(WUNIT,608)VERT,VERT,(ICHAR(J),J=1,NX)
155           GO TO 302
156       308 MJ3=MJ3+1
157           IF(NX.LT.100)GO TO 309
158           WRITE(WUNIT,609)VERT,YLAB(MJ3),VERT,(ICHAR(J),J=1,NX),VERT,
159          &YLAB(MJ3),VERT
160           GO TO 302
161       309 WRITE(WUNIT,609)VERT,YLAB(MJ3),VERT,(ICHAR(J),J=1,NX)
162       302 CONTINUE
163           WRITE(WUNIT,602)(XLAB(I),I=MB,ME)
164           MB=ME
165           ME=ME+ME-1
166       300 CONTINUE
167           WRITE(WUNIT,606)(XLABEL(J),J=1,10)
168           NPAGE=IPAGE
169           RETURN
170           END
```

```
1     C#################################################################
2     C#################################################################
3     C##                                                            ##
4     C##                                                            ##
5     C##        $ $ $ $ $  S I Z E  $ $ $ $ $                       ##
6     C##                                                            ##
7     C##                                                            ##
8     C#################################################################
9     C#################################################################
10          SUBROUTINE SIZE(DATA,DMAX,DMIN,NPTS,NSET,NCOMP6,NWVLS)
11    C
12    C       THIS ROUTINE FINDS THE MAXIMUM AND MINIMUM ELEMENT OF THE
13    C       NSET VECTOR OF THE ARRAY DATA.  THE VALUES ARE RETURNED IN
14    C       DMAX AND DMIN.
15    C
```

```
16          REAL*8 DATA(NCOMP6,NWVLS)
17          DMIN=DATA(NSET,1)
18          DMAX=DMIN
19          DO 100 I=2,NPTS
20          D=DATA(NSET,I)
21          IF(D.GT.DMAX)DMAX=D
22          IF(D.LT.DMIN)DMIN=D
23    100 CONTINUE
24    103 RETURN
25          END
```

```
1    C##################################################################
2    C##################################################################
3    C##                                                              ##
4    C##                                                              ##
5    C##        $ $ $ $ $   L A B E L   $ $ $ $ $                     ##
6    C##                                                              ##
7    C##                                                              ##
8    C##################################################################
9    C##################################################################
10          SUBROUTINE LABEL(AMAX,AMIN,AFACT,NA,ALAB,LINES,NN)
11    C
12    C      THIS ROUTINE DETERMINES THE SCALING FACTOR FOR THE X- AND
13    C      Y-AXES.  LABELS FOR EACH AXIS ARE LOADED INTO THE ALAB VECTOR.
14    C
15    C      USE OF THIS ROUTINE IS AS FOLLOWS:
16    C
17    C      INPUT REQUIREMENTS:-
18    C      AMAX,AMIN - THE MAX. AND MIN. VALUES OF THE AXIS
19    C                  IN QUESTION.  AMAX AND AMIN COME FROM S/R SIZE.
20    C      NA - TOTAL NUMBER OF PRINTER LINES OR CHARACTER POSITIONS
21    C           REQUIRED FOR THE PLOT.
22    C      LINES - NUMBER OF LINES OR POSITIONS BETWEEN EACH LABEL.
23    C
24    C      OUTPUT WILL BE:
25    C      ALAB - VECTOR CONTAINING THE X- OR Y-AXIS LABELS.
26    C      AFACT - A FACTOR TO CONVERT RAW DATA POINTS TO COORDINATES
27    C              RELATIVE TO THE PLOTTING SURFACE.
28    C
29          DIMENSION ALAB(13)
30          AFACT=ABS(AMAX-AMIN)/FLOAT(NA-1)
31          AST=AMIN
32          NN=NA/LINES
33          ALINE=LINES*AFACT
34          NN=NN+1
35          DO 200 I=1,NN
36          ALAB(I)=AST
37    200 AST=AST+ALINE
38          RETURN
39          END
```

```
1    C##################################################################
2    C##################################################################
3    C##                                                              ##
4    C##                                                              ##
5    C##        $ $ $ $ $   S C A T E R   $ $ $ $ $                   ##
6    C##                                                              ##
7    C##                                                              ##
8    C##################################################################
9    C##################################################################
10          SUBROUTINESCATER(R,ITN)
11    C
12    C      THIS ROUTINE PRODUCES A SCATTER PLOT OF THE RESIDUALS
13    C      FOR THE CALCULATION ABSORBANCE(CALC.) - ABSORBANCE(OBS.)
14    C      THE SCATTER PLOT HAS A SELF  EXPANDING Y-AXIS.
15    C      THE SCATTER PLOT IS USEFUL FOR DETERMINING WHETHER THE
16    C      THE FITTED MODEL REASONABLY DESCRIBES THE DATA.
17    C
18    C
19          COMMON/UNIT/RUNIT,WUNIT,PUNIT
20          INTEGERLINE(100),BLANK/' '/,STAR/'*'/,POINT/'.'/,CROSS/'+'/,
```

```
21              &II(9),RUNIT,WUNIT,PUNIT
22              DATAII/10,20,30,40,50,60,70,80,90/
23              REAL*8R(100)
24       600 FORMAT(2X,'.',2X,1PE12.2,'+',100A1,'+')
25       601 FORMAT(2X,25('.'),9('+(',I2,').....'))
26       602 FORMAT('0')
27              RMAX=0.0
28              DO11K=1,ITN
29              IF(ABS(R(K)).GT.RMAX)RMAX=ABS(R(K))
30       11   CONTINUE
31              ICOUNT=0
32              IITN=ITN/10
33              TEST=7.5E-04
34       10   IF(RMAX.LT.TEST)GOTO20
35              RMAX=RMAX/2.0
36              ICOUNT=ICOUNT+1
37              GOTO10
38       20   RANGE=5.0E-05*(2.0**ICOUNT)*15.0
39              RINC=RANGE/15.0
40              WRITE(WUNIT,602)
41              WRITE(WUNIT,601)(II(I),I=1,IITN)
42              ISET=ITN/100
43              IST=-99
44              IF(ISET.EQ.0)ISET=1
45              DO50ILINE=1,ISET
46              IST=IST+100
47              IFIN=IST+99
48              IF(IFIN.GT.ITN)IFIN=ITN
49              DO30I=1,31
50              IF(I.NE.16)GOTO31
51              DO42K=IST,IFIN
52       42   LINE(K)=POINT
53              GOTO32
54       31   DO40K=IST,IFIN
55       40   LINE(K)=BLANK
56       32   CONTINUE
57              DO43J=1,IITN
58              JJ=II(J)
59       43   LINE(JJ)=CROSS
60              DO41K=1,ITN
61              IF(R(K).LT.RANGE)GOTO41
62              LINE(K)=STAR
63              R(K)=-99.9
64       41   CONTINUE
65              IF(I.EQ.16)RANGE=0.E0
66              WRITE(WUNIT,600)RANGE,(LINE(K),K=IST,IFIN)
67       30   RANGE=RANGE-RINC
68              WRITE(WUNIT,601)(II(I),I=1,IITN)
69              WRITE(WUNIT,602)
70       50   CONTINUE
71              RETURN
72              END
```

5.2. Input Data

The following is a listing of a typical data set for SQUAD. The data are taken from a nonaqueous spectrophotometric titration of rhodium butyrate with imidazole and DMSO.

```
¶<-- Column 1

   TITRATION OF RHODIUM BUTYRATE/IMID WITH DMSO

                    DATA SET 111279

DICTIONARY:

MTL1=RHBU;LIG1=IMID;LIG2=DMSO:
```

```
END:
SPECIES:
RHBU(1)IMID(1);6.15;VB;VE:
RHBU(1)IMID(2);11.2;VB;VE:
RHBU(1)IMID(1)DMSO(1);7.75;VB;VE:
RHBU(1)DMSO(1);3.8411;FB;FE:
RHBU(1)DMSO(2);5.8612;FB;FE:
END:
DATA:
  500.0       670.0       10.0
LOGB
PRIN
NOCD
NNLS
PLOT 1
CRT
  15
1.0
MOL.ABS.:
RHBU;RHBU(1)DMSO(1);RHBU(1)DMSO(2):
END:
```

18.04	68.05	319.0
15.32	89.43	304.5
14.73	119.8	272.5
15.31	152.5	242.6
16.59	185.2	207.8
19.98	211.2	173.7
25.35	229.1	131.1
36.33	229.8	120.4
52.40	222.3	97.11
75.01	206.5	79.64
104.3	185.5	61.85
139.4	161.5	47.60
177.6	137.1	36.90
213.4	114.8	27.97
241.5	94.84	21.74
256.1	77.29	17.11
256.9	62.14	13.02
243.9	49.07	11.06

BASELINE:

.012	.013	.013	.013	.014	.014	.012	.011
.010	.009	.009	.009	.008	.007	.005	.004
.003	.002						

SPECTRA:

2.51859E-4   0.01.21621E-41.42871E-2   0.0   10.0

.465	.489	.506	.517	.522	.519	.502	.473
.428	.377	.324	.273	.223	.181	.145	.117
.093	.074						

2.51859E-4   0.01.21621E-42.85743E-2   0.0   10.0

.569	.581	.576	.555	.529	.496	.459	.416
.365	.313	.266	.220	.178	.144	.114	.092
.074	.057						

2.51859E-4   0.01.21621E-47.14358E-2   0.0   10.0

.668	.668	.640	.592	.533	.473	.413	.357
.301	.251	.208	.170	.136	.109	.087	.068
.053	.044						

2.51932E-4   0.02.52056E-41.42871E-2   0.0   10.0

.425	.464	.496	.522	.542	.548	.538	.509
.464	.408	.349	.292	.238	.193	.154	.123
.101	.083						

2.51932E-4   0.02.52056E-42.85743E-2   0.0   10.0

.521	.550	.559	.555	.540	.516	.481	.437
.387	.333	.283	.234	.190	.153	.120	.098
.078	.062						

2.51932E-4   0.02.52056E-47.14358E-2   0.0   10.0

.632	.648	.633	.596	.544	.487	.428	.370
.313	.262	.218	.178	.144	.114	.091	.074
.059	.049						

2.51932E-4   0.02.52056E-41.42872E-1   0.0   10.0

.690	.698	.669	.612	.543	.470	.399	.334
.277	.225	.182	.148	.118	.096	.076	.059
.048	.039						

2.52365E-4   0.05.08225E-41.42871E-2   0.0   10.0

.405	.458	.506	.544	.569	.575	.560	.527
.477	.417	.356	.297	.242	.198	.159	.130
.107	.091						

2.52365E-4   0.05.08225E-42.85743E-2   0.0   10.0

.489	.539	.570	.580	.574	.553	.520	.474

| .419 | .362 | .305 | .253 | .209 | .170 | .138 | .114 |
| .095 | .079 | | | | | | |

2.52365E-4   0.05.08225E-47.14358E-2   0.0   10.0

.593	.632	.638	.613	.570	.516	.454	.395
.336	.281	.234	.192	.157	.128	.105	.087
.072	.059						

2.52365E-4   0.05.08225E-41.42872E-1   0.0   10.0

.659	.689	.679	.637	.570	.497	.422	.353
.293	.241	.198	.161	.131	.107	.087	.072
.059	.050						

2.52221E-4   0.02.54700E-31.42871E-2   0.0   10.0

.514	.585	.638	.662	.651	.613	.554	.489
.422	.359	.303	.258	.218	.185	.159	.141
.124	.113						

2.52221E-4   0.02.54700E-32.85743E-2   0.0   10.0

.542	.608	.654	.664	.643	.596	.530	.461
.393	.332	.279	.235	.199	.169	.145	.126
.112	.100						

2.52221E-4   0.02.54700E-37.14358E-2   0.0   10.0

.590	.649	.677	.669	.628	.566	.493	.422
.355	.298	.248	.210	.177	.151	.129	.113
.101	.091						

2.52221E-4   0.02.54700E-31.42872E-1   0.0   10.0

.629	.682	.697	.670	.612	.537	.457	.381
.313	.257	.213	.176	.147	.122	.104	.091
.080	.071						

-1.0

7

General Computer Programs for the Determination of Formation Constants from Various Types of Data

JOSEF HAVEL and MILAN MELOUN

1. MULTIPARAMETRIC CURVE FITTING USING ABLET

1.1. Introduction

In chemical equilibrium studies it is often necessary to find an adequate model description, represented by the function $f(x; \beta_1, \ldots, \beta_m)$. Good agreement between the experimental data and the values calculated from that model is taken as evidence of having selected the best model. It should be noted immediately that this is not the only criterion by which the correctness of the model is determined, but it is certainly one of the most important indicators.

Consider an observable dependent variable, y_i, $i = 1, \ldots, n$, representing n instrument readings for a set of independent variables, x_i. Let us denote the vector of m unknowns for the model by $\boldsymbol{\beta}$ (β_j, $j = 1, \ldots, m$), i.e., assuming that the objective function of the model is known analytically as $\Psi = \Psi(x; \boldsymbol{\beta})$. Consequently each observed y_i, for a given x_i is related to the expected value of y_i, $F(y_i, x_i) = \Psi_i$, by the equation

$$y_i = \Psi_i + \epsilon_i = \Psi(x_i, \boldsymbol{\beta}) + \epsilon_i \qquad (1)$$

where ϵ_i is the random error for the ith point. To obtain the estimates of $\boldsymbol{\beta}$ it is assumed that the values of y_i are random observations from the distributions of y_i

JOSEF HAVEL and MILAN MELOUN • Department of Analytical Chemistry, J. E. Purkyne, University, Kotlarska 2, 611 37 Brno; Department of Analytical Chemistry, University of Chemical Technology, Leninovo nam. 565, 532 10, Pardubice, Czechoslovakia.

about Ψ. Using the method of least squares the sum of squared residuals, U, is minimized to obtain the desired parameters:

$$U = \sum_i w_i [y_{\exp,i} - \Psi(x_i\beta_1,\ldots,\beta_m)]^2 = \min \tag{2}$$

where w_i represents a statistical weighting factor and $y_{\exp,i}$ is a single observation made at x_i.

Ideally the error sum square will be at a minimum and therefore its use is justified if

    a. the correct form of the objective function Ψ is used as a function of β;
    b. the errors, ϵ_i, are random errors in y_i only, and in particular contain no systematic errors;
    c. the random errors follow a Gaussian distribution;
    d. the weighting factor w_i precisely reflects the inherent accuracy of y_i.

Rarely are all the above conditions fulfilled exactly; nonetheless, the location of a minimum value of U is a widely used and accepted criterion for model correctness.

The general program ABLET[1] contains a suite of subprograms that embody the nonlinear least-squares algorithm found in the LETAGROP approach pioneered by Sillen.[2] Additional routines are supplied by the user that describe the particular problem and input the data. ABLET performs the following general tasks:

    i. estimates the nonlinear parameters, β, for the model;
    ii. tests agreement between the data and the proposed model, and the refined parameters, using appropriate statistical criteria;
    iii. generates synthetic data for preselected values of β.

1.2. The "Pit-Mapping" Technique

In essence the problem is reduced to finding the set of parameters, β, that minimize the function $U(\beta)$. The LETAG[1] minimization procedure finds this minimum by assuming that in the neighborhood of the minimum $U(\beta)$ approaches a second degree surface, i.e., a parabola, paraboloid, or hyperparaboloid, in $(m + 1)$-dimensional space.

Starting with an initial set of parameters, β_c, an error square sum U_c^0 is calculated for a sufficient number of surrounding sets where some elements of β_c have been changed by $\pm\Delta\beta$. For m parameters, β_i to be varied there will be $\frac{1}{2}(m + 1)(m + 2)$ evaluations of U. This is shown in Figure 1. The coefficients of the equation for a second degree hypersurface are calculated from these values of $U(\beta)$ thereby enabling the coordinates of $U_{\min}^i$, that is $\beta_{\min}^i$, for that surface. This procedure of locating $\beta_{\min}^i$ is called a "shot."

If $U_{\min}^i$ is smaller than $U_{\min}^{i-1}$, $\beta_{\min}^i$ is used for the next round of calculations and in many situations the final minimum value for $U_{\min}^k$ for $\beta_{\min}^k$ is rapidly found. However, if $U_{\min}^i$ is larger than a previously calculated $U_{\min}$, $U_{\min}^*$, the algorithm returns

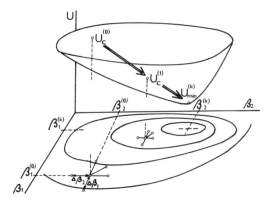

FIGURE 1. Graphical representation of a systematic search of the error surface for U_{min} in $(m + 1)$-dimensional space. The search commences at $U_c^{(0)}$ for the initial guesses of the sought parameters. Block STEG provides the parametric steps; block LETAG performs the systematic "shooting" around the central point, $U_c^{(i)}$; a new position $U_c^{(1)}$ is calcuated. U_{min} is located iteratively.

to β_{min}^*, and adjusts these parameters that led to U_{min}^i since these are likely to be the parameters that are furthest removed from the "best" values. Once the minimization procedure has been brought back on track the normal procedure of selecting shots is continued.

A commonly encountered problem in the refinement of $\boldsymbol{\beta}$ is that two or more parameters are strongly correlated, leading to a skewed and/or narrowed pit. To counteract this the twist matrix algorithm is envoked. This procedure twists parametric space so that the parameters β_i are varied along the principal axis of the error surface rather than parallel to the coordinate axes.

The square of the standard deviation for y may be calculated from $U(\boldsymbol{\beta})$ by dividing U_{min} by the number of degrees of freedom:

$$\sigma^2(y) = u_{min}/(n - m) \tag{3}$$

The "D boundary" is the hypercurve on which U is equal to $U_{min} + \sigma^2(y)$, as shown in Figure 2.

The standard deviation of each parameter is defined as the maximum difference between the value for β_i at any point on the D boundary and the value for β_i at the minimum:

$$\sigma(\beta_i) = \max [(\beta_D - \beta_{min})_i] \tag{4}$$

Should any parameter take a negative value, which would have no physical significance (concentration, molar absorptivity, ion-size parameter), the program sets this value to zero and the algorithm continues to search for the "reduced pit." This procedure is invoked only close to the minimum to prevent premature elimination of such parameters from $\boldsymbol{\beta}$.

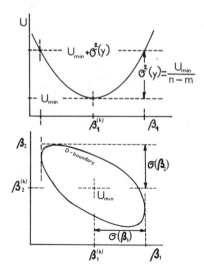

FIGURE 2. Graphical representation for the determination of standard deviations for a two-parameter model. The standard deviation $\sigma(y)$ is calculated from the "D-boundary." Standard deviations for the parameters are obtained from the maximum distance between the value of β_i at any point on the D-boundary and the coordinates of β_i at the minimum.

1.3. Description of the Program ABLET

ABLET comprises a total of 16 subroutines. Twelve are unchanged for any of the family of programs that extract β from various types of data. Two (for data I/0) are given in recommended form but may be changed to suit the user's preference. The last two are related to the specific member of the family of programs and therefore are changed for each different type of problem. Execution-time dimensioning has been used wherever practical. Currently ABLET is set for 50 data points and for 8 parameters in the model. A block diagram showing the inter-relationships of each routine is shown in Figure 3.

1.3.1. Routines Specific to the Model

Two routines, specific to the model and problem at hand, are supplied by the user. DATA reads in experimental data and performs any necessary preliminary calculations. This routine must consist of the following coding:

```
SUBROUTINE DATA(IOU,NB,N)
COMMON/FUNC/XEXP(N),YEXP(N),ERR(N),YCALC(N),W(N)
COMMON/KANAL/NI
CALL READ(NB,1)
        :
        :
        :
```

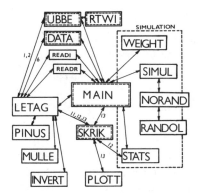

FIGURE 3. Block diagram for the ABLET suite of programs. User-supplied routines are shown by dou-
bled outlined boxes; DATA for input; UBBE for calculation of the error sum square; SKRIK for output;
MAIN to control the logic flow. UBBE may require additional routines to solve equations relating to the
calculation of the error sum square.

```
        User-supplied code for reading in data
        .
        .
        RETURN
        END
```

where NI and IOU are input and output unit numbers, respectively; NB is the number
of points; N is the maximum number of points; XEXP, YEXP are the vectors for the
data. Subroutine UBBE calculates the error sum square, according to equation (2).
This routine, supplied by the user, must consist of the following:

```
        SUBROUTINE UBBE(U,NK,XK,NB)
        DIMENSION XK(NK)
        COMMON/FUNC/XEXP(N),YEXP(N),ERR(N),YCALC(N),W(N)
        U=0.0
        DO 100 I=1,NB
        .
        .
        User-supplied code for reading in data
        .
        .
        YCALC(I)= .... User-supplied model equation ....
100     U=U+W(I) * (YEXP(I)-YCALC(I))**2
        RETURN
        END
```

where NK is the number of parameters, XK is the parameter vector, and W is the
weighting vector.

1.3.2. Basic Routines in the ABLET Family

The main program reads in a series of refinement keys, refinement termination criteria, and either synthetic data (SINST > 0) or experimental data (SINST < 0).

For simulated data the values, standard deviations, and weighting factors of the parameters, the standard deviation of the dependent variable, and a set of exact data points are read in. To these data points are added a generated random error having a normal Gaussian distribution.

Subroutine SKRIK (NB,NK,IOU,IRUR,XK,SIGXK) provides the output of the estimated parameters (XK) together with their standard deviations (SIGXK), a statistical analysis of the residuals from subroutine STATS, and printer plots of the experimental and calculated dependent variables (subroutine PLOTT[3]). This routine may be extended by the user to provide additional output. IRUR is an integer that controls the content of the output. This integer is also used extensively to indicate several options during data input.

Subroutine LETAG (IOU, NAUT, NK, NB, XK, UMIN, ISSW, DARK2, DATA, UBBE,SKRIK) is by Cermak and Meloun[1] derived from Sillen's LETAGROP VRID.[2] This routine contains the FORTRAN coding of the pit-mapping procedure described in Section 1.2. The routine is divided into logical units which are encountered in an order determined by the particular value of IRUR. ISSW is an integer vector that controls the type and extent of the output and the minimization strategy. Other principal arrays are DARK2, which holds the standard deviations of the parameters, and XK, which holds the current values of the parameters.

Subroutine SIMUL(X,YC,YM,NB,SIG,ISTART) is used in the simulation option of ABLET. YM is the dependent variable with applied random error, EPS:

$$YM(I) = YC(I) + SIG*EPS(I)$$

EPS is generated by subroutine NORAND, a standard routine[4] for the generation of random numbers.

Subroutine STATS(IOU,NK,NB,X,Y,EPS) is adapted from MINIQUAD[5] and performs a statistical analysis of residuals to determine how closely they follow a Gaussian distribution. The Pearson test is also applied.

Subroutine WEIGHT(UBBE,NB,XK,SIGXK,WEI) calculates the weighting for the simulated data, held in W. SIGXK and WEI are the standard deviations and weights of the parameters. The values of YC are calculated by subroutine UBBE according to

$$^0YC(I) = f(X(I);\ XK(1),\ XK(2),\dots XK(NK))$$

$$^1YC(I) = f(X(I);\ XK(1)+SIGXK(1),XK(2),\dots XK(NK))$$

$$^{NC}YC(I) = f(X(I);\ XK(1),XK(2),\dots XK(NK)+SIGXK(NK))$$

the weights are then calculated according to

$$W(I) = \sum_{J=1}^{NK} OMEGA(J)*WEI(I)*ABS(^JYC)(I) - {}^0YC(I))$$

where

$$\text{OMEGA}(J) = 1. \bigg/ \sum_{I=1}^{NB} [{}^J\text{YC}(I) - {}^0\text{YC}(I)]^2$$

This scheme permits the investigation of the variation of each parameter on the variation of the dependent variable. OMEGA compensates for the influence of a change in parameter value on the change in U. If the key ISSW(5) equals unity, WEI increases the effect of ill-conditioned parameters on some of the data points during the minimization process and thus increases the sensitivity of these parameters in the model. If ISSW(5) equals 2 the weighting factor for each datum point is given by the reciprocal value of W(I).

Subroutine PLOTT (XX, YY, NDATA, NDMAX, ISYMBL, NF, XLINE, MX, YLINE,MY,NLS,NCL,MM, LL, AREA, YSCALE)[3] provides a printer plot of the fit of the model to the data. The data, held in XX and YY, are scaled to the plotting surface range 0 to NCL and 0 to NLS, respectively. These values are used to locate and print an appropriate Hollerith character on the plotting surface. NDMAX is the maximum number of points in the largest data set; NDATA(J) holds the number of points per data set; ISYMBL(J) is the Hollerith character for the Jth curve; NF is the number of curves; MX is the number of grid lines (can be zero); NLS is the number of lines for the plot (NLS = 56 for one page); NCL is the width of the plot (NCL = 122 for full page width). The type of display is controlled by MM(=1, across the page; =2, down the page) and LL (=1, linear scale; =2, semilog scale, =3, semilog root scale). [Note: This routine contains coding that is IBM-type machine specific. Testing of this program required the replacement of this code with a subroutine (PRTPLT) that is machine independent. Subroutine PLOTT has been included in Section 1.5 after the listing for ABLET (*Editor*).]

Subroutine READI(I,N) and subroutine READR(A,N) allow for format-free input of data.[1] Integers are read in through READI, real numbers through READR. A number of data points, N, are to be loaded into the particular real array, A, or the integer array, I. Repeated data points may be read in using the format $n*$value, where n is the number of repeated values. A blank for 'value' implies that the following n values will be ignored by READR(READI).

LETAG uses three utility routines: INVERT[6] for matrix inversion; PINUS[2] for vector–matrix multiplication; MULLE[2] for matrix–matrix multiplication.

1.4. Data Input Instructions

1.4.1. Introduction

ABLET, and its relatives, makes use of the value of IRUR to organize and control the minimization process. The description of the permitted values of IRUR and their significance is shown in Table 1. The use of the terms Process Card, Step Card, etc. will be described in Section 1.4.2. A second parameter is the vector ISSW, known as the Key Card. The allowed values for the seven elements of ISSW are shown in Table 2.

TABLE 1.   Permitted Values of IRUR and Their Significance

IRUR	Significance
1	Subroutine UBBE is called. Value is included on Process Card
2	As for IRUR = 1 with U and XK being printed for each cycle.
3	Minimization steps, STEK(I), for each parameter are read in. Value included on the Step Card.
4	Minimization steps, STEK(I), are recalculated from the standard deviations of each parameter, DARK(I). Value included on the Step Card.
5	Block LETA is accessed and one iteration is performed.
6	Data are provided to the program via a BLOCK DATA and are also read in. Value is written on the Data Card(s).
7	Three possible variations are possible for IRUR = 7: (A) Block LASK is accessed and the initial guesses of the parameters, XK, and twist matrix elements, S, read in. Value is included on the Initial Guess and Matrix Cards. (B) Block LASK is accessed and NBYK parameters are changed. Value is included on the Initial Guess Card. (C) Block LASK is accessed and NEGK parameters are allowed to be negative. Posk protection is removed by including IRUR on the Posk Removal Card. If used this card must follow an Initial Card. *Note:* Parameters read in for the three meanings of IRUR=7 are: (A) NK,NBYK,(XK(I),I=1,NBYK),ISKIN,((I(K),J(K),S(I(K),J(K)),K=1,ISKIN) (B) NK,NBYK,((IK(K),XK(K)))K=1,NBYK),ISKIN,((I(K),J(K),S(I(K),J(K)), K=1,ISKIN) (C) NK,−1,NEGK,(IK(I)I=1,NEGK)
8	Termination of the minimization process. Value is included on the Process Card.
9	As IRUR = 8.
10	The LETAG algorithmic strategy is used. Value is included on the Process Card.
11	Subroutine SKRIK is called. Value is included on the Process Card.
12 and 13	As IRUR = 11.

TABLE 2.   Allowed Values for the Key Card, ISSW

Key	Equal to 0	Equal to 1
ISSW(1)	No print.	Print XK and U for each call to UBBE.
ISSW(2)	No print.	Print out the twist matrix for each iteration.
ISSW(3)	No print.	Print out the values of the determinant.
ISSW(4)	STEK(I) = PSI*STEK(I) (See item 3)	STEK(I) = PSI*DARK(I) (See item 3 and IRUR=4)
ISSW(5)	Equal to 1 implies improvement of sensitivity for ill-conditioned parameters during simulation.	Equal to 2 implies that statistical weights for simulated data are calculated.
ISSW(6)	Algorithmic Strategy. (See Section 1.4.3)	Heuristic Strategy. (See Section 1.4.3)
ISSW(7)	Used for special purposes.	Used for special purposes.

1.4.2. Data Input Requirements

All data are read in format free. A blank or a comma may be used to separate each datum. Since data processing options are, in several instances, controlled by the particular use of the program, there is no unique setup for the data deck. The following descriptions cover all possibilities, some of which are optional.

1. Title Card: **TITLE**
   A descriptive title for the system under consideration. A maximum of 80 characters.
2. Key Card: **(ISSW(I),I=1,7)**
   ISSW controls the extent of output during the minimization process. (See Table 2)
3. Termination Card: **EPS,ITMAX,PSI,NK,MY,MX,NLS,NCL,SINST**
   EPS: Termination criterion for convergence, defined as

$$(|U_i - U_{i-1}|)/U_i \leq \text{EPS}$$

   ITMAX: Maximum number of iterations
   PSI: Controls the change in the step for the parameters after each iteration. Recommended values are from 0.1 to 0.5
   NK: Number of parameters.
   MY: Number of grid lines for the y axis.
   MX: Number of grid lines for the x axis.
   NLS: Number of lines for the printer plot. NLS = 50 for one page.
   NCL: Number of columns for the printer plot. NCL = 122 for full page width.
   Note: If the user chooses the alternative (non-machine-dependent) plotter routine these parameters become redundant. See listing for details of use.
   SINST: A preselected value for the standard deviation of the dependent variable, y. If experimental data are to be processed set SINST < 0; for simulations set SINST > 0.
4. Simulation Card: **(XK(I),I=1,NK),(SIGXK(I),I=1,NK),(WEI(I), I=1,NK)**
   XK(I): NK preselected parameter value, for simulation of data.
   SIGXK(I): Standard deviations of the XK(I) parameters.
   WEI(I): Weighting factors for the XK(I) parameters. The NK values of WEI should sum to unity.
5. Data Card(s):
   The number of parameters and format specifications will be established by the user in the design of subroutine DATA. Each card is started with IRUR(=6).
6. Initial Guess Card
   This card is formulated in one of three different ways:
a. Initial Guess Card: **IRUR,NK,NBYK,(XK(I),I=1,NBYK)**
   IRUR: Set equal to 7.

NK: Number of parameters.

NBYK: Set equal to NK. The number of parameters that must remain positive.

XK(I): Initial estimates for the parameters.

Note: A matrix card must follow this card.

b. Initial Guess Card: **IRUR,NK,NBYK,((IK(I),X(IK(I))),I=1,NBYK)**

IRUR: Set equal to 7.

NK: Number of parameters.

NBYK: Number of parameters whose initial value is to be changed after the first iteration.

IK(I): Index for the particular parameters estimate to be changed.

XK(IK(I)): Parameter estimate to be changed.

Note: This card must be followed by a Matrix Card.

c. Posk Removal Card: **IRUR,NK,−1,NEGK,(IK(I),I=1,NEGK)**

IRUR: Set to 7.

NK: Number of parameters

NBYK: Set to −1. This implies that parameters in the minimization process may take on both positive and negative values.

IK(I): indices of parameters that may take a negative value.

7. Matrix Card: **ISKIN,((I(K),J(K),S(I(K),J(K)),K=1,ISKIN)**

S(I(K),J(K)): A unit diagonal symmetric matrix that provides the coefficients for parameter "twisting." These values may be obtained from the output of the SIK block in LETAG.

ISKIN: Number of elements in S. If ISKIN = 0 no values are available.

Either

8. Step Card: **IRUR,N,((I(K),STEK(I(K)),K=1,N)**

IRUR: Set equal to 3.

N: Number of parameters to be varied during a "shot."

STEK(I(K)): Step values for the minimization of the $I(K)$th parameter.

Note: If, during minimization, any of the values become negative the new step value is set to the product to STEK and the standard deviation, DARK2, for the particular parameters, obtained from the previous cycle.

or

8. Step Card: **IRUR,H**

IRUR: Set equal to 4.

H: A factor, between 0.1 and 0.5, used to control the step size. The new step is H∗DARK2(I).

Note: Start with IRUR = 3 and optionally follow with second card with IRUR = 4.

9. Process Card: **IRUR,IRUR,...**

IRUR: Set to a series of values depending upon the type and progress of minimization desired. For example, the series 2, 5, 13 gives

a. IRUR=2: the error sum square for a set of parameters;

b. IRUR=5: causes control to pass to LETAG to perform the "shots";

c. IRUR=13: prints of parameters and their standard deviations followed by a printer-plot of the observed vs. calculated. Alternatively IRUR=12 gives a statistical analysis of residuals together with a histogram; IRUR=11 gives a table of observed and calculated together with (observed–calculated) values. Both 11 and 12 also provided the parameters and standard deviations. All permitted values for IRUR that can be used in this context are described in Table 1.

1.4.3. Algorithmic and Heuristic Strategies

Algorithmic Strategy. When ISSW(4) = 0 the minimization steps for all parameters are calculated automatically as a multiple of PSI from the previous step, i.e., $STEK(I)_i = PSI*STEK(I)_{i-1}$. The initial values are obtained from the step card.

When ISSW(4) = 1 minimization steps are calculated using the standard deviation of each parameter from the previous step, i.e., $STEK(I)_i = PSI*DARK2(I)_{i-1}$

Example: ISSW(6)=0, ISSW(4)=0.
Initial Guess Card: 7,3,3,4.8,5.0,0.3, (Three parameters.)
Matrix Card: 0 (No matrix elements available.)
Step Card: 3,3,1,0.12,2,0.5,3,0.15

All parameters will be varied; 0.12,0.5 and 0.15 are the steps for the parameters 4.8, 5.0, and 0.3, respectively. For the second and subsequent iterations steps are calculated from $STEK(I)_i = PSI*STEK(I)_{i-1}$.
Process Card: 2,5,13. (Minimization terminates when EPS or ITMAX criteria are fulfilled.)

Heuristic Strategy. The minimization steps are supplied by the user and consequently minimization is controlled by values obtained from the step and Process Cards. This is achieved as follows: After reading in the steps (not all are necessary) in STEG(IRUR=3), a central U, U_c is calculated by subroutine UBBE(IRUR=2). The parameters are varied systematically in LETAG (IRUR=5). The user organizes several iterations using Process Card in the order 3, . . . , 2, 5. Termination is achieved using IRUR=8 or 9, or alternatively IRUR = −1. An additional option may be that after a few iterations the step size factor H (IRUR=4) may be employed.

Example: ISSW(6)=1, ISSW(4)=0.
Initial Guess Card: 7,3,3,4.8,5.0,0.3 (Three parameters)
Matrix Card: 0
Step and Process Cards:
   3,3,1,0.2,2,0.5,3,0.3,2,5 (All parameters to be varied and minimization steps are read in here.)
   3,2,1,0.05,2,0.2,2,5 (First and second parameters will be varied using the steps read in here.)
   3,2,2,0.01,3,0.01,2,5 (Second and third parameters will be varied using the steps read in here.)
   3,3,1,0.01,2,0.01,3,0.01,2,5 (All parameters will be varied using the steps read in here.)
   −1 (Return to the algorithm strategy and minimization proceeds until criteria EPS or ITMAX are fulfilled.)
If IRUR = −1 is omitted calculations terminate here.

1.5. Listing of ABLET

The following listing contains the subroutines common to the ABLET family of programs, namely, STATS, SIMUL, NORAND, RANDOL, PRTPLT, LETAG, MULLE, PINUS, INVER, WEIGHT, READR, READI. Subroutines SKRIK, UBBE, and DATA and the main program, which are specific to the type of problem at hand, are shown in Sections 2.5, 3.5, and 4.5.

Note: The listing of PLOTT found at the end of the listing of ABLET is the original version of the printer plot.

```
 1          SUBROUTINE STATS(IOU,NK,NB,X,Y,EPS)
 2      C
 3      C---  STATISTICAL ANALYSIS OF RESIDUALS: CALCULATION OF MEAN  MEAN DEV.,
 4      C     STAND.DEV., VARIANCE, COEFFICIENTS OF SKEWNESS AND KURTOSIS, RFAC
 5      C
 6            DIMENSION X(NB),Y(NB),EPS(NB),JPOP(8),CLIM(8),EXFR(8)
 7            INTEGER PLUS,BLANK,STAR,PT(116)
 8            DATA PLUS,BLANK,STAR/'+',' ','*'/
 9            WAR=0.
10            W=FLOAT(NB)
11            DO 100 I=1,NB
12        100 WAR=WAR+EPS(I)*EPS(I)
13            SD=SQRT(WAR/W)
14            NR=8
15            CLIM(1)=-1.150*SD
16            CLIM(2)=-0.675*SD
17            CLIM(3)=-0.319*SD
18            CLIM(4)=0.
19            CLIM(5)=-CLIM(3)
20            CLIM(6)=-CLIM(2)
21            CLIM(7)=-CLIM(1)
22            CLIM(8)=1.E37
23            DO 108 I=1,NR
24        108 JPOP(I)=0
25            AM=0.
26            DM=0.
27            COSQ=0.
28            COKU=0.
29            RDEN=0.
30            DO 130 I=1,NB
31            DO 119 K=1,NR
32            IF(EPS(I)-CLIM(K))118,118,119
33        118 JPOP(K)=JPOP(K)+1
34            GO TO 120
35        119 CONTINUE
36            JPOP(NR)=JPOP(NR)+1
37        120 AM=AM+EPS(I)
38            RDEN=RDEN+Y(I)*Y(I)
39            DM=DM+ABS(EPS(I))
40            COSQ=COSQ+EPS(I)**3
41        130 COKU=COKU+EPS(I)**4
42            WRITE(IOU,24)
43            DO 260 IP=1,NB
44            DO 210 I=1,116
45        210 PT(I)=BLANK
46            DO 220 I=2,116,19
47        220 PT(I)=PLUS
48            J=EPS(IP)/SD*19.+59.5
49            IF(J.LE.1)J=2
50            IF(J.GT.116)J=116
51            PT(J)=STAR
52        260 WRITE(IOU,21)IP,PT
53            AM=AM/W
54            DM=DM/W
55            COSQ=COSQ/(WAR*SD)
56            COKU=COKU*W/(WAR*WAR)
57            WAR=SD*SD
58            WRITE(IOU,10)AM,DM,SD,WAR,COSQ,COKU
```

```
59             DO 270 I=1,NR
60         270 EXFR(I)=.125
61             OBSCH=0.
62             WRITE(IOU,12)
63             R1=-CLIM(NR)
64             DO 350 K=1,NR
65             EXPOP=EXFR(K)*W
66             OBFR=FLOAT(JPOP(K))/W
67             RAPP=((FLOAT(JPOP(K))-EXPOP)**2)/EXPOP
68             WRITE(IOU,17)K,R1,CLIM(K),EXFR(K),OBFR,EXPOP,JPOP(K),RAPP
69             R1=CLIM(K)
70         350 OBSCH=OBSCH+RAPP
71             EXPCH=12.6
72             WRITE(IOU,18)OBSCH,EXPCH
73             RFACT=SQRT(WAR*W/RDEN)
74             WRITE(IOU,4)RFACT
75           4 FORMAT (//' R FACTOR=',F12.6)
76          10 FORMAT(//'    ARITHMETIC MEAN                = ',1PE15.5/
77             &          '    MEAN DEVIATION               = ',1PE15.5/
78             &          '    STANDARD DEVIATION           = ',1PE15.5/
79             &          '    VARIANCE                     = ',1PE15.5/
80             &          '    MOMENT COEFF.OF SKEWNESS (=0) = ',1PE15.5/
81             &          '    MOMENT COEFF.OF KURTOSIS (=3) = ',1PE15.5//)
82          12 FORMAT (16X,'CLASS LIMITS',13X,'PROBABILITY',10X,'FREQUENCY',8X,
83             &'PARTIAL'/12X,'LOWER',10X,'HIGHER',7X,'CALC',7X,'OBS',7X,'CALC',
84             &5X,'OBS',5X,'CHI-SQUARE'//)
85          17 FORMAT(I5,2(1PE15.5),2F10.4,F10.1,I8,F12.3)
86          18 FORMAT(//'    OBSERVED CHI SQUARE IS ',F10.2/
87             &          '    CHI SQUARE (6,0.95) SHOULD BE ',F10.2/)
88          21 FORMAT (I4,116A1)
89          24 FORMAT(/'  RESIDUALS PLOT '/)
90             RETURN
91             END
```

```
1              SUBROUTINE LETAG (IOU,NAUT,NK,NB,XK,UMIN,ISSW,DARK2,DATA,UBBE,
2             &SKRIK)
3      C
4      C
5      C--- M I N I M I Z A T I O N    S U B R O U T I N E    L E T A G
6      C     (A MODIFICATION OF SILLEN'S PROGRAM LETAGROP VRID BY
7      C     J. CERMAK AND M. MELOUN)
8      C
9      C
10             DIMENSION DARK(8),DARK2(NK),IPOSK(8),XK(NK),ISSW(7),
11            &ABOM(8),AC(8),ACSP(8),AMIN(8),AV(8),PINA(8),DIA1(8),DIA2(8),
12            &PINNE(8),RUCKA(8,8),RUTA(8,8),RUT1(8,8),RUT3(8,8),RUTIN
13            &(8,8),S(8,8),SH(8,8),SHINV(8,8),SK(8,8),STEK(8),STEKG
14            &(8),V(8),VBOM(8),VRI(8,8),IMI(8),IVAR(8),IVARG(8),DARR1(8),
15            &DARR2(8)
16             COMMON/KANAL/NI
17             NI=5
18             ISSW6=ISSW(6)
19             NDIM=8
20             IF(NAUT)3,106,104
21           3 W0=UMIN
22             DO 2 I=1,NDIM
23           2 DARK(I)=-1.
24             GO TO 106
25     C
26     C KNUT
27     C
28         100 IF(IRUR.EQ.9)GOTO 111
29             IF(NAUT)106,107,108
30         106 CALL READI(IRUR,1)
31             IF(IRUR)5111,5111,5100
32        5100 WRITE(IOU,1602)IRUR
33             GO TO(1200,102,800,103,104,105,400,110,111,112,700,700,700),IRUR
34         107 NAUT=1
35             RETURN
36         102 WRITE(IOU,1603)
37             GO TO 1200
```

```
38      108 IRUR=4
39          NAUT=0
40          IF(ISSW(4)) 4,4,800
41        4 DO 5 I=1,N
42        5 STEK(I)=W0*STEK(I)
43          GO TO 100
44      103 CALL READR(W0,1)
45          GO TO 800
46      104 WRITE(IOU,1604)
47          NAUT=1
48          IF(ISSW6)5004,5004,5003
49     5003 NAUT=-1
50     5004 NORV=0
51          IRUR=5
52          GO TO 300
53 C
54 C DATA
55 C
56      105 CALL DATA(IOU,NB)
57          NORV=0
58          GO TO 100
59      110 NAUT=2
60          RETURN
61      111 NAUT=3
62          RETURN
63      112 NAUT=1
64          RETURN
65     5111 ISSW(6)=0
66          RETURN
67 C
68 C GROP
69 C
70      200 CALL PINUS(PINNE,RUTIN,N,VBOM,1)
71          UNO=UC
72          DO 201 I=1,N
73      201 UNO=UNO-PINNE(I)*VBOM(I)
74          IF(UNO)202,202,210
75      202 WRITE(IOU,2001)
76          GO TO 220
77      210 SIG2Y=UNO/FLOAT(NB-N)
78          SIGY=SQRT(SIG2Y)
79          WRITE(IOU,2002)SIGY
80      220 WRITE(IOU,2003)
81          CALL PINUS(VBOM,SH,N,ABOM,-1)
82          CALL MULLE(SH,RUTIN,N,N,N,RUT1,1)
83          DO 221 I=1,N
84      221 ABOM(I)=ABOM(I)+AC(I)
85          DO 222 I=1,N
86          DIA1(I) = RUTIN(I,I)
87          DIA2(I) = 0
88          DO 2210 M=1,N
89     2210 DIA2(I) = DIA2(I) + RUT1(I,M)*SH(I,M)
90          IK=IVAR(I)
91          XK(IK)=ABOM(I)
92          IF(DIA1(I)) 223,223,224
93      224 IF(UNO)223,223,225
94      225 IF(IRUR-15)226,223,226
95      226 W = ABS(SQRT(SIG2Y*DIA1(I))*SH(I,I))
96          DARR1(I) = W
97          GO TO 227
98      223 W=-1.
99      227 IF(IRUR-15)228,229,228
100     228 DARK(IK)=DARR1(I)
101     229 IF(DIA2(I)) 230,230,231
102     231 IF(UNO)230,230,232
103     232 DARR2(I) = SQRT(SIG2Y*DIA2(I))
104         W1 = DARR2(I)
105         GO TO 233
106     230 W1=-1.
107     233 WRITE(IOU,2004)IK,ABOM(I),W,W1
108         DARK2(IK)=DARR2(I)
109     222 CONTINUE
110         IF(N-1)234,235,234
111     235 IF(W)236,234,234
112     236 W=SQRT(ABS(0.01*UC/PINNE(1)))
113         IK=IVAR(1)
```

```
114              DARK(IK) = ABS(W*SH(1,1))
115         234 IF(IRUR-15)237,500,237
116         237 IF(N-1)1000,500,1000
117     C
118     C LETA
119     C
120         300 IF(NORV*(NORV+2))301,1200,301
121         301 DO 302 I=1,N
122         302 V(I)=0.
123             IF(NORV-1)303,310,303
124         303 IF(NORV +1)304,305,304
125         305 IF(U-UMIN*1.5)307,307,306
126         306 NORV=-3
127             GO TO 800
128         307 NORV=1
129         304 IF(U-UMIN)310,310,308
130         308 GO TO (320,330,340,350),NORV
131         310 UMIN=U
132             DO 311 I=1,N
133         311 AMIN(I)=AV(I)
134             GO TO (320,330,340,350),NORV
135         320 UC=U
136             DO 321 I=1,N
137             DO 321 J=1,N
138         321 SH(I,J)=S(I,J)*STEK(J)
139             IR=1
140             JR=0
141             V(1)=1.
142             NORV=2
143             GOTO 3000
144         330 U1=U
145             V(IR)=-1.
146             NORV=3
147             GO TO 3000
148         340 U2=U
149             IF(U2-U1)341,341,360
150         341 U2=U1
151             U1=U
152             STEK(IR)=-STEK(IR)
153         360 PINNE(IR)=(U2-U1)*.25
154             RUTA(IR,IR)=(U2+U1)*.5-UC
155             IR=IR+1
156             IF(IR-N)361,361,370
157         361 V(IR)=1.
158             NORV=2
159             GO TO 3000
160         350 W=.5*(U-UC)+PINNE(IR)+PINNE(JR)-.5*(RUTA(IR,IR)+RUTA(JR,JR))
161             RUTA(IR,JR)=W
162             RUTA(JR,IR)=W
163             JR=JR+1
164             IF(JR-N)390,390,380
165         370 IR=0
166             NORV=4
167             DO 371 I=1,N
168             DO 371 J=1,N
169         371 SH(I,J)=S(I,J)*STEK(J)
170         372 STEK(I)=ABS(STEK(I))
171         380 IR=IR+1
172             IF(IR-N)381,3010,381
173         381 JR=IR+1
174         390 V(IR)=1.
175             V(JR)=1.
176        3000 CALL PINUS(V,SH,N,AV,-1)
177             DO 3001 I=1,N
178             IK=IVAR(I)
179             AV(I)=AV(I)+AC(I)
180        3001 XK(IK)=AV(I)
181        3020 IF(NORV-1)1200,900,1200
182        3010 IR=0
183             JR=0
184             NORV=1
185             IRV=N
186             GO TO 900
187     C
188     C LASK
189     C
```

```
190      400 NORV=0
191          WRITE(IOU,4000)
192          CALL READI(NK,1)
193          CALL READI(NBYK,1)
194          WRITE(IOU,4001)NK,NBYK
195          IF(NBYK+1)401,460,401
196      401 IF(NBYK-NK)410,402,402
197      402 CALL READR(XK,NK)
198          WRITE(IOU,4002)(XK(I),I=1,NK)
199          DO 403 I=1,NDIM
200          DO 404 J=1,NDIM
201      404 SK(I,J)=0.
202          SK(I,I)=1.
203          DARK(I)=-1.
204          DARK2(I)=-1.
205      403 IPOSK(I)=1
206          DO 405 I=1,NK
207      405 DARK2(I)=-1.0
208      450 CALL READI(ISKIN,1)
209          WRITE(IOU,4003)ISKIN
210          IF(ISKIN)470,470,14
211       14 DO 12 I1=1,ISKIN
212          CALL READI(I,1)
213          CALL READI(J,1)
214          CALL READR(SK(I,J),1)
215       12 WRITE(IOU,4004)I,J,SK(I,J)
216          GO TO 470
217      410 DO 411 M=1,NBYK
218          CALL READI(IK,1)
219          CALL READR(XK(IK),1)
220          WRITE(IOU,4005)IK,XK(IK)
221          DARK(IK)=-1.
222      411 DARK2(IK)=-1.
223          GO TO 450
224      460 CALL READI(NEGK,1)
225          WRITE(IOU,4006)NEGK
226          DO 461 I=1,NEGK
227          CALL READI(IK,1)
228          WRITE(IOU,4007)IK
229      461 IPOSK(IK)=0
230      470 GO TO 100
231  C
232  C        MIKO
233  C
234      500 NIMI=0
235          DO 501 I=1,N
236          IK=IVAR(I)
237          IF(ABOM(I))502,503,503
238      502 IF(IPOSK(IK))503,503,504
239      504 NIMI=NIMI+1
240          IMI(NIMI)=I
241          XK(IK)=0.
242          GO TO 501
243      503 XK(IK)=ABOM(I)
244      501 CONTINUE
245          IF(NIMI)505,505,510
246      505 IF(IRUR-15)600,520,600
247      510 WRITE(IOU,5101)
248          IF(IRUR-15)511,512,511
249      511 UCSP=UC
250          DO 513 I=1,N
251          ACSP(I)=AC(I)
252      513 AC(I)=ABOM(I)
253      512 IF(NIMI-N)514,520,514
254      514 DO 515 I=1,N
255          DO 515 J=1,N
256      515 SHINV(I,J)=SH(I,J)
257          CALL INVER(N,SHINV,1.E-35,DET,ING)
258          IF(ISSW(3)) 5001,5001,5000
259     5000 WRITE(IOU,5102)DET
260     5001 IF(ING)516,516,517
261      517 IRUR=9
262          GO TO 611
263      516 CALL MULLE(SHINV,RUTA,N,N,N,RUT3,-1)
264          CALL MULLE(RUT3,SHINV,N,N,N,RUCKA,1)
265          DO 518 IR=1,NIMI
```

```
266              I=IMI(IR)
267              DO 518 JR=1,NIMI
268              J=IMI(JR)
269          518 UNO=UNO+AC(I)*AC(J)*RUCKA(I,J)
270              UC=UNO
271              DO 519 I=1,N
272              PINA(I)=0.
273              DO 519 JR=1,NIMI
274              J=IMI(JR)
275          519 PINA(I)=PINA(I)+AC(J)*RUCKA(J,I)
276              DO 521 IR=1,NIMI
277              I=IMI(IR)-1
278          522 I=I+1
279              IF(I-N+IR)523,523,521
280          523 IVAR(I)=IVAR(I+1)
281              AC(I)=AC(I+1)
282              PINA(I)=PINA(I+1)
283              NIR1=N-IR+1
284              DO 524 J=1,NIR1
285              RUCKA(I,J)=RUCKA(I+1,J)
286          524 SH(I,J)=SH(I+1,J)
287              NIR1=NIR1-1
288              DO 525 J=1,NIR1
289              SH(J,I)=SH(J,I+1)
290          525 RUCKA(J,I)=RUCKA(J,I+1)
291              GO TO 522
292          521 CONTINUE
293              N=N-NIMI
294              CALL PINUS(PINA,SH,N,PINNE,1)
295              CALL MULLE(SH,RUCKA,N,N,N,RUT3,-1)
296              CALL MULLE(RUT3,SH,N,N,N,RUTA,1)
297              IRUR=15
298              IRV=N
299              GO TO 900
300          520 N=NGE
301              UC=UCSP
302              DO 526 I=1,N
303              IVAR(I)=IVARG(I)
304          526 AC(I)=ACSP(I)
305    C
306    C        PROVA
307    C
308          600 IF(IRUR)601,610,601
309          601 IRUR=0
310              WRITE(IOU,6001)
311              GO TO 1200
312          610 IF(U-UMIN)630,611,611
313          611 IF(UC-UMIN)612,612,620
314          612 WRITE(IOU,6002)
315              U=UC
316              DO 613 I=1,N
317              IK=IVAR(I)
318          613 XK(IK)=AC(I)
319              GO TO 100
320          620 WRITE(IOU,6003)
321              U=UMIN
322              DO 621 I=1,N
323              IK=IVAR(I)
324              XK(IK)=AMIN(I)
325          621 AC(I)=AMIN(I)
326              GO TO 100
327          630 DO 631 I=1,N
328              IK=IVAR(I)
329              AC(I)=XK(IK)
330          631 AMIN(I)=XK(IK)
331              UMIN=U
332              GO TO 100
333    C        SKRIK
334          700 CALL SKRIK(NB,NK,IOU,XK,IRUR,DARK2)
335              GO TO 100
336    C
337    C        STEG
338    C
339          800 IF(NORV+3)801,810,801
340          801 IF(IRUR-4)802,840,802
341          802 CALL READI(N,1)
```

```
342          WRITE(IOU,8026)
343          DO 803 I=1,N
344          CALL READI(IK,1)
345          CALL READR(W,1)
346          WRITE(IOU,8025)   IK,W
347          IVAR(I)=IK
348          AC(I)=XK(IK)
349          AV(I)=XK(IK)
350          IF(W)804,804,805
351      805 STEK(I)=W
352          GO TO 803
353      804 IF(DARK(IK))806,806,807
354      807 STEK(I)=-W*DARK(IK)
355          GO TO 803
356      806 STEK(I)=0.1
357      803 CONTINUE
358      810 DO 811 I=1,N
359          IK=IVAR(I)
360          DO 811 J=1,N
361          JK=IVAR(J)
362      811 S(I,J)=SK(IK,JK)
363          IF(NORV+3)813,812,813
364      812 W=.1/SQRT(U/UMIN-1.)
365          WRITE(IOU,8001)
366          DO 814 I=1,N
367          IK=IVAR(I)
368          IF(AMIN(I)-XK(IK))815,814,814
369      815 AM=AMIN(I)
370          XK(IK)=AM
371          AC(I)=AM
372          AV(I)=AM
373          AMAX=0.
374          DO 816 J=1,N
375          IF(SLASK-AMAX)816,816,817
376      817 AMAX=SLASK
377          JMAX=J
378      816 CONTINUE
379          STEK(JMAX)=STEK(JMAX)*W
380      814 CONTINUE
381      813 CONTINUE
382          DO 818 I=1,N
383          IK=IVAR(I)
384          IF(IPOSK(IK))818,818,819
385      819 AMAX=0.
386          DO 820 J=1,N
387      820 AMAX=AMAX+ABS(S(I,J)*STEK(J))
388          IF(AMAX-XK(IK))818,818,821
389      821 XK(IK)=AMAX
390          AV(I)=AMAX
391          AC(I)=AMAX
392          IF(NORV-1)818,822,818
393      822 NORV=-2
394      818 CONTINUE
395          NGE=N
396          DO 823 I=1,N
397          IVARG(I)=IVAR(I)
398      823 STEKG(I)=STEK(I)
399          IF(NORV+3)100,825,100
400      825 NORV=-2
401          GO TO 1200
402      840 DO 841 I=1,N
403          IK=IVAR(I)
404          AV(I)=XK(IK)
405          AC(I)=AV(I)
406          STEK(I)=STEKG(I)
407          IF(DARK(IK))841,841,842
408      842 STEK(I)=W0*DARK(IK)
409      841 CONTINUE
410          GO TO 810
411  C
412  C        VAND
413  C
414      900 DO 901 I=1,IRV
415          DO 901 J=1,IRV
416      901 RUTIN(I,J)=RUTA(I,J)
417          CALL INVER(IRV,RUTIN,1.0E-35,DET,ING)
```

```
418            IF(ISSW(3)) 930,930,920
419       920 WRITE(IOU,9001)DET
420       930 IF(ING)902,902,903
421       903 IRUR=9
422            GO TO 611
423       902 IF(IRV-N)1000,200,1000
424     C
425     C VRID
426     C
427      1000 IF(N-IRV)1001,1030,1001
428      1001 DO 1002 I=1,IRV
429            W=0.
430            DO 1003 M=1,IRV
431      1003 W=W-RUTIN(I,M)*RUTA(M,IRV+1)
432      1002 VRI(I,IRV+1)=W
433      1010 IRV=IRV-1
434            IF(IRV-1)900,1020,900
435      1030 DO 1031 I=1,N
436            DO 1032 J=1,N
437      1032 VRI(I,J)=0.
438      1031 VRI(I,I)=1.
439            GO TO 1010
440      1020 VRI(1,2)=-RUTA(1,2)/RUTA(1,1)
441            DO 1021 I=1,N
442            DO 1021 J=1,N
443      1021 RUT1(I,J)=VRI(I,J)/SH(J,J)
444     C
445     C     SIK
446     C
447            CALL MULLE(SH,RUT1,N,N,N,S,1)
448            N1=N-1
449            IF(ISSW(2))1111,1111,1100
450      1100 WRITE(IOU,1101)
451      1111 DO 1104 I=1,N1
452            I1=I+1
453            IK=IVAR(I)
454            DO 1102 J=I1,N
455            JK=IVAR(J)
456            SK(IK,JK)=S(I,J)
457            IF(ISSW(2)) 1102,1102,1105
458      1105 WRITE(IOU,1103)IK,JK,SK(IK,JK)
459      1102 CONTINUE
460      1104 CONTINUE
461            GO TO 500
462     C
463     C     UBBE
464     C
465      1200 IF(NORV)1201,1202,1201
466      1201 IF(NORV+2)1203,1202,1203
467      1202 NORV=NORV+1
468      1203 IRUR1=IRUR+1
469            CALL UBBE(U,NK,XK,NB)
470            IF(ISSW(1).EQ.1.OR.IRUR.EQ.2)WRITE(IOU,1204)U,(XK(I),I=1,NK)
471      1206 GO TO (600,100,100,300,300,300),IRUR1
472        10 FORMAT(I2)
473      1101 FORMAT('   SIK(TWIST MATRIX)')
474      1103 FORMAT(2I3,2X,1PE12.5)
475      1204 FORMAT(' U=',1PE12.5,' PARAM.:  ',8(1PE12.5)/)
476      1602 FORMAT(/10X,' RURIK=',I2,/,11X,8(1H*))
477      1603 FORMAT(' ***** U T T A G  (ERROR-SQUARE-SUM) *****',/1X,41(1H*))
478      1604 FORMAT(' ***** S K O T T  (SHOT) *****',/,1X,29(1H*))
479      2001 FORMAT('   MINUSGROP (MINUS PIT)')
480      2002 FORMAT('   SIGY (STANDARD DEVIATION IN Y):',1PE12.4)
481      2003 FORMAT('   KBOM (PARAMETERS) NUMBER  VALUE      DARR1          DARR2
482           &')
483      2004 FORMAT(20X,I4,2X,3(1PE12.5))
484      4000 FORMAT(' ***** L A S K  (INITIAL GUESS) *****',/,1X,36(1H*))
485      4001 FORMAT(' NUMBER OF ESTIMATED PARAMETERS:',I5,' NUMBER OF POSITIVE
486           &PARAMETERS:',I5)
487      4002 FORMAT(' INITIAL GUESS OF PARAMETERS:',8(1PE10.3))
488      4003 FORMAT(' NUMBER OF MATRIX ELEMENTS:',I3,'   INDEX AND VALUE:')
489      4004 FORMAT(50X,2I5,F12.7)
490      4005 FORMAT(50X,I5,F12.7)
491      4006 FORMAT(' NUMBER OF NEGATIVE PARAMETERS TO BE ESTIMATED:',I5)
492      4007 FORMAT(5X,I5,' -TH PARAMETER')
493      5101 FORMAT('   MIKO(MINUS PARAMETER)')
```

```
494     5102 FORMAT('   DET=',1PE12.5)
495     6001 FORMAT('     PROVA(TESTING)')
496     6002 FORMAT('     GAMLA KONSTANTER(OLD PARAMETERS)')
497     6003 FORMAT('     SLUMPSKOTT(HIT BY ACCIDENT)')
498     8001 FORMAT('     KOMNER(COME DOWN)')
499     8002 FORMAT(' ***** P L U S K A  (TEST,CONTROL)   *****',/,1X,40(1H*))
500     8025 FORMAT(6X,I4,F12.5)
501     8026 FORMAT(' ***** S T E G  (STEPS OF PARAMETERS) *****',/,1X,42(1H*))
502     9001 FORMAT('   DET=',1PE12.5)
503          RETURN
504          END
```

```
1            SUBROUTINE WEIGHT( UBBE,NB,NK,XK,SIGXK,WEI,ISSW)
2     C
3     C      THE CALCULATION OF A STATISTICAL WEIGHT ACCORDING TO THE VALUE OF
4     C      PARAMETERS' WEIGHT
5     C
6            COMMON/FUNC/X(50),YEXP(50),DAF(50),YCAL(50),W(50)
7            DIMENSION SIGXK(NK),XX(10),XK(NK),WEI(NK)
8            DIMENSION S1(50),DIF(50),ISSW(7)
9            ISSW5=ISSW(5)
10           W1=NB
11           DO 10 I=1,NK
12     10    XX(I)=XK(I)
13           CALL UBBE (U,NK,XX,NB)
14           DO 20 I=1,NB
15           W(I)=0.0
16     20    S1(I)=YCAL(I)
17           DO 35 IK=1,NK
18           XX(IK)=XX(IK)+SIGXK(IK)
19           CALL UBBE (U,NK,XX,NB)
20           S=0.0
21           DO 25 I=1,NB
22           DIF(I)=(1./(ABS(YCAL(I)-S1(I))+1.E-10))+1.E-37
23           GO TO(24,25),ISSW5
24     24    DIF(I)=1./DIF(I)
25     25    S=S+DIF(I)
26           S=W1*WEI(IK)/S
27           DO 30 I=1,NB
28     30    W(I)=W(I)+DIF(I)*S
29     35    XX(IK)=XX(IK)-SIGXK(IK)
30           RETURN
31           END
```

```
1            SUBROUTINE INVER(N,A,EPS,DET,ING)
2     C
3     C--- EMINV, MATRIX INVERSION SUBROUTINE
4     C
5     C
6            REAL A(8,8),PIVOT(8)
7            INTEGER IPVOT(8),INDEX(8,2)
8            EQUIVALENCE(IROW,JROW),(ICOL,JCOL)
9            ING=0
10     57    DET=1.
11           DO 17 J=1,N
12     17    IPVOT(J)=0
13           DO 135 I=1,N
14           T=0.
15           DO 9 J=1,N
16           IF(IPVOT(J)-1)13,9,13
17     13    DO 23 K=1,N
18           IF(IPVOT(K)-1)43,23,81
19     43    IF( ABS(T)- ABS(A(J,K)))83,23,23
20     83    IROW=J
21           ICOL=K
22           T=A(J,K)
23     23    CONTINUE
24     9     CONTINUE
25           IPVOT(ICOL)=IPVOT(ICOL)+1
```

```
26              IF(IROW-ICOL)73,109,73
27        73 DET=-DET
28              DO 12 L=1,N
29              T=A(IROW,L)
30              A(IROW,L)=A(ICOL,L)
31        12 A(ICOL,L)=T
32       109 INDEX(I,1)=IROW
33              INDEX(I,2)=ICOL
34              PIVOT(I)=A(ICOL,ICOL)
35              DET=DET*PIVOT(I)
36              A(ICOL,ICOL)=1.
37              IF(ABS(PIVOT(I)).GT.EPS)GO TO 206
38              ING=1
39              RETURN
40       206 DO 205 L=1,N
41       205 A(ICOL,L)=A(ICOL,L)/PIVOT(I)
42              DO 135 LI=1,N
43              IF(LI-ICOL)21,135,21
44        21 T=A(LI,ICOL)
45              A(LI,ICOL)=0.
46              DO 89 L=1,N
47        89 A(LI,L)=A(LI,L)-A(ICOL,L)*T
48       135 CONTINUE
49       222 DO 3 I=1,N
50              L=N-I+1
51              IF(INDEX(L,1)-INDEX(L,2))19,3,19
52        19 JROW=INDEX(L,1)
53              JCOL=INDEX(L,2)
54              DO 549 K=1,N
55              T=A(K,JROW)
56              A(K,JROW)=A(K,JCOL)
57              A(K,JCOL)=T
58       549 CONTINUE
59         3 CONTINUE
60        81 RETURN
61              END
```

```
1              SUBROUTINE MULLE(RAT1,RAT2,NRAD,NMEL,NKOL,RAT3,IFR)
2     C
3     C
4              DIMENSION RAT1(8,8),RAT2(8,8),RAT3(8,8)
5              DO 10 I=1,NRAD
6              DO 10 J=1,NKOL
7              W=0.
8              DO 20 M=1,NMEL
9              IF(IFR-1)30,40,30
10       40 W=W+RAT1(I,M)*RAT2(M,J)
11              GO TO 20
12       30 W=W+RAT1(M,I)*RAT2(M,J)
13       20 CONTINUE
14       10 RAT3(I,J)=W
15              RETURN
16              END
```

```
1              SUBROUTINE PINUS(PIN1,RAT,N,PIN2,IRAM)
2     C
3     C
4              DIMENSION RAT(8,8),PIN1(8),PIN2(8)
5              DO 10 I=1,N
6              W=0.
7              DO 20 J=1,N
8              IF(IRAM-1)40,30,40
9        30 W=W+PIN1(J)*RAT(J,I)
10              GO TO 20
11       40 W=W+RAT(I,J)*PIN1(J)
12       20 CONTINUE
13       10 PIN2(I)=W
14              RETURN
15              END
```

```
1           SUBROUTINE READR(R1,J)
2     C     FORMAT-FREE READING OF REALS
3           DIMENSION R1(1),KNUM(10)
4           LOGICAL TECKA,NUM
5           COMMON /KANAL/NI/READ/IST(80),NS,IA,NA,IHV,RA
6           DATA KZN,KHV,KT,KE,KNUM/1H-,1H*,1H.,1HE,1H0,1H1,1H2,1H3,1H4,1H5,
7          &1H6,1H7,1H8,1H9/
8           EQUIVALENCE(NUL,KNUM(1)),(IDEV,KNUM(10))
9           NUM(I)=I.LT.NUL.OR.I.GT.IDEV
10          N=IABS(J)
11          IF(J.GT.0 .AND.NS.LE.80)GO TO 10
12          READ(NI,101,END=170)IST
13      101 FORMAT(80A1)
14          IHV=0
15          NA=0
16          NS=0
17       10 DO 165 K=1,N
18          IF(IHV.NE.0)GO TO 150
19          TECKA=.FALSE.
20          IEZ=1
21          IE=0
22          NZN=1
23          NZ=0
24       20 NS=NS+1
25          IF(NS.LE.80)GO TO 30
26          READ(NI,101,END=170)IST
27          NS=1
28       30 IS=IST(NS)
29          IF(IS.NE.KZN)GO TO 40
30          NZN=-1
31          GO TO 20
32       40 IF(IS.EQ.KT) TECKA=.TRUE.
33          IF(NUM(IS))GO TO 20
34          L=0
35       50 L=L+1
36          IF(KNUM(L).NE.IS)GO TO 50
37          I=L-1
38          IF(TECKA)NZ=NZ+1
39       60 NS=NS+1
40          IF(NS.GT.80)GO TO 90
41          IS=IST(NS)
42          IF(IS.NE.KT)GO TO 70
43          TECKA=.TRUE.
44          GO TO 60
45       70 IF(NUM(IS))GO TO 90
46          L=0
47       80 L=L+1
48          IF(KNUM(L).NE.IS)GO TO 80
49          I=I*10+L-1
50          IF(TECKA)NZ=NZ+1
51          GO TO 60
52       90 IF(IS.NE.KE)GO TO 130
53          NS=NS+1
54          IS=IST(NS)
55          IF(IS.NE.KZN)GO TO 110
56          IEZ=-1
57      100 NS=NS+1
58          IS=IST(NS)
59      110 IF(NUM(IS))GO TO 130
60          L=0
61      120 L=L+1
62          IF(KNUM(L).NE.IS)GO TO 120
63          IE=IE*10+L-1
64          GO TO 100
65      130 RA    =FLOAT(I*NZN)*10.**(IE*IEZ-NZ)
66          IF(IHV.NE.0)GO TO 150
67          IF(IS.NE.KHV)GO TO 160
68          IHV=1
69          NA=I
70          IS=IST(NS+1)
71          IF(IS.NE.KZN.AND.IS.NE.KT.AND.NUM(IS)) GOTO 150
72          IHV=2
73          GO TO 20
74      150 NA=NA-1
75      160 IF(IHV.NE.1)R1(K)=RA
76      165 IF(NA.LE.0)IHV=0
```

```
77              RETURN
78        170   WRITE(6,102)NI
79        102   FORMAT(1X,'  DATA ARE MISSING  ',I6)
80              RETURN
81              END
```

```
1               SUBROUTINE READI(I1,J)
2          C    FORMAT-FREE READING OF INTEGERS
3               DIMENSION I1(1),KNUM(10)
4               LOGICAL NUM
5               COMMON /KANAL/NI/READ/IST(80),NS,IA,NA,IHV,RA
6               DATA KZN,KHV,KNUM/1H-,1H*,1H0,1H1,1H2,1H3,1H4,1H5,1H6,1H7,1H8,1H9/
7               EQUIVALENCE (NUL,KNUM(1)),(IDEV,KNUM(10))
8               NUM(I)=I.LT.NUL.OR.I.GT.IDEV
9               N=IABS(J)
10              IF(J.GT.0 .AND.NS.LE.80)GO TO 10
11              READ(NI,101,END=120)IST
12              NS=0
13        101   FORMAT(80A1)
14              IHV=0
15              NA=0
16        10    DO 115 K=1,N
17              IF(IHV.NE.0)GO TO 100
18              IZN=1
19        20    NS=NS+1
20              IF(NS.LE.80)GO TO 30
21              READ(NI,101,END=120)IST
22              NS=1
23        30    IS=IST(NS)
24              IF(IS.NE.KZN)GO TO 40
25              IZN=-1
26              GO TO 20
27        40    IF(NUM(IS))GO TO 20
28              L=0
29        50    L=L+1
30              IF(KNUM(L).NE.IS)GO TO 50
31              I=L-1
32        60    NS=NS+1
33              IF(NS.GT.80)GO TO 80
34              IS=IST(NS)
35              IF(NUM(IS))GO TO 80
36              L=0
37        70    L=L+1
38              IF(KNUM(L).NE.IS)GO TO 70
39              I=I*10+L-1
40              GO TO 60
41        80    IA=IZN*I
42              IF(IHV.NE.0)GO TO 100
43              IF(IS.NE.KHV)GO TO 110
44              IHV=1
45              NA=I
46              IS=IST(NS+1)
47              IF(IS.NE.KZN.AND.NUM(IS))GO TO 100
48              IHV=2
49              GO TO 20
50        100   NA=NA-1
51        110   IF(IHV.NE.1)I1(K)=IA
52        115   IF(NA.LE.0)IHV=0
53              RETURN
54        120   WRITE(6,102) NI
55        102   FORMAT(1X,'  DATA ARE MISSING  ',I6)
56              RETURN
57              END
```

```
1               SUBROUTINE NORAND(D1,D2,IS)
2          C    GENERATION OF RANDOM NUMBERS
3         10    X=RANDOL(IS)
4               Y=2.0*RANDOL(IS)-1.0
5               XX=X*X
```

```
 6              YY=Y*Y
 7              S=XX+YY
 8              IF(S-1.0) 20,20,10
 9           20 A=SQRT(-2.0*ALOG(RANDOL(IS)))/S
10              D1=(XX-YY)*A
11              D2=2.0*X*Y*A
12              RETURN
13              END
```

```
 1              FUNCTION RANDOL(IS)
 2              K=67108864
 3              IS= MOD(5*MOD(25*MOD(IS,K),K),K)
 4              RANDOL=FLOAT(IS)/FLOAT(K)
 5              RETURN
 6              END
```

```
 1              SUBROUTINE SIMUL(X,YC,YM,NB,SIG,ISTART)
 2      C
 3      C   LOADING ACCURATE VALUE OF THE Y BY SIMULATED ERROR
 4      C   (NORMAL GAUSSIAN DISTRIBUTION, STANDARD DEVIATION EQUALS TO SINST)
 5      C
 6              DIMENSION X(NB),YC(NB),YM(NB)
 7              YC(NB+1)=1.
 8              DO 10 I=1,NB,2
 9              CALL NORAND(D1,D2,ISTART)
10              YM(I)=YC(I)+D1*SIG
11           10 YM(I+1)=YC(I+1)+D2*SIG
12              RETURN
13              END
```

```
 1              SUBROUTINE PRTPLT(X,Y,NPTSET,N,NPAGE,NDMAX)
 2      C
 3      C       SUBROUTINE PRTPLT PRODUCES A PRINTER PLOT OF USER CONTROLLED
 4      C       SIZE FOR DATA THAT ARE NOT EVENLY SPACED IN THE X OR Y DIRECTION.
 5      C
 6              DIMENSION ICHAR(100),X(NDMAX,2),XLAB(100),YLAB(100),Y(NDMAX,2),
 7             &LINE(10),NSAVE(5),LINE1(10),LINE2(10),NPTSET(2),XMN(5),
 8             &YMN(5),XMX(5),YMX(5)
 9              INTEGER CHAR(9),VERT,S
10              DATA LINE1/'I',9*'-'/,MINUS/'-'/,LINE2/'I',9*' '/,VERT/'I'/,
11             &CHAR/'1','2','3','4','5','6','7','8','9'/,IBLAN/' '/,S/'S'/
12          602 FORMAT(3X,11F10.2)
13          608 FORMAT(2X,A1,6X,102A1,6X,A1)
14          609 FORMAT(2X,A1,F6.2,102A1,F6.2,A1)
15      C
16      C       LOCATE THE SINGLE LARGEST AND SMALLEST X AND Y VALUES
17      C       FOR ALL DATA SETS TO BE PROCESSED.
18      C
19              DO 100 J=1,N
20              NPTS=NPTSET(J)
21              JPT=J
22              CALL SIZE(X,XMAX,XMIN,NPTS,JPT,NDMAX)
23              CALL SIZE(Y,YMAX,YMIN,NPTS,JPT,NDMAX)
24              XMX(J)=XMAX
25              XMN(J)=XMIN
26              YMX(J)=YMAX
27              YMN(J)=YMIN
28          100 CONTINUE
29              XMAX=XMX(1)
30              XMIN=XMN(1)
31              YMAX=YMX(1)
32              YMIN=YMN(1)
33              DO 102 J=2,N
34              IF(XMX(J).GT.XMAX)XMAX=XMX(J)
35              IF(XMN(J).LT.XMIN)XMIN=XMN(J)
```

```
 36              IF(YMX(J).GT.YMAX)YMAX=YMX(J)
 37              IF(YMN(J).LT.YMIN)YMIN=YMN(J)
 38        102 CONTINUE
 39    C
 40    C        READJUST THE ABSOLUTE MAXIMUM AND MINIMUM VALUES OF
 41    C        OF THE COMPLETE DATA SET TO BE PLOTTED SO AS TO GIVE
 42    C        PORTIONS OF THE PLOT ON EITHER SIDE OF THE START AND
 43    C        FINISH OF THE EXPERIMENTAL DATA.
 44    C
 45              DIFF=(YMAX-YMIN)*0.05
 46              YMAX=YMAX+DIFF
 47              YMIN=YMIN-DIFF
 48              DIFF=(XMAX-XMIN)*0.05
 49              XMAX=XMAX+DIFF
 50              XMIN=XMIN-DIFF
 51    C
 52    C        SET UP PARAMETERS DEPENDING ON THE PAGE SIZE NEEDED.
 53    C
 54              IPAGE=NPAGE
 55              IF(NPAGE.EQ.0)GO TO 400
 56              NX=100
 57              NY=61
 58              GO TO 401
 59        400 NX=68
 60              NY=41
 61              NPAGE=0
 62        401 CALL LABEL(XMAX,XMIN,XFACT,NX,XLAB,10,NN)
 63              CALL LABEL(YMAX,YMIN,YFACT,NY,YLAB,5,NN)
 64              NY=NY-1
 65    C
 66    C        RELOAD THE Y AXIS TO READ FROM THE BOTTOM UP
 67    C
 68              NND2=NN/2
 69              DO 402 J=1,NND2
 70              IBOT=NN+1-J
 71              YSAVE=YLAB(J)
 72              YLAB(J)=YLAB(IBOT)
 73        402 YLAB(IBOT)=YSAVE
 74    C
 75    C        DUMP THE PLOT LINE-BY-LINE
 76    C
 77              MJ1=0
 78              MJ2=0
 79              MJ3=1
 80              MB=1
 81              ME=11
 82              IF(NX.LT.100)ME=7
 83              DO 300 KJ=1,NPAGE
 84              WRITE(6,602)(XLAB(I),I=MB,ME)
 85    C
 86    C        SET UP THE GRID FOR THE PLOTTING SURFACE.
 87    C
 88              J1=0
 89        212 DO 211 JJ=1,10
 90              J1=J1+1
 91        211 ICHAR(J1)=LINE1(JJ)
 92              IF(J1.LT.NX)GO TO 212
 93              ICHAR(NX)=VERT
 94              IF(NX.LT.100)GO TO 319
 95              WRITE(6,609)VERT,YLAB(MJ3),VERT,(ICHAR(J),J=1,NX),VERT,
 96             &YLAB(MJ3),VERT
 97              GO TO 312
 98        319 WRITE(6,609)VERT,YLAB(MJ3),VERT,(ICHAR(J),J=1,NX)
 99        312 DO 302 K=1,NY
100              DO 203 J=1,10
101        203 LINE(J)=LINE1(J)
102              KK=K-K/5*5
103              IF(KK.EQ.0)GO TO 205
104              DO 204 J=1,10
105        204 LINE(J)=LINE2(J)
106        205 J1=0
107        202 DO 201 JJ=1,10
108              J1=J1+1
109        201 ICHAR(J1)=LINE(JJ)
110              IF(J1.LT.NX)GO TO 202
111              ICHAR(NX)=VERT
```

```
112    C
113    C          START PLOTTING THE DATA.
114    C
115               DO 110 J=1,N
116               NPTS=NPTSET(J)
117               DO 110 I=1,NPTS
118               XII=X(I,J)
119               YY=Y(I,J)
120               IF(YY.LT.-1.0E35)GO TO 110
121               IY=IFIX((YY-YMIN)/YFACT)
122               KTEST=NY-K+1
123               IF(IY.NE.KTEST)GO TO 110
124               IX=IFIX((XII-XMIN)/XFACT)
125               IF(IX.EQ.0)IX=1
126               L=ICHAR(IX)
127               IF(L.EQ.S)GO TO 126
128               IF(L.EQ.IBLAN.OR.L.EQ.VERT.OR.L.EQ.MINUS)GO TO 127
129        128 ICHAR(IX)=S
130               GO TO 126
131        127 ICHAR(IX)=CHAR(J)
132        126 Y(I,J)=-1.0E35
133        110 CONTINUE
134    C
135    C          SET UP TO PRINT THE PLOT AS DICTATED BY
136    C          THE PAGE PARAMETER, NPAGE.
137    C
138        301 IF(KK.EQ.0)GO TO 308
139               IF(NX.LT.100)GO TO 320
140               WRITE(6,608)VERT,VERT,(ICHAR(J),J=1,NX),VERT,VERT
141               GO TO 302
142        320 WRITE(6,608)VERT,VERT,(ICHAR(J),J=1,NX)
143               GO TO 302
144        308 MJ3=MJ3+1
145               IF(NX.LT.100)GO TO 309
146               WRITE(6,609)VERT,YLAB(MJ3),VERT,(ICHAR(J),J=1,NX),VERT,
147             &YLAB(MJ3),VERT
148               GO TO 302
149        309 WRITE(6,609)VERT,YLAB(MJ3),VERT,(ICHAR(J),J=1,NX)
150        302 CONTINUE
151               WRITE(6,602)(XLAB(I),I=MB,ME)
152               MB=ME
153               ME=ME+ME-1
154        300 CONTINUE
155               NPAGE=IPAGE
156               RETURN
157               END
```

```
  1               SUBROUTINE SIZE(DATA,DMAX,DMIN,NPTS,NSET,NDMAX)
  2    C
  3    C          THIS ROUTINE FINDS THE MAXIMUM AND MINIMUM ELEMENT OF THE
  4    C          NSET VECTOR OF THE ARRAY DATA.  THE VALUES ARE RETURNED IN
  5    C          DMAX AND DMIN.
  6    C
  7               DIMENSION DATA(NDMAX,2)
  8               DMIN=DATA(1,NSET)
  9               DMAX=DATA(1,NSET)
 10               DO 100 I=2,NPTS
 11               D=DATA(I,NSET)
 12               IF(D.GT.DMAX)DMAX=D
 13               IF(D.LT.DMIN)DMIN=D
 14        100 CONTINUE
 15        103 RETURN
 16               END
```

```
  1               SUBROUTINE LABEL(AMAX,AMIN,AFACT,NA,ALAB,LINES,NN)
  2    C
  3    C          THIS ROUTINE DETERMINES THE SCALING FACTOR FOR THE X- AND
  4    C          Y-AXES.  LABELS FOR EACH AXIS ARE LOADED INTO THE ALAB VECTOR.
  5    C
```

```
 6           DIMENSION ALAB(100)
 7           AFACT=ABS(AMAX-AMIN)/FLOAT(NA-1)
 8           AST=AMIN
 9           NN=NA/LINES
10           ALINE=LINES*AFACT
11           NN=NN+1
12           DO 200 I=1,NN
13           ALAB(I)=AST
14      200  AST=AST+ALINE
15           RETURN
16           END
```

```
 1           SUBROUTINE PLOTT(XX,YY,NDATA,NDMAX,ISYMBL,NF,XLINE,MX,YLINE,MY,
 2          &NLS,NCL,MM,LL,AREA,YSCALE)
 3    C
 4    C        * * ADD * * DIMENSION STATEMENT IN CALLING ROUTINE * *
 5    C
 6    C        DIMENSION XX(NDMAX,EXP1),YY(NDMAX,EXP1),NDATA(EXP1),XLINE(EXP2),
 7    C       1YLINE(EXP3),AREA(13,EXP4),ISYMBL(EXP1),YSCALE(EXP4)
 8    C        EXP1    EQUAL TO OR GREATER THAN  NF
 9    C        EXP2    EQUAL TO OR GREATER THAN  MX
10    C        EXP3    EQUAL TO OR GREATER THAN  MY
11    C        EXP4    EQUAL TO OR GREATER THAN  NLS
12    C
13    C        FIELD DEFINITIONS  ( * MUST BE DEFINED IN CALLING ROUTINE)
14    C
15    C  * XX(J,I) AND YY(J,I) ARE THE COORDINATES OF THE JTH POINT OF THE
16    C        ITH ARRAY TO BE PLOTTED.
17    C  * NDATA(J) IS THE NUMBER OF POINTS IN THE JTH ARRAY.
18    C  * NDMAX  IS THE MAXIMUM NUMBER OF ELEMENTS WHICH CAN BE PLOTTED IN
19    C        ANY ONE ARRAY AND IS EQUAL TO, J, IN THE DIMENSION STATEMENT
20    C        FOR XX(J,I) AND YY(J,I) IN THE ROUTINE WHICH CALLS PLOTT.
21           DIMENSION AMASK(4),AREA(31,1),BORDER(31),NDATA(1),ISYMBL(1),
22          &XLINE(1),XSCALE(31),XX(NDMAX,1),YLINE(1),YSCALE(1),YY(NDMAX,1),
23          &ZZ(2),JBRDR(5),BMASK(4)
24           EQUIVALENCE (NTEST,ITEST),(NTUBE,ITUBE)
25    C    ISYMBL(J) IS THE SYMBOL FOR THE JTH ARRAY  (EG  ISYMBL(1)=4H****)
26    C    NF  IS THE NUMBER OF ARRAYS TO BE PLOTTED.
27    C    XLINE(I) IS THE VALUE FOR THE ITH X REFERENCE LINE.
28    C    MX  IS THE NUMBER OF X REFERENCE LINES  (EG  MX=0, IF NONE DESIRED)
29    C    YLINE(I) IS THE VALUE FOR THE ITH Y REFERENCE LINE.
30    C    MY  IS THE NUMBER OF Y REFERENCE LINES  (EG  MY=0, IF NONE DESIRED)
31    C    NLS  IS THE NUMBER OF LINES TO BE USED IN PLOT  (EG  NLS=55)
32    C    NCL  IS THE NUMBER OF COLUMNS TO BE USED IN PLOT  (EG  NCL=122)
33    C    MM=1  PLOTS ACROSS THE PAGE        MM=2  PLOTS DOWN THE PAGE
34    C    LL=1  LINEAR     LL=2  SEMI-LOG      LL=3  SQUARE ROOT
35    C    AREA(13,J) IS THE ARRAY WHICH SPANS THE PLOT SURFACE.
36    C    YSCALE(J) IS THE SCALE SCALAR RUNNING DOWN THE PAGE.
37    C    AREA AND YSCALE APPEAR IN THE CALL LIST IN ORDER THAT THEY CAN BE
38    C        DIMENSIONED EXTERNAL TO THE PLOTT SUBROUTINE IN ACCORDANCE WITH
39    C        THE NUMBER OF DESIRED LINES  (NLS)
40    C
41    C    JSIZE IS THE MAXIMUM HOLLERITH WORD SIZE = 4
42    C    AMASK(J) AND BMASK(J) ARE FOR MASKING CONVIENCE.
43    C    NCLMAX  IS THE MAXIMUM NUMBER OF COLUMNS OF PRINTER CAPACITY = 132,
44    C        MINUS THE NUMBER OF COLUMNS USED FOR YSCALE(J) = 14, NCLMAX = 118
45    C    JACRSS IS THE NUMBER OF WORDS NECESSARY TO SPAN A LINE = 31
46    C    AREA(J,NLS), XSCALE(J) AND BORDER(J) HAVE DIMENSION J = JACRSS
47    C    AMASK(J) MUST HAVE DIMENSION J EQUAL TO JSIZE.
48    C
49           LOGICAL AMASK,BMASK,AREA,BLANK,BORDER,DASH,EQUAL,SIDE,SYM,
50          &UPLINE,ITEST,ITUBE,ISYMBL,IYM,INF,IMX,IMY,INLS,INCL,IMM,
51          &ILL,ML,JBRDR
52           DATA AMASK(1)/ZFF000000/
53           DATA AMASK(2)/Z00FF0000/
54           DATA AMASK(3)/Z0000FF00/
55           DATA AMASK(4)/Z000000FF/
56           DATA BMASK(1)/Z00FFFFFF/
57           DATA BMASK(2)/ZFF00FFFF/
58           DATA BMASK(3)/ZFFFF00FF/
59           DATA BMASK(4)/ZFFFFFF00/
60           DATA JBRDR(1)/4HI---/
61           DATA JBRDR(2)/4H-I--/
```

```
 62            DATA JBRDR(3)/4H--I-/
 63            DATA JBRDR(4)/4H---I/
 64            DATA JBRDR(5)/4H----/
 65            DATA SIDE/4HHHHHH/
 66            DATA BLANK/4H    /
 67            DATA DASH/4H----/
 68            DATA UPLINE/4HIIII/
 69            DATA EQUAL/4H====/
 70            DATA IYM/4H****/
 71            DATA INF/2HNF/
 72            DATA IMX/2HMX/
 73            DATA IMY/2HMY/
 74            DATA INLS/3HNLS/
 75            DATA INCL/3HNCL/
 76            DATA IMM/2HMM/
 77            DATA ILL/2HLL/
 78            JSIZE=4
 79            NCLMAX=118
 80            JACRSS=NCLMAX/JSIZE+1
 81      C     THE FOLLOWING SEVEN RESTRICTIONS PREVENT SERIOUS PROGRAMMER ERRORS.
 82            IF(NF.GT.0 .AND. NF.LT.33)GO TO 1
 83            ML=INF
 84            LM=1
 85            ISYMBL(1)=IYM
 86            WRITE(6,82)ML,LM,XX(1,1),YY(1,1),NDATA(1),NF,MX,MY,
 87           &NLS,NCL,MM,LL
 88            NF=LM
 89      1     IF(MX.GT.-1 .AND. MX.LT.33)GO TO 2
 90            ML=IMX
 91            LM=0
 92            WRITE(6,82)ML,LM,XX(1,1),YY(1,1),NDATA(1),NF,MX,MY,
 93           &NLS,NCL,MM,LL
 94            MX=LM
 95      2     IF(MY.GT.-1 .AND. MY.LT.33)GO TO 3
 96            ML=IMY
 97            LM=0
 98            WRITE(6,82)ML,LM,XX(1,1),YY(1,1),NDATA(1),NF,MX,MY,
 99           &NLS,NCL,MM,LL
100            MY=LM
101      3     IF(NLS.GT.0)GO TO 4
102            ML=INLS
103            LM=56
104            WRITE(6,82ML,LM,XX(1,1),YY(1,1),NDATA(1),NF,MX,MY,
105           &NLS,NCL,MM,LL
106            NLS=LM
107      4     IF(NCL.GT.0 .AND. NCL.LT.NCLMAX+1)GO TO 5
108            ML=INCL
109            LM=NCLMAX
110            WRITE(6,82)ML,LM,XX(1,1),YY(1,1),NDATA(1),NF,MX,MY,
111           &NLS,NCL,MM,LL
112            NCL=LM
113      5     IF(MM.GT.0 .AND. MM.LT.3)GO TO 6
114            ML=IMM
115            LM=1
116            WRITE(6,82)ML,LM,XX(1,1),YY(1,1),NDATA(1),NF,MX,MY,
117           &NLS,NCL,MM,LL
118            MM=LM
119      6     IF(LL.GT.0 .AND. LL.LT.4)GO TO 7
120            ML=ILL
121            LM=1
122            WRIT (6,82)ML,LM,XX(1,1),YY(1,1),NDATA(1),NF,MX,MY,
123           &NLS,NCL,MM,LL
124            LL=LM
125      7     CONTINUE
126      C     THE ABOVE SEVEN RESTRICTIONS ARE SUPERFLUOUS AND MAY BE OMITTED.
127            DO 9 I=1,JACRSS,5
128            BORDER(I  )=JBRDR(1)
129            BORDER(I+1)=JBRDR(2)
130            BORDER(I+2)=JBRDR(3)
131            BORDER(I+3)=JBRDR(4)
132      9     BORDER(I+4)=JBRDR(5)
133            DO 10 I=1,JACRSS
134            DO 10 J=1,NLS
135     10     AREA(I,J)=BLANK
136            NCL1=NCL+1
137            NCLP=NCL1/10
```

```
138              JWD=NCL1/JSIZE+1
139              JWP=JWD+1
140              JPS= MOD (NCL1,JSIZE)+1
141              IF (JWP .GT. JACRSS) GO TO 12
142              DO 11 I=JWP,JACRSS
143       11     BORDER(I)=BLANK
144       12     DO 13 I=JPS,JSIZE
145       13     BORDER(JWD)=BORDER(JWD).AND.BMASK(I) .OR. BLANK.AND.AMASK(I)
146              GO TO (14,15),MM
147       14     LX=1
148              LY=2
149              NX=MX
150              NY=MY
151              GO TO 16
152       15     LX=2
153              LY=1
154              NX=MY
155              NY=MX
156       16     IF (LL.EQ.1) GO TO 24
157              LLL=1
158              DO 19 I=1,NF
159              NDF=NDATA(I)
160              DO 19 J=1,NDF
161              YYY=YY(J,I)
162              IF (YYY) 18,17,19
163       17     IF (LL.EQ.1 .OR. LL.EQ.3)GO TO 19
164       18     WRITE(6,83)J,I,YYY
165              LLL=2
166       19     CONTINUE
167              IF (NY.LE.0)GO TO 23
168              DO 22 I=1,NY
169              YLINEG=YLINE(I)
170              IF(YLINEG)21,20,22
171       20     IF (LL.EQ.1 .OR. LL.EQ.3)GO TO 22
172       21     WRITE(6,84)I,YLINEG
173              LLL=2
174       22     CONTINUE
175       23     IF(LLL.EQ.1)GO TO 24
176              WRITE(6,85)
177              WRITE(6,86)
178              LL=1
179       24     ZZ(LX)=XX(1,1)
180              ZZ(LY)=YY(1,1)
181              XMAX=ZZ(1)
182              XMIN=XMAX
183              YMAX=ZZ(2)
184              YMIN=YMAX
185              DO 25 I=1,NF
186              NDF=NDATA(I)
187              DO 25 J=1,NDF
188              ZZ(LX)=XX(J,I)
189              ZZ(LY)=YY(J,I)
190              XXX=ZZ(1)
191              YYY=ZZ(2)
192              XMAX=AMAX1(XMAX,XXX)
193              XMIN=AMIN1(XMIN,XXX)
194              YMAX=AMAX1(YMAX,YYY)
195       25     YMIN=AMIN1(YMIN,YYY)
196              IF(NX.LE.0)GO TO 27
197              DO 26 J=1,NX
198              ZZ(LX)=XLINE(J)
199              ZZ(LY)=YLINE(J)
200              XLINEG=ZZ(1)
201              XMAX=AMAX1(XMAX,XLINEG)
202       26     XMIN=AMIN1(XMIN,XLINEG)
203       27     IF(NY.LE.0)GO TO 29
204              DO 28 J=1,NY
205              ZZ(LX)=XLINE(J)
206              ZZ(LY)=YLINE(J)
207              YLINEG=ZZ(2)
208              YMAX=AMAX1(YMAX,YLINEG)
209       28     YMIN=AMIN1(YMIN,YLINEG)
210       29     GO TO(34,30,32),LL
211       30     IF(MM.EQ.2)GO TO 31
212              YMIN=ALOG(YMIN)
213              YMAX=ALOG(YMAX)
```

```
214            GO TO 34
215   31       XMIN=ALOG(XMIN)
216            XMAX=ALOG(XMAX)
217            GO TO 34
218   32       IF(MM.EQ.2)GO TO 33
219            YMIN=SQRT(YMIN)
220            YMAX=SQRT(YMAX)
221            GO TO 34
222   33       XMIN=SQRT(XMIN)
223            XMAX=SQRT(XMAX)
224   34       XDIFX=XMAX-XMIN
225            YDIFY=YMAX-YMIN
226            IF(XDIFX.NE.0.)GO TO 35
227            XMIN=XMIN-.5*ABS(XMIN)-1.0E-77
228            XMAX=XMAX+.5*ABS(XMAX)+1.0E-77
229            XDIFX=XMAX-XMIN
230   35       IF(YDIFY.NE.0.)GO TO 36
231            YMIN=YMIN-.5*ABS(YMIN)-1.0E-77
232            YMAX=YMAX+.5*ABS(YMAX)+1.0E-77
233            YDIFY=YMAX-YMIN
234   36       XSC=(NCL-1.0)/XDIFX
235            YSC=(NLS-1.0)/YDIFY
236            IF(LL.EQ.1)GO TO 37
237            IF(MM.EQ.2)GO TO 39
238   37       IF(XMAX*XMIN.GE.0.0)GO TO 39
239            JX=-XMIN*XSC+JSIZE+1.5
240            JWORD=JX/JSIZE
241            JPOS=MOD(JX,JSIZE)+1
242            DO 38 J=1,NLS
243   38       AREA(JWORD,J)=AREA(JWORD,J).AND.BMASK(JPOS) .OR. UPLINE.AND.AMAS
244            &K(JPOS)
245   39       IF(LL.EQ.1)GO TO 40
246            IF(MM.EQ.1)GO TO 43
247   40       IF(YMAX*YMIN.GE.0.0)GO TO 43
248            JY=-YMIN*YSC+1.5
249            DO 41 I=1,JWD
250   41       AREA(I,JY)=DASH
251            DO 42 I=JPS,JSIZE
252   42       AREA(JWD,JY)=AREA(JWD,JY).AND.BMASK(I) .OR. BLANK.AND.AMASK(I)
253   43       IF(NX.LE.0)GO TO 48
254            DO 47 I=1,NX
255            ZZ(LX)=XLINE(I)
256            ZZ(LY)=YLINE(I)
257            XLINEG=ZZ(1)
258            GO TO(46,44,45),LL
259   44       IF(MM.EQ.1)GO TO 46
260            XLINEG=ALOG(XLINEG)
261            GO TO 46
262   45       IF(MM.EQ.1)GO TO 46
263            XLINEG=SQRT(XLINEG)
264   46       JX=(XLINEG-XMIN)*XSC+JSIZE+1.5
265            JWORD=JX/JSIZE
266            JPOS=MOD(JX,JSIZE)+1
267            DO 47 J=1,NLS
268   47       AREA(JWORD,J)=AREA(JWORD,J).AND.BMASK(JPOS) .OR. UPLINE.AND.AMAS
269            &K(JPOS)
270   48       IF(NY.LE.0)GO TO 54
271            DO 53 J=1,NY
272            ZZ(LX)=XLINE(J)
273            ZZ(LY)=YLINE(J)
274            YLINEG=ZZ(2)
275            GO TO(51,49,50),LL
276   49       IF(MM.EQ.2)GO TO 51
277            YLINEG=ALOG(YLINEG)
278            GO TO 51
279   50       IF(MM.EQ.2)GO TO 51
280            YLINEG=SQRT(YLINEG)
281   51       JY=(YLINEG-YMIN)*YSC+1.5
282            DO 52 I=1,JWD
283   52       AREA(I,JY)=DASH
284            DO 53 I=JPS,JSIZE
285   53       AREA(JWD,JY)=AREA(JWD,JY).AND.BMASK(I) .OR. BLANK.AND.AMASK(I)
286   54       DO 61 I=1,NF
287            NDF=NDATA(I)
288            DO 61 J=1,NDF
289            ZZ(LX)=XX(J,I)
```

```
290              ZZ(LY)=YY(J,I)
291              XXX=ZZ(1)
292              YYY=ZZ(2)
293              GO TO(59,55,57),LL
294      55      IF(MM.EQ.2)GO TO 56
295              YYY=ALOG(YYY)
296              GO TO 59
297      56      XXX=ALOG(XXX)
298              GO TO 59
299      57      IF(MM.EQ.2)GO TO 58
300              YYY=SQRT (YYY)
301              GO TO 59
302      58      XXX=SQRT (XXX)
303      59      JX=(XXX-XMIN)*XSC+JSIZE+1.5
304              JY=(YYY-YMIN)*YSC+1.5
305              JWORD=JX/JSIZE
306              JPOS=MOD(JX,JSIZE)+1
307              ITEST=AREA(JWORD,JY).AND.AMASK(JPOS)
308              ITUBE=EQUAL.AND.AMASK(JPOS)
309              IF(NTEST.EQ.NTUBE)GO TO 61
310              SYM=ISYMBL(I)
311              DO 60 II=1,NF
312              ITUBE=ISYMBL(II).AND.AMASK(JPOS)
313              IF(NTEST.EQ.NTUBE)SYM=EQUAL
314      60      CONTINUE
315              AREA(JWORD,JY)=AREA(JWORD,JY).AND.BMASK(JPOS) .OR. SYM.AND.AMASK
316             &(JPOS)
317      61      CONTINUE
318              DO 62 J=1,NLS
319              AREA(1,J)=AREA(1,J).AND.BMASK(1) .OR. SIDE.AND.AMASK(1)
320      62      AREA(JWD,J)=AREA(JWD,J).AND.BMASK(JPS) .OR. SIDE.AND.AMASK(JPS)
321              DO 63 J=1,NLS
322      63      YSCALE(J)=YMIN+(J-1.0)/YSC
323              DO 64 J=1,NCLP
324      64      XSCALE(J)=XMIN+10.0*(J-1.0)/XSC
325              GO TO(73,65,69),LL
326      65      IF(MM.EQ.2)GO TO 67
327              DO 66 J=1,NLS
328              SAVE=YSCALE(J)
329      66      YSCALE(J)=EXP(SAVE)
330              GO TO 73
331      67      DO 68 J=1,NCLP
332              SAVE=XSCALE(J)
333      68      XSCALE(J)=EXP(SAVE)
334              GO TO 73
335      69      IF(MM.EQ.2)GO TO 71
336              DO 70 J=1,NLS
337      70      YSCALE(J)=YSCALE(J)*YSCALE(J)
338              GO TO 73
339      71      DO 72 J=1,NCLP
340      72      XSCALE(J)=XSCALE(J)*XSCALE(J)
341      73      XUNIT=1.0/XSC
342              YUNIT=1.0/YSC
343              GO TO(74,76,75),LL
344      74      WRITE(6,87)XUNIT,YUNIT
345              GO TO 77
346      75      WRITE(6,88)XUNIT
347              GO TO 77
348      76      WRITE(6,89)XUNIT
349      77      IF(MM.EQ.2)GO TO 79
350              WRITE(6,90)(BORDER(I),I=1,JACRSS)
351              NLS1=NLS+1
352              DO 78 JC=1,NLS
353              J=NLS1-JC
354      78      WRITE(6,91)YSCALE(J),(AREA(I,J),I=1,JACRSS)
355              WRITE(6,90)(BORDER(I),I=1,JACRSS)
356              WRITE(6,92)(XSCALE(J),J=1,NCLP,2)
357              WRITE(6,93)(XSCALE(J),J=2,NCLP,2)
358              GO TO 81
359      79      WRITE(6,92)(XSCALE(J),J=1,NCLP,2)
360              WRITE(6,93)(XSCALE(J),J=2,NCLP,2)
361              WRITE(6,90)(BORDER(I),I=1,JACRSS)
362              DO 80 J=1,NLS
363      80      WRITE(6,91)YSCALE(J),(AREA(I,J),I=1,JACRSS)
364              WRITE(6,90)(BORDER(I),I=1,JACRSS)
365      81      WRITE(6,94)
```

```
366          RETURN
367    82    FORMAT(5X,'RESTRICTIVE STATEMENT LIMITATION FOR,',2X,A3,I4,9X,
368         &'XX(1,1)',E12.3,5X,'YY(1,1)',E12.3,5X,'NDATA(1)',I4/5X,'NF',I4,9X
369         &,'MX',I4,9X,'MY',I4,9X,'NLS',I4,9X,'NCL',I4,9X,'MM',I4,9X,'LL',I4)
370    83    FORMAT(2X,'Y(',I3,1H,,I2,') =',E16.9)
371    84    FORMAT(2X,'YLINE(',I2,') =',E16.9)
372    85    FORMAT2X,'SEMILOG PLOTT OR SQUARE ROOT PLOTT WILL NOT ACCEPT',
373         &' A NEGATIVE ARGUMENT, AND SEMILOG PLOTT WILL NOT ACCEPT A ',
374         &'ZERO ARGUMENT.')
375    86    FORMAT(2X,'ERRORS FOUND AND LISTED ABOVE.    (LL=1, LINEAR; ',
376         &'LL=2, SEMI-LOG;   LL=3, SQUARE ROOT)')
377    C     1H   INHIBITS LINE PRINTER FROM SKIPPING SPACES BETWEEN PAGES
378    87    FORMAT(2X,'Q= INDICATES MORE THAN ONE POINT IS PLOTTED IN THAT ',
379         &'SPACE.',5X,'HORIZONTAL UNIT =',E10.3,5X,'VERTICAL UNIT =',E10.3)
380    88    FORMAT(2X,'Q= INDICATES MORE THAN ONE POINT IS PLOTTED IN THAT ',
381         &'SPACE.',5X,'HORIZONTAL UNIT =',E10.3,8X,'SQUARE ROOT VERTICAL ',
382         &'SCALE')
383    89    FORMAT(2X,'Q= INDICATES MORE THAN ONE POINT IS PLOTTED IN THAT ',
384         &'SPACE.',5X,'HORIZONTAL UNIT =',E10.3,8X,'SEMILOG VERTICAL ',
385         &'SCALE')
386    90    FORMAT(1H,12X,29A4,A3)
387    91    FORMAT(1H,1PE10.3,1X,30A4)
388    92    FORMAT(1H,8X,F10.5,4X,1HI,5(5X,F10.5,4X,1HI))
389    93    FORMAT(1H,8X,6(4X,1HI,5X,F10.5))
390    94    FORMAT(1HR)
391          RETURN
392          END
```

2. MRLET: FORMATION CONSTANTS OF M_mL_n FROM MOLE RATIO DATA

2.1. Model Formulation

The mole ratio method is based on plotting the absorbance of solutions against their mole ratio C_M/C_L, where C_L is held constant. If the system forms a stable complex which does not show appreciable dissociation the plot gives a sharp break indicating the composition of the complex. If, however, a weak complex is formed a curved plot results and deductions made concerning the stoichiometry of the complex may be unreliable. For very weak complexes the stoichiometry cannot be determined graphically.

Consider a reaction between a nonabsorbing metal ion and a light absorbing ligand giving rise to the complex M_mL_n. For the reaction

$$mM + nL \rightleftharpoons M_mL_n$$

the conditional formation constant K'_{mn} is given by

$$K'_{mn} = [M_mL'_n]/[M]^m[L']^n \tag{5}$$

the mass balances for M and L are

$$C_M = [M] + m[M_mL'_n] = C_M^0 \cdot V_M/V_0 \tag{6}$$

$$C_L = [L'] + n[M_mL'_n] = C_L^* \cdot f_L \tag{7}$$

where C_M^0 is the concentration of the metal ion in the buret; V_M is the volume of the metal ion added to the initial volume V_0 in the cuvet. The apparent concentration of the ligand, C_L^*, allows for the possibility of an impurity, and f_L is an effective concentration factor. Let the mole ratio, q_M, be defined as

$$q_M = C_M/C_L = C_M^0 \cdot V_M/(C_L^* \cdot f_L \cdot V_0) \tag{8}$$

and the normalized absorbance, α, be defined as

$$\alpha = [A(1 + V_M/V_0) - A_0]/(A_{ext} - A_0) \tag{9}$$

where A_0 is the absorbance of the ligand solution for $V_M = 0$; A_{ext} is the absorbance for the complex $M_m L_n$ (i.e., when A for V_M^i equals A for V_M^{i-1}) and A is the observed absorbance for any V_M. For a constant ligand concentration, C_L, q_M may be expressed implicitly as a function of C_M:

$$q_M = \frac{m}{n} + \frac{1/m}{(n \cdot K'_{mn})^{1/m} \cdot C_L^{(m+n-1)/m} \cdot (1 - \alpha)^{n/m}} \tag{10}$$

This function describes the experimental dependence $A = f(q_M)$ or $Z = f(V_M)$ and contains the independent variable V_M and the dependent variable A and four unknown parameters to be estimated $(A_{ext}, f_L, K'_{mn}, n)$

The experimental reading error for V_M is usually quite small compared to the instrumental error for A. Since the explicit expression for the absorbance, $A = f(V_M)$, is not possible A_{calc} may be evaluated using the secant method due to Wegstein.[7] MRLET uses the least-squares procedure, described in Section 1, to minimize

$$U = \sum_{i=1}^{N} W_i (A_{calc}^i - A_{exp}^i)^2 = \text{minimum} \tag{11}$$

where W_i is a statistical weighting factor, taken as unity for experimental data and calculated by MRLET for simulation purposes.

2.2. Description of MRLET

MRLET is written for 50 absorbance values at a constant wavelength and will determine the best values for A_{ext}, f_L, log K'_{mn}, and n. As described in Section 1.3 the user supplies appropriate MAIN, SHRIK, DATA, and UBBE routines. Subroutine SKRIK performs most of the output functions and conducts a statistical analysis of the estimated parameter. Subroutine DATA reads in the experimental data comprising the mole ratio titration. Subroutine UBBE calculates NB dependent variables. The absorbance A_{calc}^i is obtained using Wegstein's algorithm, which is coded in RIWI. The algorithm determines the real root of the nonlinear equation (10) iteratively; EPSL is a user supplied convergence criterion.

2.3. Data Input Requirements

The general layout of the data cards has been described in Section 1.4. There are some requirements that are specific to MRLET.

1. Title Card: **TITLE**
   A suitable title (80 characters)
2. Key Card: **(ISSW(I),I=1, 7)**
   Simulated data use ISSW(7) = 1 or 3
   Experimental data use ISSW(7) = 2 or 4
   See Table 2 for other values.
3. Termination Card: **EPS,ITMAX,PSI,NK,MY,NLS,NCL,SINST**
   EPS: $\Big\}$ As Section 1.4.2, item 3.
   SINST:
4. Simulation Card: **(XK(I),I=1,NK),(SIGX(I),I=1,NK),(WEI(I),I=1,NK)**
   XK(I):
   SIGX(I): $\Big\}$ As Section 1.4.2, item 4.
   WEI(I):
5. Data Card: **IRUR,NBI,CMO,VO,AL,CL,CORM,ALFMIN,ALFMAX,**
   B,C,X2,X1,EPSL,(V(I),AEX(I),I=1,NBI)
   IRUR: Set equal to 6 for experimental data. *Omit* for simulation runs.
   NBI: Number of volume, absorbance pairs of data.
   CMO: Concentration of metal ion solution in buret (M).
   VO: Initial volume of solution to be titrated.
   AL: Absorbance of ligand solution before addition of metal.
   CL: Total concentration of ligand solution (M).
   CORM: Correction factor for buret reading:
   $\qquad$ True volume = Buret reading $*$ CORM
   ALFMIN: Lower limit for normalized absorbance.
   ALFMAX: Upper limit for normalized absorbance.
   B: Stoichiometric coefficient m in M_mL_n.
   C: Stoichiometric coefficient n.
   X2: Estimated value for the concentration factor f_L.
   X1: Estimated value for A_{exp}.
   EPSL: Convergence criterion for Wegstein's algorithm.
   V(I),AEX(I): Volume of added metal solution and measured absorbance, respectively, for the ith point. There will be NBI values.
   Note: If NBI is set to a negative value transmittances may be entered.
6. Initial Guess Card: **IRUR,NK,NBYK(XK(I),i=1,NBYK)**
   IRUR:
   $\Big\}$ As for Section 1.4.2, item 6a.
   XK(I):

Note: For MRLET
$$XK(1) = A_{ext}$$
$$XK(2) = f_L$$
$$XK(3) = \log K'_{mn}$$
$$XK(4) = n$$

7. Matrix Card: **ISKIN,((I(K),J(K),S(I(K),J(K)),K=1,ISKIN)**

   ISKIN: $\left.\begin{array}{l} \vdots \\ S(\ ): \end{array}\right\}$ As for Section 1.4.2, item 7.

8. Step Card: **IRUR,N,((I(K),STEK(I(K)),K=1,N)**

   IRUR: $\left.\begin{array}{l} \vdots \\ STEK(\ ): \end{array}\right\}$ As for Section 1.4.2, item 8.

9. Process Card: **IRUR,IRUR,...**

   IRUR: As for Section 1.4.2, item 9. For MRLET set to 2, 5, 13.

The above set of cards represent one data block; any number of blocks of cards for separate experiments may follow.

2.4. Presentation of MRLET

2.4.1. Listing of MRLET

There follows a listing of the MAIN program and subroutines SKRIK, UBBE, and DATA. Those routines listed in Section 1.5 are added to form the complete program MRLET.

```
1     *#RUN= #05;06
2     C
3     C      M R L E T    P R O G R A M
4     C      (MAIN PROGRAM)
5     C      (MILAN MELOUN:TALANTA (IN PRESS))
6     C
7     C
8            DIMENSION XK(4),SIGXK(4),WEI(4)
9            COMMON/FUNC/XEXP(50),YEXP(50),ERR(50),YCAL(50),W(50)/KANAL/INP
10           COMMON/TIT/ TITLE(20)/PLOT/ MY,MX,NLS,NCL,DUMMY(1930)/ISW/ISSW(7)
11           EQUIVALENCE (UMIN,PSI)
12           EXTERNAL DATA,UBBE,SKRIK
13        10 INP=5
14           ISSW(1)=0
15           ISSW(2)=0
16           ISSW(3)=0
17           ISSW(4)=0
18           ISSW(5)=1
19           ISSW(6)=0
20           EPS=1.0E-6
21           ITMAX=40
22           PSI=0.3
23           NK=4
24           MY=0
```

```
25          MX=0
26          NLS=50
27          NCL=100
28          SINST=-1.0
29          IOU=6
30          IR=1225
31          DO 60 I=1,50
32          W(I)= 1.0
33       60 CONTINUE
34          READ(INP,130,END=999)TITLE
35          WRITE(IOU,131)TITLE
36          CALL READI(ISSW,-7)
37          CALL READR(EPS,1)
38          CALL READI(ITMAX,1)
39          CALL READR(PSI,1)
40          CALL READI(NK,1)
41          CALL READI(MY,4)
42          CALL READR(SINST,1)
43          IF(SINST)30,30,20
44       20 CALL READR(XK,NK)
45          CALL READR(SIGXK,NK)
46          CALL READR(WEI,NK)
47          continue
48          CALL DATA(IOU,NB)
49          WRITE(IOU,100)SINST,(XK(I),SIGXK(I),WEI(I),I=1,NK)
50          DO 21 I=1,NB
51       21 W(I)=1.0
52          CALL UBBE(U,NK,XK,NB)
53          CALL SIMUL (XEXP,YCAL,YEXP,NB,SINST,IR)
54          DO 11  I=1,NB
55       11 ERR(I)=YEXP(I)-YCAL(I)
56          CALL WEIGHT(UBBE,NB,NK,XK,SIGXK,WEI,ISSW)
57          WRITE(IOU,110)
58          DO 12 I=1,NB
59       12 YCAL(I)=YEXP(I)-ERR(I)
60          WRITE(IOU,111)(I,XEXP(I),YCAL(I),ERR(I),YEXP(I),W(I),I=1,NB)
61          CALL STATS(IOU,NK,NB,XEXP,YEXP,ERR)
62          GO TO 330
63       30 WRITE(IOU,112)
64      330 U2=1.0E30
65          U1=U2
66          NAUT=-1
67          DO 50 IT=1,ITMAX
68          CALL LETAG(IOU,NAUT,NK,NB,XK,UMIN,ISSW,SIGXK,DATA,UBBE,SKRIK)
69          WRITE(IOU,120)IT,UMIN,(XK(I),I=1,NK)
70          NNAUT=IABS(NAUT)
71          GO TO (40,40,70),NNAUT
72       40 IF ((ABS((U2-UMIN)/U2)).LE.EPS)GO TO 70
73          U2=U1
74          NAUT=1
75       50 U1=UMIN
76       70 CALL SKRIK(NB,NK,IOU,XK,13,SIGXK)
77          GO TO 10
78      100 FORMAT(//31X,' T H E   S I M U L A T I O N   O F   M O D E L
79         & C U R V E ',//' INSTRUMENTAL ERROR=',F9.6//6X,' PARAMETERS +-
80         & SIGMA        WEIGHT OF P.',(7(/1X,F14.5,2F14.5))/)
81      110 FORMAT(13X,' V(EXP)     A(ACCURATE) + ERROR =   A(LOADED)     WEI
82         &GHT '/)
83      111 FORMAT(2X,I4,6X,F9.6,4X,F9.6,3X,F9.6,3X,F9.6,F13.6)
84      112 FORMAT(//31X,' T H E   E X P E R I M E N T A L   D A T A '//)
85      120 FORMAT(1X,I3,' ITER.  U= ',1PE11.5,' PARAM.=',
86         &7(0PF12.3))
87      130 FORMAT (20A4)
88      131 FORMAT(1H1,120(1H*),//46X,' M R L E T    P R O G R A M ',//
89         &1X,120(1H*),//2X,20A4,//1X,120(1H*)//)
90      999 STOP
91          END



1           SUBROUTINE DATA(IOU,NB)
2    C
3    C --- D A T A M R ---
4    C     INPUT OF CONSTANS,PRELIMINARY CALCULATIONS WITH DATA
5    C
6           DIMENSION V(50),ALF(50),RAT(50)
```

```
 7          COMMON/FUNC/VOL(50),AEX(50),ERROR(50),AC(50),W(50)/KANAL/NI
 8          COMMON/MR1/CMO,VO,AL,ALFMIN,ALFMAX,B,C,X2,X1,EPSL
 9          COMMON/MR2/ CME(50),G(50),GG(50)
10          CALL READI(NBI,1)
11          CALL READR(CMO,11)
12          NBA=IABS(NBI)
13          CALL READR(EPSL,1)
14          WRITE(IOU,5)CMO,CL,AL,X2,VO,CORM,ALFMIN,ALFMAX,B,C,X1
15          J=0
16          WRITE(IOU,6)
17          NB=0
18          DO 3 I=1,NBA
19          CALL READR(V(I),1)
20          CALL READR(AEX(I),1)
21          IF(NBI-1)1,2,2
22        1 AEX(I)=-.43429448*ALOG(AEX(I))
23        2 VI=V(I)*CORM
24          GI=1.0+VI/VO
25          ACI=GI*AEX(I)-AL
26          ALFI=ACI/(X1-AL)
27          WRITE(IOU,7)I,V(I),AEX(I)
28          IF((ALFI-ALFMIN)*(ALFMAX-ALFI))3,3,4
29        4 J=J+1
30          G(J)=GI
31          ERROR(J)=0.0
32          VOL(J)=VI
33          AEX(J)=AEX(I)
34          NB=NB+1
35          AC(J)=ACI
36          CMEE=VI*CMO/VO
37          CME(J)=CMEE
38          ALF(J)=ALFI
39          RAT(J)=CMEE/(CL*X2)
40          WRITE(IOU,8)J,CME(J),AEX(J),RAT(J),ALF(J)
41        3 CONTINUE
42        5 FORMAT(1X,' *****  D A T A  ***** '/2X,21(1H*)/1X,'EXPERIMENTAL ',
43         &'CONDITIONS: '/1X,'METAL : TOTAL CONCENTRATION IN BURETTE :',
44         &F10.6,'(MOL/L),'/1X,'LIGAND : TOTAL CONCENTRATION IN VESSEL :',
45         &F10.6,'(MOL/L),   INITIAL ABSORBANCE:',F6.4,' GUESSED FACTOR :',
46         &F7.5/1X,'INITIAL VOLUME :',F7.3,' (ML), FACTOR OF MICROBURETTE :',
47         &F7.5/1X,'LIMITS OF ALPHA   ALPHAMIN :',F6.4,'   ALPHAMAX :',
48         &F6.4/1X,'STOICHIOMETRIC COEFFICIENTS M=',F3.1,' N=',F3.1,
49         &' GUESSED EXTRAPOL. ABSORB. OF COMPLEX =',F6.4///)
50        6 FORMAT(  1X,'MEASURED POINTS OF A-V CURVE',10X,'SELECTED POINTS OF
51         & A-V CURVE BEING IN (ALFMIN,ALFMAX) RANGE'//3X,'I  VOLUME(ML)  ABS
52         &ORBANCE',12X,'J  METAL CONC.  ABSORBANCE        RATIO     ALPHA'/7
53         &X,'/......READ VALUES.../',14X,'/...ACCURATE VALUES.../',4X,'/..AP
54         &PROXIMATE VALUES../'/)
55        7 FORMAT(1X,I4,F10.5,F12.4)
56        8 FORMAT(1H+,37X,I3,F13.8,F10.4,F14.3,F10.3)
57          RETURN
58          END


 1          SUBROUTINE UBBE(U,NK,XK,NB)
 2  C --- U B B E M R ---
 3  C         CALCULATION OF ERROR SQUARE SUM FUNCTION
 4  C
 5          DIMENSION XK(NK)
 6          COMMON/MR1/CMO,VO,AL,CL,CORM,ALFMIN,ALFMAX,B,C,X2,X1,EPSL
 7          COMMON /FUNC/VOL(50),AEX (50),ERROR(50),ACAL(50),W(50)
 8          COMMON/MR2/ CME(50),G(50),GG(50)
 9          C=AINT(XK(4)+0.5)
10          AK=EXP(2.3025851*XK(3))
11          AEXT=XK(1)-AL
12          CCL=CL*XK(2)
13          HELP1=1/B
14          HELP2=(C-1.0)/B
15          HELP3=C/B
16          HELP4=(C*AK)**HELP1*CCL**HELP2/HELP3*CCL
17          U=0.
18          DO 8 I=1,NB
19          V=VOL(I)*CMO*C/(CCL*VO*B)
20          ALFA=(AEX(I)*G(I)-AL)/AEXT
21          CALL RTWI(ALFC,VAL,FCT,ALFA,EPSL,100,IER,HELP1,HELP3,HELP4,V)
```

```
22              IF(IER.NE.0)WRITE(6,1)IER,I,V,VAL,ALFC
23            1 FORMAT(1X,'WARNING*POINT*X-F(X)*V*ALFAC***   ',2I5,3(1PE15.7))
24              IF(ALFC.LE.0.0)ALFC=1.0E-10
25              IF(ALFC.GT.1.0)ALFC=1.0-1.0E-10
26              ACAL(I)=ALFC*AEXT+AL
27              U=U+(AEX(I)-ACAL(I))**2*W(I)
28            8 CONTINUE
29              RETURN
30              END
```

```
1               SUBROUTINE RTWI(X,VAL,FOT,XST,EPSL,IEND,IER,HELP1,HELP3,HELP4,V)
2               FCT(X)=V-(ABS(X))**HELP1/(HELP4*(ABS(1.0-X))**HELP3)
3               IER=0
4               TOL=XST
5               X=FCT(TOL)
6               A=X-XST
7               B=-A
8               TOL=X
9               VAL=X-FCT(TOL)
10              DO 6 I=1,IEND
11              IF(VAL)1,7,1
12            1 B=B/VAL-1.
13              IF(B)2,8,2
14            2 A=A/B
15              X=X+A
16              B=VAL
17              TOL=X
18              VAL=X-FCT(TOL)
19              TOL=EPSL
20              D=ABS(X)
21              IF(D-1.)4,4,3
22            3 TOL=TOL*D
23            4 IF(ABS(A)-TOL)5,5,6
24            5 IF(ABS(VAL)-10.*TOL)7,7,6
25            6 CONTINUE
26              IER=1
27            7 FOT=FCT(X)
28              RETURN
29            8 IER=2
30              RETURN
31              END
```

```
1               SUBROUTINE SKRIK(NB,NK,IOU,XK,IRUR,SIGXK)
2     C
3     C --- S K R I K ---
4     C       OUTPUT OF PARAMETERS AND THEIR STANDARD DEVIATIONS
5     C       STATISTICAL TEST OF CURVE FITTING......ANALYSIS OF REZIDUALS
6     C
7               DIMENSION DIF(50),SIGXK(NK),XK(NK)
8               DIMENSION XX(50,2),YY(50,2),NDATA(2)
9               COMMON /FUNC/XEXP(50),YEXP(50),ERR(50),YCAL(50),W(50)
10              COMMON/TIT/ TITLE(20) /PLOT/ MY,MX,NLS,NCL,XLINE(5),YLINE(5),
11             &AREA(31,60),YSCALE(60)
12              LOGICAL ISYMBL(2)
13              DATA NDMAX/50/
14              DATA ISYMBL(1)/4H****/
15              DATA ISYMBL(2)/4H0000/
16              WRITE(IOU,99)
17              WRITE(IOU,100)(XK(IK),SIGXK(IK),IK=1,NK)
18              IF(IRUR.LT.11)RETURN
19              CALL UBBE(U,NK,XK,NB)
20              DO 10 I=1,NB
21           10 DIF(I)=YEXP(I)-YCAL(I)
22              WRITE(IOU,101)
23              WRITE(IOU,102)(I,XEXP(I),YEXP(I),YCAL(I),DIF(I),I=1,NB)
24              IF(IRUR.LT.12)RETURN
25              CALL STATS(IOU,NK,NB,XEXP,YEXP,DIF)
26              IF(IRUR.LT.13)RETURN
27              NDATA(1)=NB
28              NDATA(2)=NB
29              DO 5 I=1,NB
```

```
30          XX(I,1)=XEXP(I)
31          YY(I,1)=YEXP(I)
32          YY(I,2)=YCAL(I)
33        5 XX(I,2)=XEXP(I)
34          WRITE(IOU,2001)TITLE
35   C    3 CALL PLOTT(XX,YY,NDATA,NDMAX,ISYMBL, 2,XLINE,MX,YLINE,MY,NLS,
36   C      &NCL, 1, 1,AREA,YSCALE)
37          NPAGE=0
38          N=2
39          CALL PRTPLT(XX,YY,NDATA,N,NPAGE,NDMAX)
40       99 FORMAT(' *****  S K R I K  (OUTPUT)  *****',/,1X,33(1H*))
41      100 FORMAT(/1X,' PARAMETERS AND THEIR STANDARD DEVIATIONS : '/(7(4X,
42          &F14.5, ' +- ',F14.5/)))/)
43      101 FORMAT(' I     V(EXP)       A(EXP)       A(CAL)       RESIDUAL'/)
44      102 FORMAT(1X,I2,2X,F8.5,F13.4,F13.4,F13.4)
45     2001 FORMAT(1H1, '      GRAPH IS PRINTED FOR SYSTEM:',6X,20A4,//)
46          RETURN
47          END
```

2.4.2. Input Data

There follows a listing of a typical data set for MRLET.

```
¶<-- Column 1

MRLET DATA(1):SIMUL.,ALGORITMIC PROCESS, FULL OUTPUT, 15.11.1981 ***

1 5* , 1, 8* , 0.003

1.   0.7   6.0   1.0  0.0  0.0  0.0  0.0  0.25  0.25  0.25  0.25

33   0.001  20.  0.0001  0.000015  1.  0.05  0.955  1  1  0.7  1.0  0.0001

0.0222,   0.1,  0.0335,  0.15,  0.045,  0.2,  0.0567,  0.249,

0.0686,  0.299,  0.0808,  0.349,  0.0934,  0.398,  0.1064,  0.447,

0.12,  0.497,  0.1344,  0.546,  0.1405,  0.566,  0.15,  0.596,

0.1533,  0.605,  0.1566,  0.615,  0.1601,  0.625,  0.1636,  0.635

0.1671,  0.645,  0.1708,  0.655,  0.1746,  0.664,  0.1785,  0.674,

0.1825,  0.684,  0.1866,  0.694,  0.191,  0.703,  0.1954,  0.713,

0.2001,  0.723,  0.2049,  0.733,  0.21,  0.742,  0.2153,  0.752,

0.221,  0.762,  0.2269,  0.771,  0.2332,  0.781,  0.24,  0.791,

0.2834,  0.838,

7  4  4  1.1  0.75  6.2  1.0  0  3  3  1  0.08  2  0.08  3  0.3  2  5
```

3. POLET: FORMATION CONSTANTS FROM POTENTIOMETRIC DATA

3.1. Model Formulation

POLET has been designed to search for the best set of formation constants, β_{pqrs}, for the general complex $A_pB_qC_rD_s$ that fit the supplied potentiometric data. The independent variable is taken to be the free concentration of A, [A], measured either by the glass electrode, i.e., A, H^+, or by an ion-selective electrode. The function $Z = f(\text{pH})$ is the dependent variable defined as

$$Z = 1/C_B, \sum_{j=1}^{NK} p\beta_{pqrs}[A]^p[B]^q[C]^r[D]^s \tag{12}$$

The variable Z represents the average number of A ions (e.g., H^+) bound per mole of B (if key KV is 1) *or* bound per mole of C [if KV is 2 and $1/C_B$ is replaced by $1/C_c$ in equation (12)]. The values of Z may be obtained from the titration data in the conventional manner. NK is the number of species (formation constants) needed to describe the equilibrium system. The components A, B, C, and D may be conventionally formulated as H^+, M^{m+}, L^{n-}, or as less fundamental combinations such an inert complex MX reacting with a second ligand Y.

For the data Z, $(-\log [A])$ the best estimates of the NK constants are found by minimizing

$$U = \sum_i W_i(Z_{calc}^i - Z_{exp}^i)^2 \tag{13}$$

POLET can treat equilibrium data where polynuclear and/or mixed ligand complexes may be formed, in addition to simple binuclear complexes, from two metals and one ligand *or* one metal and two ligands. Although component A is normally H^+ (pH measurements being made) it may be any component for which an ion-selective electrode is available. Hydroxo species $(OH)_pM_qL_r$ are treated as species with negative numbers of protons $(H)_{-p}M_qL_r$. POLET does not possess the capability to distinguish between OH being bound and H^+ being abstracted from bound H_2O). C may represent any protonated form of ligand, say H_4L. However, it is advantageous to choose L (rather than HL, H_2L, H_3L or H_4L) since the protonation constants can be supplied as input data for POLET. POLET has been used in the study of germanium equilibria with aminocarboxylate and aminohydroxy chelates,[8] EDTA complexes of Mo(VI),[9] and tartrate complexes of Mo(VI) and W(VI).[10]

3.2. Description of POLET

POLET is designed to process up to 99 titration points, four components, and will refine up to eight formation constants. The potentiometric data comprise the volume of titrant and Z, derived from the free concentration of component A together with total concentrations and parameters that control the progress of the minimization process.

The MAIN program fulfills an organizational role together with the printout of a few parameters. Subroutine DATA reads in the {Z,pA} data together with parameters relating to the proposed equilibrium model. The pK_a values for component C are also read in at this point and are used to calculate the coefficient of side reaction for C, α_{H_4L}:

$$\alpha_{H_4L} = 1 + \sum_{j=1}^{4} \left(\prod_{j=1}^{4} K_{a,j} \right) \cdot [H]^{-j} \tag{14}$$

Subroutine UBBE calculates values of Z_{calc} for the current set of β_{pqrs}. Subroutine COGS calculates the free concentrations of each component necessary to obtain Z_{calc}. Subroutine COGS is derived from the original COGS written by Perrin and Sayce.[11] Subroutine SKRIK prints out the progress and result of the minimization process, performs a statistical analysis of the refined parameters, the species distributions α and β:

$$\alpha = \frac{q[A_pB_qC_rD_s]}{C_B}; \ \beta = \frac{r[A_pB_qC_rD_s]}{C_C} \tag{15}$$

and the species distribution for the first ligand (i.e., component C) defined as $[H_iL]/C_C$.

3.3. Data Input Requirements

The general setup of the data cards has been described in Section 1.4. There are some requirements that are specific to POLET. The program may be used to either process experimental data or simulate titration data. All data entry is format-free.

1. Title Card: **TITLE**
   A suitable title (80 characters)
2. Key Card: **(ISSW(I), I = 1, 7)**
   Values are chosen according to Table 2. ISSW(7) is not used by POLET.
3. Termination Card: **EPS,ITMAX,PSI,NK,MY,MX,NLS,NCL,SINST**
   EPS: ⎱
   ⋮   ⎰ As Section 1.4.2, item 3.
   SINST: ⎰
4. Simulation Card: **(XK(I),I=1,NK),(SIGXK(I),I=1,NK),(WEI(I),I=1,NK)**
   XK(I): ⎱
   SIGXK(I): ⎰ As Section 1.4.2, item 4.
   WEI (I): ⎰
   Note: Values required for simulation runs; *omit* this card for experimental data.
5. Data Card: **IRUR, KV, NA, NB, (P(I), Q(I), R(I), S(I), I=1, NK), (MDE-SCR (I), I = 1, NK), (PH(I), CBTOT(I), CCTOT(I), CDTOT(I), ZEXP(I), I=1,NB),PK1,PK2,PK3,PK4**
   IRUR: Set equal to six for experimental data. *Omit* for simulation runs.
   KV: Key for calculation of Z_{calc}. Set to one for average number of moles of A bound per mole of B; set to two for A bound to C.
   NA: Number of species.
   NB: Number of titration data pairs.

P(I),Q(I),R(I),S(I): Stoichiometric coefficients for the i the complex A_pB_q-C_rD_s. There will be NK sets.

MDESCR(I): Character description of the ith reaction. Format is 15A4. Each reaction is placed on a separate card for a total of NK cards.

PH(I),(BTOT(I),CCTOT(I),CDTOT(I),ZEXP(I): Experimental data are read in, one point per card, for $-\log$ [A] (here pH), C_B, C_C, C_D and Z_{exp}. All values, except for $-\log$ [A] are precalculated for the ith point. There will be NP points.

PK1,PK2,PK2,PK4: The pK_a values for component C. If any or none of the pK_a's are known set values to -1.0.

6. Initial Guess Card: **IRUR,NK,NBYK,(XK(I),I=1,NBYK)**

   IRUR: $\left.\begin{array}{l} \\ \vdots \\ \\ XK(I) \end{array}\right\}$ As for Section 1.4.2, item 6a.

7. Matrix Card: **ISKIN,((I(K),J(K),S(I(K),J(K)),K=1,ISKIN)**

   ISKIN: $\left.\begin{array}{l} \\ \vdots \\ \\ S(\ ): \end{array}\right\}$ As for Section 1.4.2, item 7.

   Note for MRLET set to zero.

6. Posk Removal Card: **IRUR,NK,$-$1,NEGK,(IK(I),I=1,NEGK)**

   IRUR: $\left.\begin{array}{l} \\ \vdots \\ \\ IK(\ ): \end{array}\right\}$ As for Section 1.4.2, item 6c.

   Note: This card is included to permit selected formation constants to take on negative values.

8. Step Card: **IRUR,N,((I(K),STEK(I(K)),K=1,N)**

   IRUR: $\left.\begin{array}{l} \\ \vdots \\ \\ STEK(\ ): \end{array}\right\}$ As for Section 1.4.2, item 8.

9. Process Card: **IRUR,IRUR,...**

   IRUR: As for Section 1.4.2, item 9. For POLET set to 2,5,13.

3.4. Presentation of POLET

3.4.1. Listing of POLET

There follows a listing of the MAIN program and subroutines, SKRIK, UBBE, COGS, and DATA. Those routines listed in Section 1.5 are added to form the complete program POLET.

```
 1      *#RUN=(NWARN) #Ø5;Ø6
 2      C       P O L E T    P R O G R A M
 3      C       (MAIN PROGRAM)
 4      C PUBLICATION: E.MIKANOVA,M.BARTUSEK,J.HAVEL, SCRIPTA FAC SCI.NAT.
 5      C               UNIV.PURK.BRUN.1Ø,NO.1-2(CHEMIA),3-12(198Ø)
 6      C
 7      C  AUTHORS OF THIS REVISED VERSION: JOSEF HAVEL, PAVLA HAVLOVA
 8      C                                  AND MILAN MELOUN
 9      C
1Ø      C
11      C
12              DIMENSION  XK(8),SIGXK(8),WEI(8)
13              COMMON /FUNC/XEXP(99),YEXP(99),ERR(99),YCAL(99),W(99)/KANAL/INP
14             &/TIT/TITLE(2Ø)/PLOT/MY,MX,NLS,NCL,DUMMY(193Ø)/ISW/ISSW(7)
15             &/POT/NA,NB,NK,DUMMA(1Ø51),KV
16              COMMON/OUT/IOUT
17              EQUIVALENCE (UMIN,PSI)
18              EXTERNAL DATA,UBBE,SKRIK
19        1Ø INP=5
2Ø              ISSW(1)=Ø
21              ISSW(2)=Ø
22              ISSW(3)=Ø
23              ISSW(4)=Ø
24              ISSW(5)=1
25              ISSW(6)=Ø
26              ISSW(7)=2
27              EPS=1.ØE-6
28              ITMAX=4Ø
29              PSI=Ø.3
3Ø              NK=8
31              MY=Ø
32              MX=Ø
33              NLS=5Ø
34              NCL=1ØØ
35              SINST=-1.Ø
36              DO 6Ø I=1,99
37              W(I)= 1.Ø
38        6Ø CONTINUE
39              IOU=6
4Ø              IOUT=IOU
41              IR=1225
42              READ(INP,13Ø,END=999)TITLE
43              WRITE(IOU,131)TITLE
44              CALL READI(ISSW,-7)
45              CALL READR(EPS,1)
46              CALL READI(ITMAX,1)
47              CALL READR(PSI,1)
48              CALL READI(NK,1)
49              CALL READI(MY,4)
5Ø              CALL READR(SINST,1)
51              IF(SINST)3Ø,3Ø,2Ø
52        2Ø CALL READR(XK,NK)
53              CALL READR(SIGXK,NK)
54              CALL READR(WEI,NK)
55              CALL DATA(IOU,NB)
56              WRITE(IOU,1ØØ)SINST,(XK(I),SIGXK(I),WEI(I),I=1,NK)
57              DO 21 I=1,NB
58              W(I)=1.Ø
59        21 CONTINUE
6Ø              CALL UBBE(U,NK,XK,NB)
61              CALL SIMUL (XEXP,YCAL,YEXP,NB,SINST,IR)
62              DO  11  I=1,NB
63        11 ERR(I)=YEXP(I)-YCAL(I)
64              CALL WEIGHT(UBBE,NB,NK,XK,SIGXK,WEI,ISSW)
65              WRITE(IOU,11Ø)
66              DO 12 I=1,NB
67        12 YCAL(I)=YEXP(I)-ERR(I)
68              WRITE(IOU,111)(I,XEXP(I),YCAL(I),ERR(I),YEXP(I),W(I),I=1,NB)
69              CALL STATS(IOU,NK,NB,XEXP,YEXP,ERR)
7Ø              GO TO 33Ø
71        3Ø  WRITE(IOU,112)
72        33Ø U2=1.ØE3Ø
73              U1=U2
74              NAUT=-1
```

```
75          DO 50 IT=1,ITMAX
76          CALL LETAG(IOU,NAUT,NK,NB,XK,UMIN,ISSW,SIGXK,DATA,UBBE,SKRIK)
77          WRITE(IOU,120)IT,UMIN,(XK(I),I=1,NK)
78          NNAUT=IABS(NAUT)
79          GO TO (40,40,70),NNAUT
80       40 IF ((ABS((U2-UMIN)/U2)).LE.EPS)GO TO 70
81          U2=U1
82          NAUT=1
83       50 U1=UMIN
84       70 CALL SKRIK(NB,NK,IOU,XK,13,SIGXK)
85          GO TO 10
86      100 FORMAT(//31X,' T H E     S I M U L A T I O N     O F     M O D E L
87        & C U R V E ',//' INSTRUMENTAL ERROR=',F9.6//6X,' PARAMETERS +-
88        & SIGMA       WEIGHT OF P.',(7(/1X,F14.5,2F14.5)))/)
89      110 FORMAT(13X,'PH(EXP)     Z(ACCURATE) + ERROR =  Z(LOADED)    WEI
90        &GHT '/)
91      111 FORMAT(2X,I4,6X,F9.6,4X,F9.6,3X,F9.6,3X,F9.6,F13.6)
92      112 FORMAT(//31X,' T H E     E X P E R I M E N T A L     D A T A '//)
93      120 FORMAT(1X,I3,' ITER.  U= ',1PE11.5,' PARAM.=',
94        &7(0PF12.3))
95      130 FORMAT (20A4)
96      131 FORMAT(1H1,120(1H*),//46X,' P O L E T     P R O G R A M ',//
97        &1X,120(1H*),//2X,20A4,//1X,120(1H*)//)
98      999 STOP
99          END
```

```
1           SUBROUTINE DATA(IOU,NB)
2    C
3    C --- D A T A P O ---
4    C
5    C       PRELIMINARY CALCULATIONS WITH DATA
6    C
7            INTEGER P,Q,R,S,MDESCR(15)
8            COMMON/FUNC/PH(99),ZEXP(99),ERROR(99),ZVYP(99),W(99)
9          &/POT/NA,NP,NK,CA(99),CB(99),CC(99),CD(99),CBTOT(99),CCTOT(99),
10         &CDTOT(99),P(8),Q(8),R(8),S(8),B(12),DELTA(99),ALFA(99),
11         &BETA(8),GAMA(9),KV
12          COMMON/KANAL/INP
13          CALL READI(KV,1)
14          CALL READI(NA,2)
15          WRITE(IOU,32)
16          DO 3 I=1,NK
17          CALL READI(P(I),1)
18          CALL READI(Q(I),1)
19          CALL READI(R(I),1)
20          CALL READI(S(I),1)
21        3 WRITE(IOU,31) I,P(I),Q(I),R(I),S(I)
22          WRITE(IOU,34)
23          DO 6 I=1,NK
24          READ(INP,37) MDESCR
25        6 WRITE(IOU,36) I,MDESCR
26          NB=NP
27          WRITE(IOU,38)
28          DO 4 I=1,NP
29          CALL READR(PH(I),1)
30          CALL READR(CBTOT(I),1)
31          CALL READR(CCTOT(I),1)
32          CALL READR(CDTOT(I),1)
33          CALL READR(ZEXP(I),1)
34        4 WRITE(IOU,33) PH(I),CBTOT(I),CCTOT(I),CDTOT(I),ZEXP(I)
35          CALL READR(PK1,1)
36          CALL READR(PK2,1)
37          CALL READR(PK3,1)
38          CALL READR(PK4,1)
39          WRITE(IOU,35) PK1,PK2,PK3,PK4
40          DO 10 II=1,4
41       10 B(II)=0.0
42          IF(PK1.LT.0.0) GOTO 5
43          B(1)=10.**(-PK1)
44          IF(PK2.LT.0.) GOTO 5
45          B(2)=10.**(-PK1-PK2)
46          IF(PK3.LT.0.) GOTO 5
```

```
47              B(3)=10.**(-PK1-PK2-PK3)
48              IF(PK4.LT.0.) GOTO 5
49              B(4)=10.**(-PK1-PK2-PK3-PK4)
50          5 DO 18 I=1,NP
51              CA(I)=EXP(-2.302585*PH(I))
52              ALFA(I)=(B(4)*1./CA(I)+B(3))*1./(CA(I)**3)
53         18 ALFA(I)=ALFA(I)+(B(2)*1./CA(I)+B(1))*1./CA(I)+1.
54         31 FORMAT(5I3)
55         32 FORMAT(/,' I  P  Q  R  S',/)
56         33 FORMAT (5F10.6)
57         34 FORMAT(/,' I SPECIES DESCRIPTION',/)
58         35 FORMAT(4F10.4,'    PK VALUES OF LIGAND(3 RD COMP.)')
59         36 FORMAT(I3,2X,15A4)
60         37 FORMAT(15A4)
61         38 FORMAT(//,' T H E  E X P E R I M E N T A L  D A T A  '/,
62             &4X,'PH',6X,'CBTOT',5X,'CCTOT',5X,'CDTOT',5X,'Z(EXP)'/)
63              RETURN
64              END
```

```
1               SUBROUTINE UBBE(U,NK,XK,NB)
2       C
3       C --- U B B E P O ---
4       C       CALCULATION OF ERROR SQUARE SUM FUNCTION U
5       C       UBBE CALCULATES FOR GIVEN SERIES OF EQUILIBRIUM CONSTANTS THE SUM
6       C       OF SQUARES OF DEVIATIONS (CALCULATED AND EXP. VALUES Z )
7       C       CA (CATOT)    FREE (TOTAL)CONCN.OF THE FIRST COMPONENT
8       C       CB (CBTOT)    FREE (TOTAL) METAL CONCENTRATION
9       C       CC (CCTOT)    FREE (TOTAL)CONCN.OF THE FIRST REAGENT
10      C       CD (CDTOT)    FREE (TOTAL)CONCN.OF THE SECOND REAGENT
11      C
12              INTEGER P,Q,R,S
13              DIMENSION XK(NK)
14              COMMON/FUNC/PH(99),ZEXP(99),ERROR(99),ZVYP(99),W(99)
15             &/POT/NA,NP,IN,CA(99),CB(99),CC(99),CD(99),CBTOT(99),CCTOT(99),
16             &CDTOT(99),P(8),Q(8),R(8),S(8),B(12),DELTA(99),ALFA(99),
17             &BETA(8),GAMA(9),KV
18              NB=NP
19              U=0.
20              DO 1 I=1,NP
21              SUM=0.
22              IF(CBTOT(I).EQ.0.) CBTOT(I)=1.
23              IF(I.EQ.1) GOTO 3
24              J=I-1
25              CB(I)=CB(J)
26              GOTO 4
27          3 CONTINUE
28              CB(I)=CBTOT(I)
29          4 CONTINUE
30              IF(CCTOT(I).GT.0.) GOTO 10
31              CCTOT(I)=1.
32              CC(I)=CCTOT(I)
33              GOTO 11
34         10 IF(I.EQ.1) GOTO 12
35              J=I-1
36              CC(I)=CC(J)
37              GOTO 13
38         12 CC(I)=CCTOT(I)/ALFA(I)
39         13 CONTINUE
40         11 IF(CDTOT(I).EQ.0.)GOTO 14
41              IF(I.EQ.1) GOTO 15
42              J=I-1
43              CD(I)=CD(J)
44              GOTO 16
45         14 CDTOT(I)=1.
46         15 CD(I)=CDTOT(I)
47         16 CONTINUE
48              DO 20 K=1,NK
49              KK=4+K
50      20      B(KK)=EXP(2.302585*XK(K))
51              CALL COGS(I)
52              SUM=SUM+((-4.)*B(4)/CA(I)+(-3.)*B(3))*CC(I)/CA(I)**3
53              SUM=SUM+((-2.)*B(2)/CA(I)+(-1.)*B(1))*CC(I)/CA(I)
```

```
54              DO 2 K=1,NK
55              KK=4+K
56              SUM=SUM+FLOAT(P(K))*B(KK)*CA(I)**P(K)*CB(I)**Q(K)*
57           &           CC(I)**R(K)*CD(I)**S(K)
58            2 CONTINUE
59              GOTO (8,7),KV
60            7 ZVYP(I) =SUM/CCTOT(I)
61              GO TO 9
62            8 ZVYP(I)=SUM/CBTOT(I)
63            9 DELTA(I)=ZEXP(I)-ZVYP(I)
64              U=U+(DELTA(I)*DELTA(I)*W(I))
65            1 CONTINUE
66              RETURN
67              END
```

```
1               SUBROUTINE COGS(I)
2     C         THE PROCEDURE FOR THE CALCN. OF EQUIL. CONCENTRATIONS OF COMPLEXES
3     C         DESCRIPTION OF PROCEDURE   GOGS
4     C         VX - EQUIL.CONCN. OF THE METAL (SECOND COMPONENT)
5     C         TX - EQUIL.CONCN. OF THE FIRST LIGAND (THIRD COMP.)
6     C         UX - EQUIL.CONCN. OF THE SECOND LIGAND (FOURTH COMP.)
7     C
8               DIMENSION    C(8),TERM(8)
9               COMMON/FUNC/PH(99),ZEXP(99),ERROR(99),ZVYP(99),W(99)
10            &/POT/NA,NP,IN,AH(99),VX(99),TX(99),UX(99),CM(99),CR(99),
11            &CX(99),MN(8),MM(8),ML(8),MS(8),B(12),DELTA(99),ALFA(99),
12            &BETA(8),GAMA(9),KV
13              COMMON/OUT/IOUT
14              Y1=CM(I)*0.001
15              IF(NA.EQ.2) GOTO55
16              IF(NA.EQ.3)GOTO 7
17              Y5=CX(I)*0.001
18              Y3=CR(I)*0.001
19              GOTO 9
20            7 Y5 = 1.
21           55 CONTINUE
22              Y3=CR(I)*0.001
23            9 NIT=0
24              DO 222 K =1,IN
25              II=4+K
26          222 TERM(K)=B(II)*AH(I)**MN(K)
27            2 DO 3 K=1,IN
28            3 C(K)=TERM(K)
29              DO 4 K=1,IN
30            4 C(K)=C(K)*VX(I)**MM(K)
31              DO 5 K=1,IN
32            5 C(K)=C(K)*TX(I)**ML(K)*UX(I)**MS(K)
33              NIT=NIT+1
34              IF(NA.EQ.2)GOTO 6
35              BO=VX(I)
36              DO 8 K=1,IN
37              BO=BO+C(K)*FLOAT(MM(K))
38            8 CONTINUE
39              RATIO=BO/CM(I)
40              VY=VX(I)/SQRT(RATIO)
41              Y2=ABS(BO-CM(I))
42              AXO=UX(I)
43              DO 111 K=1,IN
44          111 AXO=AXO+C(K)*FLOAT(MS(K))
45              RATIO=AXO/CX(I)
46              UY=UX(I)/SQRT(RATIO)
47              Y6=ABS(AXO-CX(I))
48              ALO=TX(I)
49              DO 88 K=1,4
50           88 ALO=ALO+ALO*B(K)/(AH(I)**K)
51              DO 71 K=1,IN
52           71 ALO=ALO+C(K)*FLOAT(ML(K))
53              RATIO=ALO/CR(I)
54              TY=TX(I)/SQRT(RATIO)
55              Y4=ABS(ALO-CR(I))
56              GOTO 99
57            6 ALO=TX(I)
```

```
58            DO 10 K=1,IN
59      10 ALO=ALO+C(K)*FLOAT(ML(K))
60            RATIO=ALO/CR(1)
61            TY=TX(I)/SQRT(RATIO)
62            Y4=ABS(ALO-CR(1))
63            GOTO 77
64      99 IF(NIT-9999)11,11,999
65      77 IF(NIT-9999)22,22,999
66      22 IF(Y3-Y4)25,23,23
67      23 TX(I)=TY
68            RETURN
69      25 TX(I)=TY
70            GOTO 2
71      11 IF(Y1-Y2) 15,12,12
72      12 IF(Y3-Y4) 15,13,13
73      13 IF(Y5-Y6) 15,14,14
74      14 TX(I)=TY
75            VX(I)=VY
76            UX(I)=UY
77            RETURN
78      15 TX(I)=TY
79            VX(I)=VY
80            UX(I)=UY
81            GOTO 2
82     999 WRITE(IOUT,998) I
83     998 FORMAT(1X,' NO CONVERGENCE ',I2,'TH SOLUTION ')
84            RETURN
85            END
```

```
1            SUBROUTINE SKRIK(NB,NK,IOU,XK,IRUR,SIGXK)
2     C
3     C --- S K R I K ---
4     C      OUTPUT OF PARAMETERS AND THEIR STANDARD DEVIATIONS
5     C      STATISTICAL TEST OF CURVE FITTING......ANALYSIS OF REZIDUALS
6     C
7            INTEGER P,Q,R,S
8            DIMENSION XK(NK),SIGXK(NK)
9            DIMENSION XX(99,2),YY(99,2),NDATA(2)
10           COMMON/FUNC/ PH(99),ZEXP(99),ERROR(99),ZCAL(99),W(99)
11          &/POT/NA,NP,NL,CA(99),CB(99),CC(99),CD(99),CBTOT(99),CCTOT(99),
12          &CDTOT(99),P(8),Q(8),R(8),S(8),B(12),  DIF(99),ALFA(99),AH(99),
13          &BETA(8),GAMA(9),KV
14           COMMON/TIT/ TITLE(20) /PLOT/ MY,MX,NLS,NCL,XLINE(5),YLINE(5),
15          &AREA(31,60),YSCALE(60)
16           LOGICAL ISYMBL(2)
17           DATA NDMAX/99/
18           DATA ISYMBL(1)/4H****/
19           DATA ISYMBL(2)/4H0000/
20           WRITE(IOU,99)
21           WRITE(IOU,100)(XK(IK),SIGXK(IK),IK=1,NK)
22           IF(IRUR.GE.12)GO TO 9
23           WRITE (IOU,30)
24           DO 13 I=1,NB
25      13 WRITE (IOU,32) PH(I),CBTOT(I),CB(I),CCTOT(I),CC(I),CDTOT(I),CD(I),
26          &ZEXP(I),ZCAL(I),DIF(I)
27           DO 134 I=1,8
28           GAMA(I)=0.
29     134 BETA(I)=0.
30           GAMA(9)=0.
31           WRITE(IOU,36)
32           DO 130 I=1,NB
33           DO 131 K=1,NK
34           JJ=4+K
35           GAMA(K)=FLOAT(Q(K))*
36          &B(JJ)*CA(I)**P(K)*CB(I)**Q(K)*CC(I)**R(K)*CD(I)**S(K)
37          &/CBTOT(I)
38     131 BETA(K)=FLOAT(R(K))*
39          &B(JJ)*CA(I)**P(K)*CB(I)**Q(K)*CC (I)**R(K)*CD(I)**S(K)
40          &/CCTOT(I)
41     130  WRITE(IOU,37) PH(I),(GAMA(K),K=1,8),(BETA(J),J=1,8)
42           WRITE(IOU,38)
43           DO 132 I=1,NB
```

```
44              GAMA(1)=CC(I)/CCTOT(I)
45              DO 133 K=1,4
46              J=K+1
47              GAMA(J)=CC(I)*B(K)/CCTOT(I)/CA(I)**K
48          133 CONTINUE
49              WRITE(IOU,37)PH(I),(GAMA(K),K=1,5)
50          132 CONTINUE
51              IF(IRUR.LT.11)RETURN
52            9 CALL UBBE(U,NK,XK,NB)
53              DO 10 I=1,NB
54           10 DIF(I)=ZEXP(I)-ZCAL(I)
55              WRITE(IOU,101)
56              WRITE(IOU,102)(I,   PH(I),ZEXP(I),ZCAL(I),DIF(I),I=1,NB)
57              IF(IRUR.LT.12)RETURN
58              CALL STATS(IOU,NK,NB,   PH,ZEXP,DIF)
59              IF(IRUR.LT.13)RETURN
60              NDATA(1)=NB
61              NDATA(2)=NB
62              DO 5 I=1,NB
63              XX(I,1)=PH(I)
64              YY(I,1)=ZEXP(I)
65              YY(I,2)=ZCAL(I)
66            5 XX(I,2)=PH(I)
67              WRITE(IOU,2001)TITLE
68    C       3 CALL PLOTT(XX,YY,NDATA,NDMAX,ISYMBL,  2,XLINE,MX,YLINE,MY,NLS,
69    C         &NCL,  1,  1,AREA,YSCALE)
70              NPAGE=0
71              N=2
72              CALL PRTPLT(XX,YY,NDATA,N,NPAGE,NDMAX)
73           30 FORMAT (//////3X,14('*')/4X,'* FUNCTION Z *'/4X,14('*')///51X,
74              &'TABLE OF VALUES'/50X,17(1H*)///4X,'PH',8X,'CBTOT',13X,'CB',12X,
75              &'CCTOT',13X,'CC',12X,'CDTOT',13X,'CD',8X,'ZEXP',4X,'ZVYP',6X,
76              & 'RESID'/2X,127(1H*)//)
77           32 FORMAT(2X,F5.2,6(E16.5),2(F8.3),F10.3)
78           36 FORMAT(//4X,'DISTRIBUTION OF INDIVIDUAL COMPLEXES VS. TOTAL METAL
79              &(SECOND COMPONENT)'/4X,'OR VS. TOTAL LIGAND(THIRD COMPONENT)'
80              &,'CONCENTRATIONS'
81              & ,///2X,'PH',5X,'ALFA 1 ALFA 2 ALFA 3 ALFA 4 ALFA 5 ALFA 6 ALFA 7
82              &ALFA 8  BETA 1 BETA 2 BETA 3 BETA 4 BETA 5 BETA 6 BETA 7 BETA 8'//
83              &/)
84           37 FORMAT(1X,17F7.4)
85           38 FORMAT(//' DISTRIBUTION OF LIGAND (THIRD COMPONENT) SPECIES'//3X,
86              &'PH',6X,'LH4',4X,'LH3',4X,'LH2',4X,'LH',5X,'L')
87           99 FORMAT(' *****  S K R I K  (OUTPUT)  *****',/,1X,33(1H*))
88          100 FORMAT(/1X,'  PARAMETERS AND THEIR STANDARD DEVIATIONS : '/(7(4X,
89              &F14.5, ' +- ',F14.5))/)
90          101 FORMAT(' I     PH(EXP)       Z(EXP)      Z(CAL)      RESIDUAL'/)
91          102 FORMAT(1X,I2,2X,F8.5,F13.4,F13.4,F13.4)
92         2001 FORMAT(1H1, '     GRAPH IS PRINTED FOR SYSTEM:',6X,20A4,//)
93              RETURN
94              END
```

3.4.2. Input Data

There follows a listing of a typical data set for POLET. The data are taken from the molybdate, oxalate, catechol ternary system.[12]

¶<-- Column 1   1 1 0 2

POLET EXPERIMENTAL DATA FOR H+,MOLYBDATE,PYROCATECHOL,OXALATE SYSTEM   1 1 0 1

0 1 1 7* , 6 5* ,   2 1 1 0

6 1 4 53   2 1 1 1

 1 0 1 0   PROTONATION OF THE FIRST LIGAND

 0 1 0 2   COMPLEX 1:2 MOLYBDATE:OXALATE

PROTONATED 1:2 MOLYBDATE:OXALATE COMPLEX					5.05	.01	.028	.036	1.590
PROTONATED 1:1 MOLYBDATE OXALATE COMPLEX					4.95	.01	.028	.036	1.715
PYROCATECHOL MOLYBDATE COMPLEX 1:1					6.03	.01	.028	.048	.175
QUATERNARY H-MOLYBDATE-PYROCATECHOL-OXALATE COMPLEX					5.93	.01	.028	.048	.244
6.31	.01	.028	.03	.175	5.83	.01	.028	.048	.331
6.23	.01	.028	.03	.244	5.73	.01	.028	.048	.433
6.13	.01	.028	.03	.324	5.65	.01	.028	.048	.563
6.03	.01	.028	.03	.419	5.55	.01	.028	.048	.702
5.93	.01	.028	.03	.526	5.45	.01	.028	.048	.858
5.83	.01	.028	.03	.643	5.35	.01	.028	.048	1.02
5.73	.01	.028	.03	.779	5.25	.01	.028	.048	1.19
5.65	.01	.028	.03	.916	5.15	.01	.028	.048	1.35
5.55	.01	.028	.03	1.062	5.05	.01	.028	.048	1.51
5.45	.01	.028	.03	1.208	4.95	.01	.028	.048	1.66
5.35	.01	.028	.03	1.355	4.85	.01	.028	.048	1.79
5.25	.01	.028	.03	1.501	4.75	.01	.028	.048	1.92
5.15	.01	.028	.03	1.627	5.93	.01	.028	.06	.156
5.05	.01	.028	.03	1.745	5.65	.01	.028	.06	.399
6.23	.01	.028	.036	.148	5.55	.01	.028	.06	.526
6.13	.01	.028	.036	.206	5.45	.01	.028	.06	.663
6.03	.01	.028	.036	.292	5.35	.01	.028	.06	.828
5.93	.01	.028	.036	.370	5.28	.01	.028	.06	.994
5.83	.01	.028	.036	.487	5.15	.01	.028	.06	1.179
5.73	.01	.028	.036	.604	5.05	.01	.028	.06	1.364
5.65	.01	.028	.036	.741	4.95	.01	.028	.06	1.52
5.55	.01	.028	.036	.877	4.85	.01	.028	.06	1.676
5.45	.01	.028	⸎036	1.014	4.75	.01	.028	.06	1.822
5.35	.01	.028	.036	1.169	-1. -1. -1. -1.				
5.25	.01	.028	.036	1.322	7 6 6 4.02 5.0 9.3592 8.3206 14.083 14.5 0				
5.15	.01	.028	.036	1.462	3 2 5 0.1 6 0.5 2 5 12				

4. EXLET: FORMATION CONSTANTS AND RELATED PARAMETERS FROM SPECTROPHOTOMETRIC DATA

4.1. Model Formulation

EXLET, derived from DISTR-LETAG[13] and EXT-DISTR-LETAG[14], has been set up to process distribution and extraction spectrophotometric data obtained from absorbance measurements of the organic phase.

Extraction equilibria are between two phases, aqueous and organic, and four components H, M, R, and X. The free concentration of the first component, not necessarily the proton, is measured potentiometrically. The distribution between the two phases is followed by (i) measurement of the distribution ratio D of a second component (the metal) via its concentration using an ion-selective electrode, atomic absorption spectrophotometry, or other appropriate methods; (ii) a spectrophotometric measurement of the organic phase.

For the general reaction:

$$pH + qM + rR + sX \rightleftharpoons H_pM_qR_rS_x \tag{16}$$

occurring in either the aqueous or organic phase, let C_M, C_R, C_X be the total concentrations of the components, and β_{pqrs} the formation (or extraction) constant for $H_pM_qR_rS_x$. Depending on whether a complex exists in the aqueous or organic phase an index f is assigned equal to zero or one, respectively. Two extraction equilibria relationships, describing the distribution ratio, D, and the absorbance, A, are

$$D = \frac{C_{M(org)}}{C_{M(aq)}} = \frac{\sum\limits_{j=1}^{NK} q_j \cdot \beta[H]^{p_j} [M]^{q_j} [R]^{r_j} [X]^{s_j} \cdot r_v^{f_j} \cdot f_j}{[M] + \sum\limits_{j=1}^{NK} q_j \cdot \beta_j[H]^{p_j} [M]^{q_j} [R]^{r_j} [X]^{s_j} \cdot r_v^{f_j} \cdot (1 - f_j)} \tag{17}$$

$$A = \sum\limits_{j=1}^{NK} \epsilon_j\beta_j[H]^{p_j} [M]^{q_j} [R]^{r_j} [X]^{s_j} \cdot f_j \tag{18}$$

where ϵ_j is the molar absorptivity of the jth species. The symbol β is used to indicate a formation, extraction, or distribution constant, depending on the nature of the components and the indices of the species. The ratio of the volumes of organic to aqueous phases is denoted by r_v. The total concentrations are given in moles per liter and are related to the aqueous phase, irrespective of the phase in which the component is originally found. The sum of squares of residuals, U, is minimized:

$$U = \sum\limits_{i=1}^{NB} w_i(Y^i_{calc} - Y^i_{exp})^2 \tag{19}$$

where Y_i is log D_i for distribution ratio data or A_i for absorbance data. The statistical weighting w_i is usually set to unity.

4.2. Description of EXLET

EXLET considers any set of complexes as defined by equation (17). The nature of the components H, M, R, and X is left largely to the user. R may be up to a tetraprotic acid with the pK_a's declared separately. The concentration of the first component, usually H^+, is measured using a glass electrode. However, it is possible that

H may be a second ligand provided that a suitable electrode is available. A total of eight equilibria, excluding the four pK_a's for R, may be investigated.

Absorbance data must be measured at a single wavelength. For every species with a nonzero molar absorptivity the maximum allowable number of reactions considered is lowered by one, since the molar absorptivities of the species will be determined by EXLET.

The MAIN program performs an organizational role and prints out a few parameters. Subroutine DATA reads in the experimental data and the equilibrium model description. The pK_a's for component R are read in, and H_4L is calculated, if appropriate. Subroutine UBBE calculates values of the dependent variable D_{calc} (or A_{calc}). Subroutine COGS[8] is used to calculate free concentrations in order that equations (17) and (18) may be evaluated. Subroutine SKRIK performs the usual functions of major printout and statistical analysis.

4.3. Data Input Requirements

The general setup of the data cards has been described in Section 1.4. There are some requirements that are specific to EXLET:

1. Title Card: **TITLE**
   A suitable title (80 characters).
2. Key Card: **(ISSW(I),I=1,7)**
   Value chosen according to Table 2.
   ISSW(7) is not used by EXLET.
3. Termination Card: **EPS,ITMAX,PSI,NX,MY,MX,NLS,NCL,SINST**
   EPS: ⎫

   ⋮   ⎬ As for Section 1.4.2, item 3.

   SINST: ⎭
4. Simulation Card: **(XK(I),I=1,NK),(SIGX(I),I=1,NK),(WEI(I),I=1,NK)**
   XK(I):   ⎫
   SIGX(I): ⎬ As for Section 1.4.2, item 4.
   WEI(I):   ⎭
   Note: (1) The order of the XK parameters is arbitary except that they must be in the order dictated by the order of the stoichiometric coefficients that appear on the Data Card.
   (2) If spectrophotometric data is supplied (KV=2) the molar absorptivities of the complex(es) being extracted ($f_j = 1$) must follow the particular XK value. This card is omitted for experimental data.
5. Data Card: **IRUR,KV,NA,NB**
   IRUR: Set equal to 6 for experimental; otherwise omit.
   KV: Set equal to 1 for distribution data;
       set equal to 2 for absorbance data.
   NA: Number of components in the model.
   NB: Number of data points.

Note: The following data are required, depending on the value of NA:

NA=2: Implies equilibria between H and R only.

Hence the following are needed:

CR(I),D(I),PH(I),RV(I): CR is the analytical concentration of R; D is the distribution ratio or absorbance value; PH is the measured pH; RV is volume ratio, organic:aqueous.

NA=3: Implies equilibria between H, M, and R.

NA=4: Implies equilibria between H, M, R, and X.

For either NA=3 or NA=4 the following parameters are required:

D(I),PH(I),CM(I),CR(I),CX(I),RV(I),PK1,PK2,PK3,PK4:

CM is the analytical concentration of M; CX is the analytical concentration of X, set to zero for NA=3; PK1 ... PK4 are the pK_a's for R, set to -1.0 for NA=3 and if, for NA=4, any or all are unknown.

IN: Number of complexes (NK) plus the number of molar absorptivities of complexes extracted into the organic phase ($f_j = 1$), if spectrophotometric data are read in.

P(J),Q(J),R(J),S(J),F(J):Stoichiometric coefficients for the jth complex $H_p M_q R_r X_s$, and the key $F(J)$ used to indicate whether the complex appears in the aqueous (=0) or organic (=1) phase. If a complex appears in both phases two sets of PQRS coefficients are given, one with $F(J) = 1$ and the other with $F(J) = 0$. Following the stoichiometric coefficients a set of coefficients are read in for the molar absorptivity of the complex, if absorbance data have been supplied.

6. Initial Guess Card: **IRUR,NK,NK,(XK(I),I=1,NK)**

   IRUR:

   $\left.\begin{array}{c} \vdots \\ \vdots \\ \end{array}\right\}$ As for Section 1.4.2, item 6(a).

   XK(I):

7. Matrix Card: **ISKIN,((I(K),J(I),S(I(K),J(K)),K=1,ISKIN)**

   ISKIN:

   $\left.\begin{array}{c} \vdots \\ \vdots \\ \end{array}\right\}$ As for Section 1.4.2, item 7.

   S( ):

8. Posk Removal Card: **IRUR,NK,$-$1,NEGK,(IK(I),I=1,NEGK)**

   IRUR:

   $\left.\begin{array}{c} \vdots \\ \vdots \\ \end{array}\right\}$ As for Section 1.4.2, item 6(c).

   IK(I):

9. Step Card: **IRUR,N,((I(K),STEK(I(K)),K=1,ISKIN)**

   IRUR:

   $\left.\begin{array}{c} \vdots \\ \vdots \\ \end{array}\right\}$ As for Section 1.4.2, item 8.

   STEK

10. Process Card: **IRUR,IRUR, . . .**

    IRUR: As for Section 1.4.2. For EXLET set to 2,5,13 or 2,5,11,12,13

4.4. Presentation of EXLET

4.4.1. Listing of EXLET

There follows a listing of the MAIN program and the subroutines SKRIK, UBBE, COGS, and DATA. Those routines listed in Section 1.5 are added to form the complete program EXLET.

```
 1    *#RUN=(NWARN) #05
 2    C   E X L E T   P R O G R A M
 3    C   INTEGRATED  DISTR-LETAG AND EXT-DISTR-LETAG  PROGRAMS
 4    C
 5    C   AUTHORS OF THIS REVISED VERSION: JOSEF HAVEL AND MILAN MELOUN
 6    C
 7    C
 8    C   PUBLICATIONS  1. J.HAVEL,M.VRCHLABSKY,J.SEKANINOVA AND J.KOMAREK
 9    C                    SCRIPTA FAC.SCI.NAT.UJEP BRUN.,CHEMIA 2,9:51-66
10    C                    (1979)   (DISTR-LETAG)
11    C   2. L.MULLEROVA,M.VRCHLABSKY AND J.HAVEL,SCRIPTA FAC.SCI.NAT.UNIV.
12    C   PURK.BRUN. 10,NO.1-2(CHEMIA),13-22(1980)  (EXT-DISTR-LETAG)
13    C
14    C   (MAIN PROGRAM)
15    C
16          DIMENSION XK(8),SIGXK(8),WEI(8)
17          INTEGER F
18          COMMON/TIT/ TITLE(20)/PLOT/ MY,MX,NLS,NCL,DUMMY(1930)/ISW/ISSW(7)
19          COMMON/FUNC/XEXP(100),YEXP(100),ERR(100),YCAL(100),W(100)
20         &/EX/ALFA(100),CM(100),CR(100),CX(100),AH(100),DELTA(100),P(100),
21         &TX(100),VX(100),UX(100),RV(100),MM(8),MN(8),ML(8),MS(8),F(8),
22         &B(12),AEPS(12),IN,KV,NA,NB
23          COMMON/KANAL/ INP,IOUT
24          EQUIVALENCE (UMIN,PSI)
25          EXTERNAL DATA,UBBE,SKRIK
26       10 INP=5
27          ISSW(1)=0
28          ISSW(2)=0
29          ISSW(3)=0
30          ISSW(4)=0
31          ISSW(5)=1
32          ISSW(6)=0
33          ISSW(7)=2
34          EPS=1.0E-6
35          ITMAX=40
36          PSI=0.3
37          NK=8
38          MY=0
39          MX=0
40          NLS=50
41          NCL=100
42          SINST=-1.0
43          IOU=6
44          IOUT=IOU
45          IR=1225
46          DO 60 I=1,100
47          W(I)= 1.0
48       60 CONTINUE
49          READ(INP,130,END=999)TITLE
50          WRITE(IOU,131)TITLE
51          CALL READI(ISSW,-7)
52          CALL READR(EPS,1)
53          CALL READI(ITMAX,1)
54          CALL READR(PSI,1)
55          CALL READI(NK,1)
56          CALL READI(MY,4)
57          CALL READR(SINST,1)
58          IF(SINST)30,30,20
59       20 CONTINUE
60          CALL READR(XK,NK)
```

```
61            CALL READR(SIGXK,NK)
62            CALL READR(WEI,NK)
63            CALL DATA(IOU,NB)
64            WRITE(IOU,100)SINST,(XK(I),SIGXK(I),WEI(I),I=1,NK)
65            DO 21 I=1,NB
66         21 W(I)=1.0
67            CALL UBBE(U,NK,XK,NB)
68            CALL SIMUL (XEXP,YCAL,YEXP,NB,SINST,IR)
69            DO  11 I=1,NB
70         11 ERR(I)=YEXP(I)-YCAL(I)
71            CALL WEIGHT(UBBE,NB,NK,XK,SIGXK,WEI,ISSW)
72            WRITE(IOU,110)
73            DO 12 I=1,NB
74         12 YCAL(I)=YEXP(I)-ERR(I)
75            WRITE(IOU,111)(I,XEXP(I),YCAL(I),ERR(I),YEXP(I),W(I),I=1,NB)
76            CALL STATS(IOU,NK,NB,XEXP,YEXP,ERR)
77            GO TO 330
78         30 WRITE(IOU,112)
79        330 U2=1.0E30
80            U1=U2
81            NAUT=-1
82            DO 50 IT=1,ITMAX
83            CALL LETAG(IOU,NAUT,NK,NB,XK,UMIN,ISSW,SIGXK,DATA,UBBE,SKRIK)
84            WRITE(IOU,120)IT,UMIN,(XK(I),I=1,NK)
85            NNAUT=IABS(NAUT)
86            GO TO (40,40,70),NNAUT
87         40 IF ((ABS((U2-UMIN)/U2)).LE.EPS)GO TO 70
88            U2=U1
89            NAUT=1
90         50 U1=UMIN
91         70 CALL SKRIK(NB,NK,IOU,XK,13,SIGXK)
92            GO TO 10
93        100 FORMAT(//31X,' T H E     S I M U L A T I O N     O F     M O D E L
94           & C U R V E  ',//' INSTRUMENTAL ERROR=',F9.6//6X,' PARAMETERS +-
95           & SIGMA        WEIGHT OF P.',(7(/1X,F14.5,2F14.5))/)
96        110 FORMAT(13X,'PH(EXP)      Y(ACCURATE) + ERROR =  Y(LOADED)   WEI
97           &GHT  '/,20X,' Y IS EITHER LOG D (LOG OF DISTRIBUTION RATIO) OR A'/
98           &,20X,'        A (ABSORBANCE OF THE ORG. PHASE)'/)
99        111 FORMAT(2X,I4,6X,F9.6,4X,F9.6,3X,F9.6,3X,F9.6,F13.6)
100       112 FORMAT(//31X,' T H E     E X P E R I M E N T A L     D A T A '//)
101       120 FORMAT(1X,I3,' ITER. U= ',1PE11.5,' PARAM.=',
102           &7(0PF12.3))
103       130 FORMAT (20A4)
104       131 FORMAT(1H1,120(1H*),//46X,' E X L E T    P R O G R A M ',//
105           &1X,120(1H*),//2X,20A4,//1X,120(1H*)//)
106       999 STOP
107           END



  1            SUBROUTINE DATA(IOU,N)
  2  C
  3  C --- D A T A E X ---
  4
  5            INTEGER F,MDESCR(15)
  6            COMMON/FUNC/PH(100),D(100),ERR(100),DV(100),W(100)
  7           &/EX/ALFA(100),CM(100),CR(100),CX(100),AH(100),DELTA(100),P(100),
  8           &TX(100),VX(100),UX(100),RV(100),MM(8),MN(8),ML(8),MS(8),F(8),
  9           &B(12),AEPS(12),IN,KV,NA,NP
 10           &/KANAL/INP
 11            DO 20 II=1,4
 12         20 B(II)=0.0
 13            CALL READI(KV,1)
 14            CALL READI(NA,1)
 15            CALL READI(N,1)
 16            NP=N
 17            IF(NA.GT.2.OR.KV.GT.1)GOTO 23
 18            CALL READR(CR(1),1)
 19  C   NA = 2
 20            WRITE(IOU,32)
 21            DO 21 I=1,N
 22            CALL READR(D(I),1)
 23            CALL READR(PH(I),1)
 24            CALL READR(RV(I),1)
```

```
25        21 AH(I)=EXP(-2.302585*PH(I))
26           DO 11 I=1,N
27        11 WRITE(IOU,34) D(I),PH(I),CR(1),RV(I)
28           GOTO 18
29        23 CONTINUE
30    C    NA GREATER THAN 2
31           WRITE(IOU,63)
32           DO 17 I=1,N
33           CALL READR(D(I),1)
34           CALL READR(PH(I),1)
35           CALL READR(CM(I),1)
36           CALL READR(CR(I),1)
37           CALL READR(CX(I),1)
38           CALL READR(RV(I),1)
39           WRITE(IOU,34) D(I),PH(I),CM(I),CR(I),CX(I),RV(I)
40        17 AH(I) = EXP(-2.302585*PH(I))
41        18 CONTINUE
42           CALL READR(PK1,1)
43           CALL READR(PK2,1)
44           CALL READR(PK3,1)
45           CALL READR(PK4,1)
46           WRITE(IOU,35) PK1,PK2,PK3,PK4
47           CALL READI(IN,1)
48           WRITE(IOU,36)
49           DO 33 J=1,IN
50           CALL READI(MN(J),1)
51           CALL READI(MM(J),1)
52           CALL READI(ML(J),1)
53           CALL READI(MS(J),1)
54           CALL READI(F(J),1)
55        33 WRITE(IOU,31) J,MN(J),MM(J),ML(J),MS(J),F(J)
56           WRITE(IOU,38)
57           DO 330 I=1,IN
58           READ(INP,37) MDESCR
59       330   WRITE(IOU,39) I,MDESCR
60        31 FORMAT(6I3)
61        32 FORMAT(1X,' T H E   D A T A '/,1X,'LOG D(OR A)',2X,'PH',10X,'CBTOT'
62           &,6X,'VOL.RATIO',/)
63        63 FORMAT('   T H E         D A T A    '/,2X,'D(OR A)',5X,
64           &'PH',6X,'CBTOT',10X,'CCTOT',10X,'CDTOT',10X,'VOL.RATIO',/)
65        34 FORMAT(1X,F6.3,6X,F6.3,2X,E12.5,3X,E12.5,3X,E12.5,3X,F6.3)
66        35 FORMAT(4F10.4,6X,'PK VALUES OF LIGAND(3-RD COMPONENT)')
67        36 FORMAT(/,' I P Q R S PHASE',/)
68        37 FORMAT(15A4)
69        38 FORMAT(/,' I    SPECIES DESCRIPTION',/)
70        39 FORMAT(I3,2X,15A4)
71           IF(PK1.LT.0.) GOTO 7
72           B(1)=10.**(-PK1)
73           IF(PK2.LT.0.) GOTO 7
74           B(2)=10.**(-PK1-PK2)
75           IF(PK3.LT.0.) GOTO 7
76           B(3)=10.**(-PK1-PK2-PK3)
77           IF(PK4.LT.0.) GOTO 7
78           B(4)=10.**(-PK1-PK2-PK3-PK4)
79         7 DO 10 I=1,N
80           AH(I) = EXP(-2.302585*PH(I))
81           ALFA(I)=(B(4)*1./AH(I)+B(3))*1./(AH(I)**3)
82        10 ALFA(I)=ALFA(I)+(B(2)*1./AH(I)+B(1))*1./AH(I)+1.
83           RETURN
84           END



1           SUBROUTINE UBBE(U,NK,XK,N)
2     C
3     C --- U B B E E X ---
4     C    CALCULATION OF ERROR SQUARE SUM FUNCTION U
5     C    UBBE CALCULATES FOR GIVEN SERIES OF EQUILIBRIUM CONSTANS THE SUM
6     C    OF SQUARES OF THE DEVIATIONS OF CALCULATED AND EXPERIMENTAL
7     C    VALUES D, I.E. D EXP - D CALC OR THE SUM OF SQUARES
8     C    OF THE DEVIATIONS OF CALCULATED AND EXP.ABSORBANCE VALUES A,
9     C    I.E. A EXP - A CALC
10    C    VX - EQUIL.CONCN. OF THE METAL (SECOND COMPONENT)
11    C    TX - EQUIL.CONCN. OF THE FIRST LIGAND (THIRD COMPONENT)
```

```
12  C      UX - EQUIL.CONCN. OF THE SECOND LIGAND (FOURTH COMPONENT)
13  C
14         DIMENSION  XK(NK)
15         INTEGER F
16         COMMON/FUNC/PH(100),D(100),ERR(100),DV(100),W(100)
17        &/EX/ALFA(100),CM(100),CR(100),CX(100),AH(100),DELTA(100),P(100),
18        &TX(100),VX(100),UX(100),RV(100),MM(8),MN(8),ML(8),MS(8),F(8),
19        &B(12),AEPS(12),IN,KV,NA,NP
20         N=NP
21         NB=N
22         IF(KV.EQ.2.OR.KV.EQ.3) GO TO 70
23         IF(NA.EQ.2) GOTO 50
24  C      H+-METAL-LIGAND EQUIL.
25  C      H+-METAL-1ST LIGAND-2ND LIGAND EQUIL.
26  C
27         U = 0.
28         DO 4 K=1,IN
29         II=4+K
30         IF(XK(K).GT.50.) GO TO 1
31         IF(XK(K).LT.(-50.)) GO TO 2
32         GOTO 3
33       1 B(II)=10.**37
34         GOTO 4
35       2 B(II)=0.
36         GOTO 4
37       3 B(II)=10.**XK(K)
38       4 CONTINUE
39         DO 10 I=1,N
40         IF(I.EQ.1) GOTO 100
41         JJ=I-1
42         TX(I)=TX(JJ)
43         VX(I)=VX(JJ)
44         UX(I)=UX(JJ)
45         GOTO 101
46     100 TX(I)=CR(I)/ALFA(I)
47         VX(I)=CM(I)
48         UX(I)=CX(I)
49     101 CONTINUE
50         CALL COGS(I)
51         DD=0.0
52         D3=VX(I)
53         DO 20 K=1,IN
54         II=4+K
55         D1=FLOAT(MM(K))*B(II)*AH(I)**MN(K)*VX(I)**MM(K)*TX(I)**ML(K)*UX(I)
56        &**MS(K)
57         D2=D1*(1.-FLOAT(F(K)))
58         D1=D1*FLOAT(F(K))
59         DD=DD+D1
60         D3 = D3+D2
61      20 CONTINUE
62         DV(I)=DD/D3
63         DV(I)=ALOG10(DV(I))
64      10 U=U+((D(I)-DV(I))**2)*W(I)
65         RETURN
66      50 CONTINUE
67  C      H+-LIGAND EQUIL.
68         U = 0.0
69         DO 22 K=1,IN
70      22 B(K)=10.**XK(K)
71         DO 60 I=1,N
72         TX(I)=CR(1)
73         VX(I)=1.
74         UX(I)=1.
75         CALL COGS(I)
76         DD=0.0
77         D3=TX(I)
78         DO 30 K=1,IN
79         D1=FLOAT(ML(K))*B(K)*AH(I)**MN(K)*TX(I)**ML(K)
80         D2=D1*(1.-FLOAT(F(K)))
81         D1=D1*FLOAT(F(K))
82         DD=DD+D1
83         D3= D3+D2
84      30 CONTINUE
85         DV(I)=DD/D3
86         DV(I)=ALOG10(DV(I))
87      60 U=U+((D(I)-DV(I))**2)*W(I)
```

```
 88            RETURN
 89    C
 90    C       SPECTROPHOTOMETRY
 91       70 DO 9 K=1,IN
 92            II=4+K
 93            B(II)=0.0
 94        9 AEPS(II)=0.0
 95            U=0.0
 96            DO 5 K=1,IN
 97            II=4+K
 98            IF(K.EQ.1) GO TO 7
 99            IF(F(K).EQ.0.AND.F(K-1).EQ.1) GO TO 6
100            GO TO 7
101        6 IE=II-1
102            AEPS(IE)=XK(K)
103            GO TO 5
104        7 B(II)=10.**XK(K)
105        5 CONTINUE
106            DO 41 I=1,N
107       41 DV(I)=0.0
108            DO 40 I=1,N
109            IF(I.EQ.1) GOTO 200
110            JJ=I-1
111            TX(I)=CR(JJ)
112            VX(I)=CM(JJ)
113            UX(I)=CX(JJ)
114            GOTO 201
115      200 TX(I)=CR(I)/ALFA(I)
116            VX(I)=CM(I)
117            UX(I)=CX(I)
118      201 CONTINUE
119            CALL COGS(I)
120            DO 31 K=1,IN
121            II=4+K
122            DV(I)=DV(I)+
123        &      AEPS(II)*B(II)*AH(I)**MN(K)*VX(I)**MM(K)*TX(I)**ML(K)*UX(I)
124        &**MS(K) *FLOAT(F(K))
125       31 CONTINUE
126       40 U=U+((D(I)-DV(I))**2)*W(I)
127            RETURN
128            END


  1            SUBROUTINE COGS(I)
  2    C       VX - EQUIL.CONCN. OF THE METAL (SECOND COMPONENT)
  3    C       TX - EQUIL.CONCN. OF THE FIRST  LIGAND (THIRD COMP.)
  4    C       UX - EQUIL.CONCN. OF THE SECOND LIGAND (FOURTH COMP.)
  5    C
  6            DIMENSION C(8),TERM(8)
  7            INTEGER F
  8            DIMENSION XK(8),SIGXK(8),WEI(8)
  9            COMMON/FUNC/PH(100),D(100),ERR(100),DV(100),W(100)
 10        &/EX/ALFA(100),CM(100),CR(100),CX(100),AH(100),DELTA(100),P(100),
 11        &TX(100),VX(100),UX(100),RV(100),MM(8),MN(8),ML(8),MS(8),F(8),
 12        &B(12),AEPS(12),IN,KV,NA,NP
 13            COMMON/KANAL/ INP,IOUT
 14            Y1= CM(I)*0.001
 15            IF(NA.EQ.2) GOTO 55
 16            Y3= CR(I)*0.001
 17            IF(NA.EQ.3) GOTO 7
 18            Y5= CX(I)*0.001
 19            GOTO 9
 20        7 Y5 = 1.
 21            GOTO 9
 22       55 Y3=CR(1)*0.001
 23            NIT  = 0
 24            DO 1 K=1,IN
 25        1 TERM(K) = B(K)*AH(I)**MN(K)
 26            GOTO 2
 27        9 CONTINUE
 28            NIT=0
 29            DO 222 K=1,IN
 30            II=4+K
```

```
31          222 TERM(K)=B(II)*AH(I)**MN(K)
32            2 DO 3 K=1,IN
33            3 C(K) = TERM(K)
34              DO 4 K=1,IN
35            4 C(K) = C(K)*VX(I)**MM(K)
36              DO 5 K=1,IN
37            5 C(K) = C(K)*TX(I)**ML(K)*UX(I)**MS(K)*RV(I)**F(K)
38              NIT = NIT +1
39              IF(NA.EQ.2) GOTO 6
40              BO = VX(I)
41              DO 8 K=1,IN
42            8 BO = BO+C(K)*FLOAT(MM(K))
43              RATIO = BO/CM(I)
44              VY = VX(I)/SQRT(RATIO)
45              Y2 =   ABS(BO-CM(I))
46              AXO = UX(I)
47              DO 111 K=1,IN
48          111 AXO = AXO+C(K)*FLOAT(MS(K))
49              RATIO = AXO/CX(I)
50              UY = UX(I)/SQRT(RATIO)
51              Y6 = ABS(AXO-CX(I))
52              ALO = TX(I)
53              DO 88 K=1,4
54           88 ALO = ALO+ALO*B(K)/(AH(I)**K)
55              DO 71 K=1,IN
56           71 ALO = ALO+C(K)*FLOAT(ML(K))
57              RATIO = ALO/CR(I)
58              TY = TX(I)/SQRT(RATIO)
59              Y4 = ABS(ALO-CR(I))
60              GOTO 99
61            6 ALO=TX(I)
62              DO 10 K=1,IN
63           10 ALO = ALO+C(K)*FLOAT(ML(K))
64              RATIO=ALO/CR(1)
65              TY=TX(I)/SQRT(RATIO)
66              Y4=ABS(ALO-CR(1))
67              GOTO 77
68           99 IF(NIT-9999)11,11,999
69           77 IF(NIT-9999)22,22,999
70           22 IF(Y3-Y4)25,23,23
71           23 TX(I)=TY
72              RETURN
73           25 TX(I)=TY
74              GOTO 2
75           11 IF(Y1-Y2) 15,12,12
76           12 IF(Y3-Y4) 15,13,13
77           13 IF(Y5-Y6) 15,14,14
78           14 TX(I)=TY
79              VX(I)=VY
80              UX(I)=UY
81              RETURN
82           15 TX(I)=TY
83              VX(I)=VY
84              UX(I)=UY
85              GOTO 2
86          999 WRITE(IOUT,998) I
87          998 FORMAT(' ITERATIONS DID NOT CONVERGE IN',I2,'-TH SOLUTION')
88              RETURN
89              END




 1              SUBROUTINE SKRIK(NB,NK,IOU,XK,IRUR,SIGXK)
 2      C
 3      C --- S K R I K  ---
 4      C       OUTPUT OF PARAMETERS AND THEIR STANDARD DEVIATIONS
 5      C       STATISTICAL TEST OF CURVE FITTING......ANALYSIS OF REZIDUALS
 6      C
 7              DIMENSION XK(NK),SIGXK(NK),DIF(100)
 8              INTEGER F
 9              DIMENSION XX(100,2),YY(100,2),NDATA(2)
10              COMMON/FUNC/PH(100),D(100),ERR(100),DV(100),W(100)
11             &/EX/ALFA(100),CM(100),CR(100),CX(100),AH(100),DELTA(100),P(100),
12             &TX(100),VX(100),UX(100),RV(100),MM(8),MN(8),ML(8),MS(8),F(8),
```

```
13          &B(12),AEPS(12),IN,KV,NA,NP
14          &/ISW/ISSW(7)
15           COMMON /TEXT/ IA(35)
16           COMMON/TIT/ TITLE(20) /PLOT/ MY,MX,NLS,NCL,XLINE(5),YLINE(5),
17          &AREA(31,60),YSCALE(60)
18           LOGICAL ISYMBL(2)
19           DATA NDMAX/100/
20           DATA ISYMBL(1)/4H****/
21           DATA ISYMBL(2)/4H0000/
22           N=NB
23           WRITE(IOU,99)
24           WRITE(IOU,109)(XK(IK),SIGXK(IK),IK=1,NK)
25           IF(IRUR.LT.11)RETURN
26           CALL UBBE(U,NK,XK,NB)
27       56  IF(KV.EQ.0.OR.KV.EQ.1) GO TO 65
28           IF(NA.EQ.4) WRITE(IOU,161)
29           IF(NA.LT.4) WRITE(IOU,164)
30      161  FORMAT(3X,'PH',5X,'RV',8X,'CM',10X,'VX',10X,'CR',10X,'TX',10X,'CX'
31          &,10X,'UX',8X,'A EXP',5X,'A CALC',5X,'DELTA',7X,'P')
32      164  FORMAT(3X,'PH',5X,'RV',8X,'CM',10X,'VX',10X,'CR',10X,'TX',
33          &8X,'A EXP',5X,'A CALC',5X,'DELTA',7X,'P')
34           DO 300 I=1,N
35      300  DELTA(I)=D(I)-DV(I)
36           DO 142 I=1,N
37           DD=0.0
38           DO 41 K=1,IN
39           II=4+K
40           D1=FLOAT(MM(K))*B(II)*AH(I)**MN(K)*VX(I)**MM(K)*TX(I)**ML(K)*UX(I)
41          &**MS(K)*RV(I)**F(K)
42           D1=D1*FLOAT(F(K))
43       41  DD=DD+D1
44           P(I)=DD*100./CM(I)
45           IF(NA.LT.4) GOTO 45
46           WRITE(IOU,131)PH(I),RV(I),CM(I),VX(I),CR(I),TX(I),CX(I),UX(I),
47          &D(I),DV(I),DELTA(I),P(I)
48           GOTO 142
49       45  WRITE(IOU,130)PH(I),RV(I),CM(I),VX(I),CR(I),TX(I),
50          &D(I),DV(I),DELTA(I),P(I)
51      142  CONTINUE
52           GO TO 5
53       65  CONTINUE
54           IF(NA.EQ.4) WRITE(IOU,160)
55           IF(NA.LT.4) WRITE(IOU,165)
56           IF(NA.EQ.2)GOTO 2
57           DO 146 I=1,N
58           DELTA(I) = D(I)-DV(I)
59           DD=0.0
60           DO 20K=1,IN
61           II=4+K
62           D1=FLOAT(MM(K))*B(II)*AH(I)**MN(K)*VX(I)**MM(K)*TX(I)**
63          &ML(K)*UX(I)**MS(K)*RV(I)**F(K)
64           D1=D1*FLOAT(F(K))
65       20  DD=DD+D1
66           P(I)=DD*100./CM(I)
67           IF(NA.LT.4) GOTO 145
68      140  WRITE(IOU,131)PH(I),RV(I),CM(I),VX(I),CR(I),TX(I),CX(I),UX(I),
69          &D(I),DV(I),DELTA(I),P(I)
70           GOTO 146
71      145  WRITE(IOU,130)PH(I),RV(I),CM(I),VX(I),CR(I),TX(I),
72          &D(I),DV(I),DELTA(I),P(I)
73      146  CONTINUE
74           GOTO 5
75        2  DO 141 I=1,N
76           DELTA(I)=D(I)-DV(I)
77           DD=0.0
78           DO 31 K=1,IN
79           D1=FLOAT(ML(K))*B(K)*AH(I)**MN(K)*TX(I)**ML(K)*RV(I)**F(K)
80           D1=D1*FLOAT(F(K))
81       31  DD=DD+D1
82           P(I)=DD*100./CR(1)
83      141  WRITE(IOU,130)PH(I),RV(I),CM(I),VX(I),CR(I),TX(I),
84          &D(I),DV(I),DELTA(I),P(I)
85      160  FORMAT(3X,'PH',5X,'RV',8X,'CM',10X,'VX',10X,'CR',10X,'TX',10X,
86          &'CX',10X,'UX',8X,'LOG D',5X,'LOG DV',5X,'DELTA',7X,'P')
87      165  FORMAT(3X,'PH',5X,'RV',8X,'CM',10X,'VX',10X,'CR',10X,'TX',
88          &8X,'LOG D',5X,'LOG DV',5X,'DELTA',7X,'P')
```

```
89      130 FORMAT(1X,F5.2,1X,F7.3,4E12.5,1X,F10.7,1X,F10.7,1X,F10.7,1X,F6.2)
90      131 FORMAT(1X,F5.2,1X,F7.3,6E12.5,1X,F10.7,1X,F10.7,1X,F10.7,1X,F6.2)
91        5 CONTINUE
92      100 FORMAT(2F10.0,7I2)
93          DO 10 I=1,NB
94       10 DIF(I)=D(I)-DV(I)
95          WRITE(IOU,101)
96          WRITE(IOU,102)(I,PH(I),D(I),DV(I),DIF(I),I=1,NB)
97          IF(IRUR.LT.12)RETURN
98          CALL STATS(IOU,NK,NB,PH,D,DIF)
99          IF(IRUR.LT.13)RETURN
100         NDATA(1)=NB
101         NDATA(2)=NB
102         DO 50 I=1,NB
103         XX(I,1)=PH(I)
104         YY(I,1)=D(I)
105         YY(I,2)=DV(I)
106      50 XX(I,2)=PH(I)
107         WRITE(IOU,2001)TITLE
108    C  3 CALL PLOTT(XX,YY,NDATA,NDMAX,ISYMBL, 2,XLINE,MX,YLINE,MY,NLS,
109    C     &NCL, 1, 1,AREA,YSCALE)
110         NPAGE=0
111         N=2
112         CALL PRTPLT(XX,YY,NDATA,N,NPAGE,NDMAX)
113      99 FORMAT(' ***** S K R I K  (OUTPUT)  *****',/,1X,33(1H*))
114     109 FORMAT(/1X,' PARAMETERS AND THEIR STANDARD DEVIATIONS : '/(7(4X,
115         &F14.5, ' +- ',F14.5)))/)
116     101 FORMAT('  I   PH(EXP)      A(EXP)        A(CAL)       RESIDUAL',
117         &/13X,'OR LOG D(EXP)   LOG D(CAL)')
118     102 FORMAT(1X,I2,2X,F8.5,F13.4,F13.4,F13.4)
119    2001 FORMAT(1H1, '       GRAPH IS PRINTED FOR SYSTEM:',6X,20A4,//)
120         RETURN
121         END
```

4.4.2. Input Data

There follows a listing of a typical data set for EXLET. The data are taken form the extraction of 1-(2-pyridylazo)-2-naphthol (PAN) into methyl iso-butyl ketone.[15] The initial values for the pK_a's of PAN [XK(1)] and for the K_D [XK(2)] were obtained from grahical analysis. Data are also included for the reaction of cadmium with PAN under the same conditions.

```
¶<-- Column 1

EXLET: TEST DATA NO.3 EXTRACTION OF 1-PAN INTO METHYL ISOBUTYL KETONE

 1 1 1 0 1 0 1

1.0E-6  40  0.3  2 0 0 50 100

  -1.0

  6

 1 2

43

0.001

1.999    0.0    1.0

1.999    1.0    1.0

1.999    2.0    1.0

1.998    2.4    1.0
```

1.997	2.8	1.0
1.995	3.0	1.0
1.993	3.2	1.0
1.989	3.4	1.0
1.983	3.6	1.0
1.973	3.8	1.0
1.958	4.0	1.0
1.936	4.2	1.0
1.902	4.4	1.0
1.854	4.6	1.0
1.787	4.8	1.0
1.698	5.0	1.0
1.587	5.2	1.0
1.454	5.4	1.0
1.302	5.6	1.0
1.136	5.8	1.0
0.958	6.0	1.0
0.773	6.2	1.0
0.583	6.4	1.0
0.389	6.6	1.0
0.291	6.7	1.0
0.193	6.8	1.0
0.094	6.9	1.0
-0.004	7.0	1.0
-0.202	7.2	1.0
-0.401	7.4	1.0
-0.601	7.6	1.0
-0.800	7.8	1.0
-1.000	8.0	1.0
-1.200	8.2	1.0
-1.400	8.4	1.0
-1.600	8.6	1.0
-1.800	8.8	1.0
-2.000	9.0	1.0
-2.200	9.2	1.0
-2.400	9.4	1.0
-2.600	9.6	1.0
-2.800	9.8	1.0
-3.000	10.0	1.0

```
-1.0 -1.0 -1.0 -1.0
2
-1 0 1 0 0
0 0 1 0 1
DISSOCN. OF THE LIGAND
DISTRIBUTION COEF. OF THE LIGAND
7 2 2 -5.0 2.0 0
7 2 -1  1 1
3 2 1 0.1 2 0.1 2     5  10 11 13
EXLET: TEST DATA NO.4  CD(II) EXTN WITH 1-PAN INTO METHYL ISOBUTYL KETONE
 1 0 0 0 1 0 1
0.0001 20 0.3  5 0 0 50 100
 -1.0
6   1   3   43
```

-1.34	4.60	0.00004	0.002	1.0	1.0
-0.95	4.84	0.00004	0.002	1.0	1.0
-0.59	4.98	0.00004	0.002	1.0	1.0
-0.48	5.09	0.00004	0.002	1.0	1.0
-0.15	5.27	0.00004	0.002	1.0	1.0
-0.05	5.33	0.00004	0.002	1.0	1.0
0.35	5.50	0.00004	0.002	1.0	1.0
0.47	5.53	0.00004	0.002	1.0	1.0
0.65	5.64	0.00004	0.002	1.0	1.0
0.80	5.78	0.00004	0.002	1.0	1.0
1.15	5.96	0.00004	0.002	1.0	1.0
1.48	6.15	0.00004	0.002	1.0	1.0
1.79	6.34	0.00004	0.002	1.0	1.0
2.10	6.55	0.00004	0.002	1.0	1.0
2.51	7.00	0.00004	0.002	1.0	1.0
2.75	7.41	0.00004	0.002	1.0	1.0
2.82	7.70	0.00004	0.002	1.0	1.0
2.82	8.60	0.00004	0.002	1.0	1.0
2.82	9.55	0.00004	0.002	1.0	1.0
-0.48	5.37	0.00004	0.001	1.0	1.0
0.11	5.69	0.00004	0.001	1.0	1.0
0.59	5.95	0.00004	0.001	1.0	1.0
0.96	6.14	0.00004	0.001	1.0	1.0
1.39	6.41	0.00004	0.001	1.0	1.0
1.57	6.49	0.00004	0.001	1.0	1.0

2.21	7.00	0.00004	0.001	1.0	1.0
2.66	7.56	0.00004	0.001	1.0	1.0
2.82	7.98	0.00004	0.001	1.0	1.0
2.82	9.07	0.00004	0.001	1.0	1.0
0.00	5.93	0.00001	0.0005	1.0	1.0
0.50	6.20	0.00001	0.0005	1.0	1.0
0.80	6.36	0.00001	0.0005	1.0	1.0
1.28	6.63	0.00001	0.0005	1.0	1.0
1.77	6.93	0.00001	0.0005	1.0	1.0
-0.08	5.59	0.00001	0.001	1.0	1.0
-0.08	5.59	0.00002	0.001	1.0	1.0
-0.08	5.59	0.00003	0.001	1.0	1.0
0.52	5.59	0.00004	0.002	1.0	1.0
0.26	5.59	0.00004	0.0015	1.0	1.0
-0.08	5.59	0.00004	0.001	1.0	1.0
-0.42	5.59	0.00004	0.0007	1.0	1.0
-0.73	5.59	0.00004	0.0005	1.0	1.0
-1.38	5.59	0.00004	0.000223	1.0	1.0
10.18	-1.0	-1.0	-1.0		

```
 5
 1 0 1 0 0
 0 0 1 0 1
-1 1 1 0 0
-2 1 2 0 0
-2 1 2 0 1
PROTONATION OF LIGAND
DISTRIBUTION COEF. OF THE LIGAND
CD(L) COMPLEX IN AQ.PHASE
CD(L)(L) COMPLEX IN AQ.PHASE
CD(L)(L) COMPLEX IN ORG. PHASE
7 5 5 2.08 2.82 -1.18 -2.36 0.46 0 7 5 -1 3  3 4 5
3 3 3 0.2 4 0.2 5 0.2  2 5 13
```

5. OTHER PROGRAMS OF THE ABLET FAMILY

Three other programs have been written that perform the analysis of solution equilibrium related data. They are designed in a manner very similar to the described in Sections 1–4. The pattern and build-up to the ABLET family has been established in these earlier sections and consequently only the objective functions for each of the remaining three programs will be described.

5.1. DHLET: Estimation of Thermodynamic Dissociation Constants and Extended Debye–Huckel Parameters

The dissociation equilibrium of an ion with charge Z, $HL^z \rightleftharpoons L^{z-1} + H^+$, is characterized by the thermodynamic constant $K_a^T = a_H \cdot a_L / a_{HL}$. The dependence of the mixed dissociation constant $K_a = a_H \cdot [L]/[HL]$ on ionic strength is given by the extended Debye–Huckel equation:

$$pK_a = pK_a^T - A \cdot I^{1/2} \left[\frac{(Z - 1)^2}{1 + B\mathring{a}_L I^{1/2}} - \frac{Z^2}{1 + B\mathring{a}_{HL} I^{1/2}} \right] + (C_{HL} - C_L)I \quad (20)$$

where the constants A and B have the values 0.5115 mole$^{-1/2}$ $l^{1/2}$ K$^{3/2}$ and 0.3291 m^{-10} mole$^{-1/2}$ $l^{1/2}$ K$^{1/2}$, respectively, for aqueous solutions at 25°C. Assume that the ion-size parameter, $\mathring{a}$, for HL^z and L^{z-1} equals $1 - 2Z$. Introducing a salting out coefficient, C, defined as $C_{HL} - C_L$, equation (20) may be approximated as

$$pK_a = pK_a^T - A \cdot I^{1/2}(1 - 2Z)/(1 + B \cdot \mathring{a}I^{1/2}) + C \cdot I \quad (21)$$

The independent variable is I and the measured pK_a is the dependent variable. The three parameters estimated by DHLET are pK_a^T, $\mathring{a}$, and C.[16] The function minimized is

$$U = \sum_{i=1} w_i (pK_{a,\text{calc}}^i - pK_{a,\text{exp}}^i)^2 \quad (22)$$

5.2. DCLET: Estimation of Successive Dissociation Constants from pH–Absorbance Data

For the weak acid H_jL that absorbs radiation the absorbance of a solution $C_{H_jL}M$ is given by

$$A = d \cdot C_{H_jL} \cdot \frac{\epsilon_L + \sum_{j=1}^{J} \epsilon_{H_jL} \cdot 10^{(\log \beta_{H_jL} - j \cdot pH)}}{1 + \sum_{j=1}^{J} 10^{(\log \beta_{H_jL} - j \cdot pH)}} \quad (23)$$

where d is the path length, ϵ_L, ϵ_{HL}, $\cdots$, ϵ_{H_jL} are the molar absorptivities of H_jL at the selected wavelength and β_{H_jL} is the overall formation constant for the jth protonation step. DCLET[17] minimizes U:

$$U = \sum_i w_i (A_{\text{calc}}^i - A_{\text{exp}}^i)^2 \quad (24)$$

by finding the best values of the pK_a's and hence β_{H_jL}, and the molar absorptivities for each protonated species and for the fully deprotonated species. A^i_{calc} is obtained from equation (23) using various pH's. The program can handle 50 absorbances and up to three pK_a's with four molar absorptivities. The program has been employed to evaluate all seven parameters for 7-(3-carboxy-phenylazo)-8-hydroxyquinoline-5-sulfonic acid.[18]

5.3. NCLET: Evaluation of Competitive Equilibria in a Chelatometric Titration

The end point of a chelatometric titration, using a metallochromic indicator, is based on the displacement reaction between the indicator complex, $MInd_n$, and the titrant Y:

$$MInd_n + Y \rightleftharpoons MY + nInd$$

Assuming that the uncomplexed indicator is predominately in one protonated form H_jInd, the absorbance A, at a particular wavelength and pH, is given by

$$A = A_{MInd} - (A_{MInd} - A_{Ind})\alpha \tag{25}$$

where α is the fraction of the indicator in the form H_jInd not complexed to the metal. Hence when $\alpha = 0$ all the indicator is bound to the metal. The equation for the chelatometric titration curve, expressed as $v = f(A)$, is given by

$$v_i = R\,\alpha_i^n/(1 - \alpha_i) - S \cdot \alpha_i^n - T(1 - \alpha_i)/\alpha_i^n + Q\alpha_i + W \tag{26}$$

where

$$R = K'_{MInd}\,C_{Ind}^{n-1} \cdot n \cdot V_{eq}/K'_{MY} \tag{27a}$$

$$S = K'_{MInd} \cdot C_{Ind}^n \cdot n \cdot V/(K'_{MY} \cdot f_Y) \tag{27b}$$

$$T = 1/(K'_{MInd} \cdot n \cdot C_{Ind}^{n-1} \cdot f_Y) \tag{27c}$$

$$Q = C_{Ind} \cdot V/N \cdot f_Y) \tag{27d}$$

$$W = V_{eq} - V/f_Y \cdot (C_{Ind}/N + 1/K'_{MY}) \tag{27e}$$

$$\alpha = (A_{MInd} - A(1 + V/V_o))/(A_{MInd} - A_{Ind}) \tag{27f}$$

$$K'_{MInd} = [MInd_n]/[M] \cdot [Ind]^n \tag{28}$$

$$K'_{MY} = [MY']/[M'] \cdot [Y'] \tag{29}$$

$$C_Y = [Y'] + [MY'] = a \cdot C_M \tag{30}$$

$$C_M = [M'] + [MY'] + [MInd_n] \tag{31}$$

5.3.1. Listing of MAIN, DATA, UBBE, and SKRIK for NCLET

Since the derivation of the function to be minimized and to calculation of A^i_{calc} is not straightforward, listings of DATA and UBBE have been provided, together with the MAIN program and SKRIK, for completeness. To these are added the subroutines listed in Section 1.5.

```
  1     *#RUN=(NWARN) #Ø5
  2     C
  3     C        N C L E T    P R G R A M
  4     C        (MAIN PROGRAM)
  5     C        (MILAN MELOUN, JOSEF CERMAK,, TALANTA 23,15 (1976))
  6     C
  7     C
  8             DIMENSION XK(5),SIGXK(5),WEI(5)
  9             COMMON/TIT/ TITLE(2Ø)/PLOT/ MY,MX,NLS,NCL,DUMMY(193Ø)/ISW/ISSW(7)
 1Ø             COMMON/FUNC/XEXP(5Ø),YEXP(5Ø),ERR(5Ø),YCAL(5Ø),W(5Ø)
 11             COMMON/KANAL/ INP,IOUT
 12             EQUIVALENCE (UMIN,PSI)
 13             EXTERNAL DATA,UBBE,SKRIK
 14        1Ø INP=5
 15             ISSW(1)=Ø
 16             ISSW(2)=Ø
 17             ISSW(3)=Ø
 18             ISSW(4)=Ø
 19             ISSW(5)=1
 2Ø             ISSW(6)=Ø
 21             ISSW(7)=2
 22             EPS=1.ØE-6
 23             ITMAX=4Ø
 24             PSI=Ø.3
 25             NK=5
 26             MY=Ø
 27             MX=Ø
 28             NLS=5Ø
 29             NCL=1ØØ
 3Ø             SINST=-1.Ø
 31             IOU=6
 32             IOUT=IOU
 33             IR=1225
 34             DO 6Ø I=1,5Ø
 35        6Ø W(I)= 1.Ø
 36             READ(INP,13Ø,END=999)TITLE
 37             WRITE(IOU,131)TITLE
 38             CALL READI(ISSW,-7)
 39             CALL READR(EPS,1)
 4Ø             CALL READI(ITMAX,1)
 41             CALL READR(PSI,1)
 42             CALL READI(NK,1)
 43             CALL READI(MY,4)
 44             CALL READR(SINST,1)
 45             IF(SINST)3Ø,3Ø,2Ø
 46        2Ø CALL READR(XK,NK)
 47             CALL READR(SIGXK,NK)
 48             CALL READR(WEI,NK)
 49             CALL DATA(IOU,NB)
 5Ø             WRITE(IOU,1ØØ)SINST,(XK(I),SIGXK(I),WEI(I),I=1,NK)
 51             DO 21 I=1,NB
 52        21 W(I)=1.Ø
 53             CALL UBBE(U,NK,XK,NB)
 54             CALL SIMUL (XEXP,YCAL,YEXP,NB,SINST,IR)
 55             DO  11  I=1,NB
 56        11 ERR(I)=YEXP(I)-YCAL(I)
 57             CALL WEIGHT(UBBE,NB,NK,XK,SIGXK,WEI,ISSW)
 58             WRITE(IOU,11Ø)
 59             DO 12 I=1,NB
 6Ø        12 YCAL(I)=YEXP(I)-ERR(I)
 61             WRITE(IOU,111)(I,XEXP(I),YCAL(I),ERR(I),YEXP(I),W(I),I=1,NB)
 62             CALL STATS(IOU,NK,NB,XEXP,YEXP,ERR)
 63             GO TO 33Ø
```

```
64        30  WRITE(IOU,112)
65       330  U2=1.0E30
66            U1=U2
67            NAUT=-1
68            DO 50 IT=1,ITMAX
69            CALL LETAG(IOU,NAUT,NK,NB,XK,UMIN,ISSW,SIGXK,DATA,UBBE,SKRIK)
70            WRITE(IOU,120)IT,UMIN,(XK(I),I=1,NK)
71            NNAUT=IABS(NAUT)
72            GO TO (40,40,70),NNAUT
73        40  IF ((ABS((U2-UMIN)/U2)).LE.EPS)GO TO 70
74            U2=U1
75            NAUT=1
76        50  U1=UMIN
77        70  CALL SKRIK(NB,NK,IOU,XK,13,SIGXK)
78            GO TO 10
79       100  FORMAT(//31X,' T H E     S I M U L A T I.O N     O F     M O D E L
80           & C U R V E ',//' INSTRUMENTAL ERROR=',F9.6//6X,' PARAMETERS +-
81           & SIGMA     WEIGHT OF P.',(7(/1X,F14.5,2F14.5))/)
82       110  FORMAT(13X,' V(EXP)     A(ACCURATE) + ERROR = A(LOADED)    WEI
83           &GHT '/)
84       111  FORMAT(2X,I4,6X,F9.6,4X,F9.6,3X,F9.6,3X,F9.6,F13.6)
85       112  FORMAT(//31X,' T H E     E X P E R I M E N T A L     D A T A '//)
86       120  FORMAT(1X,I3,' ITER. U= ',1PE11.5,' PARAM.=',
87           &7(0PF12.3))
88       130  FORMAT (20A4)
89       131  FORMAT(1H1,120(1H*),//46X,' N C L E T     P R O G R A M ',//
90           &1X,120(1H*),//2X,20A4,//1X,120(1H*)//)
91       999  STOP
92            END
```

```
1             SUBROUTINE DATA(IOU,NB)
2     C
3     C --- D A T A N C ---
4     C     INPUT OF CONSTANTS AND SOME PRELIMINARY CALCULATIONS WITH DATA
5     C
6             DIMENSION VA(100),AA(2,4)
7             COMMON/FUNC/V(50),A(50),ERROR(50),ACAL(50),W(50)/KANAL/NI
8            &/NC1/N,NBI,VO,FY,VM,AMIN,AMAX,ASB,AI,H,X5,ESP,GMI,GMY,AN
9            &/NC2/DELT(50),G(50),AL(50)
10        1  CALL READI(N,2)
11           CALL READR(VO,10)
12           WRITE(IOU,97)
13           WRITE(IOU,98)N,VO,FY,VM,AMIN,AMAX,H,AI
14           NB=IABS(NBI)
15           WRITE(IOU,100)
16           DO 3 I=1,2
17           CALL READR(A,4)
18           S=1.
19           S1=0.
20           DO 4 J=1,4
21           S1=S1+A(J)-H
22           S=S+EXP(S1*2.302585)
23        4  AA(I,J)=A(J)
24        3  V(I)=S
25           WRITE(IOU,101)(((AA(I,J),J=1,4),V(I)),I=1,2)
26           WRITE(IOU,102)
27           GMI=V(1)
28           GMY=V(2)
29           NB2=NB+NB
30           CALL READR(VA,NB2)
31           DO 6 I=2,NB2,2
32           IF(NBI)7,7,6
33        7  VA(I)=-0.43429448*ALOG(VA(I))
34        6  VA(I)=VA(I)+ASB
35           J=1
36           DO 8 I=1,NB
37           G(I)=1.+VA(J)/VO
38        8  J=J+2
39           VO=VO/FY
40           AN=FLOAT(N)
41        2  NB2=0
42           J=2
```

```
43            AI5=1./(AI-X5)
44            DO 10 I=1,NB
45            F=(G(I)*VA(J)-X5)*AI5
46            WRITE(IOU,103)VA(J-1),VA(J),F
47            IF((F-AMIN)*(AMAX-F))10,10,9
48          9 NB2=NB2+1
49            V(NB2)=VA(J-1)
50            A(NB2)=VA(J)
51            G(NB2)=G(I)
52         10 J=J+2
53            WRITE(IOU,104)NB,NB2
54            NB=NB2
55         97 FORMAT(1X,'CONDITION OF TITRATION:',/2X,'N',4X,'VO',5X,'FY',8X,
56           &'AW',6X,'AMIN',3X,'AMAX',5X,'PH',6X,'AI')
57         98 FORMAT(1X,I2,F7.1,F9.5,F9.3,F8.2,F7.2,F9.3,F8.4)
58        100 FORMAT(/2X,'INDICATOR',51X,'TITRAT.AGENT',/2X,'PK11',5X,'PK21',
59           &6X,'PK31',6X,'PK41',3X,'LOG ALPHA(IN)',11X,'PK11',6X,'PK21',6X,
60           &'PK31',6X,'PK41',5X,'LOG ALPHA(Y)'/)
61        101 FORMAT(1X,F6.3,F9.3,2(F10.3),E15.4,7X,4(F10.3),E15.4)
62        102 FORMAT(/7X,' T I T R A T I O N   C U R V E ',/
63           &7X,'V(ML.)',7X,'ABS',9X,'ALPHA')
64        103 FORMAT(1X,5F12.4)
65        104 FORMAT(/1X,' NUMBER OF OBSERVED POINTS = ',I7,
66           &'    NUMBER OF USED POINTS = ',I7/)
67            RETURN
68            END
```

```
 1            SUBROUTINE UBBE(U,NK,XK,NB)
 2    C
 3    C --- U B B E N C ---
 4    C       CALCULATION OF ERROR SQUARE SUM FUNCTION U
 5    C
 6            DIMENSION  XK(NK)
 7            COMMON/FUNC/V(50),A(50),ERROR(50),ACAL(50),W(50)
 8           &/NC1/N,NBI,VO,FY,VM,AMIN,AMAX,ASB,AI,H,X5,ESP,GMI,GMY,AN
 9           &/NC2/DELT(50),G(50),AL(50)
10            AMY=(10.**XK(3)/(GMY+1.0E-20)+1.0E-20)
11            AMI=(10.**XK(2)/(GMI+1.0E-20)+1.0E-20)
12            CI=10.**(-ABS(XK(4)))
13            X5=XK(5)
14            AI5=AI-X5
15            X1=XK(1)
16            U=0.
17            Z=X1+VO*(1./AMY-CI/AN)
18            R=AMI*CI**(N-1)+1.0E-30
19            T=VO/(AN*R)
20            S=VO*CI*R/AMY
21            R=R*AN*X1/AMY
22            Q=VO*CI/AN
23            DO 1 I=1,NB
24            EP=1.
25         50 EP=EP*0.9
26            AL2=1.0E-20
27            A1=A(I)
28            IT=0
29            GI=G(I)
30         10 ALFA=((GI*A1-X5)/AI5)+1.0E-20
31            AL(I)=ALFA
32            ALN=(ALFA**N)+1.0E-20
33            AL1=(1.0-ALFA)+1.0E-20
34            AL1=(V(I)-R*ALN/AL1+S*ALN+T*AL1/ALN-Q*ALFA-Z)/((R*ALN*(AN*AL1+ALFA
35           & )/(AL1*AL1*ALFA)-S*AN*ALN/ALFA+T*(ALFA*(1.-AN)+AN)/(ALN*ALFA)+Q)
36           &*G(I)/AI5)
37            IT=IT+1
38            A1=A1+AL1*EP
39            IF(ABS(AL1)-ESP)40,40,20
40         20 IF(IT-10)10,10,50
41         40 AL2=(A(I)-A1)+1.0E-20
42            U=U+AL2*AL2*W(I)
43            DELT(I)=AL2
44          1 ACAL(I)=A1
45            RETURN
46            END
```

```
 1          SUBROUTINE SKRIK(NB,NK,IOU,XK,IRUR,SIGXK)
 2    C
 3    C --- S K R I K ---
 4    C     OUTPUT OF PARAMETERS AND THEIR STANDARD DEVIATIONS
 5    C        STATISTICAL TEST OF CURVE FITTING.......ANALYSIS OF REZIDUALS
 6    C
 7          DIMENSION DIF(50),SIGXK(NK),XK(NK)
 8          DIMENSION XX(50,2),YY(50,2),NDATA(2)
 9          COMMON/FUNC/XEXP(50),YEXP(50),ERR(50),YCAL(50),W(50)
10         &/NC1/N,NBI,VO,FY,VM,AMIN,AMAX,ASB,AI,H,X5,ESP,GMI,GMY,AN
11          COMMON/TIT/ TITLE(20) /PLOT/ MY,MX,NLS,NCL,XLINE(5),YLINE(5),
12         &AREA(31,60),YSCALE(60)
13          LOGICAL ISYMBL(2)
14          DATA NDMAX/50/
15          DATA ISYMBL(1)/4H****/
16          DATA ISYMBL(2)/4H0000/
17          RX=XK(1)*VM*FY
18          RRX=XK(1)/VO
19          WRITE(IOU,99)
20          WRITE(IOU,100)(XK(IK),SIGXK(IK),IK=1,NK)
21          WRITE(IOU,103)XK(1),RX,RRX
22          IF(IRUR.LT.11)RETURN
23          CALL UBBE(U,NK,XK,NB)
24          DO 10 I=1,NB
25       10 DIF(I)=YEXP(I)-YCAL(I)
26          WRITE(IOU,101)
27          WRITE(IOU,102)(I,XEXP(I),YEXP(I),YCAL(I),DIF(I),I=1,NB)
28          IF(IRUR.LT.12)RETURN
29          CALL STATS(IOU,NK,NB,XEXP,YEXP,DIF)
30          IF(IRUR.LT.13)RETURN
31          NDATA(1)=NB
32          NDATA(2)=NB
33          DO 50 I=1,NB
34          XX(I,1)=XEXP(I)
35          YY(I,1)=YEXP(I)
36          YY(I,2)=YCAL(I)
37       50 XX(I,2)=XEXP(I)
38          WRITE(IOU,2001)TITLE
39    C    3 CALL PLOTT(XX,YY,NDATA,NDMAX,ISYMBL, 2,XLINE,MX,YLINE,MY,NLS,
40    C       &NCL, 1, 1,AREA,YSCALE)
41          NPAGE=0
42          N=2
43          CALL PRTPLT(XX,YY,NDATA,N,NPAGE,NDMAX)
44       99 FORMAT(' *****  S K R I K  (OUTPUT)  *****',/,1X,33(1H*))
45      100 FORMAT(/1X,'  PARAMETERS AND THEIR STANDARD DEVIATIONS : '/(7(4X,
46         &F14.5, ' +- ',F14.5/))/1X,'V(EKV.)=',F10.5,5X,'MG METAL =',F10.5,
47         &5X,'CM=',E12.5,'( MOL/L)',/)
48      101 FORMAT(' I    V(EXP)        A(EXP)        A(CAL)       RESIDUAL'/)
49      102 FORMAT(1X,I2,2X,F8.5,F13.4,F13.4,F13.4)
50      103 FORMAT(/1X,' V(EKV. IN MILLILITRES)= ',F10.5,5X,
51         &' MILLIGRAMMES (METAL)= ',F10.5,5X,' MOLARITY =',E12.5,
52         &'( MOL/L)',/)
53     2001 FORMAT(1H1, '        GRAPH IS PRINTED FOR SYSTEM:',6X,20A4,//)
54          RETURN
55          END
```

6. REFERENCES

1. J. Cermak and M. Meloun, LETAG—Procedure for Minimization of Nonlinear Models, *2nd Symposium on Algorithms in Computation Technique*, High Tatras 1974, Kniznica algoritmov, Vol. 2, SVTS, Bratislava (1974).

2. N. Ingri and L. G. Sillen, High Speed Computers as a Supplement to Graphical Methods. IV. An ALGOL Version of LETAGROP VRID, *Ark. Kemi* **23**, 97–121 (1964).

3. C. F. Moore, PLOTT—Printer-Plotting Subroutine, *Comput. Phys. Commun.* **2**, 97–121 (1964).

4. J. R. Bell, Normal Random Deviate, *Commun. Assoc. Comp. Mach.*, Algorithm 334 **11**, 498–498 (1968).

5. A. Sabatini, A. Vacca, and P. Gans, MINIQUAD—A General Computer Programme for the Computation of Formation Constants from Potentiometric Data, *Talanta* **21**, 53–77 (1974).

6. S. Kuo, *Numerical Methods and Computers*, Addison-Wesley, Reading, Massachusetts (1965).

7. J. Wegstein, Root Finder, *Commun. Assoc. Comp. Mach.*, Algorithm 2 **3**, 74–74 (1960).

8. E. Mikanova, M. Mikesova, and M. Bartusek, Reactions of Ge(IV) with Aminocarboxylic Chelones and Aminohydroxylic Chelones, *Coll. Czech. Chem. Commun.* **46**, 701–707 (1980).

9. A. Mikan, J. Havel, and M. Bartusek, Molybdate Chelates of EDTA, *Scr. Fac. Sci. Nat. Univ. Purk. Brunensis* **10**, (Chemia) 23–28 (1980).

10. P. Havlova, J. Havel, and M. Bartusek, Tartrate Complexes of Molybdenum (VI), *Coll. Czech. Chem. Commun.* **47**, 1570–1579 (1982).

11. D. D. Perrin and I. G. Sayce, Computer Calculations of Equilibrium Concentrations in Mixtures of Metal Ions and Complexing Species, *Talanta* **14**, 833–842 (1967).

12. P. Havlova, J. Havel. S. Koch, and M. Bartusek, Über die Bildung des Ternaren Komplexes Mo(VI)-Oxalat-Brenzkatechin, *Scr. Fac. Sci. Nat. Univ. Purk. Brunensis* **12**, (Chemia) 257–264 (1982).

13. J. Havel, M. Vyrchlabsky, J. Sekaninova, and J. Komarek, Computer Applications in Chemistry. III. Minimizing Program DISTR-LETAG for the Evaluation of Equilibrium Constants from Liquid–Liquid Distribution Data, *Scr. Fac. Sci. Nat. Univ. Purk. Brunensis* **9**, 51–66 (1979).

14. L. Mullerova, M. Vrchlabsky, and J. Havel, Computer Applications in Chemistry. V. EXT-DISTR-LETAG Computer Program for Treatment of Extraction Photometric Data, *Scr. Fac. Sci. Nat. Univ. Purk. Brunensis* **10**, (Chemia) 13–22 (1980).

15. J. Komarek, J. Havel, and L. Sommer, The Use of Chelates of Copper, Nickel, Cobalt, Cadmium and Zinc with Heterocyclic Azodyes in the AAS Determination of These Elements, *Coll. Czech. Chem. Commun.* **44**, 3241–3255 (1979).

16. M. Meloun and S. Kotrly, Multiparametric Curve Fitting. II. Determation of Thermodynamic Dissociation Constant and Parameters of the Extended Debye–Huckel Expression. Application for Some Sulphonephthalein Indicators, *Coll. Czech. Chem. Commun.* **42**, 2115–2125 (1977).

17. M. Meloun and J. Cermak, Multiparametric Curve Fitting. IV. Computer-Assisted Estimation of Successive Dissociation Constants and of Molar Absorptivities from Absorbance–pH Curves by the DCLET Program, *Talanta* **26**, 569–575 (1979).

18. M. Meloun and J. Chylkova, Complexation Equilibria of Some Azo Derivatives of 8-Hydroxquinoline-5-Sulphonic Acid. IV. Determination of Dissociation Constants of 7-(Carboxyphenylazo)-8-Hydroxyquinoline-5-Sulphonic Acids by Non-Linear Regression of Spectrophotometric Data, *Coll. Czech. Chem. Commun.* **44**, 2815–2827 (1979).

8

PSEQUAD

A Comprehensive Program for the Evaluation of Potentiometric and/or Spectrophotometric Equilibrium Data Using Analytical Derivatives

L. ZEKANY and I. NAGYPAL

1. INTRODUCTION

Potentiometry and spectrophotometry are the most frequently used experimental methods to study equilibrium systems in solutions; these methods may be used in many different types of experimental arrangements. The equilibrium system can be described through the mass-balance equations as

$$C_1 = \sum_{j=1}^{n} \alpha_{j1} [S_j] = \sum_{j=1}^{n} \alpha_{j1} \beta_j \prod_{i=1}^{k} [c_i]^{\alpha_{ji}}$$

$$\vdots$$

$$C_k = \sum_{j=1}^{n} \alpha_{jk} [S_j] = \sum_{j=1}^{n} \alpha_{jk} \beta_j \prod_{i=1}^{k} [c_i]^{\alpha_{ji}}$$

(1)

where n is the number of species in the system, including the components; S_j is the jth species present in the system; k is the number of components in the system; $[c_1] \cdots [c_k]$ is the equilibrium (free) concentration of the components; $\beta_j = [S_j]/(\prod_{i=1}^{k} [c_i]^{\alpha_{ji}})$, the formation constant of the species, (the formation constants of the components are unity); α_{ji} are stoichiometric numbers, giving the number of the ith component in the jth species.

L. ZEKANY and I. NAGYPAL • Computer Center, Institute of Inorganic and Analytical Chemistry, Lajos Kossuth University, H-4010, Debrecen, Hungary.

   The stoichiometric numbers are arranged in k columns and n rows forming the composition or α matrix of the system. It is expedient to arrange the species in a special order, such that the first k rows of the composition matrix form a unit matrix. The stoichiometric numbers are positive integers, apart from those relating to the component taking part in the self-dissociation of the solvent. Change of the solvent concentration is normally disregarded in solution equilibrium studies. Therefore, since only one of the ions taking part in the self-dissociation process is considered as a component, the stoichiometric numbers of the species containing the counter-ion or deficient in the selected component are negative integers. This situation most frequently occurs in aqueous solutions, where, with the hydrogen ion taken as a component, the stoichiometric number of the OH^- ion is -1 for hydrogen ion and zero for the other components. Thus it follows from this general definition that the formation constant of the OH^- ion is $K_w = [H^+][OH^-]$. The appropriate stoichiometric number of the species containing OH^- ion, or having fewer protons than that form of the ligand selected to be a component, is also a negative integer for protons.

   It follows from the above definition that the total concentration of the proton in aqueous solution may be negative; this special case has some consequences on the solution of the mass-balance equations, as we shall see later.

   The experiments are frequently carried out as titrations, which means that the total concentration of one of the components is changing step by step over a relatively wide range, while the total concentrations of the other components are changing owing to dilution. Occasionally there will be no concentration change since their concentrations are the same in the two solutions being mixed. In this situation the volume of the titrant, or the total concentration of one of the components, may be regarded as an experimental datum, denoted by X_1^v in the following.

   For potentiometry, the directly measured experimental data (X_l^P) can be expressed by the free concentration of one or more of the components:

$$X_l^P = A_l + M_l \log [c_l] \qquad (l = 2 \ldots m) \tag{2}$$

where $m - 1$ is the number of potentials measured in the system $(m - 1 \leqslant k)$; A_l is an additive term, for example, E_0 if emf is measured, or an additive term to convert the directly measured pH into $-\log [H^+]$; and, M_l is multiplicative coefficient to be calculated from the Nernst equation in case of emf measurements, or -1 for pH measurements.

   For spectrophotometry, the following general relation is valid:

$$X_l^A = \sum_{j=1}^{n} \epsilon_{jl} \beta_j \prod_{j=1}^{k} [c_i]^{\alpha_{ji}} \tag{3}$$

$$(l = m + 1 \ldots p)$$

where $p - m$ is the number of wavelengths studied; X_l^A is the measured absorbance at the lth wavelength; and, ϵ_{jl} is the molar absorptivity of the jth species at the lth wavelength.

   The program PSEQUAD solves equation (1) for the unknown free concentrations and obtains the unknown formation constants and/or molar absorbancies by minimizing

$$F = \sum_{q=1}^{n_d} F_q$$

where

$$F_q = w_1 \sum_{i=1}^{r} (\Delta X_1^V)^2 + \sum_{i=1}^{r} \sum_{l=2}^{m} w_l (\Delta X_l^P)^2 + w_A \sum_{i=1}^{r} \sum_{l=m+1}^{p} (\Delta X_l^A)^2 \qquad (4)$$

and w_1 is the weighting factor of the volume of the titrant or total concentrations; w_l ($l = 2 \ldots m$) is the weighting factor for the potential measurements; and w_A is the weighting factor for the absorbance measurements. F_q is any function that is "correctly composed" from the experimental data. (The meaning of *correctly composed* will be explained later in detail.)

The program has the capacity to evaluate simultaneously different types of measurements carried out in solutions having varying compositions, or even in groups of solutions in which the number of the components differ from one subset to another, for example, the proton/ligand A; proton/ligand B, proton/ligand A/metal ion; proton/ligand B/metal ion; and proton/ligand A/ligand B/metal ion subsets. Titration curves may be evaluated simultaneously for the pK_a's of the ligands and for the formation constants of $M_zA_yH_z$, $M_xB_wH_z$ and $M_xA_yB_wH_z$ complexes. In this way the effect of error accumulation, which may occur if the titrations are evaluated step by step, can be avoided. If, for example, these measurements are supplemented with spectrophotometric measurements in the same or separate samples, these may also be evaluated simultaneously. This possibility is implied by equation (4) by $\sum_{q=1}^{n_d} F_q$, where n_d is the number of sets of measurements derived from different types of primary experimental data.

2. DESCRIPTION OF THE PROGRAM PSEQUAD

There are two main steps in the calculations: the solution of equation (1) for the unknown free concentrations and the refinement of the formation constants and/or molar absorptivities.

2.1. Calculation of the Free Concentrations

The calculation of the unknown free concentration is based on the standard Newton–Raphson procedure, by solving the equations

$$\left(C_i^{calc} - C_i^{exp} \right) = 0 \qquad (5)$$

for C_i^{calc}. The Cholesky[1] algorithm is used to solve the linear equations in the New-ton–Raphson procedure.

The free concentrations are calculated on a log $[c_i]$ scale, thereby preventing the occurrence of negative concentrations during the course of the iterative procedure. The starting $[c_i]_0$ values are $C_i/2$ if $C_i \geq 10^{-6}$ or 10^{-6} if $C_i \leq 10^{-6}$. This restriction is necessary because, as we have seen, the total concentration for the protons may be negative. There is a step-size control in the algorithm that limits a change in $[c_i]$ to two orders of magnitude; the step sizes of the other $[c_i]$ values are proportionally decreased on a logarithmic scale. The iteration terminates when

$$|(C_i^{exp} - C_i^{calc})/C_i^{exp}| \leq 5 \times 10^{-4}$$

for all of the concentrations. If the standard Newton–Rapson method fails to converge, then the equation

$$\ln \frac{C_i^{calc}}{C_i^{exp}} = 0 \tag{6}$$

is solved for log $[c_i]$ by the Newton–Raphson method. Four orders of magnitude of change for $[c_i]$ are allowed in this iteration.

As we have seen, the total concentration for the hydrogen ion may be negative; thus equation (6) is modified for the hydrogen ion as follows:

$$\ln \left\{ \frac{C_{+H}^{calc} + \dfrac{\text{sign}(C_H^{exp}) - 1}{2} C_H^{exp}}{C_{-H}^{calc} + \dfrac{\text{sign}(C_{-H}^{exp}) + 1}{2} C_H^{exp}} \right\} = 0 \tag{7}$$

where C_{-H}^{calc} and C_{+H}^{calc} are the calculated total hydrogen concentration with negative and positive stoichiometric numbers, respectively.

Iteration terminates when equations (6) and (7) are valid to within 5×10^{-4}. The final iterative cycle is always a standard Newton–Raphson one, because the Cho-lesky-type triangular matrix calculated in this cycle will be used for further calcula-tions.

The above procedure is used only for the first iterative cycle for the formation constants and/or molar absorptivities. In the second and subsequent cycles, the pre-vious $[c_i]$ values are used as $[c_i]_0$ for the current cycle. For the evaluation of titration curves, the $[c_i]_0$ values defined above are used only at the first titration point. For the second and subsequent titration points, they are calculated by extrapolation from the previous point using the method of analytical derivatives of the implicit function sys-tems.

It is worth noting that although the method of analytical differentiation of the implicit function systems, such as equation (1), may be found in many mathematical analysis texts, it had been overlooked by most workers in the field of solution equilibria. However, groups studying the fundamental aspects of equilibrium systems have been employing this method for some time. The most recent advances can be found in a review by Smith.[2] Bugaevsky et al.[3-5] have used the method of analytical differentiation in their calculations on solution equilibria and recently we have drawn attention to this possibility.[6] The method is illustrated here only for the extrapolation mechanism. Essentially the same method is used throughout the program for the calculation of the different $dx/d \log \beta_j$ values used in the minimization of equation (4).

Let us suppose that the $\ln [c_1]$ and $C_2 \ldots C_k$ values are known for each point in the experiment. The $\ln [c_2] \ldots \ln [c_k]$ values are also known at a given experimental point. The concentrations at this point are denoted by superscript 1. The extrapolated $\ln [c_i]$ ($i = 2 \ldots k$) values at the next point (superscript 2) are as follows:

$$\ln [c_i]^{(2)} = \ln [c_i]^{(1)} + \left(\frac{d \ln [c_i]}{d \ln [c_1]}\right)^{(1)} \cdot (\ln [c_1]^{(2)} - \ln [c_1]^{(1)})$$

$$+ \sum_{j=2}^{k} \left(\frac{d \ln [c_i]}{dC_j}\right)^{(1)} \cdot (C_j^2 - C_j^1) \tag{8}$$

Use of equation (8) requires the derivatives $d \ln [c_i]/d \ln [c_1]$ and $d \ln [c_i]/dC_j$. The derivatives $d \ln [c_i]/d \ln [c_1]$ may be calculated from the system of equations $d(C_i^{\text{calc}} - C_i^{\text{exp}})/d \ln [c_1] = 0$, i.e.,

$$\sum_{i=2}^{k} \left(\sum_{j=1}^{n} \alpha_{j2} \alpha_{ji} \beta_j \prod_{i=1}^{k} [c_i]^{\alpha_{ji}}\right) \frac{d \ln [c_i]}{d \ln [c_1]} = -\sum_{j=1}^{n} \alpha_{j2} \alpha_{j1} \beta_j \prod_{i=1}^{k} [c_i]^{\alpha_{ji}}$$

$$\vdots \qquad\qquad\qquad\qquad\qquad \vdots \tag{9}$$

$$\sum_{i=2}^{k} \left(\sum_{j=1}^{n} \alpha_{jk} \alpha_{ji} \beta_j \prod_{i=1}^{k} [c_i]^{\alpha_{ji}}\right) \frac{d \ln [c_i]}{d \ln [c_1]} = -\sum_{j=1}^{n} \alpha_{jk} \alpha_{j1} \beta_j \prod_{i=1}^{k} [c_i]^{\alpha_{ji}}$$

If the $d \ln [c_i]/dC_j$ derivatives are calculated, then the coefficients on the left-hand side are the same, and the right-hand sides are the jth unit vector. By solving these systems of linear equations, equation (8) yields the starting $[c_i]$ values for the next titration point.

It is easy to realize that the coefficients on the left-hand side are those quantities which form the Jacobi matrix in the Newton–Raphson procedure, based on equation (5). As the Jacobi matrix is transformed to the Cholesky-type triangular matrix in the Newton–Raphson procedure, the calculation of all necessary derivatives require only the calculation of the appropriate vectors and the use of the final steps in the Cholesky algorithm.

In case of titration curves it may happen that the free concentrations are changing by some considerable magnitude between adjacent points. In this situation, in addition to the extrapolation procedure utilizing equation (8), the so-called continuation method[7] is adopted. The allowed maximum step in the $[c_i]$ values is three orders of magnitude; the steps of the other $[c_i]$ values are proportionally decreased on a logarithmic scale.

It is possible for the Jacobi matrix to become ill conditioned. In this situation the diagonal elements are multiplied by increasing powers of 2, until the appropriate linear equations can be solved.

2.2. Refinement of the Formation Constants and/or the Molar Absorptivities

The Gauss–Newton method is used for the refinement of the parameters by minimizing any function correctly composed from equation (4). However, since some of the primary data included in equation (4) are interdependent, inconclusive results may follow. Allowed combinations of the X_l values are as follows.

(a) If X_1^V (volume of the titrant or the total concentration of the first component) is not included in the experimental data to be fitted, then all possible combinations of the remaining X_l values are allowed and all the $C_1 \ldots C_k$ values are fixed data.

(b) If some of the measured potentials are fixed values, then all potential values must be arranged so that the fixed potential(s) are the first X_l ($l = 2, \ldots, m1$) values, and the potential(s) to be fitted are the remaining X_l ($l = m1 + 1, \ldots, m$) values. In this instance the volume of the titrant containing the first component, or the C_1 values, as well as the other X_l values ($l = m1 + 1, \ldots, p$) may be optionally combined in equation (4). The $C_2 \ldots C_{m1}$ data are then not used either as fixed parameters, or as variables to be fitted. This enables the processing of photometric measurements at fixed pH values.

(c) It may be expedient, for pH titrations, to fit the parameters by a combined minimization of the volume $-$ pH data (orthogonal regression).[8] The ratio of the variance of the volume and pH will be required for this situation.

The unknown formation constants are nonlinear, whereas the molar absorptivities are linear parameters; thus the equation relating to the molar absorptivities (Beer's law) are solved first, and the β values are refined subsequently. This is essentially the same procedure as used in the LETAGROP program, "third strategy."[9] A detailed description of this approach may be found in the paper by Barham and Drane.[10]

The molar absorptivities are calculated directly, no restriction being imposed on their values. There is, however, a step-size control for the refinement of the formation constants. The allowed maximum step is one logarithmic unit for one cycle.

Refinement is complete when the change in the error square sum is less than $5 \times 10^{-2}\%$, or the maximum number of iterations is exceeded. If the error square sum in a subsequent iteration cycle is found to be increasing, then half the calculated step sizes are used for the shift of the formation constants in that cycle.

2.3. Model Selection by PSEQUAD

A model-selection algorithm has been included in PSEQUAD. The formation constants are classified into four groups, by assigning an integer number or zero to each of them. A zero is assigned to the formation constant if it is not to be refined. An example would be a formation constant determined in a previous experiment. Unity is assigned to those constants which are to be calculated and must not be excluded from the equilibrium model assumed. Positive integer numbers (other than one) are assigned to those constants which may be *excluded* from the equilibrium model; negative numbers are assigned, less than minus one, to those constants which may be *included* into the equilibrium model. Initially, the program refines those constants which are marked by positive integer numbers.

Once convergence has been achieved, or the maximum number of iterative cycles has been exceeded, the results of the refinement process are examined by PSEQUAD. If there are formation constants marked by an integer greater than one for which the decrease would be higher than one logarithmic unit even for the final cycle, then the constant having the largest integer marker is eliminated from the model and the refinement process is reinitiated. This procedure of selectively removing one constant at a time is repeated until there are no constants remaining for which the decrease is higher than one logarithmic unit is the last cycle. The second step in model selection is then begun by including constants marked be negative integers, in the order of increasing absolute value of the marker. Constants having -1 assigned to them will not be included in the model during that particular run of the program.

2.4. Estimation of Parameter Correlation

A number of statistical parameters are calculated by the program. Since most of the formulas are given in the ouput, they are not described here. However the meaning of the partial, multiple, and total correlation coefficients will be discussed. These are calculated from the elements of the matrix

$$\mathbf{B} = \mathbf{J}^T \mathbf{w} \mathbf{J} \tag{10}$$

where $\mathbf{J}$ is the Jacobian, $\mathbf{J}^T$ is its transpose, and $\mathbf{w}$ is the diagonal weighting matrix used in the Gauss–Newton method. The partial correlation coefficients, r_{ij}, give the measure of interdependence between two constants β_i and β_j assuming that the other constants have fixed values:

$$r_{ij} = \frac{-B_{ij}}{(B_{ii} B_{jj})^{1/2}} \tag{11}$$

The total correlation coefficients, S_{ij}, also provide a measure of interdependence between two constants, the other constants being regarded as fitted parameters:

$$S_{ij} = \frac{C_{ij}}{(C_{ii} C_{jj})^{1/2}} \tag{12}$$

The multiple correlation coefficients, R_i, give the measure of the independence of a given constant from that of all the others:

$$R_i = \left(1 - \frac{1}{B_{ii}\,C_{ii}}\right)^{1/2} \tag{13}$$

[The appropriate elements of $\mathbf{C} = \mathbf{B}^{-1}$ are denoted by C_{ij} in equations (12) and 13).]

Each of these correlation coefficients may take values between zero and ± 1. Zero implies the total independence of the species, $+1$ or -1 means a complete correlation, and consequently the two species in question should not be refined simultaneously. The correlation coefficients are arranged in matrix form in the output. The diagonal elements contain the multiple correlation coefficients, the upper triangle contains the partial, while the lower triangle contains the total correlation coefficients.

2.5. Program Limits

There are only a few limitations to the program:

(a) The computer used should have a line-printer of at least 120 characters.

(b) A maximum of five different types of potentials can be simultaneously measured.

(c) Most of the calculations are carried through in two common blocks. Their sizes are 1000 for the integer and 5000 for the real variables, in the present form of the program. These sizes may be easily changed as shown in the program.

Several parameters (the working precision, convergence criteria for the Newton–Raphson or the Gauss–Newton procedure, step sizes, etc.) that are given in the text as definite numbers may be changed, if necessary, via the input data. Moreover, the program may be used to simulate experimental data, or to calculate the concentration distribution of species in a known system. For this latter option the maximum number of Gauss–Newton iterations and the allowed step size of the formation constants is set to zero.

As can be seen, the program is very flexible and capable of handling almost all potentiometric and/or spectrophotometric types of measurements. As we have seen, however, no numerical constraints are applied to the molar absorptivities; thus negative values may be calculated. This flexibility, on the other hand, means that the program is rather large, mainly due to the necessary data-organization procedures. This also increases the run time, but it is still considerably lower than if numerical differentiation had been employed. We have no comparable run-time data with other general programs, thus we quote the statement of Züberbuhler and Kaden,[11] who have also used analytical differentiation in their TITFIT program:

> In contrast to other programs which also minimize the titre difference, analytical derivatives are used for the calculation. No numerical differentiation, with all of its problems of step-size control and loss of significant figures, is needed in the entire program. Analytical derivatives are also used in MINIQUAD,[12] which, however, employs a completely different approach, since it performs a least squares fit on all three mass-balance equations, thus mixing independent and dependent variables. ... The analytical derivatives can be calculated using the derivatives, ... which have already been computed to solve for the concentrations. The advantages of the analytical derivatives are higher speed of calculation and better convergence because of the greater number of significant figures. In fact the most time-consuming part in such a problem is the calculation of the species concentrations at

each titration point. By using analytical derivatives the additional time needed for a complete iteration cycle is only 30–60%, whereas by using numerical derivatives, the time for a cycle is proportional to the number of parameters refined.

3. REFERENCES

1. G. E. Forsythe and C. B. Moler, *Computer Solution of Linear Algebraic Systems*, Prentice-Hall, Englewood Cliffs, New Jersey [Hungarian translation, Müszaki Könyvkiadó, Budapest, (1976)].
2. W. R. Smith, The Computation of Chemical Equilibria in Complex System, *Ind. Eng. Chem. Fundam.* **19**, 1–10 (1980).
3. A. A. Bugaevsky and B. A. Dunai, Calculation of the Equilibrium Composition and Buffer Properties of Solutions with the Use of Computers (in Russian), *Zh. Anal. Khim.* **26**, 205–209 (1971).
4. A. A. Bugaevsky, L. E. Rudnaya, and T. P. Mukhina, Calculation of Equilibria in Complex Systems (in Russian), *Zh. Anal. Khim.* **27**, 1675–1679 (1972).
5. A. A. Bugaevsky and L. E. Nikishina, On the Analytical Method of Calculating the Derivatives of Equilibrium Concentrations, *Talanta* **28**, 977 (1981).
6. I. Nagypál, I. Páka, and L. Zékány, Analytical Evaluation of the Derivatives Used in Equilibrium Calculations, *Talanta* **25**, 549–550 (1978).
7. J. M. Ortega and W. C. Rheinboldt, *Iterative Solution of Nonlinear Equations in Several Variables*, Academic Press, New York (1970).
8. Yu. V. Linnik, *Method of Least Squares and Principles of the Theory of Observations*, Pergamon Press, Oxford (1961).
9. L. G. Sillén and B. Warnqvist, High-speed Computers as a Supplement to Graphical Methods. 6. A Strategy for Two-level LETAGROP Adjustment of Common and "Group" Parameters. Some features that avoid divergence. *Ark. Kemi* **31**, 315–339 (1969).
10. R. H. Barham and W. Drane, An Algorithm for Least Squares Estimation of Nonlinear Parameters When Some of the Parameters are Linear, *Technometrics* **14**, 757–766 (1972).
11. A. D. Zuberbuhler and Th. A. Kaden, TITFIT, A Comprehensive Program for Numerical Treatment of Potentiometric Data using Analytical Derivatives and Automatically Optimized Subroutines with the Gauss–Newton–Marquardt Algorithm, *Talanta* **29**, 201–206 (1982).
12. A. Sabatini, A. Vacca, and P. Gans, MINIQUAD—A General Computer Programme for the Computation of Formation Constants from Potentiometric Data, *Talanta* **21**, 53–77 (1974).
13. H. M. Irving, M. G. Miles, and L. D. Pettit, A Study of Some Problems in Determining the Stoicheiometric Proton Dissociation Constants of Complexes By Potentiometric Titrations Using a Glass Electrode, *Anal. Chim. Acta* **38**, 475–488 (1967).

4. PROGRAM INSTRUCTIONS

4.1. Introduction

PSEQUAD is a very flexible program that can handle pH-metric, potentiometric, and photometric data separately or in any combination; refine all or part of the equilibrium model; and perform data conversion and various types of data weighting and statistical analysis. Moreover, data pertaining to pK_w, pK_a, and $\log \beta$ calculations may be processed within a single run of PSEQUAD. The data for the Cu^{2+}–NTA–proton system are listed in Section 5.2. It comprises two major sections. The first section contains the experimental data: pH, potential and absorbance data, solution compositions, and the total equilibrium model description. The second section contains program directives as to how the data in the first section are to be processed.

Only three types (formats) of data card are employed by PSEQUAD. They are as follows:

Type A: FORMAT (I2,A8,10I2,5F10.0)
Type B: FORMAT (8F10.0)
Type C: FORMAT (A80)

Type A cards are used to indicate specific options or instructions for data processing. Type B cards are for experimental data, and Type C cards are used for descriptive titles.

The following descriptions of the requirements for the input data will be focused mainly on the Type A card. This card has four sections:

a. Key [I2]: an integer that determines the role of card and, in many instances, one or more cards that follow.
b. Text [A8]: an eight-character label, used to identify the purpose of the particular card. The keywords used in the examples below are of our own making and may be changed to suit the user. PSEQUAD uses the key integer rather than the keyword.
c. i1, i2, i3, $\cdots$, i10 [10I2]: ten integers, used for a variety of purposes, that are positioned from column 11 through column 30. In the following examples designations such as
      i. i1 i2 $\cdots$ i10
     ii. i1 i2 i3
    iii. i1 i3 i5 i7
will be used. This means that (i) a value must be supplied for each integer; (ii) a value is needed only for the first three integers; (iii) a value is required only for the first, third, fifth, and seventh integers.
d. f1 f2 f3 f4 f5 [5F10.0]: five floating point numbers, used for a variety of purposes, that are positioned from column 31 through column 80. The designation f1 f3 f5 or f1 f2 has the same implication as for the integers above.

The input protocol, to be described, will use as an example data derived from a study of the copper (II)–NTA equilibrium system. Data from a complete set of experiments have been processed during one run of PSEQUAD. The separate experiments are as follows:

A. Determination of pK_w, one titration, pH measured.
B. Determination of NTA pK_a's, one titration, pH measured.
C. Titration of copper (II) and NTA two titrations, pH and pCu (potential) measured for each.
D. Titration of copper (II) and NTA with KOH, two titrations, pH and pCu (potential) measured for each.
E. Photometric titration of copper (II) and NTA (1:2) with KOH, one titration, 11 absorbances per spectrum, 15 spectra.
F. Photometric titration of copper (II) with NTA (1:1.25) with KOH, one titration, 11 absorbances per spectrum, 12 spectra.

The data from these experiments, together with type A cards and appropriate title cards (type C), comprise the first section of the data deck. The setup of this section will be described in detail before elaborating upon the second section.

4.2. Input of Potentiometric and/or Photometric Data

As indicated earlier in this chapter the values of a number of parameters, related to the various least-squares algorithms employed by PSEQUAD, may be reset. This is achieved using type A cards with key equal to 91 or 92. These cards may be placed at the beginning of the first or second section. They are supervisor cards, hence the key word SUPERVIS. The format of the cards is as follows:

91SUPERVISi1 i2   f1 f2 f3

i1: The number of subsequent iterations for which the convergence criterion should be fulfilled before terminating the Gauss–Newton (GN) algorithm. Usually i1 = 1 but may be set to 2, 3, etc. for ill-conditioned systems.

i2: The maximum number of half-steps permitted in the GN algorithm.

f1: Convergence criterion for GN algorithm.

f2: Maximum step size in log β values for those species that cannot be excluded during the course of the model selection process.

f3: The starting value for $[c_i]_o$, Section 2.1.

92SUPERVISi1 i2   f1 f2 f3 f4 f5

i1: The maximum number of iterations allowed for the standard Newton–Raphson (NR) algorithm.

i2: The maximum number of iterations allowed for the NR algorithm minimizing on equations (6) and (7), Section 2.1.

f1: Convergence criterion for standard NR.

f2: Convergence criterion for NR algorithm minimizing equations (6) and (7).

f3: Maximum step size for log $[c_i]$ in standard NR.

f4: Maximum step size for log $[c_i]$ in NR for equations (6) and (7).

f5: Maximum step size for the continuation method, Section 2.1.

PSEQUAD uses the following preset values for each of the above parameters:

Key	i1	i2	f1	f2	f3	f4	f5
91	1	1	5.0×10^{-4}	1.0	1.0×10^{-6}	—	—
92	25	75	5.0×10^{-4}	5.0×10^{-4}	2.0	4.0	3.0

Note that even if only one parameter on 91 and 92 need be altered, all of the other parameters must be "reset" to the default values. If all values are acceptable then the 91 and/or 92 cards are omitted.

Having established the convergence criteria, explicitly or implicitly, the experimental data are now read in. Each experiment is preceded *and* followed by type A cards that describe the type of data to be expected. Referring directly to the data listed in Section 4.1, the following arrangement of cards is seen for experiment A:

10 DATAIN

Key = 10 indicates that either an experimental data set begins or that all information relating to a particular experiment has been entered.

01 SET   i1

Key = 01 indicates that data-type description cards follow. Integer i1 is used to indicate the number of titrations (pH, potentiometric, or photometric) of the same type that follow. For batch titrations, one data set per solution, i1 indicates the number of solutions.

02 TOTAL   i1   i3   i5   i7   i9   f1   f2   f3   f4   f5

Key = 02 indicates that this is a total concentration(s) card, for titrant. The integers i1, i3, . . . are the i column numbers of the matrix (Section 1). The floating point numbers f1, f2 are the total concentrations of the component i1, component i3, respectively, *in the buret*. A negative value is used for total hydrogen ion concentration when titrant is a strong base. For more than five components, a second 02 card is used. In this example, Section 4.1, DETER-MINATION OF PKW, i1 = 1 (Hydrogen ion is titrant) and f1 = −0.2324 (Strong base, 0.2324 M).

03 TITR

Key = 03 implies that titration-type data are to be evaluated, i.e., several data points relate to a single solution. Batch-type titration (one datum per solution) can be processed by PSEQUAD by omitting this card.

04 POT   i1   i3   i5   i7   i9

Key = 04 signifies that potentiometric measurements have been made. Integers i1, i3 . . . have the same meaning as for card 02. Thus for pH measurements [hydrogen (component 1) ion measured] i1 = 1 and i3–i9 are set to zero. If no potential measurements have been made then this card is omitted.

05 ABS   i1   i2

Key = 05 indicates that absorbance data will follow. Integer i1 is the marker for the first, while i2 is the marker for the last wavelength for the absorbance data that are read in.

10 ENDSET   i1

Key = 10, in this context, signals the completion of this particular set of type A cards. Thus for every 01 card there must be a 10 card. PSEQUAD now expects a title card (type C) if i1 = 1, or a type B card if i1 = 0.

At this point data for the particular experiment will follow using the type B cards (preceded, optionally by a title card, which is good practice). The arrangement of data on the type B cards will be determined by the presence (or absence) of cards 02, 03, 04, and/or 05.

Titration data. The first card contains the initial volume followed by the initial total concentrations of components *in the solution to be titrated*. The order of the total concentrations is determined by the order established on the 02 card. The following cards will contain the volume of titrant for that point and then the measured potentials (or pH), followed by the absorbances, if the potentials and absorbances are measured in the same solutions. [This is not the case for the Cu(II)-NTA example.] Consequently either potentials or absorbances may be absent. Again the ordering of the potential (pH) measurements follows the order specified by the 04 card while the ordering of the absorbances follows the order specified by the 05 card. If more than seven data are to be entered a new card is used. However, data for a new point must begin on a new card. For pointwise measurements the type B cards are set up as follows: total concentrations, ordered according to the 02 card; measured potentials in the 04 card order; measured absorbances in the 05 card order. The end of the titration data is indicated by -1.0 in columns 1–10. There is no need for the card for pointwise measurements.

It should be noted that if more than one card is needed for each point, then, when terminating with -1.0, add as many blank cards as necessary after the termination card as continuation cards, see Section 5.2.2, PHOTOMETRIC TITRATION . . .

If, for the 01 card, i1 is greater than unity, and there is an 03 card, then a second (third, . . .) set of titration data will follow, terminated by -1.0.

Once all the data for the particular experiment have been entered then either

01 SET   i1

or

10 END DATA   i1 i2

should appear. The first possibility indicates the beginning of a new experiment and 01, 02, 03, 04, 05, 10 cards are used as indicated above. The second possibility means that all the data pertaining to all the experiments have been entered. For this situation i1 and i2 have the following significance:

    i1 = 1 a title card follows
    i1 = 0 no title card
    i2 = 0 no data to be printed
    i2 = 1 print out solution compositions and other relevant input data
    i2 = 2 print out all input data.

In this context the 10 END DATA card pairs with the 10 DATAIN card.

One further option exists for inputting experimental data. All of the data between 10 DATAIN and 10 END DATA may be stored on tape or disk. If this is the situation the following cards are required:

10 DATAIN
−2 INTAPE i1
10 END DATA

where i1 is the logical unit number for the mass storage device.

4.3. Data Processing Options and Instructions

The second section of the data deck comprises the instructions to PSEQUAD for the several data processing options relating to the various experiments, including model selection. Each set of calculations begins with an 11 card and ends with a 10 card. We shall refer to these calculation sets as Tasks. Refering to Section 5.2.2, immediately following 10 END DATA the overall title for second section for the data deck, i.e., PH-METRIC, POTENTIOMETRIC . . . , is found. The first task to be performed is the calculation of the pK_w. The second task will be the refinement of the pK_a's for NTA, and subsequent tasks will be the refinement of all or some of the formation constants using the model selection option. Notice that the tasks are performed in a logical sequence so that the value the pK_w from the first task will be incorporated into the equilibrium model for the second task. Consider the data deck for the first task:

11 Task   i1 i2 i3 i4 i5 i6 i7 i8 i9 i10
   Key = 11 indicates the beginning of a task. The integers i1–i10 are used to indicate number of iterations, number of colored species, *etc*. The omission of certain values, i.e., iN set to zero, also provides PSEQUAD with information concerning the type of experimental data to be included in this task. The significances of the integer values are as follows:
   - i1: Number of components, i.e., number of columns in α matrix.
   - i2: Number of complexes *plus* components, i.e., number of rows in α matrix.
   - i3: Maximum number of iterations in GN algorithm.
   - i4: Maximum number of additional calculations in the model selection procedure. If i4 = 0 there is no model selection.
   - i5: Not used, set to zero.
   - i6: Not used, set to zero.
   - i7: Number of species with unknown molar absorptivities.
   - i8: Total number of absorbing species.
   - i9: The marking number of the first wavelength used in the calculation.
   - i10: The marking number of the last wavelength used in the calculation.

   These two integers, i9 and i10, are not necessarily the same as i1 and i2 for the 05 card. Any part of the absorbance data that has been read in may be selected for calculation. This is the purpose of i9 and i10. This option was not used for the Cu(II)–NTA system.

00 Text   i1 i2 i3 i4 i5 i6 i7 i8 i9 i10 f1
   Key = 00 implies initial equilibrium model specification. There will be (i2 − i1, on card 11) cards, i.e., one per complex. It is at this point in the data set that the total equilibrium model is defined, including probable and not so probable complexes, in preparation for model selection. "Text" (eight characters maximum) may be used for the formula of the complex, HL, ML_2, H, OH−, for example. The integers i1–i10 have the following meanings:
   i1: Serial number for the complex. This should be started at (number of components + 1).
   i2: A value assigned to each complex in accord with the model selection rules, Section 2.3.
   i3 . . . i10: The elements of the α matrix, one row per card.

f1: Value for log β. If i2 = 0 a predetermined value from previous experiments is employed; if i2 $\neq$ 0 the estimated value is supplied. If there are more than eight components only i1, i2, and f1 are entered on this card and the elements of the α-matrix are punched on consecutive cards.

Note that the model definition cards need only appear once. In the example, Section 4.1, they occur in the first task, calculation of pK_w. All i2 values, except for species 4, are set to -1, meaning that they will not be included in the model to be refined. The complexes to be included for refinement in the subsequent tasks are specified through the 13 or 14 cards.

13 SPECIES i1 i2 i3 i4 i5 i6 i7 i8 i9 i10 f1 f2 f3 f4 f5
  i1: Serial number for the complex in the α-matrix.
  i2: As for 00 card, i2.
  f1: Value for log β for complex i1.
  i3, i4, f2: As for i1, i2, f1 but for the second complex.
  i5, i6, f3: For the third complex, *etc.*
14 SPECIES i1 i2 i3 i4 i5 i6 i7 i8 i9 i10
  i1, i2: As for i1, i2 in the 13 card.
  i3, i4: As for i3, i4 in the 13 cards, *etc.*

The 13 and 14 cards are used in conjunction with the 00 cards. Their purpose is to reset the i2 value of the 00 cards (the 14 cards), or to reset the i2 and f1 values of the 00 cards (the 13 cards), in subsequent tasks. For example, the input data (Section 4.1) require the calculation of pK_w initially, followed by the calculation of pK_a's for NTA. This is accomplished by using a 14 card in the second task having the values

14 SPECIES 4 0 5 1 6 1 7 1

which implies that the constant for species 4 (pK_w), obtained in the first task, will be used as a fixed value, i2 = 0, replacing the value assigned by the original 00 card for species 4. Additionally species 5, 6, and 7 will have their constants refined (i4, i6, i8 = 1). It can also be seen from Section 4.1 that the third task involves the refinement of the full equilibrium model with the constants for species 4–7 held fixed at values obtained from the previous tasks. Note that a second 14 card is necessary in this third task. When using the model selection option i2, i4 . . . integers may take on values 0, 1, 2, . . . , -1, -2, ... The significance of these values has been described in Section 2.3.

The remaining type A cards relate to various data preprocessing options. Two cards, 15 and 16, are used to provide statistical weighting information to be applied to the potentiometric and/or absorbance data:

15 STD DEV i1 i3 i5 i7 i9 f1 f2 f3 f4 f5
  i1 . . . i9: Indices relating to the experimental sets of measurements that use the same weighting procedure.
  f1: Standard deviation of the volume of titrant or of the first total concentration.

f2: The ratio of the standard deviations of the volume and of the $-\log [c_1]$ data for orthogonal regression, Section 2.2.

f3: Standard deviation of the first measured potential.

f4: Standard deviation of the second measured potential.

f5: Standard deviation of absorbance measurements. One of the f1, f3, f4, and f5 should be 1.0, and the others are the estimated ratios of the standard deviations; see Section 5.3.1.

If more than two potentials are measured, then a 16 card is necessary:

16 STD DEV   i1 i3 i5 i7 i9   f3 f4 f5

where i1 . . . i9 have the same meaning as in the 15 card.

f3, f4, f5 are the standard deviations of the subsequent potentials.

If more than five measurements are evaluated in the same manner, then a second 15 card (and 16 card) is necessary. If the standard deviation of the particular type of data is positive, then this is included in equation (4); if zero then it is fixed; and if negative then the given data are desregarded in the calculation. The weighting factors in equation (4) are the reciprocal values of the square of the standard deviations.

Since PSEQUAD accepts emf data, a mechanism is required whereby the potentials may be directly converted into $\log [c_i]$ values. The Nernst equation, in the form given by equation (2), serves this purpose. A 17 card provides the appropriate additive terms A_l and an 18 card gives the multiplicative factors M_l for the lth component:

17 ADDITIVE   i1 i3 i5 i7 i9 f1 f2 f3 f4 f5

i1 . . . i9: Have the same meaning as in the 15 card.

f1 . . . f5: The additive terms to convert potentials into $\log [c_i]$ values according to equation (2). They are ordered in the same manner as the potentials on the 04 card.

18 MULTIPL   i1 i3 i5 i7 i9 f1 f2 f3 f4 f5

This card contains the multiplicative coefficients in the same sense as for the 17 card.

These cards are necessary only if the additive terms differ from zero and the multiplicative terms differ from -1. The values used in the 17 and 18 cards remain in force from one task to next, until new 17 and/or 18 cards are encountered.

The end of the task is indicated by a 10 card. The values of the integers i1–i10 on this card give rise to a variety of further processes.

10 ENDTASK   i1 i2 i3 i4 i5 i6 i7 i8 i9 i10

i1 = 1: The next card will by a type C.

i1 = 0: (need not be punched): There is no following title card

i2 = 0: The program lists only the calculated parameters together with their standard deviations.

i2 = 1: The program lists the back-calculated data and the deviation between the measured and calculated data.

i2 = 2: The program gives a complete output, including the concentration distribution of each species.

i2 = 3: The program provides only the concentration distribution. The commands given by i2 may be supervised by i3.

i3 = 0: There is no supervision.

i3 = 1: The program provides the information only at the end of model—
   selection *and* if the sum of squares minimum is reached.

i3 = 4: There is no output prior to the elimination of the species during
   model-selection.

i4 . . . i10: are the codes of many specific tasks; these positions may be left
empty. The most important possibilities are as follows:

The continuation method is used, when i5 = 1, both for titration curves and for
pointwise measurements. If back-calculations are carried out on the experimental
data for a given model and i6 = −2, then these calculated data are modified by
random errors corresponding to the given standard deviations. The calculated values are regarded as "experimental" data in the subsequent task(s).

When fitting titration curves, it is expedient to use the i7 = 1 code. In this
situation the total concentrations are calculated from the experimental volumes
only for the first Gauss–Newton iteration. The previously calculated volumes are
used for the second and subsequent iteration. For the back-calculation of titration
curves, if i7 = 1 and i8 > 0, then the procedure is carried out i8 times, using
previously calculated volumes to obtain total concentrations. This is the recommended procedure.

Some additional points need to be made concerning photometric titrations evaluated
in a given task. The serial numbers of the colored species are read in, following the
title card, in the i1 . . . i10 positions of type A cards ordered so that the species with
unknown molar absorptivities occupy the first positions.

These cards may then be followed by type B cards, containing the concentration
of a species, with measured molar absorptivities in the first ten columns. The subsequent F10.0 positions are for the absorbances measured in the wavelength order determined by i9 and i10 on the 11 card. If more than seven wavelengths are used then
the absorbance readings are continued on subsequent cards, columns 1–80. Data for
each new species must begin on a new card. If 1.0 is written into the first F10.0
positions of the first card of a species, then the molar absorptivities should be read
in, otherwise the program calculates the molar absorptivities from the given concentrations and measured absorbances.

This completes the description of the second section of the data deck. One or
more additional cards may be present. First, a new task may be initiated with an 11
card, as described above. In the second and subsequent tasks, however, the i1 and i2
positions are set to zero. It should be recalled that many of the parameter values
established in previous tasks may be left unchanged or reset within the new task.
Second, a 10 card in the DATAIN context may be present which would indicate a
second complete data comprising the two major sections described above. Third, the
experimental data and log $[c_i]$ values may be written out to a storage device (tape or
disk) using

10 DATAIN

−3 OUTTAPE   i1
   i1:   logical unit number of storage device

10 ENDDATA   i1 i2

Fourth, and always present as the final card;

82 STOP

The physical end of the data deck, permitting an orderly completion to data processing.

5. PRESENTATION OF THE PROGRAM

5.1. Listing of PSEQUAD

```
 1    C --------------------------------------------------------------------------
 2    C --
 3    C --              P S E Q U A D
 4    C --    A COMPREHENSIVE PROGRAM FOR THE EVALUATION OF POTENTIOMETRIC AND
 5    C --    SPECTROPHOTOMETRIC EQUILIBRIUM DATA USING ANALYTICAL DERIVATIVES
 6    C --
 7    C --------------------------------------------------------------------------
 8          DIMENSION IBUF(10)
 9          DOUBLE PRECISION
10    C***  REAL
11         &H0,H0INP,DFIT,DALGB1,FREED,FIT,ALN10,ALGE,CMIN,XCMAX,FNRMAX,
12         &FHMAX,XMAX,XHMAX,TITLD,CREAL,BUF(7),TITLE(20),BLANK
13          DATA BLANK/4H     /
14          COMMON/REPAR/H0,H0INP,DFIT,DALGB1,FREED,FIT,ALN10,ALGE,CMIN,
15         &XCMAX,FNRMAX,FHMAX,XMAX,XHMAX
16          COMMON/INTPAR/KRAND,MAINIT,MAXIT,MODEL,MODELN,MITFIT,MXHALV,
17         &MINIT,MMM,ITNSUM,ITHSUM,MXITNR,MXITH
18          COMMON/IPERO/IO(10)
19          COMMON/MEAS/MXI,MXR,NTIP,NI,NR,NC,MCOMP,
20         &NINT(1000)
21          COMMON TITLD(20),CREAL(5000)
22              MXI=1000
23              MXR=5000
24              DO 5 K=1,10
25      5       IO(K)=K
26              IO(1)=5
27              IO(3)=6
28    C --
29    C --------------------------------------------------------------------------
30    C --
31    C --        A D A P T A T I O N
32    C --
33    C --    A LINE PRINTER OF AT LEAST 120 CHARACTERS SHOULD BE USED. THE IO
34    C --    ARRAY IS BE FILLED WITH THE SERIAL NUMBERS OF THE INPUT/OUTPUT
35    C --    UNITS.  IO(1) SHOULD CONTAIN THE SERIAL NUMBER OF CARD-READER,
36    C --    IO(3) SHOULD CONTAIN THE SERIAL NUMBER OF LINE-PRINTER.
37    C --    DEFAULT VALUES ARE 1 FOR READER AND 3 FOR PRINTER, SEE ABOVE.
38    C --
39    C --    IF THE DATA ARE READ IN FROM A UNIT DIFFERENT FROM THAT OF IO(1),
40    C --    OR  IF THE DATA ARE ON CARDS DIFFERENT FROM THAT OF B,
41    C --    OR  THE STRUCTURE OF THE DATA ARE DIFFERENT FROM THAT OF GIVEN IN
42    C --    THE INSTRUCTIONS, THEN THE SUBROUTINE PSDINP SHOULD BE MODIFIED.
43    C --
44    C --    IF MORE THAN 1000 MEMORY ARE TO BE USED FOR INTEGER AND/OR MORE
45    C --    THAN 5000 MEMORY ARE TO BE USED FOR REAL NUMBERS, THEN
46    C --    THE 'NINT(1000)' CARD SHOULD BE CHANGED IN THE PSDATA, PSTASK AND
47    C --    PSABS SUBROUTINE  ALSO.
48    C --
49    C --    THE PRESENT VERSION WORKS WITH DOUBLE PRECISION. IF THERE IS NO
50    C --    NEED FOR THAT, THEN THE COMMENT CARD DENOTED BY 'C***' SHOULD
51    C --    REPLACE THE CARD IMMEDIATELY FOLLOWING IT.
52    C --
53    C --    THE 'RANDOM' SUBROUTINE MAY BE CHANGED FOR AN APPROPRIATE
54    C --    SUBROUTINE   WHICH GENERATES RANDOM NUMBERS WITH
55    C --    STANDARD NORMAL DISTRIBUTION.
56    C --
57    C --------------------------------------------------------------------------
```

```
 58    C --
 59                  NTIP=5
 60                  NG=Ø
 61                  NI=Ø
 62                  NR=Ø
 63                  NC=Ø
 64                  HØ=Ø.DØ
 65    C***         HØ=Ø.
 66                  ALN1Ø=1Ø.
 67                  ALN1Ø=DLOG(ALN1Ø)
 68    C***         ALN1Ø=ALOG(ALN1Ø)
 69                  ALGE=1./ALN1Ø
 70                  MITFIT=1
 71                  MXHALV=1
 72                  DFIT=5D-4
 73                  DALGB1=1.
 74                  CMIN=1.D-6
 75                  MXITNR=25
 76                  MXITH=75
 77                  FNRMAX=5.D-4
 78                  FHMAX=5.D-4
 79                  XMAX=ALN1Ø*2.Ø
 80                  XHMAX=ALN1Ø*4.Ø
 81                  XCMAX=ALN1Ø*3.Ø
 82                  I=IO(3)
 83                  WRITE(I,40)(K,K=1,9),(K,K=1,5)
 84        40       FORMAT(10X,8HKEY TEXT,4X,9(2H I,I1),4H I1Ø,5(3X,1HF,I1,7X)/)
 85                  IERR=3
 86                  KERR=2
 87        90       KEDIT=-1
 88       100       KOD=KEYPSD(IBUF,BUF,TITLE)
 89                  IF(KOD.EQ.11)GO TO 200
 90                  IF(KOD-90)102,102,900
 91       102       IF(KOD-10)199,104,199
 92       104       HØINP=BUF(5)
 93                  CALL PSDATA(NG,NCOMP,MXPOT)
 94                  IF(NG.LE.Ø)GO TO 90
 95                  MCOMP=Ø
 96                  MXPOT=Ø
 97                  I=Ø
 98                  DO 140 KG=1,NG
 99                     J1=NINT(I+12)
100                     J=NINT(I+10)
101                     IF(J.GT.J1)GO TO 190
102                     IF(J.GT.MXPOT)MXPOT=J
103                     J2=I+10+NTIP
104                     J3=J2+NINT(I+13)
105                     DO 130 K=1,J1
106                        J2=J2+1
107                        J3=J3+1
108                        J4=NINT(J2)
109                        IF(J4.LT.1)GO TO 189
110                        IF(J4.GT.MCOMP)MCOMP=J4
111                        IF(K-J)120,120,130
112       120             J5=NINT(J3)
113                        IF(J4-J5)190,130,190
114       130          CONTINUE
115                     I=I+NINT(I+1)
116       140       CONTINUE
117                  IF(MXPOT.GT.5)GO TO 190
118                  I=MXPOT*NG
119                  J1=NR+1
120                  J2=J1+I
121                  J3=J2+I
122                  J4=J3+I
123                  J5=J4+NG
124                  J6=J5+NG
125                  J7=J6+NG
126                  J8=J7+I
127                  J9=J8+NG
128                  J1Ø=J9+NG
129                  I1=NI+1
130                  I2=I1+NG
131                  I3=I2+NG
132                  I4=I3+5*NG
133                  IF(I4.GT.MXI .OR. J1Ø.GT.MXR)GO TO 192
```

```
134                IF(KEDIT.LT.Ø)KEDIT=Ø
135                IF(MXPOT)100,100,150
136        150     J=J2-1
137                DO 160 K=J1,J
138        160     CREAL(K)=HØ
139                J=J3-1
140                DO 170 K=J2,J
141        170     CREAL(K)=-1.
142                GO TO 100
143        192     IERR=IERR+2
144        191     IERR=IERR+3
145        188     KERR=KERR+7
146        190     KERR=KERR+2
147        189     CALL PSEXIT(IERR,KERR)
148        199         CALL PSEXIT(1,Ø)
149        200     IF(IBUF(1))300,300,210
150        210     NCOMP=IBUF(1)
151                MXSPEC=IBUF(2)
152                IF(KEDIT)191,215,215
153        215     IF(MXSPEC.LE.NCOMP .OR. NCOMP.LT.MCOMP)GO TO 188
154                JNAME=MXR-2*MXSPEC+1
155                JM1=JNAME-MXSPEC
156                IM1=MXI-NCOMP*MXSPEC+1
157                IM2=IM1+NCOMP*NCOMP-MXSPEC
158                IF(I4.GT.IM2 .OR. J1Ø.GT.JM1)GO TO 192
159                J=IM2+MXSPEC-1
160                DO 220 K=IM2,J
161        220     NINT(K)=-1
162                DO 225 K=JNAME,MXR
163        225     CREAL(K)=BLANK
164                KEDIT=1
165                DO 230 K=1,20
166        230     TITLE(K)=TITLD(K)
167    C
168        300     IF(KEDIT)188,188,310
169        310     J=J7-1
170                DO 315 K=J3,J
171        315     CREAL(K)=-1.
172                MAXIT=IBUF(3)
173                MAINIT=-1
174                MODEL=IBUF(4)
175                MODELN=Ø
176                NSPEC=MXSPEC
177                NSPCOL=IBUF(8)
178                NWAVE1=IBUF(9)
179                IF(NWAVE1.GT.Ø .AND. IBUF(1Ø).EQ.Ø)IBUF(1Ø)=NWAVE1
180                NWAVE=IBUF(1Ø)-NWAVE1+1
181                IF(NWAVE1.EQ.Ø)NWAVE=Ø
182                IF(Ø.GT.IBUF(7) .OR. IBUF(7).GT.NSPCOL. OR. NSPCOL.GT.MXSPEC
183        &        .OR. Ø.GT.NWAVE1 .OR. Ø.GT.NWAVE .OR. NWAVE+NSPCOL.GT.Ø
184        &        .AND. NWAVE*NSPCOL*NWAVE1.EQ.Ø)GO TO 188
185                NGCOL=Ø
186                IF(NWAVE)34Ø,34Ø,32Ø
187        32Ø     NGCOL=NG
188                I=IBUF(6)
189                IF(I.GT.Ø .AND. I.LT.NG)NGCOL=I
190        33Ø     NRFCOL=IBUF(7)
191                IF(NSPCOL)34Ø,34Ø,35Ø
192        34Ø     NGCOL=Ø
193                NRFCOL=Ø
194                NSPCOL=Ø
195                NWAVE=Ø
196        35Ø     I=IBUF(5)
197                IF(I.GT.NCOMP)NSPEC=I
198                J=J1Ø-1
199                DO 355 K=J7,J
200        355     CREAL(K)=HØ
201                I5=I4+NSPEC*NG
202                I6=I5+NSPEC*NG
203                I8=I6+NSPEC
204                I9=I8+NSPCOL*NGCOL
205                IØ=I9+NGCOL
206                JØ=J1Ø+NWAVE*NSPCOL
207                JM2=JM1+NCOMP-NWAVE*NGCOL
208                I7=IM2-1Ø*((NSPCOL+9)/1Ø)
209                I=I7-IØ-66
```

```
210              IF(NSPEC.GT.66)I=I+66-NSPEC
211              IF(I.LT.0 .OR. J0.GT.JM2)GO TO 192
212              CALL PSTASK(NINT(IM1),NINT(IM2),NINT(I4),NINT(I5),NINT(I6),
213        &     NINT(I1),NINT(I2),NINT(I7),NINT(I8),NINT(I9),NINT(I0),
214        &     IBUF,BUF,TITLE,I0,J0,JM2,IFAULT,CREAL(JM1),CREAL(J3),
215        &     CREAL(J4),CREAL(J5),CREAL(J6),CREAL(J2),CREAL(J1),
216        &     CREAL(J7),CREAL(J8),CREAL(J9),CREAL(JM2),CREAL(J10),
217        &     NWAVE1,NINT(I3),CREAL(JNAME),NG,NCOMP,MXPOT,
218        &     NWAVE,NSPCOL,NSPREF,NSPEC,MXSPEC,NRFCOL,NGCOL,
219        &     I13,I15,NCURVE,NPOINT,KODCOL,KODPOT,KCOMP,KSPEC,KODREF)
220              IF(MAINIT+4)330,330,360
221        360   IF(IFAULT)100,100,199
222        900   K=KOD-90
223              GO TO(910,920,930,100,100,100,100,100,100),K
224        910   MITFIT=IBUF(1)
225              MXHALV=IBUF(2)
226              DFIT= BUF(1)
227              DALGB1= BUF(2)
228              CMIN= BUF(3)
229              GO TO 100
230        920   MXITNR=IBUF(1)
231              MXITH=IBUF(2)
232              FNRMAX= BUF(1)
233              FHMAX= BUF(2)
234              XMAX= BUF(3)*ALN10
235              XHMAX= BUF(4)*ALN10
236              XCMAX= BUF(5)*ALN10
237              GO TO 100
238        930   KRAND=0
239              DO 935 I=1,4
240        935   KRAND=100*KRAND+IBUF(K)
241          GO TO 100
242          END
```

```
 1    C -------------------------------------------------------------------------
 2    C --
 3    C --   S U B R O U T I N E     P S D A T A
 4    C --
 5    C --              INPUT AND CHECK OF THE DATA CARDS
 6    C --
 7    C -------------------------------------------------------------------------
 8          SUBROUTINE  PSDATA(NG,NCOMP,MXPOT)
 9          COMMON/MEAS/MXI,MXR,NTIP,NI,NR,NC,MCOMP,
10         &NINT(1000)
11          COMMON TITLD(20),CREAL(5000)
12          DOUBLE PRECISION
13    C***  REAL
14         &TITLD,CREAL,BUF(7),TITLE(20)
15          DIMENSION IBUF(10)
16              IFL=3
17              GO TO 15
18    10      NG=0
19              NI=0
20              NR=0
21              NC=0
22    15      KEDIT=0
23    20      KOD=KEYPSD(IBUF,BUF,TITLE)
24              IF(KOD.EQ.10)GO TO 60
25              IF(KOD)80,99,30
26    30      IF(KOD.GT.NTIP .OR. KEDIT.GT.KOD)GO TO 99
27              IF(KOD.EQ.3)GO TO 40
28              IF(KOD.GT.1)GO TO 50
29              KEDIT=1
30              NG=NG+1
31              K1=NI+1
32              K10=NI+10
33              K15=K10+NTIP
34              DO 35  K=K1,K15
35    35      NINT(K)=0
36              I11=IBUF(1)
37              IF(I11.LE.0)GO TO 98
38              KB1=IBUF(2)
```

```
39                      GO TO 20
40          40          IF(KEDIT.NE.2)GO TO 99
41                      KEDIT=3
42                      K=NINT(K10+2)+1
43                      NINT(K10+3)= K
44                      NINT(K15+K)= 0
45                      GO TO 20
46          50          IF(IBUF(1).LE.0)GO TO 98
47          52          IF(KOD.EQ.KEDIT)GO TO 54
48                      KEDIT=KEDIT+1
49                      K=K10 +KEDIT
50                      NINT(K)=NINT(K-1)
51                      GO TO 52
52          54          KB2=K10+KEDIT
53                      KB2=NINT(KB2)
54                      DO 56  K=1,5
55                         KB5= IBUF(2*K-1)
56                         KB6=IABS(IBUF(2*K))
57                         IF(KB6.EQ.0)KB6=KB5
58                         IF(KB5.LT.0 .OR. KB6.LT.KB5)GO TO 98
59                         IF(KB5.EQ.0)GO TO 58
60                         DO 55  KB=KB5,KB6
61                            KB2=KB2+1
62                            KK=K15+KB2
63                            NINT(KK)=KB
64                            KK=NR+KB2
65                            IF(KEDIT.EQ.2)CREAL(KK)=BUF(K)
66          55          CONTINUE
67          56          CONTINUE
68          58          KK=K10+KEDIT
69                      NINT(KK)=KB2
70                      GO TO 20
71          60          IF(KEDIT.EQ.0)GO TO 70
72          62          IF(KEDIT.EQ.NTIP)GO TO 64
73                         KEDIT=KEDIT+1
74                         K=K10+KEDIT
75                         NINT(K)=NINT(K-1)
76                         GO TO 62
77          64          I12=NINT(NI+12)
78                      I13=NINT(NI+13)
79                      I7=I13-I12
80                      KK=K10+NTIP
81                      I15=NINT(KK)
82                      I8=I15-I12
83                      KB2=K15+I15+1
84                      KEDIT=0
85                      IF(I12.GT.0 .AND. I15.GT.I13)GO TO 65
86                         KOD=4
87                         IF(I12.LE.0)KOD=2
88                         GO TO 99
89          65          IF(MXI.LE.KB2+I7*I11)GO TO 993
90                      KB5=NR+I7*I12+1
91                      KB6=KB5+I13*I11
92                      KB=(MXR-KB6+1)/I15
93                      KB6=KB6+I12
94                      IF(I7)99,66,68
95          66          IF(I11-KB)67,67,993
96          67          KB=0
97          68          CALL PSDINP(CREAL(KB5),I13,I11,CREAL(KB6),I15,
98           &          I8,KB,NINT(KB2),KB1,IBUF,BUF)
99                      I9=I11
100                     KB=K15+I15+I11*I7
101                     IF(I7.EQ.1)I9=NINT(KB)
102                     KB6=KB6-I12
103                     NINT(NI+1)=KB-NI
104                     NINT(NI+2)=I7*I12+I13*I11+I15*I9
105                     NINT(NI+4)=NR+1
106                     NINT(NI+5)=NINT(NI+4)+I7*I12
107                     NINT(NI+6)=NINT(NI+5)+I13*I11
108                     NINT(NI+7)=I7
109                     NINT(NI+8)=I8
110                     NINT(NI+9)=I9
111                     NINT(NI+10)=NINT(NI+14)-I13
112                     NINT(NI+11)=I11
113                     IF(IBUF(2).GT.0)
114          &          CALL PSDOUT(CREAL(KB5),I13,I11,CREAL(KB6),I15,I8,I9,
```

```
115        &    NINT(KB2),CREAL(NR+1),NINT(K1),NINT(K15+1),TITLE,
116        &    IBUF(1),IBUF(2),NG,NTIP)
117             NR=NR+NINT(NI+2)
118             NI=NI+NINT(NI+1)
119             GO TO 20
120   70        IF(NG.LE.0)GO TO 996
121             IF(IBUF(1)-1)74,71,74
122   71        DO 72 K=1,20
123   72        TITLD(K)=TITLE(K)
124   74        CALL PSDOUT(CREAL(1),NG,NI,CREAL(1),NR,NC,NC,NINT(1),
125        &    CREAL(1),NINT(1),NINT(1),TITLD,1,IBUF(2),0,NTIP)
126             IF(IBUF(2).LE.0)GO TO 78
127             K1=1
128             DO 76 KG=1,NG
129             I8=NINT(K1+7)
130             I9=NINT(K1+8)
131             I11=NINT(K1+10)
132             I13=NINT(K1+12)
133             K15=K1-1+10+NTIP
134             I15=NINT(K15)
135             KB=K15+I15+1
136             I4=NINT(K1+3)
137             KB1=I4+NINT(K1+6)*NINT(K1+11)
138             KB2=KB1+I11*I13
139             CALL PSDOUT(CREAL(KB1),I13,I11,CREAL(KB2),I15,I8,I9,NINT(KB),
140        &    CREAL(I4),NINT(K1),NINT(K15+1),TITLE,0,IBUF(2),KG,NTIP)
141             K1=K1+NINT(K1)
142   76        CONTINUE
143   78   RETURN
144   80        IF(KOD.EQ.-1)GO TO 10
145             IF(KEDIT.GT.0)GO TO 99
146             KB1=IBUF(1)
147             IF(KB1.LE.0)GO TO 997
148             IF(KOD+3)99,100,90
149   90        KB=1
150   92        CALL PSDTAP(IBUF(1),IBUF(3),IBUF(5),IBUF(7),IBUF(9),
151        &    NINT(NI+1),TITLD,CREAL(NR+1),KB1,KB,NNI,NNR)
152             IBUF(3)=NNI
153             IBUF(5)=NNR
154             IF(KB.GT.1)GO TO 96
155             IF(IBUF(1).LE.0 .OR. IBUF(7).LT.0 .OR. IBUF(9).NE.NTIP)
156        &    GO TO 998
157             IF(NI+IBUF(3).GT.MXI .OR. NR+IBUF(5).GT.MXR)GO TO 993
158             KB=2
159             GO TO 92
160   96        K1=NG+1
161             KB=NG+IBUF(1)
162             DO 97 KG=K1,KB
163             NG=NG+1
164             NINT(NI+4)=NR+1
165             NR=NR+NINT(NI+2)
166             NC=NC+NINT(NI+3)
167             NI=NI+NINT(NI+1)
168   97        CONTINUE
169             GO TO 20
170   100       IF(IBUF(1).GT.0)CALL PSDTAP(NG,NI,NR,NC,NTIP,NINT,TITLD,
171        &    CREAL,IBUF(1),0,NNI,NNR)
172             GO TO 20
173   993       IFL=IFL+1
174   997       IFL=IFL+1
175   996       IFL=IFL+1
176   998       IFL=IFL+1
177   99        IFL=IFL+1
178   98        CALL PSEXIT(IFL,KOD)
179             CALL PSEXIT(1,0)
180             RETURN
181             END
```

```
1    C -------------------------------------------------------------------------
2    C --
3    C --      S U B R O U T I N E   P S D I N P
4    C --
5    C --             INPUT OF THE DATA OF ONE SET FROM TYPE B CARDS
```

```
 6    C --
 7    C --    THE VARIABLE KEYINP MAY BE USED TO SELECT THE METHOD FOR DATA
 8    C --    INPUT. THIS IS READ FROM THE I2 POSITION OF '01 SET' CARD.
 9    C --    THE OTHER NECESSARY DATA MAY BE GIVEN ON THE I3,I4,...,I10  AND
10    C --    F1,F2,...,F5  POSITIONS OF '10 ENDSET' CARD, WHICH WILL BE GIVEN
11    C --    TO THE KPRVAT(3) ,...,KPRVAT(10) AND PRIVAT(1),...,PRIVAT(5)
12    C --
13    C ------------------------------------------------------------------------
14          SUBROUTINE  PSDINP(PT,KT,KSUB,PM,KD,KM,KTITR,KPOINT,KEYINP,
15         &          KPRVAT,      PRIVAT)
16          DOUBLE PRECISION
17    C***  REAL
18         &PT,PM,PRIVAT
19          DIMENSION PT(KT,KSUB),PM(KD,1),KPOINT(KSUB),KPRVAT(10),PRIVAT(5)
20          COMMON/IPERO/IO(10)
21    101   FORMAT(8F10.0)
22          KINP=IO(1)
23          KTM1=KT-1
24          IF(KTITR .GT. 0)GO TO 103
25    C --
26    C --       INPUT FOR POINTWISE MEASUREMENTS (KTITR.EQ.0)
27    C --
28    C --  KSUB .........NUMBER OF EXPERIMENTAL POINTS
29    C --  KT ...........NUMBER OF COMPONENTS
30    C --  PT(KT,KSUB)....TOTAL CONCENTRATIONS
31    C --  KM ...........NUMBER OF MEASURED POTENTIALS AND ABSORBANCIES
32    C --             KD=KM+KT
33    C --  PM(KD,KSUB)...POTENTIALS, ABSORBANCIES (AND THE LOGARITHM  OF
34    C --             COMPONENT CONCENTRATIONS)
35    C --  THE TOTAL CONCENTRATIONS, POTENTIALS AND ABSORBANCIES BELONGING TO
36    C --  POINT K1 ARE IN THE  PT(1,K1),...,PT(KT,K1),PM(1,K1),...,PM(KM,K1)
37    C --  VARIABLES.
38    C --
39          DO 102  K1=1,KSUB
40          READ(KINP,101)(PT(K,K1),K=1,KT),(PM(K,K1),K=1,KM)
41    102   CONTINUE
42          RETURN
43    C --
44    C --         IN CASE OF TITRATION CURVES(KTITR.GT.0)
45    C --  KTITR ...ONLY  KTITR  POINT MAY BE READ IN. THIS IS CALCULATED
46    C --          BY THE PROGRAM.
47    C --  KSUB ....NUMBER OF TITRATION CURVES
48    C --  KT-1 ....NUMBER OF COMPONENTS
49    C --  KM-1 ....NUMBER OF MEASURED POTENTIALS AND ABSORBANCIES
50    C --  THE TOTAL CONCENTRATIONS OF K1-ST STARTING SOLUTION IS IN
51    C --  PT(1,K1),...,PT(KT-1,K1)VARIABLES. ITS VOLUME IS IN PT(KT,K1).
52    C --  THE SERIAL NUMBER OF THE LAST POINT OF K1-ST TITRATION CURVE IS
53    C --  IN  KPOINT(K1).
54    C --  THE VOLUME OF K2-ND TITRATION POINT IS IN PM(1,K2), THE MEASURED
55    C --  POTENTIALS AND ABSORBANCIES ARE IN  PM(2,K2),...,PM(KM,K2).
56    C --
57    103   K2=0
58          DO 107  K1=1,KSUB
59          READ(KINP,101)PT(KT,K1),(PT(K,K1),K=1,KTM1)
60    104   K2=K2+1
61          IF(K2-KTITR)106,106,105
62    105     CALL PSEXIT(8,0)
63          CALL PSEXIT(1,0)
64    106   READ(KINP,101)(PM(K,K2),K=1,KM)
65          IF(PM(1,K2).GE. 0.)GO TO 104
66          K2=K2-1
67          KPOINT(K1)=K2
68    107   CONTINUE
69          RETURN
70          END



 1    C ------------------------------------------------------------------------
 2    C --
 3    C --    F U N C T I O N  R A N D O M
 4    C --
 5    C --  COMPUTES A NORMALLY DISTRIBUTED RANDOM NUMBER (MEAN=0. , STANDARD
 6    C --  DEVIATION=1.) FOR THE IBM SYSTEM/360. FROM SUBROUTINES 'RANDU'
```

```
 7      C --    'GAUSS' OF THE SSP PACKAGE. STARTING VALUE OF 'K' CAN READ IN
 8      C --    FROM I1...,I4 POSITIONS OF CARD KEY:93.
 9      C --
10      C -------------------------------------------------------------------
11              DOUBLE PRECISION
12      C       REAL
13            & FUNCTION    RANDOM(K)
14              SH=-6.
15              DO 30 I=1,12
16                 K=K*65539
17                 IF(K)10,20,20
18         10      K=K+2147483647+1
19         20      H=K
20         30      SH=SH+H*.4656613E-9
21              RANDOM=SH
22              RETURN
23              END



 1      C -------------------------------------------------------------------
 2      C --
 3      C --     S U B R O U T I N E    P S D O U T
 4      C --
 5      C --                LISTING OF THE MEASURED DATA
 6      C --
 7      C -------------------------------------------------------------------
 8              SUBROUTINE  PSDOUT(TOTV,I13,I11,XVPA,I15,I8,I9,IP,TOTB,IM,IDENT,
 9            &TITLE,KTITLE,KPR,KG,NTIP)
10              COMMON/IPERO/IO(10)
11              DIMENSION TOTV(I13,I11),XVPA(I15,I9),TOTB(1),IM(1),IDENT(I15),
12            &TITLE(20),CHAR(10),IP(1),CH(4)
13              DOUBLE PRECISION
14      C***    REAL
15            &TOTV,XVPA,TOTB,TITLE,CHAR,CH
16              DATA CHAR(2),CHAR(3),CHAR(4),CHAR(5)/3HTOT,3HVOL,3HPOT,3HABS/
17              DATA CH(1),CH(2),CH(3),CH(4)/3HCUR,3HVE ,3H NO,3H.  /
18         10 FORMAT(1H1)
19         20 FORMAT(1X)
20         30 FORMAT(  /2X,2(2H* ),14HP S E Q U A D ,3(2H* ),14HP S E Q U A D ,
21            &4(2H* ),27HM E A S U R E D    D A T A ,4(2H* ),14HP S E Q U A D ,
22            &3(2H* ),14HP S E Q U A D ,2(2H* ))
23         40 FORMAT(17X,22(4HXXXX))
24         50 FORMAT(17X,1HX,86X,1HX)
25         60 FORMAT(17X,1HX,3X,20A4,3X,1HX)
26         70 FORMAT(/18X,I3,5H   SET,8X,16HSTORAGE(INTEGER=,I5,7H, REAL=,I5,1H))
27         71 FORMAT(1H+,25X,1HS)
28         80 FORMAT(17X,5(3H***)/17X,8H*     SET,I3,22H  *  STORAGE(INTEGER=,I5
29            &,7H, REAL=,I5,1H))
30         90 FORMAT(1H+,80X,16HSTORED LN(FREEC))
31        100 FORMAT(17X,5(3H***)/18X,I3,7H POINTS)
32        104 FORMAT(1H+,28X,I4,7H CURVE )
33        105 FORMAT(1H+,38X,1HS)
34        106 FORMAT(18X,I3,6H DATA:)
35        110 FORMAT(18X,A3,  25I4/(21X,25I4))
36        130 FORMAT(/4X,2A3,3X,9(2X,A3,7X))
37        135 FORMAT(/(9X,  8HBURETTE:,8X,8(2X,A3,I2,5X)))
38        140 FORMAT(1H+,12X,9(5X,I2,5X))
39        150 FORMAT(3X,I4,6X,1P9G12.5)
40        155 FORMAT(1X,24X,1P8G12.5)
41              J= IO(3)
42              IF(KG.GT.0 .AND. KPR.EQ.2)WRITE(J,10)
43              WRITE(J,30)
44              IF(KTITLE.LE.0)GO TO 200
45              WRITE(J,20)
46              WRITE(J,40)
47              WRITE(J,50)
48              WRITE(J,60)(TITLE(K),K=1,20)
49              WRITE(J,50)
50              WRITE(J,40)
51        200   IF(KG.GT.0)GO TO 210
52              WRITE(J,70)I13,I11,I15
53              IF(I13.GT.1)WRITE(J,71)
54              RETURN
```

```
 55   210      WRITE(J,80)KG,IM(1),IM(2)
 56              IF(IM(3).GT.0)WRITE(J,90)
 57              WRITE(J,100)I9
 58              IF(I9.GT.I11)WRITE(J,104)I11
 59              IF(I9.GT.I11 .AND. I11.GT.1)WRITE(J,105)
 60              WRITE(J,106)I15
 61              II=10+NTIP
 62              K2=0
 63              DO 220 I=12,II
 64                 K1= K2+1
 65                 K2= IM(I)
 66                 IF(K2.GE.K1)WRITE(J,110)CHAR(I-10),(IDENT(K),K=K1,K2)
 67   220      CONTINUE
 68              I12=IM(12)
 69              IF(KPR.EQ.1)RETURN
 70              K2=0
 71              I7=IM(7)
 72              IF(I7.EQ.1)GO TO 240
 73   225      K1=K2+1
 74              K2=K2+9
 75              IF(K2.GT.I12)K2=I12
 76              WRITE(J,130)CH(3),CH(4), (CHAR(2),K=K1,K2)
 77              WRITE(J,140)(IDENT(K),K=K1,K2)
 78              WRITE(J,20)
 79              DO 230 I=1,I11
 80   230      WRITE(J,150)I,(TOTV(K,I),K=K1,K2)
 81              IF(K2.GT.I12)K2=I12
 82              IF(K2.LT.I12)GO TO 225
 83              IF(I15.EQ.I12)RETURN
 84              I=14
 85              GO TO 260
 86   240      K1=K2+1
 87              K2=K2+8
 88              IF(K2.GT.I12)K2=I12
 89              WRITE(J,135)(CHAR(2),IDENT(K),K=K1,K2)
 90              WRITE(J,155)(TOTB(K),K=K1,K2)
 91              IF(K2.LT.I12)GO TO 240
 92              I=13
 93   260      K1=K2+1
 94              IF(K1.GT.I15)RETURN
 95              K2=K2+9
 96              IF(K2.GT.I15)K2=I15
 97              DO 300 K=K1,K2
 98   270         IF(K-IM(I))290,290,280
 99   280         I=I+1
100              GO TO 270
101   290         II=K+1-K1
102                TITLE(II)=CHAR(I-10)
103   300      CONTINUE
104              IF(I7.EQ.1)GO TO 320
105              II=K2-K1+1
106              WRITE(J,130)CH(3),CH(4),(TITLE(K),K=1,II)
107              WRITE(J,140)(IDENT(K),K=K1,K2)
108              WRITE(J,20)
109              DO 310 II=1,I11
110   310      WRITE(J,150)II,(XVPA(K,II),K=K1,K2)
111              GO TO 260
112   320      I2=0
113              DO 340 IS=1,I11
114                 I1= I2+1
115                 I2= IP(IS)
116                 L2=0
117   325         L1=L2+1
118                 L2=L2+8
119                 IF(L2.GT.I12)L2=I12
120                 WRITE(J,130)CH(1),CH(2),CHAR(3),(CHAR(2),K=L1,L2)
121                 WRITE(J,140)IDENT(I13),(IDENT(K),K=L1,L2)
122                 WRITE(J,150)IS,TOTV(I13,IS),(TOTV(K,IS),K=L1,L2)
123                 IF(L2.LT.I12)GO TO 325
124                 II=K2+1-K1
125                 WRITE(J,130)CH(3),CH(4),(TITLE(K),K=1,II)
126                 WRITE(J,140)(IDENT(K),K=K1,K2)
127                 WRITE(J,20)
128                 DO 330 II=I1,I2
129   330            WRITE(J,150)II,(XVPA(K,II),K=K1,K2)
130                 WRITE(J,20)
```

```
131      34Ø      CONTINUE
132              WRITE(J,2Ø)
133              GO TO 26Ø
134          END




  1   C ------------------------------------------------------------------------
  2   C --
  3   C --    S U B R O U T I N E    P S T A S K
  4   C --
  5   C --            ORGANIZATION OF THE CALCULATION
  6   C --
  7   C ------------------------------------------------------------------------
  8          SUBROUTINE PSTASK(ISTCOF,MODREF,ISORT,ISELEC,NISORT,IREFIN,ISPEC,
  9         &NICOL,ICOL,KW1,IH,IBUF,BUF,TITLE,IØ,JØ,JM,IFAULT,
 10         &ALGB,VPOT,VVOL,VORT,VABS,POTMLT,POTADD,VARPOT,VARVOL,
 11         &VARORT,VARABS,ABCOEF,NWAVE1,KEYCLC,SPNAME,NG,NCOMP,MXPOT,
 12         &NWAVE,NSPCOL,NSPREF,NSPEC,MXSPEC,NRFCOL,NGCOL,
 13         &I13,I15,NCURVE,NPOINT,KODCOL,KODPOT,KCOMP,KSPEC,KODREF)
 14          DOUBLE PRECISION
 15   C***   REAL
 16         &      BUF(7),TITLE(2Ø),ALGB(MXSPEC),VABS(NG),VVOL(NG),VORT(NG),
 17         &      VPOT(MXPOT,NG),POTMLT(MXPOT,NG),POTADD(MXPOT,NG),VARVOL(NG),
 18         &      VARPOT(MXPOT,NG),VARORT(NG),VARABS(NWAVE,NGCOL),  TITLD,CREAL,
 19         &      ABCOEF(NWAVE,NSPCOL),HØ,HØINP,DFIT,DALGB1,FREED,FIT,ALN1Ø,
 20         &      ALGE,CMIN,XCMAX,FNRMAX,FHMAX,XMAX,XHMAX,SPNAME(2,MXSPEC),
 21         &      H,SH,REDUCE
 22          DIMENSION ISTCOF(NCOMP,MXSPEC),MODREF(MXSPEC),ISORT(NSPEC,NG),
 23         &          ISELEC(NSPEC,NG),NISORT(NSPEC),IREFIN(NG),ISPEC(NG),
 24         &          NICOL(NSPCOL),ICOL(NSPCOL,NGCOL),KW1(NGCOL),
 25         &          IH(NSPEC),IBUF(1Ø),KEYCLC(5,NG)
 26          COMMON/REPAR/HØ,HØINP,DFIT,DALGB1,FREED,FIT,ALN1Ø,ALGE,CMIN,
 27         &             XCMAX,FNRMAX,FHMAX,XMAX,XHMAX
 28          COMMON/INTPAR/KRAND,MAINIT,MAXIT,MODEL,MODELN,MITFIT,MXHALV,
 29         &MINIT,MMM,ITNSUM,ITHSUM,MXITNR,MXITH
 30          COMMON/LABGRP/KODVOL,KFIX,K1,KODTOT,KCOMP1,KODORT,KD,KREFIN,
 31         &LB,LB1,LB2,LB3,LB4
 32          COMMON/IPERO/IO(1Ø)
 33          COMMON/MEAS/MXI,MXR,NTIP,NI,NR,NC,MCOMP,
 34         &NINT(1ØØØ)
 35          COMMON TITLD(2Ø),CREAL(5ØØØ)
 36          DATA JBLANK/1H /,JSTAR/1H*/
 37        3 FORMAT(8F1Ø.Ø)
 38          IFAULT=Ø
 39          INP=IO(1)
 40          IPR=IO(3)
 41          IF(MAINIT+4)1Ø5,1Ø5,1
 42        1 KOD=KEYPSD(IBUF,BUF,TITLE)
 43          IF(KOD.EQ.Ø)GO TO 21
 44          IF(KOD.LT.1Ø .OR. KOD.GT.18)GO TO 99
 45          K=KOD-9
 46          GO TO(4Ø,99,21,3Ø,3Ø,5,5,7,7),K
 47        5      DO 6 K=1,5
 48        6          IF(BUF(K).EQ.HØINP)BUF(K)=HØ
 49                CONTINUE
 50        7      DO 19 I=1,9,2
 51          N1=IBUF(I)
 52          IF(N1)1,1,8
 53        8      N2=IABS(IBUF(I+1))
 54          IF(N2.EQ.Ø)N2=N1
 55          IF(N2.LT.N1 .OR. N2.GT.NG)GO TO 97
 56          DO 18 KG=N1,N2
 57              IF(KOD.GT.15)GO TO 9
 58              VVOL(KG)=BUF(1)
 59              VORT(KG)=BUF(2)
 60              IF(MXPOT.GE.1)VPOT(1,KG)=BUF(3)
 61              IF(MXPOT.GE.2)VPOT(2,KG)=BUF(4)
 62              IF(KG.LE.NGCOL)VABS(KG)=BUF(5)
 63              GO TO 18
 64        9      IF(KOD-17)1Ø,12,15
 65       1Ø      IF(MXPOT.LT.3)GO TO 99
 66                DO 11 K=3,MXPOT
 67       11          VPOT(K,KG)=BUF(K)
```

```
68                              GO TO 18
69      12              IF(MXPOT)99,99,13
70      13                  DO 14 K=1,MXPOT
71      14              POTADD(K,KG)=BUF(K)
72                          GO TO 18
73      15              IF(MXPOT)99,99,16
74      16                  DO 17 K=1,MXPOT
75      17              POTMLT(K,KG)=BUF(K)
76      18          CONTINUE
77      19          CONTINUE
78              GO TO 1
79      21      I=IBUF(1)
80              IF(I.LE.Ø .OR. I.GT.MXSPEC)GO TO 97
81                  SPNAME(1,I)=BUF(6)
82                  SPNAME(2,I)=BUF(7)
83                  IF(I.LE.NCOMP)GO TO 1
84                  MODREF(I)=IBUF(2)
85                  ALGB(I)=BUF(1)
86                  IF(NCOMP-8)22,22,24
87      22          DO 23 K=1,NCOMP
88      23          ISTCOF(K,I)=IBUF(K+2)
89                  GO TO 27
90      24          J=1
91                  N2=(NCOMP+9)/1Ø
92                  DO 25 II=1,N2
93                  KOD=KEYPSD(IH(J),BUF,CREAL(JØ))
94                  IF(KOD)99,25,99
95      25          J=J+1Ø
96                  DO 26 K=1,NCOMP
97      26          ISTCOF(K,I)=IH(K)
98      27          J=Ø
99                  DO 28 K=1,NCOMP
100     28          J=J+IABS(ISTCOF(K,I))
101                 IF(J)97,97,1
102     30          DO 35 K=1,5
103                     I=IBUF(2*K-1)
104                     IF(I)35,35,32
105     32              IF(I.LE.NCOMP .OR. I.GT.MXSPEC)GO TO 97
106                     MODREF(I)=IBUF(2*K)
107                     IF(KOD.EQ.13)ALGB(I)=BUF(K)
108     35          CONTINUE
109             GO TO 1
110     40      NRFCL1=NRFCOL
111             IF(NGCOL)100,100,41
112     41      J=1
113             N2=(NSPCOL+9)/1Ø
114             DO 42 II=1,N2
115             KOD=KEYPSD(NICOL(J),BUF,CREAL(JØ))
116             IF(KOD)99,42,99
117     42      J=J+1Ø
118             IF(NRFCOL)5Ø,5Ø,43
119     43          DO 45 I=1,NRFCOL
120                 DO 45 K=1,NWAVE
121     45          ABCOEF(K,I)=Ø.
122     50      IF(NSPCOL-NRFCOL)61,61,55
123     55      N1=NRFCOL+1
124             DO 6Ø I=N1,NSPCOL
125             READ(INP,3)H,(ABCOEF(K,I),K=1,NWAVE)
126             IF(H)6Ø,6Ø,56
127     56      DO 57 K=1,NWAVE
128     57      ABCOEF(K,I)=ABCOEF(K,I)/H
129     60      CONTINUE
130     61      IF(NRFCOL.EQ.Ø)GO TO 100
131             NRFCOL=Ø
132             II=NRFCL1
133             DO 65 KK=1,NSPCOL
134     62      I=NICOL(KK)
135             IF(I.LE.Ø .OR. I.GT.MXSPEC)GO TO 97
136             IF(KK.GT.II)GO TO 65
137             IF(I.LE.NCOMP)GO TO 64
138             IF(MODREF(I))63,64,64
139     63          NICOL(KK)=NICOL(II)
140                 NICOL(II)=I
141                 II=II-1
142             GO TO 62
143     64          NRFCOL=NRFCOL+1
```

```
144      65          CONTINUE
145      70 FORMAT(1X)
146      71 FORMAT(17X,22(4HXXXX))
147      72 FORMAT(17X,1HX,86X,1HX)
148      73 FORMAT(17X,1HX,3X,20A4,3X,1HX)
149      74 FORMAT(1H1)
150      75 FORMAT(/ 13X,16HWEIGHT=1/(SD*SD),8X,
151         &31HPOTENTIAL=ADD+MULT*ALOG10(CONC))
152      76 FORMAT(2X,12HSET  SD(VOL),12X,7HSD(POT),4X,4HPOT.)
153      77 FORMAT(1H+,15X,7HVOL/POT,21X,3HADD,10X,4HMULT,10X,7HSD(ABS),5X,
154         &8HSPECIES:)
155      78 FORMAT(/1X,1H*,I2,2H* )
156      87 FORMAT(1H+,5X, 1PG10.3)
157      88 FORMAT(1H+,15X,1PG10.3)
158      79 FORMAT(1H+,68X,1PG10.3)
159      80 FORMAT(1H+,25X,1PG10.3,2H( ,I2,2H ))
160      81 FORMAT(1H+,41X,2(1PG12.5,1X))
161      82 FORMAT(1H+,79X,10I4)
162      83 FORMAT(1H+,80X,10(3X,A1))
163      84 FORMAT(/4X,22HNUMBER OF FITTED DATA=,I5,5X,18HNUMBER OF REFINED ,
164         &11HPARAMETERS=,I5,5X,18HDEGREE OF FREEDOM=,F7.0/ 4X,8HFITTING ,
165         &13HPARAMETER =  ,1PG12.5,32H = SQRT(SUM(WEIGHT*RESIDUAL**2)/,
166         &12HDEGR.FREED.)/40H SD=ESTIMATED STANDARD DEVIATION IN SET=,
167         &53HSQRT(SUM IN SET(RESIDUAL**2)/NFD IN SET)*SQRT(NFD/DF))
168      86 FORMAT(1H+,15X,7HSD(ORT))
169      89 FORMAT(/5H ABS.,11(A1,4H SET,I3,2X))
170      90 FORMAT(1X,I3,1X,1P11G10.3)
171      91 FORMAT(/11X,16HSTORAGE(INTEGER=,I5,10H  ,  REAL= ,I5,1H)/)
172      92 FORMAT(//50H ***FAULT IN THE CALCULATION OF CONCENTRATIONS***
173         &/3X,5H  SET,I3,5X,5HPOINT,I4,14X,6HLN(C): /(10F10.3))
174     100          KLIST=IBUF(2)
175                  KLISTT=IBUF(3)
176                  KEYCNT=IBUF(5)
177                  KWRITE=IBUF(6)
178                  KEYVOL=IBUF(7)
179                  MINIT=IBUF(8)
180                  KEYCOR=IBUF(9)
181                  KEYFLT=IBUF(10)
182     105          NSPREF=0
183                  JVOLC=0
184                  JORT=0
185                  JCOLC=0
186                  JCOLD=0
187                  JHELP=0
188                  NDATA=0
189                  KCOMP1=NCOMP+1
190                  DO 120 I=KCOMP1,MXSPEC
191                      IF(MODREF(I))120,120,110
192     110              NSPREF=NSPREF+1
193                      NISORT(NSPREF)=I
194     120          CONTINUE
195                  IF(MAINIT.LE.-4)WRITE(IPR,74)
196                  WRITE(IPR,71)
197                  WRITE(IPR,72)
198                  WRITE(IPR,73)(TITLD(K),K=1,20)
199                  WRITE(IPR,72)
200                  WRITE(IPR,73)(TITLE(K),K=1,20)
201                  WRITE(IPR,72)
202                  WRITE(IPR,71)
203                  WRITE(IPR,75)
204                  WRITE(IPR,76)
205                  WRITE(IPR,77)
206                  N1=0
207                  DO 390 KG=1,NG
208                      NPOINT=NINT(N1+9)
209                      KEYCLC(1,KG)=0
210                      KCOMP=NINT(N1+12)
211                      II=N1+10+NTIP
212                      DO 130 I=1,NCOMP
213     130              IH(I)=0
214                      DO 140 K=1,KCOMP
215                          II=II+1
216                          I=NINT(II)
217                          ISORT(K,KG)=I
218                          IH(I)=K
219     140              CONTINUE
```

```
220                      KODREF=Ø
221                      KREFIN=KCOMP
222                      KD=NCOMP
223                      K1=Ø
224                      DO 18Ø I=KCOMP1,MXSPEC
225                        N2=MODREF(I)
226                        IF(N2)18Ø,15Ø,15Ø
227       15Ø            II=Ø
228                      DO 155 K=1,NCOMP
229                        J=ISTCOF(K,I)
230                        IF(IH(K))151,151,152
231       151            IF(J)18Ø,155,18Ø
232       152            IF(J.NE.Ø)II=1
233       155            CONTINUE
234                      IF(II)18Ø,18Ø,16Ø
235       16Ø            IF(N2)18Ø,165,17Ø
236       165            KD=KD+1
237                      IH(KD)=I
238                      GO TO 18Ø
239       17Ø            KODREF=KODREF+1
240                      KREFIN=KREFIN+1
241                      ISORT(KREFIN,KG)=I
242                      K1=K1+1
243       171            IF(I-NISORT(K1))18Ø,173,172
244       172            K1=K1+1
245                      GO TO 171
246       173            ISELEC(KODREF,KG)=K1
247       18Ø            CONTINUE
248                      II=KREFIN-NCOMP
249                      IREFIN(KG)=KREFIN
250                      KSPEC=KD+II
251                      ISPEC(KG)=KSPEC
252                      IF(KD-NCOMP)19Ø,19Ø,183
253       183            DO 185 K=KCOMP1,KD
254                        I=II+K
255                        ISORT(I,KG)=IH(K)
256       185            CONTINUE
257       19Ø            IF(KG-NGCOL)2ØØ,2ØØ,24Ø
258       2ØØ            IF(VABS(KG))24Ø,24Ø,21Ø
259       21Ø            II=N1+1Ø+NTIP
260                      N2=II+NINT(N1+14)+1
261                      N3=II+NINT(N1+15)
262                      KW1(KG)=Ø
263                      DO 22Ø I=N2,N3
264                        J=NINT(I)
265                        IF(J-NWAVE1)22Ø,215,22Ø
266       215            IF(N3.LT.I-1+NWAVE)GO TO 98
267                      KW1(KG)=I-II
268                      KWW1=I
269                      GO TO 221
270       22Ø            CONTINUE
271       221            IF(KW1(KG))98,98,222
272       222            II=Ø
273                      DO 23Ø I=1,NSPCOL
274                        J=NICOL(I)
275                        ICOL(I,KG)=Ø
276                        DO 228 K=1,KSPEC
277                        IF(J-ISORT(K,KG))228,225,228
278       225            ICOL(I,KG)=K
279                      II=II+1
280                      GO TO 23Ø
281       228            CONTINUE
282       23Ø            CONTINUE
283                      IF(II)235,235,25Ø
284       235            IF(MAINIT+4)24Ø,98,98
285       24Ø            VABS(KG)=-1.
286       25Ø            KD=Ø
287                      KFIX=Ø
288                      KODORT=Ø
289                      KODPOT=Ø
290                      K1=NINT(N1+1Ø)
291                      II=N1+1Ø+NTIP+NINT(N1+13)
292                      N3=Ø
293                      I=1
294                      H=-1.
295                      IF(K1.GT.Ø)H=VPOT(1,KG)
```

```
296                  KODTOT=-1
297                  IF(VVOL(KG).GE.HØ .OR. H.GE.HØ)WRITE(IPR,78)KG
298                  IF(VVOL(KG))260,270,280
299     260             IF(H)385,300,98
300     270             KODTOT=Ø
301                     WRITE(IPR,87)VVOL(KG)
302                     IF(H)300,98,290
303     280             KODTOT=1
304                     WRITE(IPR,87)VVOL(KG)
305                     KD=KD+1
306                     JVOLC=JVOLC+NPOINT
307                     IF(H)98,290,98
308     290             IF(VORT(KG))300,300,295
309     295             KODORT=1
310                     WRITE(IPR,88)VORT(KG)
311                     JORT=JORT+NPOINT
312     300             IF(VABS(KG))320,320,310
313     310             WRITE(IPR,79)VABS(KG)
314                     KD=KD+NWAVE
315                     JCOLC=JCOLC+NSPCOL*NPOINT
316                     JCOLD=JCOLD+NSPCOL*NPOINT*KODREF
317     320             IF(I-K1)325,325,350
318     325             II=II+1
319                     WRITE(IPR,80)VPOT(I,KG),NINT(II)
320                     WRITE(IPR,81)POTADD(I,KG),POTMLT(I,KG)
321                     IF(VPOT(I,KG))340,330,335
322     330                IF(KODTOT.EQ.Ø .OR. KFIX.LT.I-1)GO TO 98
323                        KFIX=I
324                        GO TO 340
325     335             KODPOT=I
326                     KD=KD+1
327     340             I=I+1
328     350             IF(N3.GE.KSPEC)GO TO 370
329                     N2=N3+1
330                     N3=N3+10
331                     IF(N3.GT.KSPEC)N3=KSPEC
332                     WRITE(IPR,82)(ISORT(K,KG),K=N2,N3)
333                     J=1
334                     DO 360 K=N2,N3
335                        IBUF(J)=JBLANK
336                        IF(K.GT.KCOMP .AND. K.LE.KREFIN)IBUF(J)=JSTAR
337                        J=J+1
338     360             CONTINUE
339                     J=N3-N2+1
340                     WRITE(IPR,83)(IBUF(K),K=1,J)
341     370          IF(N3.GE.KSPEC .AND. I.GT.K1)GO TO 375
342                  WRITE(IPR,70)
343                  GO TO 320
344     375          IF(KD)385,385,380
345     380          IF(KSPEC-KCOMP)98,384,384
346     384             NDATA=NDATA+KD*NPOINT
347                     KEYCLC(1,KG)=NPOINT
348                     KEYCLC(2,KG)=KODTOT
349                     KEYCLC(3,KG)=KODORT
350                     KEYCLC(4,KG)=KODPOT
351                     KEYCLC(5,KG)=KFIX
352                     J=KCOMP*(KSPEC+6)+2*KSPEC+KODREF*(KODPOT+1)
353                     IF(J.GT.JHELP)JHELP=J
354     385          N1=N1+NINT(N1+1)
355     390       CONTINUE
356               I=Ø
357               DO 395 KG=1,NG
358     395       I= I+KEYCLC(1,KG)
359               IF(NSPCOL.GT.Ø .AND. JCOLC.EQ.Ø .OR. I.LE.Ø)GO TO 98
360               H=NDATA
361               FREED=NDATA-(NSPREF+NRFCOL*NWAVE)
362               IF(FREED.LE.HØ)FREED=H
363               REDUCE=DSQRT(H/FREED)
364   C           REDUCE= SQRT(H/FREED)
365               KEYCOL=JCOLC
366               J1=J0+NSPREF*NSPREF
367               J2=J1+NSPREF
368               J3=J2+NSPREF
369               J4=J3+NSPREF
370               J5=J4+NRFCOL*NRFCOL
371               J6=J5+NRFCOL
```

```
372                J7=J6+JVOLC
373                J8=J7+JORT
374                JH0=J8+JCOLC
375                JH1=JH0+NSPREF
376                JH2=JH1+NRFCOL
377                JH7=JH2+NSPREF*NRFCOL
378                JCD0=JM-JCOLD
379                JM1=JCD0-NWAVE*NRFCOL
380                JM2=JM1-NWAVE*NRFCOL
381                K=NSPREF*NSPREF
382                IF(KEYCOR.NE.0)K=0
383                JC=JCD0-K
384                K=JCD0-JHELP
385                J=JC-JH1
386                IF(K.LT.J)J=K
387                K=JM2-(JH7+20)
388                IF(K.LT.J)J=K
389                J=MXR-J
390                I=I0+65+(NCOMP+1)* MXSPEC-NCOMP*NCOMP+10*((NSPCOL+9)/10)
391                IF(NSPEC.GT.66)I=I-66+NSPEC
392                WRITE(IPR,91)I,J
393                IF(J.GT.MXR)GO TO 95
394      500       CALL PSGAUS(NISORT,MODREF,ISTCOF,NINT(I0),NCOMP,KEYCOR,
395        &       CREAL(J0),CREAL(JC),CREAL(J1),CREAL(J2),CREAL(JH0),
396        &       ALGB,CREAL(J3),SPNAME,
397        &       NWAVE,NSPCOL,NSPREF,NSPEC,MXSPEC,NRFCOL,NGCOL)
398                IF(MAINIT)600,800,800
399      600       IF(KEYCOL)700,700,610
400      610       DO 620 K=1,NWAVE
401                DO 620 KG=1,NGCOL
402      620       VARABS(K,KG)=H0
403                CALL PSABS(NICOL,KW1,IREFIN,ISELEC,KWW1,JCD0,IFAULT,
404        &       CREAL(J0),CREAL(J2),CREAL(JH0),CREAL(JH2),CREAL(J4),
405        &       CREAL(J5),CREAL(JH1),ABCOEF,CREAL(J8),VABS,VARABS,
406        &       CREAL(JM1),CREAL(JM2),SPNAME,CREAL(JH7),NG,NCOMP,MXPOT,
407        &       NWAVE,NSPCOL,NSPREF,NSPEC,MXSPEC,NRFCOL,NGCOL)
408      700       K=NSPREF+NWAVE*NRFCOL
409                WRITE(IPR,84)NDATA,K,FREED,FIT
410                WRITE(IPR,76)
411                WRITE(IPR,86)
412                N1=0
413                DO 730 KG=1,NG
414                IF(KEYCLC(1,KG))730,730,705
415      705          WRITE(IPR,78)KG
416                   VARVOL(KG)=VARVOL(KG)*REDUCE
417                   IF(VVOL(KG).GT.H0)WRITE(IPR,87)VARVOL(KG)
418                   VARORT(KG)=VARORT(KG)*REDUCE
419                   IF(VORT(KG).GT.H0)WRITE(IPR,88)VARORT(KG)
420                   KODPOT=NINT(N1+10)
421                   IF(KODPOT)730,730,710
422      710          KK=N1+10+NTIP+NINT(N1+13)
423                   II=0
424                   DO 720 I=1,KODPOT
425                      KK=KK+1
426                      IF(VPOT(I,KG))720,720,715
427      715             IF(II.GT.0)WRITE(IPR,70)
428                      II=1
429                      VARPOT(I,KG)=VARPOT(I,KG)*REDUCE
430                      WRITE(IPR,80)VARPOT(I,KG),NINT(KK)
431      720          CONTINUE
432      730       N1=N1+NINT(N1+1)
433                IF(KEYCOL)790,790,740
434      740       DO 750 K=1,NWAVE
435                DO 750 KG=1,NGCOL
436      750       VARABS(K,KG)=VARABS(K,KG)*REDUCE
437                N2=0
438      760       N1=N2+1
439                N2=N2+10
440                IF(N2.GT.NGCOL)N2=NGCOL
441                I=KWW1
442                WRITE(IPR,89)(JBLANK,KG,KG=N1,N2)
443                DO 770 K=1,NWAVE
444                WRITE(IPR,90)NINT(I),(VARABS(K,KG),KG=N1,N2)
445      770       I=I+1
446                IF(N2.LT.NGCOL)GO TO 760
447      790       IF(MAINIT.EQ.-3 .AND. NSPREF.GT.0)IFAULT=10
```

```
448            IF(KLIST.LE.0 .OR. KLISTT*(KLISTT+MAINIT).LT.0)GO TO 1000
449    800     JVOLC=J6
450            JORT=J7
451            JCOLC=J8
452            JCOLD=JCD0
453            N1=0
454            DO 900 KG=1,NG
455                NPOINT=KEYCLC(1,KG)
456                IF(NPOINT)890,890,810
457    810         KODTOT=KEYCLC(2,KG)
458                KODORT=KEYCLC(3,KG)
459                KODPOT=KEYCLC(4,KG)
460                KFIX=KEYCLC(5,KG)
461                K1=KFIX+1
462                KCOMP=NINT(N1+12)
463                KCOMP1=KCOMP+1
464                KD=KCOMP-KFIX
465                KREFIN=IREFIN(KG)
466                KSPEC=ISPEC(KG)
467                KODREF=KREFIN-KCOMP
468                NCURVE=NINT(N1+11)
469                I13=NINT(N1+13)
470                I15=NINT(N1+15)
471                IP1=N1+16+I15
472                KODVOL=-1
473                IF(NINT(N1+7).GT.0)KODVOL=KEYVOL
474                IF(KODVOL.EQ.1 .AND. MAINIT.EQ.0)KODVOL=0
475                KODCOL=0
476                IF(VABS(KG).GT.H0)KODCOL=NSPCOL
477                JTOTB=NINT(N1+4)
478                JTOTV=NINT(N1+5)
479                JXVPA=NINT(N1+6)
480                JH3=JH1+KCOMP*KSPEC
481                JH4=JH3+KSPEC
482                N2=JH3+KCOMP
483                N3=JH1+KCOMP*KCOMP
484                IF(KSPEC.LE.KCOMP)GO TO 827
485                DO 825 I=KCOMP1,KSPEC
486                    II=ISORT(I,KG)
487                    DO 820 KC=1,KCOMP
488                        K=ISORT(KC,KG)
489                        CREAL(N3)=ISTCOF(K,II)
490    820             N3=N3+1
491                    CREAL(N2)=ALN10*ALGB(II)
492    825         N2=N2+1
493    827         IF(MAINIT)830,850,850
494    830         IF(IABS(KWRITE)*KODORT-2)840,832,832
495    832         IF(KODTOT)840,840,834
496    834         KODORT=0
497                VPOT(1,KG)=VVOL(KG)/VORT(KG)
498                KODPOT=1
499    840         J=N1+15+KW1(KG)
500                II=JH4
501                DO 845 KC=1,KSPEC
502                    I=ISORT(KC,KG)
503                    CREAL(II)=SPNAME(1,I)
504                    CREAL(II+1)=SPNAME(2,I)
505    845         II=II+2
506                CALL PSLIST(NINT(IP1),ISORT(1,KG),NINT(J),KW1(KG),NINT(I0),
507        &      KG,KLIST,KWRITE,CREAL(JH3),CREAL(JH1),CREAL(JTOTV),
508        &      CREAL(JXVPA),CREAL(JVOLC),CREAL(JORT),VVOL(KG),
509        &      VORT(KG),VPOT(1,KG),VABS(KG),POTMLT(1,KG),POTADD(1,KG),
510        &      VARVOL(KG),VARPOT(1,KG),ABCOEF,CREAL(JCOLC),CREAL(JH4),
511        &      NWAVE,NSPCOL,NSPREF,NSPEC,MXSPEC,NRFCOL,NGCOL,
512        &      I13,I15,NCURVE,NPOINT,KODCOL,KODPOT,KCOMP,KSPEC,KODREF)
513                GO TO 870
514    850     JH5=JH4+KODREF
515            JH6=JH5+KODREF*KODPOT
516            LB1=JH6+KCOMP
517            LB2=LB1+KSPEC
518            LB3=LB2+KCOMP
519            LB4=LB3+KCOMP
520            LB=KCOMP+KCOMP1
521            KEYCN1=-NINT(N1+3)
522            JHELP=-KEYCN1
523            IF(KEYCN1.EQ.0)KEYCN1=KEYCNT
```

```
524                  CALL PSCALC(NINT(IP1),ISELEC(1,KG),ICOL(1,KG),MXPOT,II,
525        &         KEYCN1,JCOLD,CREAL(JH3),CREAL(JH1),CREAL(JTOTV),
526        &         CREAL(JXVPA),CREAL(JVOLC),CREAL(JORT),VVOL(KG),
527        &         VORT(KG),VPOT(1,KG),POTMLT(1,KG),CREAL(JH6),
528        &         POTADD(1,KG),VARVOL(KG),VARORT(1,KG),
529        &         CREAL(JTOTB),CREAL(J0),CREAL(J1),CREAL(J2),CREAL(JH0),
530        &         CREAL(JH4),CREAL(JH5),CREAL(JCOLC),
531        &         NWAVE,NSPCOL,NSPREF,NSPEC,MXSPEC,NRFCOL,NGCOL,
532        &         I13,I15,NCURVE,NPOINT,KODCOL,KODPOT,KCOMP,KSPEC,KODREF)
533                  IF(II)870,870,855
534      855             I=JXVPA+(II-1)*I15
535                      J=I+KCOMP-1
536                      WRITE(IPR,92)KG,II,(CREAL(K),K=I,J)
537                      IFAULT=10
538                      GO TO 96
539      870         IF(KODTOT.GT.0)JVOLC=JVOLC+NPOINT
540                  IF(KODORT.GT.0)JORT=JORT +NPOINT
541                  JCOLC=JCOLC+NPOINT*KODCOL
542                  IF(JHELP)880,880,890
543      880         NINT(N1+3)=1
544                  NC=NC+1
545      890      N1=N1+NINT(N1+1)
546      900      CONTINUE
547               IF(MAINIT)1000,910,910
548      910      IF(KEYCOL)500,500,920
549      920      CALL PSABS(NICOL,KW1,IREFIN,ISELEC,KWW1,JCD0,IFAULT,
550        &         CREAL(J0),CREAL(J2),CREAL(JH0),CREAL(JH2),CREAL(J4),
551        &         CREAL(J5),CREAL(JH1),ABCOEF,CREAL(J8),VABS,VARABS,
552        &         CREAL(JM1),CREAL(JM2),SPNAME,CREAL(JH7),NG,NCOMP,MXPOT,
553        &         NWAVE,NSPCOL,NSPREF,NSPEC,MXSPEC,NRFCOL,NGCOL)
554               IF(IFAULT)500,500,96
555      1000     IF(IFAULT)1010,1010,96
556      1010     IF(MAINIT+4)1040,1020,1200
557      1020     IF(NRFCL1-NRFCOL)1040,1040,1030
558      1030     II=NRFCOL+1
559               DO 1039 I=II,NRFCL1
560                  IF(NICOL(I)-MODELN)1039,1035,1039
561      1035        NICOL(I)=NICOL(II)
562                  NICOL(II)=MODELN
563                  NRFCOL=II
564                  GO TO 1040
565      1039     CONTINUE
566      1040     IF(NRFCOL)1100,1100,1050
567      1050     II=NRFCOL
568               DO 1090 I=1,II
569      1055        IF(I.GT.NRFCOL)GO TO 1090
570                  J=NICOL(I)
571                  IF(J-NCOMP)1090,1090,1060
572      1060        IF(MODREF(J))1070,1090,1090
573      1070        NICOL(I)=NICOL(NRFCOL)
574                  NICOL(NRFCOL)=J
575                  DO 1080 K=1,NWAVE
576                     ABCOEF(K,I)=ABCOEF(K,NRFCOL)
577                     ABCOEF(K,NRFCOL)=0.
578      1080        CONTINUE
579                  NRFCOL=NRFCOL-1
580                  GO TO 1055
581      1090     CONTINUE
582      1100     II=NCOMP
583               KCOMP1=NCOMP+1
584               DO 1120 I=KCOMP1,MXSPEC
585                  IF(MODREF(I))1120,1110,1110
586      1110        II=II+1
587      1120     CONTINUE
588               IBUF(5)=II
589               IBUF(7)=NRFCOL
590      1200     CALL PSEXIT(2,0)
591               IF(MAINIT.EQ.-2 .AND. KEYFLT.GT.1)IFAULT=1
592               RETURN
593      98       IFAULT=IFAULT+1
594      95       IFAULT=IFAULT+4
595      99       IFAULT=IFAULT+1
596      97       IFAULT=IFAULT+3
597      96       CALL PSEXIT(IFAULT,KOD)
598               IF(IFAULT.GT.4 .AND. KEYFLT.EQ.0)IFAULT=0
599               MAINIT=-2
```

```
600          GO TO 1200
601          END




  1    C -------------------------------------------------------------------------
  2    C --
  3    C --    S U B R O U T I N E    P S D T A P
  4    C --
  5    C --              INPUT/OUTPUT FROM OR TO MAGNETIC TAPE
  6    C --
  7    C -------------------------------------------------------------------------
  8          SUBROUTINE  PSDTAP(NG,NI,NR,NC,NTIP,NINT,TITLE,CREAL,IBUF1,KOD,
  9         &NNI,NNR)
 10          COMMON/IPERO/IO(10)
 11          DOUBLE PRECISION
 12    C***  REAL
 13         &TITLE(20),CREAL(NR)
 14          DIMENSION NINT(NI)
 15             KTAPE=IBUF1
 16             IF(KOD-1)10,20,30
 17      10     REWIND  KTAPE
 18             WRITE(KTAPE)NG,NI,NR,NC,NTIP
 19             WRITE(KTAPE)(TITLE(K),K=1,20),CREAL,NINT
 20             ENDFILE KTAPE
 21          RETURN
 22      20     REWIND  KTAPE
 23             READ(KTAPE)NG,NNI,NNR,NC,NTIP
 24          RETURN
 25      30     READ(KTAPE)(TITLE(K),K=1,20),CREAL,NINT
 26          RETURN
 27          END




  1    C -------------------------------------------------------------------------
  2    C --
  3    C --  S U B R O U T I N E    P S G A U S
  4    C --
  5    C --         ORGANIZATION AND EXECUTION OF GAUSS-NEWTON PROCEDURE
  6    C --
  7    C -------------------------------------------------------------------------
  8          SUBROUTINE  PSGAUS(NISORT,MODREF,ISTCOF,JCH20,NCOMP,KEYCOR,
  9         &A,C,ADIAG,B,BH,ALGB,DB,SPNAME,
 10         &NWAVE,NSPCOL,NSPREF,NSPEC,MXSPEC,NRFCOL,NGCOL)
 11          DOUBLE PRECISION
 12    C***  REAL
 13         &A(NSPREF,NSPREF),ADIAG(NSPREF),B(NSPREF),BH(NSPREF),
 14         &C(NSPREF,NSPREF),ALGB(MXSPEC),DB(NSPREF),H,SH,OLDFIT,H0,
 15         &H0INP,DFIT,DALGB1,FREED,FIT,ALN10,ALGE,REP,SPNAME(2,MXSPEC)
 16          COMMON/REPAR/H0,H0INP,DFIT,DALGB1,FREED,FIT,ALN10,ALGE,REP(6)
 17          DIMENSION NISORT(NSPREF),MODREF(MXSPEC),ISTCOF(NCOMP,MXSPEC),
 18         &JCH20(20),JCH8(8)
 19          COMMON/INTPAR/KRAND,MAINIT,MAXIT,MODEL,MODELN,MITFIT,MXHALV,
 20         &MINIT,MMM,ITNSUM,ITHSUM,MXITNR,MXITH
 21          COMMON/IPERO/IO(10)
 22          DATA JBLANK/1H /,JEND/1H]/,JSTAR/1H*/
 23       10 FORMAT(1H1)
 24       20 FORMAT(1X)
 25       30 FORMAT(  /2X,2(2H* ),14HP S E Q U A D ,3(2H* ),14HP S E Q U A D ,
 26         &4(2H* ),27HM A I N   I T E R A T I O N ,4(2H* ),14HP S E Q U A D ,
 27         &3(2H* ),14HP S E Q U A D ,2(2H* ))
 28       52 FORMAT(1X,45HNO. LG (BETA) MAX.STEP  FORMULA  COMPOSITION /)
 29       53 FORMAT(5X,3HNEW/23H NO. LG(BETA) G-N.STEP /)
 30       55 FORMAT(1X,I2,F10.4,1X,1PG9.2)
 31       56 FORMAT(1H+,24X,2A4,2H( ,8(I2,A1))
 32       57 FORMAT(/ 1X,9HITERATION,I3,3X,18HFITTING PARAMETER=,1PG12.5,
 33         &30X,16H(NEWTON-RAPHSON= ,I5,7H, HELP= ,I5,1H))
 34       58 FORMAT(1H+,58X,10F6.2)
 35       59 FORMAT(1H+,54X,10(5X,A1))
 36       60 FORMAT(1H+,51X,19H***** MINIMUM *****)
 37       61 FORMAT(/39X,12H*** SPECIES(,I2,2H),2A4,21H TO BE ELIMINATED ***)
 38       62 FORMAT(/39X,12H*** SPECIES(,I2,2H),2A4,21H TO BE  ADMITTED  ***)
```

```
39        63 FORMAT(1H+,46X,27H*** HALF STEP BACKWARDS ***)
40        64 FORMAT(15X,16HSTANDARD          ,27X,A1,24HCORRELATION COEFFICIENTS)
41        65 FORMAT(46H NO. LG(BETA)DEVIATION  FORMULA  COMPOSITION ,12X,A1,
42        &53HLOWER: TOTAL     *DIAGONAL: MULTIPLE    UPPER: PARTIAL)
43        66 FORMAT(57X,10I6)
44        76 FORMAT(1H+, 24X,2A4,1X,29I3,(33X,29I3))
45        77 FORMAT(/31H NO.   CORRELATION COEFFICIENTS /5X,19I6)
46        78 FORMAT(1X,I2,4X,19F6.2)
47        79 FORMAT(1H+,2X,19(5X,A1))
48              IPR=IO(3)
49              IF(MAINIT)100,300,300
50   100        MAINIT=0
51              IF(NCOMP-8)110,110,130
52   110        DO 120 K=1,NCOMP
53   120        JCH8(K)=JBLANK
54              JCH8(NCOMP)=JEND
55   130        ITFIT=0
56              WRITE(IPR,52)
57              IF(NSPREF)200,200,140
58   140        DO 190 KK=1,NSPREF
59                 I=NISORT(KK)
60                 H=MODREF(I)
61                 H=DALGB1*H
62                 WRITE(IPR,55)I,ALGB(I),H
63                 IF(NCOMP-8)180,180,170
64   170            WRITE(IPR,76)SPNAME(1,I),SPNAME(2,I),
65        &            (ISTCOF(K,I),K=1,NCOMP)
66                 GO TO 190
67   180            WRITE(IPR,56)SPNAME(1,I),SPNAME(2,I),
68        &            (ISTCOF(K,I),JCH8(K),K=1,NCOMP)
69   190        CONTINUE
70   200        KK=NCOMP+1
71              DO 250 I=KK,MXSPEC
72                 IF(MODREF(I))250,210,250
73   210            WRITE(IPR,55)I,ALGB(I)
74                 IF(NCOMP-8)230,230,220
75   220               WRITE(IPR,76)SPNAME(1,I),SPNAME(2,I),
76        &               (ISTCOF(K,I),K=1,NCOMP)
77                 GO TO 250
78   230               WRITE(IPR,56)SPNAME(1,I),SPNAME(2,I),
79        &               (ISTCOF(K,I),JCH8(K),K=1,NCOMP)
80   250        CONTINUE
81              GO TO 600
82   300        FIT=DSQRT(FIT/ FREED)
83 C 300        FIT= SQRT(FIT/ FREED)
84              WRITE(IPR,57)MAINIT,FIT,ITNSUM,ITHSUM
85              MAINIT=MAINIT+1
86              IF(NSPREF)305,305,320
87   305          IF(MAINIT-MINIT)800,800,310
88   310          MAINIT=-3
89              WRITE(IPR,30)
90              RETURN
91   320        IF(MAINIT.EQ.1)GO TO 520
92              H=FIT*DFIT
93              IF(FIT.GT.OLDFIT)H=OLDFIT*DFIT
94              SH=FIT-OLDFIT
95              IF(DABS(SH)-H)500,500,330
96 C            IF( ABS(SH)-H)500,500,330
97   330        ITFIT=0
98              IF(SH)520,520,400
99   400          IF(ITHALV.GE.MXHALV .OR. MAINIT.GT.MAXIT)GO TO 520
100                WRITE(IPR,63)
101                WRITE(IPR,20)
102                WRITE(IPR,53)
103                DO 410 K=1,NSPREF
104                   I=NISORT(K)
105                   DB(K)=DB(K)*0.5
106                   ALGB(I)=ALGB(I)-DB(K)
107                   WRITE(IPR,55)I,ALGB(I)
108   410          CONTINUE
109                ITHALV=ITHALV+1
110                GO TO 600
111   500        IF(ITHALV.EQ.0)ITFIT=ITFIT+1
112              IF(ITFIT-MITFIT)520,510,510
113   510        WRITE(IPR,60)
114              IF(MAINIT.GT.MINIT)MAINIT=-1
```

```
115    520     DO 530 K=1,NSPREF
116               H=ADIAG(K)+A(K,K)
117               A(K,K)=H
118               ADIAG(K)=H
119    530     CONTINUE
120            CALL FACTOR(A,NSPREF,NSPREF,KK)
121            IF(KK)540,540,310
122    540     CALL SUBST(A,NSPREF,NSPREF,BH,B)
123            ITHALV=0
124            IF(MAINIT.LT.0)GO TO 710
125            IF(MAINIT.GT.MAXIT)GO TO 700
126            IF(MAINIT.EQ.1)WRITE(IPR,10)
127            WRITE(IPR,30)
128            WRITE(IPR,53)
129            DO 550 K=1,NSPREF
130              I=NISORT(K)
131              SH=MODREF(I)
132              SH=DALGB1*SH
133              H=ALGE*BH(K)
134              IF(DABS(H).GT.SH)H=(H/DABS(H))*SH
135    C***      IF(ABS(H).GT.SH)H=(H/ABS(H))*SH
136              ALGB(I)=ALGB(I)+H
137              DB(K)=H
138              H=ALGE*BH(K)
139              WRITE(IPR,55)I,ALGB(I),H
140    550     CONTINUE
141            OLDFIT=FIT
142    600     FIT=0.
143            ITNSUM=0
144            ITHSUM=0
145            IF(NSPREF.LE.0)RETURN
146            DO 620 I=1,NSPREF
147              ADIAG(I)=0.
148              B(I)=0.
149              BH(I)=0.
150              DO 610 K=1,NSPREF
151    610        A(I,K)=0.
152    620     CONTINUE
153            RETURN
154    700     MAINIT=-2
155    710     IF(MODEL.LE.0)GO TO 800
156    715     II=0
157            KK=0
158            DO 720 K=1,NSPREF
159              I=NISORT(K)
160              N1=MODREF(I)
161              SH=N1
162              IF(ALGE*BH(K).GT.-SH *DALGB1 .OR. N1.LT.KK)GO TO 720
163                KK=MODREF(I)
164                II=I
165    720     CONTINUE
166            IF(KK-1)750,750,725
167    725       WRITE(IPR,30)
168              WRITE(IPR,61)II,SPNAME(1,II),SPNAME(2,II)
169              MODREF(II)=-1
170              IF(MODELN-II)740,730,740
171    730         MODELN=0
172                GO TO 715
173    740       MODEL=MODEL-1
174              MODELN=0
175              MAINIT=-5
176              GO TO 800
177    750     MODELN=0
178            II=0
179            KK=-100
180            N1=NCOMP+1
181            DO 760 I=N1,MXSPEC
182              K=MODREF(I)
183              IF(K+1)753,760,760
184    753       IF(K-KK)760,760,755
185    755         KK=K
186                II=I
187    760     CONTINUE
188            IF(II)800,800,770
189    770       WRITE(IPR,30)
190              WRITE(IPR,62)II,SPNAME(1,II),SPNAME(2,II)
```

```
191                       MODREF(II)=-KK
192                       MODELN=II
193                       MODEL=MODEL-1
194                       MAINIT=-4
195          800         WRITE(IPR,10)
196                       WRITE(IPR,30)
197                       WRITE(IPR,20)
198                       N2=NSPREF
199                       IF(N2.GT.10)N2=10
200                       IF(N2.LT.2 .OR. KEYCOR.NE.0 .OR. NCOMP.GT.8)N2=0
201                       IF(NSPREF-1)850,810,820
202          810          BH(1)=1./DSQRT(ADIAG(1))
203   C***810             BH(1)=1./SQRT(ADIAG(1))
204                       GO TO 850
205          820         IF(KEYCOR)850,830,850
206          830         DO 835 I=1,NSPREF
207                       DO 832 K=1,NSPREF
208          832          BH(K)=0.
209                       BH(I)=1.
210                       CALL SUBST(A,NSPREF,NSPREF,C(1,I),BH)
211          835         CONTINUE
212                       DO 839 I=1,NSPREF
213                       BH(I)=DSQRT(C(I,I))
214   C                   BH(I)= SQRT(C(I,I))
215                       C(I,I)=DSQRT(1.-1./(C(I,I)*ADIAG(I)))
216   C                   C(I,I)= SQRT(1.-1./(C(I,I)*ADIAG(I)))
217                       ADIAG(I)=DSQRT(ADIAG(I))
218   C                   ADIAG(I)= SQRT(ADIAG(I))
219                       IF(I.EQ.1)GO TO 839
220                       KK=I-1
221                       DO 838 K=1,KK
222                       C(I,K)=C(I,K)/(BH(K)*BH(I))
223          838          C(K,I)=-A(I,K)/(ADIAG(I)*ADIAG(K))
224          839         CONTINUE
225                       IF(N2)850,850,842
226          842         DO 845 K=2,N2
227          845         JCH20(K)=JBLANK
228                       JCH20(1)=JSTAR
229                       WRITE(IPR,64)JBLANK
230                       WRITE(IPR,65)JBLANK
231                       WRITE(IPR,66)(NISORT(I),I=1,N2)
232                       GO TO 860
233          850         WRITE(IPR,64)
234                       WRITE(IPR,65)
235                       WRITE(IPR,20)
236                       IF(NSPREF)870,870,860
237          860         DO 869 KK=1,NSPREF
238                       I=NISORT(KK)
239                       H=BH(KK)*FIT*ALGE
240                       WRITE(IPR,55)I,ALGB(I),H
241                       IF(NCOMP-8)864,864,862
242          862             WRITE(IPR,76)SPNAME(1,I),SPNAME(2,I),
243          &               (ISTCOF(K,I),K=1,NCOMP)
244                         GO TO 869
245          864         WRITE(IPR,56)SPNAME(1,I),SPNAME(2,I),
246          &           (ISTCOF(K,I),JCH8(K),K=1,NCOMP)
247                       IF(N2)869,869,866
248          866         WRITE(IPR,58)(C(KK,K),K=1,N2)
249                       IF(KK.GT.N2)GO TO 869
250                         WRITE(IPR,59)(JCH20(K),K=1,N2)
251                       JCH20(KK)=JBLANK
252                       JCH20(KK+1)=JSTAR
253          869         CONTINUE
254          870         KK=NCOMP+1
255                       DO 879 I=KK,MXSPEC
256                       IF(MODREF(I))879,872,879
257          872         WRITE(IPR,55)I,ALGB(I)
258                       IF(NCOMP-8)877,877,875
259          875             WRITE(IPR,76)SPNAME(1,I),SPNAME(2,I),
260          &               (ISTCOF(K,I),K=1,NCOMP)
261                         GO TO 879
262          877         WRITE(IPR,56)SPNAME(1,I),SPNAME(2,I),
263          &           (ISTCOF(K,I),JCH8(K),K=1,NCOMP)
264          879         CONTINUE
265          880         IF(KEYCOR.NE.0 .OR. NSPREF.LT.2 .OR. N2.GE.NSPREF)RETURN
266                       N1=N2+1
```

```
267                     N2=N2+19
268                     IF(N2.GT.NSPREF)N2=NSPREF
269                     WRITE(IPR,77)(NISORT(K),K=N1,N2)
270                     DO 885 K=1,19
271        885          JCH20(K)=JBLANK
272                     JCH20(1)=JSTAR
273                     I=1
274                     DO 889 KK=1,NSPREF
275                        WRITE(IPR,78)NISORT(KK),(C(KK,K),K=N1,N2)
276                        IF(KK.LT.N1 .OR. KK.GT.N2)GO TO 889
277                           WRITE(IPR,79)(JCH20(K),K=1,19)
278                           JCH20(I)=JBLANK
279                           JCH20(I+1)=JSTAR
280                           I=I+1
281        889          CONTINUE
282             GO TO 880
283             END
```

```
  1    C -------------------------------------------------------------------------
  2    C --
  3    C --    S U B R O U T I N E    P S A B S
  4    C --
  5    C --              EXECUTION OF PHOTOMETRIC CALCULATIONS
  6    C --
  7    C -------------------------------------------------------------------------
  8             SUBROUTINE PSABS(NICOL,KW1,IREFIN,ISELEC,KWW1,ICD0,IFAULT,
  9            &A,B,XH,AMIX,AA,AADIAG,BB,ABCOEF,COLC,VABS,VARABS,
 10            &VAR,COR,SPNAME,TITLE,NG,NCOMP,MXPOT,
 11            &NWAVE,NSPCOL,NSPREF,NSPEC,MXSPEC,NRFCOL,NGCOL)
 12             DIMENSION NICOL(NSPCOL),KW1(NGCOL),IREFIN(NG),ISELEC(NSPEC,NG)
 13             DOUBLE PRECISION
 14    C***    REAL
 15            &A(NSPREF,NSPREF),B(NSPREF),XH(NSPREF),AMIX(NRFCOL,NSPREF),
 16            &AA(NRFCOL,NRFCOL),AADIAG(NRFCOL),BB(NRFCOL),COLC(NSPCOL,1),
 17            &VARABS(NWAVE,NGCOL),VAR(NWAVE,NRFCOL),COR(NWAVE,NRFCOL),
 18            &ABCOEF(NWAVE,NSPCOL),VABS(NG),SPNAME(2,MXSPEC),TITLE(2,10),
 19            &TITLD,CREAL,REP,FIT,REP2,H,SH,W,REZ
 20             COMMON TITLD(20),CREAL(5000 )
 21             COMMON/MEAS/MXI,MXR,NTIP,MI,NR,NC,MCOMP,
 22            &NINT(1000)
 23             COMMON/REPAR/REP(5),FIT,REP2(8)
 24             COMMON/INTPAR/KRAND,MAINIT,MAXIT,MODEL,MODELN,MITFIT,MXHALV,
 25            &MINIT,MMM,ITNSUM,ITHSUM,MXITNR,MXITH
 26             COMMON/IPERO/IO(10)
 27         10 FORMAT(1X)
 28         30 FORMAT(/7X,40HMOLAR ABSORPTIVITIES OF COLOURED SPECIES )
 29         31 FORMAT(//2X,4HNO. ,2A4,8(5X,2A4))
 30         32 FORMAT(1X,4HABS.,A1,5HSPEC(,I2,1H),8(4X,A1,5HSPEC(,I2,1H)))
 31         33 FORMAT(1X,I3,9(1X,1PG12.5))
 32         41 FORMAT(//2X,3HNO.,4(4X,2A4,3X,14HSTANDARD MULT. ))
 33         42 FORMAT(1X,4HABS.,4(A1,3X,5HSPEC(,I2,1H),2X,15HDEVIATION CORR. ))
 34         43 FORMAT(1X,I3,1X,4(2X,1PG12.5,   G9.2,1H*,0PF5.2))
 35             DATA JBL/1H /
 36             IPR=IO(3)
 37             N4=NRFCOL
 38             IF(MAINIT)300,80,90
 39         80 IF(NSPCOL-NRFCOL)100,100,85
 40         85 N4=NSPCOL
 41             IF(NRFCOL.LE.0)GO TO 800
 42         90 IF(NRFCOL)300,300,100
 43        100 DO 110 I=1,NRFCOL
 44             DO 110 J=1,I
 45        110 AA(I,J)=0.
 46             NI=0
 47             N2=0
 48             DO 160 KG=1,NGCOL
 49             IF(VABS(KG))160,160,120
 50        120 W=1./(VABS(KG)*VABS(KG))
 51             N1=N2+1
 52             N2=N2+NINT(NI+9)
 53             DO 140 I=1,NRFCOL
 54             DO 140 J=1,I
```

```
55                         SH=Ø.
56                         DO 130 K=N1,N2
57      130                SH=SH +COLC(I,K)*COLC(J,K)
58      140              AA(I,J)=AA(I,J)+W*SH
59      160            NI=NI+NINT(NI+1)
60                     DO 170 K=1,NRFCOL
61      170            AADIAG(K)=AA(K,K)
62                     CALL FACTOR(AA,NRFCOL,NRFCOL,I)
63                     IF(I)200,200,190
64      190            IFAULT=10
65                     RETURN
66      200          DO 290 KW=1,NWAVE
67                     DO 210 I=1,NRFCOL
68      210            BB(I)=Ø.
69                     NI=Ø
70                     N2=Ø
71                     DO 260 KG=1,NGCOL
72                       IF(VABS(KG))260,260,220
73      220              W=1./(VABS(KG)*VABS(KG))
74                       N1=N2+1
75                       N2=N2+NINT(NI+9)
76                       L=NINT(NI+6)+KW1(KG)+KW-2
77                       I15=NINT(NI+15)
78                       DO 250 K=N1,N2
79                         REZ=CREAL(L)
80                         DO 230 I=1,NSPCOL
81      230                REZ=REZ-ABCOEF(KW,I)*COLC(I,K)
82                         DO 240 I=1,NRFCOL
83      240                BB(I)=BB(I)+W*COLC(I,K)*REZ
84      250              L=L+I15
85      260            NI=NI+NINT(NI+1)
86                     CALL SUBST(AA,NRFCOL,NRFCOL,BB,BB)
87                     DO 270 I=1,NRFCOL
88      270            ABCOEF(KW,I)=ABCOEF(KW,I)+BB(I)
89      290          CONTINUE
90      800        WRITE(IPR,30)
91                 DO 820 N1=1,N4,9
92                   N2=N1+8
93                   IF(N2.GT.N4)N2=N4
94                   I=Ø
95                   DO 815 K=N1,N2
96                     I=I+1
97                     J=NICOL(K)
98                     TITLE(1,I)=SPNAME(1,J)
99                     TITLE(2,I)=SPNAME(2,J)
100     815          CONTINUE
101                  WRITE(IPR,31)(TITLE(1,K),TITLE(2,K),K=1,I)
102                  WRITE(IPR,32)(JBL,NICOL(K),K=N1,N2)
103                  I=KWW1
104                  DO 810 KW=1,NWAVE
105                    WRITE(IPR,33)NINT(I),(ABCOEF(KW,K),K=N1,N2)
106     810          I=I+1
107     820        CONTINUE
108     300      DO 500 KW=1,NWAVE
109                 IF(NRFCOL)315,315,305
110     305          DO 310 I=1,NRFCOL
111                    DO 310 J=1,NSPREF
112     310            AMIX(I,J)=Ø.
113     315          NI=Ø
114                  N2=Ø
115                  ICD=ICDØ
116                  DO 390 KG=1,NGCOL
117                    IF(VABS(KG))390,390,320
118     320            W=1./(VABS(KG)*VABS(KG))
119                    N1=N2+1
120                    N2=N2+NINT(NI+9)
121                    L=NINT(NI+6)+KW1(KG)+KW-2
122                    I15=NINT(NI+15)
123                    KODREF=IREFIN(KG)-NINT(NI+12)
124                    SH=Ø.
125                    DO 380 K=N1,N2
126                      REZ=CREAL(L)
127                      DO 330 I=1,NSPCOL
128     330              REZ=REZ-ABCOEF(KW,I)*COLC(I,K)
129                      SH=SH+REZ*REZ
130                      IF(KODREF)380,380,335
```

```
131   335           DO 370 J=1,KODREF
132                 JJ=ISELEC(J,KG)
133                 H=0.
134                   DO 340 I=1,NSPCOL
135                   H=H+ABCOEF(KW,I)*CREAL(ICD)
136   340           ICD=ICD+1
137               IF(NRFCOL)355,355,345
138   345           DO 350 I=1,NRFCOL
139   350           AMIX(I,JJ)=AMIX(I,JJ)+W*H*COLC(I,K)
140   355         IF(MAINIT.LT.0)GO TO 370
141               B(JJ)=B(JJ)+ W*REZ*H
142               XH(J)=H
143                 DO 360 I=1,J
144                 II=ISELEC(I,KG)
145   360           A(JJ,II)=A(JJ,II)+W*H*XH(I)
146   370         CONTINUE
147   380       L=L+I15
148             IF(MAINIT)382,385,385
149   382       H=N2+1-N1
150             VARABS(KW,KG)=DSQRT(SH/H)
151   C         VARABS(KW,KG)= SQRT(SH/H)
152             GO TO 390
153   385       FIT=FIT+W*SH
154   390     NI=NI+NINT(NI+1)
155           IF(NRFCOL.LE.0)GO TO 500
156           IF(MAINIT)400,391,391
157   391     IF(NSPREF.LE.0)GO TO 500
158           DO 395 J=1,NSPREF
159             CALL SUBST(AA,NRFCOL,NRFCOL,BB,AMIX(1,J))
160             DO 395 JJ=1,J
161             SH=0.
162               DO 392 I=1,NRFCOL
163   392           SH=SH +AMIX(I,JJ)*BB(I)
164             A(J,JJ)=A(J,JJ)-SH
165   395     CONTINUE
166           GO TO 500
167   400     IF(NSPREF)430,430,410
168   410     IF(MAINIT.EQ.-3)GO TO 500
169           DO 420 J=1,NSPREF
170   420     CALL SUBST(AA,NRFCOL,NRFCOL,AMIX(1,J),AMIX(1,J))
171   430     DO 490 I=1,NRFCOL
172             DO 440 K=1,NRFCOL
173   440         BB(K)=0.
174             BB(I)=1.
175             CALL SUBST(AA,NRFCOL,NRFCOL,BB,BB)
176             SH=BB(I)
177             IF(NSPREF)480,480,450
178   450         DO 460 J=1,NSPREF
179   460         B(J)=AMIX(I,J)
180             CALL SUBST(A,NSPREF,NSPREF,B,B)
181             DO 470 J=1,NSPREF
182   470         SH=SH+AMIX(I,J)*B(J)
183   480       VAR(KW,I)=DSQRT(SH)*FIT
184   C 480     VAR(KW,I)= SQRT(SH)*FIT
185             COR(KW,I)=DSQRT(1.-1./(SH*AADIAG(I)))
186   C         COR(KW,I)= SQRT(1.-1./(SH*AADIAG(I)))
187   490     CONTINUE
188   500   CONTINUE
189         IF(MAINIT.GE.0 .OR. NRFCOL.LE.0)GO TO 700
190         IF(MAINIT.EQ.-3 .AND. NSPREF.GT.0)GO TO 700
191         WRITE(IPR,30)
192         DO 650 N1=1,NRFCOL,4
193         N2=N1+3
194         IF(N2.GT.NRFCOL)N2=NRFCOL
195         I=0
196         DO 610 K=N1,N2
197           I=I+1
198           J=NICOL(K)
199           TITLE(1,I)=SPNAME(1,J)
200           TITLE(2,I)=SPNAME(2,J)
201   610   CONTINUE
202         WRITE(IPR,41)(TITLE(1,K),TITLE(2,K),K=1,I)
203         WRITE(IPR,42)(JBL,NICOL(K),K=N1,N2)
204         I=KWW1
205         DO 620 KW=1,NWAVE
206         WRITE(IPR,43)NINT(I),(ABCOEF(KW,J),VAR(KW,J),COR(KW,J),
```

```
207            &                    J=N1,N2)
208      620            I=I+1
209      650      CONTINUE
210      700      RETURN
211           END
```

```
  1   C ---------------------------------------------------------------------------
  2   C --
  3   C --   S U B R O U T I N E     P S C A L C
  4   C --
  5   C --        CALCULATIONS BASED ON THE POTENTIOMETRIC MEASUREMENTS
  6   C --
  7   C ---------------------------------------------------------------------------
  8            SUBROUTINE   PSCALC(IP,ISELEC,ICOL,MXPOT,IFAULT,KEYCN1,ICD,ALNB,
  9           &STCOEF,TOTV,XVPA,VOLC,REZORT,VVOL,VORT,VPOT,POTMLT,TOT,POTADD,
 10           &VARVOL,VARORT,VARPOT,TOTB,A,ADIAG,B,BH,DERTOT,DERPOT,COLC,
 11           &NWAVE,NSPCOL,NSPREF,NSPEC,MXSPEC,NRFCOL,NGCOL,
 12           &I13,I15,NCURVE,NPOINT,KODCOL,KODPOT,KCOMP,KSPEC,KODREF)
 13            DIMENSION IP(NCURVE),ISELEC(KODREF),ICOL(KODCOL)
 14            DOUBLE PRECISION
 15   C***    REAL
 16           &ALNB(KSPEC),STCOEF(KCOMP,KSPEC),TOTV(I13,NCURVE),VPOT(KODPOT),
 17           &XVPA(I15,NPOINT),VOLC(NPOINT),REZORT(NPOINT),VVOL,VORT,VABS,
 18           &POTMLT(MXPOT),POTADD(MXPOT),VARVOL,VARORT,VARPOT(KODPOT),
 19           &A(NSPREF,NSPREF),ADIAG(NSPREF),B(NSPREF),BH(KODREF),
 20           &TOT(KCOMP),DERTOT(KODREF),DERPOT(KODREF,KODPOT),TOTB(KCOMP),
 21           &COLC(KODCOL,NPOINT),H,HH,SH,VOL0,VPT,WVOL,W,ORT,REZID,
 22           &TITLD,CREAL,REP,FIT,ALN10,ALGE,CMIN,XCMAX,FNRMAX,REP2,DERORT,
 23            COMMON TITLD(20),CREAL(5000)
 24            COMMON/REPAR/REP(5),FIT,ALN10,ALGE,CMIN,XCMAX,FNRMAX,REP2(3)
 25            COMMON/INTPAR/KRAND,MAINIT,MAXIT,MODEL,MODELN,MITFIT,MXHALV,
 26           &MINIT,MMM,ITNSUM,ITHSUM,MXITNR,MXITH
 27            COMMON/LABGRP/KODVOL,KFIX,K1,KODTOT,KCOMP1,KODORT,KD,KREFIN,
 28           &LB,LB1,LB2,LB3,LB4
 29            IFAULT=0
 30            KODCNT=-1
 31            IF(KEYCN1.GT.0)KODCNT=0
 32            VARVOL=0.
 33            VARORT=0.
 34            IF(KODPOT)112,112,110
 35      110   DO 111 K=K1,KODPOT
 36      111   VARPOT(K)=0.
 37      112   IF(KODTOT.GT.0)WVOL=1./(VVOL*VVOL)
 38            IF(KODORT.GT.0)ORT=VORT*ALGE*POTMLT(1)
 39            N2=0
 40            DO 199 KCURVE=1,NCURVE
 41               IF(KEYCN1.EQ.0)KODCNT=0
 42               N1= N2+1
 43               IF(KODVOL)120,130,130
 44      120      N2=N1
 45               GO TO 200
 46      130      N2=IP(KCURVE)
 47               VOL0=TOTV(KCOMP1,KCURVE)
 48               IF(KODTOT+KODORT.GT.0)VPT=VOL0/(TOTB(1)-TOTV(1,KCURVE))
 49      200      DO 299 KPOINT=N1,N2
 50                  IF(KFIX.LE.0 .OR. MAINIT.NE.0)GO TO 209
 51                     DO 201 K=1,KFIX
 52                     II=I13+K
 53                     XVPA(K,KPOINT)=(XVPA(II,KPOINT)-POTADD(K))
 54           &            *ALN10/POTMLT(K)
 55      201            CONTINUE
 56      209         IF(KODVOL)225,210,220
 57      210         H=XVPA(KCOMP1,KPOINT)
 58               GO TO 225
 59      220         H=VOLC(KPOINT)
 60      225         IF(KFIX-KCOMP)230,240,240
 61      230         DO 239 K=K1,KCOMP
 62                  SH= TOTV(K,KCURVE)
 63                  IF(KODVOL.GE.0)SH=(VOL0*SH+H*TOTB(K))/(VOL0+H)
 64                  TOT(K)=SH
 65                  IF(KSPEC-KCOMP)231,231,232
 66      231            XVPA(K,KPOINT)=DLOG(SH)
```

```
 67    C***231              XVPA(K,KPOINT)=ALOG(SH)
 68                         GO TO 239
 69    232          IF(SH-CMIN)233,238,238
 70    233          SH=DABS(SH)
 71    C 233        SH= ABS(SH)
 72                 IF(KODVOL.GE.0)SH=SH+DABS(TOTB(K)*H/VOL0)
 73    C            IF(KODVOL.GE.0)SH=SH+ ABS(TOTB(K)*H/VOL0)
 74                 IF(SH.LT.CMIN)SH=CMIN
 75    238          ALNB(K)=SH*FNRMAX
 76                 IF(KODCNT .EQ. 0  )XVPA(K,KPOINT)=DLOG(SH)-0.7
 77    C***         IF(KODCNT .EQ. 0  )XVPA(K,KPOINT)=ALOG(SH)-0.7
 78    239       CONTINUE
 79    240       CALL PSDER(STCOEF,ALNB,TOT,XVPA(1,KPOINT),DERTOT,DERPOT,
 80        &        DERORT,XCMAX,KODCNT,ICOL,COLC(1,KPOINT),CREAL(ICD),
 81        &        CREAL(LB1),CREAL(LB2),CREAL(LB3),CREAL(LB4),II,
 82        &        I13,I15,NCURVE,NPOINT,KODCOL,KODPOT,KCOMP,KSPEC,KODREF)
 83              IF(KODCOL.GT.0)ICD=ICD+KODCOL*KODREF
 84              IF(II)300,300,99
 85    99        IFAULT=KPOINT
 86              RETURN
 87    300       IF(KODTOT)400,400,310
 88    310       H=1.
 89              SH=TOT(1)
 90              REZID=TOTV(1,KCURVE)-SH
 91              IF(KODVOL)330,320,320
 92    320          SH=REZID*VOL0/(SH-TOTB(1))
 93                 REZID=XVPA(KCOMP1,KPOINT)-SH
 94                 H=H+SH/VOL0
 95                 H=VPT*H*H
 96    330       VOLC(KPOINT)=SH
 97              VARVOL=VARVOL+REZID*REZID
 98              IF(KODORT)350,350,340
 99    340          SH=H*DERORT/ORT
100                 REZID=REZID/DSQRT(1.+SH*SH)
101    C            REZID=REZID/ SQRT(1.+SH*SH)
102                 VARORT=VARORT+REZID*REZID
103                 REZORT(KPOINT)=REZID
104    350       IF(KODREF.GT.0)CALL PSADD(A,BH,NSPREF,KODREF,REZID,
105        &            DERTOT,WVOL,H)
106    400       IF(KODPOT)299,299,410
107    410       DO 499 K=K1,KODPOT
108                 W=VPOT(K)
109                 IF(W)499,499,420
110    420             W=1./(W*W)
111                    H=POTMLT(K)*ALGE
112                    II=I13+K
113                    REZID=XVPA(II,KPOINT)-(H*XVPA(K,KPOINT)+POTADD(K))
114                    VARPOT(K)=VARPOT(K)+REZID*REZID
115                    IF(K-KODORT)460,430,460
116    430                SH=ORT*DERORT
117                       IF(KODVOL)450,440,440
118    440                   HH=1.+XVPA(KCOMP1,KPOINT)/VOL0
119                          SH=SH/(VPT*HH*HH)
120    450                SH=1.+SH*SH
121                       REZID=REZID/DSQRT(SH)
122    C                  REZID=REZID/ SQRT(SH)
123                       VARORT=VARORT+REZID*REZID
124                       REZORT(KPOINT)=REZID
125    460             IF(KODREF.GT.0)CALL PSADD(A,BH,NSPREF,KODREF,REZID,
126        &                DERPOT(1,K),W,H)
127    499       CONTINUE
128    299       CONTINUE
129    199    CONTINUE
130           H=NPOINT
131           IF(KODTOT)550,550,510
132    510    IF(KODORT)530,530,520
133    520       FIT=FIT+WVOL*VARORT
134              VARORT=DSQRT(VARORT/H)
135    C         VARORT= SQRT(VARORT/H)
136              GO TO 540
137    530       FIT=FIT+WVOL*VARVOL
138    540       VARVOL=DSQRT(VARVOL/H)
139    C 540     VARVOL= SQRT(VARVOL/H)
140    550    IF(KODPOT)600,600,560
141    560       DO 569 K=K1,KODPOT
142                 W=VPOT(K)
```

```
143                      IF(W)569,569,561
144      561            IF(K-KODORT)565,562,565
145      562        FIT=FIT+VARORT/(W*W)
146                  VARORT=DSQRT(VARORT/H)
147  C               VARORT= SQRT(VARORT/H)
148                  GO TO 566
149      565            FIT=FIT+VARPOT(K)/(W*W)
150      566        VARPOT(K)=DSQRT(VARPOT(K)/H)
151  C 566          VARPOT(K)= SQRT(VARPOT(K)/H)
152      569        CONTINUE
153      600    IF(KODREF)900,900,610
154      610    DO 699 KR=1,KODREF
155                  I=ISELEC(KR)
156                  B(I)=B(I)+BH(KR)
157                  BH(KR)=0.
158                  ADIAG(I)=ADIAG(I)+A(KR,KR)
159                  A(KR,KR)=0.
160                  IF(KR.EQ.KODREF)GO TO 699
161                  K=KR+1
162                  DO 620 KC=K,KODREF
163                      II= ISELEC(KC)
164                      A(II,I)=A(II,I)+A(KR,KC)
165                      A(KR,KC)=0.
166      620        CONTINUE
167      699    CONTINUE
168      900 RETURN
169          END
```

```
1   C ----------------------------------------------------------------------
2   C --
3   C --   S U B R O U T I N E   P S D E R
4   C --
5   C --          CALCULATION OF ANALYTICAL DERIVATES AT ONE POINT
6   C --
7   C ----------------------------------------------------------------------
8          SUBROUTINE  PSDER(STCOEF,ALNB,TOT,X,DERTOT,DERPOT,DERORT,
9         &XCMAX,KODCNT,ICOL,COLC,DERC,CONC,XH,TOTH,DX,IFAULT,
10        &I13,I15,NCURVE,NPOINT,KODCOL,KODPOT,KCOMP,KSPEC,KODREF)
11         DIMENSION ICOL(KODCOL)
12         DOUBLE PRECISION
13  C***   REAL
14        &STCOEF(KCOMP,KSPEC),ALNB(KSPEC),TOT(KCOMP),X(KCOMP),
15        &COLC(KODCOL),DERC(KODCOL,KODREF),CONC(KSPEC),DERTOT(KODREF),
16        &DERPOT(KODREF,KODPOT),XH(KCOMP),TOTH(KCOMP),DX(KCOMP),
17        &H,SH,DERORT,XCMAX
18         COMMON/LABGRP/KODVOL,KFIX,K1,KODTOT,KCOMP1,KODORT,KD,KREFIN,
19        &LB,LB1,LB2,LB3,LB4
20             IFAULT=0
21             IF(KFIX.LT.KCOMP .AND. KCOMP.LT.KSPEC)GO TO 90
22             DO 10 K=1,KCOMP
23      10         CONC(K)=DEXP(X(K))
24  C   10         CONC(K)= EXP(X(K))
25             DERORT=CONC(1)
26             IF(KODTOT.GT.0)TOT(1)=CONC(1)
27             IF(KSPEC-KCOMP)800,800,20
28      20     DO 50 I=KCOMP1,KSPEC
29                 SH=ALNB(I)
30                 DO 30 K=1,KCOMP
31      30             SH=SH+STCOEF(K,I)*X(K)
32                 SH=DEXP(SH)
33  C               SH= EXP(SH)
34                 CONC(I)=SH
35                 H=STCOEF(1,I)
36                 DERORT=DERORT+H*H*SH
37                 IF(KODTOT)50,50,40
38      40             TOT(1)=TOT(1)+H*SH
39                     K=I-KCOMP
40                     IF(K.LE.KODREF)DERTOT(K)=H*SH
41      50     CONTINUE
42             IF(KODCOL)950,950,60
43      60     IF(KODREF)800,800,70
44      70     J=0
```

```
 45                      DO 80 I=KCOMP1,KREFIN
 46                         J=J+1
 47                         DO 80 IC=1,KODCOL
 48                         DERC(IC,J)=0.
 49                         IF(I.EQ.ICOL(IC))DERC(IC,J)=CONC(I)
 50        80             CONTINUE
 51                      GO TO 950
 52        90         IF(KODCNT)200,200,100
 53       100         DO 150 KC=K1,KCOMP
 54                      SH=0.
 55                      IF(KFIX)140,140,110
 56       110             DO 130 KR=1,KFIX
 57                         H=0.
 58                         DO 120 I=KCOMP1,KSPEC
 59       120             H= H-STCOEF(KC,I)*STCOEF(KR,I)*CONC(I)
 60                         SH=SH+H*(X(KR)-XH(KR))
 61       130             CONTINUE
 62       140             DX(KC)=SH+TOT(KC)-TOTH(KC)
 63       150         CONTINUE
 64                   CALL SUBST(STCOEF(K1,K1),KCOMP,KD,DX(K1),DX(K1))
 65                   H=0.
 66                   DO 160 K=K1,KCOMP
 67                      SH= DABS(DX(K))
 68    C               SH=   ABS(DX(K))
 69                      IF(SH.GT.H)H=SH
 70       160         CONTINUE
 71                   IF(H.LE.XCMAX)GO TO 190
 72                   H=XCMAX/H
 73                   SH=1.-H
 74                   IF(KFIX)180,180,170
 75       170         DO 175 K=1,KFIX
 76                      XH(K)=SH*XH(K)+H*X(K)
 77                      CONC(K)=DEXP(XH(K))
 78    C               CONC(K)= EXP(XH(K))
 79       175         CONTINUE
 80       180         DO 185 K=K1,KCOMP
 81                      XH(K)=XH(K)+H*DX(K)
 82                      TOTH(K)=SH*TOTH(K)+H*TOT(K)
 83       185         CONTINUE
 84                   CALL PSCONC(STCOEF,STCOEF,ALNB,ALNB,TOTH,XH,CONC,
 85          &         DX,DX(KCOMP1),DX(LB),II,
 86          &         I13,I15,NCURVE,NPOINT,KODCOL,KODPOT,KCOMP,KSPEC,KODREF)
 87                   IF(II)100,100,990
 88       190         DO 195 K=K1,KCOMP
 89       195         X(K)=XH(K)+DX(K)
 90       200         IF(KFIX)220,220,210
 91       210         DO 215 K=1,KFIX
 92       215         CONC(K)=DEXP(X(K))
 93    C*** 215       CONC(K)=EXP(X(K))
 94       220         CONTINUE
 95                   CALL PSCONC(STCOEF,STCOEF,ALNB,ALNB,TOT,X,CONC,
 96          &         DX,DX(KCOMP1),DX(LB),II,
 97          &         NWAVE,NSPCOL,NSPREF,NSPEC,MXSPEC,NRFCOL,NGCOL,
 98          &         I13,I15,NCURVE,NPOINT,KODCOL,KODPOT,KCOMP,KSPEC,KODREF)
 99                   IF(II.GT.0)GO TO 990
100                   IF(KODTOT)400,400,300
101       300         SH=CONC(1)
102                   DO 310 I=KCOMP1,KSPEC
103       310         SH=SH+STCOEF(1,I)*CONC(I)
104                   TOT(1)=SH
105                   IF(KODORT+KODREF)500,500,320
106       320         DO 340 K=K1,KCOMP
107                      SH= 0.
108                      DO 330 I=KCOMP1,KSPEC
109       330             SH=SH +STCOEF(1,I)*STCOEF(K,I)*CONC(I)
110                      TOTH(K)=SH
111       340         CONTINUE
112                   IF(KODORT)500,500,350
113       350         CALL SUBST(STCOEF(K1,K1),KCOMP,KD,DX(K1),TOTH(K1))
114                   SH=CONC(1)
115                   DO 360 I=KCOMP1,KSPEC
116                      H=STCOEF(1,I)
117                      SH=SH+H*H*CONC(I)
118       360         CONTINUE
```

```
119                    DO 370 K=K1,KCOMP
120        370         SH=SH -DX(K)*TOTH(K)
121                    DERORT=SH
122                    GO TO 500
123        400         IF(KODORT)500,500,410
124        410         DO 420 K=1,KCOMP
125        420         DX(K)=0.
126                    DX(1)=1.
127                    CALL SUBST(STCOEF,KCOMP,KCOMP,DX,DX)
128                    DERORT=DX(1)
129        500         IF(KODREF.LE.0)GO TO 800
130                    J=0
131                    DO 600 I=KCOMP1,KREFIN
132                    J=J+1
133                    H=CONC(I)
134                    DO 510 K=K1,KCOMP
135        510         DX(K)=-H*STCOEF(K,I)
136                    CALL SUBST(STCOEF(K1,K1),KCOMP,KD,DX(K1),DX(K1))
137                    IF(KODTOT)540,540,520
138        520         SH=H*STCOEF(1,I)
139                    DO 530 K=K1,KCOMP
140        530         SH=SH +TOTH(K)*DX(K)
141                    DERTOT(J)=SH
142        540         IF(KODPOT)570,570,550
143        550         DO 560 K=K1,KODPOT
144        560         DERPOT(J,K)=DX(K)
145        570         IF(KODCOL)600,600,580
146        580         DO 590 IC=1,KODCOL
147                    II=ICOL(IC)
148                    SH=0
149                    IF(II-KCOMP)582,582,584
150        582         IF(II-K1)588,583,583
151        583         SH=DX(II)
152                    GO TO 587
153        584         IF(II.EQ.I)SH=1.
154                    DO 586 K=K1,KCOMP
155        586         SH=SH +STCOEF(K,II)*DX(K)
156        587         SH=SH*CONC(II)
157        588         DERC(IC,J)=SH
158        590         CONTINUE
159        600         CONTINUE
160        800         IF(KODCOL)900,900,810
161        810         DO 820 IC=1,KODCOL
162                    II=ICOL(IC)
163                    IF(II)812,812,814
164        812         COLC(IC)=0.
165                    GO TO 820
166        814         COLC(IC)=CONC(II)
167        820         CONTINUE
168        900         IF(KODCNT)950,910,920
169        910         KODCNT=1
170        920         DO 930 K=1,KCOMP
171                    TOTH(K)=TOT(K)
172                    XH(K)=X(K)
173        930         CONTINUE
174        950         RETURN
175        990         IFAULT=II
176                RETURN
177                END




1      C ------------------------------------------------------------------------
2      C --
3      C --   S U B R O U T I N E    P S C O N C
4      C --
5      C --        CALCULATION OF SPECIES CONCENTRATIONS AT ONE POINT
6      C --
7      C ------------------------------------------------------------------------
8             SUBROUTINE PSCONC(FDERIV,STCOEF,ALNB,FMAX,TOTM,X,CONC,TOTR,FP,FM,
9           &IFAULT,
10          &I13,I15,NCURVE,NPOINT,KODCOL,KODPOT,KCOMP,KSPEC,KODREF)
11             DOUBLE PRECISION
12     C***    REAL
13          &FDERIV(KCOMP,KCOMP),STCOEF(KCOMP,KSPEC),ALNB(KSPEC),
```

```
14        &FMAX(KCOMP),TOTM(KCOMP),X(KCOMP),CONC(KSPEC),TOTR(KCOMP),
15        &FP(KCOMP),FM(KCOMP),H,SH,SHP,SHM,REP,FHMAX,XMAX,XHMAX
16         COMMON/REPAR/REP(11),FHMAX,XMAX,XHMAX
17         COMMON/INTPAR/KRAND,MAINIT,MAXIT,MODEL,MODELN,MITFIT,MXHALV,
18        &MINIT,MMM,ITNSUM,ITHSUM,MXITNR,MXITH
19         COMMON/LABGRP/KODVOL,KFIX,K1,KODTOT,KCOMP1,KODORT,KD,KREFIN,
20        &LB,LB1,LB2,LB3,LB4
21   C --
22   C --      CALCULATION OF SPECIES CONCENTRATIONS IN ONE POINT BASED ON
23   C --      EQUATION(5)
24   C --
25              IFAULT=0
26              ITH=0
27        20    ITNR=0
28        30    DO 32 I=KCOMP1,KSPEC
29              SH=ALNB(I)
30              DO 31 K=1,KCOMP
31        31    SH=SH+STCOEF(K,I)*X(K)
32              CONC(I)=DEXP(SH)
33   C***       CONC(I)=EXP(SH)
34        32    CONTINUE
35              II=0
36              DO 36 K=K1,KCOMP
37              SH=DEXP(X(K))
38   C          SH=EXP(X(K))
39              CONC(K)=SH
40              SH=TOTM(K)-SH
41              DO 35 I=KCOMP1,KSPEC
42        35    SH=SH-STCOEF(K,I)*CONC(I)
43              TOTR(K)=SH
44              IF(DABS(SH).GT.FMAX(K))II=1
45   C***       IF(ABS(SH).GT.FMAX(K))II=1
46        36    CONTINUE
47              IF(ITH.GT.0)II=0
48              IF(II.EQ.1 .OR. ITNR.EQ.0)GO TO 37
49              ITNSUM=ITNSUM+ITNR
50              ITHSUM=ITHSUM+ITH
51              RETURN
52        37    IF(MXITNR-ITNR)100,100,40
53        40    DO 43 KR=K1,KCOMP
54              DO 42 KC=K1,KR
55              SH=0.
56              DO 41 I=KCOMP1,KSPEC
57        41    SH=SH +STCOEF(KR,I)*STCOEF(KC,I)*CONC(I)
58        42    FDERIV(KR,KC)=SH
59              FDERIV(KR,KR)=FDERIV(KR,KR)+CONC(KR)
60        43    CONTINUE
61              DO 450 K=K1,KCOMP
62        450   FM(K)=FDERIV(K,K)
63              SH=1.
64        451   CALL FACTOR(FDERIV(K1,K1),KCOMP,KD,II)
65              ITNR=ITNR+1
66              IF(II)460,460,452
67        452   DO 453 K=K1,KCOMP
68        453   FDERIV(K,K)=(1.+SH)*FM(K)
69              SH=SH* 2.
70              IF(MXITNR-ITNR)100,100,451
71        460   CONTINUE
72              CALL SUBST(FDERIV(K1,K1),KCOMP,KD,TOTR(K1),TOTR(K1))
73              H=0
74              DO 45 K=K1,KCOMP
75              SH=DABS(TOTR(K))
76   C          SH=ABS(TOTR(K))
77              IF(SH.GT.H)H=SH
78        45    CONTINUE
79              SH=1.0
80              IF(H.GT.XMAX)SH=XMAX/H
81              DO 47 K=K1,KCOMP
82        47    X(K)=X(K)+SH*TOTR(K)
83              GO TO 30
84   C --
85   C --      CALCULATION OF SPECIES CONCENTRATIONS IN ONE POINT BASED ON
86   C --      EQUATIONS(5,6)
87   C --
88        100   ITNSUM=ITNSUM+ITNR
89        101   II=0
```

```
 90                    DO 120 K=K1,KCOMP
 91                       SHP=CONC(K)
 92                       SHM=0.
 93                       IF(TOTM(K))111,113,112
 94       111               SHP=SHP-TOTM(K)
 95                         GO TO 113
 96       112               SHM=SHM+TOTM(K)
 97       113             DO 118 I=KCOMP1,KSPEC
 98                          H=STCOEF(K,I)
 99                          IF(H)116,118,117
100       116                  SHM=SHM-H*CONC(I)
101                            GO TO 118
102       117                  SHP=SHP+H*CONC(I)
103       118             CONTINUE
104                       FP(K)=SHP
105                       FM(K)=SHM
106                       H=DLOG(SHM)-DLOG(SHP)
107     C                 H=ALOG(SHM)-ALOG(SHP)
108                       TOTR(K)=H
109                       IF(DABS(H).GT.FHMAX)II=1
110     C                 IF( ABS(H).GT.FHMAX)II=1
111       120           CONTINUE
112                     IF(ITH.EQ.0)GO TO 140
113                     IF(II .EQ.0)GO TO 20
114                     IF(ITH.GE.MXITH)GO TO 200
115       140           DO 145 KR=K1,KCOMP
116                       DO 145 KC=K1,KCOMP
117                         SHP=0.
118                         SHM=0.
119                         IF(KR.EQ.KC)SHP=SHP+CONC(KR)
120                         DO 143 I=KCOMP1,KSPEC
121                           H=STCOEF(KR,I)
122                           IF(H)141,143,142
123       141                   SHM=SHM+H*STCOEF(KC,I)*CONC(I)
124                             GO TO 143
125       142                   SHP=SHP+H*STCOEF(KC,I)*CONC(I)
126       143                 CONTINUE
127                         FDERIV(KR,KC)=SHP/FP(KR)+SHM/FM(KR)
128       145           CONTINUE
129                     DO 149 KC=K1,KCOMP
130                       SH=0.
131                       DO 146 K=K1,KCOMP
132                         FM(K)=FDERIV(K,KC)
133                         SH=SH+FM(K)*TOTR(K)
134       146             CONTINUE
135                       FP(KC)=SH
136                       DO 148 KR=KC,KCOMP
137                         SH=0.
138                         DO 147 K=K1,KCOMP
139       147                 SH=SH +FDERIV(K,KR)*FM(K)
140                         FDERIV(KR,KC)=SH
141       148             CONTINUE
142       149           CONTINUE
143                     DO 150 K=K1,KCOMP
144       150           FM(K)=FDERIV(K,K)
145                     SH=1.
146       151           CALL FACTOR(FDERIV(K1,K1),KCOMP,KD,II)
147                     ITH=ITH+1
148                     IF(5*(ITH/5)+1 .EQ. ITH)GO TO 152
149                     IF(II)160,160,152
150       152           DO 153 K=K1,KCOMP
151       153           FDERIV(K,K)=(1.+SH)*FM(K)
152                     SH=SH* 2.
153                     IF(MXITH-ITH)200,200,151
154       160           CALL SUBST(FDERIV(K1,K1),KCOMP,KD,FP(K1),FP(K1))
155                     H=0.
156                     DO 161 K=K1,KCOMP
157                       SH=DABS(FP(K))
158     C                 SH=ABS(FP(K))
159                       IF(SH.GT.H)H=SH
160       161           CONTINUE
161                     IF(H.LE.XHMAX)GO TO170
162                     SH=XHMAX/H
163                     DO 162 K=K1,KCOMP
164       162           FP(K)=SH*FP(K)
165       170           DO 171 K=K1,KCOMP
```

```
166                 X(K)=X(K)+FP(K)
167                 CONC(K)=DEXP(X(K))
168   C             CONC(K)=EXP(X(K))
169      171     CONTINUE
170              DO 173 I=KCOMP1,KSPEC
171                 SH=ALNB(I)
172                 DO 172 K=1,KCOMP
173      172        SH=SH +STCOEF(K,I)*X(K)
174                 CONC(I)=DEXP(SH)
175   C             CONC(I)=EXP(SH)
176      173     CONTINUE
177              GO TO 101
178      200     IFAULT=ITH
179            RETURN
180            END
```

```
1     C ---------------------------------------------------------------------
2     C --
3     C --    S U B R O U T I N E    P S L I S T
4     C --
5     C --            LISTING OF THE FINAL RESULTS
6     C --
7     C ---------------------------------------------------------------------
8           SUBROUTINE PSLIST(IP,ISORT,IDENT,KW1,JCH,KG,KLIST,KWRITE,ALNB,
9          &STCOEF,TOTV,XVPA,VOLC,REZORT,VVOL,VORT,VPOT,VABS,POTMLT,
10         &POTADD,VARVOL,VARPOT,ABCOEF,COLC,SPNAME,
11         &NWAVE,NSPCOL,NSPREF,NSPEC,MXSPEC,NRFCOL,NGCOL,
12         &I13,I15,NCURVE,NPOINT,KODCOL,KODPOT,KCOMP,KSPEC,KODREF)
13          DIMENSION IP(NCURVE),ISORT(KSPEC),IDENT(NWAVE),JCH(66),
14         &JCM1(10),JC1(5),JC3(5),IK(5),JK(5)
15          DOUBLE PRECISION
16    C***  REAL
17         &ALNB(KSPEC),STCOEF(KCOMP,KSPEC),TOTV(I13,NCURVE),VPOT(KODPOT),
18         &XVPA(I15,NPOINT),VOLC(NPOINT),REZORT(NPOINT),VVOL,VORT,VABS,
19         &POTMLT(KODPOT),POTADD(KODPOT),VARVOL,ABCOEF(NWAVE,KODCOL),
20         &VARPOT(KODPOT),COLC(KODCOL,NPOINT),VV(5),X(10),CALC(5),
21         &REZ(5),SM,SC,SR,H,SH,W,SPNAME(2,KSPEC),REP,ALN10,ALGE,REP2
22          COMMON/REPAR/REP(6),ALN10,ALGE,REP2(6)
23          COMMON/INTPAR/KRAND,MAINIT,MAXIT,MODEL,MODELN,MITFIT,MXHALV,
24         &MINIT,MMM,ITNSUM,ITHSUM,MXITNR,MXITH
25          COMMON/LABGRP/KODVOL,KFIX,K1,KODTOT,KCOMP1,KODORT,KD,KREFIN,
26         &LB,LB1,LB2,LB3,LB4
27          COMMON/IPERO/IO(10)
28          DATA JCM1(1),JCM1(2),JCM1(3),JCM1(4),JCM1(5),JCM1(6),JCM1(7),
29         &JCM1(8),JCM1(9),JCM1(10)/1H1,1H2,1H3,1H4,1H5,1H6,1H7,1H8,1H9,
30         &1HA/,J0,JBL,JPL,JST/1H0,1H ,1H+,1H*/,JTOT,JVOL,JPOT,JABS
31         &/3HTOT,3HVOL,3HPOT,3HABS/
32        1 FORMAT(1X)
33        2 FORMAT(1H1,16X,5(3H***)/17X,8H*    SET,I3,4H    */17X,5(3H***))
34        3 FORMAT(// 41X,27HC O N C E N T R A T I O N S )
35        4 FORMAT(3X,9HNO.    PC(,I2,3H)   ,8(1X,A1,5HSPEC(,I2,4H)    ))
36        5 FORMAT(1X,I4,2X,F8.4,2X,8(1X,1PG12.5))
37        6 FORMAT(1H+,5X,8(12X,A1))
38        7 FORMAT(// 41X,17HR E S I D U A L S )
39        8 FORMAT(// 41X,19HS I M U L A T I O N /
40         &31X,19HNEW MEASURED DATE = )
41        9 FORMAT(// 6X,5(A1,2X,6H CALC.,5X,6HREZID.,3X))
42       10 FORMAT(3X,3HNO.,5(4X,A3,I2,5X,A3,I2,4X))
43       11 FORMAT(1H+,40X,2(15X,A1, 7H SDSET(,A3,I2,2H)=,1PG10.3))
44       12 FORMAT(1H+,56X,4H-3SD,6X,4H-2SD,7X,3H-SD,8X,1H0,8X,3H SD,6X,
45         &4H 2SD,6X,4H 3SD)
46       13 FORMAT(1X,I4,1X ,5(1X,1PG12.5,G10.3))
47       14 FORMAT(1H+,52X,3A1,1HX,1X63A1)
48       15 FORMAT(18X,1PG11.4,4(12X,G11.4))
49       16 FORMAT(17H+SQRT(SUM(R*R)/N))
50       17 FORMAT(17H+MEAN = SUM(R)/N )
51       18 FORMAT(17H+SQRT(SQ*SQ-M*M) )
52       19 FORMAT(1H1)
53       22 FORMAT(17X,8H*  CURVE,I3,3X,1H*/17X,5(3H***))
54       24 FORMAT(// 17X,8(2X,2A4,3X))
55       81 FORMAT(1H+,50X,8HMEASURED)
56       82 FORMAT(1H+,49X,10HCALCULATED )
```

```
57        83 FORMAT(1H+,59X,11H+ SD*RANDOM )
58           IPR=IO(3)
59           WRITE(IPR,2)KG
60           SM=1.
61           SR=0.
62           IF(KWRITE.LT.0)SM=0.
63           SC=1.-SM
64           IF(IABS(KWRITE).GE.2)SR=1.
65           KPRC=KLIST-1
66           KPREZ=KLIST
67           IF(KLIST.GE.3)KPREZ=0
68           JMAX=0
69           IF(KPREZ)180,180,130
70       130 IF(KODTOT)140,140,132
71       132    JMAX=1
72              IK(1)=0
73              IF(KWRITE.NE.0)VARVOL=VVOL
74              VV(1)=VARVOL
75              IF(KODVOL)134,136,136
76       134       K=ISORT(1)
77                 JK(1)=K
78                 IF(K.GT.9)K=K-(K/10)*10
79                 IF(K.EQ.0)K=10
80                 JC1(1)=JCM1(K)
81                 JC3(1)=JTOT
82                 GO TO 140
83       136       JK(1)=0
84                 JC1(1)=J0
85                 JC3(1)=JVOL
86       140    IF(KODPOT)160,160,142
87       142       DO 150 K=1,KODPOT
88                    IF(VPOT(K))150,150,144
89       144             JMAX=JMAX+1
90                       IK( JMAX)=K
91                       IF(KWRITE.NE.0)VARPOT(K)=VPOT(K)
92                       VV( JMAX)=VARPOT(K)
93                       I=ISORT(K)
94                       JK( JMAX)=I
95                       IF(I.GT.9)I=I-(I/10)*10
96                       IF(I.EQ.0)I=10
97                       JC1(JMAX)=JCM1(I)
98                       JC3(JMAX)=JPOT
99       150       CONTINUE
100      160    IF(JMAX.GT.0 .AND. JMAX.LE.2 .AND. KODORT.EQ.1)JC1(JMAX+1)=JST
101      180    N3=1
102             IF(KODVOL.GE.0)N3=NCURVE
103             N2=0
104             DO 800 KCURVE=1,N3
105             IF(KCURVE.GT.1)WRITE(IPR,19)
106             IF(KODVOL.GE.0)WRITE(IPR,22)KCURVE
107             N1=N2+1
108             IF(KODVOL.LT.0)N2=NCURVE
109             IF(KODVOL.GE.0)N2=IP(KCURVE)
110             H=N2+1-N1
111             IF(KPRC)300,300,200
112      200    WRITE(IPR,3)
113             KPC1=ISORT(1)
114             DO 290 L1=1,KSPEC,8
115                L2=L1+7
116                IF(L2.GT.KSPEC)L2=KSPEC
117                L3=L2+1-L1
118                WRITE(IPR,24)(SPNAME(1,K),SPNAME(2,K),K=L1,L2)
119                WRITE(IPR,4)KPC1,(JBL,ISORT(K),K=L1,L2)
120                WRITE(IPR,1)
121                IF(KODVOL)206,202,202
122      202       DO 204 I=1,L3
123                   JCH(I)=JPL
124                   JCH(10+I)=1
125      204       CONTINUE
126      206       DO 260 KPOINT=N1,N2
127                   J=L1
128                   DO 250 I=1,L3
129                      IF(J-KCOMP)210,210,220
130      210             SH=XVPA(J,KPOINT)
131                      GO TO 240
132      220             SH=ALNB(J)
```

```
133                          DO 230 K=1,KCOMP
134        230               SH=SH+ STCOEF(K,J)*XVPA(K,KPOINT)
135        240               SH=DEXP(SH)
136      C 240               SH=EXP(SH)
137                          IF(KPOINT.EQ.N1 .OR. KODVOL.LT.0)GO TO 249
138                          K=JCH(10+I)
139                          IF(SH-X(I))242,248,248
140        242               IF(K)249,244,246
141        244               JCH(10+I)=-1
142                          JCH(I)=JBL
143                          GO TO 249
144        246               JCH(10+I)=0
145                          JCH(I)=JST
146            GO TO 249
147        248               IF(K.GT.0)GO TO 249
148                          JCH(10+I)=1
149                          JCH(I)=JPL
150        249               X(I)=SH
151                          J=J+1
152        250               CONTINUE
153                          IF(KODVOL.GE.0 .AND. KPOINT.GT.N1)WRITE(IPR,6)
154          &               (JCH(I),I=1,L3)
155                          SH=-ALGE*XVPA(1,KPOINT)
156        260             WRITE(IPR,5)KPOINT,SH,(X(I),I=1,L3)
157                        IF(KODVOL.LT.0 .OR. N1.EQ.N2)GO TO 290
158                        DO 270 I=1,L3
159                          IF(JCH(10+I).GT.0)JCH(I)=JST
160        270             CONTINUE
161                        WRITE(IPR,6)(JCH(I),I=1,L3)
162        290           CONTINUE
163        300           IF(JMAX)500,500,302
164        302           IF(KWRITE)303,310,303
165        303             WRITE(IPR,8)
166                        IF(SM-0.5)305,307,306
167        305             WRITE(IPR,82)
168                        GO TO 307
169        306             WRITE(IPR,81)
170        307             IF(SR-0.5)315,315,308
171        308             WRITE(IPR,83)
172        310           WRITE(IPR,7)
173        315           WRITE(IPR,9)(JBL,K=1,JMAX)
174                      IF(JMAX.LE.2)WRITE(IPR,11)(JBL,JC3(K),JK(K),VV(K),K=1,JMAX)
175                      WRITE(IPR,10)(JC3(K),JK(K),JC3(K),JK(K),K=1,JMAX)
176                      IF(JMAX.LE.2)WRITE(IPR,12)
177                      WRITE(IPR,1)
178                      DO 320 K=1,JMAX
179                        X(K)=0.
180                        X(K+5)=0.
181        320           CONTINUE
182                      DO 490 KPOINT=N1,N2
183                        DO 390 II=1,JMAX
184                          K=IK(II)
185                          IF(K)340,340,330
186        330               SH=ALGE*POTMLT(K)*XVPA(K,KPOINT)+POTADD(K)
187                          J=I13+K
188                          IF(KWRITE.NE.0)XVPA(J,KPOINT)=SM*XVPA(J,KPOINT)+
189          &               SC*SH+SR*VV(II)*RANDOM(KRAND)
190                          W=XVPA(J,KPOINT)-SH
191                          GO TO 370
192        340               SH=VOLC(KPOINT)
193                          IF(KODVOL)350,360,360
194        350               IF(KWRITE.NE.0)TOTV(1,KPOINT)=SM*TOTV(1,KPOINT)+
195          &             SC*SH+SR*VV(II)*RANDOM(KRAND)
196                          W=TOTV(1,KPOINT)-SH
197                          GO TO 370
198        360               IF(KWRITE.NE.0)XVPA(I13,KPOINT)=SM*XVPA(I13,KPOINT)+
199          &             SC*SH+SR*VV(II)*RANDOM(KRAND)
200                          W=XVPA(I13,KPOINT)-SH
201        370               CALC(II)=SH
202                          REZ(II)=W
203                          X(II)=X(II)+W
204                          X(II+5)=X(II+5)+W*W
205        390             CONTINUE
206                        WRITE(IPR,13)KPOINT,(CALC(K),REZ(K),K=1,JMAX)
207                        IF(JMAX-2)400,400,490
208        400             DO 410 K=1,66
```

```
209    410        JCH(K)=JBL
210               DO 420 K=5,66,10
211    420        JCH(K)=JPL
212               II=JMAX
213               DO 430 K=1,JMAX
214    430        REZ(K)=REZ(K)/VV(K)
215               IF(KODORT)450,450,440
216    440        II=JMAX+1
217               REZ(II)=REZORT(KPOINT)/VV(1)
218    450        DO 480 K=1,II
219                   J=1
220                   SH=REZ(K)*10.0
221                   IF(SH)460,465,465
222    460            J=-1
223    465            L=DABS(SH)+0.49999
224  C 465            L=ABS(SH)+0.49999
225                   IF(L-30)475,475,470
226    470            JCH(K)=JC1(K)
227                   L=31
228    475            L=35+J*L
229    480            JCH(L)=JC1(K)
230               WRITE(IPR,14)(JCH(K),K=1,66)
231    490        CONTINUE
232               IF(KWRITE.EQ.-1)GO TO 500
233               DO 495 I=1,JMAX
234                   SH=X(I)/H
235                   X(I)=SH
236                   W=X(I+5)/H
237                   CALC(I)=DSQRT(W)
238  C                CALC(I)=SQRT(W)
239                   REZ(I)=DSQRT(W-SH*SH)
240  C                REZ(I)=SQRT(W-SH*SH)
241    495        CONTINUE
242               WRITE(IPR,1)
243               WRITE(IPR,15)(CALC(I),I=1,JMAX)
244               WRITE(IPR,16)
245               WRITE(IPR,15)(X(I),I=1,JMAX)
246               WRITE(IPR,17)
247               WRITE(IPR,15)(REZ(I),I=1,JMAX)
248               WRITE(IPR,18)
249    500        IF(KODCOL*KPREZ)800,800,502
250    502        IF(KWRITE)503,510,503
251    503            WRITE(IPR,8)
252                   IF(SM-0.5)505,507,506
253    505            WRITE(IPR,82)
254                   GO TO 507
255    506            WRITE(IPR,81)
256    507            IF(SR-0.5)515,515,508
257    508            WRITE(IPR,83)
258    510        WRITE(IPR,7)
259    515        DO 590 L1=1,NWAVE,5
260                   L2=L1+4
261                   IF(L2.GT.NWAVE)L2=NWAVE
262                   L3=L2+1-L1
263                   DO 520 K=1,5
264                       X(K)=0.
265                       X(K+5)=0.
266    520            CONTINUE
267               WRITE(IPR,9)(JBL,K=L1,L2)
268               WRITE(IPR,10)(JABS,IDENT(K),JABS,IDENT(K),K=L1,L2)
269               WRITE(IPR,1)
270               DO 550 KPOINT=N1,N2
271                   II=1
272                   L=KW1+L1-1
273                   DO 540 I=L1,L2
274                       SH=0.
275                       DO 530 J=1,KODCOL
276    530                SH=SH +ABCOEF(I,J)*COLC(J,KPOINT)
277                       CALC(II)=SH
278                       IF(KWRITE.NE.0)XVPA(L,KPOINT)=SM*XVPA(L,KPOINT)+
279        &             SC*SH+SR*VABS*RANDOM(KRAND)
280                       SH=XVPA(L,KPOINT)-SH
281                       REZ(II)=SH
282                       X(II)=X(II)+SH
283                       X(II+5)=X(II+5)+SH*SH
284                       II=II+1
```

```
285                        L=L +1
286     540       CONTINUE
287     550       WRITE(IPR,13)KPOINT,(CALC(II),REZ(II),II=1,L3)
288               IF(KWRITE.EQ.-1)GO TO 590
289               DO 560 I=1,L3
290                  SH=X(I)/H
291                  X(I)=SH
292                  W=X(I+5)/H
293                  CALC(I)=DSQRT(W)
294     C            CALC(I)= SQRT(W)
295                  REZ(I)=DSQRT(W-SH*SH)
296     C            REZ(I)= SQRT(W-SH*SH)
297     560       CONTINUE
298               WRITE(IPR,1)
299               WRITE(IPR,15)(CALC(I),I=1,L3)
300               WRITE(IPR,16)
301               WRITE(IPR,15)(X(I),I=1,L3)
302               WRITE(IPR,17)
303               WRITE(IPR,15)(REZ(I),I=1,L3)
304               WRITE(IPR,18)
305     590       CONTINUE
306     800       CONTINUE
307           RETURN
308           END
```

```
1     C ----------------------------------------------------------------------
2     C --
3     C --   F U N C T I O N   K E Y P S D
4     C --
5     C --  INPUT OF TYPE A AND TYPE C CARDS
6     C --
7     C ----------------------------------------------------------------------
8           FUNCTION KEYPSD(IBUF,BUF,TITLE)
9           COMMON/IPERO/IO(10)
10          DOUBLE PRECISION
11    C***    REAL
12         &BUF(7),TITLE(20)
13          DIMENSION IBUF(10)
14      10 FORMAT(I2,2A4,10I2,5F10.0)
15      20 FORMAT(20A4)
16      30 FORMAT(11X,I2,1X,2A4,10I3,1X,1P5D12.4)
17      40 FORMAT(1H1,9X,8HKEY TEXT,4X,9(2H I,I1),4H I10,5(3X,1HF,I1,7X)/)
18      50 FORMAT(21X,20A4)
19          JINP=IO(1)
20          JOUT=IO(3)
21          READ(JINP,10)K1,BUF(6),BUF(7),IBUF,(BUF(K),K=1,5)
22          IF(K1.EQ.11)WRITE(JOUT,40)(K,K=1,9),(K,K=1,5)
23          WRITE(JOUT,30)K1,BUF(6),BUF(7),(IBUF(K),K=1,10),(BUF(K),K=1,5)
24          KEYPSD=K1
25          IF(K1.NE.10 .OR. IBUF(1).NE.1)RETURN
26          READ(JINP,20)(TITLE(K),K=1,20)
27          WRITE(JOUT,50)(TITLE(K),K=1,20)
28          RETURN
29          END
```

```
1     C ----------------------------------------------------------------------
2     C --
3     C --   S U B R O U T I N E   P S E X I T
4     C --
5     C --   ERROR MESSAGES AND STOP
6     C --
7     C ----------------------------------------------------------------------
8           SUBROUTINE  PSEXIT(K1,K2)
9           COMMON     /IPERO/ IO(10)
10          DIMENSION  A(2,3),B(6,10)
11          DATA A(1,1),A(2,1),B(1,4),B(6,5),B(5,6),B(6,6),B(6,7),B(5,8),
12         &B(6,8),B(5,9),B(6,9),B(5,10),B(6,10)/13*4H****/,A(1,2),A(2,2),
13         &B(1,1),B(1,2),B(2,1),B(2,2),B(3,1),B(4,1),B(5,1),B(6,1),B(3,2),
14         &B(6,2),B(4,2),B(5,2)/4H  E ,4HN D ,2*4H  O,2*4H F ,4H  P ,
```

```
15          &4HS E ,4HQ U ,4HA D ,2*4H      ,4HT A ,4HS K /,A(1,3),A(2,3),
16          &B(1,3),B(2,4),B(2,3),B(3,4),B(3,3),B(4,3),B(4,4),B(5,3),B(5,4),
17          &B(6,3),B(6,4)/4H INT,4HEGER,2*4H  ER,2*4HROR ,4HON C,4HARD ,
18          &4H ON ,2*4HKEY(,2*4H  ) /,B(1,5),B(2,5),B(3,5),B(4,5),B(5,5)
19          &/4HERRO,4HR ON,4H THE,4H TAP,4HE***/,B(1,6),B(2,6),B(3,6),
20          &B(4,6)/4H MI,4HSSIN,4HG DA,4HTA  /, B(1,7),B(2,7),B(3,7),
21          &B(4,7),B(5,7)/4HINVA,4HLID ,4HLOGI,4HCAL ,4HUNIT/,B(1,8),
22          &B(2,8),B(3,8),B(4,8)/4HWANT,4HING ,4H STO,4HRAGE/,
23          &B(1,9),B(2,9),B(3,9),B(4,9)/4H SEN,4HSELE,4HSS T,4HASK /,
24          &B(1,10),B(2,10),B(3,10),B(4,10)/4H MIS,4HCALC,4HULAT,4HION /
25       10 FORMAT(/1X,11(4H****),32X,11(4H****))
26       20 FORMAT(1H+,72X,I2)
27       30 FORMAT(1H+,44X,8A4)
28             IPR=IO(3)
29             WRITE(IPR,10)
30             IF(K1.EQ.3 .OR. K1.EQ.4)WRITE(IPR,20)K2
31             I=1
32             IF(K1-3)110,100,120
33      100    I=2
34      110    I=I+1
35      120    WRITE(IPR,30)A(1,I),A(2,I),(B(K,K1),K=1,6)
36             IF(K1.EQ.1)STOP
37          RETURN
38          END
```

```
 1   C -----------------------------------------------------------------------
 2   C --
 3   C --   S U B R O U T I N E   F A C T O R
 4   C --
 5   C --     CHOLESKY FACTORIZATION
 6   C --
 7   C -----------------------------------------------------------------------
 8          SUBROUTINE  FACTOR(A,NDCL,NAKT,IERR)
 9          DOUBLE PRECISION
10   C***   REAL
11          &A,SH,H
12          DIMENSION   A(NDCL,NAKT)
13             IERR=0
14             DO 7 I=1,NAKT
15               SH=A(I,I)
16               IM1=I-1
17               IP1=I+1
18               IF(I.EQ.1)GO TO 2
19               DO 1 K=1,IM1
20                  H=A(K,I)
21                  SH=SH-H*H
22       1         CONTINUE
23       2         IF(SH)3,3,4
24       3         IERR=I
25                 RETURN
26       4         SH=DSQRT(SH)
27   C   4         SH= SQRT(SH)
28                 A(I,I)=SH
29                 IF(I.EQ.NAKT)GO TO 7
30                 DO 6 II=IP1,NAKT
31                   H=0.
32                   IF(I.EQ.1)GO TO 6
33                   DO 5 K=1,IM1
34       5           H=H-A(K,I)*A(K,II)
35       6           A(I,II)=(A(II,I)+H)/SH
36       7   CONTINUE
37          RETURN
38          END
```

```
 1   C -----------------------------------------------------------------------
 2   C --
 3   C --   S U B R O U T I N E   S U B S T
 4   C --
 5   C --                  FINAL STEPS OF THE CHOLESKY ALGORITHM
```

```
 6    C --
 7    C ------------------------------------------------------------------
 8              SUBROUTINE  SUBST(A,NDCL,NAKT,X,B)
 9              DOUBLE PRECISION
10    C***    REAL
11            &A,X,B,SH
12            DIMENSION A(NDCL,NAKT),X(NAKT),B(NAKT)
13                DO 20 I=1,NAKT
14                    SH=B(I)
15                    IF(I.EQ.1)GO TO  20
16                    IM1=I-1
17                    DO  10  K=1,IM1
18    10              SH=SH-A(K,I)*X(K)
19    20        X(I)=SH/A(I,I)
20              DO 40 II=1,NAKT
21                    I=NAKT+1-II
22                    IP1=I+1
23                    SH=X(I)
24                    IF(I.EQ.NAKT)GO TO  40
25                    DO  30  K=IP1,NAKT
26    30              SH=SH-A(I,K)*X(K)
27    40        X(I)=SH/A(I,I)
28              RETURN
29              END
```

```
 1    C ------------------------------------------------------------------
 2    C --
 3    C --    S U B R O U T I N E   P S A D D
 4    C --
 5    C --    THIS SUBROUTINE BUILDS UP THE LINEAR EQUATIONS FOR GAUSS-NEWTON
 6    C --    PROCEDURE
 7    C --
 8    C ------------------------------------------------------------------
 9              SUBROUTINE  PSADD(A,B,NDCL,NAKT,REZID,DERIV,W,H)
10              DOUBLE PRECISION
11    C***    REAL
12            &A,B,REZID,DERIV,W,H,SH
13            DIMENSION A(NDCL,NDCL),B(NAKT),DERIV(NAKT)
14                DO 10 KC=1,NAKT
15                    SH=H*DERIV(KC)
16                    B(KC)=B(KC)+SH*W*REZID
17                    DO 10 KR=1,KC
18                    A(KR,KC)=A(KR,KC)+SH*W*H*DERIV(KR)
19    10        CONTINUE
20              RETURN
21              END
```

5.2. Input Data

5.2.1. Experimental Details

The potentiometric data have been collected by a titrimetric technique. The pH measurements were made using a Radiometer PHM-52 pH meter and a GK-2322C combination electrode. The electrode system was calibrated following Irving et al.,[13] the calibration constant, K, being found to be 0.05 for the equation.

$$pH_{obs.} = K - \log [H^+]$$

Copper ion measurements were performed using a Radelkis OP 208 pH meter equipped with a Radelkis OP 8303 calomel electrode and a Radiometer F 3002 Cu Selectrode. The copper ion selective electrode was calibrated from data obtained from copper(II)–glycine titrations, measuring pH and emf simultaneously. The titrations were evaluated using the pH measurements and the copper electrode subsequently calibrated from emf. vs. $-\log [Cu^{2+}]$ plots which provided E^0 (313.8 mV) and Nerst slope (28.935).

Absorbance measurements were obtained using a Beckman ACTA M4 spectrophotometer. Solution from the titration vessel was pumped through a 1-cm pathlength flow-cell using a peristaltic pump. Absorbance values were digitized at 20-nm intervals starting at 800 nm.

The pH, pCu titrations were performed as follows: Initial solutions of $Cu(NO_3)_2$ (5.551×10^{-3} M, curve 1; 4.683×10^{-3} M, curve 2; 0.10M KNO_3) were titrated with a solution of H_3NTA (9.92×10^{-3} M for both curves; 0.10 M KNO_3). The data from each titration comprise set 3. After an appropriate amount of H_3NTA had been added the titrations were completed using 0.2324 M KOH, comprising set 4. The data were collected in the order set 3, curve 1 followed by set 4, curve 1 followed by set 3, curve 2, set 4 curve 2. However, PSEQUAD processes the curves for sets 3 and 4 by set.

5.2.2. Data for the System Copper(II)–Nitriloacetic Acid

The following is a listing of the data obtained from a combined potentiometric/spectrophotometric study of the copper(II)–nitriloacetic acid system. The data are from the six experiments A–F mentioned in Section 4.1.

```
¶<-- Column 1

91SUPERVIS 1 2            0.001     1.0       0.000001

10 DATAIN

01 SET      1

02 TOTAL    1                 -0.2324

03 TITR

04 POT      1

10 ENDSET  1

          DETERMINATION  OF  PKW

   25.0      0.010431

  1.4       11.210

  1.5       11.343

  1.6       11.444

  1.7       11.526

  1.8       11.595

  1.9       11.657

  -1.0

01 SET      1

02 TOTAL    1    3            -0.2324

03 TITR
```

```
04 POT      1

10 ENDSET   1

            DETERMINATION  OF    PROTONATION    CONSTANTS

   25.0     0.017856    0.005952

  0.2       2.348

  0.4       2.460

  0.5       2.525

  0.6       2.600

  0.7       2.684

  0.8       2.780

  0.9       2.893

  1.0       3.045

  1.1       3.270

  1.4       9.139

  1.5       9.481

  1.6       9.750

  1.7       9.995

  -1.0

01 SET      2

02 TOT      1   2   3         0.02976     0.0      0.00992

03 TITR

04 POT      1   2

10 ENDSET   1

            TITRATION  OF    CU    WITH   NTA

  18.0      0.0       0.005551    0.0

  11.0      2.091       199.0

  12.0      2.088       192.0

  13.0      2.087       188.3

  14.0      2.085       184.0

  15.0      2.084       181.2

  -1.0

  16.0      0.0       0.0046838   0.0

   8.0      2.138       199.8

   9.0      2.130       190.4

  10.0      2.127       184.5

  11.0      2.125       180.3

  12.0      2.124       177.2

  13.0      2.123       174.6
```

14.0	2.122	172.4
15.0	2.122	170.5
-1.0		

```
01 SET     2
02 TOTAL  1   2   3            -0.2324
03 TITR
04 POT    1   2
10 ENDSET  1
```

TITRATION OF CU - NTA WITH KOH

33.0	0.013527	0.0030279	0.0045091
0.0	2.084	181.2	
0.2	2.155	178.2	
0.4	2.225	174.8	
0.6	2.303	171.2	
0.8	2.407	166.9	
1.0	2.521	161.4	
1.2	2.676	155.1	
1.3	2.779	150.5	
1.4	2.912	145.4	
1.6	3.359	131.5	
1.75	7.285	17.6	
1.80	7.790	4.8	
1.85	8.170	-3.8	
1.90	8.500	-11.5	
1.95	8.807	-18.4	
2.00	9.100	-26.1	
2.10	9.689	-46.7	
2.20	10.257	-71.9	
2.30	10.650	-87.9	
2.40	10.875	-97.1	
-1.00			
31.0	0.0144	0.0024174	0.0048
0.4	2.270	163.5	
0.6	2.365	159.2	
1.0	2.620	148.4	
1.2	2.815	141.0	
1.3	2.943	136.3	
1.4	3.125	131.0	

1.5	3.410	121.0
1.65	7.290	5.3
1.70	7.760	-7.1
1.75	8.120	-17.9
1.80	8.468	-28.8
1.85	8.839	-40.5
1.90	9.249	-55.0
1.95	9.665	-70.5
2.0	10.000	-82.9
2.10	10.465	-98.6
2.20	10.745	-107.5
2.3	10.935	-113.5
2.4	11.075	-117.9

```
-1.00

01 SET      1

02 TOTAL    1   2   3        -0.997

03TITR

05 ABS      111

10 ENDSET  1
```

         PHOTOMETRIC TITRATION AT 2-1 NTA-CU RATIO

50.0	0.02234	0.0104	0.01984				
0.65	0.640	0.631	0.606	0.567	0.511	0.443	0.378
0.292	0.220	0.155	0.102				
0.75	0.559	0.554	0.540	0.515	0.477	0.428	0.370
0.310	0.250	0.190	0.138				
0.85	0.478	0.479	0.477	0.465	0.444	0.412	0.370
0.326	0.278	0.224	0.172				
0.95	0.400	0.406	0.411	0.413	0.410	0.396	0.372
0.342	0.306	0.260	0.210				
1.05	0.330	0.343	0.359	0.370	0.380	0.381	0.377
0.360	0.332	0.293	0.240				
1.150	0.286	0.303	0.326	0.349	0.366	0.376	0.379
0.371	0.351	0.314	0.262				
1.25	0.302	0.319	0.343	0.367	0.382	0.390	0.388
0.364	0.349	0.308	0.256				
1.30	0.330	0.346	0.369	0.389	0.400	0.401	0.394
0.373	0.341	0.300	0.245				
1.40	0.380	0.398	0.416	0.430	0.434	0.427	0.402

```
0.370    0.330    0.280    0.222
1.50     0.436    0.446    0.462    0.471    0.470    0.449    0.410
0.365    0.313    0.260    0.202
1.653    0.491    0.500    0.512    0.518    0.504    0.469    0.418
0.359    0.297    0.236    0.178
1.90     0.542    0.551    0.560    0.559    0.535    0.490    0.422
0.351    0.280    0.212    0.155
2.2      0.572    0.580    0.588    0.584    0.555    0.500    0.429
0.350    0.271    0.200    0.140
3.0      0.598    0.601    0.608    0.600    0.565    0.502    0.424
0.340    0.257    0.183    0.122
4.0      0.598    0.600    0.606    0.596    0.560    0.500    0.418
0.332    0.250    0.176    0.117
-1.0
0.0
01 SET      1
02 TOTAL    1    2    3          -0.997
03TITR
05 ABS      111
10 ENDSET   1
```

PHOTOMETRIC TITRATION AT 1.25-1 NTA-CU RATIO

```
50.0     0.01396   0.0104   0.01240
0.65     0.610     0.603    0.582    0.550    0.500    0.440    0.372
0.301    0.234     0.172    0.121
0.70     0.580     0.577    0.560    0.530    0.490    0.439    0.375
0.310    0.248     0.189    0.136
0.80     0.561     0.560    0.551    0.531    0.498    0.450    0.390
0.330    0.266     0.204    0.150
0.90     0.562     0.565    0.563    0.550    0.520    0.472    0.410
0.344    0.278     0.211    0.153
0.95     0.570     0.571    0.572    0.561    0.530    0.482    0.420
0.352    0.281     0.215    0.156
1.05     0.580     0.584    0.588    0.580    0.552    0.500    0.432
0.360    0.285     0.217    0.154
1.10     0.589     0.592    0.596    0.591    0.560    0.509    0.437
0.360    0.281     0.211    0.151
1.20     0.602     0.609    0.615    0.610    0.578    0.515    0.441
0.360    0.278     0.205    0.144
```

1.25	0.619	0.621	0.628	0.621	0.586	0.526	0.445
0.360	0.275	0.202	0.141				
1.45	0.638	0.643	0.647	0.640	0.602	0.538	0.450
0.360	0.274	0.200	0.139				
2.00	0.650	0.654	0.658	0.649	0.608	0.540	0.450
0.360	0.270	0.195	0.132				
4.00	0.620	0.622	0.628	0.620	0.580	0.513	0.432
0.345	0.261	0.188	0.130				

```
-1.0

0.0

10 ENDDATA 1 2

        PH-METRIC,POTENTIOMETRIC AND PHOTOMETRIC STUDY OF CU-NTA SYSTEM

11 TASK     31210

     OH      4 1-1 0 0       -14.0

     HA      5-1 1 0 1        10.0

    H2A      6-1 2 0 1        12.0

    H3A      7-1 3 0 1        14.0

     MA      8-1 0 1 1        13.0

    MA2      9-1 0 1 2        17.0

   MAH-1 10-1-1 1 1            4.0

   MAH-2 11-1-2 1 1           -8.0

    MAH    12-1 1 1 1         15.0

17ADDITIVE 1    2             0.05

17 ADD     3    4             0.05     313.8

18MULTIPL  1    2            -1.0

18MULTIPL  3    4            -1.0      28.935

15 STDEV   1                                  1.0

10ENDTASK  1 2

                    CALCULATION  OF  PKW

11 TASK        10

14 SPECIES 4 0 5 1 6 1 7 1

15 STDEV   2                                  1.0

10ENDTASK  1 2

                CALCULATION  OF  PROTONATION  CONSTANTS

11 TASK        15 5    8 4 °4 111

14 SPECIES 4 0 5 0 6 0 7 0 8 1

14 SPECIES 9 110 111-312-2

15 STDDEV  1    2                             1.0
```

```
15 STDDEV  3   4                                        1.0        57.87
15 STDDEV  5   6                         0.0                                      0.3
10 ENDTASK 1 2
     EVALUATION  OF  ALL EXPERIMENTS   WITH  MODEL - SELECTION
          8 91011
11 TASK        15 0   8 4 4 111
14 SPECIES 4 1 5 1 6 1 7 1
15 STDDEV  1   2                                        1.0
15 STDDEV  3   4                                        1.0        57.87
15 STDDEV  5   6                         0.0                                      0.3
10 ENDTASK 1 2
               EVALUATION  OF  ALL EXPERIMENTS
          8 91011
82 STOP
```

5.3. Output from PSEQUAD

After all the experimental data have been printed out together with processing options required, the various parameters are refined together with the estimation of pK_w. The evaluation is continued with the refinement of the pK_a's of NTA using set 2, minimizing pH. Note that the first task provided the value of pK_w (13.7745) which is now used in the second and subsequent tasks. The next task is the calculation of the formation constants using sets 3–6. Only MA, MA_2, and MAH_{-1} are considered at this point in the calculations. The species MAH and MAH_{-2} are included only in subsequent tasks (-2 and -3 are assigned to these species on the 14 card). The absorbance data are included in these calculations. by using appropiate values for i6–i10 on the task 11 card. This task is entitled "Evaluation of all experiments with Model Selection."

It is our experience that the pCu measurements are about half as precise as the pH data absorbance measurements are about three times as precise as the pH data. This information is relayed to PSEQUAD using the 15 Card (STEDV) so that the function, F, to be minimized becomes

$$F = (\Delta pH)^2 + \frac{1}{57.87^2} (\Delta emf)^2 + \frac{1}{0.3^2} (\Delta A)^2$$

where

$$\frac{1}{57.87^2} (\Delta emf)^2 = \frac{1}{2^2} (\Delta pCu)^2$$

This function is defined by the second and third 15 card. It will be seen from the output that the inclusion MAH decreases F significantly from 0.00353 to 0.00213.

However, the inclusion of MAH_{-2} leads to only a slight decrease in the sum of squares from 0.00213 to 0.0021. The standard deviation for this latter constant is an order of magnitude higher compared to the other constants.

Finally all six sets of data are included in the refinement and F is further decreased to 0.00203. It should be noted that all the refined values of the constants are changed slightly in this final task as there is now no error accumulation from one task to the next.

9

STBLTY
Methods for Construction and Refinement of Equilibrium Models

ALEX AVDEEF

1. INTRODUCTION

STBLTY is an integrated collection of nine major FORTRAN programs suited for the construction, refinement, and assessment of general equilibrium models derived from potentiometric data. The extensive modular library of programs has been under active development since 1975, and has been tested on the CDC7600, IBM370/155, and IBM4341 large-mainframe computers. Applications of STBLTY have been varied and have included studies of ligands having as many as eight dissociable protons, Fe^{3+}–catechol[1] and –enterobactin[2] complexes, Pu^{4+}– and Th^{4+}–cathechol complexes,[3] polynuclear Cd^{2+}–penicillamine[4] and –cysteamine[5] complexes, metal-ion–cellulose interactions, and humic acid protonation reactions. Both ion-selective electrode and glass electrode data have been used in STBLTY calculations.

2. PROGRAM DESCRIPTION

Many of the technical terms used in this presentation are defined in Table 1. To illustrate some of the features of STBLTY, we have chosen the sufficiently general Cu^{2+}–diethylenetriamine (dien) system as a case study. The equilibrium reactions of interest consist of the sequences shown overleaf.

Parallel equlibria involving the protonation of the dien ligand are also present. The equilibrium quotients chosen in the present study are not in a commonly encountered form, a feature STBLTY can cope with:

$$K_1' = [Cu(dien\ H)^{3+}]\ [H^+]^2/[Cu^{2+}]\ [dien\ H_3^{3+}]$$

$$K_2' = [Cu(dien)^{2+}]\ [H^+]/[Cu(dien\ H)^{3+}]$$

ALEX AVDEEF • Syracuse University, Department of Chemistry, Syracuse, New York 13210.

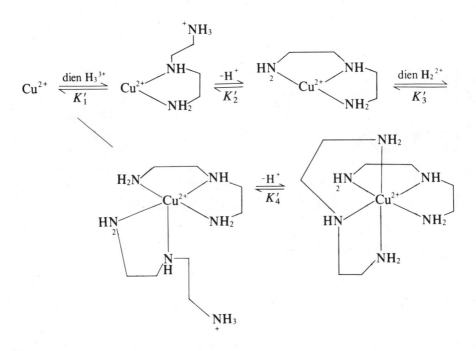

TABLE 1. Glossary of Terms

C_j	Concentration of the jth associated species: $C_j = [M_{e_{mj}}L_{e_{lj}}H_{e_{hj}}] = m^{e_{mj}}l^{e_{lj}}h^{e_{hj}}\beta_{e_{mj}e_{lj}e_{hj}}$
e_{kj}	Stoichiometric coefficient, referring to the number of kth type of reactant in the jth associated species. For example, for the jth species Cd(pen)$_2$ (OH), $e_{mj} = 1$, $e_{lj} = 2$, $e_{hj} = -1$. The value of e_{hj} is negative to signify a hydroxide. Positive values refer to protons.
$\bar{e}'_k$	Average stoichiometric coefficient of the kth reactant at a particular pH in *metal-containing* species.
H	Total hydrogen excess, defined as A − B + 3L, where A = [HClO$_4$], B = [NaOH]. The factor in front of L refers to the three dissociable protons introduced by the ligand.
h	Free hydrogen ion concentration, [H$^+$].
oh	Free hydroxide ion concentration, [OH$^-$].
$K_{w'}$	[H$^+$][OH$^-$] = $10^{-13.787}$ at 25°C and 0.1 M ionic strength.[26]
L	Total concentration of the ligand.
l	Concentration of the unassociated (and deprotonated) ligand.
M	Total concentration of the metal.
m	Concentration of the free metal.
pH,pL,pM	Negative of the logarithm, base 10, of the corresponding free concentrations.
pL$_0$,pM$_0$	Integration constants.[9–14]
V	Volume of titrant, ml.
β	Cumulative formation constant of the jth associated species, referring to the equilibrium expression $$e_{mj}M + e_{lj}L + e_{hj}H \rightleftharpoons M_{e_{mj}}L_{e_{lj}}H_{e_{hj}}$$

$$K_3' = [Cu(dien)\,(dien\,H)^{3+}]\,[H^+]/[Cu(dien)^{2+}][dien\,H_2^{2+}]$$

$$K_4' = [Cu(dien)_2^{2+}]\,[H^+]/[Cu(dien)\,(dien\,H)^{3+}]$$

The user may choose any form of constants as long as information is provided for the conversion of the input constants to the so-called "log β" form, which STBLTY uses for internal, user-transparent calculations:

$$\beta_{111} = [Cu(dien\,H)^{3+}]/[Cu^{2+}][dien][H^+] \qquad (\text{cf. } K_1')$$

$$\beta_{110} = [Cu(dien)^{2+}]/[Cu^{2+}][dien] \qquad (\text{cf. } K_2')$$

$$\beta_{121} = [Cu(dien)\,(dien\,H)^{3+}]/[Cu^{2+}][dien]^2[H^+] \qquad (\text{cf. } K_3')$$

$$\beta_{120} = [Cu(dien)_2^{2+}]/[Cu^{2+}][dien]^2 \qquad (\text{cf. } K_4')$$

The triplet-index subscript in the above expressions refers to the stoichiometric coefficients (Table 1) of the associated species, in the order metal–ligand–hydrogen. The matrix (called A in Section 4), which converts constants in the log K form to the log β form, is supplied by the user (or a default matrix may be evoked). For the Cu^{2+}–dien system A has the form

$$
\begin{bmatrix}
\log \beta_{011} \\
\log \beta_{012} \\
\log \beta_{013} \\
\\
\log \beta_{111} \\
\log \beta_{110} \\
\log \beta_{121} \\
\log \beta_{120}
\end{bmatrix}
=
\begin{bmatrix}
1\ 0\ 0 & 0\ 0\ 0\ 0 \\
1\ 1\ 0 & 0\ 0\ 0\ 0 \\
1\ 1\ 1 & 0\ 0\ 0\ 0 \\
\\
1\ 1\ 1 & 1\ 0\ 0\ 0 \\
1\ 1\ 1 & 1\ 1\ 0\ 0 \\
2\ 2\ 1 & 1\ 1\ 1\ 0 \\
2\ 2\ 1 & 1\ 1\ 1\ 1
\end{bmatrix}
\begin{bmatrix}
\log K_1^H \\
\log K_2^H \\
\log K_3^H \\
\\
\log K_1' \\
\log K_2' \\
\log K_3' \\
\log K_4'
\end{bmatrix}
$$

How has this matrix been derived? Each element of any row indicates the numbers of each log K values that need to be added together to give the equivalent log β value. The examples will illustrate the method.

By definition

$$\beta_{011} = [HL]/[H][L], \qquad \beta_{013} = [H_3L]/[H]^3[L]$$

$$K_1^H = [HL]/[H][L], \qquad K_2^H = [H_2L]/[HL][H], \qquad K_3^H = [H_3L]/[H_2L][H]$$

Therefore

$$\beta_{011} = K_1^H$$

or

$$\log \beta_{011} = 1 \cdot \log K_1^H$$

and

$$\beta_{013} = K_1^H \cdot K_2^H \cdot K_3^H$$

or

$$\log \beta_{013} = 1 \cdot \log K_1^H + 1 \cdot \log K_2^H + 1 \cdot \log K_3^H$$

Each column of A relates to a specific K-type constant and thus the value of log β is obtained by employing the matrix–vector multiplication rule. Hence for log β_{013}, the third row of A contains the entries 1, 1, 1, 0, 0, 0, 0, implying the right-hand side of the definition for log β_{013}.

As a second example consider the generation of a value for log β_{121}. By definition

$$\beta_{121} = [ML_2H]/[M][L]^2[H]$$

and

$$K_1' = [MLH][H]^2/[M][H_3L]$$

$$K_2' = [ML][H]/[MLH]$$

$$K_3' = [ML_2H][H]/[ML][H_2L]$$

Seeking a relationship between β_{121} and the K-type constants, in concentration terms, we have

$$\frac{[ML_2H]}{[M][L]^2[H]} = \frac{[ML_2H][H]}{[ML][H_2L]} \frac{[ML][H]}{[MLH]} \frac{[MLH][H]^2}{[M][H_3L]}$$

$$\times \left(\frac{[HL]}{[H][L]}\right)^2 \left(\frac{[H_2L]}{[HL][H]}\right)^2 \frac{[H_3L]}{[H_2L]\,[H]}$$

Or in terms of β_{121} and K's

$$\log \beta_{121} = 2 \cdot \log K_1^H + 2 \cdot \log K_2^H + 1 \cdot \log K_3^H + 1 \cdot \log K'$$
$$+ 1 \cdot \log K_2' + 1 \cdot \log K_3'$$

The relationship between K-form and β-form constants may appear awkward at first. However, we have encountered examples where refinement of constants directly

TABLE 2. Equilibrium Constants

		Correct constants[a]	Linear LS (Section 2.6)	Nonlinear LS: FICS-Data (Section 2.8)	Nonlinear LS: pH data (Section 2.7)
K-form	$\log K_1'$	−3.74	$(-3.87)^b$	−3.69 ± 0.02	−3.7460 ± 0.0009
	$\log K_2'$	−3.20	(−3.00)	−3.15 ± 0.02	−3.1897 ± 0.0007
	$\log K_3'$	−5.70	(−5.63)	−5.67 ± 0.02	−5.7010 ± 0.0008
	$\log K_4'$	−8.20	(−8.18)	−8.28 ± 0.02	−8.1975 ± 0.0008
	$\log K_1^H$	9.80			
	$\log K_2^H$	8.96			
	$\log K_3^H$	4.20			
β-form	$\log \beta_{111}$	19.22	19.09 ± 0.02		
	$\log \beta_{110}$	16.02	16.10 ± 0.01		
	$\log \beta_{121}$	29.08	29.32 ± 0.02		
	$\log \beta_{120}$	20.88	21.05 ± 0.01		
	$\log \beta_{011}$	9.80			
	$\log \beta_{012}$	18.76			
	$\log \beta_{013}$	22.96			

[a] "Correct" in the sense that these constants were used to generate the data (in Section 2.1). The values were determined[6] at 0.1 M ionic strength at 25°C; $K_j^H = [H_jL]/(H_{j-1}L][H]$.
[b] *K*-form constants obtained by multiplying refined *β*-form constants by $\mathbf{A}^{-1}$ (Section 2).

in the *β*-form was unsuccessful, owing to high correlations between the *β* constants. Redefining the constants to the appropriate *K*-form can improve markedly the refinement procedure in such cases.

The numerical values of the constants are listed in the first column of Table 2.

2.1. Computer Generation of Potentiometric Data and Distribution of Species

Ten potentiometric titration sets $(V, pH)_{M,L}$ were generated using the subprogram FANTASY CALCULATION; the "experimental" conditions assumed are listed in Table 3. Constants in the *K*-form were used in the simulation calculation.

It is worth noting that the input constants were automatically adjusted to accommodate minor, unavoidable changes in the ionic strength due to the presence of ionic

TABLE 3. Total Compositions[a]

Set no.	$[Cu(ClO_4)_2]$ (*m*)	[dien · 3HCl] (*m*)	Set no.	$[Cu(ClO_4)_2]$ (*m*)	[dien · 3HCl] (*m*)
M1	0.001	0.010	L1	0.003	0.006
M2	0.002	0.010	L2	0.003	0.008
M3[b]	0.003	0.010	L3[b]	0.003	0.010
M4	0.004	0.010	L4	0.003	0.012
M5	0.005	0.010	L5	0.003	0.014

[a] $[HClO_4] = 0.02\ M$; initial total solution volume 25 mL; $[NaClO_4] = 0.07\ M$; 1.0 M NaOH titrant.
[b] "Common point" (see Section 2.3).

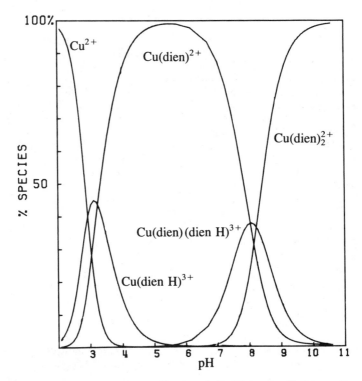

FIGURE 1. Distribution of species as a function of pH. "Common point" connections used.

species other than the supporting electrolyte. A Davies-modified Debye–Huckel equation is used in the adjustment.[1] In addition, the data were also corrected for the dilution effects introduced by the addition of titrant.

Additionally, after the equilibrium model is fully developed and refined, FANTASY CALCULATION can be used to calculate and plot distribution-of-species curves as a function of pH. Such a plot is shown in Figure 1 for our test case.

2.2. Data Reduction and Bjerrum Analysis

The data generated in Section 2.1 are treated by the subprogram DATA REDUCTION as though they were actually measured. The pH values are converted from pH-meter readings to values based on true concentrations of H^+. The pertinent calibration parameters for the above conversions are derived in the subprogram CALIBRATE ELECTRODE; we need not detail the algorithm, since it has already been described in the literature.[7] The total concentrations (Table 3) are corrected for dilution at each observation pair $(V, pH)_{M, L}$. The titration curves for the metal-varied series (sets M1–M5, Table 3) are drawn (Calcomp plotter) in Figure 2a. The ligand series (L1–L5) titration curves are drawn in Figure 2b.

A Bjerrum analysis is also performed.[8] The function $\bar{n}_L$, equation (1), (see Table 1)

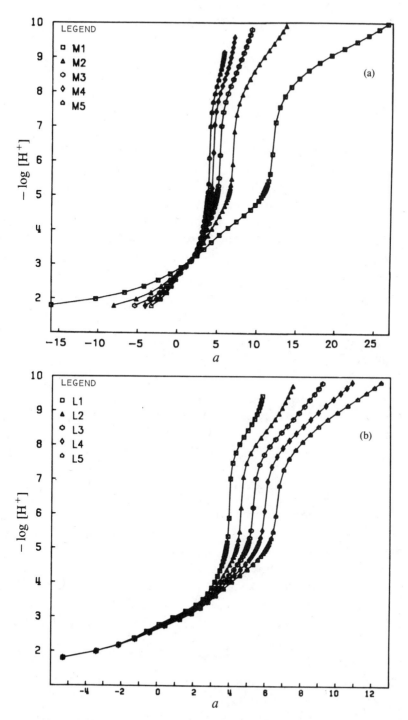

FIGURE 2. Titration curves: $(a, pH)_{M,L}$ where a = moles of base added per mole of metal ion present. (a) Metal-varied series of titrations (M1–M5); (b) ligand-varied series of titrations (L1–L5).

$$\bar{n}_L = [L - (H - (h - oh))/\bar{n}_H]/M \tag{1}$$

$$\bar{n}_H = \sum_{j=1}^{3} j\beta_{01j}h^j \Big/ \left(1 + \sum_{j=1}^{3} \beta_{01j}h^j\right) \tag{2}$$

is calculated. The free ligand concentration is derived[8]:

$$pL = -\log\left[(L - \bar{n}_L M)\Big/\left(1 + \sum_{j=1}^{3} \beta_{01j}h^j\right)\right] \tag{3}$$

If ternary complexes are not present, then $\bar{n}_L$ refers to the average number of metal-bound ligands per metal ion. If protonated metal–ligand complexes are present, then the meaning of $\bar{n}_L$ is complicated and pL calculated from it is not precisely equal to the free ligand concentration. However, the Bjerrum analysis is still a useful guide to the nature of the species present in solution, as is suggested by the following discussion.

The various Bjerrum plots $(\bar{n}_L, pL)$ and $(\bar{n}_L, pH)$ are shown in Figures 3a–3d. Due to the presence of the ternary complexes $[Cu(dienH)]^{3+}$ and $[Cu(dien)(dienH)]^{3+}$,

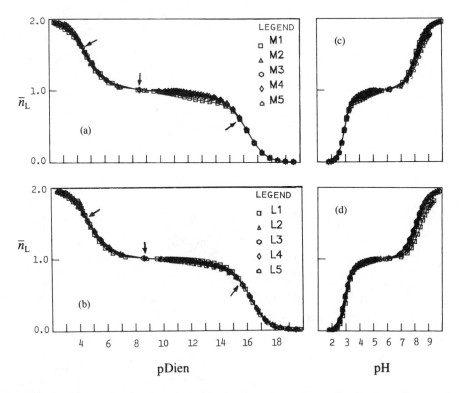

FIGURE 3. Bjerrum function [equation (1)] plotted against pL or pH. (a) $(n_L, pL)_L$, metal-varied series: (b) $(\bar{n}_L, pL)_M$, ligand-varied series; (c) $(\bar{n}_L, pH)_L$; (d) $(\bar{n}_L, pH)_M$.

the formation curves in Figures 3a and 3b depend systematically on total metal and total ligand concentrations. For pL > 16, all of the formation curves superimpose. (Stable polynuclear complexes would produce nonsuperimposable curves in this region in Figure 3a.[4]) At pL ~ 16 (pH ~ 3.2) the curves begin to spread. This is where the concentration of $[Cu(dienH)]^{3+}$ maximizes (~ 45% of the metal is in this form). The maximum spread is reached at pL ~ 13 (pH ~ 4.5), where both $[Cu(dienH)]^{3+}$ and $[Cu(dien)]^{2+}$ are present. The spreading appears to be proportional to L/M. When $[Cu(dien)]^{2+}$ is the predominant complex (pL ~ 8.5, pH ~ 5.6), the formation curves superimpose once more. In the interval pL ~ 8.5–4.5 (pH ~ 5.6–8.1) the curves again spread out, but to a lesser extent and in the *reverse* order of dependence on the total concentrations, since protonation of $[Cu(dien)]^{2+}$ is taking place with the addition of $dienH^+$, in contrast to the preceding deprotonation of $[Cu(dienH)]^{3+}$. When the concentration $[Cu(dien)(dien\,H)]^{3+}$ reaches a maximum (~ 39% of the metal is in this form at pL ~ 4.5, pH ~ 8.1), the curves again superimpose. For pL < 4.5, the curves diverge slightly as $[Cu(dien)_2]^{2+}$ begins to form. When 100% of the metal is in the latter form, the formation curves superimpose. This is quite typical for protonated complexes and the overall picture given by these plots thus provides information that there are more species than simply ML and ML_2.[4,5]

2.3. Free-Ion Concentration in Solution (FICS) Methodology

The FICS method (subprograms VARIATION and PX FROM VARIATION) has been described in the literature.[9–14] It can be used to derive pM and pL values *without any prior knowledge of the equilibrium model*, using only pH-based measurements. The pertinent equations are

$$pM = pM_0 - \int_{pH_0}^{pH} (\partial H/\partial M)_{pH,L} \, dpH|_M \qquad (4)$$

$$pL = pL_0 - \int_{pH_0}^{pH} (\partial H/\partial L)_{pH,M} \, dpH|_L \qquad (5)$$

The titration curves in Figure 2 were converted to the form $(H, pH)_L$ and $(H, pH)_M$ and are displayed in Figures 4a and 4b, respectively. Interpolation (using natural spline functions[15]) of the above titration curves at constant pH values produces the transformed data of the form $(H, M)_{L,H}$, Figure 5a, and $(H, L)_{M,h}$, Figure 5b. The discrete points in Figure 5 are fitted to deBoor's smoothing (p-factor) splines, assuming $\sigma(H) = 0.00005$.[16] The slopes of the smooth functions at concentrations corresponding to the titration sets M3 and L3 (the so-called "common points"[14]) are evaluated and plotted in Figure 6. These slope functions, when integrated, yield values of pM and pL, according to equations (4) and (5).

A plot of the FICS-derived pM and pL values is shown (discrete points) in Figure 7. The solid curves in the figure correspond to the correct pM and pL values (obtained from Section 2.1 calculation). The FICS-pM values are less precise for pH > 5. This kind of error has been discussed in the literature.[12–14] It undoubtedly is related to

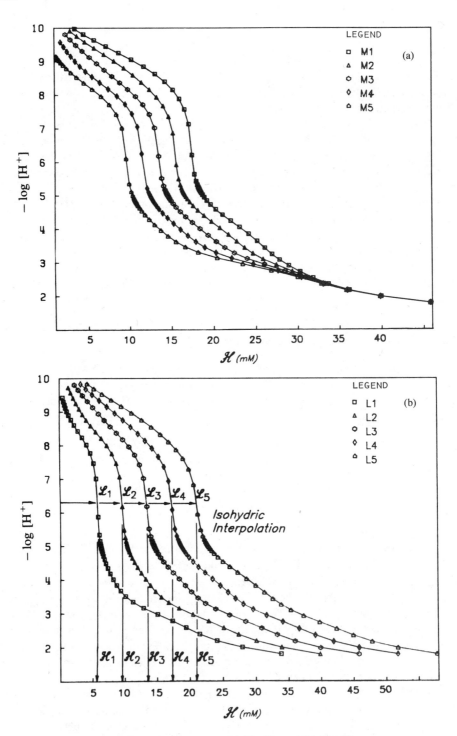

FIGURE 4. Titration curves: (a) $(H, pH)_L$ and (b) $(H, pH)_M$.

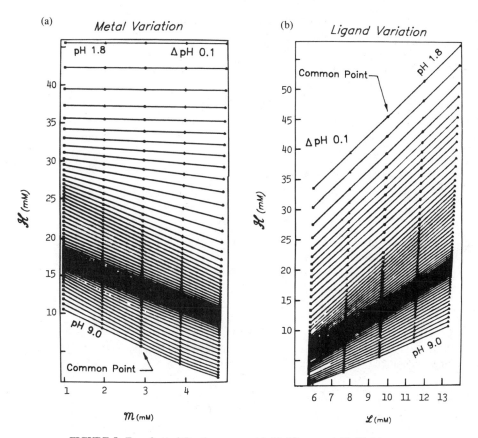

FIGURE 5. Transformed titration curves: (a) $(H, M)_{L,pH}$ and (b) $(H, L)_{M,pH}$.

interpolation difficulties, the nonadherence to constant ionic strength (as required by the FICS method), and dilution effects (invalidating the required constancies of certain total concentrations). Nonetheless, as will be demonstrated below, FICS-derived pM and pL data can be an extremely valuable aid in the construction of the equilibrium model.

2.4. Average Composition of Species in Solution (ACSS) Methodology

Having performed the FICS analysis in the preceding section, it is possible to use the pM and pL values to derive (using the subprogram STOICHIOMETRIC COEFS) the average stoichiometric coefficients of the metal-containing species, as is described elsewhere.[17, 18] Figure 8 shows the results for the Cu^{2+}–dien system. One can see that "noise" is a prominent feature, in spite of the "error-free" origin of the primary data. The ACSS analysis seems to be inescapably prone to the accumulation of errors in the various pretreatments. Still, the method has proved to be a valuable tool in

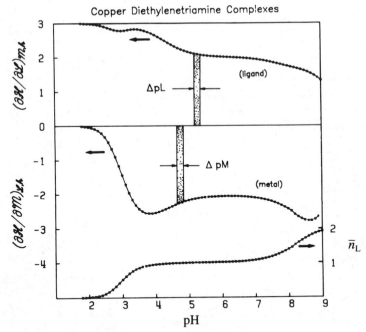

FIGURE 6. FICS slopes at the "common point" concentrations: $(\partial H/\partial L)_{M,pH}$, $(\partial H/\partial M)_{L,pH}$; the resultant McBryde[11] formation curve: $\bar{n}_L = -(\partial H/\partial M)_{L,pH}/(\partial H/\partial L)_{M,pH}$.

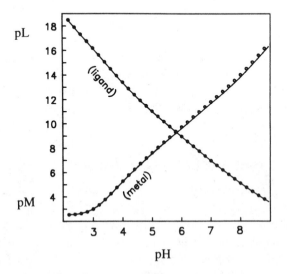

FIGURE 7. FICS-derived values of pM and pL (discrete points) compared to those derived in Section 2.1 (solid curves).

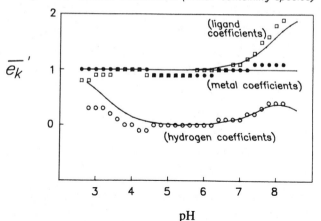

FIGURE 8. Average stoichiometric coefficients for metal-containing species. The solid curves are the correct coefficients (derived in Section 2.1) and the discrete points are those derived by the ACSS method (Section 2.4).

identifying the polynuclear complexes present in the Cd^{2+}–penicillamine system,[4] and the ternary complexes present in the Zn^{2+}– and Cd^{2+}–cysteamine systems.[5]

2.5. Linear Least-Squares Refinement of Constants

With knowledge of pM and pL values (FICS derived), the mass balance equations, equations (6–8), are linear (for an assumed set of stoichiometric coefficients) in

$$M = m + \sum_{j=1} e_{mj} C_j \tag{6}$$

$$L = l + \sum e_{lj} c_j \tag{7}$$

$$H = h - K'_w/h + \sum e_{hj} C_j \tag{8}$$

$$C_j = m^{e_{mj}} l^{e_{lj}} h^{e_{hj}} \beta_{e_{mj} e_{lj} e_{jh}} \tag{9}$$

the unknown β constants.[13] These constants are derived in the subprogram LINEAR LS REFINE. The least-squares function minimized is

$$R = \sum_{i}^{N_0} [(M_i^{obs} - M_i^{calc})^2 + (L_i^{obs} - L_i^{calc})^2 + (H_i^{obs} - H_i^{calc})^2] \tag{10}$$

where N_0 is the number of pH measurements in the "common point" sets M3 and L3. An option exists to use nonunit weights in equation (10). The algorithm NNLS[19] is used to prevent the calculation of negative β constants. The constants obtained by the linear procedure agree reasonably well with the correct values (Table 2).

2.6. Nonlinear Least-Squares Refinement of Constants: pH Data

The subprogram NONLINEAR LS REFINE is the most commonly used refinement program in STBLTY. Its use does not depend on FICS-processed data. One or
multiple titration sets may be processed in the same calculation step. Constants may
be refined in any so-called K-form. This is a particularly useful feature if refinement
of constants in β-form leads to near-unity correlation coefficients. The algorithm in
the nonlinear subprogram[1] differs significantly from those of SCOGS[20] and MINI-
QUAD.[21]

For the primary data $(V, \text{pH})_{M,L}$, the dependent variable is assumed to be pH,[22]
rather than the volume of titrant, in contrast to the approach used in the two above-
mentioned programs. Regression analysis entails the fundamental assumption that the
independent variables are error free but that the dependent variables are drawn from
a population having normally distributed (Gaussian) random errors. With a high-precision glass-piston buret, the regression effect of the errors in the volume of titrant
(< 0.001 ml) is much smaller than the effect of the uncertainties in the measured pH.
Hence, V is treated as an independent variable, as are the total concentrations. Thus
the function minimized is

$$R = \sum_{i}^{N_0} (\text{pH}^{\text{obs}} - \text{pH}^{\text{calc}})^2 / \sigma^2(\text{pH}) \tag{11}$$

where N_0 is the number of pH measurements and the variances are defined as[1]

$$\sigma^2(\text{pH}) = \sigma_c^2 + (\sigma_v \, d\text{pH}/dV)^2 \tag{12}$$

with typical values of $\sigma_c = 0.02$ and $\sigma_v = 0.001$ ml. The second variance component
in equation (12) depends on the slope of the titration curve and ascribes higher errors
to the measured pH values near end points, where titration curves have large values
of $d\text{pH}/dV$.

Since the refinement is nonlinear, one needs to assume initial estimates of the
quilibrium constants. Given their values and the accurate knowledge of the total concentrations, the values of pM, pL, and pH are iteratively calculated,[1] in much the
same way as in SCOGS and MINIQUAD. Noteworthy is that STBLTY calculates the
required derivatives $(d\text{pH}/d \log K)$ by an analytical (rather than numerical) procedure,[14,23] by utilizing the inverse Jacobian matrix used to calculate pM, pL, and pH.
This enables the program to operate 5–10 times faster, compared to the numerical
approach.

The calculated pH's are compared to those which are observed and the weighted
differences are the basis of the normal equations in the standard Gauss–Newton iterative least-squares procedure. Each cycle of refinement produces shifts in the estimated constants. The improved constants lower the value of R, equation (11). The
goodness-of-fit,

$$\text{GOF} = [R/(N_0 - N_c)]^{1/2} \tag{13}$$

where N_c is the number of varied constants, is used as an index to characterize the refinement. Usually the calculation converges after 2–3 cycles and the GOF is near zero if the equilibrium model is sufficiently valid and the observed data have the ascribed errors [equation (12)].

The case study calculation utilized all of the 10 sets of titration data ($N_0 = 429$). In the final cycle the average difference between pH^{obs} and pH^{calc} was 0.001. This is abnormally low, but then we were using "perfect" (Section 2.1) pH^{obs} data. For the same reason, GOF = 0.06. The constants (K-form) refined to values which are nearly identical to those assumed in the calculation in Section 2.1 (Table 2). The average total ionic strength was calculated to be 0.138 ± 0.013 M, with extrema being 0.18 and 0.11 M. During the calculations the equilibrium constants were adjusted slightly when the total ionic strength was calculated to be different from the reference value of 0.1 M.[1]

2.7. Nonlinear Least-Squares Refinement of Constants: pM, pL, and pH Data

Rather than using all of the 10 data sets and refining the constants on the basis of pH data, it is possible to use the FICS-derived data. (Ion-selective electrode data may also be the basis of the following approach.) The function minimized is

$$R = \sum_{}^{N_0} [(pM^{FICS} - pM^{calc})^2/\sigma^2(pM) + (pL^{FICS} - pL^{calc})^2/\sigma^2(pL) \\ + (pH^{obs} - pH^{calc})^2/\sigma^2(pH)] \tag{14}$$

where

$$\sigma^2(pM) = \sigma_c^2 + (\sigma_v \, dpM/dV)^2 \tag{15}$$

and

$$\sigma^2(pL) = \sigma_c^2 + (\sigma_v \, dpL/dV)^2 \tag{16}$$

with

$$\sigma_c = 0.05 \quad \text{and} \quad \sigma_v = 0.001 \text{ ml}$$

The refined (K-form) constants converged to the values listed in Table 2 and are quite close to the correct values. The differences result from the errors in FICS-generated pM and pL values. The GOF value of 0.83 suggests that FICS errors are not very different, on the average, from those encountered in experimentally measured data. It is interesting to note that the above constants are also slightly different from those determined by the linear procedure (Section 2.6). Insight into this difference may be gained from the discussion in Ref. 22. We are, in both situations, minimizing a *different* function [equations (10) and (14)].

2.8. Nonlinear Least-Squares Refinement of the Total Concentrations

NONLINEAR LS REFINE may be used to refine the scale factors associated with the total concentrations. The necessary derivatives are analytically calculated.[24] The approach has practical consequences when "real" data are considered.[22] In our present test case, the correlations between the total concentration scale factors and the equilibrium constants were very high and led to large errors in the refined quantities.

2.9. Analysis of Variance

The subprogram ANALYSIS OF VARIANCE may be executed after a LINEAR LS REFINE or NONLINEAR LS REFINE calculation. The object of the analysis is to search for any systematic dependencies between the nonlinear statistics $(pX^{obs} - pX^{calc})/\sigma(pX)$, $X = M, L, H$, or the linear statistics $(X^{obs} - X^{obs})/\sigma(X)$, $X = M, L, H$, and any observables (dependent or independent variables), such as ionic strength, pH, pM, pL, $(h + oh)$, m, l, dpH/dV, and dpL/dV. If the weights were properly assigned, there should be no systematic dependencies on any of these variables. If dependencies are found, means of changing the weighting scheme may be suggested. Numerous Calcomp plots are produced to aid the user in the process.

An Abrahams–Keve[25] normal probability plot is produced. This involves a comparison of the ordered distribution of the above-mentioned statistics to that expected from a normal, Gaussian distribution of the same sample size. If the weighting scheme is proper, such a plot should have a well-defined slope of unity and an intercept of zero.

The calculation in Section 2.8 was so analyzed. Figure 9 shows a plot of the

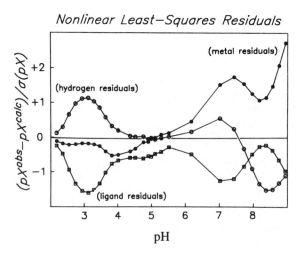

FIGURE 9. Weighted residuals as a function of pH resulting from the least-squares calculation described in Section 2.8, X = M, L, or H.

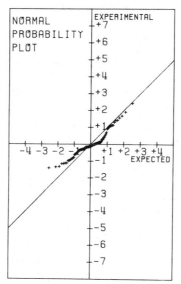

FIGURE 10. Abrahams–Keve normal probability plot for the weighted residuals resulting from the least-squares calculation described in Section 2.8.

statistics as a function of pH. Figure 10 shows the corresponding normal probability plot.

2.10. Acknowledgments

Useful discussions with K. N. Raymond, W. L. Smith, S. R. Sofen, and G. Christoph are gratefully acknowledged. I am especially grateful to Professor Raymond for his criticism, numerous specific suggestions, and encouragement which led to the initial coding of STBLTY. The project at Syracuse University was supported, in part, by the Petroleum Research Fund under Grant No. 11609-G3, administered by the American Chemical Society, and by the Syracuse University Research and Equipment Fund, for which I am grateful.

3. REFERENCES

1. A. Avdeef, S. R. Sofen, T. L. Bregante, and K. N. Raymond, Coordination Chemistry of Microbial Iron Transport Compounds. 9. Stability Constants for Catechol Models of Enterobactin, *J. Am. Chem. Soc.* **100**, 5362–5370 (1978).
2. W. R. Harris, C. J. Carrano, S. R. Cooper, S. R. Sofen, A. Avdeef, J. V. McArdle, and K. N. Raymond, Coordination Chemistry of Microbial Iron Transport Compounds. 19. Stability Constants and Electrochemical Behavior of Ferric Enterobactin and Model Complexes, *J. Am. Chem. Soc.* **101**, 6097–6104 (1979).
3. A. Avdeef, T. L. Bregante, and K. N. Raymond, Specific Sequestering Agents for the Actinides. Equilibrium Studies of Thorium (IV) 4-Nitrocatechol Complexes in Aqueous Solutions of Low Ionic Strengths, to be published.
4. A. Avdeef and D. L. Kearney, Cadmium Binding by Biological Ligands. 1. Formation of Protonated

Polynuclear Complexes Between Cadmium and D-Penicillamine in Aqueous Solution, *J. Am. Chem. Soc.* **104**, 7212–7219 (1982).

5. J. Brown, M.S. dissertation, Zinc and Cadmium Interactions in Aqueous Solution with β-Mercaptoethylamine: Model For Metallothionein, Syracuse University, (1981).

6. H. Haver, E. J. Billo, and D. W. Margerum, Ethylenediamine and Diethylenetriamine Reactions with Copper (II)-Triglycine, *J. Am. Chem. Soc.* **93**, 4173–4178 (1971).

7. A. Avdeef and J. J. Bucher, Accurate Measurements of the Concentration of Hydrogen Ions with a Glass Electrode: Calibrations Using the Prideaux and other Universal Buffer Solutions and a Computer-Controlled Automatic Titrator, *Anal. Chem.* **50**, 2137–2142 (1978).

8. H. M. Irving and H. S. Rossotti, The Calculation of Formation Curves of Metal Complexes from pH Titration Curves in Mixed Solvents, *J. Chem. Soc.* **1954**, 2904–2910.

9. R. Osterberg, The Copper(II) Complexity of O-Phosphorylethanolamine, *Acta Chem. Scand.* **14**, 471–485 (1960).

10. B. Sarkar and T. P. A. Kruck, Theoretical Considerations and Equilibrium Conditions in Analytical Potentiometry. Computer Facilitated Mathematical Analysis of Equilibria in a Multicomponent System, *Can. J. Chem.* **51**, 3541–3548 (1973).

11. W. A. E. McBryde, On an Extension of the Use of pH-Titrations for Determination of Free Metal and Free Ligand Concentrations During Metal Complex Formation, *Can. J. Chem.* **51**, 3572–3576 (1973).

12. R. Guevremond and D. L. Rabenstein, A Study of the Osterberg–Sarkar–Kruck Method for Evaluating Free Metal and Free Ligand Concentrations in Solutions of Complex Equilibria, *Can. J. Chem.* **55**, 4211–4221 (1977).

13. T. B. Field and W. A. E. McBryde, Determinations of Stability Constants by pH Titrations: A Critical Examination of Data Handling, *Can. J. Chem.* **56**, 1201–1211 (1978).

14. A. Avdeef and K. N. Raymond, Free Metal and Free Ligand Concentrations Determined from Titrations Using Only a pH Electrode. Partial Derivatives in Equilibrium Studies, *Inorg. Chem.* **18**, 1605–1611 (1979).

15. R. L. Burden, J. D. Faires, and A. C. Reynolds, *Numerical Analysis*, Prindle, Weber and Schmidt, Boston (1978), pp. 116–128.

16. C. deBoor, *A Practical Guide to Splines*, Springer-Verlag, New York (1978), pp. 235–249.

17. A. Avdeef, Composition of Ternary (Metal–Ligand–Hydrogen) Complexes Experimentally Determined from Titrations Using Only a pH Electrode, *Inorg. Chem.* **19**, 3081–3086 (1980).

18. L. C. VanPoucke, J. Yperman, and J. P. Francois, Experimental Determination of the Complexity Sum in Investigating Metal Ion Complex Formation, *Inorg. Chem.* **19**, 3078–3081 (1980).

19. C. L. Lawson and R.J. Hanson, *Solving Least Squares Problems*, Prentice Hall, Englewood Cliffs, New Jersey (1974), pp. 269–275, 304–311.

20. I. G. Sayce, Computer Calculation of Equilibrium Constants of Species Present in Mixtures of Metal Ions and Complexing Agents, *Talanta* **15**, 1397–1411 (1968).

21. A. Sabatini, A. Vacca, and P. Gans, MINIQUAD—A General Computer Programme for the Computation of Formation Constants from Potentiometric Data, *Talanta* **21**, 53–77 (1974).

22. L. Meites, J. E. Stuehr, and T. N. Briggs, Simultaneous Determination of Precise Equivalence Points and pK Values from Potentiometric Data: Single pK Systems, *Anal. Chem.* **47**, 1485–1486 (1975).

23. I. Nagypal, I. Paka, and L. Zekany, Analytical Evaluation of the Derivatives Used in Equilibrium Calculations, *Talanta* **25**, 549–550 (1978).

24. A. Avdeef, unpublished results (1981).

25. S. C. Abrahams and E. T. Keve, Normal Probability Plot Analysis of Error in Measured and Derived Quantities and Standard Deviations, *Acta Crystallogr.* **A27**, 157 (1971).

26. F. H. Sweeton, R. E. Mesmer, and C. F. Bases, Jr., Acidity Measurements at Elevated Temperatures. 7. Dissociation of Water, *J. Solution Chem.* **3**, 191–214 (1974).

4. DESCRIPTION OF BATCH INPUT PROTOCOL

Since STBLTY is a batch-oriented program, the complete input deck must be prepared before the programs can be executed. The explanations of input requirements are given in Sections 4.1–4.11. Within each section there are a number of items designated 1, 2, etc. In most situations one item corresponds to one card. Obvious exceptions to this would include an item comprising the titration data, one point per card. The format requirements are given in square brackets, e.g., [20A4], and followed by the variable name list. Only those variable names printed in boldface type require input values for that section. However, the full variable list has been included so that field locations can be determined. The location of each READ statement (routine and line number) has been included at the end of each section. Sample sets of data have been included for the more commonly used features of STBLTY.

Molar units of concentration are used throughout; volumes are assumed to be in milliliters. The current dimension limitations allow up to 600 pH measurements to be treated simultaneously. Up to four different types of reactants (including the hydrogen ion) may be considered. Primary data cards must be ordered in ascending pH within each titration set. A titration set is defined as a collection of pH (and/or pM) measurements with the same total nonhydrogen reactant concentrations—for example, data one obtains in a single alkalimetric titration.

Reactants are entered and addressed in the order metal (if present), before ligand (if present), before hydrogen. We may term this the standard "stoichiometric" ordering.

The default logical unit numbers for input/output devices are 5 = card reader, 6 = line printer, and 7 = card punch. The user may redefine these assignments by altering the values of ICRD, IPCH, and IPRT in the main program.

The descriptions of the input decks below have been abridged in order to minimize the length of the current presentation. There are many options which are not explicitly described. The Calcomp plotting subroutines, and the subprograms CALIBRATE ELECTRODE (7) and GRAPHICAL CONSTANTS (which consists of a specialized application of LINEAR LS REFINE) are not in this listing of the program. An 80-page manual describing the complete protocol may be obtained from us. It is not likely, however, that most prospective users of STBLTY would require the complete protocol manual.

4.1. FANTASY Calculation

4.1.1. Instructions for Data Input

The input file will consist of the following lines:
1. 1 line [20A4]: **TITLE**
   A descriptive title for the system under consideration.
2. 1 line [20A4]: **FMT**
   FMT: Not used in this section. Include a blank card.
3. 1 line [3A4]: **(YNAME (I), I = 1, NC)**
   Names of reactants (in the standard "stoichiometric" order).
4. 1 line [5A4,2F5.0,4I1,3F2.0,1X,4I1,1X,4A1,1X,4A1,4X,I2,8I1,1X,2F5.0]:
   (CODEN(I),I=1,5),TEMP,UCORR,NM,NL,IACT,ISORT,

(QC(J),J = 1,3),(NH(I),I = 1,4),(IHM(I),I = 1,4),(LIGS(I),I = 1.4),
NSET,**IPUN,LONG,IPLOT,**IDAT,**IPAR,IGUESS,**TAPIN,TAPOUT,
PH1,PH2

CODEN: Code word(s) used to determine data processing option.
In this instance use FANTASY.

TEMP: Temperature of the titration cell.

UCORR: Reference ionic strength.

NM: Number of reactants that are metals.

NL: Number of ligands.

IACT: ⎫
  ·   ⎪
  ·   ⎬   Not used in this section.
  ·   ⎪
ISORT: ⎭

QC():   Charge on unassociated ("free") form of first, second, and third
reactant. There will be (NM + NL) values.

NH(1),NH(2), . . .:   Maximum number of dissociable protons on first
ligand, on second ligand, *etc*. There will be NL values.

IHM(1),IHM(2), . . .:   Number of protons on ligand in the form in which
it was introduced into solution. There will be NL values.

LIGS(1),LIGS(2), . . .:   Maximum number of ligands bound to the metal
ion. There will be NL values.

NSET: Number of titration sets to be considered. Not used in this section.

IPUN: Control for punched output. IPUN = 1 for a deck of V,pH cards;
IPUN = 0 otherwise.

LONG: Signal which controls the quantity of printed output. LONG = 5
usually.

IPLOT: Control for plotted output. IPLOT = 5 for Calcomp plots;
IPLOT = 1 for line-printer plots.

IDAT: Not used in this section.

IGUESS: Control integer used to indicate the presence (= 1) or absence (= 0) of initial values for free component concentrations. If equal
to one, item 7 is required. If equal to zero, omit item 7.

PH1:   Minimum pH value for data generation.

PH2:   Maximum pH value for data generation.

5. 1 line   [6F5.0]:   **SCALEX, SCALEY, SCALEN, DELPH,** SHRINK, DM

SCALEX: Scale factor (in./volume unit) for titration curve plot in x
direction.

SCALEY: Scale factor (in./pH) for titration curve plot in y direction.

SCALEN: Scale factor (in./pL) for Bjerrum plot in x direction.

Note: SCALEX, SCALEY, and SCALEN are only used when IPLOT = 5.

DELPH: pH interval between generated data.

6. 1 line   [5F10.0]:   **VO, ACID, BASE, SE,** (TOTX(I),I = 1,NC)

VO: Initial volume of solution.

ACID: Concentration of mineral acid already in solution.

BASE: Titrant concentration (mineral base).

SE: Concentration of supporting electrolyte.

TOTX(I): Total concentration of Ith reactant. There will be NC($=$ NM + NL) values. They will be in the same order as dictated by the third item, YNAME(1), YNAME(2) . . . .

7. 1 line [4F10.0]: (PCF(J,1),J=1,NC)

NC values for uncomplexed (free) concentrations for the first reactant, second reactant . . . . There will be NC values.

This card is present *only* if IGUESS $\neq$ 0.

8. 1 line [I1,A1,4I2,F10.0,3F5.0,7A4]: **IS, KI(I), (IX(J,I),J=1,4), BL(I),SL(I),DP(I),SHFT(I),(LABP(I,J),J=1,7).**

IS: For last constant (K-form) in set, IS$=$9. Otherwise IS$=$0.

IX(1,I),IX(2,I), . . .: Number of molecules of reactant 1 in complex I, number of molecules of reactant 2 in complex I . . .

BL(I): Log_{10} of the input K-form constant.

LABP(I,J): A short title (28 characters) that may be used to describe the Ith constant.

Note there will be N cards where the Nth card will contain a 9 in column 1.

9. 1 line [2012]:**(ICB(I,J),J=1,NSP)**

ICB will hold the conversion mark A, where $\log \beta_i = \sum_j A_{ij} \log K'_j$. One row of the matrix is entered per card; there are as many cards as constants, in item 8. If all of the constants are entered in β-form then ICB is a unit matrix. See Section 2.

10. 1 line [5A4, . . .]: **(CODEN(I), I=1,5),** . . .

This represents the final card for the particular data processing option. It may be END or another key word to initiate another type of data processing. END will determine execution of STBLTY.

READ statements for this section are located as follows:

Item	Routine	Line No.
1	MAIN	92
2	MAIN	92
3	MAIN	93
4	MAIN	94–96
5	FANTASY	39
6	FANTASY	41
7[a]	FANTASY	42
8	READP	34
9	READP	53
10	MAIN	94–96

[a]Read only if IGUESS $\neq$ 0.

4.1.2. Input data for FANTASY CALCULATION

There follows a typical data set for a fantasy calculation for the copper(II)–dien system.

```
¶<-- Column 1

CU2+ DIETHYLENETRIAMINE EQUILIBRIUM REACTIONS

A Blank Card Goes Here

CU2+DIENH+

FANTASY CALCULATION  25.   .1  115 +2 0+1 3000 3000 2000      055100    2.1 10.9

   .2   .75   .2  .1    .3  1.

   25.        .02       1.        .07        .003      .01

     0 1 1   9.8                    LOG K1 DIEN

     0 1 2   8.96                   LOG K2 DIEN

     0 1 3   4.2                    LOG K3 DIEN

     1 1 1   -3.74                  100+013=111+2(001)

     1 1 0   -3.2                   111=110+001

     1 2 1   -5.7                   110+012=121+001

9  1 2 0   -8.2                     121=120+001

   1 0 0 0 0 0 0

   1 1 0 0 0 0 0

   1 1 1 0 0 0 0

   1 1 1 1 0 0 0

   1 1 1 1 1 0 0

   2 2 1 1 1 1 0

   2 2 1 1 1 1 1

END
```

4.2. DATA REDUCTION

4.2.1. Instructions for Data Input

1–3.   As Section 4.1, items 1–3.
4. 1 line [As Section 4.1, item 4]:
              (CODEN(I),I=1,5),TEMP,UCORR,NM,NL,IACT,ISORT,

(QC(J),J=1,3),(NH(I),I=1,4),(IHM(I),I=1,4),
(LIGS(I),I=1,4), NSET, IPUN, LONG, IPLOT, IDAT,
IPAR, IGUESS, TAPIN, TAPOUT, PH1, PH2
CODEN: DATA REDUCTION

TEMP:⎫
  ⋮   ⎬ As for Section 4.1, item 4.
  ⋮   ⎪
NL: ⎭

IACT: Signal for choice of pH, pM calibration scheme. Choose IACT=5.

ISORT:⎫
  ⋮   ⎬ As for Section 4.1, item 4.
  ⋮   ⎪
IHM( ):⎭

LIGS(1),LIGS(2),LIGS(3),LIGS(4): 0,0,5,1

NSET: Number of titrations to be processed.

IPUN: SET IPUN = 1 to product (V,pH) cards. Useful for Section 4.6. Otherwise set to zero.

LONG: LONG=5 produces normal printed output; LONG=0 produces a shortened version of the output.

IPLOT: Setting IPLOT=5 produces Bjerrum function Calcomp. IPLOT=0 suppresses plotting.

IGUESS: Enter 1 for normal $\bar{n}$ calculation.

TAPIN: and TAPOUT: Not used in this section.

PH1: Minimum pH value to be considered in the refinement.

PH2: Maximum pH value to be considered in the refinement.

The following items, 5–9, are entered NSET times.

5. 1 line [2I1,4X,A4,7F10.0]:
ITITR,MV,LABEL,VO,ACID,BASE,STOCK,SE,VCOR,CAB(N)

ITITR: For base titrant, ITITR=0. For acid titrant ITITR=1.

MV: For metal ISE used, MV=1. Otherwise MV=0.

LABEL: A four-character identification label for the titration set.

VO: Initial volume of solution.

ACID: If ITITR=0 then ACID is initial mineral acid concentration. If ITITR=1 then ACID is titrant concentration.

BASE: If ITITR=0 then BASE is titrant concentration. Otherwise BASE is initial mineral base concentration.

SE: Initial concentration of supporting electrolyte insolution.

6. 1 line [4F10.6]: (TOTX(I),I=1,NC)
TOTX(1),TOTX(2), ...: total concentration of reactant 1, total concentration of reactant 2, ... There will be NC values.

7. 1 line [F10.6]: DELPH
As for Section 4.1, item 5.

8. 1 line [8F10.6]: (BO(I),I=1,4),(DO(I),I=1,4)

BO(I),BO(I),BI(3),BO(4): Calibration parameters for pH meter readings converting them to p[H$^+$],[7] where

$$pH_{meter} = BO(1) + BO(2)*p[H^+] + BO\ (3)*[H^+]$$
$$+ BO\ (4)*[OH^-]$$

DO(1),DO(2),DO(3),DO(4): Calibration parameters for metal ion selective electrode, if used.

$$pM_{meter} = DO(1) + DO(2)*p[M]$$

At present DO(3) and DO(4) have not been required.

9. 1 line   [I1,F9.0,F10.0,7F7.3]:
IS,VT,PCF(NC1,I),EO,SLO,PHO,PCF(1,I),EOM,SLOM,PROM
Primary data are read in at this point, one point per card.
IS: Control integer to indicate end of data (=1). Otherwise IS=0.
VT: Titrant volume.
PCF(NC1,I):   pH meter reading, for Ith point.
PCF(1,I):   pM meter reading, if used.

10. 1 line   [I1,A1,4I2,F10.0,3F5.0,7A4]:
IS,KI(I),(IX(J,I),J=1,4),BL(I),SL(I),DP(I),SHFT(I),
(LABP(I,J),J=1,7)
Ligand constants are read in here, *only* if there are at least three reactants and the *first* reactant is a metal ion.
IS: Control integer to signal final constant card (IS=1).
Otherwise IS=0.
IX(1,I),IX(2,I),IX(3,I),IX(4,I):   Set IX(1,I) and IX(4,I) to zero.
Set IX(2,I) to 1 (implies one ligand) and IX(3,I) to j for H$_j$L.
BL(I):   Value of log K_j^H in order log K_1^H, log K_2^H, ...
LABP(I,J):   A suitable label for each constant.

11. 1 line   [5A4, ...]:   **(CODEN(I),I=1,5), ...**
Final card, i.e., END (See Section 4.1, item 9).

READ statements for this section are located as follows:

Item	Routine	Line No.
1	MAIN	92
2	MAIN	92
3	MAIN	93
4	MAIN	94–96
5	REDUCE	86
6	REDUCE	90
7	REDUCE	91
8[a]	REDUCE	102

Item	Routine	Line No.
—[a]	REDUCE	108
9	REDUCE	127
10[b]	READP	34
11	MAIN	94–96

[a]Read depending on value of IACT.
[b]READP called only if NM > 0.

4.2.2. Input Data for DATA REDUCTION

There follows a data set derived from copper(II)–dien titrations. The first half of the data set is for the variation of the metal concentration.

```
¶<--  Column 1

CU2+ DIEN       METAL VARIED

Blank Card Goes Here.

CU2+DIENH+

DATA REDUCTION      25.  .1  1150+2 0+1 3000 3000   21    5053 0112  1.7  10.

      M1   25.      .02      1.                   .07

.001      .01

.01

.000      1.000     0.000     0.000

  0.10040   1.79673-15.984  3.005 19.649  1.7970.001000.009960.04580

  0.24263   1.98787-10.295  3.011 19.078  1.9880.000990.009900.03991

  0.33471   2.17479 -6.612  3.023 18.520  2.1750.000990.009870.03613

  0.39628   2.35733 -4.149  3.047 17.977  2.3570.000980.009840.03362

  0.44146   2.53934 -2.342  3.099 17.438  2.5390.000980.009830.03178

  0.47956   2.72531 -0.818  3.207 16.892  2.7250.000980.009810.03024

  0.51318   2.90869  0.527  3.399 16.358  2.9090.000980.009800.02888

  0.54074   3.08208  1.629  3.671 15.855  3.0820.000980.009790.02777

  0.56332   3.25413  2.532  4.014 15.357  3.2540.000980.009780.02686

  0.58454   3.44233  3.381  4.444 14.815  3.4420.000980.009770.02601

  0.60722   3.64826  4.288  4.947 14.232  3.6480.000980.009760.02510

  0.63235   3.85766  5.294  5.472 13.654  3.8580.000980.009750.02410

  0.65922   4.06004  6.368  5.978 13.115  4.0600.000970.009740.02302

  0.68584   4.25138  7.433  6.446 12.626  4.2510.000970.009730.02196

  0.70988   4.42913  8.395  6.868 12.191  4.4290.000970.009720.02101

  0.72973   4.59090  9.189  7.241 11.811  4.5910.000970.009720.02022

  0.74491   4.73359  9.796  7.560 11.486  4.7340.000970.009710.01962
```

```
      0.75585    4.85440 10.234  7.825 11.218  4.8540.000970.009710.01919

      0.76341    4.95220 10.537  8.037 11.005  4.9520.000970.009700.01889

      0.76852    5.02777 10.740  8.198 10.843  5.0280.000970.009700.01869

      0.77196    5.08492 10.878  8.319 10.721  5.0850.000970.009700.01855

      0.77433    5.12772 10.973  8.409 10.630  5.1280.000970.009700.01846

      0.77607    5.16147 11.043  8.480 10.559  5.1610.000970.009700.01839

      0.77754    5.19157 11.101  8.543 10.496  5.1920.000970.009700.01833

      0.77903    5.22417 11.161  8.611 10.428  5.2240.000970.009700.01827

      0.78084    5.26665 11.234  8.700 10.339  5.2670.000970.009700.01820

      0.78329    5.32979 11.331  8.831 10.208  5.3300.000970.009700.01810

      0.78677    5.43559 11.470  9.049  9.989  5.4360.000970.009690.01796

      0.79175    5.64103 11.670  9.469  9.569  5.6410.000970.009690.01777

      0.79875    6.21027 11.949 10.626  8.420  6.2100.000970.009690.01749

      0.80829    7.13155 12.331 12.544  6.588  7.1320.000970.009690.01712

      0.82079    7.57118 12.831 13.547  5.730  7.5710.000970.009680.01662

      0.83669    7.88205 13.467 14.331  5.136  7.8820.000970.009680.01600

      0.85647    8.14873 14.258 15.065  4.638  8.1490.000970.009670.01522

      0.88082    8.39510 15.233 15.784  4.194  8.3950.000970.009660.01426

      0.91026    8.62744 16.410 16.471  3.798  8.6270.000960.009650.01311

      0.94462    8.84737 17.784 17.098  3.452  8.8470.000960.009640.01177

      0.98277    9.05663 19.311 17.648  3.158  9.0570.000960.009620.01028

      1.02283    9.25695 20.912 18.114  2.913  9.2570.000960.009610.00873

      1.06256    9.44939 22.502 18.496  2.715  9.4490.000960.009590.00719

      1.09979    9.63345 23.992 18.797  2.560  9.6330.000960.009580.00576

      1.13280    9.80757 25.311 19.025  2.443  9.8080.000960.009570.00449

1   1.16075    9.97113 26.429 19.194  2.357  9.9710.000960.009560.00341

          M2   25.      .02    1.                 .07

   .002     .01

   .01

   .000     1.000    0.000    0.000

      0.10075    1.79673 -7.985  2.704 19.653  1.7970.001990.009960.04579

      0.24349    1.98792 -5.130  2.710 19.082  1.9880.001980.009900.03987

      0.33686    2.17523 -3.263  2.722 18.524  2.1750.001970.009870.03604

      0.40158    2.35993 -1.969  2.746 17.976  2.3600.001970.009840.03340

      0.45342    2.54797 -0.932  2.799 17.424  2.5480.001960.009820.03130

      0.50289    2.74184  0.058  2.913 16.864  2.7420.001960.009800.02930

      0.55048    2.93133  1.010  3.111 16.323  2.9310.001960.009780.02738

      0.58992    3.10786  1.798  3.386 15.821  3.1080.001950.009770.02580
```

0.61971	3.27483	2.394	3.717	15.344	3.2750.001950.009760.02460
0.64431	3.44804	2.886	4.109	14.849	3.4480.001950.009750.02362
0.66798	3.63873	3.359	4.573	14.310	3.6390.001950.009740.02268
0.69261	3.83985	3.852	5.075	13.755	3.8400.001950.009730.02170
0.71801	4.03850	4.360	5.572	13.224	4.0390.001940.009720.02069
0.74268	4.22721	4.854	6.035	12.739	4.2270.001940.009710.01971
0.76473	4.40184	5.294	6.451	12.310	4.4020.001940.009700.01883
0.78283	4.55913	5.656	6.815	11.937	4.5590.001940.009700.01812
0.79663	4.69596	5.932	7.124	11.624	4.6960.001940.009690.01758
0.80657	4.81032	6.131	7.376	11.368	4.8100.001940.009690.01718
0.81346	4.90147	6.269	7.574	11.168	4.9010.001940.009680.01691
0.81814	4.97163	6.363	7.725	11.016	4.9720.001940.009680.01673
0.82134	5.02463	6.427	7.838	10.902	5.0250.001940.009680.01660
0.82362	5.06535	6.472	7.925	10.816	5.0650.001940.009680.01651
0.82542	5.09975	6.508	7.997	10.742	5.1000.001940.009680.01644
0.82708	5.13355	6.541	8.068	10.671	5.1340.001940.009680.01638
0.82896	5.17432	6.579	8.154	10.585	5.1740.001940.009680.01630
0.83139	5.23198	6.628	8.274	10.464	5.2320.001940.009680.01621
0.83480	5.32543	6.696	8.468	10.270	5.3250.001940.009680.01607
0.83971	5.49819	6.794	8.824	9.914	5.4980.001940.009680.01588
0.84672	5.92187	6.934	9.686	9.054	5.9220.001930.009670.01560
0.85647	6.86097	7.129	11.601	7.181	6.8610.001930.009670.01522
0.86945	7.35523	7.389	12.654	6.218	7.3550.001930.009660.01471
0.88584	7.67394	7.717	13.380	5.612	7.6740.001930.009660.01407
0.90529	7.93070	8.106	14.010	5.136	7.9310.001930.009650.01331
0.92727	8.15954	8.545	14.614	4.721	8.1600.001930.009640.01245
0.95145	8.37715	9.029	15.223	4.338	8.3770.001930.009630.01150
0.97794	8.59158	9.558	15.838	3.975	8.5920.001920.009620.01047
1.00677	8.80365	10.135	16.438	3.640	8.8040.001920.009610.00935
1.03744	9.01019	10.749	16.987	3.343	9.0100.001920.009600.00816
1.06888	9.20934	11.378	17.462	3.092	9.2090.001920.009590.00695
1.09968	9.40002	11.993	17.855	2.887	9.4000.001920.009580.00576
1.12840	9.58116	12.568	18.168	2.726	9.5810.001910.009570.00465
1.15389	9.75131	13.078	18.407	2.603	9.7510.001910.009560.00367
1 1.17571	9.91031	13.514	18.586	2.512	9.9100.001910.009550.00284

```
        M3  25.     .02    1.           .07
 .003      .01
 .01
```

```
 .000      1.000     0.000     0.000

0.10106   1.79672 -5.319   2.528 19.656   1.7970.002990.009960.04577

0.24434   1.98798 -3.409   2.534 19.086   1.9880.002970.009900.03984

0.33899   2.17568 -2.147   2.546 18.527   2.1760.002960.009870.03595

0.40673   2.36210 -1.244   2.570 17.977   2.3620.002950.009840.03319

0.46490   2.55401 -0.468   2.624 17.418   2.5540.002950.009820.03083

0.52492   2.75122  0.332   2.737 16.857   2.7510.002940.009790.02841

0.58578   2.94346  1.144   2.933 16.320   2.9430.002930.009770.02596

0.63741   3.12324  1.832   3.205 15.820   3.1230.002930.009750.02389

0.67525   3.28997  2.337   3.528 15.352   3.2900.002920.009740.02239

0.70376   3.45464  2.717   3.896 14.886   3.4550.002920.009730.02125

0.72865   3.63186  3.049   4.323 14.388   3.6320.002920.009720.02026

0.75285   3.82238  3.371   4.798 13.861   3.8220.002910.009710.01931

0.77681   4.01490  3.691   5.279 13.345   4.0150.002910.009700.01836

0.79954   4.19947  3.994   5.733 12.869   4.1990.002910.009690.01746

0.81960   4.36984  4.261   6.141 12.447   4.3700.002900.009680.01667

0.83594   4.52185  4.479   6.495 12.084   4.5220.002900.009680.01603

0.84836   4.65241  4.645   6.792 11.782   4.6520.002900.009670.01554

0.85729   4.75981  4.764   7.030 11.540   4.7600.002900.009670.01519

0.86350   4.84442  4.847   7.216 11.353   4.8440.002900.009670.01494

0.86774   4.90858  4.903   7.355 11.212   4.9090.002900.009660.01478

0.87067   4.95737  4.942   7.460 11.107   4.9570.002900.009660.01466

0.87283   4.99561  4.971   7.541 11.024   4.9960.002900.009660.01458

0.87462   5.02929  4.995   7.613 10.952   5.0290.002900.009660.01451

0.87639   5.06507  5.019   7.689 10.876   5.0650.002900.009660.01444

0.87851   5.11034  5.047   7.784 10.780   5.1100.002900.009660.01436

0.88136   5.17813  5.085   7.927 10.637   5.1780.002900.009660.01424

0.88544   5.29216  5.139   8.164 10.399   5.2920.002900.009660.01408

0.89134   5.51872  5.218   8.630  9.932   5.5190.002900.009660.01385

0.89979   6.17909  5.330   9.970  8.598   6.1790.002900.009650.01352

0.91154   7.01755  5.487  11.676  6.941   7.0180.002890.009650.01306

0.92720   7.42591  5.696  12.535  6.161   7.4260.002890.009640.01245

0.94688   7.71768  5.958  13.176  5.623   7.7180.002890.009640.01168

0.96986   7.95957  6.265  13.738  5.191   7.9600.002890.009630.01079

0.99462   8.17347  6.595  14.264  4.820   8.1730.002890.009620.00982

1.01962   8.37085  6.928  14.776  4.488   8.3710.002880.009610.00885

1.04410   8.56103  7.255  15.288  4.178   8.5610.002880.009600.00791

1.06808   8.75054  7.574  15.803  3.883   8.7510.002880.009590.00698
```

```
  1.09168    8.94130  7.889 16.305  3.606  8.9410.002870.009580.00607
  1.11472    9.13043  8.196 16.767  3.359  9.1300.002870.009570.00518
  1.13669    9.31392  8.489 17.167  3.149  9.3140.002870.009570.00434
  1.15694    9.48870  8.759 17.494  2.978  9.4890.002870.009560.00356
  1.17496    9.65218  8.999 17.750  2.846  9.6520.002870.009550.00287
1 1.19064    9.80530  9.209 17.948  2.745  9.8050.002860.009550.00227
        M4  25.       .02     1.                  .07
 .004       .01
 .01
 .000    1.000    0.000     0.000
  0.10141    1.79672 -3.986  2.403 19.659  1.7970.003980.009960.04576
  0.24518    1.98800 -2.548  2.409 19.089  1.9880.003960.009900.03980
  0.34107    2.17602 -1.589  2.421 18.531  2.1760.003950.009870.03587
  0.41176    2.36395 -0.882  2.445 17.978  2.3640.003940.009840.03299
  0.47590    2.55827 -0.241  2.499 17.417  2.5580.003930.009810.03039
  0.54567    2.75707  0.457  2.610 16.860  2.7570.003910.009790.02757
  0.61894    2.95089  1.189  2.799 16.332  2.9510.003900.009760.02463
  0.68286    3.13357  1.828  3.064 15.837  3.1340.003890.009730.02208
  0.72962    3.30208  2.296  3.380 15.374  3.3020.003890.009720.02023
  0.76275    3.46220  2.627  3.730 14.928  3.4620.003880.009700.01891
  0.78917    3.62797  2.892  4.125 14.465  3.6280.003880.009690.01787
  0.81305    3.80594  3.130  4.566 13.974  3.8060.003870.009690.01693
  0.83561    3.98961  3.356  5.024 13.480  3.9900.003870.009680.01604
  0.85643    4.16785  3.564  5.464 13.018  4.1680.003870.009670.01522
  0.87448    4.33249  3.745  5.860 12.606  4.3320.003860.009660.01451
  0.88907    4.47814  3.891  6.202 12.256  4.4780.003860.009660.01394
  0.90010    4.60133  4.001  6.484 11.968  4.6010.003860.009650.01351
  0.90802    4.70107  4.080  6.708 11.741  4.7010.003860.009650.01320
  0.91352    4.77856  4.135  6.879 11.567  4.7790.003860.009650.01298
  0.91730    4.83703  4.173  7.007 11.438  4.8370.003860.009650.01284
  0.91995    4.88099  4.199  7.102 11.341  4.8810.003860.009650.01273
  0.92195    4.91638  4.219  7.179 11.264  4.9160.003860.009640.01266
  0.92367    4.94860  4.237  7.248 11.194  4.9490.003860.009640.01259
  0.92547    4.98416  4.255  7.324 11.118  4.9840.003860.009640.01252
  0.92770    5.03127  4.277  7.424 11.017  5.0310.003860.009640.01243
  0.93076    5.10300  4.308  7.576 10.864  5.1030.003860.009640.01231
  0.93519    5.22576  4.352  7.833 10.606  5.2260.003860.009640.01214
  0.95086    6.25226  4.509  9.922  8.522  6.2520.003850.009630.01153
```

```
      0.96373    7.05244   4.637 11.537   6.951   7.052.003850.009630.01103

      0.98095    7.44497   4.809 12.340   6.214   7.445.003850.009620.01036

      1.00270    7.73204   5.027 12.938   5.701   7.732.003850.009610.00951

      1.02821    7.97181   5.282 13.451   5.294   7.972.003840.009600.00852

      1.05552    8.18296   5.555 13.915   4.953   8.183.003840.009590.00746

      1.08212    8.37278   5.821 14.347   4.661   8.373.003830.009590.00644

      1.10606    8.54551   6.061 14.750   4.405   8.546.003830.009580.00551

      1.12666    8.70532   6.267 15.130   4.177   8.705.003830.009570.00472

      1.14430    8.85717   6.443 15.491   3.971   8.857.003820.009560.00404

      1.15963    9.00585   6.596 15.837   3.779   9.006.003820.009560.00345

      1.17317    9.15337   6.732 16.163   3.602   9.153.003820.009550.00294

      1.18518    9.29850   6.852 16.460   3.444   9.299.003820.009550.00248

      1.19583    9.44025   6.958 16.721   3.307   9.440.003820.009540.00207

  1   1.20546    9.57967   7.055 16.946   3.190   9.580.003820.009540.00170

          M5   25.      .02      1.                    .07
   .005       .01

   .01

   .000      1.000     0.000      0.000

      0.10172    1.79671  -3.186   2.306 19.663   1.797.004980.009960.04575

      0.24601    1.98805  -2.032   2.312 19.093   1.988.004950.009900.03977

      0.34314    2.17637  -1.255   2.324 18.535   2.176.004930.009860.03578

      0.41667    2.36553  -0.667   2.348 17.980   2.366.004920.009840.03279

      0.48648    2.56161  -0.108   2.402 17.419   2.562.004900.009810.02996

      0.56521    2.76110   0.522   2.510 16.869   2.761.004890.009780.02679

      0.64988    2.95582   1.199   2.690 16.351   2.956.004870.009750.02340

      0.72584    3.14098   1.807   2.943 15.865   3.141.004860.009720.02038

      0.78236    3.31226   2.259   3.250 15.408   3.312.004850.009700.01814

      0.82103    3.47077   2.568   3.586 14.975   3.471.004840.009680.01661

      0.84945    3.62715   2.796   3.952 14.544   3.627.004840.009670.01550

      0.87318    3.79170   2.985   4.356 14.092   3.792.004830.009660.01456

      0.89441    3.96314   3.155   4.783 13.630   3.963.004830.009650.01373

      0.91332    4.13197   3.307   5.200 13.190   4.132.004820.009650.01299

      0.92939    4.28829   3.435   5.578 12.796   4.288.004820.009640.01237

      0.94222    4.42579   3.538   5.903 12.461   4.426.004820.009640.01186

      0.95185    4.54056   3.615   6.169 12.190   4.541.004820.009630.01149

      0.95875    4.63197   3.670   6.376 11.978   4.632.004820.009630.01122

      0.96354    4.70185   3.708   6.533 11.820   4.702.004810.009630.01103

      0.96684    4.75386   3.735   6.648 11.703   4.754.004810.009630.01091

      0.96918    4.79337   3.753   6.735 11.615   4.793.004810.009630.01081
```

```
  0.97097    4.82499   3.768   6.804 11.545   4.8250.004810.009630.01074

  0.97258    4.85499   3.781   6.869 11.479   4.8550.004810.009630.01068

  0.97431    4.88856   3.794   6.942 11.406   4.8890.004810.009620.01061

  0.97652    4.93441   3.812   7.040 11.306   4.9340.004810.009620.01053

  0.97959    5.00423   3.837   7.190 11.156   5.0040.004810.009620.01041

  0.98406    5.12345   3.872   7.442 10.902   5.1230.004810.009620.01023

  0.99057    5.36100   3.924   7.936 10.406   5.3610.004810.009620.00998

  0.99990    6.10644   3.999   9.449  8.895   6.1060.004810.009620.00962

  1.01290    7.01575   4.103  11.268  7.111   7.0160.004810.009610.00911

  1.03028    7.42675   4.242  12.083  6.349   7.4270.004800.009600.00844

  1.05224    7.72003   4.418  12.659  5.841   7.7200.004800.009600.00759

  1.07797    7.96265   4.624  13.127  5.451   7.9630.004790.009590.00660

  1.10544    8.17427   4.843  13.527  5.141   8.1740.004790.009580.00554

  1.13187    8.36211   5.055  13.870  4.890   8.3620.004780.009570.00452

  1.15487    8.52762   5.239  14.161  4.689   8.5280.004780.009560.00364

  1.17322    8.67045   5.386  14.400  4.532   8.6700.004780.009550.00293

  1.18697    8.79011   5.496  14.590  4.412   8.7900.004770.009550.00241

  1.19690    8.88766   5.575  14.737  4.321   8.8880.004770.009540.00203

  1.20408    8.96660   5.633  14.851  4.253   8.9670.004770.009540.00175

  1.20949    9.03269   5.676  14.942  4.199   9.0330.004770.009540.00155

  1.21402    9.09352   5.712  15.022  4.152   9.0940.004770.009540.00137

1 1.21848    9.15963   5.748  15.105  4.104   9.1600.004770.009540.00120

   0 1 1   9.8                     LOG K1 DIEN

   0 1 2   8.96                    LOG K2 DIEN

 1 0 1 3   4.2                     LOG K3 DIEN

TITRATION CURVE      .193 .75 11                      51  1

TITRATION CURVE      .177 .75 11                      50  1

END
```

The second half of the Data Reduction data comprises equilibrium titrations in which the ligand concentration is varied.

```
¶<-- Column 1

CU2+ DIEN     LIGAND VARIED


CU2+DIENH+

DATA REDUCTION     25.  .1  1150+2 0+1 3000 3000   21    5053 0112 1.7 10.

     L1   25.     .02     1.                  .07
```

.003   .006

.01

.000   1.000   0.000   0.000

0.10154	1.79802	-5.313	2.526	19.846	1.7980.002990.005980.03380
0.24568	1.99262	-3.391	2.532	19.266	1.9930.002970.005940.02790
0.33965	2.18476	-2.138	2.540	18.695	2.1850.002960.005920.02409
0.40440	2.37521	-1.275	2.557	18.133	2.3750.002950.005900.02148
0.45638	2.56851	-0.582	2.594	17.574	2.5690.002950.005890.01939
0.50852	2.76746	0.114	2.676	17.017	2.7670.002940.005880.01731
0.56513	2.96582	0.868	2.827	16.485	2.9660.002930.005870.01505
0.61934	3.15774	1.591	3.060	15.985	3.1580.002930.005850.01291
0.66187	3.33891	2.158	3.362	15.510	3.3390.002920.005850.01123
0.69078	3.50713	2.544	3.708	15.059	3.5070.002920.005840.01009
0.71069	3.66915	2.809	4.082	14.618	3.6690.002920.005830.00931
0.72625	3.83567	3.017	4.488	14.167	3.8360.002920.005830.00870
0.73961	4.00798	3.195	4.913	13.710	4.0080.002910.005830.00817
0.75124	4.17870	3.350	5.331	13.271	4.1790.002910.005820.00772
0.76097	4.33841	3.480	5.713	12.875	4.3380.002910.005820.00734
0.76866	4.48082	3.582	6.045	12.534	4.4810.002910.005820.00704
0.77439	4.60163	3.658	6.321	12.253	4.6020.002910.005820.00681
0.77846	4.69951	3.713	6.540	12.031	4.7000.002910.005820.00665
0.78127	4.77524	3.750	6.707	11.862	4.7750.002910.005820.00654
0.78317	4.83183	3.776	6.830	11.737	4.8320.002910.005820.00647
0.78449	4.87381	3.793	6.921	11.645	4.8740.002910.005820.00642
0.78545	4.90642	3.806	6.991	11.574	4.9060.002910.005820.00638
0.78624	4.93474	3.817	7.052	11.512	4.9350.002910.005820.00635
0.78703	4.96389	3.827	7.115	11.450	4.9640.002910.005820.00632
0.78798	5.00195	3.840	7.196	11.368	5.0020.002910.005820.00628
0.78929	5.05763	3.857	7.314	11.249	5.0580.002910.005820.00623
0.79121	5.15194	3.883	7.512	11.050	5.1520.002910.005820.00616
0.79413	5.33739	3.922	7.897	10.663	5.3370.002910.005820.00604
0.79855	5.87077	3.981	8.983	9.576	5.8710.002910.005810.00587
0.80513	7.00820	4.068	11.262	7.320	7.0080.002910.005810.00561
0.81452	7.46948	4.194	12.176	6.449	7.4690.002910.005810.00525
0.82717	7.77165	4.362	12.767	5.915	7.7720.002900.005810.00476
0.84279	8.01654	4.571	13.238	5.516	8.0170.002900.005800.00415
0.86017	8.23121	4.802	13.639	5.198	8.2310.002900.005800.00347
0.87739	8.42502	5.032	13.988	4.940	8.4250.002900.005800.00281
0.89269	8.60036	5.236	14.290	4.731	8.6000.002900.005790.00221

```
   0.90508    8.75670   5.401 14.543   4.565   8.7570.002900.005790.00173

   0.91449    8.89319   5.526 14.749   4.435   8.8930.002890.005790.00137

   0.92140    9.01048   5.619 14.915   4.334   9.0100.002890.005790.00110

   0.92651    9.11193   5.687 15.048   4.254   9.1120.002890.005790.00091

   0.93055    9.20418   5.741 15.161   4.188   9.2040.002890.005780.00075

   0.93415    9.29965   5.789 15.269   4.126   9.3000.002890.005780.00061

1  0.93799    9.42024   5.840 15.392   4.056   9.4200.002890.005780.00046

       L2   25.       .02    1.                  .07

 .003      .008

 .01

 .000      1.000     0.000     0.000

   0.10130    1.79735 -5.316   2.527 19.738   1.7970.002990.007970.03979

   0.24501    1.99028 -3.400   2.533 19.162   1.9900.002970.007920.03387

   0.33935    2.18012 -2.142   2.543 18.598   2.1800.002960.007890.03002

   0.40567    2.36843 -1.258   2.564 18.042   2.3680.002950.007870.02733

   0.46101    2.56108 -0.520   2.610 17.483   2.5610.002950.007860.02510

   0.51776    2.75942  0.237   2.709 16.923   2.7590.002940.007840.02282

   0.57741    2.95502  1.032   2.887 16.386   2.9550.002930.007820.02043

   0.63073    3.14091  1.743   3.147 15.880   3.1410.002930.007800.01831

   0.67029    3.31341  2.270   3.466 15.407   3.3130.002920.007790.01674

   0.69817    3.47756  2.642   3.824 14.950   3.4780.002920.007780.01564

   0.72015    3.64651  2.935   4.227 14.479   3.6470.002920.007780.01477

   0.73990    3.82730  3.199   4.674 13.983   3.8270.002910.007770.01399

   0.75855    4.01309  3.447   5.137 13.486   4.0130.002910.007760.01326

   0.77573    4.19343  3.676   5.580 13.022   4.1930.002910.007760.01258

   0.79062    4.36070  3.875   5.980 12.607   4.3610.002910.007750.01200

   0.80262    4.51015  4.035   6.328 12.251   4.5100.002910.007750.01153
   0.81167    4.63791  4.155   6.618 11.955   4.6380.002910.007750.01117

   0.81816    4.74247  4.242   6.851 11.719   4.7420.002900.007750.01092

   0.82265    4.82460  4.302   7.031 11.537   4.8250.002900.007750.01074

   0.82571    4.88675  4.343   7.166 11.401   4.8870.002900.007740.01062

   0.82783    4.93315  4.371   7.266 11.300   4.9330.002900.007740.01054

   0.82939    4.96983  4.392   7.344 11.221   4.9700.002900.007740.01048

   0.83068    5.00217  4.409   7.413 11.151   5.0020.002900.007740.01043

   0.83196    5.03603  4.426   7.485 11.079   5.0360.002900.007740.01038

   0.83351    5.07949  4.447   7.577 10.986   5.0790.002900.007740.01032

   0.83561    5.14491  4.475   7.715 10.848   5.1450.002900.007740.01023

   0.83867    5.25599  4.515   7.946 10.615   5.2560.002900.007740.01011

   0.84319    5.48117  4.576   8.411 10.150   5.4810.002900.007740.00994
```

0.84984	6.17662	4.664	9.820	8.745	6.1770.002900.007740.00968
0.85938	7.06181	4.792	11.606	6.998	7.0620.002900.007730.00931
0.87252	7.46835	4.967	12.439	6.228	7.4680.002900.007730.00879
0.88956	7.75754	5.194	13.046	5.705	7.7580.002900.007730.00813
0.90997	7.99773	5.466	13.566	5.292	7.9980.002890.007720.00733
0.93218	8.21012	5.762	14.042	4.944	8.2100.002890.007710.00647
0.95412	8.40303	6.055	14.491	4.642	8.4030.002890.007710.00562
0.97422	8.58164	6.323	14.919	4.373	8.5820.002890.007700.00484
0.99196	8.75114	6.559	15.333	4.129	8.7510.002890.007690.00416
1.00763	8.91679	6.768	15.733	3.902	8.9170.002880.007690.00355
1.02169	9.08206	6.956	16.117	3.692	9.0820.002880.007690.00301
1.03442	9.24707	7.125	16.472	3.502	9.2470.002880.007680.00252
1.04593	9.40987	7.279	16.785	3.337	9.4100.002880.007680.00208
1.05627	9.56780	7.417	17.048	3.200	9.5680.002880.007680.00168

1 1.06570   9.72400   7.543 17.267   3.087   9.7240.002880.007670.00132

    L3   25.   .02   1.   .07

.003   .010

.01

.000   1.000   0.000   0.000

0.10106	1.79672	-5.319	2.528	19.656	1.7970.002990.009960.04577
0.24434	1.98798	-3.409	2.534	19.086	1.9880.002970.009900.03984
0.33899	2.17568	-2.147	2.546	18.527	2.1760.002960.009870.03595
0.40673	2.36210	-1.244	2.570	17.977	2.3620.002950.009840.03319
0.46490	2.55401	-0.468	2.624	17.418	2.5540.002950.009820.03083
0.52492	2.75122	0.332	2.737	16.857	2.7510.002940.009790.02841
0.58578	2.94346	1.144	2.933	16.320	2.9430.002930.009770.02596
0.63741	3.12324	1.832	3.205	15.820	3.1230.002930.009750.02389
0.67525	3.28997	2.337	3.528	15.352	3.2900.002920.009740.02239
0.70376	3.45464	2.717	3.896	14.886	3.4550.002920.009730.02125
0.72865	3.63186	3.049	4.323	14.388	3.6320.002920.009720.02026
0.75285	3.82238	3.371	4.798	13.861	3.8220.002910.009710.01931
0.77681	4.01490	3.691	5.279	13.345	4.0150.002910.009700.01836
0.79954	4.19947	3.994	5.733	12.869	4.1990.002910.009690.01746
0.81960	4.36984	4.261	6.141	12.447	4.3700.002900.009680.01667
0.83594	4.52185	4.479	6.495	12.084	4.5220.002900.009680.01603
0.84836	4.65241	4.645	6.792	11.782	4.6520.002900.009670.01554
0.85729	4.75981	4.764	7.030	11.540	4.7600.002900.009670.01519
0.86350	4.84442	4.847	7.216	11.353	4.8440.002900.009670.01494

```
  0.86774    4.90858  4.903  7.355 11.212  4.9090.002900.009660.01478
  0.87067    4.95737  4.942  7.460 11.107  4.9570.002900.009660.01466
  0.87283    4.99561  4.971  7.541 11.024  4.9960.002900.009660.01458
  0.87462    5.02929  4.995  7.613 10.952  5.0290.002900.009660.01451
  0.87639    5.06507  5.019  7.689 10.876  5.0650.002900.009660.01444
  0.87851    5.11034  5.047  7.784 10.780  5.1100.002900.009660.01436
  0.88136    5.17813  5.085  7.927 10.637  5.1780.002900.009660.01424
  0.88544    5.29216  5.139  8.164 10.399  5.2920.002900.009660.01408
  0.89134    5.51872  5.218  8.630  9.932  5.5190.002900.009660.01385
  0.89979    6.17909  5.330  9.970  8.598  6.1790.002900.009650.01352
  0.91154    7.01755  5.487 11.676  6.941  7.0180.002890.009650.01306
  0.92720    7.42591  5.696 12.535  6.161  7.4260.002890.009640.01245
  0.94688    7.71768  5.958 13.176  5.623  7.7180.002890.009640.01168
  0.96986    7.95957  6.265 13.738  5.191  7.9600.002890.009630.01079
  0.99462    8.17347  6.595 14.264  4.820  8.1730.002890.009620.00982
  1.01962    8.37085  6.928 14.776  4.488  8.3710.002880.009610.00885
  1.04410    8.56103  7.255 15.288  4.178  8.5610.002880.009600.00791
  1.06808    8.75054  7.574 15.803  3.883  8.7510.002880.009590.00698
  1.09168    8.94130  7.889 16.305  3.606  8.9410.002870.009580.00607
  1.11472    9.13043  8.196 16.767  3.359  9.1300.002870.009570.00518
  1.13669    9.31392  8.489 17.167  3.149  9.3140.002870.009570.00434
  1.15694    9.48870  8.759 17.494  2.978  9.4890.002870.009560.00356
  1.17496    9.65218  8.999 17.750  2.846  9.6520.002870.009550.00287
1 1.19064    9.80530  9.209 17.948  2.745  9.8050.002860.009550.00227
     L4   25.      .02    1.               .07
 .003      .012
 .01
 .000     1.000    0.000    0.000
  0.10082    1.79608 -5.322  2.528 19.591  1.7960.002990.011950.05176
  0.24365    1.98571 -3.418  2.536 19.025  1.9860.002970.011880.04581
  0.33858    2.17135 -2.152  2.549 18.473  2.1710.002960.011840.04189
  0.40761    2.35603 -1.232  2.576 17.927  2.3560.002950.011810.03906
  0.46818    2.54714 -0.424  2.637 17.369  2.5470.002940.011780.03659
  0.53055    2.74288  0.407  2.761 16.809  2.7430.002940.011750.03406
  0.59163    2.93142  1.222  2.968 16.277  2.9310.002930.011720.03159
  0.64164    3.10561  1.888  3.245 15.785  3.1060.002920.011700.02958
  0.67870    3.26871  2.383  3.569 15.321  3.2690.002920.011680.02809
  0.70856    3.43620  2.781  3.947 14.844  3.4360.002920.011670.02690
```

```
  0.73660   3.62092   3.155   4.395 14.321   3.6210.002910.011660.02578
  0.76525   3.81832   3.537   4.888 13.773   3.8180.002910.011640.02464
  0.79450   4.01506   3.927   5.381 13.245   4.0150.002910.011630.02347
  0.82275   4.20175   4.303   5.841 12.762   4.2020.002900.011620.02235
  0.84796   4.37350   4.639   6.253 12.336   4.3730.002900.011610.02136
  0.86867   4.52668   4.915   6.610 11.970   4.5270.002900.011600.02054
  0.88447   4.65833   5.126   6.909 11.666   4.6580.002900.011590.01992
  0.89588   4.76677   5.278   7.150 11.421   4.7670.002900.011580.01947
  0.90382   4.85232   5.384   7.338 11.231   4.8520.002900.011580.01916
  0.90925   4.91758   5.457   7.479 11.089   4.9180.002890.011580.01894
  0.91302   4.96708   5.507   7.586 10.982   4.9670.002890.011580.01879
  0.91578   5.00603   5.544   7.669 10.898   5.0060.002890.011580.01868
  0.91807   5.04035   5.574   7.742 10.824   5.0400.002890.011570.01859
  0.92033   5.07631   5.604   7.818 10.748   5.0760.002890.011570.01851
  0.92300   5.12197   5.640   7.914 10.651   5.1220.002890.011570.01840
  0.92655   5.18911   5.687   8.055 10.509   5.1890.002890.011570.01826
  0.93155   5.30011   5.754   8.286 10.278   5.3000.002890.011570.01806
  0.93865   5.51215   5.849   8.723  9.841   5.5120.002890.011570.01779
  0.94856   6.07907   5.981   9.876  8.693   6.0790.002890.011560.01740
  0.96198   6.93450   6.160  11.624  6.997   6.9340.002890.011560.01687
  0.97939   7.36474   6.392  12.543  6.168   7.3650.002890.011550.01619
  1.00080   7.66622   6.677  13.226  5.603   7.6660.002880.011540.01535
  1.02554   7.91477   7.007  13.826  5.148   7.9150.002880.011530.01439
  1.05251   8.13557   7.367  14.397  4.754   8.1360.002880.011520.01334
  1.08090   8.34336   7.745  14.965  4.391   8.3430.002880.011500.01224
  1.11055   8.54790   8.140  15.544  4.046   8.5480.002870.011490.01109
  1.14158   8.75277   8.554  16.123  3.719   8.7530.002870.011480.00989
  1.17371   8.95539   8.983  16.670  3.421   8.9550.002870.011460.00865
  1.20609   9.15185   9.414  17.154  3.163   9.1520.002860.011450.00740
  1.23748   9.33940   9.833  17.559  2.951   9.3390.002860.011430.00619
  1.26655   9.51591  10.221  17.884  2.783   9.5160.002860.011420.00508
  1.29226   9.67960  10.563  18.134  2.654   9.6800.002850.011410.00410
1 1.31421   9.83004  10.856  18.323  2.558   9.8300.002850.011400.00326
        L5  25.       .02     1.                  .07
 .003      .014
 .01
 .000      1.000    0.000    0.000
  0.10057   1.79543  -5.326   2.529 19.536   1.7950.002990.013940.05775
```

0.24294	1.98346	-3.427	2.537	18.975	1.9830	.002970	.013870.05178
0.33814	2.16717	-2.158	2.551	18.429	2.1670	.002960	.013810.04783
0.40831	2.35020	-1.223	2.581	17.887	2.3500	.002950	.013780.04493
0.47095	2.54049	-0.387	2.648	17.331	2.5400	.002940	.013740.04236
0.53500	2.73438	0.467	2.781	16.773	2.7340	.002940	.013710.03975
0.59577	2.91902	1.277	2.996	16.248	2.9190	.002930	.013670.03728
0.64444	3.08841	1.926	3.273	15.765	3.0880	.002920	.013650.03531
0.68134	3.24948	2.418	3.597	15.304	3.2490	.002920	.013630.03382
0.71289	3.42059	2.838	3.986	14.813	3.4210	.002920	.013610.03256
0.74415	3.61205	3.255	4.451	14.269	3.6120	.002910	.013600.03130
0.77722	3.81457	3.696	4.958	13.706	3.8150	.002910	.013580.02998
0.81167	4.01395	4.155	5.458	13.169	4.0140	.002910	.013560.02860
0.84540	4.20203	4.605	5.922	12.681	4.2020	.002900	.013540.02726
0.87576	4.37441	5.010	6.336	12.253	4.3740	.002900	.013530.02606
0.90084	4.52809	5.344	6.695	11.885	4.5280	.002900	.013510.02506
0.92005	4.66006	5.601	6.995	11.580	4.6600	.002890	.013500.02430
0.93396	4.76875	5.786	7.238	11.334	4.7690	.002890	.013500.02375
0.94365	4.85460	5.915	7.426	11.144	4.8550	.002890	.013490.02337
0.95028	4.92013	6.004	7.568	11.001	4.9200	.002890	.013490.02311
0.95488	4.96983	6.065	7.675	10.893	4.9700	.002890	.013480.02293
0.95825	5.00888	6.110	7.758	10.809	5.0090	.002890	.013480.02280
0.96104	5.04309	6.147	7.831	10.736	5.0430	.002890	.013480.02269
0.96377	5.07877	6.183	7.907	10.660	5.0790	.002890	.013480.02258
0.96698	5.12391	6.226	8.002	10.564	5.1240	.002890	.013480.02245
0.97120	5.18895	6.283	8.139	10.427	5.1890	.002890	.013480.02229
0.97704	5.29380	6.360	8.357	10.208	5.2940	.002890	.013470.02206
0.98516	5.48621	6.469	8.754	9.811	5.4860	.002890	.013470.02174
0.99627	5.95260	6.617	9.705	8.864	5.9530	.002890	.013460.02130
1.01097	6.82008	6.813	11.482	7.138	6.8200	.002880	.013460.02072
1.02967	7.28914	7.062	12.490	6.226	7.2890	.002880	.013450.01999
1.05238	7.60700	7.365	13.220	5.624	7.6070	.002880	.013430.01910
1.07866	7.86679	7.715	13.863	5.141	7.8670	.002880	.013420.01807
1.10795	8.09904	8.106	14.480	4.717	8.0990	.002870	.013410.01693
1.14009	8.31987	8.534	15.102	4.324	8.3200	.002870	.013390.01568
1.17530	8.53769	9.004	15.735	3.949	8.5380	.002870	.013370.01432
1.21367	8.75274	9.515	16.355	3.601	8.7530	.002860	.013350.01283
1.25448	8.96128	10.060	16.923	3.293	8.9610	.002860	.013330.01126
1.29626	9.16048	10.617	17.416	3.032	9.1600	.002850	.013310.00965

```
    1.33707    9.34912 11.161 17.823  2.819  9.3490.002850.013290.00808

    1.37495    9.52572 11.666 18.146  2.652  9.5260.002840.013270.00664

    1.40832    9.68860 12.111 18.393  2.525  9.6890.002840.013250.00537

1   1.43648    9.83684 12.486 18.577  2.431  9.8370.002840.013240.00429

    0 1 1   9.8                        LOG K1 DIEN

    0 1 2   8.96                       LOG K2 DIEN

1   0 1 3   4.2                        LOG K3 DIEN

TITRATION CURVE        .15  .75 11                         51  1

TITRATION CURVE        .4   .75 11                         50  1

END
```

4.3. VARIATION

1–10.   As Section 4.2, items 1–10.

11. 1 line [As Section 4.1, item 4]: **(CODEN(I), I=1,5) TEMP,**UCORR, **NM, NL, IACT, ISORT, (QC(J), J=1,3), (NH(I), i=1,4), (IHM(I), I=1,4), (LIGS(I), I=1,4), NEST, IPUN, LONG, IPLOT, IDAT,** IPAR, IGUESS, TAPIN, TAPOUT, **PH1, PH2**

CODEN: VARIATION

TEMP: TEMP is used here for value of DELPH, for S/R OSTR1.

NM: ⎫
: ⎬ As for Section 4.1, item 4.
: ⎪
QC( ): ⎭

NH(1):   NH(1) is used in this section to indicate number of titration sets where total *metal* concentration is varied but total ligand is held constant (max = 7).

NH(2):   NH(2) is used to indicate the number of titration sets where total ligand is varied with constant metal.

LIGS(1):   Titration set in the metal-varied series which is the common point; see Section 2.3.

LIGS(2):   Titration set in the ligand-varied set which is the common point; see Section 2.3.

NSET: Number of titration sets. The metal-varied series *must* precede the ligand-varied series.

IPLOT: IPLOT=5 produces Calcomp plots; see Figures 5–6. Otherwise 0. Note: The Calcomp routines are not included here.

IDAT: Set equal to unity.

PH1:   Minimum pH value to be considered in the refinement.

PH2:   Maximum pH value to be considered in the refinement.

12. 1 line [6F10.6]: SCALEX,SCALEY,XTYPE,**PHA,PHB,SIGHT**

PHA: Minimum pH for FICS integration. Choose a value slightly larger than the largest minimum pH in any of the titration sets.

PHB: Maximum pH for FICS integration. Choose a value slightly lower than the lowest maximum pH in any of the titration sets.

SIGHT: Parameter for spline smoothing; try 0.00005; see Section 2.3.

13. 1 line [6F10.6]: **(CT5(I),I=1,6)**

CT5(1): Total metal concentration of the "common point" set, corrected for *dilution* at the pH specified by PHA.

CT5(2): Total ligand concentration of the "common point" set, corrected for dilution at the pH specified by PHB.

CT5(3), . . . CT5(6): Not used by this section.

14. 1 line [5A4, . . .]: **(CODEN(I),I=1,5),** . . .

Final card, i.e., END. See Section 4.1, item 9.

READ statements for this section are located as follows:

Item	Routine	Line No.
1–10	As Section 4.3 Items 1–10	
11	MAIN	94–96
12	OSTR1	41
13	OSTR1	46
14	MAIN	94–96

4.4. PX FROM VARIATION

1–13.   As Section 4.3, items 1–13.

11. 1 line [As Section 4.1, item 4]: **(CODEN(I), I=1,5),** TEMP, UCORR, **NM, NL,** IACT, ISORT, (QC(J), J=1,3), **(NH(I), I=1,4),** (IHM(I), I=1,4, (LIGS(I),I=1,4), **NSET, IPUN, LONG,** IPLOT, IDAT, IPAR, IGUESS, TAPIN, **TAPOUT,** PH1, PH2

CODEN: PX FROM VARIATION

NM and NL: As for Section 4.3, item 11.

NSET: Number of titration sets, always 1.

IPUN: Set IPUN=1 to produce "common point" deck for use in Sections 4.6 and 4.7.

LONG: Set equal to 5 for normal operation.

TAPOUT: Set equal to 2 for normal operation.

15. 1 line [I1,F9.0,F10.0]: **IS,PHS(I),DPX(I,N)**

Card deck produced by Section 4.2 for the "common point" set. A "1" must appear in column 1 of the last data card.

16. 1 line [5A4, . . .]: **(CODEN(I),I=1,5),** . . .

Final card, i.e., END. See Section 4.1, item 9.

READ statements for this section are located as follows:

Item	Routine	Line No.
1–13	As for Section 4.5, items 1–13	
14	MAIN	94–96
15	OSTR2	33
16	MAIN	94–96

4.5. STOICHIOMETRIC COEFS

4.5.1. Instructions for Data Input

1–13.   As Section 4.3, items 1–13.

14. 1 line [As Section 4.1, item 4]: **(CODEN(I), I=1,5), TEMP, UCORR, NM, NL,** IACT, ISORT, **(QC(J), J=1,3), (NH(I), I=1,4), (IHM(I), I=1,4, (LIGS(I), I=1,4),** NSET, IPUN, LONG, IPLOT, IDAT, IPAR, IGUESS, TAPIN, TAPOUT, PH1, PH2

CODEN: STOICHIOMETRIC COEFS

TEMP: ⎫
: ⎪
: ⎪
NL: ⎬ As for Section 4.1, item 4.
QC( ): ⎪
NH( ): ⎪
IHM( ): ⎪
LIGS( ): ⎭

15. 1 line [8F10.6]: (PX1(I), I=1, 3), S1, **(CT5(J), J=1, NC)**

CT5(1):   Total metal concentration in "common point" set, *undiluted*.

CT5(2):   Total ligand concentration in "common point" set, *undiluted*.

16. 1 line [20X, 3F10.6]: **ACID, BASE,** STOCK

ACID: Acid concentration added or titrant; see Section 4.2, item 5.

BASE: Base concentration added or titrant; see Section 4.2, item 5.

17. 1 line [8F10.6]: (FS(J), J=1, NC2), SF

FS(J): ⎫
: ⎬ Not used in this section. Blank card required.
: ⎪
SF: ⎭

18. 1 line [I1, A1, 4I2, F10.0, 3F5.0, 7A4]: **IS, KI(I), (IX(J,I), J=1, 4), BL(I),** SL(I), DP(I), SHFT(I), **(LABP(I, J), J=1, 7)**

Ligand log K_j^H constants. Same as Section 4.2, item 10.

19. 1 line [5A4,...]: **(CODEN(I), I=1, 5),** . . .

Final card, i.e., END. (See Section 4.1, item 9)

READ statements for this section are located as follows:

Item	Routine	Line No.
1–13	As for Section 4.5, items 1–13	
14	MAIN	94–96
15	STOICH	39

Item	Routine	Line No.
16	STOICH	40
17	STOICH	41
18	READP	34
19	MAIN	94–96

4.5.2. Input Data for VARIATION, PX FROM VARIATION, and STOICHIOMETRIC COEFS

Data for these three sections are usually combined. Consequently Section 4.3, item 14 and Section 4.4, item 17 would be omitted from the combined data set.

```
¶<--  Column 1

CU2+ DIEN

A Blank Card Goes Here

CU2+DIENH+

VARIATION            .1      11 0+2 0+1 5500      3300     100 50  12

                             1.8       9.0      .00005

   0.0029877 .009959   .04567

0    -15.984  1.797   0.0    1.797 0.000996 0.009960 0.045800 0.073720  M1

0    -10.295  1.988   0.0    1.988 0.000990 0.009904 0.039908 0.078939  M1

0     -6.612  2.175   0.0    2.175 0.000987 0.009868 0.036128 0.082287  M1

0     -4.149  2.357   0.0    2.357 0.000984 0.009844 0.033616 0.084512  M1

0     -2.342  2.539   0.0    2.539 0.000983 0.009826 0.031780 0.086137  M1

0     -0.818  2.725   0.0    2.725 0.000981 0.009812 0.030238 0.087504  M1

0      0.527  2.909   0.0    2.909 0.000980 0.009799 0.028880 0.088706  M1

0      1.630  3.082   0.0    3.082 0.000979 0.009788 0.027770 0.089690  M1

0      2.533  3.254   0.0    3.254 0.000978 0.009780 0.026862 0.090494  M1

0      3.382  3.442   0.0    3.442 0.000977 0.009772 0.026010 0.091248  M1

0      4.289  3.648   0.0    3.648 0.000976 0.009763 0.025102 0.092053  M1

0      5.294  3.858   0.0    3.858 0.000975 0.009753 0.024096 0.092943  M1

0      6.369  4.060   0.0    4.060 0.000974 0.009743 0.023024 0.093893  M1

0      7.434  4.251   0.0    4.251 0.000973 0.009733 0.021964 0.094832  M1

0      8.395  4.429   0.0    4.429 0.000972 0.009724 0.021008 0.095678  M1

0      9.796  4.734   0.0    4.734 0.000971 0.009711 0.019619 0.096909  M1

0     10.234  4.854   0.0    4.854 0.000971 0.009707 0.019186 0.097292  M1

0     10.536  4.952   0.0    4.952 0.000970 0.009704 0.018887 0.097557  M1

0     10.741  5.028   0.0    5.028 0.000970 0.009702 0.018685 0.097736  M1

0     10.878  5.085   0.0    5.085 0.000970 0.009700 0.018549 0.097857  M1
```

0	10.973	5.128	0.0	5.128	0.000970	0.009700	0.018455	0.097940	M1
0	11.043	5.161	0.0	5.161	0.000970	0.009699	0.018386	0.098001	M1
0	11.102	5.192	0.0	5.192	0.000970	0.009698	0.018328	0.098052	M1
0	11.161	5.224	0.0	5.224	0.000970	0.009698	0.018269	0.098104	M1
0	11.234	5.267	0.0	5.267	0.000970	0.009697	0.018198	0.098167	M1
0	11.332	5.330	0.0	5.330	0.000970	0.009696	0.018101	0.098253	M1
0	11.471	5.436	0.0	5.436	0.000969	0.009695	0.017964	0.098375	M1
0	11.670	5.641	0.0	5.641	0.000969	0.009693	0.017767	0.098549	M1
0	11.950	6.210	0.0	6.210	0.000969	0.009690	0.017491	0.098794	M1
0	12.332	7.132	0.0	7.132	0.000969	0.009687	0.017115	0.099127	M1
0	12.832	7.571	0.0	7.571	0.000968	0.009682	0.016623	0.099563	M1
0	13.468	7.882	0.0	7.882	0.000968	0.009676	0.015997	0.100117	M1
0	14.259	8.149	0.0	8.149	0.000967	0.009669	0.015220	0.100805	M1
0	15.233	8.395	0.0	8.395	0.000966	0.009660	0.014265	0.101651	M1
0	16.410	8.627	0.0	8.627	0.000965	0.009649	0.013112	0.102672	M1
0	17.785	8.847	0.0	8.847	0.000964	0.009636	0.011770	0.103860	M1
0	19.311	9.057	0.0	9.057	0.000962	0.009622	0.010285	0.105176	M1
0	20.913	9.257	0.0	9.257	0.000961	0.009607	0.008730	0.106554	M1
0	22.502	9.449	0.0	9.449	0.000959	0.009592	0.007192	0.107916	M1
0	23.992	9.633	0.0	9.633	0.000958	0.009579	0.005755	0.109188	M1
0	25.312	9.808	0.0	9.808	0.000957	0.009567	0.004485	0.110313	M1
1	26.430	9.971	0.0	9.971	0.000956	0.009556	0.003412	0.111264	M1
0	-7.985	1.797	0.0	1.797	0.001992	0.009960	0.045785	0.073733	M2
0	-5.130	1.988	0.0	1.988	0.001981	0.009904	0.039872	0.078970	M2
0	-3.263	2.175	0.0	2.175	0.001973	0.009867	0.036040	0.082365	M2
0	-1.968	2.360	0.0	2.360	0.001968	0.009842	0.033400	0.084703	M2
0	-0.932	2.548	0.0	2.548	0.001964	0.009822	0.031296	0.086567	M2
0	0.058	2.742	0.0	2.742	0.001961	0.009803	0.029295	0.088339	M2
0	1.010	2.931	0.0	2.931	0.001957	0.009785	0.027378	0.090037	M2
0	1.798	3.108	0.0	3.108	0.001954	0.009769	0.025795	0.091439	M2
0	2.394	3.275	0.0	3.275	0.001952	0.009758	0.024602	0.092496	M2
0	2.886	3.448	0.0	3.448	0.001950	0.009749	0.023619	0.093366	M2
0	3.360	3.639	0.0	3.639	0.001948	0.009740	0.022675	0.094202	M2
0	3.852	3.840	0.0	3.840	0.001946	0.009730	0.021695	0.095070	M2
0	4.360	4.039	0.0	4.038	0.001944	0.009721	0.020685	0.095964	M2
0	4.854	4.227	0.0	4.227	0.001942	0.009711	0.019707	0.096831	M2
0	5.295	4.402	0.0	4.402	0.001941	0.009703	0.018835	0.097604	M2
0	5.657	4.559	0.0	4.559	0.001939	0.009696	0.018119	0.098237	M2

0	5.933	4.696	0.0	4.696	0.001938	0.009691	0.017575	0.098719	M2
0	6.131	4.810	0.0	4.810	0.001937	0.009687	0.017183	0.099067	M2
0	6.269	4.901	0.0	4.901	0.001937	0.009685	0.016911	0.099307	M2
0	6.363	4.972	0.0	4.972	0.001937	0.009683	0.016727	0.099470	M2
0	6.427	5.025	0.0	5.025	0.001936	0.009682	0.016601	0.099582	M2
0	6.472	5.065	0.0	5.065	0.001936	0.009681	0.016511	0.099661	M2
0	6.508	5.100	0.0	5.100	0.001936	0.009680	0.016440	0.099724	M2
0	6.542	5.134	0.0	5.134	0.001936	0.009680	0.016375	0.099782	M2
0	6.579	5.174	0.0	5.174	0.001936	0.009679	0.016301	0.099848	M2
0	6.628	5.232	0.0	5.232	0.001936	0.009678	0.016205	0.099932	M2
0	6.696	5.325	0.0	5.325	0.001935	0.009677	0.016071	0.100051	M2
0	6.794	5.498	0.0	5.498	0.001935	0.009675	0.015878	0.100222	M2
0	6.934	5.922	0.0	5.922	0.001934	0.009672	0.015603	0.100466	M2
0	7.129	6.861	0.0	6.861	0.001934	0.009669	0.015220	0.100805	M2
0	7.389	7.355	0.0	7.355	0.001933	0.009664	0.014710	0.101256	M2
0	7.717	7.674	0.0	7.674	0.001932	0.009658	0.014068	0.101825	M2
0	8.106	7.931	0.0	7.931	0.001930	0.009651	0.013307	0.102500	M2
0	8.545	8.160	0.0	8.160	0.001928	0.009642	0.012447	0.103261	M2
0	9.029	8.377	0.0	8.377	0.001927	0.009633	0.011504	0.104096	M2
0	9.559	8.592	0.0	8.592	0.001925	0.009624	0.010473	0.105010	M2
0	10.135	8.804	0.0	8.804	0.001923	0.009613	0.009353	0.106002	M2
0	10.749	9.010	0.0	9.010	0.001920	0.009602	0.008164	0.107055	M2
0	11.378	9.209	0.0	9.209	0.001918	0.009590	0.006948	0.108132	M2
0	11.994	9.400	0.0	9.400	0.001916	0.009579	0.005759	0.109184	M2
0	12.568	9.581	0.0	9.581	0.001914	0.009568	0.004654	0.110164	M2
0	13.078	9.751	0.0	9.751	0.001912	0.009559	0.003675	0.111031	M2
1	13.514	9.910	0.0	9.910	0.001910	0.009551	0.002838	0.111772	M2
0	-5.319	1.797	0.0	1.797	0.002988	0.009960	0.045773	0.073744	M3
0	-3.409	1.988	0.0	1.988	0.002971	0.009903	0.039837	0.079001	M3
0	-2.147	2.176	0.0	2.176	0.002960	0.009866	0.035953	0.082442	M3
0	-1.244	2.362	0.0	2.362	0.002952	0.009840	0.033191	0.084888	M3
0	-0.468	2.554	0.0	2.554	0.002945	0.009817	0.030831	0.086978	M3
0	0.332	2.751	0.0	2.751	0.002938	0.009794	0.028407	0.089125	M3
0	1.144	2.943	0.0	2.943	0.002931	0.009771	0.025961	0.091292	M3
0	1.832	3.123	0.0	3.123	0.002925	0.009751	0.023894	0.093122	M3
0	2.337	3.290	0.0	3.290	0.002921	0.009737	0.022385	0.094459	M3
0	2.717	3.455	0.0	3.455	0.002918	0.009726	0.021251	0.095463	M3
0	3.049	3.632	0.0	3.632	0.002915	0.009717	0.020263	0.096338	M3

0	3.371	3.822	0.0	3.822	0.002912	0.009708	0.019305	0.097187	M3
0	3.691	4.015	0.0	4.015	0.002910	0.009699	0.018357	0.098026	M3
0	3.994	4.199	0.0	4.199	0.002907	0.009690	0.017460	0.098821	M3
0	4.261	4.370	0.0	4.370	0.002905	0.009683	0.016670	0.099521	M3
0	4.479	4.522	0.0	4.522	0.002903	0.009676	0.016026	0.100091	M3
0	4.645	4.652	0.0	4.652	0.002902	0.009672	0.015538	0.100523	M3
0	4.764	4.760	0.0	4.760	0.002901	0.009668	0.015188	0.100834	M3
0	4.847	4.844	0.0	4.844	0.002900	0.009666	0.014944	0.101050	M3
0	4.903	4.909	0.0	4.909	0.002899	0.009665	0.014777	0.101197	M3
0	4.942	4.957	0.0	4.957	0.002899	0.009663	0.014663	0.101299	M3
0	4.971	4.996	0.0	4.996	0.002899	0.009663	0.014578	0.101374	M3
0	4.995	5.029	0.0	5.029	0.002899	0.009662	0.014508	0.101436	M3
0	5.019	5.065	0.0	5.065	0.002898	0.009661	0.014438	0.101497	M3
0	5.047	5.110	0.0	5.110	0.002898	0.009661	0.014355	0.101571	M3
0	5.085	5.178	0.0	5.178	0.002898	0.009659	0.014243	0.101670	M3
0	5.139	5.292	0.0	5.292	0.002897	0.009658	0.014084	0.101812	M3
0	5.218	5.519	0.0	5.519	0.002897	0.009656	0.013853	0.102016	M3
0	5.331	6.179	0.0	6.179	0.002896	0.009653	0.013522	0.102309	M3
0	5.487	7.018	0.0	7.018	0.002894	0.009648	0.013062	0.102716	M3
0	5.696	7.426	0.0	7.426	0.002893	0.009642	0.012450	0.103258	M3
0	5.958	7.718	0.0	7.718	0.002891	0.009635	0.011682	0.103938	M3
0	6.265	7.960	0.0	7.960	0.002888	0.009627	0.010787	0.104731	M3
0	6.595	8.173	0.0	8.173	0.002885	0.009617	0.009824	0.105584	M3
0	6.928	8.371	0.0	8.371	0.002882	0.009608	0.008854	0.106444	M3
0	7.255	8.561	0.0	8.561	0.002880	0.009599	0.007906	0.107283	M3
0	7.574	8.751	0.0	8.751	0.002877	0.009590	0.006979	0.108105	M3
0	7.889	8.941	0.0	8.941	0.002874	0.009582	0.006068	0.108911	M3
0	8.196	9.130	0.0	9.130	0.002872	0.009573	0.005180	0.109697	M3
0	8.489	9.314	0.0	9.314	0.002870	0.009565	0.004335	0.110446	M3
0	8.759	9.489	0.0	9.489	0.002867	0.009558	0.003558	0.111134	M3
0	8.999	9.652	0.0	9.652	0.002865	0.009551	0.002867	0.111746	M3
1	9.209	9.805	0.0	9.805	0.002864	0.009545	0.002266	0.112278	M3
0	-3.986	1.797	0.0	1.797	0.003984	0.009960	0.045758	0.073757	M4
0	-2.548	1.988	0.0	1.988	0.003961	0.009903	0.039802	0.079032	M4
0	-1.589	2.176	0.0	2.176	0.003946	0.009865	0.035868	0.082517	M4
0	-0.882	2.364	0.0	2.364	0.003935	0.009838	0.032986	0.085069	M4
0	-0.241	2.558	0.0	2.558	0.003925	0.009813	0.030386	0.087373	M4
0	0.457	2.757	0.0	2.757	0.003915	0.009786	0.027571	0.089865	M4

0	1.189	2.951	0.0	2.951	0.003903	0.009758	0.024633	0.092468	M4
0	1.829	3.134	0.0	3.134	0.003894	0.009734	0.022082	0.094727	M4
0	2.296	3.302	0.0	3.302	0.003887	0.009716	0.020225	0.096372	M4
0	2.628	3.462	0.0	3.462	0.003882	0.009704	0.018913	0.097534	M4
0	2.892	3.628	0.0	3.628	0.003878	0.009694	0.017869	0.098459	M4
0	3.131	3.806	0.0	3.806	0.003874	0.009685	0.016927	0.099293	M4
0	3.356	3.990	0.0	3.990	0.003871	0.009677	0.016039	0.100079	M4
0	3.564	4.168	0.0	4.168	0.003868	0.009669	0.015221	0.100804	M4
0	3.745	4.332	0.0	4.332	0.003865	0.009662	0.014513	0.101431	M4
0	3.891	4.478	0.0	4.478	0.003863	0.009657	0.013941	0.101938	M4
0	4.001	4.601	0.0	4.601	0.003861	0.009652	0.013510	0.102320	M4
0	4.080	4.701	0.0	4.701	0.003860	0.009650	0.013200	0.102594	M4
0	4.135	4.779	0.0	4.779	0.003859	0.009647	0.012985	0.102785	M4
0	4.173	4.837	0.0	4.837	0.003858	0.009646	0.012837	0.102916	M4
0	4.200	4.881	0.0	4.881	0.003858	0.009645	0.012733	0.103007	M4
0	4.220	4.916	0.0	4.916	0.003858	0.009644	0.012655	0.103077	M4
0	4.237	4.949	0.0	4.949	0.003857	0.009644	0.012588	0.103136	M4
0	4.255	4.984	0.0	4.984	0.003857	0.009643	0.012518	0.103198	M4
0	4.277	5.031	0.0	5.031	0.003857	0.009642	0.012431	0.103276	M4
0	4.308	5.103	0.0	5.103	0.003856	0.009641	0.012311	0.103381	M4
0	4.352	5.226	0.0	5.226	0.003856	0.009639	0.012138	0.103535	M4
0	4.509	6.252	0.0	6.252	0.003853	0.009634	0.011527	0.104076	M4
0	4.637	7.052	0.0	7.052	0.003852	0.009629	0.011026	0.104520	M4
0	4.810	7.445	0.0	7.445	0.003849	0.009622	0.010356	0.105114	M4
0	5.027	7.732	0.0	7.732	0.003846	0.009614	0.009511	0.105862	M4
0	5.282	7.972	0.0	7.972	0.003842	0.009605	0.008521	0.106738	M4
0	5.555	8.183	0.0	8.183	0.003838	0.009595	0.007464	0.107675	M4
0	5.821	8.373	0.0	8.373	0.003834	0.009585	0.006437	0.108585	M4
0	6.061	8.546	0.0	8.546	0.003831	0.009576	0.005514	0.109402	M4
0	6.267	8.705	0.0	8.705	0.003828	0.009569	0.004721	0.110104	M4
0	6.443	8.857	0.0	8.857	0.003825	0.009562	0.004043	0.110705	M4
0	6.596	9.006	0.0	9.006	0.003823	0.009557	0.003455	0.111226	M4
0	6.732	9.153	0.0	9.153	0.003821	0.009552	0.002935	0.111686	M4
0	6.852	9.299	0.0	9.298	0.003819	0.009547	0.002475	0.112093	M4
0	6.958	9.440	0.0	9.440	0.003817	0.009544	0.002068	0.112454	M4
1	7.055	9.580	0.0	9.580	0.003816	0.009540	0.001700	0.112780	M4
0	-3.186	1.797	0.0	1.797	0.004980	0.009959	0.045745	0.073769	M5
0	-2.032	1.988	0.0	1.988	0.004951	0.009903	0.039768	0.079062	M5

0	-1.255	2.176	0.0	2.176	0.004932	0.009865	0.035783	0.082592	M5
0	-0.667	2.366	0.0	2.366	0.004918	0.009836	0.032787	0.085246	M5
0	-0.108	2.562	0.0	2.562	0.004905	0.009809	0.029958	0.087752	M5
0	0.522	2.761	0.0	2.761	0.004889	0.009779	0.026786	0.090561	M5
0	1.199	2.956	0.0	2.956	0.004873	0.009747	0.023397	0.093563	M5
0	1.807	3.141	0.0	3.141	0.004859	0.009718	0.020375	0.096239	M5
0	2.259	3.312	0.0	3.312	0.004848	0.009697	0.018138	0.098221	M5
0	2.568	3.471	0.0	3.471	0.004841	0.009682	0.016613	0.099571	M5
0	2.796	3.627	0.0	3.627	0.004836	0.009671	0.015495	0.100561	M5
0	2.985	3.792	0.0	3.792	0.004831	0.009663	0.014564	0.101386	M5
0	3.155	3.963	0.0	3.963	0.004827	0.009655	0.013732	0.102123	M5
0	3.307	4.132	0.0	4.132	0.004824	0.009648	0.012993	0.102778	M5
0	3.435	4.288	0.0	4.288	0.004821	0.009642	0.012365	0.103334	M5
0	3.538	4.426	0.0	4.426	0.004818	0.009637	0.011864	0.103777	M5
0	3.615	4.541	0.0	4.541	0.004817	0.009633	0.011489	0.104110	M5
0	3.670	4.632	0.0	4.632	0.004815	0.009631	0.011220	0.104348	M5
0	3.708	4.702	0.0	4.702	0.004814	0.009629	0.011033	0.104513	M5
0	3.735	4.754	0.0	4.754	0.004814	0.009628	0.010905	0.104627	M5
0	3.753	4.793	0.0	4.793	0.004813	0.009627	0.010814	0.104708	M5
0	3.768	4.825	0.0	4.825	0.004813	0.009626	0.010744	0.104770	M5
0	3.781	4.855	0.0	4.855	0.004813	0.009626	0.010681	0.104825	M5
0	3.794	4.889	0.0	4.889	0.004812	0.009625	0.010614	0.104885	M5
0	3.812	4.934	0.0	4.934	0.004812	0.009624	0.010528	0.104961	M5
0	3.837	5.004	0.0	5.004	0.004811	0.009623	0.010409	0.105067	M5
0	3.872	5.123	0.0	5.123	0.004811	0.009621	0.010235	0.105221	M5
0	3.925	5.361	0.0	5.361	0.004809	0.009619	0.009982	0.105445	M5
0	3.999	6.106	0.0	6.106	0.004808	0.009615	0.009619	0.105766	M5
0	4.103	7.016	0.0	7.016	0.004805	0.009611	0.009115	0.106213	M5
0	4.242	7.427	0.0	7.427	0.004802	0.009604	0.008441	0.106809	M5
0	4.418	7.720	0.0	7.720	0.004798	0.009596	0.007591	0.107562	M5
0	4.624	7.963	0.0	7.963	0.004793	0.009587	0.006597	0.108443	M5
0	4.844	8.174	0.0	8.174	0.004788	0.009577	0.005538	0.109381	M5
0	5.055	8.362	0.0	8.362	0.004783	0.009567	0.004521	0.110282	M5
0	5.239	8.528	0.0	8.528	0.004779	0.009558	0.003637	0.111064	M5
0	5.386	8.670	0.0	8.670	0.004776	0.009552	0.002934	0.111687	M5
0	5.496	8.790	0.0	8.790	0.004773	0.009547	0.002407	0.112154	M5
0	5.575	8.888	0.0	8.888	0.004772	0.009543	0.002027	0.112490	M5
0	5.633	8.967	0.0	8.967	0.004770	0.009540	0.001752	0.112734	M5

0	5.676	9.033	0.0	9.033	0.004769	0.009539	0.001546	0.112917	M5
0	5.712	9.094	0.0	9.094	0.004768	0.009537	0.001373	0.113070	M5
1	5.748	9.160	0.0	9.160	0.004768	0.009535	0.001202	0.113221	M5
0	-5.313	1.798	0.0	1.798	0.002988	0.005976	0.033801	0.073762	L1
0	-3.391	1.993	0.0	1.993	0.002971	0.005942	0.027899	0.079050	L1
0	-2.138	2.185	0.0	2.185	0.002960	0.005920	0.024087	0.082466	L1
0	-1.275	2.375	0.0	2.375	0.002952	0.005904	0.021477	0.084804	L1
0	-0.582	2.569	0.0	2.569	0.002946	0.005892	0.019391	0.086673	L1
0	0.114	2.767	0.0	2.767	0.002940	0.005880	0.017307	0.088540	L1
0	0.868	2.966	0.0	2.966	0.002934	0.005867	0.015054	0.090558	L1
0	1.591	3.158	0.0	3.158	0.002927	0.005855	0.012907	0.092482	L1
0	2.158	3.339	0.0	3.339	0.002923	0.005845	0.011228	0.093986	L1
0	2.544	3.507	0.0	3.507	0.002919	0.005839	0.010090	0.095006	L1
0	2.809	3.669	0.0	3.669	0.002917	0.005834	0.009308	0.095707	L1
0	3.017	3.836	0.0	3.836	0.002915	0.005831	0.008697	0.096254	L1
0	3.195	4.008	0.0	4.008	0.002914	0.005828	0.008174	0.096723	L1
0	3.350	4.179	0.0	4.179	0.002912	0.005825	0.007718	0.097131	L1
0	3.480	4.338	0.0	4.338	0.002911	0.005823	0.007338	0.097472	L1
0	3.582	4.481	0.0	4.481	0.002911	0.005821	0.007037	0.097741	L1
0	3.659	4.602	0.0	4.602	0.002910	0.005820	0.006813	0.097942	L1
0	3.713	4.700	0.0	4.700	0.002909	0.005819	0.006654	0.098084	L1
0	3.750	4.775	0.0	4.775	0.002909	0.005818	0.006545	0.098182	L1
0	3.776	4.832	0.0	4.832	0.002909	0.005818	0.006470	0.098249	L1
0	3.793	4.874	0.0	4.874	0.002909	0.005817	0.006419	0.098295	L1
0	3.806	4.906	0.0	4.906	0.002909	0.005817	0.006381	0.098329	L1
0	3.817	4.935	0.0	4.935	0.002909	0.005817	0.006351	0.098356	L1
0	3.827	4.964	0.0	4.964	0.002908	0.005817	0.006320	0.098384	L1
0	3.840	5.002	0.0	5.002	0.002908	0.005817	0.006283	0.098417	L1
0	3.857	5.058	0.0	5.058	0.002908	0.005816	0.006232	0.098463	L1
0	3.883	5.152	0.0	5.152	0.002908	0.005816	0.006157	0.098530	L1
0	3.922	5.337	0.0	5.337	0.002908	0.005815	0.006043	0.098632	L1
0	3.981	5.871	0.0	5.871	0.002907	0.005814	0.005870	0.098787	L1
0	4.068	7.008	0.0	7.008	0.002906	0.005813	0.005614	0.099016	L1
0	4.194	7.469	0.0	7.469	0.002905	0.005811	0.005248	0.099344	L1
0	4.362	7.772	0.0	7.772	0.002904	0.005808	0.004756	0.099785	L1
0	4.571	8.017	0.0	8.017	0.002902	0.005804	0.004149	0.100329	L1
0	4.802	8.231	0.0	8.231	0.002900	0.005800	0.003474	0.100934	L1
0	5.032	8.425	0.0	8.425	0.002898	0.005797	0.002806	0.101532	L1

0	5.236	8.600	0.0	8.600 0.002897 0.005793 0.002213 0.102063	L1
0	5.401	8.757	0.0	8.757 0.002895 0.005790 0.001734 0.102493	L1
0	5.527	8.893	0.0	8.893 0.002894 0.005788 0.001370 0.102818	L1
0	5.619	9.010	0.0	9.010 0.002893 0.005787 0.001103 0.103058	L1
0	5.687	9.112	0.0	9.112 0.002893 0.005786 0.000906 0.103234	L1
0	5.741	9.204	0.0	9.204 0.002892 0.005785 0.000750 0.103374	L1
0	5.789	9.300	0.0	9.300 0.002892 0.005784 0.000611 0.103499	L1
1	5.840	9.420	0.0	9.420 0.002892 0.005783 0.000463 0.103631	L1
0	-5.316	1.797	0.0	1.797 0.002988 0.007968 0.039787 0.073753	L2
0	-3.400	1.990	0.0	1.990 0.002971 0.007922 0.033868 0.079026	L2
0	-2.142	2.180	0.0	2.180 0.002960 0.007893 0.030019 0.082455	L2
0	-1.258	2.368	0.0	2.368 0.002952 0.007872 0.027330 0.084850	L2
0	-0.520	2.561	0.0	2.561 0.002946 0.007855 0.025097 0.086839	L2
0	0.237	2.759	0.0	2.759 0.002939 0.007838 0.022817 0.088870	L2
0	1.032	2.955	0.0	2.955 0.002932 0.007819 0.020432 0.090995	L2
0	1.743	3.141	0.0	3.141 0.002926 0.007803 0.018309 0.092886	L2
0	2.271	3.313	0.0	3.313 0.002922 0.007791 0.016740 0.094284	L2
0	2.642	3.478	0.0	3.478 0.002918 0.007783 0.015637 0.095266	L2
0	2.935	3.647	0.0	3.647 0.002916 0.007776 0.014769 0.096039	L2
0	3.199	3.827	0.0	3.827 0.002914 0.007770 0.013990 0.096733	L2
0	3.447	4.013	0.0	4.013 0.002912 0.007764 0.013256 0.097387	L2
0	3.676	4.193	0.0	4.193 0.002910 0.007759 0.012580 0.097989	L2
0	3.875	4.361	0.0	4.361 0.002908 0.007755 0.011996 0.098509	L2
0	4.035	4.510	0.0	4.510 0.002907 0.007751 0.011525 0.098929	L2
0	4.156	4.638	0.0	4.638 0.002906 0.007748 0.011171 0.099245	L2
0	4.242	4.742	0.0	4.742 0.002905 0.007746 0.010916 0.099471	L2
0	4.302	4.825	0.0	4.825 0.002904 0.007745 0.010741 0.099628	L2
0	4.343	4.887	0.0	4.887 0.002904 0.007744 0.010621 0.099734	L2
0	4.371	4.933	0.0	4.933 0.002904 0.007744 0.010538 0.099808	L2
0	4.392	4.970	0.0	4.970 0.002904 0.007743 0.010477 0.099863	L2
0	4.409	5.002	0.0	5.002 0.002904 0.007743 0.010426 0.099907	L2
0	4.426	5.036	0.0	5.036 0.002903 0.007742 0.010376 0.099952	L2
0	4.447	5.079	0.0	5.079 0.002903 0.007742 0.010316 0.100006	L2
0	4.475	5.145	0.0	5.145 0.002903 0.007741 0.010234 0.100079	L2
0	4.516	5.256	0.0	5.256 0.002903 0.007740 0.010114 0.100186	L2
0	4.576	5.481	0.0	5.481 0.002902 0.007739 0.009937 0.100343	L2
0	4.665	6.177	0.0	6.177 0.002901 0.007737 0.009677 0.100575	L2
0	4.792	7.062	0.0	7.062 0.002900 0.007734 0.009305 0.100906	L2

0	4.967	7.468	0.0	7.468	0.002899	0.007730	0.008792	0.101363	L2
0	5.194	7.758	0.0	7.758	0.002897	0.007725	0.008128	0.101955	L2
0	5.466	7.998	0.0	7.998	0.002895	0.007719	0.007334	0.102662	L2
0	5.762	8.210	0.0	8.210	0.002892	0.007712	0.006471	0.103431	L2
0	6.055	8.403	0.0	8.403	0.002890	0.007706	0.005621	0.104188	L2
0	6.323	8.582	0.0	8.582	0.002887	0.007700	0.004842	0.104882	L2
0	6.559	8.751	0.0	8.751	0.002886	0.007695	0.004157	0.105493	L2
0	6.768	8.917	0.0	8.917	0.002884	0.007690	0.003552	0.106031	L2
0	6.956	9.082	0.0	9.082	0.002882	0.007686	0.003009	0.106515	L2
0	7.126	9.247	0.0	9.247	0.002881	0.007682	0.002519	0.106951	L2
0	7.279	9.410	0.0	9.410	0.002880	0.007679	0.002076	0.107346	L2
0	7.417	9.568	0.0	9.568	0.002878	0.007676	0.001678	0.107700	L2
1	7.543	9.724	0.0	9.724	0.002877	0.007673	0.001316	0.108023	L2
0	-5.319	1.797	0.0	1.797	0.002988	0.009960	0.045773	0.073744	L3
0	-3.409	1.988	0.0	1.988	0.002971	0.009903	0.039837	0.079001	L3
0	-2.147	2.176	0.0	2.176	0.002960	0.009866	0.035953	0.082442	L3
0	-1.244	2.362	0.0	2.362	0.002952	0.009840	0.033191	0.084888	L3
0	-0.468	2.554	0.0	2.554	0.002945	0.009817	0.030831	0.086978	L3
0	0.332	2.751	0.0	2.751	0.002938	0.009794	0.028407	0.089125	L3
0	1.144	2.943	0.0	2.943	0.002931	0.009771	0.025961	0.091292	L3
0	1.832	3.123	0.0	3.123	0.002925	0.009751	0.023894	0.093122	L3
0	2.337	3.290	0.0	3.290	0.002921	0.009737	0.022385	0.094459	L3
0	2.717	3.455	0.0	3.455	0.002918	0.009726	0.021251	0.095463	L3
0	3.049	3.632	0.0	3.632	0.002915	0.009717	0.020263	0.096338	L3
0	3.371	3.822	0.0	3.822	0.002912	0.009708	0.019305	0.097187	L3
0	3.691	4.015	0.0	4.015	0.002910	0.009699	0.018357	0.098026	L3
0	3.994	4.199	0.0	4.199	0.002907	0.009690	0.017460	0.098821	L3
0	4.261	4.370	0.0	4.370	0.002905	0.009683	0.016670	0.099521	L3
0	4.479	4.522	0.0	4.522	0.002903	0.009676	0.016026	0.100091	L3
0	4.645	4.652	0.0	4.652	0.002902	0.009672	0.015538	0.100523	L3
0	4.764	4.760	0.0	4.760	0.002901	0.009668	0.015188	0.100834	L3
0	4.847	4.844	0.0	4.844	0.002900	0.009666	0.014944	0.101050	L3
0	4.903	4.909	0.0	4.909	0.002899	0.009665	0.014777	0.101197	L3
0	4.942	4.957	0.0	4.957	0.002899	0.009663	0.014663	0.101299	L3
0	4.971	4.996	0.0	4.996	0.002899	0.009663	0.014578	0.101374	L3
0	4.995	5.029	0.0	5.029	0.002899	0.009662	0.014508	0.101436	L3
0	5.019	5.065	0.0	5.065	0.002898	0.009661	0.014438	0.101497	L3
0	5.047	5.110	0.0	5.110	0.002898	0.009661	0.014355	0.101571	L3

0	5.085	5.178	0.0	5.178	0.002898	0.009659	0.014243	0.101670	L3
0	5.139	5.292	0.0	5.292	0.002897	0.009658	0.014084	0.101812	L3
0	5.218	5.519	0.0	5.519	0.002897	0.009656	0.013853	0.102016	L3
0	5.331	6.179	0.0	6.179	0.002896	0.009653	0.013522	0.102309	L3
0	5.487	7.018	0.0	7.018	0.002894	0.009648	0.013062	0.102716	L3
0	5.696	7.426	0.0	7.426	0.002893	0.009642	0.012450	0.103258	L3
0	5.958	7.718	0.0	7.718	0.002891	0.009635	0.011682	0.103938	L3
0	6.265	7.960	0.0	7.960	0.002888	0.009627	0.010787	0.104731	L3
0	6.595	8.173	0.0	8.173	0.002885	0.009617	0.009824	0.105584	L3
0	6.928	8.371	0.0	8.371	0.002882	0.009608	0.008854	0.106444	L3
0	7.255	8.561	0.0	8.561	0.002880	0.009599	0.007906	0.107283	L3
0	7.574	8.751	0.0	8.751	0.002877	0.009590	0.006979	0.108105	L3
0	7.889	8.941	0.0	8.941	0.002874	0.009582	0.006068	0.108911	L3
0	8.196	9.130	0.0	9.130	0.002872	0.009573	0.005180	0.109697	L3
0	8.489	9.314	0.0	9.314	0.002870	0.009565	0.004335	0.110446	L3
0	8.759	9.489	0.0	9.489	0.002867	0.009558	0.003558	0.111134	L3
0	8.999	9.652	0.0	9.652	0.002865	0.009551	0.002867	0.111746	L3
1	9.209	9.805	0.0	9.805	0.002864	0.009545	0.002266	0.112278	L3
0	-5.322	1.796	0.0	1.796	0.002988	0.011952	0.051758	0.073735	L4
0	-3.418	1.986	0.0	1.986	0.002971	0.011884	0.045808	0.078976	L4
0	-2.152	2.171	0.0	2.171	0.002960	0.011840	0.041889	0.082427	L4
0	-1.232	2.356	0.0	2.356	0.002952	0.011807	0.039059	0.084920	L4
0	-0.424	2.547	0.0	2.547	0.002945	0.011779	0.036588	0.087096	L4
0	0.407	2.743	0.0	2.743	0.002938	0.011751	0.034055	0.089326	L4
0	1.222	2.931	0.0	2.931	0.002931	0.011723	0.031587	0.091500	L4
0	1.889	3.106	0.0	3.106	0.002925	0.011700	0.029575	0.093272	L4
0	2.383	3.269	0.0	3.269	0.002921	0.011683	0.028089	0.094580	L4
0	2.781	3.436	0.0	3.436	0.002917	0.011669	0.026895	0.095632	L4
0	3.155	3.621	0.0	3.621	0.002914	0.011657	0.025777	0.096617	L4
0	3.537	3.818	0.0	3.818	0.002911	0.011644	0.024636	0.097622	L4
0	3.927	4.015	0.0	4.015	0.002908	0.011630	0.023474	0.098645	L4
0	4.303	4.202	0.0	4.202	0.002904	0.011618	0.022354	0.099631	L4
0	4.639	4.373	0.0	4.373	0.002902	0.011606	0.021357	0.100509	L4
0	4.916	4.527	0.0	4.527	0.002899	0.011597	0.020540	0.101229	L4
0	5.126	4.658	0.0	4.658	0.002897	0.011590	0.019917	0.101778	L4
0	5.278	4.767	0.0	4.767	0.002896	0.011585	0.019467	0.102174	L4
0	5.384	4.852	0.0	4.852	0.002895	0.011581	0.019155	0.102449	L4
0	5.457	4.918	0.0	4.918	0.002895	0.011579	0.018941	0.102637	L4

0	5.507	4.967	0.0	4.967	0.002894	0.011577	0.018793	0.102768	L4
0	5.544	5.006	0.0	5.006	0.002894	0.011576	0.018684	0.102863	L4
0	5.574	5.040	0.0	5.040	0.002894	0.011575	0.018594	0.102942	L4
0	5.604	5.076	0.0	5.076	0.002893	0.011574	0.018506	0.103021	L4
0	5.640	5.122	0.0	5.122	0.002893	0.011573	0.018401	0.103113	L4
0	5.687	5.189	0.0	5.189	0.002893	0.011571	0.018261	0.103236	L4
0	5.754	5.300	0.0	5.300	0.002892	0.011569	0.018065	0.103409	L4
0	5.849	5.512	0.0	5.512	0.002891	0.011566	0.017786	0.103654	L4
0	5.981	6.079	0.0	6.079	0.002890	0.011561	0.017397	0.103997	L4
0	6.160	6.934	0.0	6.934	0.002889	0.011555	0.016872	0.104460	L4
0	6.392	7.365	0.0	7.365	0.002887	0.011548	0.016190	0.105060	L4
0	6.677	7.666	0.0	7.666	0.002885	0.011538	0.015353	0.105797	L4
0	7.007	7.915	0.0	7.915	0.002882	0.011527	0.014388	0.106647	L4
0	7.367	8.136	0.0	8.136	0.002879	0.011515	0.013338	0.107571	L4
0	7.745	8.343	0.0	8.343	0.002876	0.011503	0.012235	0.108543	L4
0	8.141	8.548	0.0	8.548	0.002872	0.011490	0.011086	0.109555	L4
0	8.554	8.753	0.0	8.753	0.002869	0.011476	0.009885	0.110612	L4
0	8.983	8.955	0.0	8.955	0.002865	0.011462	0.008646	0.111704	L4
0	9.415	9.152	0.0	9.152	0.002862	0.011448	0.007399	0.112802	L4
0	9.833	9.339	0.0	9.339	0.002859	0.011434	0.006194	0.113863	L4
0	10.221	9.516	0.0	9.516	0.002855	0.011421	0.005081	0.114844	L4
0	10.563	9.680	0.0	9.680	0.002853	0.011410	0.004098	0.115709	L4
1	10.856	9.830	0.0	9.830	0.002850	0.011401	0.003260	0.116447	L4
0	-5.326	1.795	0.0	1.795	0.002988	0.013944	0.057745	0.073726	L5
0	-3.427	1.983	0.0	1.983	0.002971	0.013865	0.051779	0.078950	L5
0	-2.158	2.167	0.0	2.167	0.002960	0.013813	0.047828	0.082411	L5
0	-1.223	2.350	0.0	2.350	0.002952	0.013775	0.044934	0.084945	L5
0	-0.387	2.540	0.0	2.540	0.002945	0.013741	0.042364	0.087195	L5
0	0.467	2.734	0.0	2.734	0.002937	0.013707	0.039749	0.089485	L5
0	1.277	2.919	0.0	2.919	0.002930	0.013674	0.037281	0.091647	L5
0	1.926	3.088	0.0	3.088	0.002925	0.013648	0.035312	0.093371	L5
0	2.418	3.249	0.0	3.249	0.002920	0.013629	0.033825	0.094673	L5
0	2.839	3.421	0.0	3.421	0.002917	0.013612	0.032556	0.095784	L5
0	3.255	3.612	0.0	3.612	0.002913	0.013595	0.031302	0.096882	L5
0	3.696	3.815	0.0	3.815	0.002910	0.013578	0.029979	0.098041	L5
0	4.156	4.014	0.0	4.014	0.002906	0.013560	0.028604	0.099245	L5
0	4.605	4.202	0.0	4.202	0.002902	0.013542	0.027262	0.100420	L5
0	5.010	4.374	0.0	4.374	0.002898	0.013526	0.026057	0.101476	L5

0	5.345	4.528	0.0	4.528	0.002896	0.013513	0.025063	0.102346	L5
0	5.601	4.660	0.0	4.660	0.002894	0.013503	0.024304	0.103011	L5
0	5.786	4.769	0.0	4.769	0.002892	0.013496	0.023754	0.103492	L5
0	5.915	4.855	0.0	4.855	0.002891	0.013491	0.023372	0.103827	L5
0	6.004	4.920	0.0	4.920	0.002890	0.013487	0.023110	0.104056	L5
0	6.065	4.970	0.0	4.970	0.002890	0.013485	0.022929	0.104215	L5
0	6.110	5.009	0.0	5.009	0.002889	0.013483	0.022796	0.104331	L5
0	6.147	5.043	0.0	5.043	0.002889	0.013482	0.022686	0.104427	L5
0	6.184	5.079	0.0	5.079	0.002889	0.013480	0.022579	0.104521	L5
0	6.226	5.124	0.0	5.124	0.002888	0.013479	0.022452	0.104632	L5
0	6.283	5.189	0.0	5.189	0.002888	0.013476	0.022286	0.104778	L5
0	6.361	5.294	0.0	5.294	0.002887	0.013473	0.022056	0.104979	L5
0	6.469	5.486	0.0	5.486	0.002886	0.013469	0.021737	0.105258	L5
0	6.617	5.953	0.0	5.953	0.002885	0.013463	0.021300	0.105641	L5
0	6.813	6.820	0.0	6.820	0.002883	0.013456	0.020723	0.106146	L5
0	7.062	7.289	0.0	7.289	0.002881	0.013446	0.019990	0.106788	L5
0	7.365	7.607	0.0	7.607	0.002879	0.013434	0.019101	0.107567	L5
0	7.715	7.867	0.0	7.867	0.002876	0.013421	0.018074	0.108466	L5
0	8.106	8.099	0.0	8.099	0.002873	0.013406	0.016932	0.109467	L5
0	8.535	8.320	0.0	8.320	0.002869	0.013389	0.015681	0.110562	L5
0	9.004	8.538	0.0	8.538	0.002865	0.013371	0.014315	0.111758	L5
0	9.516	8.753	0.0	8.753	0.002861	0.013352	0.012830	0.113058	L5
0	10.060	8.961	0.0	8.961	0.002857	0.013331	0.011256	0.114437	L5
0	10.617	9.160	0.0	9.160	0.002852	0.013310	0.009649	0.115844	L5
0	11.161	9.349	0.0	9.349	0.002848	0.013289	0.008085	0.117214	L5
0	11.666	9.526	0.0	9.526	0.002844	0.013270	0.006637	0.118482	L5
0	12.111	9.689	0.0	9.689	0.002840	0.013253	0.005365	0.119596	L5
1	12.486	9.837	0.0	9.837	0.002837	0.013239	0.004294	0.120533	L5

```
STOICHIOMETRIC COEFS 25. .1   11   +2 0+1 3000 3000 1000            00
                                      .003        .010
        M3   25.       .02    1.              .07
   1.        1.0    1.     1.       1.
     0 1 1   9.8                     LOG K1 DIEN
     0 1 2   8.98
 1  0 1 3   4.30
PX FROM VARIATION              11       5500              115 0  12
   2.528     19.656    1.797
 0    -5.319   1.797   0.0    1.797 0.002988 0.009960 0.045773 0.073744 M3
```

0	-3.409	1.988	0.0	1.988	0.002971	0.009903	0.039837	0.079001 M3
0	-2.147	2.176	0.0	2.176	0.002960	0.009866	0.035953	0.082442 M3
0	-1.244	2.362	0.0	2.362	0.002952	0.009840	0.033191	0.084888 M3
0	-0.468	2.554	0.0	2.554	0.002945	0.009817	0.030831	0.086978 M3
0	0.332	2.751	0.0	2.751	0.002938	0.009794	0.028407	0.089125 M3
0	1.144	2.943	0.0	2.943	0.002931	0.009771	0.025961	0.091292 M3
0	1.832	3.123	0.0	3.123	0.002925	0.009751	0.023894	0.093122 M3
0	2.337	3.290	0.0	3.290	0.002921	0.009737	0.022385	0.094459 M3
0	2.717	3.455	0.0	3.455	0.002918	0.009726	0.021251	0.095463 M3
0	3.049	3.632	0.0	3.632	0.002915	0.009717	0.020263	0.096338 M3
0	3.371	3.822	0.0	3.822	0.002912	0.009708	0.019305	0.097187 M3
0	3.691	4.015	0.0	4.015	0.002910	0.009699	0.018357	0.098026 M3
0	4.261	4.370	0.0	4.370	0.002905	0.009683	0.016670	0.099521 M3
0	4.479	4.522	0.0	4.522	0.002903	0.009676	0.016026	0.100091 M3
0	4.764	4.760	0.0	4.760	0.002901	0.009668	0.015188	0.100834 M3
0	4.903	4.909	0.0	4.909	0.002899	0.009665	0.014777	0.101197 M3
0	4.971	4.996	0.0	4.996	0.002899	0.009663	0.014578	0.101374 M3
0	5.047	5.110	0.0	5.110	0.002898	0.009661	0.014355	0.101571 M3
0	5.139	5.292	0.0	5.292	0.002897	0.009658	0.014084	0.101812 M3
0	5.218	5.519	0.0	5.519	0.002897	0.009656	0.013853	0.102016 M3
0	5.331	6.179	0.0	6.179	0.002896	0.009653	0.013522	0.102309 M3
0	5.487	7.018	0.0	7.018	0.002894	0.009648	0.013062	0.102716 M3
0	5.696	7.426	0.0	7.426	0.002893	0.009642	0.012450	0.103258 M3
0	5.958	7.718	0.0	7.718	0.002891	0.009635	0.011682	0.103938 M3
0	6.265	7.960	0.0	7.960	0.002888	0.009627	0.010787	0.104731 M3
0	6.595	8.173	0.0	8.173	0.002885	0.009617	0.009824	0.105584 M3
0	6.928	8.371	0.0	8.371	0.002882	0.009608	0.008854	0.106444 M3
0	7.255	8.561	0.0	8.561	0.002880	0.009599	0.007906	0.107283 M3
0	7.574	8.751	0.0	8.751	0.002877	0.009590	0.006979	0.108105 M3
1	7.889	8.941	0.0	8.941	0.002874	0.009582	0.006068	0.108911 M3

END

4.6. LINEAR LS REFINE

1-3.   As Section 4.1, items 1-3.
   4.   1 line   [As section 4.1, item 4]:   **(CODEN(I), I=5), TEMP, UCORR,
           NM, NL, IACT, ISORT, (QC(J), J=1, 3), (NH(I), I=1, 4),
           (IHM(I), I=1, 4), (LIGS(I), I=1, 4), NSET, IPUN, LONG,
           IPLOT, IDAT, IPAR, IGUESS, TAPIN, TAPOUT, PH1, PH2**

CODEN: LINEAR LS REFINE

TEMP: $\left.\begin{array}{c} \\ \vdots \\ \\ NL: \end{array}\right\}$ Same as for Section 4.1, item 3.

QC( ): $\left.\begin{array}{c} \\ \vdots \\ \\ LONG: \end{array}\right\}$ Same as for Section 4.1, item 3

5. 1 line [3I2, 5X, 4I1, 5F5.0]: NCY, **IW**, INEG, (KKI(I), I=1, 4), SOECUT, OPT1, OPT2, DELMAX, DEL

   IW: Control integer to indicate whether weight is to be used.

   IW=1 implies no weighting; IW=0 weighting used. Choose IW=1.

6. 1 line [I1, AI, 4I2, F10.0, 3F5.0, 7A4]: **IS, KI(I), (IX(J, I), J=1, 4), BL(I), SL(I), DP(I), SHFT(I), (LABP(I, J), J=7)**

   This item is used to read in equilibrium constants. See Section 4.1, item 8 for full details. *Exception:*

   KI(I): Use * to identify an unknown (to be determined) constant. There will be as many cards as constants.

7. 1 line [20I2]: **(ICB(I, J), J=1,NSP)**

   Conversion matrix, read in as shown in Section 4.1, item 9.

8. N lines, predetermined format.

   Data deck, produced in Section 4.4, is placed here.

9. 1 line [5A4, . . .]: **(CODEN(I), I=1, 5), . . .**

   Final card, i.e., END (see Section 4.1, item 9).

READ statements for this section are located as follows:

Item	Routine	Line No.
1–3	As for Section 4.1, items 1–3	
4	MAIN	94–96
5	PRELIM	29
—[a]	PRELIM	37
—[a]	PRELIM	39
6	READP	34
7	READP	53
8[b]	READD	35
9	MAIN	94–96

[a] Data read in only if IW=0.
[b] Data read in only if IDAT=0.

4.7. NONLINEAR LS REFINE

4.7.1. Instructions for Data Input

1–10. As section 4.2, items 1–10.

11. 1 line [As Section 4.1, item 4): **(CODEN(I), I=1, 5, TEMP, UCORR,**

**NM, NL, IACT, ISORT, (QC(J), J=1, 3), (NH(I), I=1, 4),
(IHM(I), I=1, 4, (LIGS(I), I=1, 4), NSET, IPUN, LONG,
IPLOT, IDAT,** IPAR, IGUESS, TAPIN, TAPOUT, PH1, PH2
CODEN: NONLINEAR LS REFINE

QC( ): ⎫
: ⎬ Same as for Section 4.1, item 3.
: ⎪
LONG: ⎭

IPLOT: IPLOT=1 produces a plot for (pH^{obs} − pH^{calc}) vs. pH. Otherwise 0.

IDAT: Set equal to unity.

PH1: Minimum pH value to be considered in the refinement.

PH2: Maximum pH value to be considered in the refinement.

12. 1 line [3I2, 5X, 4I1, 5F5.0] **NCY, IW, INEG, (KKI(I), I=1, 4),** SOECUT,
IPT, OPT2, DELMAX, DEL

NCY: Number of refinement cycles.

IW: Weighting indicator integer. Set equal to zero, thus item 3 is included.

INEG: Set to unity.

KKI( ): The significance of KKI is shown below:

KKI

1	2	3	4	
0	0	1	0	If three reactants considered; pH measured.
1	0	1	0	As above, except pH and pM measured.
0	1	0	0	If two reactants considered; pH measured.

FICS pM values may be considered as being measured, as well as those obtained from ISE metal electrodes.

13. 1 line [7F5.0]: **SOH, (SIGC(K), SIGV(K),** SIGX(K), **K=1,** NC1)
Parameters for weighting scheme as described in Section 2.6, equation (12).

SOH: Set to unity.

SIGC( ): σ_c for reactant k.

SIGV( ): σ_v for reactant k.

Note: For metal or ligand reactants, $\sigma_c = 0.05$ is typical; for hydrogen reactant, $\sigma_c = 0.02$ is typical. For all reactants $\sigma_v = 0.001$ is normal.

14. 1 line [4F10.4]: **(DEDV(J), J=1, NSET)**

DEDV: Conversion factors: da/dV = (titration concentration)/(total concentration of first reactant * total volume in milliliters)

DEDV(1), . . . DEDV(NSET): da/dV for first titration set, . . . da/dV for NSET titration set.

15. 1 line [I1, A1, 4I2, F10.0, 3F5.0, 7A4]: **IS, KI(I), (IX(J, I), J=1, 4),** BL(I),
SL(I), DP(I), SHFT(I), **(LABP(I, J), J=1, 7)**
Equilibrium constants read in as shown in Section 4.6, item 6.
Note the use of the *.

16. 1 line [20I2]: **(ICB(I, J) J=1,NSP)**
Conversion matrix; read in as shown in Section 4.1, item 9.
17. 1 line [4A1]: **(KII(I), I=1, 4)**
Alphanumeric indicators to control variation of scale factors.
Set up as follows:
***B vary scale factors of metal concentration *only*.
*LAB vary scale factor of base concentration *only*.
MLA* vary scale factor of base concentration *only*.
If none of these options is needed include a blank card.
18. 1 line [20X, 5F10.0] **ACID(J), BASE(J), STOCK(J)**
19. 1 line [8F10.4] **(TOT(J, K), K=1, NC)**
this pair of cards is included *only* if item 17 is *non*blank.
One pair of cards per titration set.
ACID( ): ⎫
　　　　　⎬ As for Section 4.2, item 5, for the *J*th set.
STOCK( ): ⎭
20. 1 line [5A4, . . .]: **(CODEN(I), I=1, 5 . . . . .**
Final card, i.e., END (see Section 4.1, item 9).

Item	Routine	Line No.
1–10	As for Section 4.2, items 1–10	
11	MAIN	94–96
12	PRELIM	29
13[a]	PRELIM	37
14[a]	PRELIM	39
15	READP	34
16	READP	53
—[b]	READD	35
17	NONLIN	96
18[c]	NONLIN	105
19[c]	NONLIN	106
—[d]	NONLIN	110
20	MAIN	94–96

[a]Data read in only if IW = 0.
[b]Data read in only if IDAT = 0.
[c]Data read in only if data are present for item 17.
[d]Data read in only if IGUESS = 1.

4.7.2. Input Data for NONLINEAR LS REFINE

The following is a data set for the nonlinear least-squares refinement option of
STBLTY. The data shown below are for items 11–20 only.

```
¶<-- Column 1

CU2+ DIEN  NONL LS  PM+PL+PH DATA

A Blank Card Goes Here
```

```
CU2+DIENH+

NONLINEAR LS REFINE  25.  .1  115 +2 0+1 3000 3000 1000     1151000 9 2.   9.
   6 0 1     1110 .001              0.50.000 .000 .00
  1.  .05  .001 .0   .05  .001 .0   .02  .001 .0
  13.33
     0 1 1   9.8                    LOG K1 DIEN
     0 1 2   8.96                   LOG K2 DIEN
     0 1 3   4.2                    LOG K3 DIEN
  * 1 1 1   -3.74                   100+013=111+2(001)
  * 1 1 0   -3.20                   111=110+001
  * 1 2 1   -5.70                   110+012=121+001
9* 1 2 0   -8.20                    121=120+001
  1 0 0 0 0 0 0
  1 1 0 0 0 0 0
  1 1 1 0 0 0 0
  1 1 1 1 0 0 0
  1 1 1 1 1 0 0
  2 2 1 1 1 1 0
  2 2 1 1 1 1 1
0    -5.319   2.528  19.656   1.797 0.002988 0.009960 0.045773 0.073744  M3
0    -3.409   2.533  19.085   1.988 0.002971 0.009903 0.039837 0.079001  M3
0    -2.147   2.543  18.525   2.176 0.002960 0.009866 0.035953 0.082442  M3
0    -1.244   2.567  17.976   2.362 0.002952 0.009840 0.033191 0.084888  M3
0    -0.468   2.621  17.416   2.554 0.002945 0.009817 0.030831 0.086978  M3
0     0.332   2.736  16.854   2.751 0.002938 0.009794 0.028407 0.089125  M3
0     1.144   2.935  16.318   2.943 0.002931 0.009771 0.025961 0.091292  M3
0     1.832   3.211  15.817   3.123 0.002925 0.009751 0.023894 0.093122  M3
0     2.337   3.540  15.349   3.290 0.002921 0.009737 0.022385 0.094459  M3
0     2.717   3.915  14.883   3.455 0.002918 0.009726 0.021251 0.095463  M3
0     3.049   4.350  14.385   3.632 0.002915 0.009717 0.020263 0.096338  M3
0     3.371   4.833  13.859   3.822 0.002912 0.009708 0.019305 0.097187  M3
0     3.691   5.325  13.340   4.015 0.002910 0.009699 0.018357 0.098026  M3
0     4.261   6.204  12.436   4.370 0.002905 0.009683 0.016670 0.099521  M3
0     4.479   6.565  12.071   4.522 0.002903 0.009676 0.016026 0.100091  M3
0     4.764   7.111  11.524   4.760 0.002901 0.009668 0.015188 0.100834  M3
0     4.903   7.441  11.194   4.909 0.002899 0.009665 0.014777 0.101197  M3
0     4.971   7.631  11.005   4.996 0.002899 0.009663 0.014578 0.101374  M3
0     5.047   7.878  10.761   5.110 0.002898 0.009661 0.014355 0.101571  M3
0     5.139   8.265  10.377   5.292 0.002897 0.009658 0.014084 0.101812  M3
```

0	5.218	8.741	9.907	5.519	0.002897	0.009656	0.013853	0.102016	M3
0	5.331	10.107	8.569	6.179	0.002896	0.009653	0.013522	0.102309	M3
0	5.487	11.846	6.907	7.018	0.002894	0.009648	0.013062	0.102716	M3
0	5.696	12.720	6.125	7.426	0.002893	0.009642	0.012450	0.103258	M3
0	5.958	13.373	5.584	7.718	0.002891	0.009635	0.011682	0.103938	M3
0	6.265	13.941	5.152	7.960	0.002888	0.009627	0.010787	0.104731	M3
0	6.595	14.469	4.784	8.173	0.002885	0.009617	0.009824	0.105584	M3
0	6.928	14.991	4.452	8.371	0.002882	0.009608	0.008854	0.106444	M3
0	7.255	15.513	4.144	8.561	0.002880	0.009599	0.007906	0.107283	M3
0	7.574	16.040	3.850	8.751	0.002877	0.009590	0.006979	0.108105	M3
1	7.889	16.552	3.577	8.941	0.002874	0.009582	0.006068	0.108911	M3

***B

    M3   25.   .02   1.   .07

  .003   .01

END

4.8. ANALYSIS OF VARIANCE

4.8.1. Instructions for Data Input

1–18.   As Section 4.7, items 1–17.

19. 1 line   [As Section 4.1, item 4]: **CODEN(I),I=1,5),**
TEMP,UCORR,NM,NL,IACT,ISORT,$(QC(J),J=1,3),(NH(I),I=1,4),$
$(IHM(I),I=1,4),$ **(LIGS(I),I=1,4), NSET,**IPUN, **LONG,IPLOT,**
IDAT,IPAR,IGUESS,TAPIN,TAPOUT, PH1,PH2
CODEN: ANALYSIS OF VARIANCE
TEMP:⎫
  · ⎪
  · ⎬ Same as for Section 4.1, item 3.
  · ⎪
NL: ⎭
IACT: Set to unity.
ISORT: Set to zero.
NSET: Number of titration sets.
LONG: Indicator for type of weighting scheme to be used for statistical
analysis. LONG=1 for unit weights; = 0 nonunit weights.
IPLOT: IPLOT=1 for Calcomp plots; else 0. Routines not included.

20. 1 or 2   [8F10.5]: **SOH,(SIGC(K),SIGV(K),**SIGX(K),K=1,NC1)
lines   SOH:⎫
SIGC:⎬As for Section 4.7, item 5. Note format change here.
SIGV:⎭
A second card will be needed for SIGC(3) and SIGV(3) if three reactants
are present.

21. 1 line [8F10.5]: (SPX(K),K=1,NC1)
   SPX: Not used for this section. Include a blank card.
22. 1 line [5A4, . . .]: **(CODEN(I),I=1,5)** . . .
   Final card, i.e., END (see Section 4.1, item 9).

READ statements for this section are located as follows:

Item	Routine	Line No.
1–18	As for Section 4.10, items 1–18	
19	MAIN	94–96
20	ERRORS	67
21	ERRORS	68
22	MAIN	94–96

4.8.2. Input Data for ANALYSIS OF VARIANCE

The following is a set of data for the analysis of variance option of STBLTY. The data shown are for items 19–21 only.

```
¶<-- Column 1

CU2+ DIEN  NONL LS ANAL VARIANCES

Blank Card Goes Here.

CU2+DIENH+

ANALYSIS OF VARIANCE 25.  .1  1110                     1 05  09

1.         .05      .001      .0      .05      .001      .000      .02

   .001     0.00

Blank Card Goes Here.

END
```

5. Source Listing of STBLTY

There follows a source listing of the program STBLTY. The user's attention is drawn to the fact that portions of the original program have been omitted from the listing shown here. These routines, the names of which are shown at the beginning of the listing, are related to the Calcomp plotting routines and to electrode calibration options. The options described in Sections 2 and 4 have all the required software presented in the following listing of the source.

```
1   C
2   C       THIS IS A REDUCED VERSION OF STBLTY WITH GRAFN, CALC4, GLASS,
3   C       PHSYMB,STATPL, TITRPL, DISTPL, FORMPL, AXES, SCALES, SIGPL,
4   C       VARNPL, TABLE, NORMPL, PLOTS, PLOT, SYMBOL, FACTOR, SYMB,
5   C       APEB, VARPL, AND STATPL REPLACED BY DUMMY CALLS.
6   C
```

```
 7    C -------*---------*---------*---------*---------*---------*---------*
 8    C
 9    C  POTENTIOMETRIC DATA. (DATA REDUCTION, DETERMINATION AND REFINEMENT
10    C  OF CONSTANTS, GRAPHICAL DISPLAY OF RESULTS, ANALYSIS OF VARIANCE)
11    C  CODED BY A.AVDEEF 1975-1978 LBL AND CHEM DEPT, UNIV OF CALIF,
12    C  BERKELEY. MAJOR MODIFICATIONS BY A.AVDEEF 1978-1982 SYRACUSE
13    C  UNIVERSITY.  PRESENT VERSION TESTED ON IBM370/155 AND IBM4341
14    C   LAST REVISION DATA:  23 JAN 1982
15    C
16    C  SUBROUTINES REQUIRED
17    C
18    C     REDUCE    READD     TITRPL    TSORT4?   NORMPL
19    C     BJERM     READP     DISTPL    MOVE1     PLOTS*
20    C     FANTSY    PRELIM    FORMPL    MOVE2     PLOT*
21    C     OSTR1     GLASS     AXES      ICOMP1    SYMBOL*
22    C     OSTR2     IONIC     SCALES    ICOMP2    FACTOR*
23    C     GRAFN     ACTIV     ERRORS    ICOMP3    DMINV
24    C     LINLSQ    WEIGHT    SIGPL     ICOMP4    SYMB
25    C     NONLIN    STOICH    LINPLT    ICOMP5    BOOT
26    C     CALC1     MINLS     FIT       RFACTR    DELTA
27    C     CALC2     CTOT      YCALC     CLEAR     APEB
28    C     CALC3     PHSYMB    MINV      SPLINE    VARPL
29    C     CALC4     SEARCH    PROB      TABLE     STATPL
30    C     MOVE3     STATPL    VARNPL    SMOOTH&   CHOLID&   PPVALU&
31    C     NNLS#     H12#      G1#       G2#       DIFF#
32    C     (# LAWSON AND HANSON 1974)
33    C     (& DE BOOR 1978)
34    C     (? UNKNOWN ORIGIN)
35    C     (* IBM370/155 ROUTINES AT SYRACUSE UNIVERSITY)
36    C
37          INTEGER TAPIN,TAPOUT
38          COMMON/LARG/STAT(600,4),SIGPX(4,600),SIGD(4,600),INDX(600)
39          COMMON/APE/UM, T, QWLOGM,CON,SQM,SQ,ITOT,ITITR,QWLOG
40          COMMON/BETA/BETA(20),BLOG(20),BL(20),B(20),SL(20),SLB(20),SB(20)
41          COMMON/CONC/CF(4),HOH,CTC(4),C(20),PC(20),TOTX(4)
42          COMMON/CONT/IPUN,IPLOT,IFPL,LONG,IDAT,IPAR,IGUESS,IDPX,ISORT,IREF
43          COMMON/CYCL/ICYCL,JTYPE,KTYPE
44          COMMON/DATA/IDATA(600),PCF(4,600),CT(4,600),EQV(600),U1(600)
45          COMMON/DERV/SD(4,4),CB(20,20),CK(20,20),ITR
46          COMMON/DETERM/DETR
47          COMMON/DIM/NO,NOS(20),NSP,NM,NL,NC,NC1,NV,NSET,IST,JST,NY,NCY,IW
48          COMMON/FANT/VO,ACID,BASE,TL,DIL
49          COMMON/INIT/IACT,AU,AH,AOH,A1,A0,B1,AKW,CKW,TEMP,UCORR
50          COMMON/IO/TAPIN,TAPOUT,INTAP(20)
51          COMMON/LIGS/HM(4),NH(4),LIGS(4),KKI(4),QC(4),QR(20),Q(20)
52          COMMON/LIMS/PH1,PH2,PHMIN,PHMAX,EMIN,EMAX,PLMIN,PLMAX,BMAX
53          COMMON/LIMS2/HTMIN,HTMAX
54          COMMON/LIMT/JMEM,JSET,MSTART,MEND,NS
55          COMMON/MISC/AL(4,200),ALPHA(20,200),U2(1200)
56          COMMON/MONT/ICRD,IPRT,IPCH,TITLE(20),FMT(20),YNAME(4),CODEN(5)
57          COMMON/PAR1/KI(20),LABP(20,7),IX(4,20),EX(4,20)
58          COMMON/PAR2/AM(20,20),V(20),DP(20),P(20),SHFT(20),SOECUT,OPT1,OPT2
59          COMMON/PREL/NCM,INDEX1(4),INDEX2(4),INDEX3(4)
60          COMMON/RFAC/SIG(50),DENOM(50),NPTS(50),SUM(50)
61          COMMON/SIG1/SIGC(4),SIGV(4),SIGX(4),SOH
62          COMMON/SIG2/SCC(4),RCC(4,4),DEL(4),RHO(20,20),SP(20),DEDV(20)
63          COMMON/SGNL/MAPEB(6)
64          DIMENSION TYPE(16),IHM(4),DEFFMT(20),DIFFMT(20)
65          DATA TYPE/4HYES ,4HEND ,4HDATA,4HVARI,4HPX F,4HGRAP,4HFANT,
66         & 4HNONL,4HLINE,4HANAL,4HTITR,4HPUBL,4HPAPE,4HCALI,4HSTOI,4H    /
67          DATA DEFFMT/4H(I1, ,4HA1,2 ,4HX,4F ,4H8.3, ,4H4F9. ,4H6,2X ,4H,
68         & A4),13*4H    /
69          DATA DIFFMT/4H(I1, ,4HA1,2 ,4HX,3F ,4H8.3, ,4H3F9. ,4H6,2X ,4H,
70         & A4),13*4H    /
71        5 FORMAT(20A4)
72        6 FORMAT(///,10X,'UNRECOGNIZED FUNCTION REQUESTED',//)
73        7 FORMAT(5A4,2F5.0,4I1,3F2.0,1X,4I1,1X,4I1,1X,4I1,4X,I2,8I1,1X,
74         & 2F5.0)
75        8 FORMAT(1H1,20A4,//,2X,5A4,/)
76       10 FORMAT(4A4,20I2)
77      408 FORMAT(' TEMPERATURE',11(1H.),F8.1,4H DEG,/,15H REF IONIC STR  ,
78         & 8(1H.),F8.2,2H M,/,8H PKW(U0),15(1H.),F10.3)
79      411 FORMAT(' DEBYE-HUCKEL CONSTANTS A=',F6.4,2X,3H B=,F6.4,2X,4H A0=,
80         &F6.4,/)
81      413 FORMAT('0DEBYE-HUCKEL ACTIVITY COEFFICIENTS FOR H+ AND OH-')
```

```
82          414 FORMAT('ØEXPERIMENTAL ACTIVITY COEFFICIENTS FOR H+ AND OH-',/
83         &' (FROM PITZER ANALYTICAL EXPRESSIONS FOR HCL AND KOH ',
84         &'ACTIVITIES IN KCL)')
85          415 FORMAT('ØEMPIRICALLY-DETERMINED RELATION BETWEEN PH(METER) AND',
86         &' P(H)')
87             IDPX=Ø
88             KTYPE=Ø
89             ICRD=5
90             IPCH=7
91             IPRT=6
92           1 READ(ICRD,5)TITLE,FMT
93             READ(ICRD,1Ø)YNAME,INTAP
94           3 READ(ICRD,7)(CODEN(I),I=1,5),TEMP,UCORR,NM,NL,IACT,ISORT,
95         &(QC(J),J=1,3),(NH(I),I=1,4),(IHM(I),I=1,4),(LIGS(I),I=1,4),
96         &NSET,IPUN,LONG,IPLOT,IDAT,IPAR,IGUESS,TAPIN,TAPOUT,PH1,PH2
97             DO 2ØØ I=1,4
98         2ØØ HM(I)=FLOAT(IHM(I))
99             IF(TAPIN.GT.ICRD)GO TO 12
100            DO 11 J=1,2Ø
101         11 INTAP(J)=ICRD
102         12 IF(TAPOUT.EQ.Ø)TAPOUT=IPRT
103            NC=NM+NL
104            NC1=NC+1
105            IF(FMT(1).NE.TYPE(16))GO TO 321
106            DO 2 I=1,2Ø
107            FMT(I)=DEFFMT(I)
108          2 IF(NC.EQ.1)FMT(I)=DIFFMT(I)
109        321 QC(NC1)=1.
110            IF(UCORR.LT.1.E-3)UCORR=.2
111            IF(TEMP.LT.1.E-3)TEMP=25.
112            IF(NSET.EQ.Ø)NSET=1
113            IF(PH1.LE.Ø.)PH1=1.5
114            IF(PH2.LE.Ø.)PH2=12.5
115            IST=NH(1)+NH(2)+NH(3)+NH(4)
116            JST=IST+LIGS(1)+LIGS(2)+LIGS(3)+LIGS(4)
117            ITYPE=Ø
118            DO 4 I=1,15
119            IF(CODEN(1).NE.TYPE(I))GO TO 4
120            ITYPE=I
121            GO TO 9
122          4 CONTINUE
123          9 WRITE(IPRT,8)TITLE,CODEN
124            IF(ITYPE.EQ.Ø)GO TO 8888
125            IF(ITYPE.LT.3)GO TO 14
126            T=273.16+TEMP
127    C  A1=.51Ø1 AT 25 DEG  (MOLAL SCALE)
128            A1=-2.97627+4.8Ø688E-2*T-2.6928E-4*T**2+7.49524E-7*T**3
129         &  -1.Ø2352E-9*T**4+5.58ØØ4E-13*T**5
130    C  B1=.3296 AT 25 DEG  (MOLAL SCALE)
131            B1=.4125*A1**.3333333333333
132            AØ=-A1*(SQRT(UCORR)/(1.+SQRT(UCORR))-.3*UCORR)
133            CALL ACTIV(UCORR,1.)
134            PKW=-ALOG1Ø(CKW)
135            GO TO (1,9999,15,1Ø4,1Ø5,15,15,15,15,11Ø,111,112,8888,15, 15),
136         & ITYPE
137         15 WRITE(IPRT,4Ø8)TEMP,UCORR,PKW
138            IF(ITYPE.EQ.15)GO TO 115
139            IF(IACT.EQ.Ø)WRITE(IPRT,414)
140            IF(IACT.EQ.1)WRITE(IPRT,413)
141            IF(IACT.GT.1)WRITE(IPRT,415)
142            WRITE(IPRT,411)A1,B1,AØ
143    C  ALLOWED CODEN LIST
144    C  -------------------
145    C  FANTASY CALCULATION
146    C  DATA REDUCTION
147    C  CALIBRATE ELECTRODE
148    C  TITRATION CURVE PLOT
149    C  VARIATION
150    C  PX FROM VARIATION
151    C  STOICHIOMETRIC COEFS
152    C  GRAPHICAL CONSTANTS
153    C  LINEAR LS REFINE
154    C  NONLINEAR LS REFINE
155    C  ANALYSIS OF VARIANCE
156    C  PUBLICATION LIST
```

```
157   C  --------------------
158        14 GO TO(1,9999,103,104,105,106,107,108,109,110,111,112,113,114,
159           & 115),ITYPE
160       103 CALL REDUCE
161           GO TO 3
162       104 IREF=0
163           CALL OSTR1
164           GO TO 3
165       105 CALL OSTR2
166           GO TO 3
167       106 CALL GRAFN
168           GO TO 3
169       107 CALL FANTSY
170           GO TO 3
171       108 CALL NONLIN
172           GO TO 3
173       109 CALL LINLSQ
174           GO TO 3
175       110 IF(TAPIN.LT.4)GO TO 1100
176           READ(TAPIN)STAT,SIGPX,SIGD,INDX,IDATA,PCF,CT,EQV,U1,U2,
177           & NO,NOS,NSP,NV,NY,IREF,SOH,DEDV,NCM,INDEX2
178      1100 CALL ERRORS
179           GO TO 3
180       111 IF(IDAT.EQ.0)CALL READD(0)
181           CALL TITRPL(TEMP,UCORR,NO,NSET,NC1)
182           GO TO 3
183       112 CALL TABLE
184       113 GO TO 3
185       114 CALL GLASS
186           GO TO 3
187       115 IF(IDPX.EQ.0)GO TO 3
188           CALL STOICH(CKW,NC,NC1,ICRD,IPRT,NS,NM,NL,UCORR)
189           GO TO 3
190      8888 WRITE(IPRT,6)
191           GO TO 3
192      9999 STOP
193           END
```

```
  1   C -------*---------*---------*---------*---------*---------*---------*
  2         SUBROUTINE FANTSY
  3   C  CALCULATION OF TITRATION, DISTRIBUTION, AND FORMATION CURVES GIVEN
  4   C  ALL THE STABILITY CONSTANTS (THIS CALCULATION REQUIRES NO DATA)
  5         EXTERNAL MOVE3,ICOMP3
  6         COMMON/LARG/ALPHA(4800),FULL(2400),INDX(600)
  7         COMMON/DATA/IDATA(600),PCF(4,600),CT(4,600),EQV(600),U1(600)
  8         COMMON/MISC/FILL(6000)
  9         COMMON/MONT/ICRD,IPRT,IPCH,TITLE(20),FMT(20),YNAME(4),CODEN(5)
 10         COMMON/DIM/NO,NOS(20),NSP,NM,NL,NC,NC1,NV,NSET,IST,JST,NY,NCY,IW
 11         COMMON/CONT/IPUN,IPLOT,IFPL,LONG,IDAT,IPAR,IGUESS,IDPX,ISORT,IREF
 12         COMMON/CONC/CF(4),HOH,CTC(4),C(20),PC(20),TOTX(4)
 13         COMMON/LIMS/PH1,PH2,PHMIN,PHMAX,EMIN,EMAX,PLMIN,PLMAX,BMAX
 14         COMMON/LIGS/HM(4),NH(4),LIGS(4),KKI(4),QC(4),QR(20),Q(20)
 15         COMMON/BETA/BETA(20),BLOG(20),BL(20),B(20),SL(20),SLB(20),SB(20)
 16         COMMON/INIT/IACT,AU,AH,AOH,A1,A0,B1,AKW,CKW,TEMP,U
 17         COMMON/FANT/VO,ACID,BASE,TL,DIL
 18         COMMON/SIG1/SIGC(12),DM
 19         COMMON/PREL/NCM,INDEX1(4),INDEX2(4),INDEX3(4)
 20         COMMON/LIMS2/HTMIN,HTMAX
 21         DIMENSION CO(4)
 22         DATA XN1,YN1,XN2,YN2/2HPX,3HBAR,2HPH,4HALFA/
 23       1 FORMAT(1H1)
 24       3 FORMAT(16F5.0)
 25       4 FORMAT(///,' INITIAL PX(NC) VALUES ARE <ITUO> ',3F8.3)
 26      93 FORMAT(/,I4,' POINTS DID NOT REFINE',/)
 27     133 FORMAT( '  PH    VOL    EQV    NBAR   HBAR   PX',/)
 28     158 FORMAT(8F10.0)
 29     408 FORMAT(5H ACID,18(1H.),F9.6,2H M,/,
 30           &      5H BASE,18(1H.),F9.6,2H M,/,
 31           &      5H SE  ,18(1H.),F9.4,2H M,/,
 32           &      5H VO  ,18(1H.),F9.4,3H ML)
 33     409 FORMAT(//,16X,4HBARE,6X,7HINITIAL,5X,23HMAX NO.  NO. PROTONS AS,
 34           &/58H REACTANTS    CHARGE  CONCENTRATION  PROTONS  INTRODUCED,/)
```

```
 35       410 FORMAT(4X,A4,I10,F14.6,I9,I12)
 36           PLMAX=0.
 37           PLMIN=30.
 38           BMAX=0.
 39           READ(ICRD,3)SCALEX,SCALEY,SCALEN,DELPH,SHRINK,DM
 40           IF(DM.LT.0.001)DM=1.
 41           READ(ICRD,158)VO,ACID,BASE,SE,(TOTX(I),I=1,4)
 42           IF(IGUESS.NE.0)READ(ICRD,158)(PCF(J,1),J=1,NC)
 43           NO=(2.+(PH2-PH1)/DELPH)
 44           IF(NO.GT.200)NO=200
 45           IF(SHRINK.LE.1.E-5)SHRINK=.3
 46           WRITE(IPRT,408)ACID,BASE,SE,VO
 47       C   READ PARAMETERS
 48           CALL READP
 49           TL=0.
 50           IF(NL.EQ.0)GO TO 301
 51           DO 302 I=1,NL
 52       302 TL=TL+TOTX(I+NM)*HM(I)
 53       301 WRITE(IPRT,409)
 54           DO 400 I=1,NC
 55           IQ=QC(I)
 56           MH=HM(I-NM)
 57           IF(I.LE.NM)WRITE(IPRT,410)YNAME(I),IQ,TOTX(I)
 58           IF(I.GT.NM)WRITE(IPRT,410)YNAME(I),IQ,TOTX(I),NH(I-NM),MH
 59       400 CONTINUE
 60       C   INITIAL FREE CONCENTRATIONS
 61           IF(IGUESS.GT.0)GO TO 147
 62           IF(NM.EQ.0)GO TO 150
 63           DO 146 K=1,NM
 64       146 CF(K)=TOTX(K)
 65       150 IF(NL.EQ.0)GO TO 59
 66           KK=1
 67           DO 149 I=1,NL
 68           NI=NH(I)
 69           DLL=1.
 70           DO 148 J=1,NI
 71           DLL=DLL+10.**(BLOG(KK)-FLOAT(J)*PH1)
 72       148 KK=KK+1
 73       149 CF(I+NM)=TOTX(I+NM)/DLL
 74        59 DO 58 J=1,NC
 75        58 PCF(J,1)=-ALOG10(CF(J))
 76       147 WRITE(IPRT,4)(PCF(J,1),J=1,NC)
 77           WRITE(IPRT,1)
 78           IF(LONG.EQ.1)WRITE(IPRT,133)
 79           DIL=1.
 80           IF(NSET.EQ.10)DIL=.95
 81           PHM=PH1-DELPH
 82           DO 88 J=1,NC1
 83        88 INDEX3(J)=J
 84           J=1
 85           ID=0
 86           KX=0
 87           DO 92 I=1,NO
 88           IDATA(I)=1
 89           INDX(I)=IDATA(I)
 90           IF(KX.EQ.0)PHM=PHM+DELPH
 91           KX=0
 92           PCF(NC1,J)=PHM
 93           KJJ=0
 94       C   SELF-CONSISTENCY ON DILUTION (3-PASS)
 95        90 KJJ=KJJ+1
 96           DO 89 L=1,NC
 97        89 CT(L,I)=TOTX(L)*DIL
 98           IF(KJJ.GT.3)GO TO 191
 99           U1(J)=(SE+BASE)*DIL
100           IF(NSET.NE.10)U1(J)=SE*DIL+(1.-DIL)*BASE
101           IF(NL.EQ.0)GO TO 199
102           DO 190 K=1,NL
103           QC1=QC(K+NM)+HM(K)
104       190 U1(J)=U1(J)+CT(K+NM,J)*(ABS(QC1)-QC1)
105       199 CALL CALC3(J,IFAIL,KJJ)
106           IF(IFAIL.NE.0)GO TO 91
107           GO TO 90
108       191 J=J+1
109           GO TO 92
110        91 PCF(NC1,J)=PCF(NC1,J)-DELPH*SHRINK
```

```
111          KX=1
112          IF(PCF(NC1,J).GT.PHM-DELPH)GO TO 90
113          ID=ID+1
114          KX=0
115       92 IF(IPLOT.EQ.1)INDX(I)=FILL(I+1200)+1.
116          NO=J-1
117          NOS(1)=NO
118          IF(ID.GT.0)WRITE(IPRT,93)ID
119          IF(NO.LT.5)RETURN
120          PHMIN=PCF(NC1,1)
121          PHMAX=PCF(NC1,J-1)
122          EMIN=EQV(1)
123          EMAX=EQV(J-1)
124          HTMIN=CT(NC1,J-1)
125          HTMAX=CT(NC1,1)
126          IF(IPLOT.GT.1)GO TO 94
127          IF(IPLOT.EQ.0)RETURN
128    C FORMATION LINEPRINTER PLOT
129          CALL TSORT4(NO,MOVE3,ICOMP3,1)
130          CALL LINPLT(FILL(1),FILL(1201),NO,CO,NORD,XN1,YN1,1,PLMIN,
131        & PLMAX,0.,BMAX)
132          DO 200 I=1,4800
133      200 FILL(I)=ALPHA(I)
134    C  DISTRIBUTION LINEPRINTER PLOT
135          DO 192 I=1,NO
136      192 U1(I)=PCF(NC1,I)
137          CALL LINPLT(U1,U1   ,NO,CO,NORD,XN2,YN2,2,PHMIN,PHMAX,0.,1.)
138          GO TO 101
139       94 IPUN=1
140          CALL TITRPL(SCALEX,SCALEY,NO,NSET,NC1)
141          CALL FORMPL(SCALEN)
142          DO 100 I=1,4800
143      100 FILL(I)=ALPHA(I)
144          CALL DISTPL
145      101 CALL SEARCH(NC,NSP,NO,ICRD,IPRT)
146          RETURN
147          END
```

```
1    C -------*---------*---------*---------*---------*---------*---------*
2           SUBROUTINE SEARCH(NC,NSP,NO,ICRD,IPRT)
3    C  SEARCH FOR PEAK MAXIMA IN THE DISTRIBUTION CURVE -- CALLED BY FANTSY
4           COMMON/DATA/III(600),PCF(4,600),XX(3600)
5           COMMON/MISC/AL(4,200),ALPHA(20,200),AF(4),AC(20),MF(4),MC(20),
6         & X(7),Y(7),CO(4),A(4,4),JX(4),XXX(1080)
7           COMMON/PAR1/KI(20),LABP(20,7),IX(4,20),EX(4,20)
8        50 FORMAT(31H1 E(K,J)      PH    MAX % COMPLEX,/)
9        60 FORMAT(1X,4I2,F8.3,F9.2,2H %)
10          NC1=NC+1
11          DO 100 K=1,NC
12      100 AF(K)=0.
13          DO 110 J=1,NSP
14      110 AC(J)=0.
15          DO 200 I=1,NO
16          DO 180 K=1,NC
17          IF(AL(K,I).LE.AF(K))GO TO 180
18          AF(K)=AL(K,I)
19          MF(K)=I
20      180 CONTINUE
21          DO 190 J=1,NSP
22          IF(ALPHA(J,I).LE.AC(J))GO TO 190
23          AC(J)=ALPHA(J,I)
24          MC(J)=I
25      190 CONTINUE
26      200 CONTINUE
27          WRITE(IPRT,50)
28          DO 300 K=1,NC
29          IF(MF(K).LT.3)GO TO 280
30          IF(MF(K).GT.NO-2)GO TO 280
31          GO TO 281
32      280 IM=MF(K)
33          PH=PCF(NC1,IM)
34          AF(K)=AF(K)*100.
```

```
35              GO TO 290
36        281 DO 282 L=1,5
37              IM=MF(K)-3+L
38              X(L)=PCF(NC1,IM)
39        282 Y(L)=AL(K,IM)*100.
40              CALL FIT(X,Y,CO,A,3,5,G,IFA)
41              PH=-.5*CO(2)/CO(3)
42              AF(K)=CO(1)+CO(2)*PH+CO(3)*PH**2
43        290 DO 295 L=1,4
44        295 JX(L)=0
45              JX(K)=1
46        300 WRITE(IPRT,60)JX,PH,AF(K)
47              DO 400 J=1,NSP
48              IF(MC(J).LT.3)GO TO 380
49              IF(MC(J).GT.NO-2)GO TO 380
50              GO TO 381
51        380 IM=MC(J)
52              PH=PCF(NC1,IM)
53              AC(J)=AC(J)*100.
54              GO TO 400
55        381 DO 382 L=1,5
56              IM=MC(J)-3+L
57              X(L)=PCF(NC1,IM)
58        382 Y(L)=ALPHA(J,IM)*100.
59              CALL FIT(X,Y,CO,A,3,5,G,IFA)
60              PH=-.5*CO(2)/CO(3)
61              AC(J)=CO(1)+CO(2)*PH+CO(3)*PH**2
62        400 WRITE(IPRT,60)(IX(K,J),K=1,4),PH,AC(J)
63              RETURN
64              END



 1  C -------*---------*---------*---------*---------*---------*---------*
 2              SUBROUTINE BJERM(II,IBIN,NBAR,HBAR,PL)
 3  C  CALCULATION OF NBAR,HBAR, AND PL ONE POINT AT A TIME
 4  C  CALLED BY REDUCE OR CALC3  OR GLASS
 5  C  IBIN=1 SIGNALS ENTRY FROM REDUCE--I.E. APPLICABLE TO BINARY MODEL
 6              REAL NBAR(4)
 7              COMMON/DATA/IDATA(600),PCF(4,600),CT(4,600),EQV(600),U1(600)
 8              COMMON/MISC/PX(1200),BAR(1200),FILL(3000),BARS(600)
 9              COMMON/DIM/NO,NOS(20),NSP,NM,NL,NC,NC1,NV,NSET,IST,JST,NY,NCY,IW
10              COMMON/LIGS/HM(4),NH(4),LIGS(4),KKI(4),QC(4),QR(20),Q(20)
11              COMMON/PAR1/KI(20),LABP(20,7),IX(4,20),EX(4,20)
12              COMMON/BETA/BETA(20),BLOG(20),BL(20),B(20),SL(20),SLB(20),SB(20)
13              COMMON/CONC/CF(4),HOH,CTC(4),C(20),PC(20),TOTX(4)
14              COMMON/INIT/IACT,AU,AH,AOH,A1,A0,B1,AKW,CKW,TEMP,UCORR
15              DIMENSION HBAR(4),DLL(3),GLL(3)
16              KNM=1+NM
17              IF(NL.EQ.0)KNM=1
18              T=CT(NC1,II)-CF(NC1)+CKW/CF(NC1)
19              IF(NL.EQ.0)GO TO 195
20              TL=0.
21              DO 180 J=1,NL
22        180 TL=TL+CT(NM+J,II)
23              IF(NM.EQ.0)GO TO 196
24        195 TM=0.
25              DO 190 J=1,NM
26        190 TM=TM+CT(J,II)
27        196 TX=CT(KNM,II)
28              IF(NL.EQ.0)TX=TM
29              IF(NM.EQ.0)TX=TL
30              DO 199  J=1,NC1
31              NBAR(J)=0.
32        199 HBAR(J)=T/TX
33              BAR(II)=HBAR(1)
34              PX(II)=-ALOG10(CF(NC1))
35              PL=0.
36              IF((NL.EQ.0).OR.(NM.EQ.0))GO TO 500
37              KL=0
38              NLX=NL
39              IF(IBIN.EQ.1)NLX=1
40              DO 200 I=1,NLX
41              NI=NH(I)
```

```
42          DLL(I)=1.
43          GLL(I)=Ø.
44          DO 201 J=1,NI
45          KL=KL+1
46          Dl=10.**(BLOG(KL)+QR(KL)*(AU-AØ)-FLOAT(J)*PCF(NC1,II))
47          DLL(I)=DLL(I)+D1
48      201 GLL(I)=GLL(I)+D1*FLOAT(J)
49          IF(IBIN.NE.1)GO TO 400
50          HBAR(1)=GLL(1)/DLL(1)
51          NBAR(1)=(CT(NM+1,II)-T/HBAR(1))/CT(1,II)
52          BAR(II)=NBAR(1)
53          PL=(CT(NM+1,II)-NBAR(1)*CT(1,II))/DLL(1)
54          PX(II)=2.
55          IF(PL.GT.Ø.)PX(II)=-ALOG10(PL)
56          PL=PX(II)
57          GO TO 5ØØ
58      400 NBAR(I)= CT(NM+I,II)-CF(NM+I)*DLL(I)
59          SAL=Ø.
60          DO 5Ø J=1,NM
61       5Ø SAL=SAL+CF(J)
62          DO 203 K=1,NSP
63    C  NBAR(J) WRT SUM METAL TYPES AND LIGAND(J)
64          IF(IX(NM+I,K).EQ.Ø)GO TO 203
65          EXX=Ø.
66          DO 2Ø2 J=1,NM
67      2Ø2 EXX=EXX+EX(J,K)
68          SAL=SAL+C(K)*EXX
69      203 CONTINUE
70          NBAR(I)=NBAR(I)/SAL
71      2ØØ HBAR(I)=GLL(I)/DLL(I)
72          BAR(II)=NBAR(1)
73          PX(II)=PCF(NM+1,II)
74      5ØØ BARM=FLOAT(NH(1))+.999
75          IF(NM.GT.Ø)BARM=.999+FLOAT(MAXØ(LIGS(1),LIGS(3)))
76          IF(BAR(II).GT.BARM)BAR(II)=Ø.
77          IF(BAR(II).LT.-.5)BAR(II)=-.5
78          IF(ABS(PX(II)-2.).LT.1.E-4)BAR(II)=Ø.
79          BARS(II)=BAR(II)
80          RETURN
81          END
```

```
 1    C -------*---------*---------*---------*---------*---------*---------*
 2          SUBROUTINE GLASS
 3          EXTERNAL MOVE1,ICOMP1
 4          DOUBLE PRECISION AWA(4,4),DET,AWAM(4,4)
 5          REAL NBAR(4)
 6    C  CALIBRATION OF GLASS ELECTRODE USING APEB UNIVERSAL BUFFER
 7    C  APEB UNI-BUFFER SOLUTION (ACETATE-PHOSPHATE-ETHYLENEDIAMINE-BORAX)
 8    C  THIS SECTION USES BUFFER REGIONS TO CALIBRATE A GLASS ELECTRODE.
 9    C  THEY MAY CONSIST OF ANY OR ALL OF THE FOLLOWING BUFFERS...
10    C  WATER(2),ACETATE(1),ETHYLENEDIAMINE(2),PHOSPHATE(3),BORATE(1)
11    C  COPPER OR CADMIUM ELECTRODE MAY BE CALIBRATED AT Ø.2 M M2+/EN
12          COMMON/LARG/DPXDV(600,4),SIGPX(4,600),A(600,5)
13          COMMON/MISC/PX(1200),BAR(1200),DILS(600),TILS(600),W(600,2),
14        &U2(6ØØ),D2(6ØØ)
15          COMMON/DATA/IDATA(600),PCF(4,600),CT(4,600),VTS(600),U1(6ØØ)
16          COMMON/BETA/BETA(20),BLOG(20),BL(20),B(20),SL(20),SLB(20),SB(20)
17          COMMON/DIM/NO,NOS(20),NSP,NM,NL,NC,NC1,NV,NSET,IST,JST,NY,NCZ,IW
18          COMMON/IO/TAPIN,TAPOUT,INTAP(20)
19          COMMON/APE/UM,T,QWLOGM,CON,SQM,SQ,ITOT,ITITR,QWLOG
20          COMMON/CONT/NCY,IPLOT,IFPL,LONG,IDAT,IPAR,ITYPE,IDPX,ISORT,IREF
21          COMMON/INIT/IACT,AU,AH,AOH,A1,AØ,B1,AKW,CKW,TEMP,UØ
22          COMMON/LIGS/HM(4),NH(4),LIGS(4),KI(4),QC(4),QR(20),Q(20)
23          COMMON/MONT/ICRD,IPRT,IPCH,TITLE(20),FMT(20),YNAME(4),CODEN(5)
24          COMMON/LIMS/PH1,PH2,ETS(7)
25          COMMON/SGNL/MAPEB(6)
26          COMMON/CONC/CF(4),HOH,CTC(4),C(20),PC(20),TOTX(4)
27          COMMON/DERV/SD(4,4),CB(8ØØ),IT
28          COMMON/SIG1/SIGC(4),SIGV(4),SIGX(4),DM
29          COMMON/PREL/NCM,INDEX1(4),INDEX2(4),INDEX3(4)
30          COMMON/PAR1/MI(20), ABP(20,7),IX(4,20),EX(4,20)
31          COMMON/CYCL/ICYCL,JTYPE,KTYPE
```

```
32          COMMON/LIMT/JMEM,JSET,MSTART,MEND,NS
33          DIMENSION CTA(4),IR(5),YNAM(7),KH(4),KM(4),BØ(4),DØ(4),S(4),
34         &KKI(4),V(4),PCFC(4),PCFS(4),SM(4),VM(4),HBAR(4),AL(4),ALPHA(20),
35         &EB(4),CTS(4,20),CAB(20)
36          DATA YNAM/4HACET,4HPHOS,4HEN  ,4HBORX,4HCU2+,4HCD2+,4HEPS /
37     1173 RETURN
38          END
```

```
1   C -------*---------*---------*---------*---------*---------*---------*
2          SUBROUTINE APEB(U)
3   C ANALYTICAL EXPRESSIONS FOR THE ACETATE, ETHYLENEDIAMINE AND PHOSPHATE
4   C AND BORIC ACID (DILUTE BORAX)
5   C CONSTANTS AS A FUNCTION OF IONIC STRENGTH AND TEMPERATURE
6   C AT Ø.2 M   CU2+/EN AND CD2+/EN CONSTANTS ALSO
7          COMMON/INIT/IACT,AU,AH,AOH,A1,AØ,B1,AKW,CKW,TEMP,UCORR
8          COMMON/DIM/NO,NOS(20),NSP,NM,NL,NC,NC1,NV,NSET,IST,JST,NY,NCY,IW
9          COMMON/LIGS/HM(4),NH(4),LIGS(4),KI(4),QC(4),QR(20),Q(20)
10         COMMON/BETA/BETA(20),BLOG(20),BL(20),B(20),SL(20),SLB(20),SB(20)
11         COMMON/APE/UM,T,QWLOGM,CON,SQM,SQ,ITOT,ITITR,QWLOG
12         COMMON/SGNL/MA,MP,ME,MB,MCU,MCD
13         DIMENSION C(6),H(6)
14         DATA C/10.537,9.134,1.000,5.620,4.950,2.090/
15         DATA H/-8.,-1.022,.25, -10.08,-1.022,.175/
16  C SEE ANALYT.CHEM.,50,2137(1978)
17         RETURN
18         END
```

```
1   C -------*---------*---------*---------*---------*---------*---------*--
2          SUBROUTINE REDUCE
3   C DATA REDUCTION
4   C PREPARES RAW DATA FOR INPUT TO SEVERAL OTHER SUBPROGRAMS.  INVOLVES
5   C CORRECTION FOR DILUTION, CALCULATION OF A PORTION OF THE IONIC
6   C STRENGTH LABELING DATA SETS, COVERSION OF VOLUMES TO EQUIVALENTS
7   C (AND CALCULATIONS NEEDED FOR WEIGHTING SCHEME),SORTING OF DATA ON
8   C PH OR EQUIV, OPTIMIZATION OF NBAR VS PL CALCULATION AND PLOT
9   C ASSUME ALL METAL COUNTER IONS ARE MONOANIONIC AND LIGAND COUNTER
10  C IONS IF PRESENT, ARE MONOCATIONIC
11         EXTERNAL MOVE1,ICOMP1,ICOMP3,MOVE3
12         INTEGER TAPIN,TAPOUT
13         REAL NBAR(4)
14         COMMON/LARG/STAT(600,4),SIGPX(4,600),SIGD(4,600),INDX(600)
15         COMMON/DATA/IDATA(600),PCF(4,600),CT(4,600),EQV(600),U1(600)
16         COMMON/MISC/PX(1200),BAR(1200),HTOTS(6,200),TOTM(6,200),
17        & TOTL(6,200)
18         COMMON/MONT/ICRD,IPRT,IPCH,TITLE(20),FMT(20),YNAME(4),CODEN(5)
19         COMMON/LIMS/PH1,PH2,PHMIN,PHMAX,EMIN,EMAX,PLMIN,PLMAX,BMAX
20         COMMON/DIM/NO,NOS(20),NSP,NM,NL,NC,NC1,NV,NSET,IST,JST,NY,NCY,IW
21         COMMON/INIT/IACT,AU,AH,AOH,A1,AØ,B1,AKW,CKW,TEMP,UCORR
22         COMMON/CONT/IPUN,IPLOT,IFPL,LONG,IDAT,IPAR,IGUESS,IDPX,ISORT,IREF
23         COMMON/CONC/CF(4),HOH,CTC(4),C(20),PC(20),TOTX(4)
24         COMMON/SIG2/AJK(4),AWA(4,4),V(4),BØ(4),SØ(4),RHO(412),DEDV(20)
25         COMMON/LIMT/JMEM,JSET,MSTART,MEND,NS
26         COMMON/BETA/BETA(20),BLOG(20),BL(20),B(20),SL(20),SLB(20),SB(20)
27         COMMON/LIGS/HM(4),NH(4),LIGS(4),KKI(4),QC(4),QR(20),Q(20)
28         COMMON/IO/TAPIN,TAPOUT,INTAP(20)
29         COMMON/LIMS2/HTMIN,HTMAX
30         DIMENSION PH( 600),XBAR( 600),DHB( 600),DNB( 600),LABL(20)
31        &,PHS(200),PLS(3),HBAR(4),BØU(6),CO(4),YBAR(600),DØ(4)
32         DATA ZO/1H /
33         DATA XN1,YN1,XN2,YN2/2HPL,4HNBAR,2HPH,4HHBAR/
34       2 FORMAT('1 BJERRUM NBAR, HBAR CALCULATION (PH-DEPENDENT METHOD)'/)
35       9 FORMAT(/)
36      12 FORMAT(16F5.0)
37      14 FORMAT(3I1,3X,A4,7F10.0)
38      15 FORMAT(1H+,80X,F7.4,F8.5,2F9.6,F10.7,F7.3)
39      16 FORMAT(8F10.3)
40      17 FORMAT(///,16H INITIAL VOLUME ,26(1H.),F8.3,3H ML,/,6H ACID ,
41        &36(1H.),F9.6,2H M,/,  6H BASE ,36(1H.),F9.6,2H M,/,  7H STOCK ,
```

```
42        &35(1H.),F10.7,2H M,/,  24H SUPPORTING ELECTROLYTE ,18(1H.),F7.4,
43        &2H M,/,33H TITRANT BURET CORRECTION FACTOR ,9(1H.),F7.4,/,
44        & 42H INITIAL REAGENT CONCENTRATIONS <ITUO> ...,3F9.6)
45     18 FORMAT(     '1 DEQV/DVOL FOR SET ',A4,' IS ',F10.3,/)
46     19 FORMAT('1 IRVING-ROSSOTTI NBAR-HBAR CALCULATION (PH-INDEPENDENT',
47        &' METHOD)',//)
48     20 FORMAT(////,10X,44HEQV, PCF(NC1), CT(NC1), U1, LABEL        <ITUO>,
49        & 31X,3HVT,5X,3HDIL,3X,4HH-OH,3X,9HA-B+STOCK,3X,2HF0,7X,4HP(H),/)
50     21 FORMAT(//,5X,4HNBAR,5X,2HPL,7X,4HHBAR,10H   -LOG(H),10X,
51        &10HHTOT(NSET)/)
52     22 FORMAT(I1,F9.0,F10.0,7F7.3)
53     36 FORMAT(4F9.3,10X,5F9.6)
54     37 FORMAT('1 BJERRUM  NBAR-HBAR CALCULATION (PH-DEPENDENT METHOD)',
55        &' FOR ',A4,/)
56     38 FORMAT(//,6X,4HNBAR,6X,2HPL,8X,4HHBAR,4X,7H  P(H) ,6X,2HPH,/)
57     39 FORMAT(//,14X,4HNBAR,4X,2HPL,3X,8H2ND DERV,11X,4HHBAR,3X,4HP(H),
58        &2X,8H2ND DERV,/)
59     40 FORMAT(2(10X,2F8.3,I6))
60     41 FORMAT(//,14X,4HHBAR,3X,4HP(H),2X,8H2ND DERV,/)
61     43 FORMAT(' B0 CONSTANTS READ IN AS ',17(1H.),4F9.5)
62   6043 FORMAT(' D0 CONSTANTS READ IN AS ',17(1H.),4F9.5)
63     44 FORMAT(' B0 CONSTANTS READ IN AS ',17(1H.),' B0(1)=',2F7.4,2H*U,
64        & F7.4,14H*U**2,  B0(2)=,2F7.4,2H*U,F7.4,21H*U**2,  B0(3)=B0(4)=0)
65     45 FORMAT(24H0DATA RECEIVED FROM UNIT,I3,7H  (UNIT,I2,' IS CARD ',
66        &'READER'))
67     48 FORMAT(//,' TOTAL NUMBER OF POINTS ACCEPTED',I4,'  COUNT BY SETS'
68        &      ,20I4)
69     49 FORMAT('1    LIST OF (PX,D2',A4,'/D',A2,'2)')
70     50 FORMAT(/,8(F9.3,F7.0))
71        ITYPE=LIGS(4)
72        IF(ITYPE.EQ.0)ITYPE=1
73        BARM=FLOAT(NH(1))+.999
74        IF(NM.GT.0)BARM=0.999+FLOAT(MAX0(LIGS(1),LIGS(3)))
75        SMALL=1.E-4
76        I=0
77        NO=0
78        EMIN=50.
79        EMAX=-EMIN
80        HTMIN=1.
81        HTMAX=-HTMIN
82        PHMAX=0.
83        PHMIN=14.
84        DO 100 N=1,NSET
85        INTP=INTAP(N)
86        READ(ICRD,14)ITITR,MV,MT,LABEL,VO,ACID,BASE,STOCK,SE,VCOR,CAB
87   C   MT=1  ==> METAL ELECTRODE USED
88        LABL(N)=LABEL
89        IF(VCOR.LT.1.E-3)VCOR=1.
90        READ(ICRD,16)(TOTX(I),I=1,4)
91        READ(ICRD,16)DELPH
92        IF(DELPH.GT.0.5)DELPH=0.5
93   C   ITITR=0 FOR BASE AND ITITR=1 FOR ACID TITRES.
94   C   MV=1 FOR MILLIVOLT DATA AND MV=0 FOR PH DATA
95        TX=TOTX(ITYPE)
96        IF(TX.LT.1.E-6)TX=CAB
97        DEDV(N)=BASE/(VO*TX)
98        IF(ITITR.EQ.1)DEDV(N)=ACID/(VO*TX)
99        WRITE(IPRT,18)LABEL,DEDV(N)
100       WRITE(IPRT,17)VO,ACID,BASE,STOCK,SE,VCOR,(TOTX(J),J=1,NC)
101       IF(IACT-5)457,458,456
102    458 READ(ICRD,16)(B0(I),I=1,4),(D0(I),I=1,4)
103       IF(B0(2).LT.SMALL)B0(2)=1.
104       IF(D0(2).LT.SMALL)D0(2)=1.
105       WRITE(IPRT,43)B0
106       IF(MT.NE.0)WRITE(IPRT,6043)D0
107       GO TO 457
108    456 READ(ICRD,16)(B0U(I),I=1,6)
109  C   B0(1)=B0U(1)+B0U(2)*U+B0U(3)*U**2
110  C   B0(2)=B0U(4)+B0U(5)*U+B0U(6)*U**2
111  C   B0(3)=B0(4)=0
112       WRITE(IPRT,44)B0U
113  C   READ OBSERVATION DATA ONE SET  TIME
114    457 WRITE(IPRT,45)INTP,ICRD
115       DO 200 II=1,200
116       NOS(N)=II
```

```
117              I=I+1
118              IDATA(I)=N
119              INDX(I)=N
120              DO 130 J=1,NC
121        130 PCF(J,I)=0.
122              IS=0
123         57 IF(IS.EQ.0)GO TO 58
124              I=I-1
125              NOS(N)=NOS(N)-1
126              GO TO 99
127         58 READ(INTP,22)IS,VT,PCF(NC1,I),E0,SL0,PH0,PCF(1,I),E0M,SL0M,PM0
128              VT=VT*VCOR
129              IF(MV.EQ.0)GO TO 60
130              IF(IS.NE.5)GO TO 150
131              PHREF=PH0
132              SLOPE=SL0
133              EREF=E0
134              PMREF=PM0
135              SLOME=SL0M
136              EREFM=E0M
137        150 PCF(NC1,I)=PHREF+(PCF(NC1,I)-EREF)/SLOPE
138              IF(MT.EQ.0)GO TO 60
139              PCF(1,I)=PMREF+(PCF(1,I)-EREFM)/SLOME
140         60 IF(PCF(NC1,I).LT.PH1)GO TO 57
141              IF(PCF(NC1,I).GT.PH2)GO TO 57
142              IF(II.EQ.1)GO TO 56
143              IF(ABS(PCF(NC1,I)-PHOLD).LT.DELPH)GO TO 57
144         56 PHOLD=PCF(NC1,I)
145              DIL=VO/(VO+VT)
146              TIL=VT/(VO+VT)
147              TL=0.
148     C   CORRECT TOTAL CONCENTRATIONS FOR DILUTION
149              DO 160 J=1,NC
150              CT(J,I)=TOTX(J)*DIL
151        160 IF(J.GT.NM)TL=TL+CT(J,I)*HM(J-NM)
152              AA=(ACID+STOCK)*DIL
153              BB=BASE*TIL
154              IF(ITITR.EQ.0)GO TO 170
155              AA=ACID*TIL+STOCK*DIL
156              BB=BASE*DIL
157        170 CT(NC1,I)=AA-BB+TL
158     C   TEMPORARY STORE
159              DNB(I)=VT
160              XBAR(I)=DIL
161              DHB(I)=AA-BB
162              TX=CT(ITYPE,I)
163              IF(TX.LT.1.E-6)TX=CAB*DIL
164              EQV(I)=(BB-AA)/TX
165              EMAX=AMAX1(EMAX,EQV(I))
166              EMIN=AMIN1(EMIN,EQV(I))
167              HTMAX=AMAX1(HTMAX,CT(NC1,I))
168              HTMIN=AMIN1(HTMIN,CT(NC1,I))
169              U1(I)=BB+(SE+ABS(STOCK))*DIL
170              IF(NL.EQ.0)GO TO 199
171              DO 190 J=1,NL
172              QC1=QC(J+NM)+HM(J)
173        190 U1(I)=U1(I)+CT(J+NM,I)*(ABS(QC1)-QC1)
174     C   NOTE THAT U1 IS ONLY A PARTIAL CONTRIBUTION TO THE IONIC STRENGTH
175     C   STANDARDIZING ELECTRODE TO READ TRUE HYDROGEN CONCENTRATION RATHER
176     C   THAN ACTIVITY-- IONIC STRENGTH MUST BE MAINTAINED CONSTANT FOR
177     C   THIS METHOD.  DATA CONVERTED PERMANENTLY FROM PH(METER) TO P(H)
178        199 HA=10.**(-PCF(NC1,I))
179              U=U1(I)+HA/.8
180              CALL ACTIV(U,HA)
181              IF(IACT.GT.1)GO TO 189
182              PCF(NC1,I)=-ALOG10(CF(NC1))
183              H=CF(NC1)
184              GO TO 191
185        189 H=CF(NC1)
186              IF(IACT.EQ.6)GO TO 198
187              DO 195 J=1,3
188              ZZZ=B0(3)*H+B0(4)*CKW/H
189              IF(ABS(7.-PCF(NC1,I)).LT.4.)ZZZ=0.
190              PHC=(PCF(NC1,I)-B0(1)-ZZZ)/B0(2)
191        195 H=10.**(-PHC)
```

```
192          PCF(NC1,I)=PHC
193          IF(MT.NE.0)PCF(1,I)=(PCF(1,I)-D0(1)-D0(3)*H-D0(4)*CKW/H)/D0(2)
194          GO TO 191
195      198 B0(1)=B0U(1)+B0U(2)*U+B0U(3)*U**2
196    C NOTE THAT B0(2)IS NEAR +1.000
197          B0(2)=B0U(4)+B0U(5)*U+B0U(6)*U**2
198          PCF(NC1,I)=(PCF(NC1,I)-B0(1))/B0(2)
199      191 PHMAX=AMAX1(PHMAX,PCF(NC1,I))
200          PHMIN=AMIN1(PHMIN,PCF(NC1,I))
201    C   TEMPORARY STORE
202          PH(I)=PCF(NC1,I)
203          PX(I)=H-CKW/H
204    C   F0 FUNCTION
205          YBAR(I)=(PX(I)-DHB(I))/XBAR(I)
206          IF(IS.EQ.1)GO TO 99
207      200 CONTINUE
208       99 MSTART=NO
209          MEND=MSTART+NOS(N)
210    C   SORT ON PH WITHIN SET
211          IF(ISORT.EQ.1)CALL TSORT4(NOS(N),MOVE1,ICOMP1,NC1)
212          IF(LONG.NE.0)WRITE(IPRT,20)
213          IS=0
214          JS=33
215          MBEGIN=MSTART+1
216          DO 98 II=MBEGIN,MEND
217          IF(LONG.EQ.0)GO TO 98
218          WRITE(IPRT,FMT)JS,ZO,EQV(II),(PCF(J,II),J=1,NC1),
219         &   (CT(J,II),J=1,NC1),U1(II),LABEL
220    C   PRINT VT,DIL,H-OH,A-B,F0,P(H)
221          WRITE(IPRT,15)DNB(II),XBAR(II),PX(II),DHB(II),YBAR(II),PH(II)
222       98 IF(IPUN.NE.0)WRITE(TAPOUT,FMT)IS,ZO,EQV(II),(PCF(J,II),J=1,NC1),
223         &   (CT(J,II),J=1,NC1),U1(II),LABEL
224      100 NO=NO+NOS(N)
225          WRITE(IPRT,48)NO,(NOS(N),N=1,NSET)
226    C   CALCULATION OF NBAR,HBAR, AND PL (1ST METAL AND 1ST LIGAND TYPES)
227    C   READ LIGAND PARAMETERS ONLY WHEN METAL IS PRESENT
228          IF(TOTX(1).LT.1.E-6)RETURN
229          IF(IGUESS.EQ.0)RETURN
230          IF(NM.NE.0)CALL READP
231          NO1=1
232          NO2=NO
233      478 PLMIN=30.
234          PLMAX=0.
235          BMAX=0.
236          IF(IGUESS-1)999,470,480
237      480 WRITE(IPRT,19)
238          WRITE(IPRT,21)
239      470 NI=NH(1)
240          IF(IGUESS.GT.1)GO TO 400
241    C   PH-DEPENDENT NBAR, HBAR CALCULATION
242          DO 300 I=NO1,NO2
243          CF(NC1)=10.**(-PCF(NC1,I))
244          U=U1(I)+CF(NC1)
245          CALL ACTIV(U,1.)
246          HOH=CF(NC1)-CKW/CF(NC1)
247          IF(NM.EQ.0)GO TO 302
248          DO 301 J=1,NI
249      301 B(J)=BETA(J)*10.**(QR(J)*(AU-A0))
250      302 CALL BJERM(I,1,NBAR,HBAR,PL)
251          IF(PX(I).GE.PLMAX)PLMAX=PX(I)
252          IF(PX(I).LT.PLMIN)PLMIN=PX(I)
253          IF(BAR(I).GT.BMAX)BMAX=BAR(I)
254          PH(I)=PCF(NC1,I)
255          XBAR(I)=HBAR(1)
256          YBAR(I)=NBAR(1)
257      300 CONTINUE
258          NSD=0
259          CALL DELTA(NSET,NOS,PH,XBAR,DHB)
260          IF((NM.EQ.0).OR.(NL.EQ.0))GO TO 350
261    C   NBAR VS PL
262          CALL DELTA(NSET,NOS,PX,YBAR,DNB)
263          DO 330 N=1,NSET
264          WRITE(IPRT,37)LABL(N)
265          WRITE(IPRT,39)
266          NDP=NOS(N)
267          DO 320 J=1,NDP
```

```
268            K=NSD+J
269            IN=DNB(K)
270            IH=DHB(K)
271        320 WRITE(IPRT,40)YBAR(K),PX(K),IN,XBAR(K),PH(K),IH
272        330 NSD=NSD+NDP
273            CALL TSORT4(NO,MOVE3,ICOMP3,1)
274            CALL LINPLT(PX,BAR,NO,CO,NORD,XN1,YN1,1,PLMIN,PLMAX,0.,BMAX)
275            GO TO 390
276    C   HBAR VS PH
277        350 DO 380 N=1,NSET
278            WRITE(IPRT,37)LABL(N)
279            WRITE(IPRT,41)
280            NDP=NOS(N)
281            DO 370 J=1,NDP
282            K=NSD+J
283            IH=DHB(K)
284        370 WRITE(IPRT,40)XBAR(K),PH(K),IH
285        380 NSD=NSD+NDP
286            YH=NH(1)+1
287            CALL TSORT4(NO,MOVE3,ICOMP3,1)
288            CALL LINPLT(PX, BAR,NO,CO,NORD,XN2,YN2,1,PHMIN,PHMAX,0.,YH)
289        390 IF(IPLOT.EQ.0)RETURN
290            SCALEN=.1*FLOAT(IPLOT)
291            CALL FORMPL(SCALEN)
292            RETURN
293    C PH-INDEPENDENT NBAR, HBAR CALCULATION ACCORDING TO IRVING-ROSSOTTI
294        400 IDAT=1
295            TEMP=.1
296            LIGS(1)=NH(1)
297            NH(1)=NSET
298            DO 401 J=2,4
299            LIGS(J)=NH(J)
300        401 NH(J)=0
301            IREF=1
302            CALL OSTR1
303            DO 402 I=1,NS
304        402 PHS(I)=PX(I)
305            I=0
306            DO 403 JJ=1,NS
307            DO 403 II=1,6
308            I=I+1
309        403 HTOTS(II,JJ)=PX(800+I)
310            NS2=NSET-2
311            PLS(1)=0.
312            DO 404 I=1,NS
313            NBAR(1)=0.
314            HBAR(1)=(HTOTS(2,I)-HTOTS(1,I))/TOTL(2,I)
315            IF(NSET.LT.3)GO TO 408
316            DO 405 J=1,NS2
317        405 NBAR(J)=(HTOTS(2,I)-HTOTS(J+2,I))/(HBAR(1)*TOTM(J+2,I))
318            DL=1.
319            DO 406 J=1,NI
320        406 DL=DL+10.**(BLOG(J)-FLOAT(J)*PHS(I))
321            DO 407 J=1,NS2
322            PLS(J)=(TOTL(2,I)-NBAR(J)*TOTM(J+2,I))/DL
323        407 IF(PLS(J).GT.0.)PLS(J)=-ALOG10(PLS(J))
324        408 WRITE(IPRT,36)NBAR(1),PLS(1),HBAR(1),PHS(I),(HTOTS(J,I),J=1,NSET)
325            IF(NSET.LT.4)GO TO 404
326            DO 409 J=2,NS2
327        409 WRITE(IPRT,36)NBAR(J),PLS(J)
328            WRITE(IPRT,9)
329        404 CONTINUE
330    C   HBAR VS PH CURVE
331            PLMIN=PHS(1)
332            PLMAX=PHS(NS)
333            NOSAVE=NO
334            NO=NS
335            NMSAVE=NM
336            NM=0
337            DO 420 I=1,NS
338            INDX(I)=2
339            BAR(I)=(HTOTS(2,I)-HTOTS(1,I))/TOTL(2,I)
340            IF(BAR(I).GT.BMAX)BMAX=BAR(I)
341        420 PX(I)=PHS(I)
342    C   HBAR 2ND DERIVATIVES
343            CALL DELTA(1,NS,PX,BAR,DHB)
```

```
344            WRITE(IPRT,49)YN2,XN2
345            WRITE(IPRT,50)(PX(J),DHB(J),J=1,NS)
346            CALL TSORT4(NS,MOVE3,ICOMP3,1)
347            CALL LINPLT(PX,BAR,NS,CO,NORD,XN2,YN2,1,PHS(1),PHS(NS),0.,BMAX)
348            SCALEN=.1*FLOAT(IPLOT)
349            IF(IPLOT.NE.0)CALL FORMPL(SCALEN)
350      C   NBAR VS PL CURVES
351            NM=NMSAVE
352            IF(NSET.LT.3)GO TO 492
353            PLMIN=30.
354            PLMAX=0.
355            BMAX=0.
356            NO=NS*NS2
357            DO 430 JJ=1,NS2
358            DO 430 I=1,NS
359            II=I+NS*(JJ-1)
360            INDX(II)=JJ+2
361            BAR(II)=(HTOTS(2,I)-HTOTS(2+JJ,I))/(HTOTS(2,I)-HTOTS(1,I))
362         &    *(TOTL(2,I)/TOTM(2+JJ,I))
363            DL=1.
364            DO 431 J=1,NI
365      431 DL=DL+10.**(BLOG(J)-FLOAT(J)*PHS(I))
366            PX(II)=2.
367            PLS(JJ)=(TOTL(2,I)-BAR(II)*TOTM(JJ+2,I))/DL
368            IF(PLS(JJ).GT.0.)PX(II)=-ALOG10(PLS(JJ))
369            IF(PX(II).GE.PLMAX)PLMAX=PX(II)
370            IF(PX(II).LT.PLMIN)PLMIN=PX(II)
371            IF(BAR(II).LT.0.)BAR(II)=0.
372            IF(BAR(II).GT.BARM)BAR(II)=BARM
373      430 IF(BAR(II).GT.BMAX)BMAX=BAR(II)
374      C   NBAR 2ND DERIVATIVES
375            CALL DELTA(1,NO,PX,BAR,DNB)
376            WRITE(IPRT,49)YN1,XN1
377            WRITE(IPRT,50)(PX(J),DNB(J),J=1,NO)
378            CALL TSORT4(NO,MOVE3,ICOMP3,1)
379            CALL LINPLT(PX,BAR,NO,CO,NORD,XN1,YN1,1,PLMIN,PLMAX,0.,BMAX)
380            IF(IPLOT.NE.0)CALL FORMPL(SCALEN)
381      492 NO=NOSAVE
382            IF(IGUESS.EQ.2)GO TO 999
383            IGUESS=1
384            NO1=NOS(1)+1
385            IF(NSET.GT.2)NO1=NO1+NOS(2)
386            NO2=NO
387            DO 493 J=1,4
388      493 NH(J)=LIGS(J)
389            GO TO 478
390      999 RETURN
391            END
```

```
1   C -------*---------*---------*---------*---------*---------*--------*
2            SUBROUTINE DELTA(NSET,NOS,X,Y,D2)
3   CALCULATION OF THE SECOND DERIVATIVES
4   CALLED BY REDUCE OR GLASS
5            DIMENSION NOS(20),X(600),Y(600),D2(600)
6            NR=0
7            DO 8500 N=1,NSET
8            N1=NR+1
9            N2=NR+NOS(N)
10           DMAX=0.
11           DO 8501 I=N1,N2
12           D2(I)=0.
13           IF((I.LT.N1+2).OR.(I.GT.N2-2))GO TO 8501
14           DY1=Y(I+2)-Y(I)
15           DX1=X(I+2)-X(I)
16           IF(ABS(DX1).LT.1.E-6)DX1=1.E-6
17           DY2=Y(I)-Y(I-2)
18           DX2=X(I)-X(I-2)
19           IF(ABS(DX2).LT.1.E-6)DX2=1.E-6
20           DX3=X(I+1)-X(I-1)
21           IF(ABS(DX3).LT.1.E-6)DX3=1.E-6
22           D2(I)=(DY1/DX1-DY2/DX2)/DX3
23           IF(ABS(D2(I)).GT.DMAX)DMAX=ABS(D2(I))
```

```
24      8501 CONTINUE
25    C SCALE TO A MAX OF 9999
26           DO 8502 J=N1,N2
27      8502 D2(J)=D2(J)*9999.9/DMAX
28      8500 NR=N2
29           RETURN
30           END


 1    C -------*---------*---------*---------*---------*---------*---------*
 2           SUBROUTINE OSTR1
 3    C HEDSTROM-OSTERBERG-SARKAR-MCBRYDE INTEGRATION METHOD FOR DETERMINING
 4    C CONCENTRATIONS IN MULTI-COMPONENT EQUILIBRIA
 5    C CT(ITYPE,I) IS THE TOTAL CONCENTRATION THAT IS VARIED WITH ALL ELSE
 6    C FIXED.   KKN = NO. TITRATION CURVES OF THE ABOVE TYPE (MUST NOT
 7    C EXCEED 6).  ENTER CURVES IN ORDER OF INCREASING TOTAL CONCENTRATION
 8    C THAT IS VARIED. IPUN=1 WILL PRODUCE A PUNCHED DECK OF PH VS DEL
 9    C PX(ITYPE), INPUT FOR PHMIN  IS  THE STARTING PH VALUE FOR THE
10    C EXTRAPOLATION CURVE HTOTAL VALUE CALLED BY MONITOR AND REDUCE
11           REAL NBAR
12           COMMON/LARG/COS(4,200),CT5(6),SLOPE(200,3),SAVE(6,199),
13          & TOT(6,200,4),XX(200),YY(200)
14           COMMON/DATA/IDATA(600),PCF(4,600),CT(4,600),EQV(600),U1(600)
15           COMMON/MONT/ICRD,IPRT,IPCH,TITLE(20),FMT(20),YNAME(4),CODEN(5)
16           COMMON/DIM/NO,NOS(20),NSP,NM,NL,NC,NC1,NV,NSET,IST,JST,NY,NCY,IW
17           COMMON/CONT/IPUN,IPLOT,IFPL,LONG ,IDAT,IPAR,NORD,IDPX,ISORT,IREF
18           COMMON/LIMS/PH1,PH2,PHMIN,PHMAX,EMIN,EMAX,PLMIN,PLMAX,BMAX
19           COMMON/INIT/IACT,AU,AH,AOH,A1,A0,B1,AKW,CKW,DELPH,UCORR
20           COMMON/LIMT/JMEM,JSET,MSTART,MEND,NS
21           COMMON/LIGS/HM(4),KKN(4),LIGS(8),Q(44)
22           COMMON/MISC/PHS(200),DPX(200,3),HTOTS(6,200),DHDCX(6,200),CO(4),
23          & CTA(6),X(8),Y(8),A(4,4),CM(6),CL(6),NBAR(146),HTS(200),
24          & TOTM(6,200),TOTL(6,200)
25           DIMENSION SY(10)
26        10 FORMAT(8F10.3)
27        29 FORMAT( '1   AVERAGE TOTAL ',A4,' IN THE KKN(',I1,')SETS ARE ',
28          & 6F9.6)
29        30 FORMAT(/4X,'PH',5X,'HTOT(KKN(',I1,')), DE BOOR P FACTOR AND GOF',
30          & /11X,'MTOT',/,11X,'LTOT',/)
31        31 FORMAT(1H1,6X,'PH',3X,'NBAR(1,NM+1)',/)
32        43 FORMAT(/,' DE BOOR SPLINE-SMOOTHING APPLIED TO HTOT VS XTOT',
33          & //,' SIGMA(HTOTAL)=',F10.6 ,/)
34        44 FORMAT(2F10.3,2X,7F10.3)
35        45 FORMAT(F7.3,10E11.4)
36        46 FORMAT(7X,6F11.7)
37        70 FORMAT(///,6X,'PH',5X,'DEL P',A4,2X,'SLOPES (LAST COLUMN USED)'/)
38       147 FORMAT(/,' REVISION DATE: 31 DEC 1980',/)
39           IF(IREF.EQ.1)GO TO 71
40           WRITE(IPRT,147)
41           READ(ICRD,10)SCALEX,SCALEY,XTYPE,PHA,PHB,SIGHT
42           IF(SIGHT.LT.1.E-9)SIGHT=1.E-6
43           WRITE(IPRT,43)SIGHT
44           IF(PHA.LT.1.)PHA=PH1
45           IF(PHB.LT.1.)PHB=PH2
46           READ(ICRD,10)(CT5(I),I=1,6)
47    C ISOHYDRIC CURVES OF HTOT AS A FUNCTION OF CX
48        71 IF(IDAT.EQ.0)CALL READD(ISORT)
49           NS=(PHB-PHA)/DELPH+1.
50           IF(NS.GT.200)NS=200
51           PHS(1)=PHA
52           DO 150 I=2,NS
53       150 PHS(I)=PHS(I-1)+DELPH
54           NR=0
55           N1=1
56           IDO=0
57           DO 900 ITYPE=1,NC
58           MS=KKN(ITYPE)
59           IF(MS.EQ.0)GO TO 955
60           IF(ITYPE.NE.1)N1=N1+KKN(ITYPE-1)
61           N2=N1+MS-1
62           DO 201 N=N1,N2
63    C SETS N1 TO N2 BELONG TO GROUPING ITYPE  AND   KKN(ITYPE)=N2-N1+1
64           M=N-N1+1
```

```
65    C   M IS AN INTRA-SET INDEX, RUNNING FROM 1 TO KKN(ITYPE)(=MS)
66            IF(N.GT.1)NR=NR+NOS(N-1)
67            NN=NOS(N)
68    C   TOTAL CONC. ISOHYDRIC INTERPOLATIONS USING NATURAL CUBIC SPLINE
69            DO 165 J=1,NC1
70            DO 160 I=1,NN
71            XX(I)=PCF(NC1,NR+I)
72    160   YY(I)=CT(J,NR+I)
73            CALL SPLINE(NN)
74            M1=1
75            DO 165 K=1,NS
76    163   M2=NR+M1+1
77            IF(PHS(K).LE.PCF(NC1,M2))GO TO 164
78            M1=M1+1
79            IF(M1.LT.NN-1)GO TO 163
80    164   IF(M1.GE.NN-1)M1=NN-1
81            H=PHS(K)-PCF(NC1,M1+NR)
82    165   TOT(M,K,J)=COS(1,M1)+COS(2,M1)*H+COS(3,M1)*H**2+COS(4,M1)*H**3
83            DO 185 K=1,NS
84            TOTM(M,K)=TOT(M,K,1)
85            TOTL(M,K)=TOT(M,K,NM+1)
86            HTOTS(M,K)=TOT(M,K,NC1)
87    185   IF(M.EQ.LIGS(ITYPE))HTS(K)=HTOTS(M,K)
88    C   AVERAGE TOTAL CONCENTRATIONS BY THE SETS
89            CTA(M)=0.
90            DO 182 I=1,NN
91    182   CTA(M)=CTA(M)+CT(ITYPE,NR+I)
92            CTA(M)=CTA(M)/FLOAT(NN)
93    201   CONTINUE
94    C   IREF=1 SIGNALS ENTRY FROM REDUCE (NBAR CALCULATION)
95            IF(IREF.EQ.1)RETURN
96            WRITE(IPRT,29)YNAME(ITYPE),ITYPE,(CTA(J),J=1,MS)
97            WRITE(IPRT,30)ITYPE
98    C   CALCULATE SLOPES
99            HTMIN=1.
100           HTMAX=-HTMIN
101           ICP=LIGS(ITYPE)
102           DO 300 I=1,NS
103           DO 298 L=1,MS
104           XX(L)=TOT(L,I,ITYPE)
105           YY(L)=HTOTS(L,I)
106           SY(L)=SIGHT
107           IF(HTOTS(L,I).GT.HTMAX)HTMAX=HTOTS(L,I)
108    298   IF(HTOTS(L,I).LT.HTMIN)HTMIN=HTOTS(L,I)
109           CALL SMOOTH(MS,XX,YY,SY,COS,P,GOFS)
110    C   SLOPES
111           DO 18 I1=1,MS
112    18    DHDCX(I1,I)=PPVALU(XX,COS,MS-1,XX(I1),1)
113           WRITE(IPRT,45)PHS(I),(HTOTS(M,I),M=1,MS),P,GOFS
114           DO 47 L=1,NC
115    47    WRITE(IPRT,46)(TOT(K,I,L),K=1,MS)
116    C   COMMON POINT HTOTAL AND SLOPE
117           IF(I.EQ.1)X0=XX(ICP)
118           DIL=XX(ICP)/X0
119           X1=DIL*CT5(ITYPE)
120           SAVE(6,I)=PPVALU(XX,COS,MS-1,X1,1)
121           IF(IDO.NE.0)GO TO 300
122           SAVE(NC1,I)=PPVALU(XX,COS,MS-1,X1,0)
123           DO 299 J=1,NC
124    299   SAVE(J,I)=DIL*CT5(J)
125    300   CONTINUE
126           IDO=1
127           IF(IPLOT.GT.1)
128          &CALL VARPL(NS,MS,ITYPE,HTMIN,HTMAX,SCALEX,SCALEY,SIGHT,XTYPE)
129    C   LIGS CONTAINS ARRAY LLN
130           M=LIGS(ITYPE)
131           WRITE(IPRT,70)YNAME(ITYPE)
132           DPX(1,ITYPE)=0.
133           SLOPE(1,ITYPE)= SAVE(6,1)
134           DO 503 I=2,NS
135           S1=SAVE(6,I-1)
136           S2=SAVE(6,I)
137           DPX(I,ITYPE)=DPX(I-1,ITYPE)-.5*DELPH*(S1+S2)
138           IF(IPUN.NE.0)WRITE(IPCH,10)PHS(I),DPX(I,ITYPE)
139           SLOPE(I,ITYPE)=S2
140    503   WRITE(IPRT,44)PHS(I),DPX(I,ITYPE),(DHDCX(J,I),J=1,MS),S2
```

```
141              GO TO 900
142         955 DO 956 I=1,NS
143         956 DPX(I,ITYPE)=0.
144         900 CONTINUE
145              IF(NM.EQ.0)GO TO 999
146              IF(KKN(1+NM).EQ.0)GO TO 999
147              IF(KKN(1).EQ.0)GO TO 999
148              WRITE(IPRT,31)
149              DO 280 I=1,NS
150      C   ADJUSTMENTS FOR STOICH
151              TOTM(1,I)=SAVE(1,I)
152              TOTM(NM+1,I)=SAVE(NM+1,I)
153              TOTM(NC1,I)=SAVE(NC1,I)
154              HTOTS(1,I)=SAVE(NC1,I)
155      C   VARIATIONAL NBAR
156              NBAR(I)=-SLOPE(I,1)/SLOPE(I,NM+1)
157         280 WRITE(IPRT,10)PHS(I),NBAR(I)
158              IF(IPLOT.GT.0)CALL VARNPL(NS,PHS,NBAR,SLOPE)
159         999 IDPX=1
160              RETURN
161              END




  1      C -------*---------*---------*---------*---------*---------*---------*
  2              SUBROUTINE SMOOTH(NPT,X,Y,S,COEF,P,GOF)
  3      C   DE BOOR 1978   PP 235-249
  4              COMMON/AV/A(10,4),V(10,7)
  5              DIMENSION DP(4),S(NPT),X(NPT),Y(NPT),COEF(4,NPT)
  6              DATA DP/.1,-.01,.001,-.0001/
  7      C   BEGIN SETUPQ
  8              NP1=NPT-1
  9              V(1,4)=X(2)-X(1)
 10              DO 11 I=2,NP1
 11              V(I,4)=X(I+1)-X(I)
 12              V(I,1)=S(I-1)/V(I-1,4)
 13              V(I,2)=-S(I)/V(I,4)-S(I)/V(I-1,4)
 14          11 V(I,3)=S(I+1)/V(I,4)
 15              V(NPT,1)=0.
 16              DO 12 I=2,NP1
 17          12 V(I,5)=V(I,1)**2+V(I,2)**2+V(I,3)**2
 18              IF(NP1.LT.3)GO TO 14
 19              DO 13 I=3,NP1
 20          13 V(I-1,6)=V(I-1,2)*V(I,1)+V(I-1,3)*V(I,2)
 21          14 V(NP1,6)=0.
 22              IF(NP1.LT.4)GO TO 16
 23              DO 15 I=4,NP1
 24          15 V(I-2,7)=V(I-2,3)*V(I,1)
 25          16 V(NP1-1,7)=0.
 26              V(NP1,7)=0.
 27              PR=(Y(2)-Y(1))/V(1,4)
 28              DO 21 I=2,NP1
 29              D=(Y(I+1)-Y(I))/V(I,4)
 30              A(I,4)=D-PR
 31          21 PR=D
 32      C   END SETUPQ
 33              L=1
 34              P=0.
 35              IF(S(1).LT.1.1E-6)P=1.
 36          30 CALL CHOLID(NPT,P)
 37              GOF=0.
 38              IF(ABS(P-1.).LT.1.E-6)GO TO 60
 39              DO 35 I=1,NPT
 40          35 GOF=GOF+(A(I,1)*S(I))**2
 41              GOF=GOF*36.*(1.-P)**2
 42              IF(GOF.GT.0.)GOF=SQRT(GOF/FLOAT(NPT))
 43              IF((DP(L).GT.0.).AND.(GOF.GT.1.))GO TO 201
 44              IF((DP(L).LT.0.).AND.(GOF.LT.1.))GO TO 201
 45              IF(P.LT.1.E-6)GO TO 60
 46              L=L+1
 47              IF(L.EQ.5)GO TO 60
 48         201 P=P+DP(L)
 49              IF(P.GT.1.)P=1.
 50              IF(P.LT.0.)P=0.
```

```
51          GO TO 30
52       60 DO 81 I=1,NPT
53       81 A(I,1)=Y(I)-6.*(1.-P)*S(I)**2*A(I,1)
54          DO 62 I=1,NPT
55       62 A(I,3)=A(I,3)*6.*P
56          DO 63 I=1,NP1
57          A(I,4)=(A(I+1,3)-A(I,3))/V(I,4)
58       63 A(I,2)=(A(I+1,1)-A(I,1))/V(I,4)-(A(I,3)+A(I,4)/3.*V(I,4))/2.
59        & *V(I,4)
60          DO 44 I=1,NP1
61          DO 44 J=1,4
62       44 COEF(J,I)=A(I,J)
63          RETURN
64          END
```

```
1   C -------*---------*---------*---------*---------*---------*---------*
2          SUBROUTINE CHOLID(NPT,P)
3   C  REF  DE BOOR 1978
4          COMMON/AV/QU(10),ZZ(10),U(10),QTY(10),V(10,7)
5          NP1=NPT-1
6          SP=6.*(1.-P)
7          DO 2 I=2,NP1
8          V(I,1)=SP*V(I,5)+2.*P*(V(I-1,4)+V(I,4))
9          V(I,2)=SP*V(I,6)+P*V(I,4)
10       2 V(I,3)=SP*V(I,7)
11         NP2=NPT-2
12         IF(NP2.GE.2)GO TO 10
13         U(1)=0.
14         U(2)=QTY(2)/V(2,1)
15         U(3)=0.
16         GO TO 41
17      10 DO 20 I=2,NP2
18         R=V(I,2)/V(I,1)
19         V(I+1,1)=V(I+1,1)-R*V(I,2)
20         V(I+1,2)=V(I+1,2)-R*V(I,3)
21         V(I,2)=R
22         R=V(I,3)/V(I,1)
23         V(I+2,1)=V(I+2,1)-R*V(I,3)
24      20 V(I,3)=R
25         U(1)=0.
26         V(1,3)=0.
27         U(2)=QTY(2)
28         DO 30 I=2,NP2
29      30 U(I+1)=QTY(I+1)-V(I,2)*U(I)-V(I-1,3)*U(I-1)
30         U(NPT)=0.
31         U(NP1)=U(NP1)/V(NP1,1)
32         I=NP1
33      40 I=I-1
34         IF(I.EQ.1)GO TO 41
35         U(I)=U(I)/V(I,1)-U(I+1)*V(I,2)-U(I+2)*V(I,3)
36         GO TO 40
37      41 PR=0.
38         DO 50 I=2,NPT
39         QU(I)=(U(I)-U(I-1))/V(I-1,4)
40         QU(I-1)=QU(I)-PR
41      50 PR=QU(I)
42         QU(NPT)=-QU(NPT)
43         RETURN
44         END
```

```
1   C -------*---------*---------*---------*---------*---------*---------*
2          FUNCTION PPVALU(X,COEF,NPT,XI,JD)
3   C  REF  DE BOOR 1978
4          DIMENSION X(NPT),COEF(4,NPT)
5          PPVALU=0.
6          F=FLOAT(4-JD)
7          IF(F.LE.0.)GO TO 99
8   C INTERV
```

```
 9           I1=1
10           I2=2
11           IF(I2.LT.NPT)GO TO 20
12           IF(XI.GE.X(NPT))GO TO 110
13           I1=NPT-1
14           I2=NPT
15        20 IF(XI.GE.X(I2))GO TO 40
16           IF(XI.GE.X(I1))GO TO 100
17           IS=1
18        31 I2=I1
19           I1=I2-IS
20           IF(I1.LE.1)GO TO 35
21           IF(XI.GE.X(I1))GO TO 50
22           IS=2*IS
23           GO TO 31
24        35 I1=1
25           IF(XI.LT.X(1))GO TO 90
26           GO TO 50
27        40 IS=1
28        41 I1=I2
29           I2=I1+IS
30           IF(I2.GE.NPT)GO TO 45
31           IF(XI.LT.X(I2))GO TO 50
32           IS=2*IS
33           GO TO 41
34        45 IF(XI.GE.X(NPT))GO TO 110
35           I2=NPT
36        50 M=(I2+I1)/2
37           IF(M.EQ.I1)GO TO 100
38           IF(XI.LT.X(M))GO TO 53
39           I1=M
40           GO TO 50
41        53 I2=M
42           GO TO 50
43        90 II=1
44           GO TO 98
45       100 II=I1
46           GO TO 98
47       110 II=NPT
48    C    END INTERV
49        98 H=XI-X(II)
50           M=5
51        10 M=M-1
52           IF(M.EQ.JD)GO TO 99
53           PPVALU=(PPVALU/F)*H+COEF(M,II)
54           F=F-1.
55           GO TO 10
56        99 RETURN
57           END
```

```
 1    C -------*---------*---------*---------*---------*---------*---------*
 2           SUBROUTINE OSTR2
 3    C  GIVEN THE RESULTS OF OSTR1 (PH VS. DEL PX) THIS PROGRAM UPDATES THE
 4    C  DATA DECK BY CALCULATING THE MISSING PX VALUES AND PRODUCING A NEW
 5    C  PUNCHED DECK.   USER MUST SUPPLY A STARTING SET OF PX VALUES AT A
 6    C  STARTING PH.   SUBPROGRAMS LINLSQ AND GRAFN REQUIRE SUCH A NEW DECK.
 7           INTEGER TAPIN,TAPOUT
 8           COMMON/LARG/COS(4,200,3),FULL(5000),X(200),Y(200)
 9           COMMON/DATA/IDATA(600),PCF(4,600),CT(4,600),EQV(600),U1(600)
10           COMMON/MONT/ICRD,IPRT,IPCH,TITLE(20),FMT(20),YNAME(4),CODEN(5)
11           COMMON/DIM/NO,NOS(20),NSP,NM,NL,NC,NC1,NV,NSET,IST,JST,NY,NCY,IW
12           COMMON/CONT/IPUN,IPLOT,IFPL,LONG,IDAT,IPAR,IGUESS,IDPX,ISORT,IREF
13           COMMON/LIMS/PH1,PH2,PHMIN,PHMAX,EMIN,EMAX,PLMIN,PLMAX,BMAX
14           COMMON/IO/TAPIN,TAPOUT,INTAP(20)
15           COMMON/LIMT/JMEM,JSET,MSTART,MEND,NS
16           COMMON/LIGS/HM(4),KKN(4),LIGS(8),Q(44)
17           COMMON/MISC/PHS(200),DPX(200,3),PXO(3),W(4,4),CO(4),CORR(3),A(4,
18          &4),FILL(5158)
19           DATA ZO/1H    /
20        10 FORMAT(8F10.3)
21        14 FORMAT(7X,'PH',9X,'DEL PX(NC)      <ITUO>',/)
22        20 FORMAT(I1,F9.0,F10.0)
23        81 FORMAT(1H1,9X,'EQV, PCF(NC1), CT(NC1), U1      <ITUO>',/)
```

```
24            IF(IDPX.NE.0)GO TO 85
25   C  IF IDPX=1 THEN   NECESSARY DEL PX VALUES ARE IN STORAGE AND NEED NOT
26   C  ENTERED.   THUS OSTR1 MUST HAVE BEEN JUST RUN FOR ALL THE NEEDED ITYPE
27            DO 170 N=1,NC
28            DO 160 I=1,200
29            IF(KKN(N).EQ.0)GO TO 150
30            NS=I
31   C  READ IN PH, DEL PM1, DEL PM2, ... , DEL PL1, DEL PL2, ...
32   C  NOTE THAT EACH OF THE NC SETS MUST HAVE THE SAME VALUES AND ORDERING
33            READ(ICRD,20)IS,PHS(I),DPX(I,N)
34            IF(IS.GT.0)GO TO 170
35            GO TO 160
36   150 DPX(I,N)=0.
37   160 CONTINUE
38   170 CONTINUE
39    85 WRITE(IPRT,14)
40            DO 180 I=1,NS
41   180 WRITE(IPRT,10)PHS(I),(DPX(I,K),K=1,NC)
42            J=NC
43   298 DO 299 I=1,NS
44            X(I)=PHS(I)
45   299 Y(I)=DPX(I,J)
46            CALL SPLINE(NS)
47            IF(J.EQ.1)GO TO 301
48            DO 300 I=1,NS
49            DO 300 K=1,4
50   300 COS(K,I,J)=COS(K,I,1)
51            J=J-1
52            GO TO 298
53   C  READ DATA ONE SET AT A TIME
54   C  DATA MUST BE ORDERED IN INCREASING PH.
55   301 I=0
56            JS=33
57            DO 270 N=1,NSET
58            INTP=INTAP(N)
59            IS=0
60            READ(ICRD,10)(PXO(I),I=1,3)
61            LIM=1
62            IF(LONG.NE.0)WRITE(IPRT,81)
63   C  INTERPOLATE DEL PX
64            II=0
65   C  II IS INTRASET INDEX AND IS RESET FOR EVERY NEW VALUE OF N
66   260 II=II+1
67            I=I+1
68            IDATA(I)=N
69            ABEL=ZO
70            IF(IDAT.NE.0)GO TO 590
71   579 IF(IS.EQ.0)GO TO 580
72            I=I-1
73            GO TO 270
74   580 READ(INTP ,FMT)IS,Z,EQV(I),(PCF(J,I),J=1,NC1),(CT(J,I),J=1,NC1),
75      & U1(I),ABEL
76            IF(PCF(NC1,I).LT.PH1)GO TO 579
77            IF(PCF(NC1,I).GT.PH2)GO TO 579
78   590 IF(PCF(NC1,I).LE.PHS(LIM+1))GO TO 591
79            LIM=LIM+1
80            IF(LIM.LT.NS-1)GO TO 590
81   591 IF(LIM.GT.NS-1)LIM=NS-1
82            DO 220 K=1,NC
83   C  PXO REFERS TO THE PCF SET, NOT THE PHS SET
84            H=PCF(NC1,I)-PHS(LIM)
85            DP=COS(1,LIM,K)+COS(2,LIM,K)*H+COS(3,LIM,K)*H*H+COS(4,LIM,K)*H**3
86            IF(II.EQ.1)CORR(K)=DP
87   220 PCF(K,I)=PXO(K)+DP-CORR(K)
88            IF(LONG.NE.0)WRITE(IPRT,FMT)JS,ZO,EQV(I),(PCF(J,I),J=1,NC1),
89      & (CT(J,I ),J=1,NC1),U1(I ), ABEL
90            IF(IPUN.NE.0)WRITE(TAPOUT,FMT)IS,ZO,EQV(I),(PCF(J,I),J=1,NC1)
91      &,(CT(J,I),J=1,NC1),U1(I),ABEL
92            IF(IDAT.NE.0)GO TO 200
93            IF(IS.EQ.1)GO TO 270
94            GO TO 260
95   200 IF(II.EQ.NOS(N))GO TO 270
96            GO TO 260
97   270 CONTINUE
98            RETURN
99            END
```

```
1    C -------*---------*---------*---------*---------*---------*---------*
2          SUBROUTINE SPLINE(NN)
3    C  CALLED FROM OSTR1,OSTR2,VARPL
4    C  NATURAL CUBIC SPLINE INTERPOLATION.  ALGORITHM 3.2, P120:
5    C  FOR X IN THE INTERVAL (XJ,XJ+1),
6    C  SJ(X)= YJ +COS(2,J)*(X-XJ)+COS(3,J)*(X-XJ)**2 +COS(4,J)*(X-XJ)**3
7          COMMON/LARG/COS(4,200),FILL(6600),X(200),Y(200)
8          DIMENSION U(200),Z(200)
9          N1=NN-1
10         U(1)=0.
11         Z(1)=0.
12         COS(1,1)=Y(1)
13         COS(1,NN)=Y(NN)
14         DO 1 I=2,N1
15         COS(1,I)=Y(I)
16         H1=X(I+1)-X(I-1)
17         H2=X(I)-X(I-1)
18         H3=X(I+1)-X(I)
19         A=3.*(Y(I+1)*H2-Y(I)*H1+Y(I-1)*H3)/(H2*H3)
20         E=2.*H1-U(I-1)*H2
21         U(I)=H3/E
22       1 Z(I)=(A-Z(I-1)*H2)/E
23         COS(3,NN)=0.
24         J=N1
25       2 COS(3,J)=Z(J)-U(J)*COS(3,J+1)
26         H1=X(J+1)-X(J)
27         COS(2,J)=(Y(J+1)-Y(J))/H1-H1*(COS(3,J+1)+2.*COS(3,J))/3.
28         COS(4,J)=(COS(3,J+1)-COS(3,J))/(3.*H1)
29         J=J-1
30         IF(J.GT.0)GO TO 2
31         RETURN
32         END
```

```
1    C--------*---------*---------*---------*---------*---------*---------*
2          SUBROUTINE STOICH(CKW,N1,N2 ,ICRD,IPRT,NS,N3,N4,UCORR)
3    C  DETERMINATION OF AVERAGE STOICHIOMETRIC COEFFICIENTS AFTER VARIATION
4    C  HAS BEEN RUN (IDPX=1).  RESULTS APPLY ONLY TO THE COMMON POINT.
5          REAL NBAR
6          COMMON/LARG/COS(800),CT5(6),SLOPE(200,3),XX(6388),PX1(3),AM(3)
7          COMMON/MISC/PHS(200),DPX(200,3),HTOTS(6,200),EXBAR(4,200),
8         & PX2(4,200),TOTX(6,200),FILL(195),FS(5),RAT(4,200),F(200)
9          COMMON/DIM/NO,NOS(20),NSP,NM,NL,NC,NC1,NV,NSET,IST,JST,NY,NCY,IW
10         COMMON/BETA/BETA(20),BLOG(20),BL(20),B(20),SL(20),SLB(20),SB(20)
11         COMMON/PAR1/KI(20),LABP(20,7),IX(4,20),EX(4,20)
12         COMMON/CONC/CF(4),HOH,CTR(4),C(20),PC(20),TO(4)
13         COMMON/CONT/IPUN,IPLOT,IFPL,LONG,IDAT,IPAR,IGUESS,IDPX,ISORT,IREF
14         COMMON/LIGS/HM(4),NH(4),LIGS(4),KKI(4),QC(4),QR(20),Q(20)
15      10 FORMAT(8F10.6)
16      11 FORMAT(20X,3F10.6)
17      18 FORMAT(' ACID,BASE,STOCK.......',3X,3F9.6)
18      19 FORMAT(///,18H    PH    SILLEN SUM,4X,4HHTOT,11X,10HEXBAR(NC1),  9X,
19         & 6HPX(NC),3X,15HDHTOT/DXTOT(NC),2X,4HHBAR,3X,4HNBAR,4X,3HEMR,4X,
20         & 3HELR,4X,3HEHR,7H SR(MM),/)
21      20 FORMAT(F6.2,2F11.6,2X,9F7.3,3F7.1,F7.2)
22      21 FORMAT(23H COMMON POINT XTOT.....,3X,3F9.6)
23      22 FORMAT(' TX,TXR ',8F9.6)
24      23 FORMAT(/)
25      24 FORMAT(/,34H PL CALCULATED FROM MCBRYDE'S NBAR,/)
26      25 FORMAT(/,' INTEGRAND: (MTOT-M)*SLOPE(M)+(LTOT-L)*SLOPE(L)-',
27         &'(HTOT-H+OH)',/)
28      26 FORMAT(/,' CONCENTRATION SCALE FACTORS',5F10.4)
29      27 FORMAT(///,5X,' PH    ELR/EMR    EHR/EMR    L*D(H)    HBAR(M)=(HT',
30         &'-HBAR*(LT-NBAR*MT)-H+OH)/MT',/)
31      28 FORMAT(F9.2,F7.2,F10.2,F13.6,F8.2)
32      29 FORMAT(/,' REVISION DATE:  6 JAN 1981',/)
33      33 FORMAT(7(2X,1H(,F5.2,1H),F9.6))
34      34 FORMAT(/,' INTEGRAND SCALE FACTOR',F10.4)
35         NM1=NM+1
36         NC2=NC+2
37    C  CT5 MUST REFER TO UNDILUTED CONCENTRATIONS. IT IS USED
38    C  TO RECOVER DILUTION FACTORS ONLY WHEN FS(NC1) AND FS(NC2) ARE ZERO.
39         READ(ICRD,10)(PX1(I),I=1,3),S1,(CT5(J),J=1,NC)
```

```
40            READ(ICRD,11)ACID,BASE,STOCK
41            READ(ICRD,10)(FS(J),J=1,NC2),SF
42            DO 48 J=1,NC2
43         48 IF(FS(J).LT.0.001)FS(J)=1.
44            IF(SF.LT.0.001)SF=1.
45            WRITE(IPRT,21)(CT5(J),J=1,NC)
46            WRITE(IPRT,18)ACID,BASE,STOCK
47            WRITE(IPRT,26)(FS(J),J=1,NC2)
48            WRITE(IPRT,34)SF
49            IF(IGUESS.NE.0)WRITE(IPRT,24)
50            WRITE(IPRT,29)
51            CALL READP
52            NI=NH(1)
53            DPH=PHS(2)-PHS(1)
54    C  PX1 REFERS TO PX0 VALUES AT THE VARIATION SPECIFIED PHA INITIAL
55    C  S1 AND PX1 VALUES ARE VALID ONLY IN REGION WITH MONONUCLEAR COMPLEXES
56            NBAR=-SLOPE(1,1)/SLOPE(1,NM1)
57            TL=0.
58            DO 112 J=1,NL
59        112 TL=TL+TOTX(NM+J,1)*FS(NM+J)
60            IF(S1.LT.1.E-6)S1=TL
61            IF(PX1(1).LT.1.E-6)PX1(1)=-ALOG10(FS(1)*(1.-NBAR)*TOTX(1,1))
62            DL=1.
63            DO 110 J=1,NI
64        110 DL=DL+10.**(BLOG(J)-FLOAT(J)*PHS(1))
65            IF(PX1(NM1).LT.1.E-6)PX1(NM1)=-ALOG10((TOTX(NM1,1)*FS(NM1)
66          & -NBAR*TOTX(1,1)*FS(1))/DL)
67            DO 46 K=1,NC
68         46 CT5(K)=CT5(K)*FS(K)
69    C  DETERMINE THE FUNCTION TO BE INTEGRATED
70            DO 50 I=1,NS
71    C  SCALING TOTAL CONCENTRATIONS
72            TL=0.
73            DO 1435 J=1,NC
74            TOTX(J,I)=TOTX(J,I)*FS(J)
75       1435 IF(J.GT.NM)TL=TL+HM(J-NM)*TOTX(J,I)
76            IF(ABS(FS(NC1)+FS(NC2)-2.).GT.1.E-6)GO TO 1237
77            DO 1239 L=1,NL
78       1239 TOTX(NC1,I)=TOTX(NC1,I)+(1.-1./FS(L+NM))*HM(L)*TOTX(L+NM,I)
79            GO TO 1238
80    C  THE ONLY USE OF CT5
81       1237 DIL=TOTX(1,I)/CT5(1)
82            TIL=1.-DIL
83            IF(ACID.LT.BASE)GO TO 1236
84            TA=TIL*ACID+DIL*STOCK
85            TB=-DIL*BASE
86            GO TO 1240
87       1236 TA=DIL*(ACID+STOCK)
88            TB=-TIL*BASE
89       1240 TOTX(NC1,I)=TA*FS(NC1)+TB*FS(NC2)+TL
90       1238 HTOTS(1,I)=TOTX(NC1,I)
91            PX2(NC1,I)=PHS(I)
92            H=10.**(-PHS(I))
93            EXBAR(NC1,I)=HTOTS(1,I)-H +CKW/H
94            F(I)=-EXBAR(NC1,I)
95            DO 47 J=1,NC
96            PX2(J,I)=PX1(J)+DPX(I,J)
97            IF(IGUESS.EQ.0)GO TO 49
98            IF(J.NE.NM1)GO TO 49
99            DL=1.
100           DO 61 K=1,NI
101        61 DL=DL+10.**(BLOG(K)-FLOAT(K)*PHS(I))
102           EL=(TOTX(J,I)+SLOPE(I,1)/SLOPE(I,J)*TOTX(1,I))/DL
103           IF(EL.GT.0.)PX2(J,I)=-ALOG10(EL)
104        49 EXBAR(J,I)=TOTX(J,I)-10.**(-PX2(J,I))
105        47 F(I)=F(I)+EXBAR(J,I)*SLOPE(I,J)
106        50 F(I)=F(I)*SF
107           WRITE(IPRT,25)
108           WRITE(IPRT,33)(PHS(I),F(I),I=1,NS)
109    C  INTEGRATION TO DETERMINE THE SILLEN SUM
110    C  S1 AND S2 REFER TO THE SILLEN SUM, AS DETERMINED FROM EXP. DATA
111           WRITE(IPRT,19)
112           S2=S1
113           DO 60 I=1,NS
114           IF(I.EQ.1)GO TO 57
```

```
115          S2=S2+2.3Ø3/2.*(F(I-1)+F(I))*DPH
116       57 DO 58 J=1,NC1
117       58 EXBAR(J,I)=EXBAR(J,I)/S2
118          DL=1.
119          GL=Ø.
120          DO 4Ø J=1,NI
121          AJ=J
122          D1=1Ø.**(BLOG(J)-AJ*PHS(I))
123          DL=DL+D1
124       4Ø GL=GL+AJ*D1
125          HBAR=GL/DL
126          NBAR =-SLOPE(I,1)/SLOPE(I,NM+1)
127     C  RESIDUAL ANALYSIS
128          S2R=S2
129          DO 3Ø K=1,NSP
130          PC(K)=-BLOG(K)
131          DO 31 J=1,NC1
132       31 PC(K)=PC(K)+EX(J,K)*PX2(J,I)
133          C(K)=1Ø.**(-PC(K))
134       3Ø S2R=S2R-C(K)
135          DO 32 J=1,NC1
136          CF(J)=1Ø.**(-PX2(J,I))
137          CTR(J)=TOTX(J,I)-CF(J)
138          DO 32 K=1,NSP
139       32 CTR(J)=CTR(J)-EX(J,K)*C(K)
140          CTR(NC1)=CTR(NC1)+CKW/CF(NC1)
141          EMR=CTR(1)/S2R
142          ELR=CTR(NM+1)/S2R
143          EHR=CTR(NC1)/S2R
144          S2R=1ØØØ.*S2R
145          RAT(1,I)=ELR/EMR
146          RAT(2,I)=EHR/EMR
147          RAT(3,I)=DL*CF(NM+1)
148          RAT(4,I)=(HTOTS(1,I)-HBAR*(TOTX(NM+1,I)-NBAR*TOTX(1,I))
149         &  -CF(NC1)+CKW/CF(NC1))/TOTX(1,I)
150          WRITE(IPRT,2Ø)PHS(I),S2,HTOTS(1,I),(EXBAR(J,I),J=1,NC1),
151         &(PX2(J,I),J=1,NC),(SLOPE(I,J),J=1,NC),HBAR,NBAR,EMR,ELR,EHR,S2R
152          WRITE(IPRT,22)(TOTX(J,I),J=1,NC1),(CTR(J),J=1,NC1)
153       6Ø WRITE(IPRT,23)
154          WRITE(IPRT,27)
155          DO 7Ø I=1,NS
156       7Ø WRITE(IPRT,28)PHS(I),(RAT(J,I),J=1,4)
157          RETURN
158          END
```

```
 1    C  -------*---------*---------*---------*---------*---------*---------*
 2          SUBROUTINE GRAFN
 3    C  INITIAL CALCULATION OF FORMATION CONSTANTS, SUCCESSIVE EXTRAPOLATION
 4    C  MUST KNOW PM,PL,PH VALUES. SPECIES HYPOTHESIES TESTING BY UNWEIGHTED
 5    C  SLOPE-INTERCEPT ANALYSIS. KNOWN INPUT BETAS CORRECTED FOR CHANGES IN
 6    C  STRENGTH.   CALCULATED BETAS CORRECTED TO THE DESIRED IONIC STRENGTH.
 7          EXTERNAL MOVE3,ICOMP3
 8          COMMON/LARG/STAT(6ØØ,4),SIGPX(4,6ØØ),SIGD(4,6ØØ),INDX(6ØØ)
 9          COMMON/DATA/IDATA(6ØØ),PCF(4,6ØØ),CT(4,6ØØ),EQV(6ØØ),U1(6ØØ)
1Ø          COMMON/MONT/ICRD,IPRT,IPCH,TITLE(2Ø),FMT(2Ø),YNAME(4),CODEN(5)
11          COMMON/DIM/NO,NOS(2Ø),NSP,NM,NL,NC,NC1,NV,NSET,IST,JST,NY,NCY,IW
12          COMMON/CONT/IPUN,IPLOT,IFPL,LONG,IDAT,IPAR,IGUESS,IDPX,ISORT,IREF
13          COMMON/INIT/IACT,AU,AH,AOH,A1,AØ,B1,AKW,CKW,TEMP,UCORR
14          COMMON/CONC/CF(4),HOH,CTC(4),C(2Ø),PC(2Ø),TOTX(4)
15          COMMON/PAR1/KI(2Ø),LABP(2Ø,7),IX(4,2Ø),EX(4,2Ø)
16          COMMON/PAR2/AM(2Ø,2Ø),V(2Ø),DP(2Ø),P(2Ø),SHFT(2Ø),SOECUT,OPT1,OPT2
17          COMMON/BETA/BETA(2Ø),BLOG(2Ø),BL(2Ø),B(2Ø),SL(2Ø),SLB(2Ø),SB(2Ø)
18          COMMON/LIMS/PH1,PH2,PHMIN,PHMAX,XMIN,XMAX,YMIN,YMAX,BMAX
19          COMMON/LIGS/HM(4),NH(4),LIGS(4),KKI(4),QC(4),QR(2Ø),Q(2Ø)
2Ø          COMMON/MISC/X(12ØØ),Y(12ØØ),CO(2),ND(2),APAI(2,2),FILL(359Ø)
21          DATA KSTAR,XN,YN,KZ/1H*,1HX,1HY,1H /
22          COMMON/CYCL/ICYCL,JTYPE
23          RETURN
24          END
```

```
1    C -------*---------*---------*---------*---------*---------*---------*
2          SUBROUTINE LINLSQ
3    C  LINEAR LEAST SQUARES REFINEMENT OF CONSTANTS
4          DOUBLE PRECISION DAM(20,20),DET
5          COMMON/LARG/STAT(600,4),SIGPX(4,600),SIGD(4,600),INDX(600)
6          COMMON/DATA/IDATA(600),PCF(4,600),CT(4,600),EQV(600),U1(600)
7          COMMON/MISC/FILL(4800),U2(1200)
8          COMMON/MONT/ICRD,IPRT,IPCH,TITLE(20),FMT(20),YNAME(4),CODEN(5)
9          COMMON/DIM/NO,NOS(20),NSP,NM,NL,NC,NC1,NV,NSET,IST,JST,NY,NCY,IW
10         COMMON/CONC/CF(4),HOH,CTC(4),C(20),PC(20),TOTX(4)
11         COMMON/CONT/IPUN,IPLOT,IFPL,LONG,IDAT,IPAR,IGUESS,IDPX,ISORT,IREF
12         COMMON/INIT/IACT,AU,AH,AOH,A1,A0,B1,AKW,CKW,TEMP,UCORR
13         COMMON/PAR1/KI(20),LABP(20,7),IX(4,20),EX(4,20)
14         COMMON/PAR2/AM(20,20),V(20),DP(20),P(20),SHFT(20),SOECUT,OPT1,OPT2
15         COMMON/BETA/BETA(20),BLOG(20),BL(20),B(20),SL(20),SLB(20),SB(20)
16         COMMON/SIG1/SIGC(4),SIGV(4),SIGX(4),SOH
17         COMMON/SIG2/SCC(4),RCC(4,4),DEL(4),RHO(20,20),SP(20),DEDV(20)
18         COMMON/LIMT/JMEM,JSET,MSTART,MEND,NS
19         COMMON/LIGS/HM(4),NH(4),LIGS(4),KKI(4),QC(4),QR(20),Q(20)
20         COMMON/RFAC/SIG(50),DENOM(50),NPTS(50),SUM(50)
21         COMMON/DERV/D(416),CK(20,20),ITR
22         COMMON/CYCL/ICYCL,JTYPE,KTYPE
23         DIMENSION AL1(4),ALPHA1(20),ISA(20),GOF(4),RW(4),DELTA(4),EB(4)
24         DIMENSION W100(20),W200(20),IW300(20),EBR(4),CTR(4)
25         DATA KSTAR,Z/1H*,1H /
26      87 FORMAT(23H0DECONVOLUTED CONSTANTS,/)
27      90 FORMAT(1X,A1,1H(,4I3,1H),F8.3,1H(,I4,1H),5X,7A4)
28      91 FORMAT(2X,2F8.3,1H(,I3,1H),2F9.6,F9.5,1H(,I3,1H),F8.2,F10.3,14X,
29        & F7.4,30X,I4)
30      94 FORMAT(10X,F8.3,1H(,I3,1H),2F9.6,F9.5,1H(,I3,1H),F8.2)
31      95 FORMAT(1H+,F17.3)
32     187 FORMAT(/,' MODE(NNLS)=',I5,/)
33     188 FORMAT( 9H1    EQUIV,7X,2HPX,7X,'TX OBS    TX CAL ',6X,3HDEL,7X,
34        &18HDEL/SIG    ION STR,16X,2HAU,33X,3HSET,/)
35     189 FORMAT(' REVISION DATE: 23 DEC 1980',//,' NNLS ALGORITHM USED ',
36        &'TO CALCULATE SCALED BETAS',/)
37     192 FORMAT(' GOODNESS-OF-FIT = SQRT(SUM(W',8H*(O-C)**,'2)/(NO-NP)) ='
38        & ,4F7.3,5X,6H<ITUO>,
39        &        /,41H RW = SQRT(SUM(W*(O-C)**2)/SUM(W*O**2)) =,4F6.3,/,
40        &' AVG ION STR =',F5.3,1H(,I3,3H) M,/,9H AU(AVG)=,F7.4,/,
41        & ' DEL/SIG (AVG) =',4F6.2)
42     198 FORMAT(/,16H SINGULAR MATRIX,/)
43     218 FORMAT(//,3X,18HCORRELATION MATRIX,/)
44     220 FORMAT(20F6.2)
45     221 FORMAT(10E12.4)
46     222 FORMAT(//,41H SCALED SYMMETRIC MATRIX BEFORE INVERSION,/)
47     223 FORMAT(//,41H SCALED SYMMETRIC MATRIX AFTERE INVERSION,/)
48     415 FORMAT( '1     (MLH/OH)    LOG BETA(SIG)',/)
49     419 FORMAT(1X,130(1H-))
50     421 FORMAT(6H ALPHA,5X, 9(F8.3,1H(,I3,1H)), /,11X,9(F8.3,1H(,I3,1H)))
51     424 FORMAT(6H LOG B,9F13.3, /,6X,9F13.3 )
52     429 FORMAT(3H PC,3X,9F13.3, /,6X,9F13.3 )
53     430 FORMAT(6H+<EX> ,5X,9(2X,A1,1H<,4I2,1H>))
54     431 FORMAT(        11X,9(2X,A1,1H<,4I2,1H>))
55    1252 FORMAT(' S',F9.6,5X,'EXBAR',4F7.3,5X,'S(R)',F9.6,5X,'EXBAR(R)',
56        & 4F6.2)
57         IREF=0
58         WRITE(IPRT,189)
59    C  READ DATA AND PARAMETERS
60         CALL PRELIM(INEG)
61         DO 19 I=1,NSP
62         IF(KI(I).NE.KSTAR)GO TO 19
63    C  SCALE FACTOR FOR SYMMETRIC MATRIX AND VECTOR IN ORDER TO PREVENT
64    C  UNDERFLOW ON AN IBM COMPUTER
65         IF(BLOG(I).LT.0.01)BLOG(I)=10.
66      19 CONTINUE
67         FREE=NO-NSP
68         ICYCL=0
69     134 ICYCL=ICYCL+1
70         IF(ICYCL.GT.NCY+1)GO TO 148
71         DO 58 I=1,NSP
72         DO 57 J=1,NSP
73         RHO(I,J)=0.
74      57 AM(I,J)=0.
75         V(I)=0.
```

```
 76        58 RHO(I,I)=1.
 77           DO 60 K=1,4
 78           DELTA(K)=0.
 79           GOF(K)=0.
 80        60 RW(K)=0.
 81           CALL CLEAR(4)
 82           UAVG=0.
 83           AUAVG=0.
 84           IF(ICYCL.NE.NCY+1)GO TO 59
 85           IF(LONG.LT.1)GO TO 59
 86           WRITE(IPRT,188)
 87        59 DO 93 I=1,NO
 88           IF((ICYCL.EQ.NCY+1).AND.(LONG.GE.1))WRITE(IPRT,419)
 89           CALL CALC2(I,U)
 90           U2(I)=U
 91           UAVG=UAVG+U
 92           AUAVG=AUAVG+AU
 93           ID=IDATA(I)
 94           DO  92 KK=1,NC1
 95           SIGDEL=SIGD(KK,I)
 96           DELT=CT(KK,I)-CTC(KK)
 97           WDY=DELT/SIGDEL
 98           DELTA(KK)= DELTA(KK)+WDY
 99           CALL RFACTR(KK,CT(KK,I),WDY,SIGDEL)
100           IF(LONG.LT.1)GO TO 92
101           IF(ICYCL.NE.NCY+1)GO TO 92
102           AL1(KK)=CF(KK)/CTC(1)
103           STAT(I,KK)=WDY
104           ISD=1.E+5*SIGDEL+.5
105           ISP=1.E+3*SIGPX(KK,I)+.5
106           IF(KK.NE.1)GO TO 88
107           WRITE(IPRT,91)EQV(I),PCF(KK,I),ISP,CT(KK,I),CTC(KK),DELT,ISD,
108          &WDY,U,AU,ID
109           GO TO 92
110        88 WRITE(IPRT,94)PCF(KK,I),ISP,CT(KK,I),CTC(KK),DELT,ISD,WDY
111           IF(KK.NE.NC1)GO TO 92
112           WRITE(IPRT,95)PCF(KK,I)
113           WRITE(IPRT,95)PCF(KK,I)
114           WRITE(IPRT,95)PCF(KK,I)
115        92 CONTINUE
116           IF(ICYCL.NE.NCY+1)GO TO 93
117           IF(LONG.LT.2)GO TO 93
118           SOM=0.
119           DO 1250 L1=1,NC1
120      1250 EB(L1)=0.
121           KV=0
122           DO 200 K=1,NSP
123           IF(KI(K).EQ.KSTAR)KV=KV+1
124           SOM=SOM+C(K)
125           DO 1251 L1=1,NC1
126      1251 EB(L1)=EB(L1)+EX(L1,K)*C(K)
127           DIMER=1.
128           IF(IX(1,K).NE.0)DIMER=EX(1,K)
129           ALPHA1(K)=DIMER*C(K)/CTC(1)
130           PC(K)=0.
131           BL(K)=BLOG(K)-QR(K)*(A0-AU)
132           IF(B(KV).LT.1.E-37)BL(K)=0.
133           IF(C(K).GT.0.)PC(K)=-ALOG10(C(K))
134    C  CALCULATION OF ERRORS IN THE DISTRIBUTION RATIO ALPHA
135           T=0.
136           DO 197 J1=1,NSP
137           DO 197 J2=1,NSP
138       197 T=T+EX(1,J1)*EX(1,J2)*C(J1)*C(J2)*SLB(J1)*SLB(J2)*RHO(J1,J2)
139           T=T/CTC(1)**2+SLB(K)**2
140           ISA(K)=9999
141       200 IF(T.GT.0.)ISA(K)=2303.*ALPHA1(K)*SQRT(T)+.5
142           DO 1253 L=1,NC1
143      1253 EB(L)=EB(L)/SOM
144           IF(NM.EQ.0)GO TO 3933
145    C  RESIDUAL ANALYSIS
146           S2R=SOM
147           DO 3930 K=1,NSP
148      3930 IF(K.LE.IST)S2R=S2R-C(K)
149           DO 3932 J=1,NC1
150           CTR(J)=CTC(J)-CF(J)
```

```
151              DO 3931 K=1,NSP
152        3931 IF(K.LE.IST)CTR(J)=CTR(J)-EX(J,K)*C(K)
153              CTR(NC1)=CTR(NC1)+CKW/CF(NC1)
154        3932 EBR(J)=CTR(J)/S2R
155        3933 NST=9
156              IF(NSP.LT.NST)NST=NSP
157              WRITE(IPRT,431)(KI(K),(IX(J,K),J=1,4),K=1,NST)
158              WRITE(IPRT,430)(KI(K),(IX(J,K),J=1,4),K=1,NST)
159              WRITE(IPRT,421)(ALPHA1(K),ISA(K),K=1,NSP)
160              WRITE(IPRT,1252)SOM,(EB(L1),L1=1,4),S2R,(EBR(L1),L1=1,4)
161              IF(LONG.LT.3)GO TO 93
162              WRITE(IPRT,429)(PC(K),K=1,NSP)
163              WRITE(IPRT,424)(BL(K),K=1,NSP)
164          93 CONTINUE
165              AVG=NO
166              UAVG=UAVG/AVG
167              AUAVG=AUAVG/AVG
168      C  CALCULATE RMS DEVIATION OF IONIC STRENGTH FROM AVERAGE
169              DU=0.
170              DO 56 I=1,NO
171          56 DU=DU+(U2(I)-UAVG)**2
172              IDU=1000.*SQRT(DU/(AVG-1.))+.5
173              SIGA=0.
174              DO 61 K=1,NC1
175              SIGA=SIGA+SIG(K)
176              DELTA(K)=DELTA(K)/AVG
177              GOF(K)=SQRT(SIG(K)/FREE)
178          61 RW(K)=SQRT(SIG(K)/DENOM(K))
179              GOFA=SQRT(SIGA/FLOAT(NC1*NO-NSP))
180              WRITE(IPRT,192)GOF,RW,UAVG,IDU,AUAVG,DELTA
181      C  INVERSION
182              IF(LONG.GT.3)WRITE(IPRT,222)
183              DO 1102 I=1,NV
184              B(I)=0.
185              IF(LONG.GT.3)WRITE(IPRT,221)(AM(I,K),K=1,NV)
186              DO 1102 J=1,NV
187        1102 DAM(I,J)=AM(I,J)
188              CALL NNLS(AM,20,NV,NV,V,B,GOFB,W100,W200,IW300,MODE)
189              GOFB=GOFB/SQRT(FLOAT(NV))
190              WRITE(IPRT,187)MODE
191              CALL DMINV(NV,20,DAM,DET,IFAIL)
192              IF(LONG.GT.3)WRITE(IPRT,223)
193              IF(IFAIL.EQ.0)GO TO 106
194              WRITE(IPRT,198)
195              GO TO 148
196         106 DO 114 I=1,NV
197              AM(I,I)=DAM(I,I)
198              SP(I)=SQRT(AM(I,I))
199              DO 113 J=1,NV
200         113 AM(I,J)=DAM(I,J)
201         114 IF(LONG.GT.3)WRITE(IPRT,221)(AM(I,K),K=1,NV)
202              WRITE(IPRT,415)
203              J=1
204              DO 122 I=1,NSP
205              IF(KI(I).NE.KSTAR)GO TO 120
206              BI=BLOG(I)
207              BLOG(I)=0.
208              SLB(I)=0.
209              IF(B(J).LT.1.E-37)GO TO 129
210              BLOG(I)=ALOG10(B(J))+QR(I)*(A0-AUAVG)+BI
211              SLB(I)=GOFA*SP(J)/(2.303*B(J))
212         129 J=J+1
213         120 ISL=1000.*SLB(I)+.5
214              WRITE(IPRT,90)KI(I),(IX(K,I),K=1,4),BLOG(I),ISL,(LABP(I,L),L=1,7)
215         122 CONTINUE
216      C  CORRELATION MATRIX
217              WRITE(IPRT,218)
218              N=0
219              DO 123 I=1,NSP
220              IF(KI(I).NE.KSTAR)GO TO 123
221              N=N+1
222              L=0
223              DO 121 J=1,NSP
224              IF(KI(J).NE.KSTAR)GO TO 121
225              L=L+1
226              RHO(I,J)=AM(N,L)/(SP(N)*SP(L))
```

```
227        121 CONTINUE
228        123 WRITE(IPRT,220)(RHO(I,J),J=1,NSP)
229            WRITE(IPRT,87)
230            DO 24 I=1,NSP
231            SL(I)=0.
232            BL(I)=0.
233            DO 23 J=1,NSP
234            BL(I)=BL(I)+CK(I,J)*BLOG(J)
235            DO 23 K=1,NSP
236         23 SL(I)=SL(I)+CK(I,J)*CK(I,K)*SLB(J)*SLB(K)*RHO(K,J)
237            IF(SL(I).GT.0.)SL(I)=SQRT(SL(I))
238            ISL=1000.*SL(I)+.5
239         24 WRITE(IPRT,90)KI(I),(IX(K,I),K=1,4),BL(I),ISL,(LABP(I,L),L=1,7)
240            GO TO 134
241        148 RETURN
242            END
```

```
1    C  -------*---------*---------*---------*---------*---------*---------*
2           SUBROUTINE NNLS (A,MDA,M,N,B,X,RNORM,W,ZZ,INDEX,MODE)
3           DIMENSION A(MDA,N), B(M), X(N), W(N), ZZ(250),INDEX(N),DUMMY(250)
4           DATA ZERO,ONE,TWO,FACTOR/0.,1.,2.,0.01/
5    C      C.L.LAWSON AND R.J.HANSON, JET PROPULSION LABORATORY, 1973 JUNE 15
6    C      TO APPEAR IN ;SOLVING LEAST SQUARES PROBLEMS;, PRENTICE-HALL, 1974
7    C      **********  NONNEGATIVE LEAST SQUARES   **********
8    C      GIVEN AN M BY N MATRIX, A, AND AN M-VECTOR, B,  COMPUTE AN
9    C      N-VECTOR, X, WHICH SOLVES THE LEAST SQUARES PROBLEM
10   C      A * X = B  SUBJECT TO X .GE. 0
11   C      A(),MDA,M,N    MDA IS THE FIRST DIMENSIONING PARAMETER FOR THE
12   C      ARRAY, A().   ON ENTRY A() CONTAINS THE M BY N
13   C      MATRIX, A.     ON EXIT A() CONTAINS
14   C      THE PRODUCT MATRIX, Q*A , WHERE Q IS AN
15   C      M BY M ORTHOGONAL MATRIX GENERATED IMPLICITY BY
16   C      THIS SUBROUTINE.
17   C      B()    ON ENTRY B() CONTAINS THE M-VECTOR, B.  ON EXIT B() CON-
18   C      TAINS Q*B.
19   C      X()    ON ENTRY X() NEED NOT BE INITIALIZED.  ON EXIT X() WILL
20   C      CONTAIN THE SOLUTION VECTOR.
21          MODE=1
22          IF(M.GT.0.AND.N.GT.0)GO TO 10
23          MODE=2
24          RETURN
25       10 ITER=0
26          ITMAX=3*N
27          DO 20 I=1,N
28          X(I)=ZERO
29       20 INDEX(I)=I
30          IZ2=N
31          IZ1=1
32          NSETP=0
33          NPP1=1
34   C      ******  MAIN LOOP BEGINS HERE   ******
35       30 CONTINUE
36          IF(IZ1.GT.IZ2.OR.NSETP.GE.M)GO TO 350
37          DO 50 IZ=IZ1,IZ2
38          J=INDEX(IZ)
39          SM=ZERO
40          DO 40 L=NPP1,M
41       40 SM=SM+A(L,J)*B(L)
42       50 W(J)=SM
43       60 WMAX=ZERO
44          DO 70 IZ=IZ1,IZ2
45          J=INDEX(IZ)
46          IF(W(J).LE.WMAX)GO TO 70
47          WMAX=W(J)
48          IZMAX=IZ
49       70 CONTINUE
50          IF(WMAX)350,350,80
51       80 IZ=IZMAX
52          J=INDEX(IZ)
53          ASAVE=A(NPP1,J)
54          CALL H12 (1,NPP1,NPP1+1,M,A(1,J),1,UP,DUMMY,1,1,0)
55          UNORM=ZERO
```

```
 56            IF(NSETP.EQ.0)GO TO 100
 57            DO 90 L=1,NSETP
 58      90 UNORM=UNORM+A(L,J)**2
 59     100 UNORM=SQRT(UNORM)
 60            IF(DIFF(UNORM+ABS(A(NPP1,J))*FACTOR,UNORM))130,130,110
 61     110 DO 120 L=1,M
 62     120 ZZ(L)=B(L)
 63            CALL H12 (2,NPP1,NPP1+1,M,A(1,J),1,UP,ZZ,1,1,1)
 64            ZTEST=ZZ(NPP1)/A(NPP1,J)
 65            IF(ZTEST)130,130,140
 66     130 A(NPP1,J)=ASAVE
 67            W(J)=ZERO
 68            GO TO 60
 69     140 DO 150 L=1,M
 70     150 B(L)=ZZ(L)
 71            INDEX(IZ)=INDEX(IZ1)
 72            INDEX(IZ1)=J
 73            IZ1=IZ1+1
 74            NSETP=NPP1
 75            NPP1=NPP1+1
 76            IF(IZ1.GT.IZ2)GO TO 170
 77            DO 160 JZ=IZ1,IZ2
 78            JJ=INDEX(JZ)
 79     160 CALL H12 (2,NSETP,NPP1,M,A(1,J),1,UP,A(1,JJ),1,MDA,1)
 80     170 CONTINUE
 81            IF(NSETP.EQ.M)GO TO 190
 82            DO 180 L=NPP1,M
 83     180 A(L,J)=ZERO
 84     190 CONTINUE
 85            W(J)=ZERO
 86            ASSIGN 200 TO NEXT
 87            GO TO 400
 88     200 CONTINUE
 89    C      ******   SECONDARY LOOP BEGINS HERE ******
 90     210 ITER=ITER+1
 91            IF(ITER.LE.ITMAX)GO TO 220
 92            MODE=3
 93            GO TO 350
 94     220 CONTINUE
 95            ALPHA=TWO
 96            DO 240  IP=1,NSETP
 97            L=INDEX(IP)
 98            IF(ZZ(IP))230,230,240
 99     230 T=-X(L)/(ZZ(IP)-X(L))
100            IF(ALPHA.LE.T)GO TO 240
101            ALPHA=T
102            JJ=IP
103     240 CONTINUE
104            IF(ALPHA.EQ.TWO)GO TO 330
105            DO 250 IP=1,NSETP
106            L=INDEX(IP)
107     250 X(L)=X(L)+ALPHA*(ZZ(IP)-X(L))
108            I=INDEX(JJ)
109     260 X(I)=ZERO
110            IF(JJ.EQ.NSETP)GO TO 290
111            JJ=JJ+1
112            DO 280 J=JJ,NSETP
113            II=INDEX(J)
114            INDEX(J-1)=II
115            CALL G1 (A(J-1,II),A(J,II),CC,SS,A(J-1,II))
116            A(J,II)=ZERO
117            DO 270 L=1,N
118            IF(L.NE.II)CALL G2 (CC,SS,A(J-1,L),A(J,L))
119     270 CONTINUE
120     280 CALL G2 (CC,SS,B(J-1),B(J))
121     290 NPP1=NSETP
122            NSETP=NSETP-1
123            IZ1=IZ1-1
124            INDEX(IZ1)=I
125            DO 300 JJ=1,NSETP
126            I=INDEX(JJ)
127            IF(X(I))260,260,300
128     300 CONTINUE
129            DO 310 I=1,M
130     310 ZZ(I)=B(I)
131            ASSIGN 320 TO NEXT
```

```
132              GO TO 400
133         320 CONTINUE
134              GO TO 210
135    C         *****  END OF SECONDARY LOOP  ******
136         330 DO 340 IP=1,NSETP
137              I=INDEX(IP)
138         340 X(I)=ZZ(IP)
139              GO TO 30
140    C         *****  END OF MAIN LOOP  ******
141         350 SM=ZERO
142              IF(NPP1.GT.M)GO TO 370
143              DO 360 I=NPP1,M
144         360 SM=SM+B(I)**2
145              GO TO 390
146         370 DO 380 J=1,N
147         380 W(J)=ZERO
148         390 RNORM=SQRT(SM)
149              RETURN
150         400 DO 430 L=1,NSETP
151              IP=NSETP+1-L
152              IF(L.EQ.1)GO TO 420
153              DO 410 II=1,IP
154         410 ZZ(II)=ZZ(II)-A(II,JJ)*ZZ(IP+1)
155         420 JJ=INDEX(IP)
156         430 ZZ(IP)=ZZ(IP)/A(IP,JJ)
157              GO TO NEXT, (200,320)
158              END
```

```
1    C -------*---------*---------*---------*---------*---------*---------*
2              SUBROUTINE H12 (MODE,LPIVOT,L1,M,U,IUE,UP,C,ICE,ICV,NCV)
3              DOUBLE PRECISION SM,B,DBLE
4              DIMENSION U(IUE,M), C(250)
5              DATA ZERO,ONE,TWO/0.,1.,2./
6    C         C.L.LAWSON AND R.J.HANSON, JET PROPULSION LABORATORY, 1973 JUN 12
7    C         IN ;SOLVING LEAST SQUARES PROBLEMS;, PRENTICE-HALL, 1974
8    C         CONSTRUCTION AND/OR APPLICATION OF A SINGLE
9    C         HOUSEHOLDER TRANSFORMATION    Q = 1 +U*(U**T)/B
10             IF(0.GE.LPIVOT.OR.LPIVOT.GE.L1.OR.L1.GT.M)RETURN
11             CL=ABS(U(1,LPIVOT))
12             IF(MODE.EQ.2)GO TO 60
13             DO 10 J=L1,M
14         10 CL=AMAX1(ABS(U(1,J)),CL)
15             IF(CL.LE.ZERO)RETURN
16             CLINV=ONE/CL
17             SM=(DBLE(U(1,LPIVOT))*CLINV)**2
18             DO 30 J=L1,M
19         30 SM=SM+(DBLE(U(1,J))*CLINV)**2
20             SM1=SM
21             CL=CL*SQRT(SM1)
22             IF(U(1,LPIVOT).GT.ZERO)CL=-CL
23             UP=U(1,LPIVOT)-CL
24             U(1,LPIVOT)=CL
25             GO TO 70
26         60 IF(CL.LE.ZERO)RETURN
27         70 IF(NCV.LE.0)RETURN
28             B=DBLE(UP)*U(1,LPIVOT)
29             IF(B)80,130,130
30         80 B=ONE/B
31             I2=1-ICV+ICE*(LPIVOT-1)
32             INCR=ICE*(L1-LPIVOT)
33             DO 120 J=1,NCV
34             I2=I2+ICV
35             I3=I2+INCR
36             I4=I3
37             SM=C(I2)*DBLE(UP)
38             DO 90 I=L1,M
39             SM=SM+C(I3)*DBLE(U(1,I))
40         90 I3=I3+ICE
41             IF(SM)100,120,100
42        100 SM=SM*B
43             C(I2)=C(I2)+SM*DBLE(UP)
44             DO 110 I=L1,M
45             C(I4)=C(I4)+SM*DBLE(U(1,I))
```

```
46        110 I4=I4+ICE
47        120 CONTINUE
48        130 RETURN
49            END




 1    C -------*---------*---------*---------*---------*---------*---------*
 2            SUBROUTINE G1 (A,B,COS,SIN,SIG)
 3            DATA ZERO,ONE,TWO/0.,1.,2./
 4    C       COMPUTE ORTHOGONAL ROTATION MATRIX..
 5            IF(ABS(A).LE.ABS(B))GO TO 10
 6            XR=B/A
 7            YR=SQRT(ONE+XR**2)
 8            COS=SIGN(ONE/YR,A)
 9            SIN=COS*XR
10            SIG=ABS(A)*YR
11            RETURN
12         10 IF(B)20,30,20
13         20 XR=A/B
14            YR=SQRT(ONE+XR**2)
15            SIN=SIGN(ONE/YR,B)
16            COS=SIN*XR
17            SIN=ABS(B)*YR
18            RETURN
19         30 SIG=ZERO
20            COS=ZERO
21            SIN=ONE
22            RETURN
23            END




 1    C -------*---------*---------*---------*---------*---------*---------*
 2            SUBROUTINE G2(COS,SIN,X,Y)
 3    C       APPLY THE ROTATION COMPUTED BY G1 TO (X,Y)
 4            XR=COS*X+SIN*Y
 5            Y=-SIN*X+COS*Y
 6            X=XR
 7            RETURN
 8            END




 1    C -------*---------*---------*---------*---------*---------*---------*
 2            FUNCTION DIFF(X,Y)
 3            DIFF=X-Y
 4            RETURN
 5            END




 1    C -------*---------*---------*---------*---------*---------*---------*
 2            SUBROUTINE NONLIN
 3    C NONLINEAR LEAST SQUARES REFINEMENT OF CONSTANTS
 4    C PRESENT VERSION CANNOT ACCOMODATE MORE THAN 15 VARIED CONSTANTS
 5            INTEGER TAPIN,TAPOUT
 6            DOUBLE PRECISION DAM(20,20),DET
 7            COMMON/LARG/STAT(600,4),SIGPX(4,600),SIGD(4,600),INDX(600)
 8            COMMON/DATA/IDATA(600),PCF(4,600),CT(4,600),EQV(600),U1(600)
 9            COMMON/MISC/FILL(4800),U2(1200)
10            COMMON/MONT/ICRD,IPRT,IPCH,TITLE(20),FMT(20),YNAME(4),CODEN(5)
11            COMMON/DIM/NO,NOS(20),NSP,NM,NL,NC,NC1,NV,NSET,IST,JST,NY,NCY,IW
12            COMMON/CONT/IPUN,IPLOT,IFPL,LONG,IDAT,IPAR,IGUESS,IDPX,ISORT,IREF
13            COMMON/LIMS/PH1,PH2,PHMIN,PHMAX,EMIN,EMAX,PLMIN,PLMAX,BMAX
14            COMMON/IO/TAPIN,TAPOUT,INTAP(20)
15            COMMON/INIT/IACT,AU,AH,AOH,A1,A0,B1,AKW,CKW,TEMP,UCORR
16            COMMON/LIGS/HM(4),NH(4),LIGS(4),KKI(4),QC(4),QR(20),Q(20)
```

```
17          COMMON/CONC/CF(4),HOH,CTC(4),C(20),PC(20),TOTX(4)
18          COMMON/PAR1/KI(20),LABP(20,7),IX(4,20),EX(4,20)
19          COMMON/PAR2/AM(20,20),V(20),DP(20),P(20),SHFT(20),SOECUT,OPT1,OPT2
20          COMMON/BETA/BETA(20),BLOG(20),BL(20),B(20),SL(20),SLB(20),SB(20)
21          COMMON/SIG1/SIGC(4),SIGV(4),SIGX(4),SOH
22          COMMON/SIG2/SCC(4),RCC(4,4),DEL(4),RHO(20,20),SP(20),DEDV(20)
23          COMMON/LIMT/JMEM,JSET,MSTART,MEND,NS
24          COMMON/PREL/NCM,INDEX1(4),INDEX2(4),INDEX3(4)
25          COMMON/RFAC/SIG(50),DENOM(50),NPTS(50),SUM(50)
26          COMMON/DERV/D(4,4),CB(20,20),CK(20,20),ITR
27          COMMON/DETERM/DETR
28          DIMENSION AL1(4),ALPHA1(20),ISA(20),ERR(2),GOF(4),SH(20),RW(4),
29        & DELTA(4),DV(20),YC1(4),YC2(4),G(3),BLU(20),EB(4),TOT(20,3)
30        & ,KII(5),TOTSC(5),TOTSCO(5),TAL(2),ACID(20),BASE(20),STOCK(20)
31          DIMENSION EBR(4),CTR(4)
32          DATA TAL/4HACID,4HBASE/
33          DATA ERR,KSTAR/1H ,1HE,1H*/
34        4 FORMAT(   /,' INITIAL PX(NC) VALUES ARE <ITUO> ',3F8.3)
35        5 FORMAT(//,' ACID,BASE,STOCK,TOTX ',6F10.6)
36       14 FORMAT(20X,6F10.0)
37       18 FORMAT(1X,130(1H-))
38       19 FORMAT('+XTOT<ITUO>',F7.5,3F13.5)
39       20 FORMAT(62X,'PX<ITUO>',4F13.3)
40       21 FORMAT(6H ALPHA,4X,9(F7.3,1H(,I4,1H) ),/,10X,9(F7.3,1H(,I4,1H) ),
41        & /,10X,9(F7.3,1H(,I4,1H) ) )
42       22 FORMAT(1H+,58X,F7.3,F10.5,F13.4,I5)
43       24 FORMAT(6H LOG B,9F13.3,3(/6X,9F13.3))
44       29 FORMAT(3H PC,3X,9F13.3,3(/6X,9F13.3))
45       91 FORMAT(1H+,F8.3,3F7.3,F8.2,3F7.3)
46      150 FORMAT(' AVG PH(OBS)-PH(CALC) =',F7.3)
47      158 FORMAT(8F10.4)
48      159 FORMAT(4H DET,E10.2,5X,I2,11H ITERATIONS,5X,4HSIGS,4F10.4)
49      186 FORMAT('1CALCULATION BASED ON PARAMETERS BEFORE CYCLE',I2,//,2X,
50        &2(7H  P(H) ),'   DEL    SIG   DEL/SIG EQV',17X,'ION STR  ',
51        &'HTOT(CALC)      AU     SET'/4X,
52        & '(OBS) (CALC) (O-C) (OBS)')
53      190 FORMAT(1X,A1,4F7.3,F8.2,3F7.3)
54      192 FORMAT('0GOODNESS-OF-FIT = SQRT(SUM(W',8H*(O-C)**,'2)/(NO-NP)) ='
55        & ,4F7.3,5X,15H(KKI(NY) ORDER),
56        &              /,41H RW = SQRT(SUM(W*(O-C)**2)/SUM(W*O**2)) =,4F6.3,/,
57        &' AVG ION STR =',F5.3,1H(,I3,3H) M,/,9H AU(AVG)=,F7.4,/,
58        & ' DEL/SIG (AVG) =',4F6.2)
59      198 FORMAT(/,16H SINGULAR MATRIX,/)
60      200 FORMAT('1PARAMETERS AFTER LEAST SQUARES CYCLE',I3,//9X,3HOLD,6X,
61        & 5HSHIFT,6X,3HNEW,6X,5HERROR,2X,'SHIFT/ERROR',5X,3HDP,  5X,
62        &'OPT SHFT FACTR=',F6.3,4X,9HGOF(OPT)=,F6.3,/)
63      202 FORMAT(1H I3,F10.5,10X,F10.5,32X,  7A4)
64      204 FORMAT(1H I3,4F10.4,F10.5,2X,  7A4)
65      218 FORMAT(//,3X,18HCORRELATION MATRIX,/)
66      220 FORMAT(18F7.3)
67      221 FORMAT(10F13.5)
68      222 FORMAT(22H0NUMERICAL DERIVATIVES,/)
69      223 FORMAT(23H0ANALYTICAL DERIVATIVES,/)
70      224 FORMAT(/,25H   PXO     PXC   DV/DP(NV),/)
71     1220 FORMAT(2F7.3,10F8.1,/,14X,10F8.1)
72     1252 FORMAT(' S',F9.6,5X,'EXBAR',4F7.3,5X,'S(R)',F9.6,5X,'EXBAR(R)',
73        & 4F6.2)
74     1332 FORMAT(I4,F10.5,10X,F10.5,32X,' SCALE FACTOR ON ',
75        & A4,' CONCENTRATION')
76     1334 FORMAT(I4,5F10.5,12X,' SCALE FACTOR ON ',A4,' CONCENTRATION')
77     1430 FORMAT(6H+<EX> ,4X,9(2X,A1,1H<,4I2,1H>))
78     1431 FORMAT(         10X,9(2X,A1,1H<,4I2,1H>))
79     1432 FORMAT(I4,F10.5,10X,F10.5,32X,' SCALE FACTOR ON TOTAL ',
80        &'CONCENTRATION OF REACTANT(',I1,')')
81     1433 FORMAT(5A1)
82     1436 FORMAT(I4,5F10.5,12X,' SCALE FACTOR ON TOTAL CONCENTRATION OF ',
83        &'REACTANT(',I1,')')
84     1438 FORMAT(' REVISION DATE: 7 JAN 1981',/)
85          WRITE(IPRT,1438)
86   C   READ DATA AND PARAMETERS
87          IREF=1
88          ISKIP=IPUN
89          CALL PRELIM(JTYPE)
90          NC2=NC1+1
91          NSPNC2=NSP+NC2
```

```
 92    C  READ SCALE FACTOR REFINEMENT INDICATORS
 93          READ(ICRD,1433)(KII(I),I=1,4)
 94          NVT=NV
 95          DO 1434 J=1,NC2
 96          IF(KII(J).EQ.KSTAR)NVT=NVT+1
 97          KI(NSP+J)=KII(J)
 98          TOTSC(J)=1.
 99    1434 TOTSCO(J)=1.
100          IF(NVT.EQ.NV)GO TO 1534
101          DO 1533 J=1,NSET
102          READ(ICRD,14)ACID(J),BASE(J),STOCK(J)
103          READ(ICRD,158)(TOT(J,K),K=1,NC)
104    1533 WRITE(IPRT,5)ACID(J),BASE(J),STOCK(J),(TOT(J,K),K=1,NC)
105    C  INITIAL FREE CONCENTRATIONS
106    1534 IF(IGUESS.EQ.0)GO TO 310
107          READ(ICRD,158)(PCF(J,1),J=1,NC)
108          GO TO 399
109     310 IF(NM.EQ.0)GO TO 330
110          DO 320 K=1,NM
111     320 CF(K)=CT(K,1)
112     330 IF(NL.EQ.0)GO TO 380
113          KK=1
114          DO 370 I=1,NL
115          NI=NH(I)
116          DLL=1.
117          DO 360 J=1,NI
118          DLL=DLL+10.**(BLOG(KK)-FLOAT(J)*PHMIN)
119     360 KK=KK+1
120     370 CF(I+NM)=CT(I+NM,1)/DLL
121     380 DO 390 J=1,NC
122     390 IF(KKI(J).EQ.0)PCF(J,1)=-ALOG10(CF(J))
123     399 WRITE(IPRT,4)(PCF(J,1),J=1,NC)
124          IF(JTYPE.EQ.0)WRITE(IPRT,222)
125          IF(JTYPE.EQ.1)WRITE(IPRT,223)
126          ICYCL=0
127          FREE=NO-NSP
128          GOFA=1000.
129     134 ICYCL=ICYCL+1
130          FREED=FREE
131          IF(ICYCL.EQ.NCY+2)GO TO 148
132          DO 58 I=1,NSPNC2
133          DO 57 J=1,NSPNC2
134          RHO(I,J)=0.
135      57 AM(I,J)=0.
136          V(I)=0.
137      58 RHO(I,I)=1.
138      62 DO 60 K=1,4
139          DELTA(K)=0.
140          GOF(K)=0.
141      60 RW(K)=0.
142          CALL CLEAR(4)
143          UAVG=0.
144          AUAVG=0.
145          AVGDEL=0.
146          IF(ICYCL.NE.NCY+1)GO TO 76
147          IF(LONG.LT.1)GO TO 76
148          WRITE(IPRT,186)ICYCL
149      76 DO 92 I=1,NO
150          ID=IDATA(I)
151          IF(NVT.EQ.NV)GO TO 1437
152    C  DILUTION FACTOR MUST BE PRESERVED
153          IF(ICYCL.EQ.1)U2(I+600)=CT(1,I)/TOT(ID,1)
154          DIL=U2(I+600)
155          TIL=1.-DIL
156          IF(ACID(ID).LT.BASE(ID))GO TO 1236
157          TA=TIL*ACID(ID)+DIL*STOCK(ID)
158          TB=-DIL*BASE(ID)
159          GO TO 1237
160    1236 TA=DIL*(ACID(ID)+STOCK(ID))
161          TB=-TIL*BASE(ID)
162    C  SCALED CONCENTRATIONS
163    1237 TL=0.
164          DO 1435 J=1,NC
165          CT(J,I)=TOT(ID,J)*DIL*TOTSC(J)
166    1435 IF(J.GT.NM)TL=TL+HM(J-NM)*CT(J,I)
```

```
167            CT(NC1,I)=TA*TOTSC(NC1)+TB*TOTSC(NC2)+TL
168      1437 IF((ICYCL.EQ.NCY+1).AND.(LONG.GE.1))WRITE(IPRT,18)
169            CALL CALC1(YC1,I,IA,U,IFAIL,0,JTYPE,ICYCL)
170            IF(IFAIL.NE.0)FREED=FREED-FLOAT(NY)
171            U2(I)=U
172            UAVG=UAVG+U
173            AUAVG=AUAVG+AU
174            ER=ERR(IFAIL+1)
175            IF(ICYCL.NE.NCY+1)GO TO 99
176            IF(LONG.LT.2)GO TO 99
177            WRITE(IPRT, 20)(PCF(J,I),J=1,NC1)
178            WRITE(IPRT,19)(CT(K,I),K=1,NC1)
179            SOM=0.
180            DO 1250 L1=1,NC1
181      1250 EB(L1)=0.
182            DO 23 K=1,NSP
183            SOM=SOM+C(K)
184            DO 1251 L1=1,NC1
185      1251 EB(L1)=EB(L1)+EX(L1,K)*C(K)
186            DIMER=1.
187            IF(IX(1,K).NE.0)DIMER=EX(1,K)
188            ALPHA1(K)=DIMER*C(K)/CTC(1)
189    C   CALCULATION OF THE ERRORS OF THE DISTRIBUTION RATIOS
190            T=0.
191            DO 25  J1=1,NSP
192            DO 25  J2=1,NSP
193         25 T=T+EX(1,J1)*EX(1,J2)*C(J1)*C(J2)*SLB(J1)*SLB(J2)*RHO(J1,J2)
194            T=T/CTC(1)**2+SLB(K)**2
195            ISA(K)=9999
196         23 IF(T.GT.0.)ISA(K)=ALPHA1(K)*SQRT(T)*2303.+.5
197            DO 127 L=1,NC1
198            EBR(L)=0.
199            EB(L)=EB(L)/SOM
200        127 AL1(L)=CF(L)/CTC(1)
201            NST=9
202            IF(NSP.LT.NST)NST=NSP
203            WRITE(IPRT,1431)(KI(K),(IX(J,K),J=1,4),K=1,NST)
204            WRITE(IPRT,1430)(KI(K),(IX(J,K),J=1,4),K=1,NST)
205            WRITE(IPRT,21)(ALPHA1(K) ,ISA(K),K=1,NSP)
206    C   RESIDUAL ANALYSIS
207            S2R=0.
208            IF(NM.EQ.0)GO TO 3933
209            S2R=SOM
210            DO 3930 K=1,NSP
211      3930 IF(K.LE.IST)S2R=S2R-C(K)
212            DO 3932 J=1,NC1
213            CTR(J)=CTC(J)-CF(J)
214            DO 3931 K=1,NSP
215      3931 IF(K.LE.IST)CTR(J)=CTR(J)-EX(J,K)*C(K)
216            CTR(NC1)=CTR(NC1)+CKW/CF(NC1)
217      3932 EBR(J)=CTR(J)/S2R
218      3933 WRITE(IPRT,1252)SOM,(EB(L1),L1=1,4),S2R,(EBR(L1),L1=1,4)
219            IF(LONG.LT.3)GO TO 99
220            DO 28 K=1,NSP
221         28 BLU(K)=BLOG(K)+QR(K)*(AU-A0)
222            WRITE(IPRT,29)(PC(M),M=1,NSP)
223            WRITE(IPRT,24)(BLU(K),K=1,NSP)
224            IF(LONG.LT.4)GO TO 99
225            WRITE(IPRT,159)DETR,ITR,(SCC(L),L=1,NC1)
226            DO 713 K=1,NC1
227        713 WRITE(IPRT,221)(RCC(K,L),L=1,NC1),(D(K,L),L=1,NC1)
228         99 DO 92 KK=1,NY
229            KX=INDEX2(KK)
230            DY=PCF(KX,I)-YC1(KK)
231    C   FOR PLOT
232            SIGD(KK,I)=ABS(DY)
233            AVGDEL=AVGDEL+ABS(DY)
234            SIGP=SIGPX(KX,I)
235            WDY=DY/SIGP
236            DELTA(KK)=DELTA(KK)+WDY
237            CALL RFACTR(KK,PCF(KX,I),WDY,SIGP)
238            STAT(I,KK)=WDY
239            IF(ICYCL.LT.NCY+1)GO TO 107
240            IF(LONG.EQ.0)GO TO 107
241            WRITE(IPRT,190)ER,PCF(KX,I),YC1(KK),DY,SIGPX(KX,I),WDY,EQV(I)
```

```
242              IF(KK.EQ.1)WRITE(IPRT,22)U,CTC(NC1),AU,IDATA(I)
243              IF(LONG.LT.2)GO TO 107
244              WRITE(IPRT,91)PCF(KX,I),YC1(KK),DY,SIGPX(KX,I),WDY,EQV(I)
245              WRITE(IPRT,91)PCF(KX,I),YC1(KK),DY,SIGPX(KX,I),WDY,EQV(I)
246       107 IF(IFAIL.GT.0)GO TO 92
247     C  DERIVATIVES
248              IF(ICYCL.EQ.NCY+1)GO TO 92
249              J=1
250              DO 82 K=1,NSP
251              IF(KI(K).NE.KSTAR)GO TO 82
252              IF(JTYPE.EQ.1)GO TO 7501
253     C  NUMERICAL
254              BL(K)=BL(K)+DP(K)
255     C  ADJUST FOR SHIFT
256              DO 7010 J1=1,NSP
257              BLU(J1)=BLOG(J1)
258              IF(ABS(CB(J1,K)).LT.0.01)GO TO 7010
259              BLOG(J1)=0.
260              DO 7009 J2=1,NSP
261      7009 BLOG(J1)=BLOG(J1)+CB(J1,J2)*BL(J2)
262      7010 CONTINUE
263              CALL CALC1(YC2,I,200,U,IFAIL,1,JTYPE,ICYCL)
264              DV(J)=1.E-5
265              IF(IFAIL.EQ.0)DV(J)=((YC2(KK)-YC1(KK))/DP(K))/SIGP
266              BL(K)=BL(K)-DP(K)
267     C  RESET
268              DO 7011 J1=1,NSP
269      7011 BLOG(J1)=BLU(J1)
270              GO TO 7801
271     C  ANALYTICAL
272      7501 R1=0.
273              DO 8001 IL=1,NSP
274              IF(ABS(CB(IL,K)).LT.0.01)GO TO 8001
275              R2=0.
276              DO 8000 IK=1,NC1
277      8000 R2=R2+D(KX,IK)*EX(IK,IL)
278              R1=R1+CB(IL,K)*C(IL)*R2
279      8001 CONTINUE
280              DV(J)= 2.303*R1/SIGP
281      7801 J=J+1
282          82 CONTINUE
283     C  ANALYTICAL DERIVATIVES FOR TOTAL CONCENTRATION VARIATION
284              DO 1082 K=1,NC2
285              IF(KII(K).NE.KSTAR)GO TO 1082
286              IF(K.GT.NC)GO TO 1075
287              IF(K.GT.NM)GO TO 1074
288     C  METAL DERIVATIVES
289              DV(J)=-D(KX,K)*CT(K,I)/SIGP
290              GO TO 1078
291     C  LIGAND DERIVATIVES
292      1074 DV(J)=(-D(KX,K)-D(KX,NC1)*HM(K-NM))*CT(K,I)/SIGP
293              GO TO 1078
294     C  ACID/BASE DERIVATIVES
295      1075 AB=TA
296              IF(K.EQ.NC2)AB=TB
297              DV(J)=-D(KX,NC1)*AB/SIGP
298      1078 J=J+1
299      1082 CONTINUE
300     C  ACCUMULATE SYMMETRIC MATRIX AND VECTOR
301              DO 90 J=1,NVT
302              DO 88 K=1,NVT
303          88 AM(J,K)=AM(J,K)+DV(J)*DV(K)
304          90 V(J)=V(J)+DV(J)*WDY
305              IF(LONG.LT.5)GO TO 92
306              IF(ICYCL.EQ.NCY)WRITE(IPRT,1220)PCF(KX,I),YC1(KK),(DV(J),J=1,NVT)
307          92 CONTINUE
308              SAVGOF=GOFA
309              IF(FREED.LT.1.)FREED=1.
310              AVG=NO
311              UAVG=UAVG/AVG
312              AUAVG=AUAVG/AVG
313              GOFA=0.
314              IF(ICYCL.GT.1)GO TO 116
315     C  CALCULATE RMS DEVIATION OF IONIC STRENGTH FROM AVERAGE
316              DU=0.
```

```
317            DO 56 I=1,NO
318      56 DU=DU+(U2(I)-UAVG)**2
319            IDU=1000.*SQRT(DU/(AVG-1.))+.5
320     116 DO 61 K=1,NY
321            DELTA(K)=DELTA(K)/AVG
322            GOF(K)=SQRT(SIG(K)/FREED)
323            RW(K)=SQRT(SIG(K)/DENOM(K))
324      61 GOFA=GOFA+SIG(K)
325            G(1)=GOFA
326            FREED=(IFIX(FREED)+NSP)*NY-NSP
327            GOFA=SQRT(GOFA/FREED)
328            WRITE(IPRT,192)GOF,RW,UAVG,IDU,AUAVG,DELTA
329   C  OPT1 SIGNALS WHEN SHIFT OPTIMIZATION IS TO BE TURNED OFF
330            IF(ICYCL.EQ.NCY+1)GO TO 148
331            IF(ABS(GOFA-SAVGOF).LT.OPT1)ISKIP=1
332            DO 1102 I=1,NVT
333            DO 1102 J=1,NVT
334    1102 DAM(I,J)=AM(I,J)
335            CALL DMINV(NVT,20,DAM,DET,IFAIL)
336            IF(IFAIL.EQ.0)GO TO 106
337            WRITE(IPRT,198)
338            GO TO 148
339     106 DO 114 I=1,NVT
340            AM(I,I)=DAM(I,I)
341            SH(I)=0.0
342            SP(I)=SQRT(AM(I,I))
343            DO 115 J=1,NVT
344            AM(I,J)=DAM(I,J)
345     115 SH(I)=SH(I)+AM(I,J)*V(J)
346     114 IF(ABS(SH(I)).GT.1.0)SH(I)=SIGN(.99999,SH(I))
347            IF(ISKIP.EQ.1)GO TO 601
348   C  SHIFT OPTIMIZATION
349            DO 600 L=2,3
350            J=1
351            DO 599 I=1,NSP
352            IF(KI(I).NE.KSTAR)GO TO 599
353            BLOG(I)=0.
354            DO 588 K=1,NSP
355     588 BLOG(I)=BLOG(I)+CB(I,K)*(BL(K)+.25*SH(J)*FLOAT(L-2))
356            J=J+1
357     599 CONTINUE
358            G(L)=0.
359            DO 600 I=1,NO
360            CALL CALC1(YC1,I,200,U,IFAIL,2,JTYPE,ICYCL)
361            DO 600 J=1,NY
362            JX=INDEX2(J)
363     600 G(L)=G(L)+((PCF(JX,I)-YC1(J))/SIGPX(JX,I))**2
364            C1=2.*(G(1)-2.*G(3))
365            C2=-3.*G(1)+4.*G(2)-G(3)
366   C  IF PARABOLA IS INVERTED (WHICH MAY BE DUE TO OSCILLATORY BEHAVIOR)
367   C  RELY ON THE ARBITRARILY SMALL SHIFT FACTOR, OPT2, UNLESS G3.LT.G1
368            OPT=OPT2
369            IF(G(3).LT.G(1))OPT=1.
370            IF(C1.GT.0.)OPT=-.5*C2/C1
371            IF(OPT.LT.0.001)OPT=OPT2
372            GMM=C1*OPT*OPT+C2*OPT+G(1)
373            IF(GMM.GT.1.E-5)GO TO 603
374            GMM=1.E-5
375            OPT=OPT2
376     603 GMM=SQRT(GMM/FREED)
377            GO TO 602
378     601 OPT=1.
379            GMM=GOFA
380     602 WRITE(IPRT,200)ICYCL,OPT,GMM
381            J=1
382            SOEMAX=0.
383            DO 122 I=1,NSP
384            IF(KI(I).EQ.KSTAR)GO TO 120
385            WRITE(IPRT,202)I,BL(I),BL(I),(LABP(I,L),L=1,7)
386            GO TO 122
387     120 POLD=BL(I)
388            SIGP=GOFA*SP(J)
389            SOE=ABS(SH(J))/SIGP
390            IF(SOE.GT.SOEMAX)SOEMAX=SOE
391            BL(I)=POLD+OPT*SH(J)*SHFT(I)
```

```
392              SL(I)=GOFA*SP(J)
393              WRITE(IPRT,204)I,POLD,SH(J),BL(I),SIGP,SOE,DP(I),(LABP(I,K),K=1,7)
394              J=J+1
395        122 CONTINUE
396    C SCALE FACTORS
397              DO 1122 K=1,NC2
398              NK=NSP+K
399              IF(KII(K).EQ.KSTAR)GO TO 1120
400              IF(K.LT.NC1)WRITE(IPRT,1432)NK,TOTSC(K),TOTSC(K),K
401              IF(K.GE.NC1)WRITE(IPRT,1332)NK,TOTSC(K),TOTSC(K),TAL(K-NC)
402              GO TO 1122
403       1120 SIGP=GOFA*SP(J)
404              SOE=ABS(SH(J))/SIGP
405              TOTSCO(K)=TOTSC(K)
406              TOTSC(K)=TOTSC(K)+SH(J)
407              IF(K.LT.NC1)
408             &WRITE(IPRT,1436)NK,TOTSCO(K),SH(J),TOTSC(K),SIGP,SOE,K
409              IF(K.GE.NC1)
410             &WRITE(IPRT,1334)NK,TOTSCO(K),SH(J),TOTSC(K),SIGP,SOE,TAL(K-NC)
411              J=J+1
412       1122 CONTINUE
413    C   SHIFT THE BLOG ALSO
414              DO 140 L=1,NSP
415              BLOG(L)=0.
416              SLB(L)=0.
417              DO 139 M=1,NSP
418              SLB(L)=SLB(L)+CB(L,M)*SL(M)**2
419        139 BLOG(L)=BLOG(L)+CB(L,M)*BL(M)
420        140 IF(SLB(L).GT.0.)SLB(L)=SQRT(SLB(L))
421              WRITE(IPRT,218)
422    C   CORRELATION MATRIX
423              N=0
424              DO 146 I=1,NSPNC2
425              IF(KI(I).NE.KSTAR)GO TO 146
426              N=N+1
427              L=0
428              DO 144 J=1,NSPNC2
429              IF(KI(J).NE.KSTAR)GO TO 144
430              L=L+1
431              RHO(I,J)=AM(N,L)/(SP(N)*SP(L))
432        144 CONTINUE
433        146 WRITE(IPRT,220)(RHO(I,J),J=1,NSPNC2)
434              IF(ICYCL.EQ.NCY-1)WRITE(IPRT,224)
435              IF(SOEMAX.GT.SOECUT)GO TO 134
436              IF(SAVGOF.GT.1.2*GOFA)GO TO 134
437              NCY=ICYCL
438              GO TO 134
439        148 IF(TAPOUT.EQ.0)GO TO 151
440              IF(TAPOUT.EQ.IPRT)GO TO 151
441    C   THIS IS FOR USE BY 'ANALYSIS OF VARIANCE'
442              WRITE(TAPOUT)STAT,SIGPX,SIGD,INDX,IDATA,PCF,CT,EQV,U1,U2,
443             & NO,NOS,NSP,NV,NY,IREF,SOH,DEDV,NCM,INDEX2
444        151 IF(IPLOT.EQ.0)RETURN
445              EMIN=0.
446              EMAX=0.
447              J=0
448              DO 149 I=1,NO
449              DO  149 K=1,NY
450              KX=INDEX2(K)
451              J=J+1
452              IF(SIGD(K,I).GT.EMAX)EMAX=SIGD(K,I)
453              FILL(J)=PCF(KX,I)
454        149 FILL(J+600)=SIGD(K,I)
455              AVGDEL=AVGDEL/FLOAT(NO)
456              WRITE(IPRT,150)AVGDEL
457    C   PLOT OF PXO-PXC VS PX
458              CALL SIGPL
459              RETURN
460              END
```

```
1     C -------*---------*---------*---------*---------*---------*---------*
2              SUBROUTINE ERRORS
3     C ANALYSIS OF VARIANCE AND LINEPRINTER PLOTS
```

```
  4    C  NORMAL PROBABILITY PLOT ON CALCOMP
  5    C  CALLED FROM MONITOR AFTER LINEAR OR NONLINEAR LS PERFORMED
  6          EXTERNAL MOVE1,ICOMP4,ICOMP5
  7          COMMON/LARG/STAT(600,4),SIGPX(4,600),SIGD(4,600),INDX(600)
  8          COMMON/DATA/IDATA(600),PCF(4,600),CT(4,600),EQV(600),U1(600)
  9          COMMON/MONT/ICRD,IPRT,IPCH,TITLE(20),FMT(20),YNAME(4),CODEN(5)
 10          COMMON/DIM/NO,NOS(20),NSP,NM,NL,NC,NC1,NV,NSET,IST,JST,NY,NCY,IW
 11          COMMON/CONC/CF(4),HOH,CTC(4),C(20),PC(20),TOTX(4)
 12          COMMON/CONT/IPUN,IPLOT,IFPL,IWT ,IDAT,IPAR,IGUESS,IDPX,ISORT,IREF
 13          COMMON/INIT/IACT,AU,AH,AOH,A1,A0,B1,AKW,CKW,TEMP,UCORR
 14          COMMON/LIMS/PH1,PH2,PHMIN,PHMAX,XMIN,XMAX,YMIN,YMAX,BMAX
 15          COMMON/RFAC/SIG(50),DENOM(50),NPTS(50),SUM(50)
 16          COMMON/PREL/NCM,INDEX1(4),INDEX2(4),INDEX3(4)
 17          COMMON/LIMT/JMEM,JSET,MSTART,MEND,NS
 18          COMMON/SIG1/SIGC(4),SIGV(4),SIGX(4),SOH
 19          COMMON/SIG2/SCC(4),RCC(4,4),DEL(4),RHO(20,20),SPP(20),DEDV(20)
 20          COMMON/ALFO/JNDX(200)
 21          COMMON/MISC/SLOPE(1200,4),U2(1200)
 22          DIMENSION A(4,4),XN(5),CO(4),C1(4),SPX(4),SC(4),SV(4),SX(4),SP(4)
 23        & ,SX1(4),SP1(4)
 24          DATA YN,XN/4HG**2,3HEQV,1HU,2HPX ,1HX ,4HPX/V/
 25        3 FORMAT(8F10.5)
 26        5 FORMAT( 7H SIG( P,A4,8H)**2 = (,F6.4,9H )**2 + (,F6.4,4H *DP,A4,
 27       &5H/DEDV,18H*DEQV/DVOL)**2 + ( ,F5.3,3H *  ,A4,4H)**2)
 28        6 FORMAT(    15H SIG(PH)**2 = (,F6.4,9H )**2 + (,F6.4,' DPH/DEQV',
 29       &'*DEQV/DVOL)**2 + (,F4.2,7H * H +,F5.2,10H * OH )**2)
 30        7 FORMAT(13X,'+ (',F6.4,' *P',A4,')**2)
 31       30 FORMAT('1ANALYSIS BY H+',F5.2,6H*OH IN,I3,' INTERVALS',/)
 32       31 FORMAT('1ANALYSIS BY EQUIVALENTS IN',I3,' INTERVALS',/)
 33       32 FORMAT('1ANALYSIS BY IONIC STRENGTH IN',I3,' INTERVALS',/)
 34       33 FORMAT('1ANALYSIS BY P',A4,3H IN,I3,' INTERVALS',/)
 35       34 FORMAT('1ANALYSIS BY  FREE ',A4,3H IN,I3,' INTERVALS',/)
 36       35 FORMAT('1ANALYSIS BY DP',A4,8H/DVOL IN,I3,' INTERVALS',/)
 37       36 FORMAT(4X,'RANGE OF VARIABLE',8X,4HMEAN,5X,4HNUMB,4X,2HRW,6X,
 38       &3HGOF,3X,6HGOF**2,4X,3HSIG,/)
 39       39 FORMAT('0LINEAR FIT OF GOF**2 VS MEAN VARIABLE HAS INTERCEPT ',
 40       & E11.3,' AND SLOPE ',E11.3,/)
 41       40 FORMAT(2H (,E10.3,4H TO ,E10.3,1H),E10.3,I4,4F8.3)
 42       41 FORMAT(1H1,6X,2HPX,8X,3HSIG,6X,5HSLOPE,3X,7HDEL/SIG,7X,3HSET,/)
 43       42 FORMAT(2F10.3,F10.1,F10.3,I10)
 44       43 FORMAT('0LINEAR FIT OF GOF**2 VS (MEAN VAR /SIG)**2 HAS ',
 45       &'INTERCEPT',E11.3,' AND SLOPE ',E11.3,/)
 46       44 FORMAT(' NEW WEIGHTING SCHEME PARAMETERS ',4F10.5,' AND + OR -('
 47       & ,F7.5,' *VARIABLE)**2',/)
 48    C  UNSORT WHOLE-DATA SORTED DECKS TO INTRA-SET SORTED FORMAT
 49          IF(IGUESS.EQ.1)CALL TSORT4(NO,MOVE1,ICOMP5,1)
 50    C  SLOPE OF TITRATION CURVE CALCULATED
 51          DO 103 J=1,NY
 52          JX=INDEX2(J)
 53          I=0
 54          DO 103 N=1,NSET
 55          NUM=NOS(N)
 56          DO 102 K=1,NUM
 57          I=I+1
 58          IF(K.NE.NUM)GO TO 101
 59          SLOPE(I,JX)=SLOPE(I-1,JX)
 60          GO TO 102
 61      101 DE=EQV(I+1)-EQV(I)
 62          IF(ABS(DE).LT.1.E-4)DE=1.E-4
 63          SLOPE(I,JX)=DEDV(N)*ABS((PCF(JX,I+1)-PCF(JX,I))/DE)
 64      102 CONTINUE
 65      103 CONTINUE
 66    C  WEIGHTS
 67          READ(ICRD,3)SOH,(SIGC(K),SIGV(K),SIGX(K),K=1,NC1)
 68          READ(ICRD,3)(SPX(K),K=1,NC1)
 69          IF(SOH.LT.1.E-4)SOH=1.
 70          AX=SIGX(NC1)*SOH
 71          DO 104 J=1,NC1
 72          SX1(J)=SIGN(1.,SIGX(J))
 73      104 SP1(J)=SIGN(1.,SPX(J))
 74          DO 150 L=1,NY
 75          K=INDEX2(L)
 76          IF(K.NE.NC1)GO TO 149
 77          WRITE(IPRT,6)SIGC(NC1),SIGV(NC1),SIGX(NC1),AX
 78          GO TO 150
 79      149 WRITE(IPRT,5)YNAME(K),SIGC(K),SIGV(K),YNAME(K),SIGX(K),YNAME(K)
```

```
 80        150 WRITE(IPRT,7)SPX(K),YNAME(K)
 81            WRITE(IPRT,41)
 82     C  IWT=0 MEANS DEL/SIG TO BE PLOTTED ON CALCOMP
 83     C  IWT=1 MEANS DEL PLOTTED ON CALCOMP
 84            YMINP=0.
 85            YMAXP=0.
 86            DO 160 I=1,NO
 87            CALL ACTIV(U2(I),1.)
 88            DO 160 J=1,NY
 89            JX=INDEX2(J)
 90            CF(JX)=10.**(-PCF(JX,I))
 91            TD=CF(JX)
 92            IF(JX.EQ.NC1)TD=TD+SOH*CKW/CF(NC1)
 93            SXX=SQRT(SIGC(JX)**2+(SIGV(JX)*SLOPE(I,JX))**2
 94          & +SX1(JX)*(SIGX(JX)*TD)**2+SP1(JX)*(SPX(JX)*PCF(JX,I))**2)
 95            STAT(I,J)=STAT(I,J)*SIGPX(JX,I)/SXX
 96            SIGPX(JX,I)=SXX
 97            IF(IWT.EQ.0)GO TO 159
 98            IF(YMAXP.LT.ABS(STAT(I,J))*SXX)YMAXP=ABS(STAT(I,J))*SXX
 99            GO TO 160
100        159 IF(YMAXP.LT.ABS(STAT(I,J)))YMAXP=ABS(STAT(I,J))
101        160 WRITE(IPRT,42)PCF(JX,I),SIGPX(JX,I),SLOPE(I,JX),STAT(I,J),
102          & IDATA(I)
103     C  THE DESIRED NUMBER OF POINTS PER INTERVAL--CHOOSE AT LEAST 10
104            NP=10*IACT+ISORT
105            NF=NY
106            IF(IREF.EQ.0)NF=NC1
107            MSTART=0
108            MEND=NO
109            INT=NO/NP
110            IF(INT*NP.LT.NO)INT=INT+1
111            DO 80 JTYPE=1,5
112            JMEM=0
113         61 JMEM=JMEM+1
114            IF(IREF.NE.1)GO TO 70
115            DO 55 J=1,NY
116            IF(JMEM.EQ.INDEX2(J))GO TO 70
117         55 CONTINUE
118            GO TO 61
119         70 GO TO (71,72,73,74,75),JTYPE
120         71 WRITE(IPRT,31)INT
121            GO TO 76
122         72 WRITE(IPRT,32)INT
123            GO TO 76
124         73 WRITE(IPRT,33)YNAME(JMEM),INT
125            GO TO 76
126         74 IF(JMEM.EQ.NC1)GO TO 77
127            WRITE(IPRT,34)YNAME(JMEM),INT
128            GO TO 76
129         77 WRITE(IPRT,30)SOH,INT
130            GO TO 76
131         75 WRITE(IPRT,35)YNAME(JMEM),INT
132         76 WRITE(IPRT,36)
133     C  ORDER USING BOOTS
134            CALL TSORT4(NO,MOVE1,ICOMP4,JTYPE)
135            XMINP=BOOT(JTYPE,1)
136            XMAXP=BOOT(JTYPE,NO)
137            M2=MSTART
138            CALL CLEAR(INT)
139            YMAX=0.
140            DO 69 I=1,INT
141            M2=M2+NP
142            M1=M2-NP+1
143            IF(I.EQ.INT)M2=MEND
144            SIGAVG=0.
145            DO 68 J=M1,M2
146            DO 68 JF=1,NF
147     C  IREF=1 FOR NONLIN AND IREF=0 FOR LINLSQ ENTRIES
148            IF(IREF.EQ.1)GO TO 66
149            SIGMA=SIGD(JF,J)
150            OBS=CT(JF,J)
151            GO TO 65
152         66 JX=INDEX2(JF)
153            SIGMA=SIGPX(JX,J)
154            OBS=PCF(JX,J)
155         65 WDY=STAT(J,JF)
```

```
156              SIGAVG=SIGAVG+SIGMA
157              CALL RFACTR(I,OBS,WDY,SIGMA)
158        68 SUM(I)=SUM(I)+BOOT(JTYPE,J)
159              FREE=NPTS(I)
160              AVG=SUM(I)/FREE
161              SIGAVG=SIGAVG/FREE
162              XMIN=BOOT(JTYPE,M1)
163              XMAX=BOOT(JTYPE,M2)
164              RW=SQRT(SIG(I)/DENOM(I))
165              GOF2=SIG(I)/FREE
166              IF(GOF2.GT.YMAX)YMAX=GOF2
167              GOF=SQRT(GOF2)
168              WRITE(IPRT,40)XMIN,XMAX,AVG,NPTS(I),RW,GOF,GOF2,SIGAVG
169              JNDX(I)=INDX(I)
170              INDX(I)=38
171              SIG(I)=AVG
172              SLOPE(I,4)=(AVG/SIGAVG)**2
173              IF(YMAX.LT.GOF2)YMAX=GOF2
174        69 DENOM(I)=GOF2
175              NORD=2
176              CALL FIT(SIG,DENOM,CO,A,NORD,INT,GOFF,IFAIL)
177              WRITE(IPRT,39)(CO(L),L=1,NORD)
178              CALL FIT(SLOPE(1,4),DENOM,C1,A,NORD,INT,GOFF,IFAIL)
179              WRITE(IPRT,43)(C1(L),L=1,NORD)
180              SC1=SQRT(ABS(C1(1)))
181              SC2=SQRT(ABS(C1(2)))
182              DO 82 J=1,NY
183              JX=INDEX2(J)
184              SC(J)=SIGC(JX)*SC1
185              SV(J)=SIGV(JX)*SC1
186              SX(J)=SIGX(JX)*SC1
187              SP(J)=SPX(JX)*SC1
188        82 WRITE(IPRT,44)SC(J),SV(J),SX(J),SP(J),SC2
189              JTY=JTYPE+3
190              CALL LINPLT(SIG,DENOM,INT,CO,NORD,XN(JTYPE),YN,JTY,
191           & SIG(1),SIG(INT),0.,YMAX)
192              DO 78 I=1,INT
193        78 INDX(I)=JNDX(I)
194              IF(IPLOT.EQ.0)GO TO 79
195              CALL STATPL(JTYPE,JMEM,IWT,IREF,XMINP,XMAXP,YMINP,YMAXP)
196        79 IF(JTYPE.LT.3)GO TO 80
197              IF(JMEM.LT.NC1)GO TO 61
198        80 CONTINUE
199     C  NORMAL PROBABILITY PLOT
200              IF(IPLOT.NE.0)CALL NORMPL(NO)
201              RETURN
202              END
```

```
  1     C -------*---------*---------*---------*---------*---------*---------*
  2              SUBROUTINE CALC4(II,LL,U,L1,L2,B1,B2)
  3     C  CALLED BY GRAFN
  4              COMMON/DATA/IDATA(600),PCF(4,600),CT(4,600),EQV(600),U1(600)
  5              COMMON/DIM/NO,NOS(20),NSP,NM,NL,NC,NC1,NV,NSET,IST,JST,NY,NCY,IW
  6              COMMON/CONC/CF(4),HOH,CTC(4),C(20),PC(20),TOTX(4)
  7              COMMON/PAR1/KI(20),LABP(20,7),IX(4,20),EX(4,20)
  8              COMMON/LIMS/PH1,PH2,PHMIN,PHMAX,EMIN,EMAX,YMIN,YMAX,BMAX
  9              COMMON/MISC/X(1200),Y(1200),FILL(3600)
 10              COMMON/INIT/IACT,AU(7),CKW,TEMP,UCORR
 11              COMMON/LIGS/Z(4),NH(12),QC(44)
 12              DATA KSTAR/1H*/
 13              RETURN
 14              END
```

```
  1     C -------*---------*---------*---------*---------*---------*---------*
  2              SUBROUTINE CALC1(PCFC,II,IA,U,IFAIL,ITYPE,JTYPE,ICYCL)
  3     C  CALLED BY NONLIN OR GLASS
  4              COMMON/LARG/STAT(600,4),    SX(4,600),SIGD(4,600),INDX(600)
  5              COMMON/DATA/IDATA(600),PCF(4,600),CT(4,600),EQV(600),U1(600)
  6              COMMON/DIM/NO,NOS(20),NSP,NM,NL,NC,NC1,NV,NSET,IST,JST,NY,NCY,IW
```

```
 7          COMMON/CONC/CF(4),HOH,CTC(4),C(20),PC(20),TOTX(4)
 8          COMMON/PREL/NCM,INDEX1(4),INDEX2(4),INDEX3(4)
 9          COMMON/INIT/IACT,AU,AH,AOH,A1,AØ,B1,AKW,CKW,TEMP,UCORR
1Ø          COMMON/SIG1/SIGC(12),DM
11          COMMON/CYCL/ICY,JTY,KTYPE
12          DIMENSION SAVE(4),SAFE(4),PCFC(4),STASH(4)
13          ICY=ICYCL
14          JTY=JTYPE
15    C  ITYPE=1 FOR DERIVATIVES AND =2 FOR SHIFT OPTIMIZATION
16          IF(ITYPE.NE.Ø)GOTO 50
17    C  REMEMBER THAT NCM IS THE NUMBER OF UNKNOWN FREE CONCENTRATIONS
18          IF(NCM.EQ.Ø)GO TO 50
19          IF(ICYCL.GT.1)GO TO 50
2Ø    C  INITIAL UNKNOWN DEPENDENT VARIABLES (FREE CONCENTRATIONS)
21          IF(II.NE.1)GO TO 22
22          LBL=1
23          GO TO 26
24       22 LBL=LBL+1
25          IF(IDATA(II).EQ.IDATA(II-1))GO TO 26
26          LBL=1
27          DO 24 J=1,NCM
28          JX=INDEX1(J)
29          PCF(JX,II)=PCF(JX,1)
3Ø       24 IF(CT(JX,II).LT.1.E-6) PCF(JX,II)=99.
31          GO TO 50
32       26 DO 31 I=1,NCM
33          LX=INDEX1(I)
34          IF(LBL-2) 50,33,35
35       33 PCF(LX,II)=PCF(LX,II-1)
36          GO TO 31
37    C  REF. TALANTA,23,43(1976).    BOOTSTRAP EXTRAPOLATION
38       35 DERV=(PCF(NC1,II-1)-PCF(NC1,II))/(PCF(NC1,II-2)-PCF(NC1,II-1))
39          PCF(LX,II)=PCF(LX,II-1)-0.85*(PCF(LX,II-2)-PCF(LX,II-1))*DERV
4Ø       31 CONTINUE
41       50 DO 52 I=1,NC1
42          SAVE(I)=PCF(I,II)
43       52 CF(I)=10.**(-PCF(I,II))
44    C  SELF-CONSISTENCY ON IONIC STRENGTH DUE TO ASSOCIATED SPECIES
45          CALL IONIC(II,U,ITYPE,IFAIL)
46          IF(IFAIL.NE.Ø) GO TO 998
47          IF(JTYPE.NE.Ø) GO TO 300
48    C  CALCULATION IN ASSOCIATION WITH NUMERICAL DERIVATIVES
49    C  CALCULATE  DEPENDENT VARIABLES -- FREE HYDROGEN CONCENTRATION FIXED
5Ø          DPX=.1
51    C  THIS IS A VERY USEFUL RELATION BETWEEN THE SHIFT AND THE SLOPE OF
52    C  THE TITRATION CURVE
53          IF(IW.NE.1) DPX=5.*SX(NC1,II)
54          HØ=CT(NC1,II)
55          H1=CTC(NC1)
56    C  NOW SHIFT EACH PX(OBS), ONE AT A TIME, TO CALC EACH TIME NEW EQV
57          DO 157 L=1,NY
58          LX=INDEX2(L)
59    C STORE IN A SAFE PLACE
6Ø          DO 152 IM=1,NC1
61          STASH(IM)=CF(IM)
62      152 SAFE(IM)=PCF(IM,II)
63    C SHIFT FROM OBSERVED VALUE
64          PCF(LX,II)=PCF(LX,II)+DPX
65          CF(LX)=10.**(-PCF(LX,II))
66          CALL IONIC(II,U,3,IFAIL)
67          IF(IFAIL.NE.Ø)GO TO 998
68          H2=CTC(NC1)
69    C RESTORE
7Ø          DO 153 IM=1,NC1
71          CF(IM)=STASH(IM)
72      153 PCF(IM,II)=SAFE(IM)
73    C  ONE USES HTOT HERE BECAUSE IT IS THE INDEPENDENT VARIABLE.  MTOT AND
74    C  ARE INDEPENDENT BUT ARE NOT VARIABLES IN ACID-BASE TITRATIONS PER SE
75      157 PCFC(L)=PCF(LX,II)-DPX*(H1-HØ)/(H2-H1)
76          IF(ABS(PCF(NC1,II)-PCFC(NY)).GT.DM)PCFC(NY)=PCF(NC1,II)+.0004
77          IF(ITYPE.EQ.Ø)RETURN
78          DO 201 J=1,NCM
79          JX=INDEX1(J)
8Ø      201 PCF(JX,II)=SAVE(JX)
81          RETURN
82    C  CALCULATION IN ASSOCIATION WITH ANALYTICAL DERIVATIVES
```

```
83    C   RESTORE OBSERVED VALUES
84        300 DO 301 J=1,NY
85            JX=INDEX2(J)
86            PCFC(J)=PCF(JX,II)
87            IF(ABS(PCFC(J)-SAVE(JX)).GT.DM)PCFC(J)=SAVE(JX)+.0004
88        301 PCF(JX,II)=SAVE(JX)
89            RETURN
90    C   NOTE THAT SHIFT-OPTIMIZED PX VALUES ARE SAVED FOR THE NEXT CYCLE
91    C   INDETERMINATE DATA HANDLING
92        998 DO 1004 I=1,NC1
93            CTC(I)=CT(I,II)
94       1004 PCF(I ,II)=SAVE(I)
95            DO 1001 J=1,NY
96            JX=INDEX2(J)
97       1001 PCFC(J)=PCF(JX,II)+.0004
98            RETURN
99            END
```

```
1     C   -------*----------*----------*----------*----------*----------*----------*
2             SUBROUTINE CALC2(II,U)
3     C   CALLED BY LINLSQ
4             COMMON/LARG/STAT(600,4),SIGPX(4,600),SIGD(4,600),INDX(600)
5             COMMON/DATA/IDATA(600),PCF(4,600),CT(4,600),EQV(600),U1(600)
6             COMMON/DIM/NO,NOS(20),NSP,NM,NL,NC,NC1,NV,NSET,IST,JST,NY,NCY,IW
7             COMMON/CONC/CF(4),HOH,CTC(4),C(20),PC(20),TOTX(4)
8             COMMON/PAR1/KI(20),LABP(20,7),IX(4,20),EX(4,20)
9             COMMON/PAR2/AM(20,20),V(20),DP(20),P(20),SHFT(20),SOECUT,OPT1,OPT2
10            COMMON/BETA/BETA(20),BLOG(20),BL(20),B(20),SL(20),SLB(20),SB(20)
11            COMMON/PREL/NCM,INDEX1(4),INDEX2(4),INDEX3(4)
12            COMMON/INIT/IA,AU(7),CKW,TE(2)
13            DIMENSION TO(4)
14            DATA KSTAR/1H*/
15    C   SELF-CONSISTENCY ON IONIC STRENGTH DUE TO ASSOCIATED SPECIES
16            CF(NC1)=10.**(-PCF(NC1,II))
17            CALL IONIC(II,U,0,IFAIL)
18    C   TOTAL CONCENTRATIONS
19            CALL CTOT(II,NC1,INDEX3,GOFC)
20    C   SYMMETRIC MATRIX AND VECTOR BUILD-UP
21            DO 150 K=1,NC1
22            TO(K)=CT(K,II)-CF(K)
23            DO 150 I=1,NSP
24        150 IF(KI(I).NE.KSTAR)TO(K)=TO(K)-EX(K,I)*C(I)
25            TO(NC1)=TO(NC1)+CKW/CF(NC1)
26            N=0
27            DO 122 I=1,NSP
28            IF(KI(I).NE.KSTAR)GO TO 122
29            S=0.
30            DO 119 K=1,NC1
31        119 S=S+EX(K,I)*TO(K)/SIGD(K,II)**2
32            N=N+1
33    C   SCALING IS IMPLICIT IN THE USE OF C(I)INSTEAD OF PI
34            V(N)=V(N)+C(I)*S
35            L=0
36            DO 121 J=1,NSP
37            IF(KI(J).NE.KSTAR)GO TO 121
38            S=0.
39            DO 120 K=1,NC1
40        120 S=S+EX(K,I)*EX(K,J)/SIGD(K,II)**2
41            L=L+1
42    C THIS IS A SUSPECT UNDERFLOW AREA WITH AN IBM MACHINE
43            AM(N,L)=AM(N,L)+C(I)*(C(J)*S)
44        121 CONTINUE
45        122 CONTINUE
46            RETURN
47            END
```

```
1     C   -------*----------*----------*----------*----------*----------*----------*
2             SUBROUTINE CALC3(II,IFAIL,KJJ)
3     C   CALLED BY FANTASY
```

```
4           REAL NBAR(4)
5           COMMON/LARG/AL(4,200),ALPHA(20,200),FULL(3000)
6           COMMON/DATA/IDATA(600),PCF(4,600),CT(4,600),EQV(600),U1(600)
7           COMMON/BETA/BETA(20),BLOG(20),BL(20),B(20),SL(20),SLB(20),SB(20)
8           COMMON/MONT/ICRD,IPRT,IPCH,TITLE(20),FMT(20),YNAME(4),CODEN(5)
9           COMMON/DIM/NO,NOS(20),NSP,NM,NL,NC,NC1,NV,NSET,IST,JST,NY,NCY,IW
10          COMMON/LIMS/PH1,PH2,PHMIN,PHMAX,EMIN,EMAX,PLMIN,PLMAX,BMAX
11          COMMON/LIGS/HM(4),NH(4),LIGS(4),KKI(4),QC(4),QR(20),Q(20)
12          COMMON/CONC/CF(4),HOH,CTC(4),C(20),PC(20),TOTX(4)
13          COMMON/CONT/IPUN,IPLOT,IFPL,LONG,IDAT,IPAR,IGUESS,IDPX,ISORT,IREF
14          COMMON/PAR1/KI(160),IX(4,20),EX(4,20)
15          COMMON/PREL/NCM,INDEX1(4),INDEX2(4),INDEX3(4)
16          COMMON/FANT/VO,ACID,BASE,TL,DIL
17          COMMON/INIT/IACT,AU,AH,AOH,A1,A0,B1,AKW,CKW,TEMP,UCORR
18          COMMON/MISC/PX(1200),BAR(1200),SAVE(4),HBAR(4),PCH(4),FILL(3588)
19          COMMON/DERV/Z(4,4),CB(20,20),CK(20,20),IT
20          DIMENSION EBR(4),CTR(4),EB(4),TEM(24)
21       13 FORMAT(6F7.3)
22       19 FORMAT(5X,6H XTOT ,6F9.6)
23       20 FORMAT(' %FREE(PX)',6(F6.1,1H(,F5.2,1H)))
24       21 FORMAT(' %COMP(PC)' ,5(F6.1,1H(,F5.2,1H)),/10X,5(F6.1,1H(,F5.2,
25          &1H)),/10X,5(F6.1,1H(,F5.2,1H)),/10X,5(F6.1,1H(,F5.2,1H))
26          &,/10X,5(F6.1,1H(,F5.2,1H)),/10X,5(F6.1,1H(,F5.2,1H)))
27       22 FORMAT(2F10.5,4F7.3,3F7.5)
28       29 FORMAT(10H PKW,LOG B,F7.3,9F7.2,2(/10X,10F7.2))
29      813 FORMAT(1X, 69(1H-),/,4H PH=,F6.3,7H    VOL=,F7.4,7H      EQV=,F8.3,
30          & 5H    U=,F6.3,I5,' ITERATIONS',
31          &/,' NBAR(HBAR)',5(F7.3,1H(,F5.3,1H)))
32     1252 FORMAT(' S',F9.6,5X,'EXBAR',4F7.3,5X,'S(R)',F9.6,5X,'EXBAR(R)',
33          & 4F6.2)
34    C   INITIAL UNKNOWN DEPENDENT VARIABLES (FREE CONCENTRATIONS)
35          DO 32 I=1,NC
36          IF(II-2)31,33,35
37       33 PCF(I,II)=PCF(I,II-1)
38          GO TO 31
39       35 DERV=(PCF(NC1,II-1)-PCF(NC1,II))/(PCF(NC1,II-2)-PCF(NC1,II-1))
40          PCF(I,II)=PCF(I,II-1)-.85*(PCF(I,II-2)-PCF(I,II-1))*DERV
41       31 SAVE(I)=PCF(I,II)
42       32 CF(I)=10.**(-PCF(I,II))
43          CF(NC1)=10.**(-PCF(NC1,II))
44    C   BEGIN REFINEMENT OF UNKNOWN FREE CONCENTRATIONS
45    C   SELF-CONSISTENCY ON IONIC STRENGTH (2-PASS)
46          KS=1
47          U=U1(II)+CF(NC1)
48      135 CALL ACTIV(U,1.)
49          CALL MINLS(II,NC,INDEX3,IFAIL)
50          IF(IFAIL.NE.0)GO TO 1000
51          IF(KS.EQ.2)GO TO 999
52          TERM=0.
53          DO 41 J=1,NSP
54       41 TERM=TERM+Q(J)*(1.+Q(J))*C(J)
55          DO 42 J=1,NC
56       42 TERM=TERM+QC(J)*(1.+QC(J))*CF(J)
57          U=U1(II)+CF(NC1)+.5*TERM
58          KS=2
59          GO TO 135
60      999 CT(NC1,II)=CTC(NC1)
61          DO 1200 KK=1,2
62    C   BASE TITRANT VOLUME
63          IF(NSET.NE.10)VT=VO*(TL*DIL+ACID-CT(NC1,II))/(CT(NC1,II)+BASE)
64    C   ACID TITRANT VOLUME
65          IF(NSET.EQ.10)VT=VO*(TL*DIL-BASE-CT(NC1,II))/(CT(NC1,II)-ACID)
66     1200 DIL=VO/(VT+VO)
67          IF(KJJ.LT.3)RETURN
68          IF(IDAT.EQ.0)IDAT=1
69    C   ADJUST PH SLIGHTLY
70          CT(NC1,II)=(ACID+TL)*DIL-BASE*(1.-DIL)
71          IF(NSET.EQ.10)CT(NC1,II)=ACID*(1.-DIL)-(BASE-TL)*DIL
72          CALL MINLS(II,NC1,INDEX3,IFAIL)
73          EQV(II)=(TL*DIL-CT(NC1,II))/CT(IDAT,II)
74    C   NBAR, HBAR
75          CALL BJERM(II,NC1,NBAR,HBAR,PL)
76          IF(PX(II).GE.PLMAX)PLMAX=PX(II)
77          IF(PX(II).LT.PLMIN)PLMIN=PX(II)
78          IF(BAR(II).GT.BMAX)BMAX=BAR(II)
79          IF(LONG.GT.1)GO TO 15
```

```
 80          WRITE(IPRT,13)PCF(NC1,II),VT,EQV(II),NBAR(1),HBAR(1),PX(II)
 81          IF(IPUN.GT.0)WRITE(IPCH,22)VT,PCF(NC1,II)
 82          RETURN
 83       15 SOM=0.
 84          S2R=0.
 85          DO 1250 L1=1,4
 86          EBR(L1)=0.
 87     1250 EB(L1)=0.
 88          DO 23 K=1,NSP
 89          SOM=SOM+C(K)
 90          DO 1251 L1=1,NC1
 91     1251 EB(L1)=EB(L1)+EX(L1,K)*C(K)
 92          DIMER=1.
 93          IF(IX(1,K).NE.0)DIMER=EX(1,K)
 94          ALPHA(K,II)=DIMER*C(K)/CTC(1)
 95       23 TEM(K)=ALPHA(K,II)*100.
 96          DO 27 I=1,NC1
 97          EB(I)=EB(I)/SOM
 98          PCH(I)=PCF(I,II)
 99          AL(I,II)=CF(I)/CTC(1)
100       27 TEM(I+20)=AL(I,II)*100.
101          IF(NM.EQ.0)GO TO 3933
102     C  RESIDUAL ANALYSIS
103          S2R=SOM
104          DO 3930 K=1,NSP
105     3930 IF(K.LE.IST)S2R=S2R-C(K)
106          DO 3932 J=1,NC1
107          CTR(J)=CTC(J)-CF(J)
108          DO 3931 K=1,NSP
109     3931 IF(K.LE.IST)CTR(J)=CTR(J)-EX(J,K)*C(K)
110          CTR(NC1)=CTR(NC1)+CKW/CF(NC1)
111     3932 EBR(J)=CTR(J)/S2R
112     3933 WRITE(IPRT,813)PCF(NC1,II),VT,EQV(II),U,IT,(NBAR(J),HBAR(J),
113        & J=1,NL)
114          WRITE(IPRT,19)(CT(I,II),I=1,NC1)
115          WRITE(IPRT,20)(TEM(K+20),PCH(K),K=1,NC1)
116          WRITE(IPRT,21)(TEM(K),PC(K),K=1,NSP)
117          PKW=-ALOG10(CKW)
118          DO 30 K=1,NSP
119       30 TEM(K)=BLOG(K)+QR(K)*(AU-A0)
120          WRITE(IPRT,29)PKW,(TEM(K),K=1,NSP)
121          WRITE(IPRT,1252)SOM,(EB(L1),L1=1,4),S2R,(EBR(L1),L1=1,4)
122          IF(IPUN.GT.0)WRITE(IPCH,22)VT,PCF(NC1,II)
123        & ,EQV(II),(PCF(JI,II),JI=1,NC1),(CT(JI,II),JI=1,NC1)
124          RETURN
125     1000 DO 1002 J=1,NC
126          PCF(J,II)=SAVE(J)
127     1002 CF(J)=10.**(-SAVE(J))
128          RETURN
129          END
```

```
  1     C -------*---------*---------*---------*---------*---------*---------*
  2          SUBROUTINE MINLS(II,NCM,INDEX,IFAIL)
  3     C LS REFINEMENT OF CONCENTRATIONS OF FREE SPECIES
  4     C  CALLED BY CALC1, CALC3, OR IONIC
  5          DOUBLE PRECISION D(4,4),DET,D0,D1,D2,D3
  6          COMMON/DATA/IDATA(600),PCF(4,600),CT(4,600),EQV(600),U1(600)
  7          COMMON/DIM/NO,NOS(20),NSP,NM,NL,NC,NC1,NV,NSET,IST,JST,NY,NCY,IW
  8          COMMON/CONC/CF(4),HOH,CTC(4),C(20),PC(20),TOTX(4)
  9          COMMON/SIG2/SCC(4),RCC(4,4),DEL(4),RHO(20,20),SP(20),DEDV(20)
 10          COMMON/PAR1/KI(20),LABP(20,7),IX(4,20),EX(4,20)
 11          COMMON/INIT/IACT,AU,AH,AOH,A1,A0,B1,AKW,CKW,TEMP,UCORR
 12          COMMON/SIG1/SIGC(12),DM
 13          COMMON/DERV/Z(4,4),CB(20,20),CK(20,20),IT
 14          COMMON/DETERM/DETR
 15          DIMENSION V(4),SH(4),A(4,4),INDEX(4)
 16          D0=2.3025851D00
 17          IF(NCM.GT.0)GO TO 11
 18          CALL CTOT(II,NCM,INDEX,GOFC)
 19          RETURN
 20       11 DO 25 ITER=1,26
 21          CALL CTOT(II,NCM,INDEX,GOFC)
 22          DMAX=1.E-9
```

```
23              DO 111 K=1,NCM
24              J=INDEX(K)
25              RDEL=ABS(1.-CTC(J)/CT(J,II))
26       111 IF(RDEL.GT.DMAX)DMAX=RDEL
27              IF(DMAX.LT.1.E-5)GO TO 888
28     C  SET UP -D(TC)/D(PX)=2.3 D(TC)/D(LN X)
29              DO 8 I=1,NCM
30              DO 7 J=1,NCM
31         7 D(I,J)=0.D00
32              LX=INDEX(I)
33              D1=CF(LX)
34         8 D(I,I)=D0*D1
35              D2=CKW
36              IF(LX .EQ.NC1)D(NCM,NCM)=D0*(D1+D2/D1)
37              DO 80 I=1,NCM
38              LX=INDEX(I)
39              DO 80 J=1,NCM
40              JX=INDEX(J)
41              DO 80 K=1,NSP
42              D1=C(K)
43              D2=EX(LX,K)
44              D3=EX(JX,K)
45        80 D(I,J)=D(I,J)+D0*D1*D2*D3
46     C   INVERSION
47              CALL DMINV(NCM,4,D,DET,IFAIL)
48              DETR=DET
49              DO 81 I=1,4
50              DO 81 J=1,4
51        81 Z(I,J)=D(I,J)
52              IF(IFAIL.NE.0)RETURN
53     C  SHIFT VECTOR
54              DO 25 I=1,NCM
55              LX=INDEX(I)
56              SH(I)=0.
57              DO 24  J=1,NCM
58        24 SH(I)=SH(I)-D(I,J)*DEL(J)
59              IF(ABS(SH(I)).GT.DM)SH(I)=SIGN(DM,SH(I))
60              PCF(LX,II)=PCF(LX,II)+SH(I)
61        25 CF(LX)=10.**(-PCF(LX,II))
62     C  COVARIANCE
63       888 IT=ITER-1
64              IF(IT.LE.0)RETURN
65              DO 160 I=1,NCM
66              DO 159 J=1,NCM
67              A(I,J)=0.
68              DO 159 K=1,NCM
69       159 A(I,J)=A(I,J)+D(I,K)*D(K,J)
70       160 SCC(I)=SQRT(A(I,I))
71              DO 32 I=1,NCM
72              DO 33 J=I,NCM
73              RCC(I,J)=A(I,J)/(SCC(I)*SCC(J))
74        33 RCC(J,I)=RCC(I,J)
75        32 SCC(I)=SCC(I)*GOFC
76              RETURN
77              END
```

```
1  C -------*---------*---------*---------*---------*---------*---------*
2              SUBROUTINE CTOT(II,NCM,INDEX,GOFC)
3  C  CALCULATION OF CONCENTRATIONS OF POSTULATED SPECIES
4  C  CALLED BY CALC2, CALC3, OR MINLS
5              COMMON/DATA/IDATA(600),PCF(4,600),CT(4,600),EQV(600),U1(600)
6              COMMON/DIM/NO,NOS(20),NSP,NM,NL,NC,NC1,NV,NSET,IST,JST,NY,NCY,IW
7              COMMON/CONC/CF(4),HOH,CTC(4),C(20),PC(20),TOTX(4)
8              COMMON/BETA/BETA(20),BLOG(20),BL(20),B(20),SL(20),SLB(20),SB(20)
9              COMMON/PAR1/KI(20),LABP(20,7),IX(4,20),EX(4,20)
10             COMMON/SIG2/SCC(4),RCC(4,4),DEL(4),RHO(20,20),SP(20),DEDV(20)
11             COMMON/LIGS/HM(4),NH(4),LIGS(4),KKI(4),QC(4),QR(20),Q(20)
12             COMMON/INIT/IACT,AU,AH,AOH,A1,A0,B1,AKW,CKW,TEMP,UCORR
13             DIMENSION INDEX(4)
14             CTC(NC1)=CF(NC1)
15             IF(NC.EQ.0)GO TO 8
16             DO 6 K=1,NSP
```

```
17            PC(K)=30.
18            IF(BLOG(K).LT.1.E-6)GO TO 6
19     C  CORRECTING FOR IONIC STRENGTH CHANGES FROM THE BASE LEVEL U0
20            BG=BLOG(K)+QR(K)*(AU-A0)
21            PC(K)=-BG
22            DO 3 J=1,NC1
23          3 PC(K)=PC(K)+EX(J,K)*PCF(J,II)
24     C  IBM OVERFLOW/UNDERFLOW PROBLEM AREA
25            IF(PC(K).GT.30.)PC(K)=30.
26          6 C(K)=10.**(-PC(K))
27            IF(NCM.EQ.0)RETURN
28            GOFC=0.
29            DO 4 J=1,NC1
30            CTC(J)=CF(J)
31            DO 4 K=1,NSP
32          4 CTC(J)=CTC(J)+EX(J,K)*C(K)
33          8 CTC(NC1)=CTC(NC1)-CKW/CF(NC1)
34            DO 5 J=1,NCM
35            JX=INDEX(J)
36            DEL(J)=CT(JX,II)-CTC(JX)
37          5 GOFC=GOFC+DEL(J)**2
38            GOFC=SQRT(GOFC/FLOAT(NCM))
39            RETURN
40            END




1      C  -------*---------*---------*---------*---------*---------*---------*
2             SUBROUTINE IONIC(II,U,ITYPE,IFAIL)
3      C  CALLED BY CALC1, CALC2, AND CALC4  (SELF-CONSISTENT IONIC STRENGTH
4      C  IREF=1 SIGNALS ENTRY FROM CALC1(NONLIN)
5      C  KTYPE=1  ==> ENTRY FROM GLASS
6             COMMON/DATA/IDATA(600),PCF(4,600),CT(4,600),EQV(600),U1(600)
7             COMMON/MISC/AL(4800),U2(1200)
8             COMMON/DIM/NO,NOS(20),NSP,NM,NL,NC,NC1,NV,NSET,IST,JST,NY,NCY,IW
9             COMMON/CONT/IPUN,IPLOT,IFPL,LONG,IDAT,IPAR,IGUESS,IDPX,ISORT,IREF
10            COMMON/PREL/NCM,INDEX1(4),INDEX2(4),INDEX3(4)
11            COMMON/CONC/CF(4),HOH,CTC(4),C(20),PC(20),TOTX(4)
12            COMMON/LIGS/HM(4),NH(4),LIGS(4),KKI(4),QC(4),QR(20),Q(20)
13            COMMON/CYCL/ICYCL  ,JTYPE,KTYPE
14            DIMENSION INDEXS(4)
15            KS=1
16            IFAIL=0
17     C  JTYPE=1 FOR ANALYTICAL DERIVATIVES;JTYPE=0 FOR NUMERICAL DERIVATIVES
18            IF(ITYPE.GT.0)GO TO 150
19     C  TWO IONIC STRENGTH ITERATIONS, PERFORMED ONLY IN FIRST CYCLE
20            IF(ICYCL.GT.1)GO TO 148
21            U=U1(II)+CF(NC1)
22            GO TO 36
23        148 U=U2(II)
24         35 KS=2
25     C  HERE ACTIV IS CALLED TO GET AN UPDATED VALUE OF U FOR CTOT CALC.
26         36 CALL ACTIV(U,1.)
27            IF(IREF.EQ.1)GO TO 150
28            DO 1 K=1,NC
29          1 CF(K)=10.**(-PCF(K,II))
30            CALL CTOT(II,0,INDEX3,GOFC)
31            GO TO 156
32        150 IF(CT(1,II).GT.1.E-6)GO TO 151
33            NCMS=NCM-1
34            IF(JTYPE.EQ.1)NCMS=NC
35            DO 152 J=1,NC
36            INDEXS(J)=INDEX3(J+1)
37        152 IF(JTYPE.EQ.0)INDEXS(J)=INDEX1(J+1)
38            PCF(1,II)=50.
39            CF(1)=1.E-37
40            CALL MINLS(II,NCMS,INDEXS,IFAIL)
41            IF(IFAIL.NE.0)RETURN
42            GO TO 153
43        151 IF(JTYPE.EQ.0)GO TO 155
44            IF(KTYPE.EQ.1)CALL APEB(U)
45            CALL MINLS(II,NC1,INDEX3,IFAIL)
46            GO TO 156
47        155 CALL MINLS(II,NCM,INDEX1,IFAIL)
```

```
48   156 IF(KS.EQ.2)GO TO 999
49       IF(IFAIL.NE.0)RETURN
50   153 IF(ITYPE.GT.0)RETURN
51   200 TERM=0.
52       DO 41 I=1,NSP
53    41 TERM=TERM+Q(I)*(1.+Q(I))*C(I)
54       DO 42 I=1,NC
55    42 TERM=TERM+QC(I)*(1.+QC(I))*CF(I)
56       U=U+.5*TERM
57       GO TO 35
58   999 RETURN
59       END
```

```
1    C -------*---------*---------*---------*---------*---------*---------*
2        SUBROUTINE WEIGHT
3    C  CALCULATION OF SIG(PX)FROM SIGC,SIGV,SIGX
4    C  CALLED ONLY BY PRELIM
5    C  IREF=1 FOR NONLIN ENTRY AND IREF=0 FOR LINLSQ ENTRY
6        COMMON/LARG/STAT(600,4),SIGPX(4,600),SIGD(4,600),INDX(600)
7        COMMON/DATA/IDATA(600),PCF(4,600),CT(4,600),EQV(600),U1(600)
8        COMMON/DIM/NO,NOS(20),NSP,NM,NL,NC,NC1,NV,NSET,IST,JST,NY,NCY,IW
9        COMMON/SIG1/SIGC(4),SIGV(4),SIGX(4),SOH
10       COMMON/SIG2/SCC(4),RCC(4,4),DEL(4),RHO(20,20),SP(20),DEDV(20)
11       COMMON/INIT/IACT,AU,AH,AOH,A1,A0,B1,AKW,CKW,TEMP,UCORR
12       COMMON/CONT/IPUN(9),IREF
13       COMMON/CONC/CF(4),HOH,CTC(4),C(20),PC(20),TOTX(4)
14       COMMON/PAR1/KI(20),LABP(20,7),IX(4,20),EX(4,20)
15       COMMON/BETA/BETA(20),BLOG(20),BL(20),B(20),SL(20),SLB(20),SB(20)
16       COMMON/MISC/D(200),IND(200),FILL(5600)
17       DATA KSTAR/1H*/
18       DO 200 K=1,NC1
19       KD=1
20    50 KC=0
21       DO 100 I=1,NO
22       ID=IDATA(I)
23       IF(ID.NE.KD)GO TO 100
24       KC=KC+1
25       IND(KC)=I
26       IF(KC.EQ.1)GO TO 100
27       J=IND(KC-1)
28       DE=EQV(I)-EQV(J)
29       IF(ABS(DE).LT.1.E-4)DE=1.E-4
30       D(KC)=(PCF(K,I)-PCF(K,J))/DE
31   100 CONTINUE
32       D(1)=D(2)
33       DO 150 I=1,KC
34       J=IND(I)
35       X=10.**(-PCF(K,J))
36       IF(K.EQ.NC1)X=X+SOH*(1.E-14)/X
37       PF=SIGV(K)*DEDV(KD)
38   C  DEDV=CTITRE/(INITIAL TOTAL VOL * TOTAL METAL)
39   150 SIGPX(K,J)=SQRT(SIGC  (K)**2+(PF*D(I))**2+(SIGX(K)*X)**2)
40       KD=KD+1
41       IF(KD.LE.NSET)GO TO 50
42   200 CONTINUE
43       IF(IREF.EQ.1)RETURN
44   C  LINEAR WEIGHTS
45       DO 118 II=1,NO
46       DO 118 J=1,NC1
47       CFJ=10.**(-PCF(J,II))
48   C  ...FROM XTOT (ASSUME  5 PERCENT ERROR IN XTOT)
49       VAR=(.0217*CT(J,II))**2
50   C  ...FROM PX
51       TLDJL=CFJ
52       IF(J.EQ.NC1)TLDJL=CFJ+(1.E-14)/CFJ
53       DO 117 L=1,NC1
54       VARPX=0.
55       IF(L.EQ.J)VARPX=TLDJL
56       DO 116 K=1,NSP
57       IF(KI(K).EQ.KSTAR)GO TO 116
58       PC(K)=-BLOG(K)
59       DO 114 KL=1,NC1
```

```
60      114 PC(K)=PC(K)+EX(KL,K)*PCF(KL,II)
61          C(K)=10.**(-PC(K))
62          VARPX=VARPX+EX(J,K)*EX(L,K)*C(K)
63      116 CONTINUE
64      117 VAR=VAR+(VARPX*SIGPX(L,II))**2
65    C ...FROM BETAS
66          DO 115 K=1,NSP
67          IF(KI(K).EQ.KSTAR)GO TO 115
68          VAR=VAR+(EX(J,K)*C(K)*SLB(K))**2
69      115 CONTINUE
70          IF(VAR.LT.1.E-14)VAR=1.E-14
71      118 SIGD(J,II)=2.3*SQRT(VAR)
72          RETURN
73          END
```

```
1     C -------*---------*---------*---------*---------*---------*---------*
2           SUBROUTINE ACTIV(U,HA)
3     C  CALCULATE ACTIVITY COEFFICIENTS OF HYDROGEN AND HYDROXIDE IONS
4     C  CALLED BY FANTASY, REDUCE, IONIC, OR BOOT
5           COMMON/DIM/NO,NOS(20),NSP,NM,NL,NC,NC1,NV,NSET,IST,JST,NY,NCY,IW
6           COMMON/INIT/IACT,AU,AH,AOH,A1,A0,B1,AKW,CKW,TEMP,UCORR
7           COMMON/CONC/CF(4),HOH,CTC(4),C(20),PC(20),TOTX(4)
8           COMMON/APE/UM,T,QWLOGM,CON,SQM,SQ,ITOT,ITITR,QWLOG
9           DIMENSION P(12)
10          DATA P/3.46916E4,105.1510,-.1075733,-2.35812E6,-6.70857E2,5.5636,
11         &-7.56934E-3,-.0626646,-8.18964E-4,-.103561,5.38147E-4,2.30258509/
12    C  MOLARITY*CON=MOLALITY AT 25 DEG    REF:- C.F.BAES,JR. AND R.E.MESMER
13    C  "THE HYDROLYSIS OF CATIONS",WILEY-INTERSCIENCE,NY,1976,P439.
14          CON=1.003+.0277*U+.002*U**2
15          UM=U*CON
16          SQM=SQRT(UM)
17    C  ANALYTICAL EXPRESSION FOR THE DISSOCIATION CONSTANT OF WATER AS A
18    C  FUNCTION OF TEMPERATURE AND IONIC STRENGTH
19    C  REF F.SWEETON,R.E.MESMER, AND C.F.BAES,JR., J.SOLUTION CHEM.,3,191
20          QWLOGM=P(1)/T+P(2)*P(12)*ALOG10(T)+P(3)*T+P(4)/T**2+P(5)
21         & +2.*A1*SQM/(1.+(P(6)+P(7)*T)*B1*SQM)+(P(8)+P(9)*T)*UM+
22         & (P(10)+P(11)*T)*UM**1.5
23    C  NOW CONVERT CONSTANT FROM MOLALITY TO MOLARITY UNITS
24          QWLOG=QWLOGM-2.*ALOG10(CON)
25    C  LOG QW = -13.787 AT 25 DEG, .1 MOLAR
26          CKW=10.**QWLOG
27          SQ=SQRT(U)
28          AU=-A1*(SQ/(1.+SQ)-.3*U)
29          IF(HA.GT.0.5)RETURN
30          IF(IACT-1)10,20,30
31       30 AH=1.
32          AOH=1.
33          GO TO 9
34    C  DEBYE-HUCKEL ACTIVITY MODEL
35       20 AH=10.**(-A1*SQ/(1.+9.*B1*SQ))
36          AOH=10.**(-A1*SQ/(1.+3.5*B1*SQ))
37          GO TO 9
38    C  EXPERIMENTAL HCL, KOH ACTIVITY COEFFICIENTS (LIT VALUES)
39       10 DELT=TEMP-25.
40    C  ACTIVITY COEFFICIENTS OF HCL AND KOH IN KCL MEDIA ACCORDING TO
41    C  K.S.PITZER AND G.MAYORGA,J.PHYS.CHEM.,77,2300(1973).  TEMP DEP DUE
42    C  TO L.F.SILVESTER AND K.S.PITZER,PREPRINT,1978.
43          BHCL0=0.1775*(1.-0.0003081*DELT)
44          BHCL1=0.2945*(1.+0.0001419*DELT)
45          CHCL=0.0008*(1.+0.00006213*DELT)
46    C  TEMP DEP FOR KOH TAKEN FROM NAOH DATA
47          BKOH0=.1298*(1.+.0007*DELT)
48          BKOH1=.32*(1.+.000134*DELT)
49          CKOH=.0041*(1.-.00001894*DELT)
50          BKCL0=.04835*(1.+.0005794*DELT)
51          BKCL1=.2122*(1.+.001071*DELT)
52          CKCL=-.00084*(1.-.00005095*DELT)
53    C  MIXING COEFFICIENTS FROM K.S.PITZER AND J.J.KIM,JACS,98,5701(1974)
54          THK=.005
55          TOHCL=-.050
56          PHKCL=-.007
57          POHKCL=-.008
```

```
58          EQ=10.**(-.86859*SQM)
59          FG=-.33333*A1*(SQM/(1.+1.2*SQM)+3.838*ALOG10(1.+1.2*SQM))
60          FI=.5*(1.-EQ*(1.+2.*SQM-2.*UM))
61    C  LOG GAMMA HCL OR KOH IN THE ABSENCE OF KCL SUPPORTING ELECTROLYTE
62          GH0=FG+(2.*UM*BHCL0+BHCL1*FI+1.5*UM**2*CHCL)/2.30259
63          GOH0=FG+(2.*UM*BKOH0+BKOH1*FI+1.5*UM**2*CKOH)/2.30259
64    C  Y IS THE MOLE FRACTION OF KCL
65          YH=1.
66          YOH=1.
67          IF(HA.LT.1.E-7)YOH=(U-CKW/HA)/U
68          IF(HA.GE.1.E-7)YH=(U-HA)/U
69          IF(YH.LT.0.)YH=0.
70          IF(YOH.LT.0.)YOH=0.
71    C  LOG GAMMA HCL OR KOH IN THE PRESENCE OF KCL SUPPORTING ELECTROLYTE
72          GHKCL=GH0 +(UM*YH/2.30259)*( (BKCL0-BHCL0)+(BKCL1-BHCL1)*EQ
73        & +THK+(CKCL-CHCL)*UM +.5*(2.-YH)*UM*PHKCL)
74          GOHKCL= GOH0 +(UM*YOH/2.30259)*( (BKCL0-BKOH0)+(BKCL1-BKOH1)*EQ
75        & +TOHCL +(CKCL-CKOH)*UM +.5*(2.-YOH)*UM*POHKCL)
76          AH=10.**GHKCL
77          AOH=10.**GOHKCL
78        9 CF(NC1)=HA/AH
79          RETURN
80          END
```

```
1     C  -------*---------*---------*---------*---------*---------*---------*
2            SUBROUTINE PRELIM(INEG)
3     C  CALLED BY LINLSQ OR NONLIN--TO READ PARAMETERS AND DATA PLUS OTHER
4     C  IREF=1 SIGNALS ENTRY FROM NONLIN
5            EXTERNAL MOVE1,ICOMP1
6            COMMON/DATA/IDATA(600),PCF(4,600),CT(4,600),EQV(600),U1(600)
7            COMMON/MONT/ICRD,IPRT,IPCH,TITLE(20),FMT(20),YNAME(4),CODEN(5)
8            COMMON/DIM/NO,NOS(20),NSP,NM,NL,NC,NC1,NV,NSET,IST,JST,NY,NCY,IW
9            COMMON/CONT/IPUN,IPLOT,IFPL,LONG,IDAT,IPAR,IGUESS,IDPX,ISORT,IREF
10           COMMON/PAR2/AM(20,20),V(20),DP(20),P(20),SHFT(20),SOECUT,OPT1,OPT2
11           COMMON/SIG1/SIGC(4),SIGV(4),SIGX(4),SOH
12           COMMON/SIG2/SCC(4),RCC(4,4),DEL(4),RHO(20,20),SP(20),DEDV(20)
13           COMMON/LIMT/JMEM,JSET,MSTART,MEND,NS
14           COMMON/PREL/NCM,INDEX1(4),INDEX2(4),INDEX3(4)
15           COMMON/LIGS/HM(4),NH(4),LIGS(4),KKI(4),QC(4),QR(20),Q(20)
16           COMMON/DERV/DI(4,4),CB(800),ITR
17         3 FORMAT(16F5.0)
18         4 FORMAT(8F10.4)
19         5 FORMAT( 7H SIG( P,A4,8H)**2 = (,F6.4,9H )**2 + (,F6.4,4H *DP,A4,
20        &5H/DEDV,18H*DEQV/DVOL)**2 + ( ,F5.3,3H * ,A4,4H)**2)
21         6 FORMAT(  15H SIG(PH)**2 = (,F6.4,9H )**2 + (,F6.4,' DPH/DEQV',
22        &'*DEQV/DVOL)**2 + (',F4.2,7H * H +,F5.2,10H * OH )**2)
23        90 FORMAT(' NONLINEAR ELECTRODE RESPONSE SCHEME CENTERS AT PH',F6.3,
24        & /    ' DEQV/DVOL FOR THE',I2,' DATA SETS ARE ',10F8.3,/35X,
25        & 10F8.3)
26       158 FORMAT(3I2,5X,4I1,10F5.0)
27       172 FORMAT(' SHIFT-OVER-ERROR CUTOFF CRITERION IS',F6.3,5X,7HOPT1 IS,
28        & F6.3,5X,7HOPT2 IS,F6.3,5X,'INITIAL ISKIP IS',I2)
29           READ(ICRD,158)NCY,IW,INEG,(KKI(I),I=1,4),SOECUT,OPT1,OPT2,
30        &DELMAX,DEL
31           IF(DELMAX.LT.0.001)DELMAX=1.
32           NY=KKI(1)+KKI(2)+KKI(3)+KKI(4)
33           IF(NY.EQ.0)NY=1
34           IF(OPT1.LT.1.E-5)OPT1=.1
35           IF(OPT2.LT.1.E-5)OPT2=.1
36           IF(SOECUT.LT.1.E-5)SOECUT=.1
37           IF(IW.EQ.1)GO TO 18
38           READ(ICRD,3)SOH,(SIGC(K),SIGV(K),SIGX(K),K=1,NC1)
39           IF(SOH.LT.1.E-4)SOH=1.
40           READ(ICRD,4)(DEDV(J),J=1,NSET)
41           AL=7.-ALOG10(SOH)
42           AX=SIGX(NC1)*SOH
43           WRITE(IPRT,5)(YNAME(K),SIGC(K),SIGV(K),YNAME(K),SIGX(K),YNAME(K),
44        & K=1,NC)
45           WRITE(IPRT,6)SIGC(NC1),SIGV(NC1),SIGX(NC1),AX
46           WRITE(IPRT,90)AL,NSET,(DEDV(J),J=1,NSET)
47     C  READ PARAMETERS
48        18 CALL READP
49     C  READ OBSERVATION DATA
```

```
50              IF(IDAT.EQ.0)CALL READD(0)
51   C  SHIFT PX VALUES
52              DO 55 I=1,NO
53              DO 55 J=1,NC1
54        55 PCF(J,I)=PCF(J,I)+DEL(J)
55              IF((IREF.EQ.1).AND.(NCY.NE.0))WRITE(IPRT,172)SOECUT,OPT1,OPT2,IPUN
56   C  WEIGHTING SCHEME -- DETERMINED ONLY ONCE
57              IF(IW.NE.1)CALL WEIGHT
58              SOH=DELMAX
59   C  AT THIS POINT DATA CAN BE SORTED AS A WHOLE (IT SHOULD ALREADY BE
60   C  SORTED BY SETS)
61   C  ORDER BY PH
62              MSTART=0
63              MEND=NO
64              IF(ISORT.EQ.1)CALL TSORT4(NO,MOVE1,ICOMP1,NC1)
65              NCM=NC1-NY
66              IND=0
67              DO 59 J=1,NC1
68              SCC(J)=0.
69              DO 60 K=1,NC1
70              RCC(J,K)=0.
71        60 DI(J,K)=0.
72              RCC(J,J)=1.
73              INDEX1(J-IND)=J
74              INDEX3(J)=J
75              IF(KKI(J).EQ.0)GO TO 59
76              IND=IND+1
77              INDEX2(IND)=J
78        59 CONTINUE
79              RETURN
80              END
```

```
1    C  -------*---------*---------*---------*---------*---------*---------*
2              SUBROUTINE READP
3    C  READ AND PRINT INPUT PARAMETERS
4    C  CALLED BY FANTASY, REDUCE, GRAFN, OR PRELIM
5              COMMON/MONT/ICRD,IPRT,IPCH,TITLE(20),FMT(20),YNAME(4),CODEN(5)
6              COMMON/DIM/NO,NOS(20),NSP,NM,NL,NC,NC1,NV,NSET,IST,JST,NY,NCY,IW
7              COMMON/CONT/IPUN,IPLOT,IFPL,LONG,IDAT,IPAR,IGUESS,IDPX,ISORT,IREF
8              COMMON/LIGS/HM(4),NH(4),LIGS(4),KKI(4),QC(4),QR(20),Q(20)
9              COMMON/PAR1/KI(20),LABP(20,7),IX(4,20),EX(4,20)
10             COMMON/PAR2/AM(20,20),V(20),DP(20),P(20),SHFT(20),SOECUT,OPT1,OPT2
11             COMMON/BETA/BETA(20),BLOG(20),BL(20),B(20),SL(20),SLB(20),SB(20)
12             COMMON/DERV/D(4,4),CB(20,20),CK(20,20),ITR
13             DIMENSION NG(8),ICB(20)
14             DATA KSTAR/1H*/
15        12 FORMAT(//,53H INPUT CONSTANT    ASSOCIATED CONSTANT         Q    DQ**
16           & ,'2   DP    SHIFT  INPUT CONSTANT IDENTIFICATION',/)
17        13 FORMAT(F8.3,1H(,I3,5H)    (,    4I3,1H),F8.3,1H(,I3,1H),I3,I6,F8.4
18           &,F6.1,3X,7A4)
19        16 FORMAT(20F2.0)
20       100 FORMAT(20I3)
21       101 FORMAT( /,51H ASSOCIATED CONSTANTS = CB MATRIX * INPUT CONSTANTS,
22           & //,10H CB MATRIX,/)
23       102 FORMAT(//,11H CB INVERSE,/)
24       174 FORMAT(I1,A1,4I2,F10.0,3F5.0,7A4)
25       414 FORMAT(/)
26       432 FORMAT('0LOG B(U0)=LOG B(U)-(AU-A0)',6H*DQ**2,8X,'LOG K(U0)=LOG K(
27           &U)-(AU-A0)', 6H*DK**2,//,68H WHERE D**2 = SUM Q**2 OF LEFT SPECIES
28           & MINUS THAT OF SPECIES ON THE,/,37H  RIGHT IN THE EQUILIBRIUM EXPR
29           &ESSION,/)
30             IF(IPAR.NE.0)GO TO 30
31             NV=0
32             DO 20 I=1,20
33             NSP=I
34             READ(ICRD,174)IS,KI(I),(IX(J,I),J=1,4),BL(I),SL(I),DP(I),SHFT(I),
35           & (LABP(I,J),J=1,7)
36             IF(SL(I).LT.0.001)SL(I)=.02
37             IF(DP(I).LT.0.002)DP(I)=.01
38             IF(SHFT(I).LT.0.001)SHFT(I)=1.
39             IF(KI(I).NE.KSTAR)GO TO 15
40             NV=NV+1
41        15 Q(I)=0.
```

```
42          QR(I)=0.
43          DO 8 J=1,NC1
44          EX(J,I)=IX(J,I)
45          Q(I)=Q(I)+EX(J,I)*QC(J)
46        8 QR(I)=QR(I)+EX(J,I)*QC(J)**2
47          QR(I)=QR(I)-Q(I)**2
48          IF(IS.NE.0)GO TO 10
49       20 CONTINUE
50       10 IF(IS.NE.9)GO TO 11
51   C  USER SUPPLIES CB MATRIX WHEN IS=9
52          DO 1010 I=1,NSP
53          READ(ICRD,16)(CB(I,J),J=1,NSP)
54          DO 1010 J=1,NSP
55     1010 CK(I,J)=CB(I,J)
56          GO TO 1007
57   C  THE FOLLOWING DEFINES THE MATRIX THAT CONVERTS BL (LOG K   OR LOG BETA
58   C  TO BLOG (LOG BETA)
59       11 DO 1006 I=1,NSP
60          DO 1004 J=1,NSP
61          CK(I,J)=0.
62     1004 CB(I,J)=0.
63          CK(I,I)=1.
64     1006 CB(I,I)=1.
65          IF(NL.EQ.0)GO TO 30
66          IT=0
67   C  NG PACKS THE FIRST NON-ZERO ELEMENTS OF NH AND LIGS
68          DO 1000 L=1,NL
69          IF(NH(L).EQ.0)GO TO 1001
70          IT=IT+1
71     1000 NG(IT)=NH(L)
72     1001 DO 1002 M=1,NL
73          IF(LIGS(M).EQ.0)GO TO 1003
74          IT=IT+1
75     1002 NG(IT)=LIGS(M)
76     1003 J2=0
77          DO 1005 I=1,IT
78          J1=J2+1
79          J2=J2+NG(I)
80          DO 1005 J=J1,J2
81          DO 1005 K=J1,J
82          CK(J,K)=1.
83     1005 CB(J,K)=1.
84     1007 CALL MINV(NSP,20,CK,IFAIL)
85   C  BLOG = CB * BL          AND            BL = CK * BLOG
86       30 WRITE(IPRT,101)
87          DO 105 I=1,NSP
88          DO 115 J=1,NSP
89      115 ICB(J)=CB(I,J)
90      105 WRITE(IPRT,100)( ICB(J),J=1,NSP)
91          WRITE(IPRT,102)
92          DO 106 I=1,NSP
93          DO 116 J=1,NSP
94      116 ICB(J)=CK(I,J)
95      106 WRITE(IPRT,100)( ICB(J),J=1,NSP)
96          WRITE(IPRT,12)
97          KM=1
98          KJ=1+IST
99          DO 9 I=1,NSP
100         BLOG(I)=0.
101         SLB(I)=0.
102         DO 14 J=1,NSP
103         BLOG(I)=BLOG(I)+CB(I,J)*BL(J)
104      14 SLB(I)=SLB(I)+(CB(I,J)*SL(J))**2
105         IF(SLB(I).GT.0.)SLB(I)=SQRT(SLB(I))
106   C  IBM OVERFLOW PROBLEM REQUIRES THE FOLLOWING LIMIT
107         BETA(I)=1.E+37
108         IF(BLOG(I).LT.70.)BETA(I)=10.**BLOG(I)
109         SB(I)=2.303*BETA(I)*SLB(I)
110         ISL=1000.*SLB(I)+.5
111         JSL=1000.*SL(I)+.5
112         IF(I.EQ.KJ)WRITE(IPRT,414)
113         KL=LIGS(KM)
114         IF(I.NE.KJ+KL)GO TO 50
115         WRITE(IPRT,414)
116         KM=KM+1
117      50 IQ=Q(I)
```

```
118            IQR=QR(I)
119          9 WRITE(IPRT,13)BL(I),JSL,(IX(J,I),J=1,4),BLOG(I),ISL,IQ,IQR,DP(I),
120          & SHFT(I),(LABP(I,J),J=1,7)
121            WRITE(IPRT,432)
122            RETURN
123            END
```

```
  1     C -------*---------*---------*---------*---------*---------*---------*
  2            SUBROUTINE READD(JSORT)
  3     C  READ OBSERVATION DATA ONE SET AT A TIME
  4     C  CALLED BY MONITOR,OSTR1,GRAFN,OR PRELIM,TABLE
  5            EXTERNAL MOVE1,ICOMP1
  6            INTEGER TAPIN,TAPOUT
  7            COMMON/LARG/STAT(600,4),SIGPX(4,600),SIGD(4,600),INDX(600)
  8            COMMON/DATA/IDATA(600),PCF(4,600),CT(4,600),EQV(600),U1(600)
  9            COMMON/MONT/ICRD,IPRT,IPCH,TITLE(20),FMT(20),YNAME(4),CODEN(5)
 10            COMMON/DIM/NO,NOS(20),NSP,NM,NL,NC,NC1,NV,NSET,IST,JST,NY,NCY,IW
 11            COMMON/CONT/IPUN,IPLOT,IFPL,LONG,IDAT,IPAR,IGUESS,IDPX,ISORT,IREF
 12            COMMON/LIMS/PH1,PH2,PHMIN,PHMAX,EMIN,EMAX,PLMIN,PLMAX,BMAX
 13            COMMON/LIMT/JMEM,JSET,MSTART,MEND,NS
 14            COMMON/IO/TAPIN,TAPOUT,INTAP(20)
 15            COMMON/LIMS2/HTMIN,HTMAX
 16            DATA STAR/1H*/
 17         10 FORMAT(' TOTAL NUMBER OF POINTS ACCEPTED IS',I5,/,' COUNT BY SETS
 18          &IS',20I4)
 19         12 FORMAT(' PHMIN,PHMAX,EQVMIN,EQVMAX',4F10.3)
 20         14 FORMAT(' DATA ACCEPTED IN THE PH RANGE',F6.3,3H TO,F7.3)
 21         45 FORMAT(25H DATA RECEIVED FROM UNITS,20I3)
 22            PHMIN=14.
 23            PHMAX=0.
 24            EMIN=50.
 25            EMAX=-50.
 26            HTMIN=1.
 27            HTMAX=-HTMIN
 28            I=1
 29            NO=0
 30            DO 22 N=1,NSET
 31            INTP=INTAP(N)
 32            NOS(N)=0
 33            IS=0
 34         17 IF(IS.EQ.1)GO TO 18
 35         16 READ(INTP ,FMT)IS,Z,EQV(I),(PCF(J,I),J=1,NC1),(CT(J,I),J=1,NC1)
 36          &, U1(I)
 37            IF(PCF(NC1,I).LT.PH1)GO TO 17
 38            IF(PCF(NC1,I).GT.PH2)GO TO 17
 39            DO 25 J=1,NC
 40         25 IF(ABS(CT(J,I)).LT.1.E-9)CT(J,I)=1.E-9
 41            IF(PHMIN.GT.PCF(NC1,I))PHMIN=PCF(NC1,I)
 42            IF(PHMAX.LT.PCF(NC1,I))PHMAX=PCF(NC1,I)
 43            IF(EMIN.GT.EQV(I))EMIN=EQV(I)
 44            IF(EMAX.LT.EQV(I))EMAX=EQV(I)
 45            IF(HTMIN.GT.CT(NC1,I))HTMIN=CT(NC1,I)
 46            IF(HTMAX.LT.CT(NC1,I))HTMAX=CT(NC1,I)
 47            INDX(I)=I
 48            DO 20 J=1,NC1
 49     C  UNIT MICROWEIGHT
 50            SIGPX(J,I)=.001
 51            SIGD(J,I)=.001
 52     C  ZERO WEIGHT
 53            IF(Z.NE.STAR)GO TO 20
 54            SIGPX(J,I)=1000.
 55            SIGD(J,I)=1000.
 56         20 CONTINUE
 57            IDATA(I)=N
 58            NOS(N)=NOS(N)+1
 59            I=I+1
 60            IF(IS.EQ.0)GO TO 16
 61         18 MSTART=NO
 62            MEND=NO+NOS(N)
 63     C  ORDER BY PH WITHIN A SET
 64            IF(JSORT.EQ.1)CALL TSORT4(NOS(N),MOVE1,ICOMP1,NC1)
 65     C  ORDER BY EQUIVALENTS  WITHIN A SET
```

```
66            IF(JSORT.EQ.2)CALL TSORT4(NOS(N),MOVE1,ICOMP2,1)
67         22 NO=NO+NOS(N)
68            WRITE(IPRT,10)NO,(NOS(J),J=1,NSET)
69            WRITE(IPRT,45)(INTAP(N),N=1,NSET)
70            WRITE(IPRT,14)PH1,PH2
71            WRITE(IPRT,12)PHMIN,PHMAX,EMIN,EMAX
72            RETURN
73            END
```

```
1     C -------*---------*---------*---------*---------*---------*---------*
2             SUBROUTINE DMINV(NORD,NDIM,A,DET,IFAIL)
3     C  MATRIX INVERSION (GAUSS-JORDAN ELIMINATION BY PIVOTAL CONDENSATION)
4     C  REF: MC CORMICK AND SALVADORI  P 306
5             DOUBLE PRECISION A(NDIM,NDIM),DET,AMAX
6             DIMENSION IN(20,3)
7             IFAIL=1
8             DET=1.D00
9             DO 20 J=1,NORD
10         20 IN(J,3)=0
11            DO 550 I=1,NORD
12            AMAX=0.D00
13            DO 110 J=1,NORD
14            IF(IN(J,3).EQ.1)GO TO 110
15            DO 100 K=1,NORD
16            IF(IN(K,3).EQ.1)GO TO 100
17            IF(IN(K,3).GT.1)RETURN
18            IF(DABS(A(J,K)).LE.AMAX)GO TO 100
19            IR=J
20            IC=K
21            AMAX=DABS(A(J,K))
22        100 CONTINUE
23        110 CONTINUE
24            IN(IC,3)=IN(IC,3)+1
25            IN(I,1)=IR
26            IN(I,2)=IC
27            IF(IR.EQ.IC)GO TO 310
28            DET=-DET
29            DO 200 L=1,NORD
30            T=A(IR,L)
31            A(IR,L)=A(IC,L)
32        200 A(IC,L)=T
33        310 PIV=A(IC,IC)
34            DET=DET*PIV
35            A(IC,IC)=1.D00
36            DO 350 L=1,NORD
37        350 A(IC,L)=A(IC,L)/PIV
38            DO 550 M=1,NORD
39            IF(M.EQ.IC)GO TO 550
40            T=A(M,IC)
41            A(M,IC)=0.D00
42            DO 450 L=1,NORD
43        450 A(M,L)=A(M,L)-T*A(IC,L)
44        550 CONTINUE
45            DO 710 I=1,NORD
46            L=NORD-I+1
47            IF(IN(L,1).EQ.IN(L,2))GO TO 710
48            IR=IN(L,1)
49            IC=IN(L,2)
50            DO 709 K=1,NORD
51            T=A(K,IR)
52            A(K,IR)=A(K,IC)
53        709 A(K,IC)=T
54        710      CONTINUE
55            DO 730 K=1,NORD
56            IF(IN(K,3).NE.1)RETURN
57        730 CONTINUE
58            IFAIL=0
59            RETURN
60            END
```

```
1     C -------*---------*---------*---------*---------*---------*---------*
2           SUBROUTINE MINV(N,NDIM,A,IFAIL)
3           DIMENSION A(NDIM,NDIM)
4           IFAIL=1
5           DO 3 K=1,N
6           C=A(K,K)
7           IF(ABS(C).LT.1.E-37)RETURN
8           A(K,K)=1.
9           DO 1 J=1,N
1Ø        1 A(K,J)=A(K,J)/C
11          DO 3 I=1,N
12          IF(I.EQ.K)GO TO 3
13          C=A(I,K)
14          A(I,K)=Ø.
15          DO 2 J=1,N
16        2 A(I,J)=A(I,J)-C*A(K,J)
17        3 CONTINUE
18          IFAIL=Ø
19          RETURN
2Ø          END
```

```
1     C -------*---------*---------*---------*---------*---------*---------*
2           SUBROUTINE FIT(X,Y,C,A,NORD,NPTS,GOF,IFAIL)
3     C  Y=CO(1)+CO(2)*X+...
4     C  CALLED BY OSTR1,OSTR2,GRAFN,ERRORS, OR NORMPL
5           DOUBLE PRECISION APA(6,6),DET,CO(6),V(6),SXP(11)
6           DIMENSION X(NPTS),Y(NPTS),C(NORD),A(NORD,NORD)
7           GOF=Ø.
8           IF(NORD-1)5ØØ,2Ø,4Ø
9        2Ø C(1)=Ø.
1Ø          DO 3Ø I=1,NPTS
11       3Ø C(1)=C(1)+Y(I)
12          C(1)=C(1)/FLOAT(NPTS)
13          RETURN
14       4Ø NMAX=2*NORD-1
15          DO 5Ø J=1,NMAX
16          IF(J.LE.NORD)V(J)=Ø.DØØ
17       5Ø SXP(J)=Ø.DØØ
18          DO 6Ø I=1,NPTS
19       6Ø V(1)=V(1)+Y(I)
2Ø          SXP(1)=NPTS
21          DO 1ØØ I=1,NPTS
22          DO 1ØØ J=2,NMAX
23          IF(J.LE.NORD)V(J)=V(J)+Y(I)*X(I)**(J-1)
24      1ØØ SXP(J)=SXP(J)+X(I)**(J-1)
25          DO 2ØØ I=1,NORD
26          DO 2ØØ J=1,I
27          IJ=I+J-1
28          APA(I,J)=SXP(IJ)
29      2ØØ APA(J,I)=SXP(IJ)
3Ø          CALL DMINV(NORD,6,APA,DET,IFAIL)
31          IF(IFAIL.NE.Ø)RETURN
32          DO 3ØØ I=1,NORD
33          CO(I)=Ø.DØØ
34          DO 25Ø J=1,NORD
35          A(I,J)=APA(I,J)
36      25Ø CO(I)=CO(I)+APA(I,J)*V(J)
37      3ØØ C(I)=CO(I)
38          DO 4ØØ I=1,NPTS
39      4ØØ GOF =GOF +(Y(I)-YCALC(NORD,X(I),C))**2
4Ø          GOF =SQRT(GOF /(FLOAT(NPTS)-1.))
41          RETURN
42      5ØØ IFAIL=2
43          RETURN
44          END
```

```
1     C -------*---------*---------*---------*---------*---------*---------*
2           FUNCTION YCALC(NORD,XX,CO)
```

```
 3   C  CALLED BY LINPLT OR FIT
 4         DIMENSION CO(NORD)
 5         YCALC=CO(1)
 6         IF(NORD.LT.2)RETURN
 7         IF(ABS(XX).LT.1.E-25)XX=1.E-25
 8         DO  1  J=2,NORD
 9       1 YCALC=YCALC+CO(J)*XX**(J-1)
10         RETURN
11         END



 1   C  -------*---------*---------*---------*---------*---------*---------*
 2         SUBROUTINE PROB(N,X)
 3         DIMENSION X(N)
 4   C  CALCULATION OF EXPECTED PROBABILITIES FOR A NORMAL DISTRIBUTION OF
 5   C  ORDERED STATISTICS--CALLED ONLY BY NORMPL
 6   C   REF. ABRAMOWITZ AND STEGUN (1964), P.933.   FOR THE APPROXIMATION TO
 7   C   VALID, N SHOULD BE GT.50, E.G., ACTA CRYST.,A28,215(1972).
 8         AN=N
 9         N2=N/2
10         AP=-.5/AN
11         DO 100 JL=1,N2
12         PQ=1./(FLOAT(JL)/AN +AP)**2
13         T=SQRT(ALOG(PQ))
14         Q=2.515517+.802853*T+.010328*T*T
15         B=1.+1.432788*T+.189269*T*T+.001308*T*T*T
16         X(JL)=Q/B-T
17         JM=N-JL+1
18     100 X(JM)=-X(JL)
19         IF((N+1)/2.GT.N2)X(N2+1)=0.
20         RETURN
21         END



 1   C  -------*---------*---------*---------*---------*---------*---------*
 2         SUBROUTINE TSORT4(N,MOVE,ICOMP,I)
 3   C   TREE-SORTING, A VERY EFFICIENT ALGORITHM
 4   C  CALLED BY REDUCE, GRAFN, READD, ERRORS, OR NORMPL
 5         NH=N/2
 6         N1=N+1
 7         IF(N-2)60,55,100
 8     100 IF((N/2)*2-N)10,1,10
 9       1 IF(ICOMP(N,NH,I))2,2,3
10       3 CALL MOVE(N,N1,I)
11         CALL MOVE(NH,N,I)
12         CALL MOVE(N1,NH,I)
13       2 NH=NH-1
14      10 NS=NH+NH
15         IF(ICOMP(NS,NS+1,I))4,5,5
16       4 NS=NS+1
17       5 IF(ICOMP(NH,NS,I))6,30,30
18       6 NHT=NS
19         CALL MOVE(NH,N1,I)
20         CALL MOVE(NS,NH,I)
21         GO TO 7
22       8 IF(ICOMP(NS,NS+1,I))11,12,12
23      11 NS=NS+1
24      12 IF(ICOMP(N1,NS,I))22,29,29
25      22 CALL MOVE(NS,NHT,I)
26         NHT=NS
27       7 NS=NHT+NHT
28         IF(NS-N)8,12,29
29      29 CALL MOVE(N1,NHT,I)
30      30 NH=NH-1
31         IF(NH)10,31,10
32      31 NO=N
33         GO TO 50
34      32 CALL MOVE(NO,N1,I)
35         CALL MOVE(1,NO,I)
36         NO=NO-1
```

```
37              IF(ICOMP(2,3,I))34,34,37
38       34     CALL MOVE(3,1,I)
39              NOS=3
40              NS=6
41              GO TO 40
42       37     CALL MOVE(2,1,I)
43              NOS=2
44              NS=4
45              GO TO 40
46       42     IF(ICOMP(NS,NS+1,I))45,43,43
47       45     NS=NS+1
48       43     CALL MOVE(NS,NOS,I)
49              NOS=NS
50              NS=NOS+NOS
51       40     IF(NS-NO)42,43,49
52       49     NS=NOS/2
53       71     IF(ICOMP(N1,NS,I))73,73,72
54       72     CALL MOVE(NS,NOS,I)
55              NOS=NS
56              NS=NS/2
57              GO TO 71
58       73     CALL MOVE(N1,NOS,I)
59       50     IF(NO-3)54,54,32
60       54     CALL MOVE(3,N1,I)
61              CALL MOVE(1,3,I)
62              CALL MOVE(N1,1,I)
63       55     IF(ICOMP(1,2,I))60,60,56
64       56     CALL MOVE(1,N1,I)
65              CALL MOVE(2,1,I)
66              CALL MOVE(N1,2,I)
67       60     RETURN
68              END
```

```
1    C -------*---------*---------*---------*---------*---------*---------*
2            SUBROUTINE MOVE1(NN,MM,I)
3    C  CALLED BY REDUCE, ERRORS, PRELIM, OR READD
4            COMMON/LARG/STAT(600,4),SIGPX(4,600),SIGD(4,600),INDX(600)
5            COMMON/DATA/IDATA(600),PCF(4,600),CT(4,600),EQV(600),U1(600)
6            COMMON/MISC/SLOPE(1200,4),U2(1200)
7            COMMON/DIM/NO,NOS(20),NSP,NM,NL,NC,NC1,NV,NSET,IST,JST,NY,NCY,IW
8            COMMON/LIMT/JMEM,JSET,MSTART,MEND,NS
9            N=NN+MSTART
10           M=MM+MSTART
11           EQV(M)=EQV(N)
12           U1(M)=U1(N)
13           U2(M)=U2(N)
14           INDX(M)=INDX(N)
15           IDATA(M)=IDATA(N)
16           DO 1 K=1,NC1
17           SLOPE(M,K)=SLOPE(N,K)
18           STAT(M,K)=STAT(N,K)
19           SIGD(K,M)=SIGD(K,N)
20           SIGPX(K,M)=SIGPX(K,N)
21           PCF(K,M)=PCF(K,N)
22         1 CT(K,M)=CT(K,N)
23           RETURN
24           END
```

```
1    C -------*---------*---------*---------*---------*---------*---------*
2            SUBROUTINE MOVE2(N,M,I)
3    C  CALLED BY GRAFN OR NORMPL
4            COMMON/MISC/X(6000)
5            X(M)=X(N)
6            RETURN
7            END
```

```
1    C -------*---------*---------*---------*---------*---------*---------*
2          SUBROUTINE MOVE3(N,M,I)
3    C  CALLED BY FANTASY AND REDUCE
4          COMMON/LARG/D(7200),INDX(600)
5          COMMON/MISC/X(6000)
6          INDX(M)=INDX(N)
7          X(M)=X(N)
8          X(M+1200)=X(N+1200)
9          RETURN
10         END
```

```
1    C -------*---------*---------*---------*---------*---------*---------*
2          FUNCTION ICOMP1(NN,MM,I)
3    C  NN,MM ARE INDICES INITIALIZED TO A PARTICULAR DATA SET
4    C  CALLED BY REDUCE, READD, OR PRELIM
5          COMMON/DATA/IDATA(600),PCF(4,600),CT(4,600),EQV(600),U1(600)
6          COMMON/LIMT/JMEM,JSET,MSTART,MEND,NS
7          N=NN+MSTART
8          M=MM+MSTART
9    C  INCREASING ORDER BY PH
10         ICOMP1=+1
11         IF(PCF(I,N).LE.PCF(I,M))ICOMP1=-1
12         RETURN
13         END
```

```
1    C -------*---------*---------*---------*---------*---------*---------*
2          FUNCTION ICOMP2(NN,MM,I)
3    C  NN,MM ARE INDICES INITIALIZED TO A PARTICULAR DATA SET
4    C  CALLED BY REDUCE, READD, OR PRELIM
5          COMMON/DATA/IDATA(600),PCF(4,600),CT(4,600),EQV(600),U1(600)
6          COMMON/LIMT/JMEM,JSET,MSTART,MEND,NS
7          N=NN+MSTART
8          M=MM+MSTART
9    C  INCREASING ORDER BY EQUIVALENTS
10         ICOMP2=+1
11         IF(EQV(N).LE.EQV(M))ICOMP2=-1
12         RETURN
13         END
```

```
1    C -------*---------*---------*---------*---------*---------*---------*
2          FUNCTION ICOMP3(N,M,I)
3    C  CALLED BY GRAFN OR NORMPL
4          COMMON/MISC/X(6000)
5    C     INCREASING ORDER
6          ICOMP3=1
7          IF(X(N).LE.X(M))ICOMP3=-1
8          RETURN
9          END
```

```
1    C -------*---------*---------*---------*---------*---------*---------*
2          FUNCTION ICOMP4(NN,MM,JTYPE)
3    C  NN,MM ARE INDICES INITIALIZED TO A PARTICULAR DATA SET
4    C  CALLED BY ERRORS ONLY
5          COMMON/LIMT/JMEM,JSET,MSTART,MEND,NS
6          ICOMP4=1
7          N=NN+MSTART
8          M=MM+MSTART
9          IF(BOOT(JTYPE,N).LE.BOOT(JTYPE,M))ICOMP4=-1
10         RETURN
11         END
```

```
 1    C  ------*---------*---------*---------*---------*---------*---------*
 2           FUNCTION ICOMP5(NN,MM,I)
 3    C  CALLED FROM ERRORS ONLY--TO UNPACK PREVIOUS SORTING
 4           COMMON/LARG/STAT(600,4),SIGPX(4,600),SIGD(4,600),INDX(600)
 5           COMMON/LIMT/JMEM,JSET,MSTART,MEND,NS
 6           N=NN+MSTART
 7           M=MM+MSTART
 8           ICOMP5=1
 9           IF(INDX(N).LE.INDX(M))ICOMP5=-1
10           RETURN
11           END
```

```
 1    C  -------*---------*---------*---------*---------*---------*---------*
 2           FUNCTION BOOT(JTYPE,I)
 3    C  CALLED BY ERRORS OR ICOMP4
 4    C  I INDEX SPANS ALL SETS, INITIALIZED ONLY TO THE FIRST SET
 5           COMMON/DATA/IDATA(600),PCF(4,600),CT(4,600),EQV(600),U1(600)
 6           COMMON/DIM/NO,NOS(20),NSP,NM,NL,NC,NC1,NV,NSET,IST,JST,NY,NCY,IW
 7           COMMON/LIMT/JMEM,JSET,MSTART,MEND,NS
 8           COMMON/INIT/IACT,AU,AH,AOH,A1,A0,B1,AKW,CKW,TEMP,UCORR
 9           COMMON/SIG1/SIGC(4),SIGV(4),SIGX(4),SOH
10           COMMON/MISC/SLOPE(1200,4),U2(1200)
11           GO TO (1,2,3,4,5),JTYPE
12         1 BOOT=EQV(I)
13           RETURN
14         2 BOOT=U2(I)
15           RETURN
16         3 BOOT=PCF(JMEM,I)
17           RETURN
18         4 BT=10.**(-PCF(JMEM,I))
19           IF(JMEM.NE.NC1)GO TO 40
20           CALL ACTIV(U2(I),1.)
21           BT=BT+SOH*CKW/BT
22        40 BOOT=BT
23           RETURN
24         5 BOOT=SLOPE(I,JMEM)
25           RETURN
26           END
```

```
 1    C  -------*---------*---------*---------*---------*---------*---------*
 2           SUBROUTINE RFACTR(ITYPE,OBS,WDEL,SIGMA)
 3    C  CALLED FROM NONLIN, LINLSQ, AND ERRORS
 4           COMMON/RFAC/SIG(50),DENOM(50),NPTS(50),SUM(50)
 5           SIG(ITYPE)=SIG(ITYPE)+WDEL**2
 6           DENOM(ITYPE)=DENOM(ITYPE)+(OBS/SIGMA)**2
 7           NPTS(ITYPE)=NPTS(ITYPE)+1
 8           RETURN
 9           END
```

```
 1    C  -------*---------*---------*---------*---------*---------*---------*
 2           SUBROUTINE CLEAR(NTYPE)
 3    C  CALLED FROM NONLIN, LINLSQ, AND ERRORS
 4           COMMON/RFAC/SIG(50),DENOM(50),NPTS(50),SUM(50)
 5           DO 1 I=1,NTYPE
 6           SIG(I)=0.
 7           DENOM(I)=0.
 8           SUM(I)=0.
 9         1 NPTS(I)=0
10           RETURN
11           END
```

```
 1   C -------*---------*---------*---------*---------*---------*---------*
 2         SUBROUTINE TABLE
 3         INTEGER TAPIN,TAPOUT
 4         COMMON/DATA/IDATA(600),PCF(4,600),CT(4,600),EQV(600),U1(600)
 5         COMMON/MONT/ICRD,IPRT,IPCH,TITLE(20),FMT(20),YNAME(4),CODEN(5)
 6         COMMON/DIM/NO,NOS(20),NSP,NM,NL,NC,NC1,NV,NSET,IST,JST,NY,NCY,IW
 7   C A LIST OF DATA IS PREPARED FOR PUBLICATION PURPOSES
 8         COMMON/CONT/IP(4),IDAT,IPAR(5)
 9         COMMON/LIGS/HM(4),NH(4),LIGS(4),KKI(4),QC(44)
10         COMMON/INIT/IACT,R(7),CKW,TEMP,UCORR
11         COMMON/IO/TAPIN,TAPOUT,INTAP(20)
12         DIMENSION TOTX(3),HEAD(20),FMT2(20)
13         RETURN
14         END



 1   C -------*---------*---------*---------*---------*---------*---------*
 2         SUBROUTINE TITRPL(SCALEX,SCALEY,NO,NSET,NC1)
 3   C TITRATION CURVE PLOT ON CALCOMP  -- CALLED BY MONITOR OR FANTASY
 4         COMMON/CONT/ITYPE,NFILL(9)
 5         COMMON/DATA/IDATA(600),PCF(4,600),CT(4,600),EQV(600),U1(600)
 6         COMMON/DIM/MO,NOS(20),NSP,NM,NL,NC,MC1,NV,MSET,IST,JST,NY,NCY,IW
 7         COMMON/LIMS/PH1,PH2,PHMIN,PHMAX,EMIN,EMAX,PLMIN,PLMAX,BMAX
 8         COMMON/LIMS2/HTMIN,HTMAX
 9         COMMON/MONT/ICRD,IPRT,IPCH,TITLE(20),FMT(20),YNAME(4),CODEN(5)
10         RETURN
11         END



 1   C -------*---------*---------*---------*---------*---------*---------*
 2         SUBROUTINE LINPLT(X,Y,NPT,CO,NORD,XN,YN,ITYPE,XMIN,XMAX,YMIN,YMAX)
 3   C LINEPRINTER PLOTS-- CALLED BY GRAFN, ERRORS, FANTASY OR REDUCE
 4   C ITYPE
 5   C    1   FORMATION CURVE
 6   C    2   DISTRIBUTION CURVE
 7   C    3   GRAFN PLOTS
 8   C    4   ERRORS (EQV,GOF**2)
 9   C    5   ERRORS   (U,G**2)
10   C    6   ERRORS  (PX,GOF**2)
11   C    7   ERRORS   (X,GOF**2)
12   C    8   ERRORS (DPX/DVOL,GOF**2)
13         INTEGER DOTS,BLANK
14         COMMON/LARG/STAT(600,4),SIGPX(4,600),SIGD(4,600),INDX(600)
15         COMMON/MISC/AL(4,200),ALPHA(20,200),U2(1200)
16         COMMON/DIM/NO,NOS(20),NSP,NM,NL,NC,NC1,NV,NSET,IST,JST,NY,NCY,IW
17         COMMON/MONT/ICRD,IPRT,IPCH,TITLE(20),FMT(20),YNAME(4),CODEN(5)
18         DIMENSION X(NPT),Y(NPT),CO(NORD),LINE(61),YAXIS(6),LIST(38)
19         DATA DOTS,BLANK/1H.,1H /
20         DATA LIST/1H1,1H2,1H3,1H4,1H5,1H6,1H7,1H8,1H9,
21        &1HA,1HB,1HC,1HD,1HE,1HF,1HG,1HH,1HI,1HJ,1HK,1HL,1HM,1HN,
22        & 1HO,1HP,1HQ,1HR,1HS,1HT,1HU,1HV,1HW,1HX,1HY,1HZ,1H0,1H+,1H*/
23   224 FORMAT(1H1,35X,A4,/10X,E9.2,5E10.2,/,3X,A4,5X,12(5H+----),1H+)
24   228 FORMAT(12X,12(5H+----),1H+,/10X,E9.2,5E10.2)
25   229 FORMAT(E10.2,1X,1H,60A1,1H)
26         DO 6 K=1,60
27      6 LINE(K)=BLANK
28         INDEX=1
29         DX=(XMAX-XMIN)/43.
30         DY=(YMAX-YMIN)/59.
31         DO223 IJ=1,6
32   223 YAXIS(IJ)=10.*FLOAT(IJ-1)*DY+YMIN
33         WRITE(IPRT,224)YN,YAXIS,XN
34         JX=(XMAX-XMIN)/DX+1.5
35         DO 225 J=1,JX
36         XX=DX*FLOAT(J-1)+XMIN
37         IF(ITYPE.LT.3)GO TO 83
38         IDEAL=(YCALC(NORD,XX,CO)-YMIN)/DY+1.5
39         IF((IDEAL.LT.1).OR.(IDEAL.GT. 60))IDEAL=61
40         LINE(IDEAL)=DOTS
41     83 NEW=INDEX
42         DO 226 K=INDEX,NPT
```

```
43              NEW=K
44              DIFF= XX -X(K)
45              IF(DIFF.LT.-.5*DX)GO TO 227
46              NEW=K+1
47              IF(ITYPE.EQ.2)GO TO 241
48              IDEX=INDX(K)
49              IY=(Y(K)-YMIN)/DY   +1.5
50              IF(IY.LT.1)IY=61
51              IF(IY.GT.61 )IY=61
52              LINE(IY)=LIST(IDEX)
53              GO TO 226
54     C  DISTRIBUTION CURVE ONLY
55        241 DO 243 KJ=1,NSP
56              KL=ALPHA(KJ,K)/DY+1.5
57              IF(KL.GT. 60)KL= 61
58        243 LINE(KL)=LIST(KJ+9)
59              DO 244 KJ=1,NC
60              KL=AL(KJ,K)/DY+1.5
61              IF(KL.GT. 60)KL= 61
62        244 LINE(KL)=LIST(KJ)
63        226 CONTINUE
64        227 INDEX=NEW
65              WRITE(IPRT,229)XX,(LINE(JJ),JJ=1, 60)
66        222 DO 225 L=1, 60
67        225 LINE(L)=BLANK
68              WRITE(IPRT,228)YAXIS
69              RETURN
70              END
```

```
1      C -------*---------*---------*---------*---------*---------*---------*
2              SUBROUTINE FORMPL(SCALEN)
3      C  FORMATION CURVE PLOT ON CALCOMP--CALLED FROM FANTASY OR REDUCE
4              COMMON/DATA/IDATA(600),PCF(4,600),CT(4,600),EQV(600),U1(600)
5              COMMON/DIM/NO,NOS(20),NSP,NM,NL,NC,NC1,NV,NSET,IST,JST,NY,NCY,IW
6              COMMON/LARG/FULL(7200),INDX(600)
7              COMMON/LIGS/HM(4),NH(4),LIGS(4),KKI(4),QC(4),QR(20),Q(20)
8              COMMON/LIMS/PH1,PH2,PHMIN,PHMAX,EMIN,EMAX,PLMIN,PLMAX,BMAX
9              COMMON/MISC/PX(1200),BAR(600),XX(20,90),YY(20,90),BARS(600)
10             COMMON/MONT/ICRD,IPRT,IPCH,TITLE(20),FMT(20),YNAME(4),CODEN(5)
11             DATA IUP,IDN/3,2/
12             RETURN
13             END
```

```
1      C -------*---------*---------*---------*---------*---------*---------*
2              SUBROUTINE NORMPL(NUM)
3      C  NORMAL PROBABILITY PLOT ON THE CALCOMP -- CALLED BY ERRORS ONLY
4              EXTERNAL MOVE2,ICOMP3
5              COMMON/LARG/STAT(600,4),SIGPX(4,600),SIGD(4,600),INDX(600)
6              COMMON/DATA/IDATA(600),PCF(4,600),CT(4,600),EQV(600),U1(600)
7              COMMON/DIM/NO,NOS(20),NSP,NM,NL,NC,NC1,NV,NSET,IST,JST,NY,NCY,IW
8              COMMON/MONT/ICRD,IPRT,IPCH,TITLE(20),FMT(20),YNAME(4),CODEN(5)
9              COMMON/CONT/IPUN,IPLOT,IFPL,LONG,IDAT,IPAR,IGUESS,IDPX,ISORT,IREF
10             COMMON/LIMT/JMEM,JSET,MSTART,MEND,NS
11             COMMON/LIMS/PH1,PH2,PHMIN,PHMAX,XMIN,XMAX,YMIN,YMAX,BMAX
12             COMMON/MISC/STATO(1200),STATC(1200),FILL(3600)
13             DIMENSION BCD(17),A(4,4),CO(4)
14             DATA IUP,IDN/3,2/
15             DATA BCD/2H-8,2H-7,2H-6,2H-5,2H-4,2H-3,2H-2,2H-1,2H  ,2H+1,2H+2,
16           & 2H+3,2H+4,2H+5,2H+6,2H+7,2H+8/
17             RETURN
18             END
```

```
1      C -------*---------*---------*---------*---------*---------*---------*
2              SUBROUTINE DISTPL
3      C  DISTRIBUTION CURVE PLOT ON CALCOMP -- CALLED BY FANTASY ONLY
```

```
 4          COMMON/DATA/IDATA(600),PCF(4,600),CT(4,600),EQV(600),U1(600)
 5          COMMON/MONT/ICRD,IPRT,IPCH,TITLE(20),FMT(20),YNAME(4),CODEN(5)
 6          COMMON/DIM/NO,NOS(20),NSP,NM,NL,NC,NC1,NV,NSET,IST,JST,NY,NCY,IW
 7          COMMON/LIMS/PH1,PH2,PHMIN,PHMAX,EMIN,EMAX,PLMIN,PLMAX,BMAX
 8          COMMON/MISC/AL(4,200),ALPHA(20,200),U2(1200)
 9          DATA IUP,IDN/3,2/
10          RETURN
11          END
```

```
 1    C    ------*---------*---------*---------*---------*---------*---------*
 2          SUBROUTINE STATPL(JTYPE,JMEM,IWT,IREF,XMINP,XMAXP,YMINP,YMAXP)
 3    C CALLED BY ERRORS ONLY: PLOT OF DEL/SIG OR DEL VS TESTED VARIABLE
 4    C EQV,U*100,PX,X*1000,SLOPE/100
 5          COMMON/DATA/IDATA(600),PCF(4,600),CT(4,600),EQV(600),U1(600)
 6          COMMON/LARG/STAT(600,4),SIGPX(4,600),SIGD(4,600),INDX(600)
 7          COMMON/DIM/NO,NOS(20),NSP,NM,NL,NC,NC1,NV,NSET,IST,JST,NY,NCY,IW
 8          COMMON/MONT/ICRD,IPRT,IPCH,TITLE(20),FMT(20),YNAME(4),CODEN(5)
 9          COMMON/PREL/NCM,INDEX1(4),INDEX2(4),INDEX3(4)
10          DIMENSION HEAD(3,5),XF1(5)
11          RETURN
12          END
```

```
 1    C    -------*---------*---------*---------*---------*---------*---------*
 2          SUBROUTINE VARPL(NS,MS,ITYPE,HTMIN,HTMAX,SCALEX,SCALEY,SIGH,XTYPE)
 3    C PLOT HTOT VERSUS XTOT THAT VARIED WITH ALL ELSE HELD CONSTANT
 4          COMMON/LARG/COS(4,200),FULL(1800),TOT(6,200,4),XX(200),YY(200)
 5          COMMON/MONT/IZ(3),TITLE(20),FMT(20),YNAME(4),CODEN(5)
 6          COMMON/MISC/PHS(800),HTOTS(6,200),DHDCX(1204),CTA(6),FILL(2790)
 7          COMMON/DIM/NO,NOS(20),NSP,NM,NL,NC,NC1,NV,NSET,IST,JST,NY,NCY,IW
 8          DIMENSION SY(10)
 9          RETURN
10          END
```

```
 1    C    -------*---------*---------*---------*---------*---------*---------*
 2          SUBROUTINE VARNPL(NS,PHS,NBAR,SLOPE)
 3          REAL NBAR
 4          COMMON/DIM/NO,NOS(20),NSP,NM,NL,NC,NC1,NV,NSET,IST,JST,NY,NCY,IW
 5          COMMON/MONT/ICRD,IPRT,IPCH,TITLE(20),FMT(20),YNAME(4),CODEN(5)
 6          DIMENSION PHS(200),NBAR(146),SLOPE(200,3)
 7          RETURN
 8          END
```

```
 1    C    ------*---------*---------*---------*---------*---------*---------*--
 2          SUBROUTINE SYMB(X,Y,L,SC)
 3    C CALLED BY TITRPL, VARPL, FORMPL
 4          DATA IUP,IDN/3,2/
 5          RETURN
 6          END
```

```
 1    C    -------*---------*---------*---------*---------*---------*---------*
 2          SUBROUTINE AXES(N1,N2,N3,N4,XSCALE,ITYPE)
 3    C GRID FOR THE CALCOMP PLOTS--CALLED BY TITRPL,DISTPL, OR FORMPL
 4          DIMENSION C(67)
 5          DATA IUP,IDN/3,2/
 6          DATA C/3H-15,3H-14,3H-13,3H-12,3H-11,3H-10,2H-9,2H-8,2H-7,2H-6,
 7         & 2H-5,2H-4,2H-3,2H-2,2H-1,
 8         &      2H 0,2H 1,2H 2,2H 3,2H 4,2H 5,2H 6,2H 7,2H 8,2H 9,2H10,
```

```
 9          & 2H11,2H12,2H13,2H14,2H15,2H16,2H17,2H18,2H19,2H20,2H21,2H22,2H23,
10          &2H24,2H25,2H26,2H27,2H28,2H29,2H30,2H31,2H32,2H33,2H34,2H35,2H36,
11          &2H37,2H38,2H39,2H40,2H41,2H42,2H43,2H44,2H45,2H46,2H47,2H48,2H49,
12          &2H50,4H100%/
13           RETURN
14           END



 1    C -------*---------*---------*---------*---------*---------*---------*
 2           SUBROUTINE PHSYMB(X,Y,V)
 3    C  DRAW THE SYMBOL "PH"
 4           DATA IUP,IDN/3,2/
 5           RETURN
 6           END
```

Index